THE FLUID MECHANICS and DYNAMICS PROBLEM SOLVER®

REGISTERED TRADEMARK

A Complete Solution Guide to Any Textbook

Staff of Research and Education Association
Dr. M. Fogiel, Director

special chapter reviews by
John M. Cimbala, Ph.D.
Assistant Professor of Mechanical Engineering
Pennsylvania State University
University Park, Pennsylvania

REA

Research and Education Association
61 Ethel Road West
Piscataway, New Jersey 08854

THE FLUID MECHANICS and
DYNAMICS PROBLEM SOLVER ®

REVISED PRINTING 1994

Printed in the United States of America

Library of Congress Card Catalog Number 91-67025

International Standard Book Number 0-87891-547-8

WHAT THIS BOOK IS FOR

For as long as fluid mechanics/dynamics has been taught in schools, students have found this subject difficult to understand and learn. Despite the publication of hundreds of textbooks in this field, each one intended to provide an improvement over previous textbooks, students continue to remain perplexed, and the subject is often taken in class only to meet school/departmental requirements for a selected course of study.

In a study of the problem, REA found the following basic reasons underlying students' difficulties with this subject as taught in schools:

(a) No systematic rules of analysis have been developed which students may follow in a step-by-step manner to solve the usual problems encountered. This results from the fact that the numerous different conditions and principles which may be involved in a problem, lead to many possible different methods of solution. To prescribe a set of rules to be followed for each of the possible variations, would involve an enormous number of rules and steps to be searched through by students, and this task would perhaps be more burdensome than solving the problem directly with some accompanying trial and error to find the correct solution route.

(b) Textbooks currently available will usually explain a given principle in a few pages written by a professional who has an insight into the subject matter that is not shared by students. The explanations are often written in abstract manners which leave the students confused as to the application of the principle. The explanations given are not sufficiently detailed and extensive to make the student aware of the wide range of applications and different aspects of the principles being studied. The numerous possible variations of principles and their applications are usually not discussed, and it is left for the students to discover these for themselves while doing the exercises. Accordingly, the average student is expected to rediscover that which has been long known and practiced, but not published or explained extensively.

(c) The illustrations usually following the explanation of a principle in fluid mechanics/dynamics are too few in number and too simple to enable the student to obtain a thorough grasp of the principle involved. The illustrations do not provide sufficient basis to enable a student to solve problems that may subsequently be assigned for homework or given on examinations.

The illustrations are presented in abbreviated form which leaves out much material between steps, and requires that students derive the omitted material themselves. As a result, students find the illustrations difficult to understand--contrary to the purpose of the illustrations.

Illustrations are, furthermore, often worded in a confusing manner. They do not state the problem and then present the solution. Instead, they pass through a general discussion, never revealing what is to be solved for.

Illustrations, also, do not always include diagrams, wherever appropriate, and students do not obtain the training to draw diagrams to simplify and organize their thinking.

(d) Students can learn the subject only by doing the exercises themselves and reviewing them in class, to obtain experience in applying the principles with their different ramifications.

In doing the exercises by themselves, students find that they are required to devote considerably more time to their courses in fluid mechanics/dynamics than to other subjects of comparable credits, because they are uncertain with regard to the selection and application of the principles involved. It is also often necessary for students to discover those "tricks" not revealed in their texts (or review books), that make it possible to solve problems easily. Students must usually resort to methods of trial-and-error to discover these "tricks," and as a result they find that they may sometimes spend several hours in solving a single problem.

(e) When reviewing the exercises in classrooms, instructors usually request students to take turns in writing solutions on the board and explaining them to the class. Students often find it difficult to explain in a manner that holds the interest of the class, and enables the remaining students to follow the material written on the board. The remaining students seated in the class are, furthermore, too occupied with copying the material from the board, to listen to the oral explanations and concentrate on the methods of solution.

This book is intended to aid students in fluid mechanics/dynamics to overcome the difficulties described, by supplying detailed illustrations of the solution methods which are usually not apparent to students. The solution methods are illustrated by problems selected from those that are most often assigned for classwork and given on examinations. The problems are arranged in order of complexity to enable students to learn and understand a particular topic by reviewing the problems in sequence. The problems are illustrated with detailed step-by-step explanations of the principles involved, to save the students the large amount of time that is often needed to fill in the information that is usually omitted between the steps of the illustrations in textbooks or review/outline books.

The staff of REA considers fluid mechanics/dynamics a subject that is best learned by allowing students to view the methods of analysis and solution techniques themselves. This approach to learning the subject matter is similar to that practiced in the medical fields, for example, and various scientific laboratories.

In using this book, students may review and study the illustrated problems at their own pace; they are not limited to the time allowed for explaining problems on the board in class.

When students want to look up a particular type of problem and solution, they can readily locate it in the book by referring to the index which has been extensively prepared. It is also possible to locate a particular type of problem by glancing at just the material within the boxed portions. To facilitate rapid scanning of the problems, each problem has a heavy border around it. Furthermore, each problem is identified with a number immediately above the problem at the right-hand margin.

To obtain maximum benefit from the book, students should familiarize themselves with the section, "HOW TO USE THIS BOOK," located in the front pages.

To meet the objectives of this book, staff members of REA have selected problems usually encountered in assignments and examinations, and have solved each problem meticulously to illustrate the steps which are usually difficult for students to comprehend. Special gratitude is expressed to them for their efforts in this area, as well as to the numerous contributors who devoted brief periods of time to this work.

Gratitude is also expressed to the many persons involved in the difficult task of typing the manuscript with its endless changes, and to the REA art staff who prepared the numerous detailed illustrations together with the layout and physical features of the book.

The difficult task of coordinating the efforts of all persons was carried out by Carl Fuchs. His conscientious work deserves much appreciation. He also trained and supervised art and production personnel in the preparation of the book for printing.

Finally, special thanks are due to Helen Kaufmann for her unique talents in rendering those difficult border-line decisions and in making constructive suggestions related to the design and organization of the book.

Max Fogiel, Ph.D.
Program Director

HOW TO USE THIS BOOK

This book can be an invaluable aid to students in fluid mechanics/dynamics as a supplement to their textbooks. The book is subdivided into 15 chapters, each dealing with a separate topic. The subject matter is developed beginning with fluid properties, fluid statics, fluid kinematics and extending through dimensional analysis, pipe flow, potential and vortex flow, channel flow, compressible flow and drag/lift. Included also are sections on flow meters, hydraulic structures, propulsion and turbomachines. An extensive number of applications have been included, since these appear to be most troublesome to students.

TO LEARN AND UNDERSTAND A TOPIC THOROUGHLY

1. Refer to your class text and read the section pertaining to the topic. You should become acquainted with the principles discussed there. These principles, however, may not be clear to you at that time.

2. Then locate the topic you are looking for by referring to the "Table of Contents" in the front of this book, "The Fluid Mechanics/Dynamics Problem Solver."

3. Turn to the page where the topic begins and review the problems under each topic, in the order given. For each topic, the problems are arranged in order of complexity, from the simplest to the more difficult. Some problems may appear similar to others, but each problem has been selected to illustrate a different point or solution method.

To learn and understand a topic thoroughly and retain its contents, it will be generally necessary for students to review the problems several times. Repeated review is essential in order to gain experience in recognizing the principles that should be applied, and in selecting the best solution technique.

TO FIND A PARTICULAR PROBLEM

To locate one or more problems related to a particular subject matter, refer to the index. In using the index, be certain to note that the numbers given there refer to problem numbers, not to page numbers. This arrangement of the index is intended to facilitate finding a problem more rapidly, since two or more problems may appear on a page.

If a particular type of problem cannot be found readily, it is recommended that the student refer to the "Table of Contents" in the front pages, and then turn to the chapter which is applicable to the problem being sought. By scanning or glancing at the material that is boxed, it will generally be possible to find problems related to the one being sought, without consuming considerable time. After the problems have been located, the solutions can be reviewed and studied in detail. For this purpose of locating problems rapidly, students should acquaint themselves with the organization of the book as found in the "Table of Contents."

In preparing for an exam, it is useful to find the topics to be covered on the exam in the "Table of Contents," and then review the problems under those topics several times. This should equip the student with what might be needed for the exam.

CONTENTS

CHAPTER 1

FLUID PROPERTIES

Basic Attacks and Strategies for Solving Problems in this Chapter. See pages 1 to 25 for step-by-step solutions to problems.

A fluid is defined as a substance which cannot resist a shear stress by static deformation. Both liquids and gases are fluids and are distinguished from solids by the above definition. There are many properties of fluids to which numerical values can be given. *Density*, ρ, is defined as the mass of a small fluid element divided by its volume. Often, *specific weight*,

$$\gamma = \rho g,$$

is more useful since density and gravitational acceleration usually occur together. Both density and specific weight are dimensional quantities. *Specific gravity*, on the other hand, is dimensionless and is defined as the ratio of a fluid's density to the density of some reference fluid. For liquids, water is the reference fluid, while for gases air is used (at a standard temperature and pressure).

Buoyancy results when an object is placed in a fluid of higher density. Archimedes' two laws of buoyancy are:

a) the vertical buoyancy force on a body immersed in a fluid is equal to the weight of the displaced fluid, and

b) a floating body displaces its own weight in the fluid in which it is floating.

In addition, for a floating body in static equilibrium, there can be no net moments; this concept can be utilized to examine the stability of floating objects.

Viscosity, another important property of fluids, is the ratio of the local shearing stress to the rate of shearing strain of a fluid element in a moving fluid. For simple shear flows where velocity component u in the x-direction is a function of only the normal coordinate y, the shear stress, τ, is equal to

$$\mu \, du/dy,$$

where μ is called the coefficient of viscosity. This linear relation applies only to

Newtonian fluids; fortunately, most common fluids, such as air, water, oil, etc., are Newtonian. The shear stress on a solid surface is equal but opposite to that applied to a fluid wetting the surface. Thus, frictional forces on surfaces can be found if the velocity gradient du/dy and the coefficient of viscosity μ are known. *Kinematic viscosity*, ν, is defined as

$$\nu = \mu/\rho.$$

Vapor pressure, p_v, is defined as the pressure at which a liquid will boil at a given temperature; p_v depends greatly on temperature, and its value can be obtained from charts. In flows of liquids, local fluid pressures typically decrease as velocity increases. If the local pressure falls below p_v, local boiling or *cavitation* may occur.

When a liquid forms an interface with a second liquid or a gas, a tensional force exists at the interface, much like the tension in the skin of a balloon. This so-called surface tension is responsible for such things as the formation of soap bubbles and capillary action. The *coefficient of surface tension*, γ, is a measure of the tensional force per unit length of the surface. The dimensions of γ are thus force divided by length.

In analyzing fluid flows, the local acceleration, **a**, of a small fluid element is given by

$$\mathbf{a} = \frac{d\mathbf{V}}{dt} = \frac{\partial \mathbf{V}}{\partial t} + (\mathbf{V} \cdot \nabla)\mathbf{V}.$$

The first term on the right is the local or unsteady acceleration, while the last term (the convective acceleration) represents acceleration which arises when the particle moves through regions of varying velocity. Note that **a** can be non-zero even in a steady flowfield.

Step-by-Step Solutions to Problems in this Chapter, "Fluid Properties"

DENSITY AND SPECIFIC WEIGHT

• PROBLEM 1-1

Calculate the density of gasoline, where S = 0.680, and T = 150°F.

Solution: From Figures 1 and 2

$$\mu = 4 \times 10^{-6} \text{ lbf - sec/ft}^2$$

$$\nu = 3.2 \times 10^{-6} \text{ ft}^2\text{/sec}$$

respectively. Since ν is defined as simply the ratio of the absolute viscosity to density, i.e.,

$$\nu = \frac{\mu}{\rho},$$

the density is found to be

$$\rho = \frac{\mu}{\nu} = \frac{4 \times 10^{-6}}{3.2 \times 10^{-6}} \text{ slugs/ft}^3 = 1.25 \text{ slugs/ft}^3 .$$

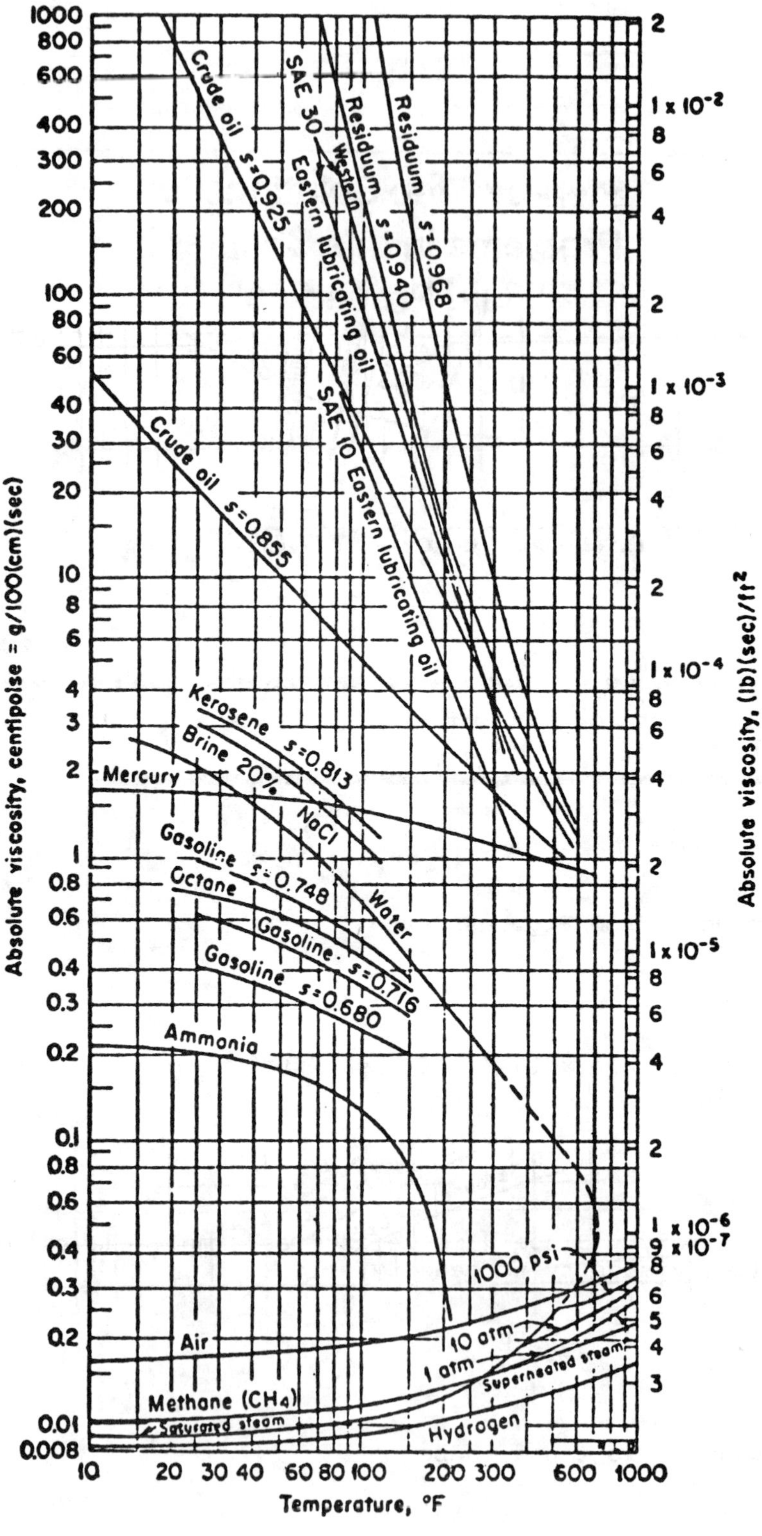
Absolute viscosity, centipoise = g/100(cm)(sec)
Absolute viscosity, (lb)(sec)/ft²
Temperature, °F
Crude oil s=0.925
SAE 30 Eastern lubricating oil
Western
Residuum s=0.940
Residuum s=0.968
SAE 10 Eastern lubricating oil
Crude oil s=0.855
Kerosene s=0.813
Brine 20% NaCl
Mercury
Gasoline s=0.748
Water
Octane
Gasoline s=0.716
Gasoline s=0.680
Ammonia
1000 psi
Air
10 atm
1 atm
Superheated steam
Methane (CH4)
Saturated steam
Hydrogen

Fig. 1

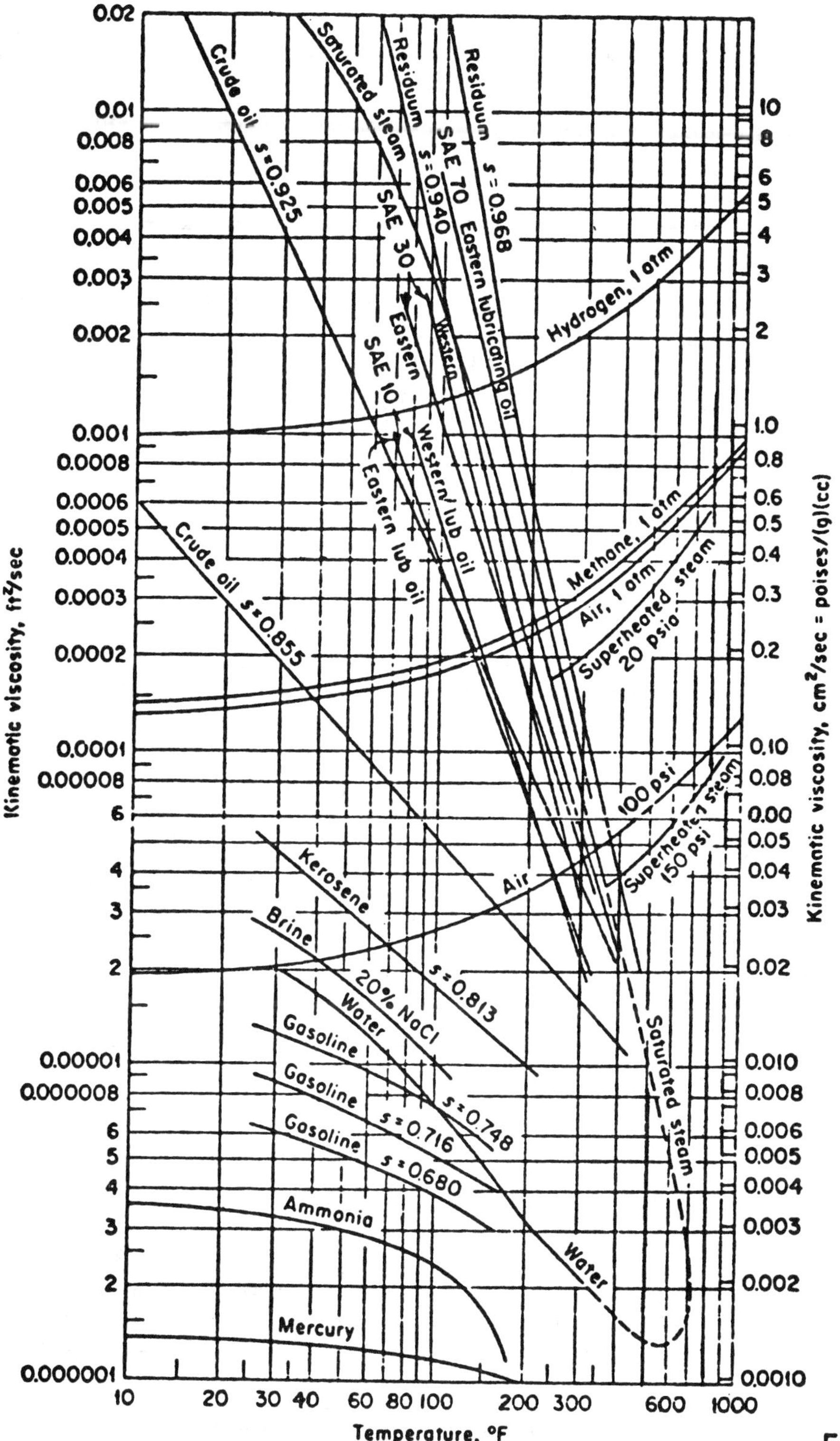
Kinematic viscosity, ft²/sec
Kinematic viscosity, cm²/sec = poises/(g)(cc)
Temperature, °F
Crude oil s=0.925
Saturated steam
Residuum s=0.940
Residuum s=0.968
SAE 70
Eastern lubricating oil
SAE 30
Eastern
Western
SAE 10
Eastern lub oil
Western lub oil
Hydrogen, 1 atm
Methane, 1 atm
Air, 1 atm
Superheated steam 20 psia
Crude oil s=0.855
100 psi
Superheated steam 150 psi
Air
Kerosene s=0.813
Brine 20% NaCl
Water
Gasoline s=0.748
Gasoline s=0.716
Gasoline s=0.680
Ammonia
Mercury
Saturated steam
Water

Fig. 2

• PROBLEM 1-2

A container weighs 3.22 lb force when empty. Filled with water at 60°F the mass of the container and its contents is 1.95 slugs. Find the weight of the water in the container and its volume in cubic feet. Assume density of water = 62.4 lb force/ft^3.

Solution: One slug is defined as 1 lb of force per 1 ft/sec^2 acceleration. By Newton's law of motion

$$1 \text{ lb force} = \frac{1 \text{ lb mass} \times 32.2 \text{ ft/sec}^2}{g_c}$$

Where

$$g_c = \frac{32.2 \text{ lb mass-ft}}{\text{lb force-sec}^2}$$

= a term serving the purpose of dimensional adjustment. From this it follows that one slug, by definition, equals 32.2 lb mass. That is;

$$1 \text{ slug} = 1 \frac{\text{lb force-sec}^2}{\text{ft}} = 32.2 \text{ lb mass}$$

Then 1.95 slug = 1.95 × 32.2 = 62.79 lb mass. This under the gravitational acceleration equals 62.79 lb force. The weight of the container equals 3.22 lb force, so that the weight of water in lb force = 62.79 - 3.22 = 59.57. Since the unit weight of water is 62.4 lb force/ft^3,

$$\text{the volume of the water in the container} = \frac{59.57 \text{ lbf}}{62.4 \text{ lbf/ft}^3}$$

$$= 0.955 \text{ ft}^3$$

• PROBLEM 1-3

The specific weight of water at ordinary pressure and temperature is 62.4 lb/ft^3 (9.81 kN/m^3). The specific gravity of mercury is 13.55. Compute the density of water and the specific weight and density of mercury.

Solution: Knowing that density and specific weight of a fluid are related as follows:

$$\rho = \frac{\gamma}{g} \quad \text{or} \quad \gamma = \rho g$$

and that specific gravity s of a liquid is the ratio of its density to that of pure water at a standard temperature, we can calculate:

$$\rho_{water} = \frac{\gamma_{water}}{g} = \frac{62.4 \text{ lb/ft}^3}{32.2 \text{ ft/s}^2} = 1.94 \text{ slugs/ft}^3 =$$

$$\frac{9.81 \text{ kN/m}^3}{9.81 \text{ m/s}^2} = 1.00 \text{ Mg/m}^3 = 1.00 \text{ g/cm}^3$$

$$\gamma_{mercury} = s_{mercury}\, \gamma_{water} = 13.55(62.4) = 846 \text{ lb/ft}^3$$

$$13.55(9.81) = 133 \text{ kN/m}^3$$

$$\rho_{mercury} = s_{mercury}\, \rho_{water} = 13.55(1.94) = 26.3 \text{ slugs/ft}^3$$

$$13.55(1.00) = 13.55 \text{ Mg/m}^3$$

• PROBLEM 1-4

A long cylindrical log of specific gravity $\frac{2}{3}$ will float in water with its axis horizontal. As its length gradually is reduced it will float in the same position. If its length is reduced so that it is only a fraction of its diameter, it will float with its axis vertical. What is its diameter-to-length ratio for which it will just float with its axis vertical?

Solution: Let the diameter be D and the length L. The required condition is when the metacentric height is zero in Eq. (1):

$$\overline{MG} = \frac{I}{\forall} - \overline{BG} = 0, \qquad (1)$$

where $I = AD^2/16$ and $\forall$ is the submerged volume and equals $\frac{2}{3}$ AL. The center of gravity is L/2 from the bottom of the short log and the center of buoyancy is L/3 from the bottom, since for a specific gravity of $\frac{2}{3}$ the log is two-thirds submerged. Then $\overline{BG} = L/6$ and setting Eq. (1) equal to zero gives

$$\frac{\frac{1}{16}AD^2}{\frac{2}{3}AL} - \frac{L}{6} = 0$$

from which $\frac{D}{L} = 1.33$.

VISCOSITY

• PROBLEM 1-5

A block weighing 100 lb and having an area of 2 ft^2 slides down an inclined plane as shown in Fig. 1, with a constant velocity. An oil gap between the block and the plane is 0.01 in. thick, the inclination of the plane is 30° to the horizontal, and the velocity of the block is 6 fps. Find the viscosity of the lubricating film.

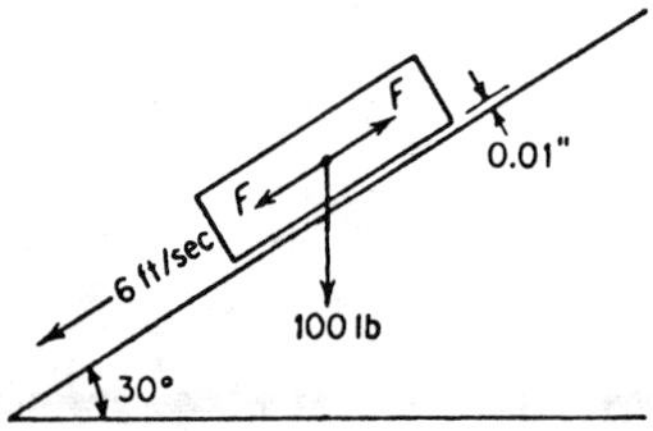

Fig. 1

Solution: Consider a fluid flowing over a smooth surface so that any fluid particle has motion parallel to the surface only (see Fig. 2). Such a flow is called laminar because the fluid moves in layers or "laminae."

Next to the surface, molecules of the fluid become embedded in the solid wall, and this layer of fluid is obviously at rest relative to the wall. Further from the wall the fluid has velocity v, which increases with distance from the wall y, giving a velocity distribution as shown in Fig. 2.

LAMINAR FLOW PROFILE CLOSE TO A BOUNDARY.

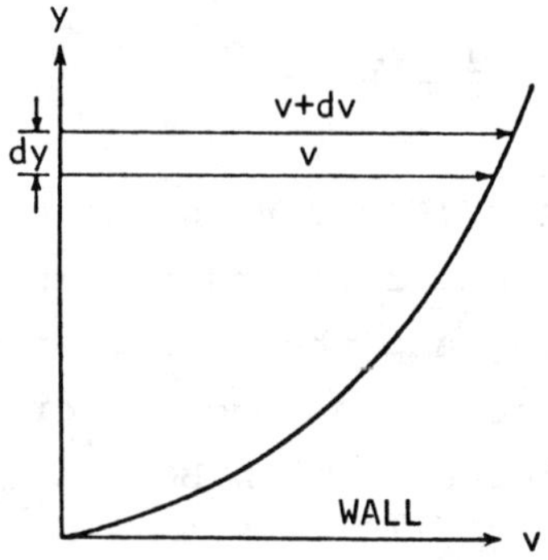

Fig. 2

Now consider two adjacent layers of fluid having velocities v and v + dv respectively and distance dy apart. The layer most remote from the wall has a velocity dv relative to the adjacent layer, and this causes a viscous or shearing stress to be present between the two layers. This stress is given the symbol τ (tau). The coefficient of viscosity μ (mu) is defined as the ratio

$$\frac{\text{Shearing stress}}{\text{Rate of shearing strain}}$$

and may be compared with the modulus of rigidity of a solid.

Rate of shearing strain is given by dv/dy, and hence

$$\mu = \frac{\tau}{dv/dy}$$

or

$$\tau = \mu \frac{dv}{dy}$$

μ is also called the absolute or dynamic viscosity and has units of lb-sec/ft^2 or slugs/ft-sec.

In this problem, the component of the weight acting down the plane is opposed by a viscous force exactly equal and opposite to it. Therefore

$$F = 100 \sin 30° = 50 \text{ lb}$$

Hence

$$\tau = \frac{F}{A} = \frac{50}{2} = 25 \text{ psf}$$

but

$$\tau = \mu \frac{dv}{dy}$$

Therefore

$$\int_0^{\frac{0.01}{12}} 25\, dy = \mu \int_0^6 dv$$

i.e.,

$$6\mu = \frac{25 \times 0.01}{12}$$

$$\mu = 0.00347 \text{ lb-sec/ft}^2$$

• PROBLEM 1-6

A shaft 15.00 cm in diameter rotates at 1800 r/min inside a bearing 15.05 cm in diameter and 30.0 cm long. The uniform space between them is filled with an oil of viscosity $\mu = 0.018$ kg/m s. What power is required to overcome viscous resistance in the bearing? Refer to Figure 1.

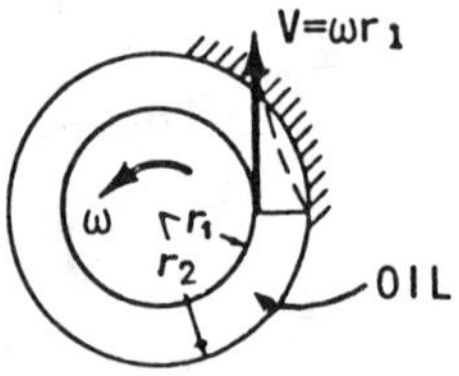

ROTATING SHAFT Fig. 1

Solution: The radial clearance is much less than the shaft diameter, and thus the velocity gradient may be considered constant at $V/\Delta r$. Thus,

power = total shear force times peripheral shaft speed

$$= \tau(A_{sheared})(V),$$

where

$$V = \omega r = 60\pi r = 4.5\pi \text{ m/s}$$

$$P = \mu(V/\Delta r)(\pi DL)(V)$$

$$= (0.018)(4.5\pi/0.00025)(\pi)(0.150)(0.300)(4.5\pi)$$

$$= 2034 \text{ W} = 2.034 \text{ kW}$$

The space between two very long parallel plates separated by a distance h is filled with fluid of constant viscosity μ. The upper plate moves steadily at a velocity V_0 relative to the lower one and the pressure is everywhere constant. Find the velocity distribution between the plates and the shear stress distribution in the fluid. This problem is to be solved by selecting an element of the fluid of some arbitrary length dx and height y above the stationary plate and considering the forces on this free body.

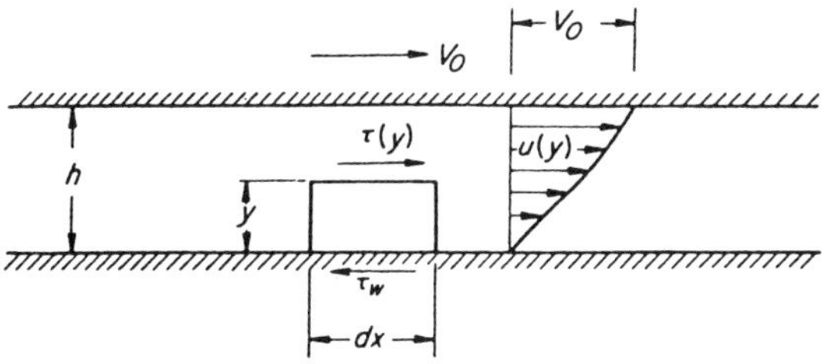

Solution: The shear stress exerted by the fluid above the element is τ while the restraining shear stress exerted by the wall on the element is τ_w in the opposite direction.

The element is not accelerating nor is there any net force due to pressure since it is presumed to be constant. Thus from statics

$$\tau = \tau_w = \text{constant}$$

But

$$\tau = \mu \frac{du}{dy} = \text{const.}$$

Hence

$$u = \frac{\tau}{\mu} y + \text{const.}$$

The no-slip condition requires that $u = 0$ at $y = 0$, and $u = V_0$ at $y = h$. Hence the integration constant is zero and

$$\tau = \frac{\mu V_0}{h} .$$

The velocity consequently is given by

$$u = V_0 \frac{y}{h} .$$

This kind of flow between two plates is often called Couette flow.

• PROBLEM 1-8

SAE 30 oil at 20°C undergoes steady shear between a fixed lower plate and an upper plate moving at speed V. The clearance between plates is h. (a) Show that the linear velocity profile in figure (1) will result if the fluid does not slip at either plate; (b) compute the shear in the oil in pascals if V = 3 m/s and h = 2 cm.

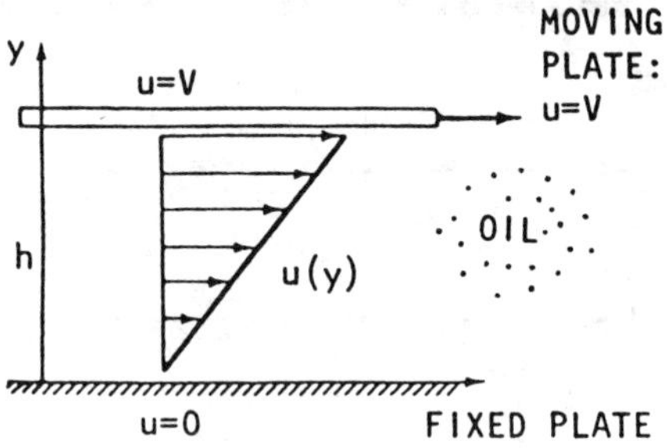

Fig. 1

Solution: (a) The shear stress τ is constant through the fluid for this geometry and motion. Assuming u = u(y) only,

$$\frac{du}{dy} = \frac{\tau}{\mu} = \text{const}$$

or

$$u = a + by \tag{1}$$

Constants a and b are evaluated from the no-slip condition at the upper and lower walls:

$$u = \begin{cases} 0 = a + b(0) & \text{at } y = 0 \\ V = a + b(h) & \text{at } y = h \end{cases}$$

Hence a = 0 and b = V/h. The velocity profile between the plates is

$$u = \frac{Vy}{h} \qquad \text{Ans. (a)} \tag{2}$$

as indicated in Fig. 2.

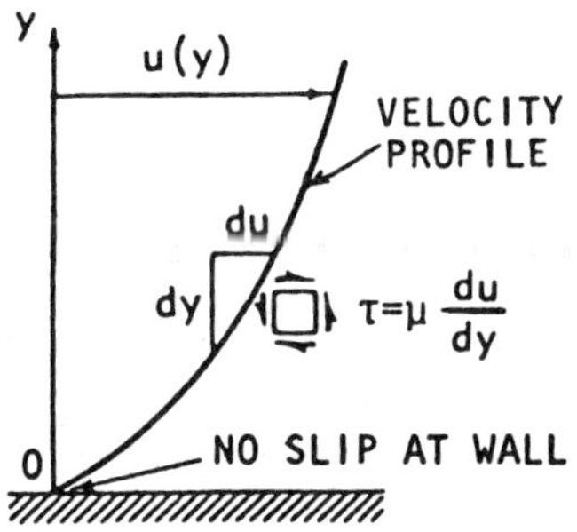

Fig. 2

GENERALIZED CASE WHERE $\frac{du}{dy} \neq$ const.

(b) From Eqs. (1) and (2) the shear stress is

$$\tau = \mu \frac{du}{dy} = \frac{\mu V}{h} \tag{3}$$

TABLE 1

VISCOSITY AND KINEMATIC VISCOSITY OF EIGHT FLUIDS AT 1 atm AND 20 C

Fluid	μ, kg/(m · s)†	*Ratio* $\mu/\mu(H_2)$	ρ, kg/m³	ν, m²/s†	*Ratio* $\nu/\nu(Hg)$
1. Hydrogen	8.9×10^{-6}	1.0	0.084	1.06×10^{-4}	910
2. Air	1.8×10^{-5}	2.1	1.20	1.51×10^{-5}	130
3. Gasoline	2.9×10^{-4}	33	680	4.27×10^{-7}	3.7
4. Water	1.0×10^{-3}	114	999	1.01×10^{-6}	8.7
5. Ethyl alcohol	1.2×10^{-3}	135	789	1.51×10^{-6}	13
6. Mercury	1.5×10^{-3}	170	13,540	1.16×10^{-7}	1.0
7. SAE 30 oil	0.26	29,700	933	2.79×10^{-4}	2,430
8. Glycerin	1.5	168,000	1,263	1.19×10^{-3}	10,200

† 1 kg/(m · s) = 0.0209 slug/(ft · s); 1 m²/s = 10.76 ft²/s.

From Table (1) for SAE 30 oil $\mu = 0.26$ kg/(m · s). Then for the given values of V and h the stress is

$$\tau = \frac{[0.26\ \text{kg/(m}\cdot\text{s)}](3\ \text{m/s})}{0.02\ \text{m}} = 39\ \text{kg/(m}\cdot\text{s}^2)$$

$$= 39\ \text{N/m}^2 = 39\ \text{Pa} \qquad \text{Ans. (b)}$$

Although oil is very viscous, this is a modest shear stress, about 3400 times less than atmospheric pressure. Viscous stresses in gases and thin liquids are even smaller.

• PROBLEM 1-9

The space between two parallel plates 1.5 cm apart is filled with an oil of viscosity $\mu = 0.050$ kg/m s. A thin 30- × 60-cm rectangular plate is pulled through the oil 0.50 cm from one plate and 1.00 cm from the other. What force is needed to pull the plate at 0.40 m/s?

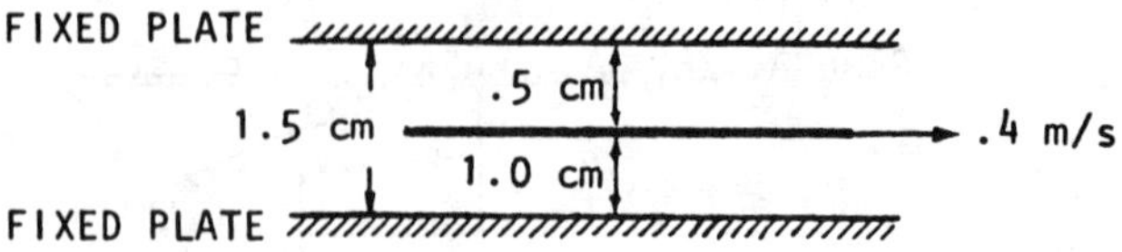

Solution: The total force overcomes the viscous shear over both the upper and the lower surface of the moving plate as indicated in Figure. Thus,

$$F_{total} = F_{upper} + F_{lower}$$

$$= (\text{upper shear stress})(\text{area}) + (\text{lower shear stress})(\text{area})$$

$$= \mu(V/h_{upper})(A) + \mu(V/h_{lower})(A)$$

$$= (0.050)(0.40/0.005)(0.180) + (0.050)(0.40/0.010)(0.180)$$

$$= 0.72 + 0.36$$

$$= 1.08 \text{ N}$$

• PROBLEM 1-10

A common type of viscosimeter for liquids consists of a small reservoir with a very slender outlet tube (all enclosed in a constant-temperature bath), the rate of outflow being determined by timing the fall in surface level. What dynamic viscosity would be represented by a drop in surface level of 0.2 in./min, if the viscosimeter has the dimensions shown and the specific gravity of the liquid is 0.8?

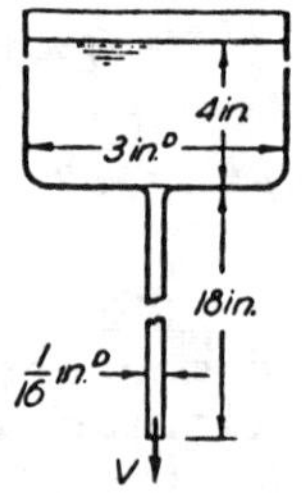

Fig. 1

Solution: From the equation of continuity

$$V = \frac{V_0 A_0}{A} = \frac{\dfrac{0.2}{12 \times 60} \times \dfrac{\pi}{4 \times 16}}{\dfrac{\pi}{4} \times \dfrac{1}{(16 \times 12)^2}} = 0.64 \text{ fps}$$

This problem involves flow through a tube. The axially symmetric counterpart of two-dimensional flow between parallel boundaries is flow through tubes of circular cross section.

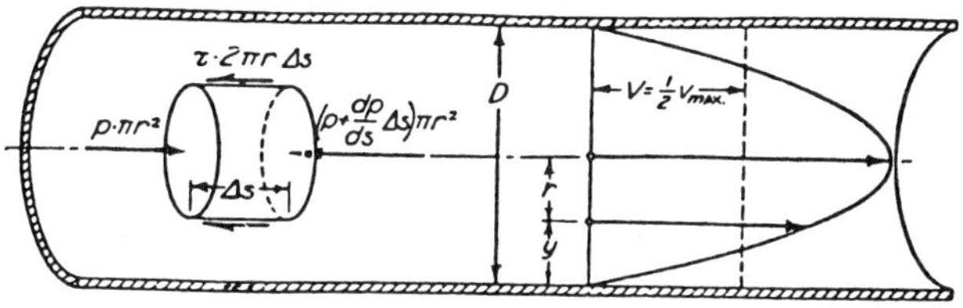

Fig. 2

VISCOUS FLOW THROUGH A UNIFORM TUBE.

With reference to Fig. 2, the corresponding equation of equilibrium between pressure and viscous shear must be based upon an element having the form of a cylinder of fluid of radius r, symmetrical about the axis of the tube. Algebraic combination of the several forces in the form

$$-\frac{dp}{ds} \Delta s \; \pi r^2 = \tau \; 2\pi r \; \Delta s$$

reduces to

$$\tau = -\frac{r}{2}\frac{dp}{ds}$$

Upon introduction of the Newtonian equation $\tau = \mu \, dv/dy = -\mu \, dv/dr$ and rearrangement of terms,

$$dv = \frac{1}{\mu}\frac{dp}{ds}\frac{r}{2} dr$$

which is readily integrated to yield

$$v = \frac{1}{\mu}\frac{dp}{ds}\frac{r^2}{4} + C$$

The constant of integration is evaluated from the condition that v = 0 when r = D/2:

$$v = -\frac{1}{\mu}\frac{dp}{ds}\left(\frac{D^2}{16} - \frac{r^2}{4}\right)$$

Evidently, the velocity distribution is parabolic in form.

Since the volume of a paraboloid of revolution is one-half that of the circumscribing cylinder, the mean velocity is now one-half the maximum, from which it follows that

$$V = \frac{1}{2} v_{max} = -\frac{D^2}{32\mu}\frac{dp}{ds}$$

After solution for dp/ds and integration in the longitudinal direction between two normal sections, the distance L apart,

$$p_1 - p_2 = \frac{32\mu VL}{D^2} \qquad (1)$$

Pressure drops in the downstream direction in proportion to μ and V and in inverse proportion to D^2. Equation (1) is commonly known as the equation of Poiseuille.

The corresponding equation for the case of liquid flow in an inclined tube is obtained by substituting for p the quantity γh:

$$h_1 - h_2 = \frac{32\mu VL}{\gamma D^2} \qquad (2)$$

Then, from Eq. (2), the velocity head being negligible,

$$h_1 - h_2 = \frac{4 + 18}{12} = \frac{32\mu VL}{\gamma D^2} = \frac{32\mu \times 0.64 \times \frac{18}{12}}{62.4 \times 0.8 \times \frac{1}{(16 \times 12)^2}}$$

$$\mu = 8.05 \times 10^{-5} \text{ lb-sec/ft}^2$$

VAPOR PRESSURE

• PROBLEM 1-11

A vertical cylinder 300 mm in diameter is fitted (at the top) with a tight but frictionless piston and is completely filled with water at 70°C. The outside of the piston is exposed to an atmospheric pressure of 100 kPa. Calculate the minimum force applied to the piston which will cause the water to boil.

Solution: The force must be applied slowly (to avoid acceleration) in a direction to withdraw the piston from the cylinder. Since the water cannot expand, a space filled with water vapor will be created beneath the piston, whereupon the water will boil. Vapor Pressure of water at 70°C is found in a table of Physical Properties of Water. The pressure on the inside of the piston will then be 31.2 kPa and the force on the piston $(100-31.2)\pi(0.3)^2/4 = 4.86$ kN. ($P = F/A$ by definition.)

• PROBLEM 1-12

At approximately what temperature will water boil if the elevation is 10,000 ft?

Solution: From Table 1, the pressure of the standard atmosphere at 10,000-ft elevation is 10.11 psia. From Table 2, the saturation pressure of water is 10.11 psia at about 193°F. Hence the water will boil at 193°F; this explains why it takes longer to cook at high elevations.

TABLE 1

Altitude, ft	Temp, °F	Pressure, psia
0	59.0	14.70
5,000	41.2	12.24
10,000	23.4	10.11
15,000	5.6	8.30
20,000	−12.3	6.76
25,000	−30.1	5.46
30,000	−47.8	4.37
35,000	−65.6	3.47
40,000	−69.7	2.73
45,000	−69.7	2.15
50,000	−69.7	1.69
60,000	−69.7	1.05
70,000	−69.7	0.65
80,000	−69.7	0.40
90,000	−57.2	0.25
100,000	−40.9	0.16

TABLE 2

Temp, °F	Specific weight γ, lb/ft³	Density ρ, slugs/ft³	Viscosity $\mu \times 10^5$, lb·s/ft²	Kinematic viscosity $\nu \times 10^5$, ft²/s	Surface tension $\sigma \times 10^2$ lb/ft	Vapor pressure p_v, psia
32	62.42	1.940	3.746	1.931	0.518	0.09
40	62.43	1.940	3.229	1.664	0.514	0.12
50	62.41	1.940	2.735	1.410	0.509	0.18
60	62.37	1.938	2.359	1.217	0.504	0.26
70	62.30	1.936	2.050	1.059	0.500	0.36
80	62.22	1.934	1.799	0.930	0.492	0.51
90	62.11	1.931	1.595	0.826	0.486	0.70
100	62.00	1.927	1.424	0.739	0.480	0.95
110	61.86	1.923	1.284	0.667	0.473	1.27
120	61.71	1.918	1.168	0.609	0.465	1.69
130	61.55	1.913	1.069	0.558	0.460	2.22
140	61.38	1.908	0.981	0.514	0.454	2.89
150	61.20	1.902	0.905	0.476	0.447	3.72
160	61.00	1.896	0.838	0.442	0.441	4.74
170	60.80	1.890	0.780	0.413	0.433	5.99
180	60.58	1.883	0.726	0.385	0.426	7.51
190	60.36	1.876	0.678	0.362	0.419	9.34
200	60.12	1.868	0.637	0.341	0.412	11.52
212	59.83	1.860	0.593	0.319	0.404	14.70

• PROBLEM 1-13

Determine the maximum theoretical height to which water at 150°F may be raised by a vacuum at sea level.

Solution: The height of a column of liquid that produces the pressure is termed the pressure head (h).

$$h = \frac{p}{\gamma}$$

The specific weight of water at 150°F is 61.2 lbs/ft^3. Under a perfect vacuum, the water will rise

$$h = \frac{14.7(144)}{61.2} = 34.59 \text{ ft.}$$

The pressure head of water at 150°F is

$$h = \frac{3.72(144)}{61.2} = 8.56 \text{ ft.}$$

Therefore the maximum height that the water will rise is

$$h_{max} = 34.59 - 8.56 = 26.03 \text{ ft.}$$

COMPRESSIBILITY

• PROBLEM 1-14

Find an expression for the coefficient of compressibility at constant temperature for an ideal gas.

Solution: The coefficient of compressibility, α, is defined by

$$\alpha = \frac{1}{\forall}\left(\frac{\partial \forall}{\partial p}\right)_T$$

For an ideal gas

$$p = \rho RT = \frac{mRT}{\forall} \quad \text{or} \quad \forall = \frac{mRT}{p} .$$

Then

$$\left(\frac{\partial V}{\partial p}\right)_T = -\frac{mRT}{p^2}$$

$$\alpha = \frac{1}{V}\left(\frac{mRT}{p^2}\right) = \frac{1}{p}$$

• PROBLEM 1-15

> One cubic foot of oxygen at 100°F and 15 psia is compressed adiabatically to 0.4 ft. What then, are the temperature and pressure of the gas. If the process had been isothermal, what would the temperature and pressure have been?

Solution: $R \approx 1540/32 = 48.1$ ft./°F

Using the perfect gas equation

$p\upsilon = RT$

yields

$$(15 \times 144)\upsilon_1, = 48.1\ (460 + 100)$$

$$\upsilon_1 = 12.5\ \text{ft}^3/\text{lb} \qquad \gamma_1 = \frac{1}{\upsilon_1} = 0.08\ \text{pcf}$$

To determine the pressure

$$P_1 \upsilon_1^K = P_2 \upsilon_2^K$$

where $K = 1.4$

$$(15 \times 144)\ (12.5)^{1.4} = (P_2 \times 144)\ \left(\frac{12.5}{1/0.4}\right)^{1.4}$$

$$P_2 = 54.0\ \text{psia}$$

$$(54 \times 144)\ \frac{12.5}{1/0.4} = 48.1\ (460 + T_2)$$

$$T_2 = 350^0\text{F}$$

If the process had been isothermal ($T_2 = 100^0$F), then

$$P\upsilon = \text{constant}$$

$$P_2 = \frac{15}{0.4} = 37.5\ \text{psia}$$

Taking the specific weight of sea-water as 64 lb. per ft.^{3}, find the pressure at a depth of 10,000 ft. in the sea, (a) neglecting compressibility, and (b) taking it into account.

Solution: (a) Pressure-intensity = 64 × 10,000 = 640,000 lb. per ft.^{2} or

$$\frac{640000}{144} = 4444 \text{ lb. per in.}^2 .$$

(b) If an increase of pressure-intensity δp reduces the volume V of a liquid to $V - \delta V$, the volumetric strain is

$$-\frac{\delta V}{V}$$

and the "Bulk Modulus"

$$K = -\frac{\delta p}{\delta V/V} = \frac{V\delta p}{\delta V} .$$

For water K is usually taken as 300,000 lb. per in.^{2}, but it actually increases with both the pressure and the temperature as shown in Table 1:

Table 1

Pressure (lb. per in.^{2})	0°C	10°C	20°C
0	283,000	301,000	319,000
2240	292,000	311,000	328,000
4480	300,000	321,000	335,000

while for sea-water K is about 9 per cent greater. These values are so high that for all ordinary purposes water may be taken as incompressible, so that $p = \gamma h$.

As the pressure-intensity is about 4480 lb. per in.^{2} we can take the figure for 2240 lb. per in.^{2} from Table 1 for 0°C as the average value of K, viz.

$$K = 1.09 \times 292{,}000 = 318{,}300 \text{ lb. per in.}^2 .$$

(Sea water is 9% greater.)

To find γ for 4444 lb. per in.^{2}, if V = volume of 1 lb.,

$$V = V_0 + \delta V = V_0\left(1 - \frac{\delta p}{K}\right);$$

$$\therefore \quad \gamma = \frac{1}{V} = \frac{\gamma_0}{1 - \frac{\delta p}{K}} = \frac{64}{1 - \frac{4444}{318300}} = 64.91 \text{ lb per ft.}^3$$

The average value of γ must be about 64.46.

$\therefore$ pressure-intensity = 644,600 lb. per ft.^{3} ,

or about 4476 lb. per in.^{2} .

SURFACE TENSION

• PROBLEM 1-17

> Of what diameter must a droplet of water (20°C) be to have the pressure within it 1.0 kPa greater than that outside?

Solution: $\sigma_{water} = 0.0728$ N/m.

Basic Relation:

$$p_i - p_0 = \frac{1}{R_i} + \frac{1}{R_2} \quad \sigma$$

where

$$R_1 = R_2 = R$$

Therefore

$$p_i - p_0 = \sigma \frac{2}{R}$$

$$p_i - p_0 = 1.0 \times 10^3 = 0.0728\left(\frac{2}{R}\right)$$

$$R = 0.00015\text{m},$$

$$d = 0.3 \text{ mm}.$$

• PROBLEM 1-18

To what height above the reservoir level will water rise in a glass tube, such as that shown in Fig. 1 if the inside diameter of the tube is $\frac{1}{16}$ in. (1.6 mm)?

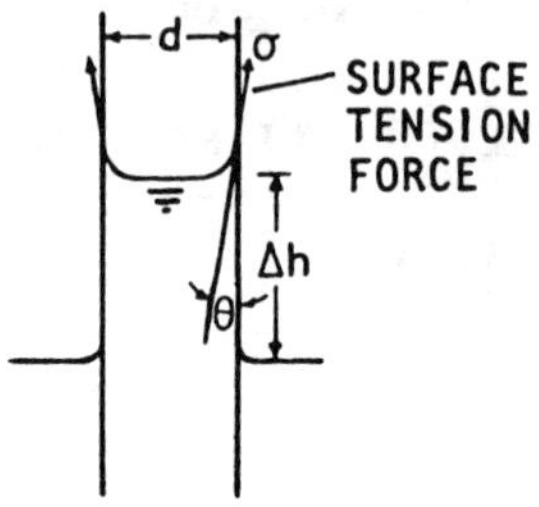

CAPILLARY ACTION IN A TUBE Fig. 1

Solution: Taking the summation of forces in the vertical direction of the water in the tube that has risen above the reservoir level, yields

$$\sigma \pi d - \gamma(\Delta h)\left(\frac{\pi d^2}{4}\right) = 0$$

Surface tension, σ, for a water-air surface is 0.005 lbf/ft (.073 N/m) at room temp.

or

$$\Delta h = \frac{4\sigma}{\gamma d}$$

$$= \frac{4 \times 0.005 \text{ lbf/ft}}{62.4 \text{ lbf/ft}^3 \times \frac{1}{16} \text{ in.} \times \frac{1}{12} \text{ ft/in.}}$$

$$= 0.062 \text{ ft}$$

or

$$\Delta h = 0.74 \text{ in. } (1.88 \text{ cm})$$

• PROBLEM 1-19

Air is introduced through a nozzle into a tank of water to form a stream of bubbles. If the bubbles are intended to have a diameter of 2 mm, calculate by how much the pressure of the air at the nozzle must exceed that of the surrounding water. Assume that $\sigma = 72{\cdot}7 \times 10^{-3}$ N m^{-1}.

Solution: The surface tension σ is measured as the force acting across unit length of a line drawn in a liquid surface.

The effect of surface tension is to reduce the surface of a free body of liquid to a minimum, since to expand the surface area,molecules have to be brought to the surface from the bulk of the liquid against the unbalanced attraction pulling the surface molecules inwards. For this reason, drops of liquid tend to take a spherical shape in order to minimize surface area. For such a small droplet, surface tension will cause an increase of internal pressure, p in order to balance the surface force.

Considering the forces acting on a diametral plane through a spherical drop of radius r,

$$\text{Force due to internal pressure} = p \times \pi r^2$$

Force due to surface tension round the perimeter =

$$2\pi r \times \sigma .$$

For equilibrium,

$$p\pi r^2 = 2\pi r\sigma$$

or

$$p = 2\sigma/r .$$

In this problem, this also gives the excess pressure

$$p = 2\sigma/r .$$

Putting $r = 1\,\text{mm} = 10^{-3}\,\text{m}$, $\sigma = 72.7 \times 10^{-3}\ \text{N m}^{-1}$:

Excess pressure,

$$p = 2 \times 72{\cdot}7 \times 10^{-3}/1 \times 10^{-3}$$

$$= 143.4\ \text{N m}^{-1} .$$

• PROBLEM 1-20

What diameter circular tube would be required to raise water at 70°F to a height of 100 ft?

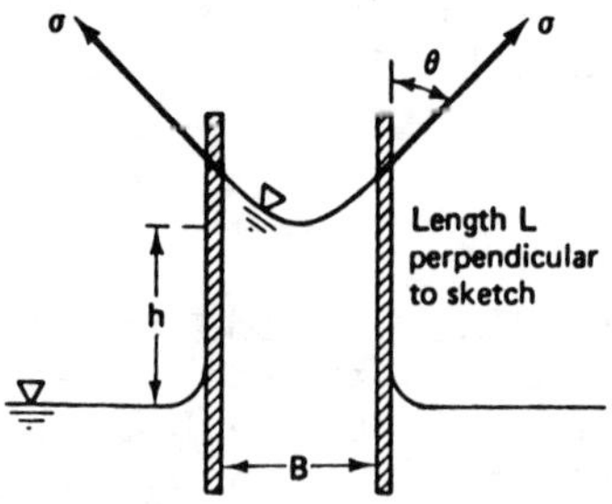

Solution: Consider the rise of a liquid between the two closely spaced flat plates in the figure. The distance h, referred to as the height of capillary rise, results from the equilibrium between the upward force due to the surface tension and the weight of water which has been lifted,

$$\sigma \cos \theta (2L) = hBl\gamma$$

This becomes

$$h = \frac{2\sigma \cos \theta}{\gamma B}$$

since the surface length L perpendicular to the figure drops out. Capillary rise in a glass tube of radius r results in the similar equation,

$$h = \frac{2\sigma \cos \theta}{\gamma r} \tag{1}$$

The angle θ is usually taken as zero for water.

TABLE

Temperature (°F)	*Specific Weight,* γ (lb/ft^3)	*Density,* ρ ($slugs/ft^3$)	*Dynamic Viscosity,* $\mu \times 10^5$ ($lb\text{-}s/ft^2$)	*Kinematic Viscosity,* $\nu \times 10^5$ (ft^2/s)	*Surface Tension,* $\sigma \times 10^2$ (lb/ft)	*Vapor Pressure,* p_v (psia)	*Modulus of Compressibility,* $E \times 10^{-5}$ (psi)
32	62.42	1.940	3.746	1.931	0.518	0.087	2.93
40	62.43	1.940	3.229	1.664	0.514	0.12	2.94
50	62.41	1.940	2.735	1.410	0.509	0.18	3.05
60	62.37	1.938	2.359	1.217	0.504	0.26	3.11
70	62.30	1.936	2.050	1.059	0.500	0.36	3.20
80	62.22	1.934	1.799	0.930	0.492	0.51	3.22
90	62.11	1.931	1.595	0.826	0.486	0.70	3.23
100	62.00	1.927	1.424	0.739	0.480	0.96	3.27
110	61.86	1.923	1.284	0.667	0.473	1.28	3.31
120	61.71	1.918	1.168	0.609	0.465	1.69	3.33
130	61.55	1.913	1.069	0.558	0.460	2.22	3.34
140	61.38	1.908	0.981	0.514	0.454	2.89	3.30
150	61.20	1.902	0.905	0.476	0.447	3.72	3.28
160	61.00	1.896	0.838	0.442	0.441	4.75	3.26
170	60.80	1.890	0.780	0.413	0.433	5.99	3.22
180	60.58	1.883	0.726	0.385	0.426	7.51	3.18
190	60.36	1.876	0.678	0.362	0.419	9.34	3.13
200	60.12	1.868	0.637	0.341	0.412	11.52	3.08
212	59.83	1.860	0.593	0.319	0.404	14.69	3.00

Using Eq. 1 with $\theta = 0°$ and $\sigma = 0.005$ lb/ft (from the table) gives

$$r = \frac{2\sigma \cos \theta}{\gamma h} = \frac{(2)(0.005)(1)}{(62.3)(100)} = 1.61 \times 10^{-6} \text{ ft}$$

or

$$d = 2r = 3.21 \times 10^{-6} \text{ ft.}$$

BULK MODULUS OF ELASTICITY

• PROBLEM 1-21

What pressure must be exerted on 1 cm^3 of water at 32°F and 15 psi to change the volume to 0.99 cm^3?

Solution: Bulk modulus of elasticity, E, is used to measure the compressibility of fluids.

$$E = -\left(\frac{\forall dp}{d\forall}\right)_{\textcircled{T}}$$

where for an increase of pressure dp, i.e., $dp > 0$, the volume $\forall$ decreases which results in the minus sign. The table below gives a few values of the modulus of elasticity E of water, in psi.

Pressure, psi	Temperature, °F				
	32	68	120	200	300
15	292,000	320,000	332,000	308,000	....
1,500	300,000	330,000	342,000	319,000	248,000
4,500	317,000	348,000	362,000	338,000	271,000
15,000	380,000	410,000	426,000	405,000	350,000

Modulus of Elasticity of Water

Then

$$dp = p_f - p_i = p_f - 0$$

$$d\forall = V_f - V_i = -0.01 \text{ cm}^3$$

$$V = V_i = 1 \text{ cm}^3$$

$$E = 292{,}000 \text{ psi}$$

Assuming an isothermal process, the equation for E becomes

$$\frac{292{,}000\ (-0.01)}{1} = -\ p_f$$

therefore,

$$p_f = 2{,}920 \text{ psi}$$

• PROBLEM 1-22

A liquid compressed in a cylinder has a volume of 1 liter (ℓ) (1000 cm^3) at 1 MN/m^2 and a volume of 995 cm^3 at 2MN/m^2. What is the bulk modulus of elasticity?

Solution: For most purposes a liquid may be considered an incompressible, but for situation involving either sudden or great changes in pressure, its compressibility becomes important, and is expressed by its bulk modulus of elasticity. If the pressure of a unit volume of liquid is increased by dp, it will cause a volume decrease -dV; the ratio -dp/dV is the bulk modulus of elasticity E. For any volume of liquid,

$$E = \frac{-dp}{dV/V}$$

where E is expressed in units of pressure.

Now, substituting the values in the equation yields

$$E = \frac{-\Delta p}{\Delta V/V} = \frac{-(2 - 1\ MN/m^2)}{(995 - 1000)/1000} = 200\ MPa$$

ACCELERATION OF A PARTICLE

• PROBLEM 1-23

A three-dimensional velocity field is given as:

$$V = (2x^2 - y)\hat{i} + (3xy + x^2)\hat{j} + (zyt^2)\hat{k}$$

Determine A) if the velocity field is steady, B) and obtain an expression for the acceleration of the particle.

Solution: A) Since the velocity vector is a function of time, the field is unsteady .

B) Since $\frac{DV}{Dt} = Vx \frac{\partial V}{\partial x} + Vy \frac{\partial V}{\partial y} + Vz \frac{\partial V}{\partial z} + \frac{\partial V}{\partial t}$ (1)

where

$$Vx \frac{\partial V}{\partial x} = (2x^2 - y)[4x\hat{i} + (3y + 2x)\hat{j}]$$

$$Vy \frac{\partial V}{\partial y} = (3xy + x^2)[-\hat{i} + 3x\hat{j} + t^2\hat{k}]$$

$$Vz \frac{\partial V}{\partial z} = (zyt^2)[(yt^2)\hat{k}]$$

and

$$\frac{\partial V}{\partial t} = (2zt)\hat{k}$$

Substituting back into equation (1),

$$\frac{DV}{Dt} = (2x^2 - y)[4x\hat{i} + (3y + 2x)\hat{j}] + (3xy + x^2)[-\hat{i} + 3xj + t^2\hat{k}] + (zyt^2)[(yt^2)\hat{k}] + (2zt)\hat{k}$$

$$\frac{DV}{Dt} = (8x^3 - 7xy - x^2)\hat{i} + (7x^3 - 2xy + 15x^2y - 3y^2)\hat{j} + (zy^2t^4 + x^2t^2 + 3xyt^2 + 2zt)\hat{k}$$

CHAPTER 2

FLUID STATICS

Basic Attacks and Strategies for Solving Problems in this Chapter. See pages 26 to 119 for step-by-step solutions to problems.

The fundamental equation that describes the pressure field in a fluid at rest is

$$\nabla p = \rho g, \tag{1}$$

where g is the acceleration vector due to gravity. If we adopt a coordinate system where z is "up" vertically, g acts downward (opposite to the direction of increasing z); Equation (1) reduces to

$$\frac{dp}{dz} = -\rho g \tag{2}$$

when applied to gases with large height differences, such as the atmosphere density in a variable. For liquid applications ρ can be assumed to be constant with negligible error. For constant ρ and g,

$$p_2 - p_1 = -\rho g(z_2 - z_1), \tag{3}$$

where 1 and 2 represent any two positions in the same fluid. At a liquid surface, the pressure must equal the pressure of the air (or other fluid) immediately above the surface. For liquids exposed to atmospheric pressure, p_{atm}, the local pressure at some depth h (measured from the surface) is thus simply

$$p = \rho g h + p_{atm}, \tag{4}$$

where ρ is the density of the liquid. Often, it is more convenient to use gage (sometimes spelled "gauge") pressure, defined as the absolute pressure minus p_{atm}. In the liquid discussed above, the gage pressure would equal $\rho g h$.

Equation (3), the basic hydrostatic equation, can be applied to columns of multiple fluids as well; since density changes abruptly at a fluid/fluid interface, Equation (3) must be applied in a piecewise fashion. In addition, Equation (3) can be integrated to find the total force and the center of pressure on solid surfaces immersed in the liquid(s).

For a plane surface of arbitrary shape inclined at an angle θ to the horizontal, let x and y be coordinates tangent to the plate, with their origin fixed at the center of gravity. Then the integration of Equation (3) yields the location of the center of pressure:

$$x_{CP} = -\rho g \sin\theta \frac{I_{xy}}{p_{CG}A},$$

$$y_{CP} = -\rho g \sin\theta \frac{I_{XX}}{p_{CG}A}, \tag{5}$$

where A is the surface area of the plate,

p_{CG} is the pressure at the location of the plate's center of gravity,

I_{XX} is the area moment of inertia about the x axis, and

I_{xy} is the product of inertia of the plate, computed in the plane of the plate.

These concepts can be extended to curved as well as plane surfaces.

When a container of fluid undergoes constant uniform linear acceleration **a** (in any direction), Equation (1) may still be applied, by substituting **g** – **a** for the vector **g**, i.e.,

$$\nabla p = \rho(\mathbf{g} - \mathbf{a}). \tag{6}$$

In other words, all the hydrostatic equations above remain valid, but with a different constant of gravity (**g** – **a** instead of **g**). The surface of an accelerating container will align itself perpendicularly to the vector **g** – **a**. The pressure increases linearly with a coordinate along the direction of **g** – **a**, rather than simply along the direction of **g** in hydrostatics.

When a container of liquid rotates at a constant angular velocity about the vertical axis, Equation (6) is still valid, with

$$\mathbf{a} = \text{centripetal acceleration} = -r\omega^2\mathbf{i}_r,$$

where r is the radial distance from the axis of rotation,

ω is the magnitude of the angular velocity, and

$\mathbf{i}_r$ is the unit coordinate in the radial direction.

Integration yields

$$p = \text{constant} - \rho g z + {}^1\!/_2\rho r^2\omega^2. \tag{7}$$

It turns out that the surface of a spinning container of liquid is shaped like a paraboloid since

$$p = \text{constant} = p_{\text{atm}}$$

at the surface, and hence

$$z_{surface} = \text{constant} + \frac{\omega^2 r^2}{2g}. \tag{8}$$

Isobars, i.e., lines of constant pressure, are everywhere parallel to this surface, with p increasing into the liquid.

Step-by-Step Solutions to Problems in this Chapter, "Fluid Statics"

HYDROSTATIC FORCES AND PRESSURE VARIATION

• PROBLEM 2-1

We can get a reasonable idea of the variation of pressure with altitude in the earth's atmosphere if we assume that the density ρ is proportional to the pressure. This would be very nearly true if the temperature of the air remained the same at all altitudes. Using this assumption, and also assuming that the variation of g with altitude is negligible, find the pressure p at an altitude y above sea level.

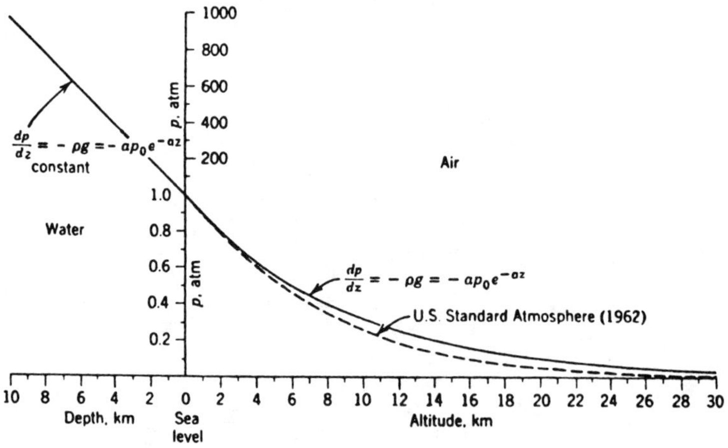

Variation of pressure with altitude in air and with depth in water, assuming p = 1 atm (exactly) at sea level. Note that the pressure scales are different for altitude and depth. The solid line for air is calculated on the assumption that the air has a constant temperature and that g does not change with altitude. The dashed line (the U.S. Standard Atmosphere-1962) is a more refined calculation in which these assumptions are not made.

Solution: We have from hydrostatics

$$\frac{dp}{dz} = -\rho g$$

Since ρ is proportional to p, we have

$$\frac{\rho}{\rho_0} = \frac{p}{p_0} ,$$

where ρ_0 and p_0 are the known values of density and pressure at sea level. Then,

$$\frac{dp}{dz} = - g\rho_0 \frac{p}{p_0} ,$$

so that

$$\frac{dp}{p} = - \frac{g\rho_0}{p_0} dz$$

Integrating this from the value p_0 at the point Z = 0 (sea level) to the value p at the point Z (above sea level), we obtain

$$\ln \frac{p}{p_0} = - \frac{g\rho_0}{p_0} Z .$$

or

$$p = p_0 e^{-g(\rho_0/p_0)Z} .$$

However,

$$g = 9.80 \text{ m/s}^2, \ \rho_0 = 1.20 \text{ kg/m}^3 \text{ (at 20°C)},$$

$$p_0 = 1.01 \times 10^5 \text{ N/m}^2 = 1.01 \times 10^5 \text{ Pa},$$

so that

$$g \frac{\rho_0}{p_0} = 1.16 \times 10^{-4} \text{ m}^{-1} = 0.116 \text{ km}^{-1} .$$

Hence,

$$p = p_0 e^{-aZ}$$

where $a = 0.116 \text{ km}^{-1}$.

Because liquids are almost incompressible the lower layers are not noticeably compressed by the weight of the upper layers superimposed on them and the density ρ is practically constant at all levels. For gases at uniform temperature the density ρ of any layer is proportional to the pressure p at that layer. The variation of pressure with distance above the bottom of the fluid for a gas is different from that for a liquid. The figure shows the pressure distribution in water and in air.

• PROBLEM 2-2

Compare the rate of change of pressure with elevation for air at sea level, p = 14.7 psia (101.3 kPa), at a temperature of 60°F (15.5°C) for fresh water. Assuming constant specific weights for air and water, determine also the total pressure change which occurs with a 10-ft decrease in elevation.

<u>Solution</u>: Determine specific weights of water and air:

$$\rho_{air} = \frac{p}{RT} = \frac{14.7 \times 144}{1716 \times 520} = 0.00237 \text{ slugs/ft}^3$$

$$\gamma_{air} = 0.00237 \times 32.2 = 0.0764 \text{ lbf/ft}^3$$

and

$$\gamma_{water} = 62.4 \text{ lbf/ft}^3$$

then

$$\frac{dp}{dz} = -\gamma$$

or

$$\left(\frac{dp}{dz}\right)_{air} = -0.0764 \text{ lbf/ft}^3 (-12.0 \text{ N/m}^3)$$

$$\left(\frac{dp}{dz}\right)_{water} = -62.4 \text{ lbf/ft}^3 (-9.81 \text{ kN/m}^3)$$

Assuming constant specific weight

$$(p_2 - p_1) = -\gamma(Z_2 - Z_1) \qquad Z_2 > Z_1$$

Total pressure change for air = (-0.0764)(-10)

= 0.764 psf (36.6 Pa) <= answer

Total pressure change for water = (-62.4)(-10)

= 624 psf (29.8 kPa) ⇐ answer

• PROBLEM 2-3

Water flows through this section of cylindrical pipe. If the static pressure at point C is 35 kPa, what are the static pressures at A and B, and where is the hydraulic grade line at this flow cross section?

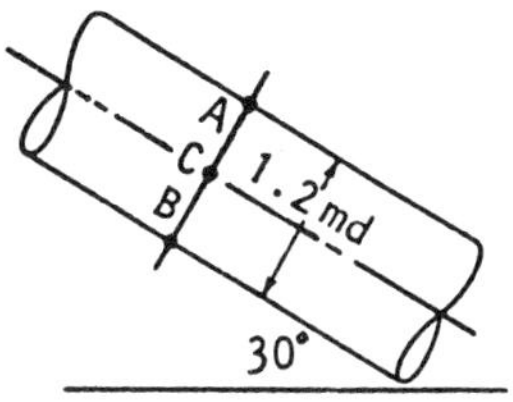

Solution: Using

$$P_A = P_C - \gamma h$$

$$P_A = P_C - \gamma(.6)\cos(30°)$$

$$P_A = 35.0 \times 10^3 - (9.8 \times 10^3)(0.866)0.6 = 29.9 \text{ kPa}$$

$$P_B = P_C + \gamma(.6)\cos(30°)$$

$$P_B = 35.0 \times 10^3 + (9.8 \times 10^3)(0.866)0.6 = 40.1 \text{ kPa}$$

The hydraulic grade line is $(35.0 \times 10^3)/9.8 \times 10^3 = 3.57\text{m}$ (vertically) above C.

• PROBLEM 2-4

A cylinder contains a fluid at a gauge pressure of 350 kN m^{-2}. Express this pressure in terms of a head of (a) water (ρ_{H_2O} = 1000 kg m^{-3}), (b) mercury (relative density 13·6).

What would be the absolute pressure in the cylinder if the atmospheric pressure is 101·3 kN m^{-2}?

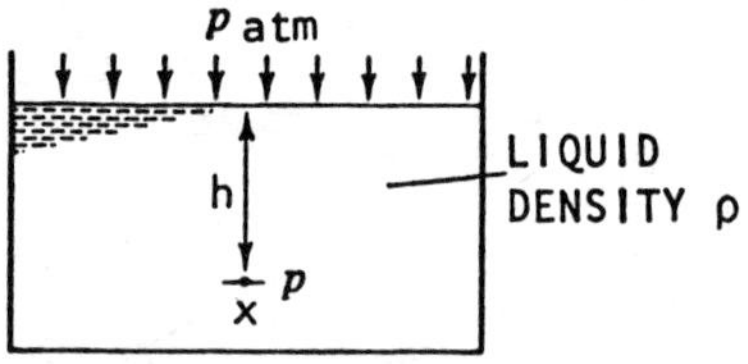

Solution: In a fluid of constant density, dp/dz = - ρg can be integrated immediately to give

$$p = -\rho g z + \text{constant}.$$

In a liquid, the pressure p at any depth z, measured downwards from the free surface so that z = -h (see figure), will be

$$p = \rho g h + \text{constant}$$

and, since, the pressure at the free surface will normally be atmospheric pressure p_{atm},

$$p = \rho g h + p_{atm} . \qquad (1)$$

It is often convenient to take atmospheric pressure as a datum. Pressures measured above atmospheric pressure are known as gauge pressures.

Since atmospheric pressure varies with atmospheric conditions, a perfect vacuum is taken as the absolute standard of pressure. Pressures measured above perfect vacuum are called absolute pressures

Absolute pressure = Gauge pressure + Atmospheric pressure.

Taking p_{atm} as zero, equation (1) becomes

$$p = \rho g h, \qquad (2)$$

which indicates that, if g is assumed constant, the gauge pressure at a point X (figure) can be defined by stating the vertical height h, called the head, of a column of a given fluid of mass density ρ which would be necessary to produce this pressure.

From equation (2), head, $h = p/\rho g$.

(a) Putting $p = 350 \times 10^3\ N\ m^{-2}$, $\rho = \rho_{H_2O} = 1000\ kg\ m^{-3}$,

$$\text{Equivalent head of water} = \frac{350 \times 10^3}{10^3 \times 9 \cdot 81} = 35 \cdot 68\ m.$$

(b) For mercury $\rho_{Hg} = \sigma\rho_{H_2O} = 13 \cdot 6 \times 1000\ kg\ m^{-3}$,

$$\text{Equivalent head of water} = \frac{350 \times 10^3}{13 \cdot 6 \times 10^3 \times 9 \cdot 81} = 2 \cdot 62\ m.$$

Absolute pressure = Gauge pressure + Atmospheric pressure

$= 350 + 101 \cdot 3 = 451 \cdot 3\ kN\ m^{-2}$.

• PROBLEM 2-5

Oil with a specific gravity of 0.80 is 3 ft (0.91 m) deep in an open tank which is otherwise filled with water. If the tank is 10 ft (3.05 m) deep, what is the pressure at the bottom of the tank?

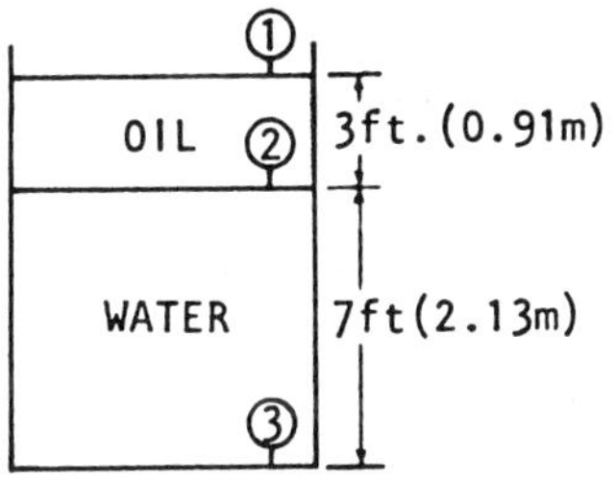

Solution: First determine the pressure at the oil-water interface staying within the oil and then calculate the pressure at the bottom.

$$\frac{p_1}{\gamma} + z_1 = \frac{p_2}{\gamma} + z_2$$

where

p_1 = pressure at free surface of oil

z_1 = elevation of free surface of oil

p_2 = pressure at interface between oil and water

z_2 = elevation at interface between oil and water

For this example, $p_1 = 0$, $\gamma = 0.80 \times 62.4$ lbf/ft^3, $z_1 = 10$ ft, and $z_2 = 7$ ft. Therefore,

$$p_2 = 3 \times 0.80 \times 62.4 = 150 \text{ psfg}$$

Now obtain p_3 from

$$\frac{p_2}{\gamma} + z_2 = \frac{p_3}{\gamma} + z_3$$

where p_2 has already been calculated and $\gamma = 62.4$ lbf/ft^3.

$$p_3 = 62.4\left(\frac{150}{62.4} + 7\right)$$

$$= 587 \text{ psfg}$$

$$= 4.07 \text{ psig} \qquad \Leftarrow \text{answer}$$

SI units $p_{3g} = 28.1$ kPa $\Leftarrow$ answer

• PROBLEM 2-6

Evaluate the pressure difference through the column of multiple fluids in the figure.

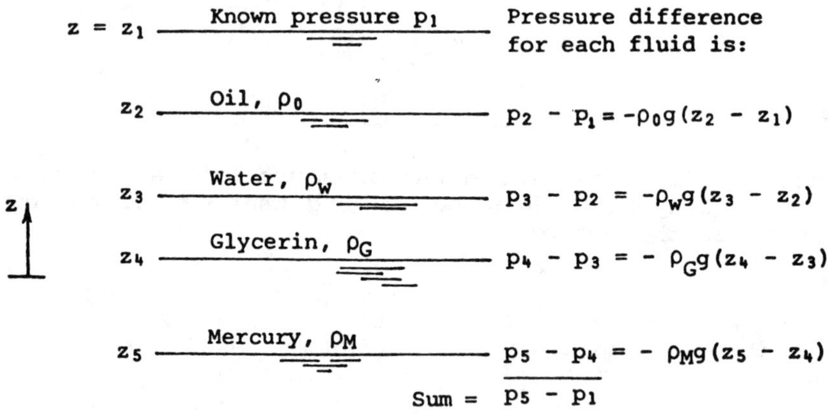

Solution: By the basic hydrostatic equation

$$p_5 - p_1 = -\rho_0 g(z_2 - z_1) - \rho_W g(z_3 - z_2) -$$

$$\rho_G g(z_4 - z_3) - \rho_M g(z_5 - z_4)$$

No additional simplification is possible on the right-hand side because of the different densities. Notice that we have placed the fluids in order from the lightest on top to the heaviest at bottom. This is the only stable configuration. If we attempt to layer them in any other manner, the fluids will overturn and seek the stable arrangement.

• PROBLEM 2-7

A tank 20 ft deep and 7 ft wide is layered with 8 ft of oil, 6 ft of water, and 4 ft of mercury. Compute (a) the total hydrostatic force and (b) the resultant center of pressure of the fluid on the right-hand side of the tank.

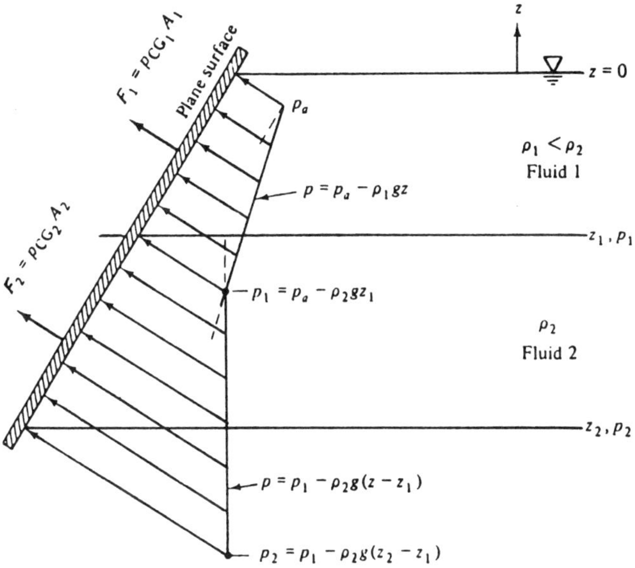

Fig. 1

Hydrostatic forces on a surface immersed in a layered fluid must be summed together in separate pieces.

Solution: If the fluid is layered with different densities, as in Fig. 1, a single formula cannot resolve the problem because the slope of the linear pressure distribution changes between layers. However, the formulas apply separately to each layer, and thus the appropriate remedy is to compute and sum the separate layer forces and moments.

Consider the slanted plane surface immersed in a two-layer fluid in Fig. 1. The slope of the pressure distribution becomes steeper as we move down into the denser second layer. Total force on the plate does not equal the pressure at the centroid times the plate area, but the plate portion in each layer does satisfy the formula, so that we can sum forces to find the total

$$F = \sum F_i = \sum p_{CG_i} A_i \qquad (1)$$

Similarly, the centroid of the plate portion in each layer can be used to locate the center of pressure on that portion

$$Y_{CP_i} = -\frac{\rho_i g \sin\theta_i \, I_{xxi}}{p_{CG_i} A_i} \qquad X_{CP_i} = -\frac{\rho_i g \sin\theta_i \, I_{xyi}}{p_{CG_i} A_i} \qquad (2)$$

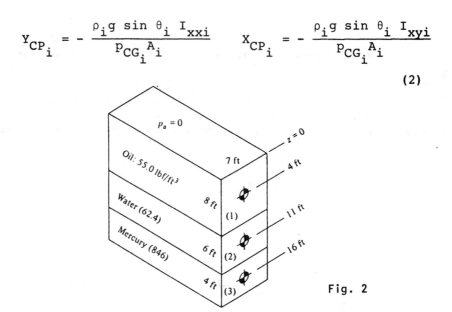

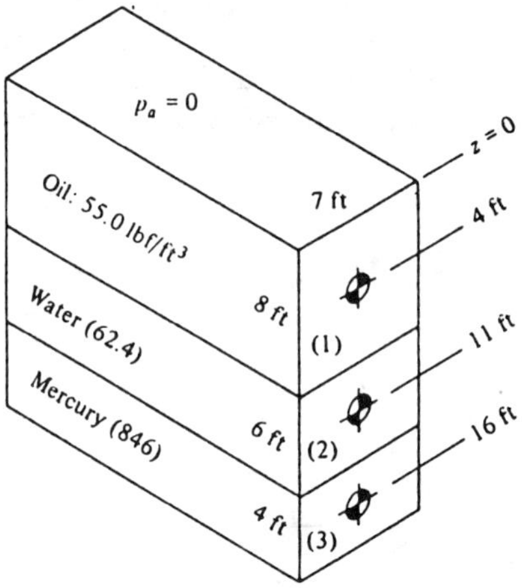

Fig. 2

(a) Divide the panel into three parts as in Fig. (2) and find the hydrostatic pressure at the centroid of each part,

$$p_{CG_1} = (55.0 \text{ lbf/ft}^3)(4 \text{ ft}) = 220 \text{ lbf/ft}^2$$

$$p_{CG_2} = (55.0)(8) + (62.4)(3) = 627 \text{ lbf/ft}^2$$

$$p_{CG_3} = (55.0)(8) + (62.4)(6) + (846)(2) = 2506 \text{ lbf/ft}^2$$

These pressures are then multiplied by the respective panel areas to find the force on each portion:

$$F_1 = p_{CG_1} A_1 = (220 \text{ lbf/ft}^2)(8 \text{ ft})(7 \text{ ft}) = 12{,}300 \text{ lbf}$$

$$F_2 = p_{CG_2} A_2 = 627(6)(7) = 26{,}300 \text{ lbf}$$

$$F_3 = p_{CG_3} A_3 = 2506(4)(7) = 70{,}200 \text{ lbf}$$

$$F = \sum F_i = 108{,}800 \text{ lbf}$$

Ans. (a)

(b) Equations (2) can be used to locate the CP of each force F_i, noting that $\theta = 90°$ and $\sin\theta = 1$ for all parts. The moments of inertia are

$$I_{xx_1} = (7\text{ ft})(8\text{ ft})^3/12 = 298.7\text{ ft}^4,$$

$$I_{xx_2} = (7)(6)^3/12 = 126.0\text{ ft}^4,$$

and

$$I_{xx_3} = (7)(4)^3/12 = 37.3\text{ ft}^4.$$

The centers of pressure are thus at

$$y_{CP_1} = -\frac{\rho_1 g I_{xx_1}}{F_1} = -\frac{(55.0\text{ lbf/ft}^3)(298.7\text{ ft}^4)}{12{,}300\text{ lbf}} = -1.33\text{ ft}$$

$$y_{CP_2} = -\frac{62.4(126.0)}{26{,}300} = -0.30\text{ ft}$$

$$y_{CP_3} = -\frac{846(37.3)}{70{,}200} = -0.45\text{ ft}$$

This locates $z_{CP_1} = -4-1.33 = -5.33$ ft, $z_{CP_2} = -11-0.30 = -11.30$ ft, and $z_{CP_3} = -16-0.45 = -16.45$ ft. Summing moments about the surface then gives

$$\sum F_i z_{CP_i} = F z_{CP}$$

or

$$12{,}300(-5.33) + 26{,}300(-11.30) + 70{,}200(-16.45) = 108{,}800 z_{CP}$$

or

$$z_{CP} = -\frac{1{,}518{,}000}{108{,}800} = -13.95\text{ ft} \qquad \text{Ans. (b)}$$

The center of pressure of the total resultant force on the right side of the tank lies 13.95 ft below the surface.

• **PROBLEM 2-8**

Determine (a) if at sea level the pressure and temperature are 14.7 psia and 59°F, respectively, what is the pressure at 5,000 ft (1,524 m) elevation, assuming that standard atmospheric conditions prevail?

(b) If the pressure and temperature are 3.28 psia (p_a = 22.6 kPa) and -67°F (-55°C) at an elevation of 36,000 ft (10,973 m), what is the pressure at 56,000 ft (17,069 m), assuming isothermal conditions over this range of elevation?

Solution: (a) This involves pressure variation in the troposphere. Let the temperature T be given by

$$T = T_0 - \alpha(z - z_0) \tag{1}$$

In this equation T_0 is the temperature at a reference level where the pressure is known and α is the lapse rate. Using the ideal gas relation (equation 2) and the basic hydrostatic equation (equation 3),

$$\rho = \frac{p}{RT} \quad \text{or} \quad \gamma = \rho g = \frac{pg}{RT} \tag{2}$$

$$\frac{dp}{dz} = -\gamma \tag{3}$$

we obtain

$$\frac{dp}{dz} = -\frac{pg}{RT} \tag{4}$$

Substituting Eq. 1 for T, we get

$$\frac{dp}{dz} = -\frac{pg}{R[T_0 - \alpha(z - z_0)]}$$

Separate the variables and integrate to obtain

$$\frac{p}{p_0} = \left[\frac{T_0 - \alpha(z - z_0)}{T_0}\right]^{g/\alpha R} \tag{5}$$

$$p = p_0\left[\frac{T_0 - \alpha(z - z_0)}{T_0}\right]^{g/\alpha R}$$

Using equation (5) where

p_0 = 14.7 **psia**

T_0 = 460 + 59 = 519°R

$\alpha = 3.566 \times 10^{-3}$ °F/ft

$g/\alpha R = 32.2/(3.566 \times 10^{-3} \times 1{,}716) = 5.261$

$z - z_0 = 5{,}000$ ft

$$p = p_0\left[\frac{519 - 3.566 \times 10^{-3} \times 5{,}000}{519}\right]^{5.261}$$

$$= 14.7 \times 0.832$$

$$= 12.2 \text{ psia}$$ ⇐ answer

SI units $p_a = 84.11$ kPa ⇐ answer

(b) This involves pressure variation in the stratosphere (isothermal conditions). In the stratosphere the temperature is assumed to be constant; therefore, when Equation (4) is integrated, we obtain

$$\ln p = -\frac{zg}{RT} + C$$

At $z = z_0$, $p = p_0$; therefore, the foregoing equation reduces to

$$\frac{p}{p_0} = e^{-(z-z_0)g/RT}$$

or

$$p = p_0 e^{-(z-z_0)g/RT} \qquad (6)$$

Thus

$$T = -67 + 460 = 393°R$$

$$p = p_0 e^{-(z-z_0)g/RT}$$

or

$$p = 3.28e^{-(20{,}000)(32.2)/(1{,}716 \times 393)}$$

$$= 3.28e^{-0.955}$$

Therefore, the pressure at 56,000 ft is

$$p = 1.26 \text{ psia}$$ ⇐ answer

SI units $p_a = 8.69$ kPa ⇐ answer

(a) An open tank of water is accelerated vertically upward at 4.5 m/s^2. Calculate the pressure at a depth of 1.5m. (Fig. 1),

(b) This open tank moves up the plane with constant acceleration. Calculate the acceleration required for the water surface to move to the position indicated. Calculate the pressure in the corner of the tank at A before and after acceleration. (Fig. 2)

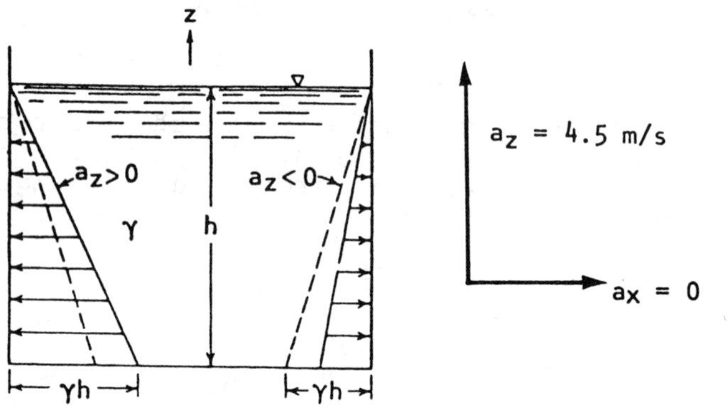

Fig. 1

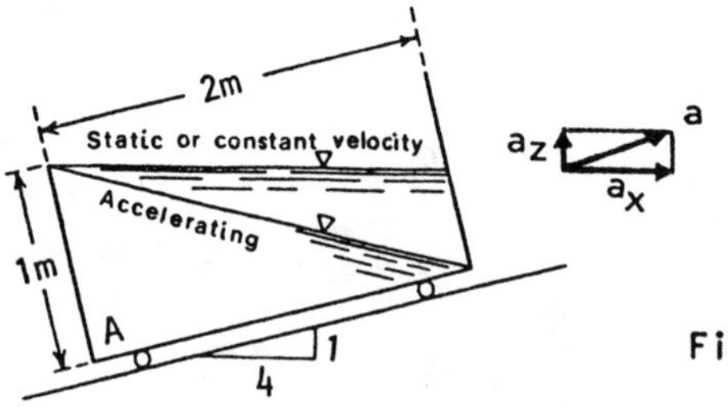

Fig. 2

Solution: (a) A generalized approach to this problem is obtained by applying Newton's second law to the fluid element of Fig. 3 which is being accelerated in such a way that its components of acceleration are a_x and a_z. The summation of force components on such an element are

$$\Sigma F_x = \left(-\frac{\partial p}{\partial x}\right) dx\, dz \tag{1}$$

$$\Sigma F_z = \left(-\frac{\partial p}{\partial z} - \gamma\right) dx\, dz \tag{2}$$

With the mass of the element equal to (γ/g_n) dx dz, the component forms of Newton's second law may be written

$$\left(-\frac{\partial p}{\partial x}\right) dx\,dz = \frac{\gamma}{g_n} a_x\,dx\,dz$$

$$\left(-\frac{\partial p}{\partial z} - \gamma\right) dx\,dz = \frac{\gamma}{g_n} a_z\,dx\,dz$$

which reduce to

$$-\frac{\partial p}{\partial x} = \frac{\gamma}{g_n}(a_x) \qquad (3)$$

$$-\frac{\partial p}{\partial z} = \frac{\gamma}{g_n}(a_z + g_n) \qquad (4)$$

These equations characterize the pressure variation through an accelerated mass of fluid, and with them specific applications can be studied.

One other generalization may be derived from the foregoing equations: this is a property of a line of constant pressure. Using the chain rule for the total differential for dp in terms of its partial derivatives,

$$dp = \frac{\partial p}{\partial x} dx + \frac{\partial p}{\partial z} dz$$

and substituting the above expressions for $\partial p/\partial x$ and $\partial p/\partial z$ give

$$dp = -\frac{\gamma}{g_n}(a_x)\,dx - \frac{\gamma}{g_n}(a_z + g_n)\,dz \qquad (5)$$

However, along a line of constant pressure dp = 0 and hence, for such a line,

$$\frac{dz}{dx} = -\left(\frac{a_x}{g_n + a_z}\right) \qquad (6)$$

Thus the slope (dz/dx) of a line of constant pressure is defined; its position must be determined from external (boundary) conditions in specific problems.

Here a container of liquid is accelerated vertically upward, $\partial p/\partial x = 0$, and with no change of pressure with x, equation (4) becomes

$$\frac{dp}{dz} = -\left(\gamma\,\frac{g_n + a_z}{g_n}\right) \qquad (7)$$

Using equation (7):

$$\frac{dp}{dz} = -\ 9.8 \times 10^3 \left(\frac{4.5 + 9.81}{9.81}\right) = -\ 14.3\ kN/m^3$$

$$p = \int_0^p dp = -\int_0^{-1.5} 14.3dz = 21.5kPa$$

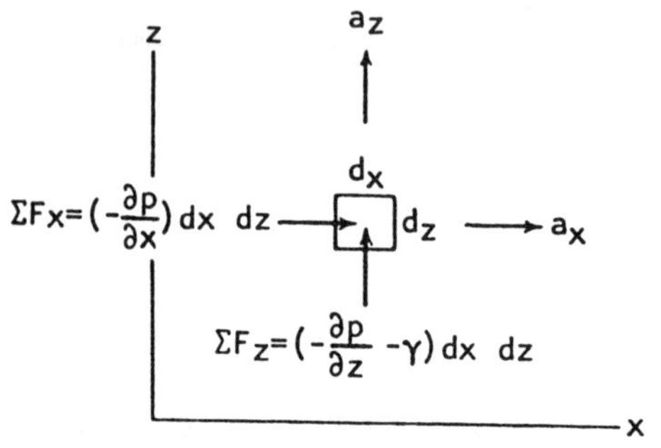

Fig. 3

(b) From geometry, the slope of the water surface during acceleration is -0.222. From the slope of the plane, $a_x = 4a_z$. Using equation (6)

$$-0.222 = \frac{-4a_z}{(a_z + 9.81)}$$

and from the foregoing

$$a_z = 0.58\ m/s^2,\ a_x = 2.3\ m/s^2,\ a = 2.38\ m/s^2$$

From geometry, the depth of water vertically above corner A before acceleration is 0.97 m; hence the pressure there is $0.97 \times 9.8 \times 10^3 = 9.51$ kPa. After acceleration this depth is 0.92 m; from equation 4,

$$\frac{\partial p}{\partial z} = -\ (9.81 + 0.58)\left(\frac{9.8 \times 10^3}{9.81}\right) = -\ 10.35\ kN/m^3$$

Therefore $p = 0.92 \times 10.35 \times 10^3 = 9.51$ kPa.

The fact that the pressures at A are the same before and after acceleration is no coincidence; general proof may be offered that p_A does not change whatever the acceleration. This means that the force exerted by the end of the tank on the water is constant for all accelerations. However, this is no violation of Newton's second law, since the mass of liquid diminishes with increased acceleration so that the product of mass and acceleration remains constant and equal to the applied force.

• PROBLEM 2-10

A fine tube of uniform bore is bent into the form of a square and filled with equal volumes of three heavy liquids of densities ρ_1, ρ_2, ρ_3 $(\rho_1<\rho_2<\rho_3)$. If the tube is placed with one side vertical, show that a side of the square will be filled with liquid of the first, third or second kind only according as ρ_2 is $>\frac{1}{3}(2\rho_3 + \rho_1)$ or $<\frac{1}{3}(\rho_3 + 2\rho_1)$ or lies between these values.

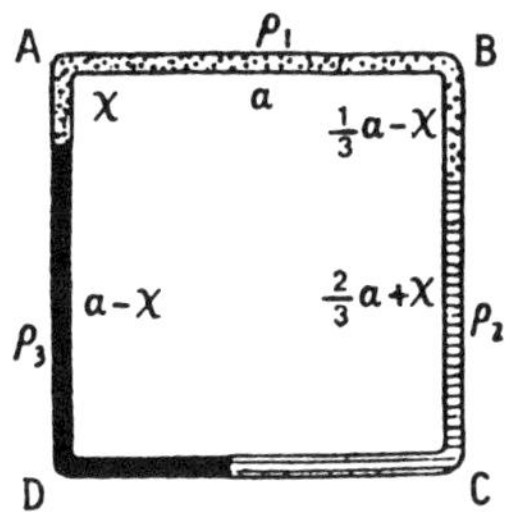

Solution: Let a be a side of the square, then the length of tube occupied by each liquid is $\frac{1}{3}a$. First let the liquid ρ_1 fill the uppermost side of the tube and lengths x and $\frac{1}{3}a - x$ of the adjacent sides. In a fluid at rest under gravity the pressure is the same at all points in the same horizontal plane. Then in the figure, the pressures at A and B are equal and the pressures at C and D are equal; so that

press. at D - press. at A = press. at C - press. at B,

or

$$g\rho_1 x + g\rho_3(a - x) = g\rho_1(\frac{1}{3}a - x) + g\rho_2(\frac{2}{3}a + x)$$

or

$$x(\rho_3 + \rho_2 - 2\rho_1) = \frac{1}{3}a(3\rho_3 - \rho_1 - 2\rho_2) \qquad (1)$$

The necessary and sufficient conditions for this arrangement of liquids to be possible are $0<x<\frac{1}{3}a$, or from (1)

$$0 < \frac{3\rho_3 - \rho_1 - 2\rho_2}{\rho_3 + \rho_2 - 2\rho_1} < 1.$$

Since $\rho_1<\rho_2<\rho_3$ the numerator and denominator are positive, so that the first inequality is satisfied, and the remaining condition is

$$3\rho_3 - \rho_1 - 2\rho_2 < \rho_3 + \rho_2 - 2\rho_1$$

or

$$\rho_2 > \frac{1}{3}(2\rho_3 + \rho_1) \qquad (2)$$

Secondly let the lowest side CD be filled with the liquid of density ρ_3, and the other liquids be as in the figure. A like argument gives

$$\rho_1(a - x) + \rho_3 x = \rho_2(\frac{2}{3}a + x) + \rho_3(\frac{1}{3}a - x)$$

or

$$x(2\rho_3 - \rho_1 - \rho_2) = \frac{1}{3}a(2\rho_2 + \rho_3 - 3\rho_1) \qquad (3)$$

and again the necessary and sufficient conditions are $0 < x < \frac{1}{3}a$, or from (3)

$$0 < \frac{2\rho_2 + \rho_3 - 3\rho_1}{2\rho_3 - \rho_1 - \rho_2} < 1.$$

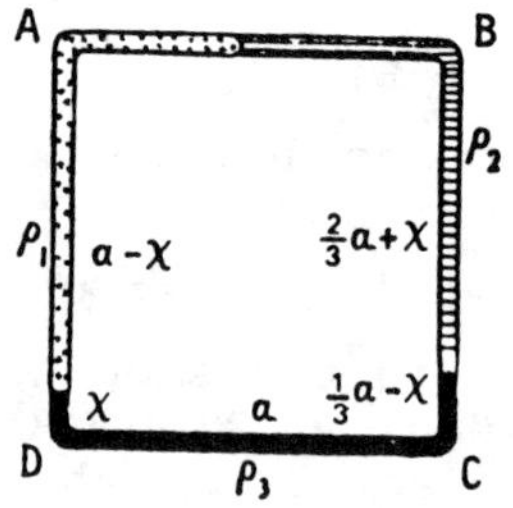

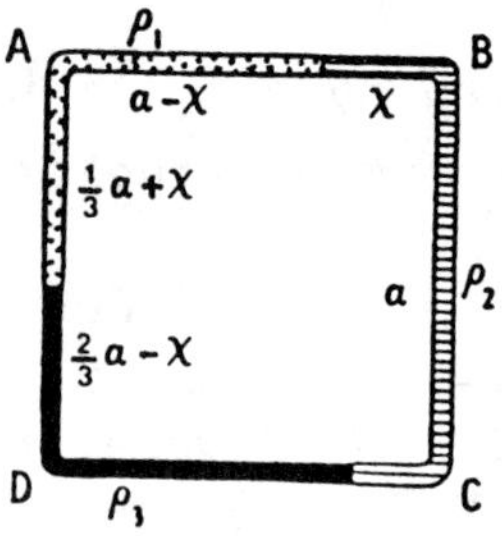

Here the numerator and denominator are positive, since $\rho_1 < \rho_2 < \rho_3$, therefore the first condition is satisfied and the second is

$$2\rho_2 + \rho_3 - 3\rho_1 < 2\rho_3 - \rho_1 - \rho_2$$

or

$$\rho_2 < \frac{1}{3}(\rho_3 + 2\rho_1) \qquad (4)$$

Finally let a vertical side BC be filled with liquid of density ρ_2, and the other liquids as in the figure. Reasoning as before we have

$$\rho_1(\frac{1}{3}a + x) + \rho_3(\frac{2}{3}a - x) = \rho_2 a$$

or

$$x(\rho_3 - \rho_1) = \frac{1}{3}a(2\rho_3 + \rho_1 - 3\rho_2) \qquad (5)$$

and the necessary and sufficient conditions for this arrangement are

$$0<x<\frac{1}{3}a$$

or

$$0 < \frac{2\rho_3 + \rho_1 - 3\rho_2}{\rho_3 - \rho_1} < 1.$$

Now the denominator is positive, hence the first inequality requires that

$$\rho_2 < \frac{1}{3}(2\rho_3 + \rho_1) \qquad (6)$$

and the second requires that

$$2\rho_3 + \rho_1 - 3\rho_2 < \rho_3 - \rho_1$$

or

$$\rho_2 > \frac{1}{3}(\rho_3 + 2\rho_1) \qquad (7);$$

and (6) and (7) are together necessary and sufficient for this arrangement.

• PROBLEM 2-11

A gate 5 ft wide is hinged at point B and rests against a smooth wall at point A. Compute (a) the force on the gate due to seawater pressure; (b) the horizontal force P exerted by the wall at point A; and (c) the reactions at the hinge B. (Fig. 1)

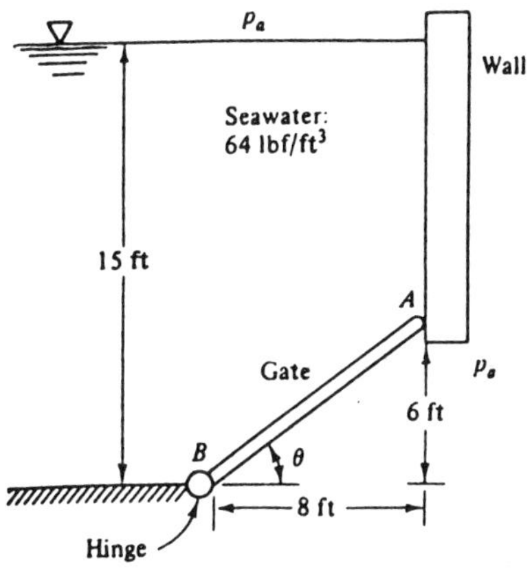

Fig. 1

Solution: Using nomenclature based on Fig. 3.

(a) By geometry the gate is 10 ft long from A to B, and its centroid is halfway between, or at elevation 3 ft above point B. The depth h_{CG} is thus 15 - 3 = 12 ft. The gate area is 5 x 10 = 50 ft^2. Neglect p_a as acting on both sides of the gate. The hydrostatic force on the gate is

$$F = p_{CG}A = \rho g h_{CG}A = (64\ lbf/ft^3)(12\ ft)(50\ ft^2) = 38{,}400\ lbf$$

Ans. (a)

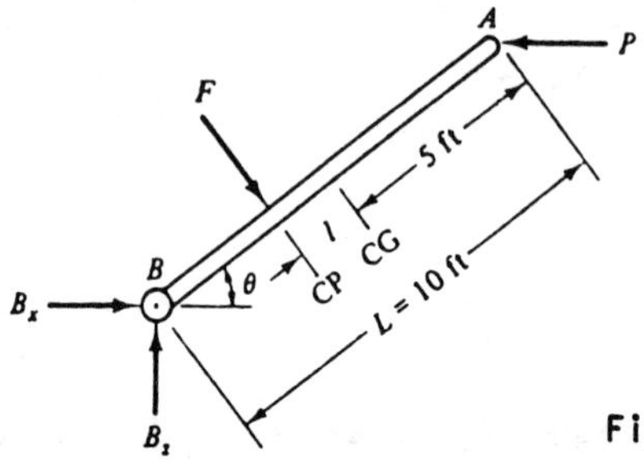

Fig. 2

(b) First we must find the center of pressure of F. A free-body diagram of the gate is shown in Fig. (2). The gate is a rectangle and hence $I_{xy} = 0$ and $I_{xx} = bL^3/12 = [(5\ ft) \times (10\ ft)^3]/12 = 417\ ft^4$. The ambient pressure p_a is neglected if it acts on both sides of the plate; e.g., the other side of the plate is inside a ship or on the dry side of a gate or dam. In this case $p_{CG} = \rho g h_{CG}$, and the center of pressure becomes independent of specific weight

$$F = \rho g h_{CG}A \Big| y_{CP} = -\frac{I_{xx}\sin\Theta}{h_{CG}A} \Big| x_{CP} = -\frac{I_{xy}\sin\Theta}{h_{CG}A} \qquad (1)$$

The distance I from the CG to the CP is given by Eq. (1), since p_a is neglected.

$$I = -y_{CP} = +\frac{I_{xx}\sin\Theta}{h_{CG}A} = \frac{(417\ ft^4)(\frac{6}{10})}{(12\ ft)(50\ ft^2)} = 0.417\ ft$$

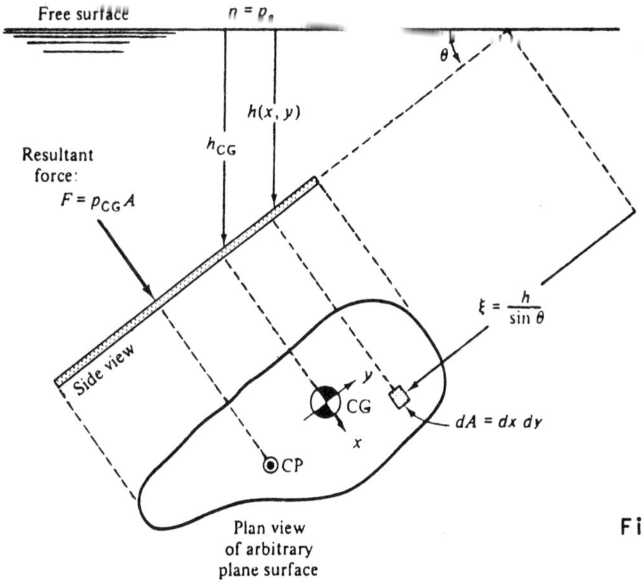

Fig. 3

Hydrostatic force and center of pressure on an arbitrary plane surface of area A inclined at an angle θ below the free surface.

The distance from point B to force F is thus 10 - I - 5 = 4.583 ft. Summing moments counterclockwise about B gives

$$PL \sin \Theta - F(5-I) = P(6 \text{ ft}) - (38{,}400 \text{ lbf})(4.583 \text{ ft}) = 0$$

or

$$P = 29{,}300 \text{ lbf}$$ Ans. (b)

(c) With F and P known, the reactions B_x and B_z are found by summing forces on the gate

$$\Sigma F_x = 0 = B_x + F \sin \theta - P = B_x + 38{,}400(0.6) - 29{,}300$$

or

$$B_x = 6300 \text{ lbf}$$

$$\Sigma F_z = 0 = B_z - F \cos \theta = B_z - 38{,}400(0.8)$$

or

$$B_z = 30{,}700 \text{ lbf}$$ Ans. (c)

Compute the atmospheric pressure at elevation 20,000 ft., considering the atmosphere as a static fluid. Assume standard atmosphere at sea level. Use four methods: (a) air of constant density; (b) constant temperature between sea level and 20,000 ft; (c) isentropic conditions; and (d) air temperature decreasing linearly with elevation at the standard lapse rate of 0.00356°F/ft. Assume T = 59°F and atmospheric pressure at sea level.

Solution: A fundamental equation for a perfect gas is

$$pv^n = p_1 v_1^n = \text{constant}$$

where p is absolute pressure, and n may have any value from zero to infinity, depending upon the process to which the gas is subjected. If the process is at constant temperature (isothermal), n = 1. If there is not heat transfer to or from the gas, the process is known as adiabatic. A frictionless adiabatic process is called an isentropic process and n is denoted by k, where $k = c_p/c_v$, the ratio of specific heat at constant pressure to that at constant volume. For air and diatomic gases at usual temperatures, k may be taken as 1.4.

Assume the pressure at the center of the element is p and that the dimensions of the element are δx, δx and δz. Since the fluid is at rest, the summation of forces acting on the element in any direction must be zero. If forces are summed up the only forces acting are the pressure forces on the vertical faces of the element. To satisfy $\Sigma F_x = 0$ and $\Sigma F_y = 0$, the pressure on the opposite vertical faces must be equal. Thus $\partial p/\partial x = \partial p/\partial y = 0$ for the case of the fluid at rest.

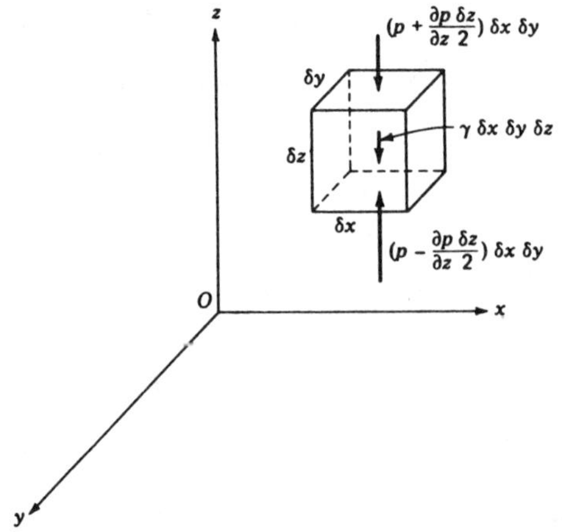

Summing up forces in the vertical direction and setting equal to zero,

$$\Sigma F_z = \left(p - \frac{\partial p}{\partial z}\frac{\partial z}{2}\right) \delta x\ \delta y - \left(p + \frac{\partial p}{\partial z}\frac{\partial z}{2}\right) \delta x\ \delta y - \gamma\ \delta x\ \delta y\ \delta z = 0$$

The results in $\partial p/\partial z = -\gamma$, which, since p is independent of x and y, can be written as

$$\frac{dp}{dz} = -\gamma$$

This is the general expression that relates variation of pressure in a static fluid to vertical position. The minus sign indicates that as z gets larger (increasing elevation), the pressure gets smaller.

TABLE

Temperature		Density	Specific weight
T, °F	T, °C	$\rho \times 10^3$, slugs/ft³	$\gamma \times 10^2$, lb/ft³
−40	−40.0	2.94	9.46
−20	−28.9	2.80	9.03
0	−17.8	2.68	8.62
10	−12.2	2.63	8.46
20	−6.7	2.57	8.27
30	−1.1	2.52	8.11
40	4.4	2.47	7.94
50	10.0	2.42	7.79
60	15.6	2.37	7.63
70	21.1	2.33	7.50
80	26.7	2.28	7.35
90	32.2	2.24	7.23
100	37.8	2.20	7.09
120	48.9	2.15	6.84
140	60.0	2.06	6.63
160	71.1	1.99	6.41
180	82.2	1.93	6.21
200	93.3	1.87	6.02
250	121.1	1.74	5.60

(a) From the Table, the conditions of the standard atmosphere at sea level are T = 59°F, p = 14.7 psia, $\gamma = 0.076\ \text{lb/ft}^3$.

For constant density:

$$\frac{dp}{dz} = -\gamma \qquad dp = -\gamma\ dz \qquad \int_{p_1}^{p} dp = -\gamma \int_{z_1}^{z} dz$$

$$p - p_1 = -\gamma(z - z_1)$$

$$p = 14.7 \times 144 - 0.076(20{,}000) = 600\ \text{lb/ft}^2,\ \text{abs} = 4.15\ \text{psia}$$

(b) For isothermal:

$pv = \text{constant, hence } \frac{p}{\gamma} = \frac{p_1}{\gamma_1}$ if g is constant

$$\frac{dp}{dz} = -\gamma \qquad \text{where} \qquad \gamma = \frac{p\gamma_1}{p_1}$$

$$\frac{dp}{p} = -\frac{\gamma_1}{p_1}\,dz$$

$$\int_{p_1}^{p} \frac{dp}{\gamma p} = \ln\frac{p}{p_1} = -\frac{\gamma_1}{p_1}\int_{z_1}^{z} dz = -\frac{\gamma_1}{p_1}(z - z_1)$$

$$\frac{p}{p_1} = \exp\left[-\frac{\gamma_1}{p_1}(z - z_1)\right]$$

$$p = 14.7 \exp\left[-\frac{0.076}{14.7 \times 144}(20{,}000)\right] = 7.18 \text{ psia}$$

(c) For isentropic:

$$pv^{1.4} = \frac{p}{\rho^{1.4}} = \text{constant} \quad \text{hence } \frac{p}{\gamma^{1.4}} = \text{constant} = \frac{p_1}{\gamma_1^{1.4}}$$

$$\frac{dp}{dz} = -\gamma \quad \text{where } \gamma = \gamma_1 \left(\frac{p}{p_1}\right)^{0.715}$$

$$\int_{p_1}^{p} p^{-0.715}\,dp = -\gamma_1 p_1^{-0.715}\int_{z_1}^{z} dz$$

$$p^{0.285} - p_1^{0.285} = -0.285\gamma_1 p_1^{-0.715}(z - z_1)$$

$$p^{0.285} = (14.7 \times 144)^{0.285} - 0.285(0.076)(14.7 \times 144)^{-0.715}(20{,}000)$$

$p = 950$ lb/ft^2, abs = 6.60 psia

(d) For temperature decreasing linearly with elevation:

$T = (460 + 59) + Kz$ where $K = -0.00356$°F/ft

$$dT = K\,dz \quad \text{hence} \quad dz = \frac{dT}{K}$$

$$\frac{pv}{T} = R = \frac{p}{\rho T} = \frac{p_1}{\rho_1 T_1}$$

$$\frac{dp}{dz} = -\gamma \quad \text{where} \quad \gamma = \frac{\gamma_1 T_1 p}{T p_1} \quad \text{if g is constant}$$

$$\frac{dp}{p} = -\frac{\gamma_1 T_1}{p_1 T} dz = -\frac{\gamma_1 T_1 dT}{p_1 \; K \; T}$$

$$\int_{p_1}^{p} \frac{dp}{p} = -\frac{\gamma_1 T_1}{p_1 K} \int_{T_1}^{T} \frac{dT}{T}$$

$$\ln \frac{p}{p_1} = \frac{\gamma_1 T_1}{p_1 K} \; n \; \frac{T}{T_1} = \ln \left(\frac{T_1}{T}\right)^{\gamma_1 T_1 / p_1 K}$$

$$p = p_1 \left(\frac{T_1}{T_1 + Kz}\right)^{\gamma_1 T_1 / p_1 K}$$

$$p = 14.7 \left(\frac{519}{447.8}\right)^{(0.076)(519)/14.7(144)(-0.00356)}$$

$$= 6.8 \text{ psia (470 mbar, abs)}$$

FORCES ON A PLANE SURFACE

• PROBLEM 2-13

A solid hemisphere floats in water as shown in Fig. 1. If the radius is 0.5 ft., determine the components of the hydrostatic force parallel and normal to the free surface on the curved surface.

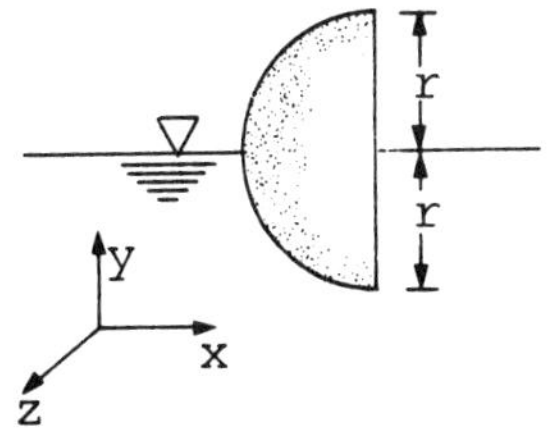

Fig. 1

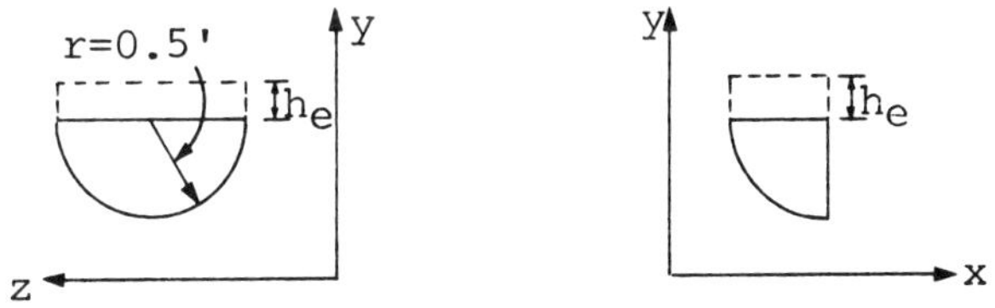

Fig. 2(a)

Fig. 2(b)

Solution: In order to calculate the parallel component F_x, we project the submerged surface onto the zy plane. A semicircle of radius r is thereby formed. Taking into account atmospheric pressure, an amount equal to the equivalent height (h_e) is raised (see Fig. 2b). That is,

$$h_e = \frac{Patm}{\gamma}$$

the parallel force F_x is given by

$$F_x = \gamma h_c A = \gamma\left(\frac{4}{3}\frac{r}{\pi} + \frac{Patm}{\gamma}\right)\frac{\pi r^2}{2}$$

$$F_x = 62.4\ \frac{lb}{ft^3}\left(\frac{4}{3}\ \frac{0.5}{\pi}\ ft + \frac{14.7}{62.4}\ \frac{lb/ft^2}{lb/ft^3}\right)\frac{\pi(0.5)^2 ft^2}{2}$$

$$F_x = 10.97\ lb.$$

The normal component Fy is evaluated by computing the weight of the water above the curved surface plus an extended amount (h_e) due to the atmospheric pressure, as shown in Fig. 2b. The cross-section of the added part is the same as the projected curved surface onto the free-surface (plane zx). Then,

$$\text{Projected area} = A_p = \frac{\pi r^2}{2} \qquad \text{(semicircle)}$$

Thus, the force will be

$$F_y = \gamma \forall = \gamma\left(\frac{1}{4}\ \frac{4}{3}\ \pi r^3 + \frac{\pi r^2}{2}\ \frac{Patm}{\gamma}\right)$$

$$F_y = 62.4\left(\frac{\pi}{3}\ (0.5)^3 + \frac{\pi(0.5)^2}{2}\ \frac{14.7}{62.4}\right)$$

$$F_y = 13.94\ lb.$$

• PROBLEM 2-14

A tank of oil has a right triangular panel near the bottom as in Fig. (1). Omitting p_a, find the (a) hydrostatic force and (b) CP on the panel.

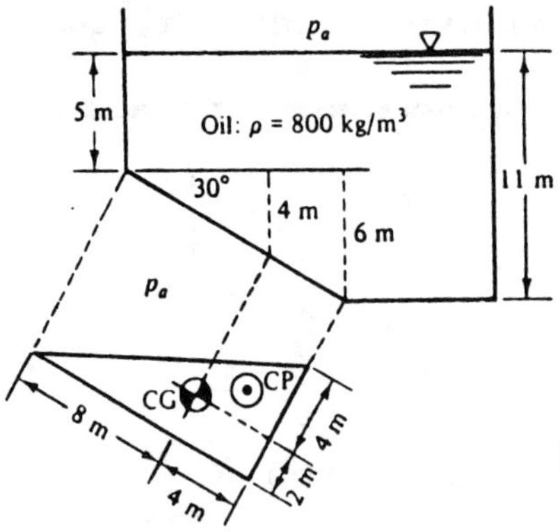

Fig. 1

Solution: (a) The triangle has the following properties. The centroid is one-third up (4 m) and one-third over (2 m) from the lower left corner, as shown. The area is ½(6 m) x (12 m) = 36 m^2. The moments of inertia are $I_{xx} = bL^3/36$ $= [(6\ m)(12\ m)^3]/36 = 288\ m^4$ and $I_{xy} = b(b - 2s)L^2/72 =$ $\{6\ m[6\ m - 2(6\ m)](12\ m)^2\}/72 = -72\ m^4$. The depth to the centroid is $h_{CG} = 5 + 4 + 9$ m; thus the hydrostatic force is $F = \rho g h_{CG} A = (800\ kg/m^3)(9.807\ m^2/s)(9\ m)(36\ m^2)$

$= 2.54 \times 10^6 \text{ (kg . m)/s}^2 = 2.54 \times 10^6 \text{ N} = 2.54 \text{ MN}$

Ans. (a)

(b) The CP is given by:

$$y_{CP} = -\frac{I_{xx} \sin \theta}{h_{CG}A} = -\frac{(288 \text{ m}^4)(\sin 30°)}{(9 \text{ m})(36 \text{ m}^2)} = -0.444 \text{ m}$$

$$x_{CP} = -\frac{I_{xy} \sin \theta}{h_{CG}A} = -\frac{(-72 \text{ m}^4)(\sin 30°)}{(9 \text{ m})(36 \text{ m}^2)} = +0.111 \text{ m}$$

Ans. (b)

The resultant force F = 2.54 MN acts through this point, which is down and to the right of the centroid.

• PROBLEM 2-15

A flat plate AB (Fig. 1) acts as a gate in a liquid, water. Atmospheric pressure acts upon the upper side of the gate and upon the water surface. Compute the normal force at B necessary to hold the gate closed.

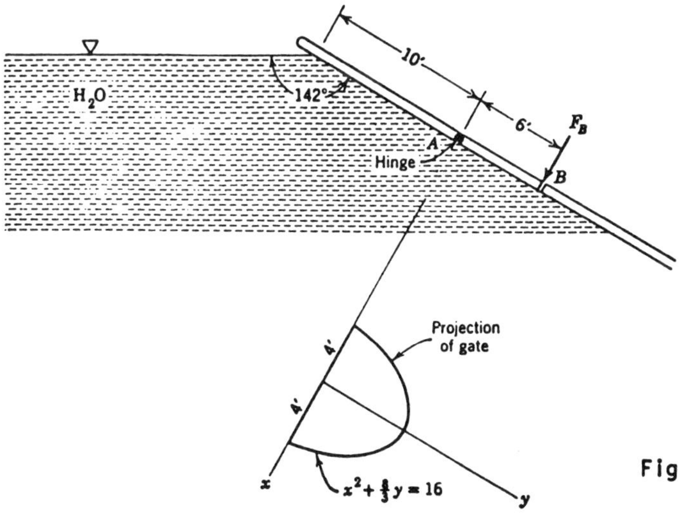

Fig. 1

Solution: Draw a free body diagram of the gate (see Fig. 2). The distributed pressure loading resolves into a resultant force F. Taking moments about the hinge A in the free body diagram (Fig. 2) allows you to solve for F_B.

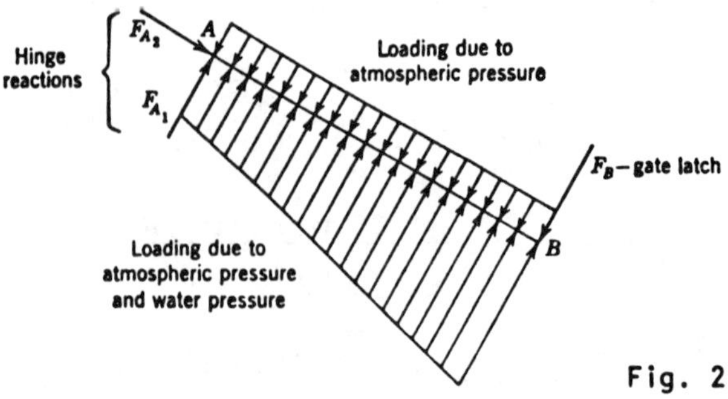

Fig. 2

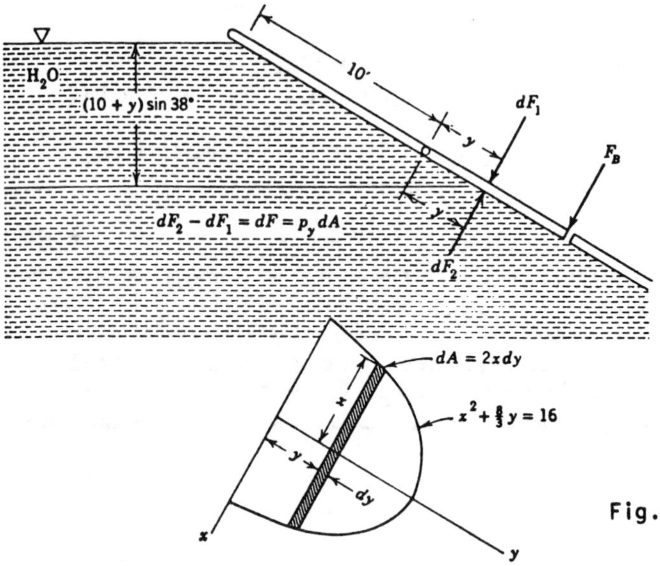

Fig. 3

Choose the elemental area dA to assure pressure p_y constant everywhere on the area. The sum of the pressures on the lower and upper sides of the gate is p_y (Fig. 3).

From the hydrostatic equation for an incompressible liquid

$$p_y = p_a + \gamma h_y - p_a = \gamma h_y$$

where γ = sp. wt. of water = 62.4 lb/ft^3

h_y = vertical distance to water surface

= (10 + y) sin (180 - 142) = (10 + y) sin 38

dF = p_ydA, where p_y is the gage pressure on the area dA and is constant over dA

= $p_y 2x\ dy = \gamma h_y 2x\ dy = 2\gamma(10 + y) \sin 38\ x\ dy$

$$F = 2\gamma \sin 38 \int_0^6 (10 + y)x\, dy$$

but $x^2 + (\frac{8}{3})y = 16$

so that $x = (16 - \frac{8}{3} y)^{1/2} = (\frac{8}{3})^{1/2}(6 - y)^{1/2}$

Therefore:

$$F = 2\gamma \sin 38 \int_0^6 (10 + y)(\frac{8}{3})^{1/2}(6 - y)^{1/2}\, dy = 15{,}220 \text{ lb}$$

From statics

$$Fy_F = \int dM_a = M_a$$

Solve for M_a

$$dM_a = y\, dF = y[2\gamma(10 + y) \sin 38x\, dy]$$

$$M_a = 2\gamma \sin 38(8/3)^{1/2} \int_0^6 (10 + y)(6 - y)^{1/2} y\, dy$$

$$= 39{,}600 \text{ ft-lb}$$

$$y_F = \frac{39{,}600}{15{,}220} = 2.6 \text{ ft in y direction}$$

By symmetry the resultant lies on line x = 0.

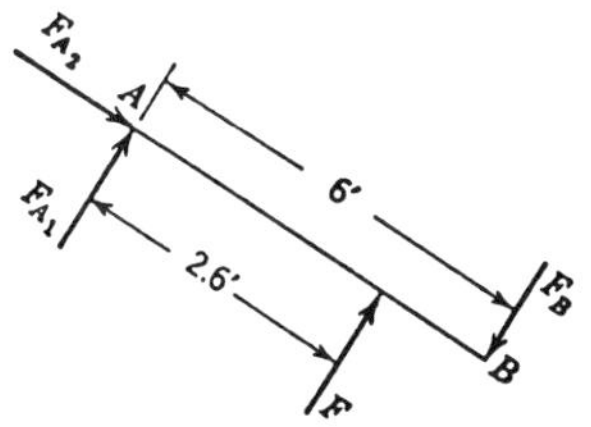

Fig. 4

Replace the distributed load in the free body by the resultant force as shown in Fig. 4.

$$\Sigma M_A = 0 \curvearrowright +$$

$$6F_B - 2.6F = 0$$

$$F_B = \frac{(2.6)F}{6} = \frac{(2.6)(15{,}220)}{6} = 6600 \text{ lb}$$

Instead of locating the resultant force of the pressures, you could compute the moment about the hinge of the pressure loading and use it in the $\Sigma M_a = 0$ without solving for the resultant pressure force.

• PROBLEM 2-16

A rectangular gate 4 by 4 ft in size is set on a 45° plane with respect to the water surface; the centroid of the gate is 10 ft below the water surface (see Figure). Compute the required location of the horizontal axis at which the gate is to be hinged if the hydrostatic pressure acting on the gate is to be balanced around the hinge so that there will be no moment causing rotation of the gate under the loading specified.

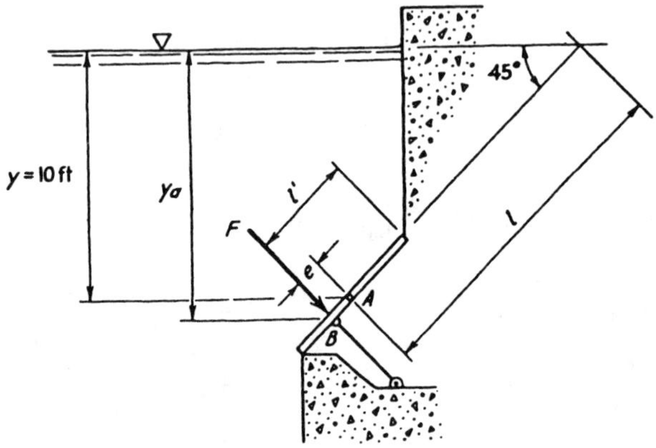

Solution: Since the centroid of the gate is 10 ft under the water surface so that point A in the plane of the gate is at a distance l from the water level. Then

$$l = \frac{10}{\sin 45°}$$

$$= \frac{10}{0.707} = 14.14 \text{ ft}$$

Under static conditions the location of hinge has to be at the point where the resultant force acts on the gate. Referring to the Figure, the distance e between centroid and the action point of the hydrostatic force is given by

$$e = \frac{I_0}{lA}$$

where I_0 is the second moment of area A with respect to the centroid, l is the distance between centroid and the line of intersection of the plane of A with the water level, and e is the distance between the centroid and the point of action of the hydrostatic force measured on the surface on which it acts.

Table

Location of Centroid, Area, and Moment of Inertia of Common Shapes

Rectangle:	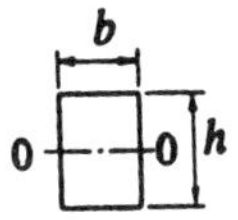	$A = b \cdot h$	$I_0 = b \cdot h^3/12$

$$I_0 = b \cdot \frac{h^3}{12} \text{ (from the Table)}$$

$$= \frac{4(4)^3}{12} = 21.33 \text{ ft}^4$$

$$A = b \cdot h = 4 \times 4 = 16 \text{ ft}^2$$

Hence

$$e = \frac{21.33}{(14.14)(16)} = 0.0943 \text{ ft} = 1.13 \text{ in.}$$

This is the distance from the centroid of the gate measured in the same plane. Hence point B is located at distance

$$l' = (2 \times 12) + 1.13 = 25.13 \text{ in.}$$

measured from the top edge of the gate.

• PROBLEM 2-17

A 1 m diameter flood gate placed vertically is 4 m below the water level at its highest point. The gate is hinged at the top. Compute the required horizontal force to be applied at the bottom of the gate in order to open it. Assume that the pressure on the other side of the gate is atmospheric pressure and neglect the weight of the gate.

<u>Solution</u>: We know

$$\gamma \text{ of water} = 1000 \text{ kg/m}^3$$

$$\text{Area of the gate} = \frac{\pi}{4}(1)^2$$

$$A = 0.7855 \text{ m}^2$$

The depth of the centroid of the gate is $l = 4.5$ m. Next determine the location of F_1 acting on the gate.

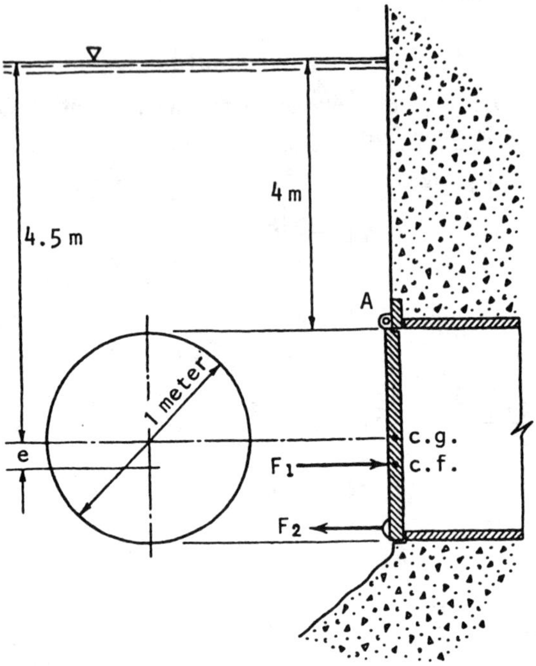

Referring to the Figure, the distance e between centroid and the action point of the hydrostatic force is given by

$$e = \frac{I_0}{l\,A}$$

where I_0 is the second moment of area A with respect to the centroid,   is the distance between centroid and the line of intersection of the plane of A with the water level, and e is the distance between the centroid and the point of action of the hydrostatic force measured on the surface on which it acts.

Table

Location of Centroid, Area, and Moment of Inertia of Common Shapes

Rectangle:	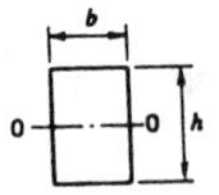	$A = b \cdot h$	$I_0 = b \cdot h^3/12$
Triangle:	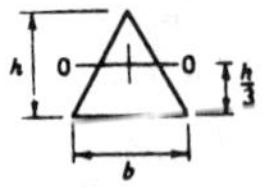	$A = b \cdot h/2$	$I_0 = b \cdot h^3/36$
Circle:	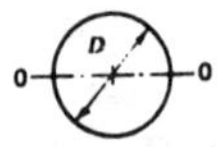	$A = \pi D^2/4$	$I_0 = \pi R^4/4$

From the table

$$I_0 = \frac{\pi R^4}{4}$$

$$= \frac{\pi(0.5)^4}{4} = 0.0491 \text{ m}^4$$

Therefore,

$$e = \frac{0.0491}{4.5(0.7855)} = 0.0139 \text{ m}$$

The force F_1 acting on the gate is located at a depth y_a = (1 + e). This equals 4.5 + 0.0139, which equals

$$y_a = 4.5139 \text{ m}$$

The magnitude of this hydrostatic force is

$$F_1 = p_a A = 1.0 \times 4.5139 \times 0.7855 = 3545.67 \text{ kg}$$

The determine F_2 we take the moment of the two forces with respect to the hinge at A:

$$F_2 = D = F_1\left(\frac{D}{2} + e\right)$$

$$F_2 = \frac{3545.67(0.5 + 0.0139)}{1}$$

$$= 1822.12 \text{ kg}$$

• PROBLEM 2-18

The rectangular gate AB is hinged at A (Fig. 1). The gate is 2m wide (w=2m) and 1.5m long (L=1.5m). Calculate

(a) The resultant force, $\vec{F}_R$, of the water on the gate AB

(b) The force $\vec{F}_B$, at point B to hold the gate closed.

Fig. 1

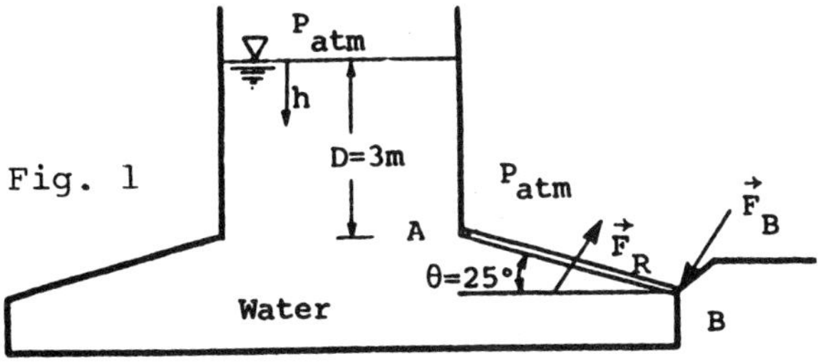

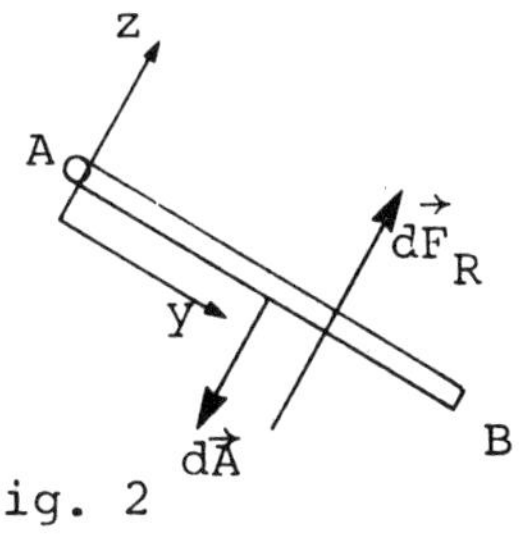

Fig. 2

Solution: (a)

In order to determine the force $\vec{F}_R$, the magnitude, the direction, and the line of action of $\vec{F}_R$ must be determined.

Consider the gate AB (Fig. 2) with coordinates as shown.

$$|\vec{F}_R| = -\int_A p\,d\vec{A} = -\int_A pw\,dy = -w\int_A p\,dy$$

To solve for the force $\vec{F}_R$ the pressure, p, as a function of y must be known.

From the basic pressure-height relation

$$\frac{dp}{dh} = \rho g \quad \text{or} \quad dp = \rho g\,dh$$

By assuming ρ = constant

$$\int_{p_{atm}}^{p} dp = \rho g \int_0^h dh \quad \text{or } p - p_{atm} = \rho g h \text{ or}$$

$$p = p_{atm} + \rho g h$$

Since atmospheric pressure acts on the top of the gate and at the free surface the above equation becomes

$$p = \gamma h$$

From the diagram (Fig. 1) $h = D + y\sin\theta$

Thus $$p = \gamma(D + y\sin\theta)$$

Hence,

$$|\vec{F}_R| = -w\int_A p\,dy$$

$$= -w\int_0^L \gamma(D + y\sin\theta)dy = -\rho g w\int_0^L (D + y\sin\theta)dy$$

$$= -\rho g w\left[Dy + \frac{y^2}{2}\sin 25^\circ\right]_0^L = -\rho g w\left[DL + \frac{L^2}{2}\sin 25^\circ\right]_0^L$$

With, $\rho = \rho_{H_2O} = 999\ \frac{kg}{m^3}$, $g = 9.81\ \frac{m}{sec^2}$, D = 5m, L = 1.5m and

w = 2m

$$|\vec{F}_R| = -(999)(9.81)(2)\left[(3)(1.5) + \frac{(1.5)^2}{2}\sin 25^\circ\right]$$

$$|\vec{F}_R| = -97,520.62\text{ N}$$

The line of action of $\vec{F}_R$ is along the positive axis z through r (Fig. 3) where

$$r = \bar{x}i + \bar{y}j$$

$$\bar{x} = \frac{1}{F_R}\int xpdA \quad \text{and} \quad \bar{y} = \frac{1}{F_R}\int ypdA$$

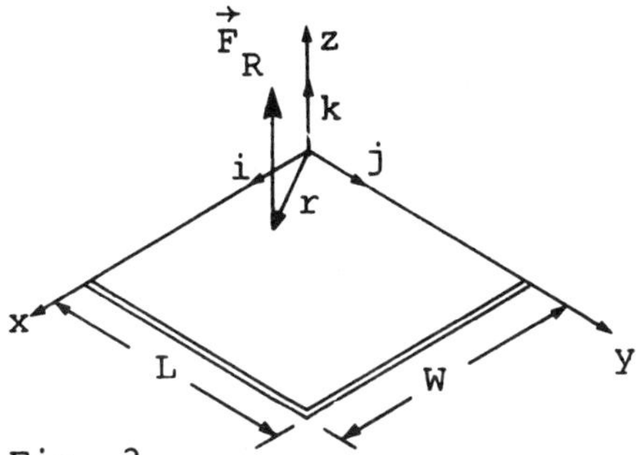

Fig. 3

then

$$\bar{y} = \frac{1}{F_R}\int ypdA = \frac{1}{F_R}\int_0^L y\rho w dy = \frac{\rho g w}{F_R}\int_0^L y(D + y\sin 25^\circ)dy$$

$$= \frac{\rho g w}{F_R}\left[\frac{D}{2}y^2 + \frac{y^3}{3}\sin 25^\circ\right]_0^L = \frac{\rho g w}{F_R}\left[\frac{DL^2}{2} + \frac{L^3}{3}\sin 25^\circ\right]$$

$$= \frac{(999)(9.81)(2)}{(97,520.62)}\left[\frac{(3)(1.5)^2}{2} + \frac{(1.5)^3}{3}\sin 25^\circ\right]$$

$$\Rightarrow \bar{y} = 0.7739\text{m}$$

and $\bar{x} = \frac{1}{F_R}\int xpdA = \frac{1}{F_R}\int \frac{w}{2}pdA = \frac{w}{2F_R}\int pdA = \frac{w}{2F_R}F_R = \frac{w}{2}$

or $\bar{x} = \frac{w}{2} = 1\text{m}$

Therefore, the resultant force $\vec{F}_R$, has a magnitude of $|\vec{F}_R| =$ 97,520.62N, direction towards the x-axis, and passing through the point $\vec{r} = 0.773\ i + 1j$.

(B) The gate will remain closed if $\Sigma M = 0$. Taking moments about the hinge at point A (see Fig. 4),

$\circlearrowleft + \Sigma M_A = 0$ or

$\vec{F}_B L - \vec{F}_R y = 0$

or $\vec{F}_B = \frac{\vec{F}_R y}{L}$

or $|\vec{F}_B| = \frac{|\vec{F}_R| y}{L}$ or

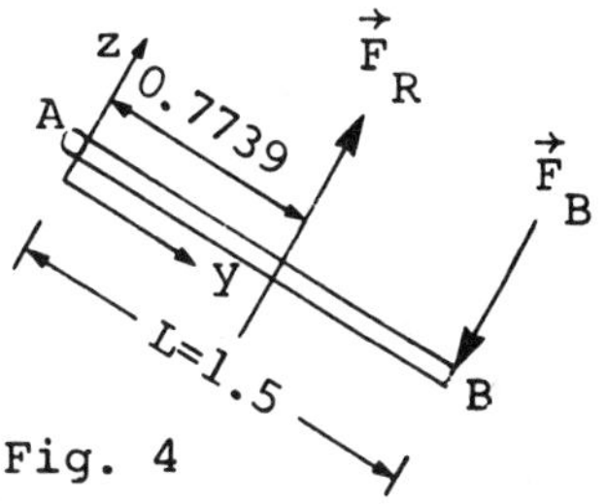

Fig. 4

$$|\vec{F}_B| = \frac{(97,520.62)(0.7739)}{(1.5)} = |\vec{F}_B| = 50,314.13\text{ N}$$

• PROBLEM 2-19

Determine the magnitude of the force on the inclined gate shown in Fig. 1. The tank of water is completely closed and the pressure gage at the lower corner reads 88,000 N/m^2. Assume $\gamma = 9800\ N/m^3$.

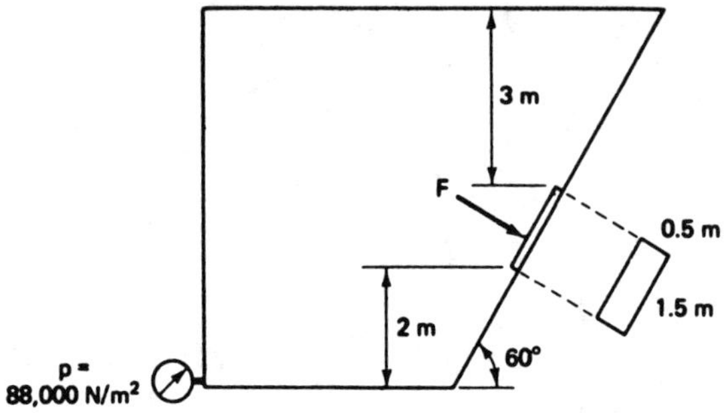

Fig. 1

Solution: Forces on horizontal surfaces may be determined. Since the surface in question is everywhere at the same depth, it is everywhere at constant pressure as well. Thus, we have,

$$F = \int_A p\, dA = pA$$

or

$$F = \gamma hA$$

where h is the distance from the free surface or plane of zero pressure down to the horizontal surface. Although derivations generally assume that the zero (i.e., atmospheric) pressure surface is a free surface, this is not always the case. If the liquid is completely contained and under sufficient pressure, then there is no plane of zero pressure within the liquid. However, by assuming that liquid of the same specific weight replaces the top of the container to a depth consistent with the pressure against the top, an imaginary free surface can be created. Specifically, if the pressure against the top of the container is p, the imaginary depth given by

$$h = \frac{p}{\gamma}$$

would be required. For purposes of calculation this surface may be treated as the free surface.

To determine a more general expression for the force on a plane surface, consider the arbitrarily shaped surface of area A shown in Fig. 2. This surface is inclined at an angle θ to the free surface. The projection shown is the true shape, that is, the shape seen by an observer looking directly at the face of the surface. Distances measured from the free surface along the incline are indicated by y and vertical distances by h. Distances to the centroid of the area are indicated by an overbar, and the subscript p indicates the distance to the center of pressure, or point of application of the resultant force.

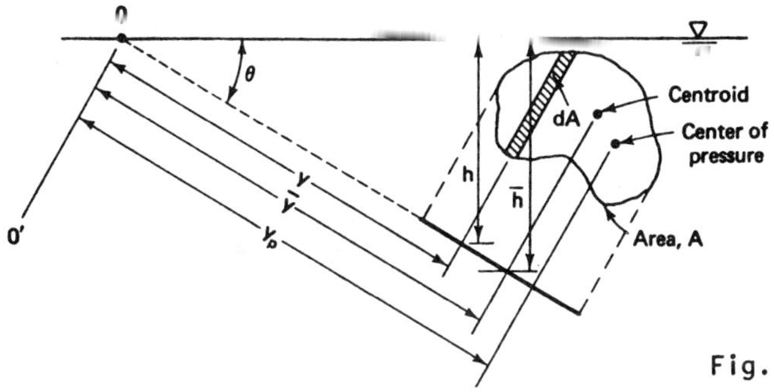

Fig. 2

The force on the differential area is

$$dF = p\ dA = \gamma h\ dA$$

and the integration over the area gives the total force,

$$F = \int_A \gamma h\ dA = \gamma \sin\theta \int_A y\ dA$$

The quantity $\int_A y\ dA$ represents the first moment of the area about the O-O' axis, or $\bar{y}A$. Thus,

$$F = \gamma \sin\theta\ \bar{y}A = \gamma\bar{h}A = p_c A \qquad (1)$$

where p_c is the pressure at the centroid.

A pressure of 88,000 N/m^2 is equivalent to that due to a depth of water of

$$h = \frac{p}{\gamma} = \frac{88{,}000\ N/m^2}{9800\ N/m^3} = 8.98\ m$$

The line of zero pressure is 8.98 m above the bottom of the tank. Since the centroid of the gate is

$$2 + \frac{(1.5)\ \sin 60°}{2} = 2.65\ m$$

above the tank bottom it also lies 8.98 - 2.65 = 6.33 m below the line of zero pressure. The force on the gate can now be calculated by Eq. 1:

$$F = \gamma\bar{h}A = (9800\ N/m^3)(6.33\ m)(1.5\ m)(0.5\ m)$$

$$= 46{,}500\ N$$

• PROBLEM 2-20

Calculate magnitude, direction, and location of the total force exerted by the water on one side of this composite area which lies in a vertical plane.

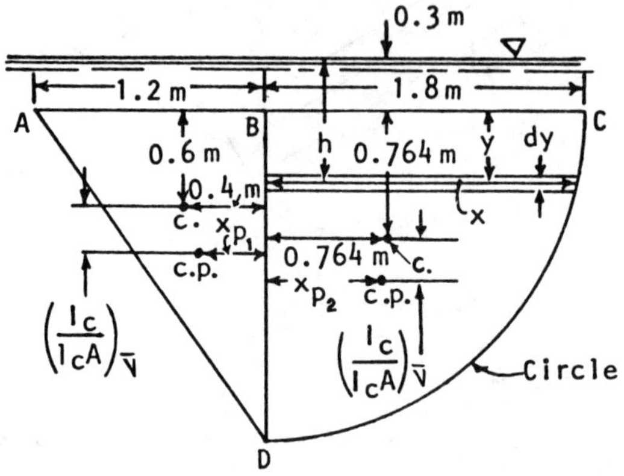

Fig. 1

Solution: For the definition of symbols see fig (2). By inspection the direction of the force is normal to the area.

Magnitude (see Table of Area and Volume Properties for area and volume, location of centroid and I or I_c .)

h_c: vertical distance from surface to the centroid of the submerged area.

Force on triangle $\gamma h_c A = 9.8 \times 10^3 \times 0.9 \times 1.08 =$ 9.53kN

Force on quadrant $\gamma h_c A = 9.8 \times 10^3 \times 1.064(3.24\pi/4$ = 26.53kN

Total force on composite area = 36.06kN

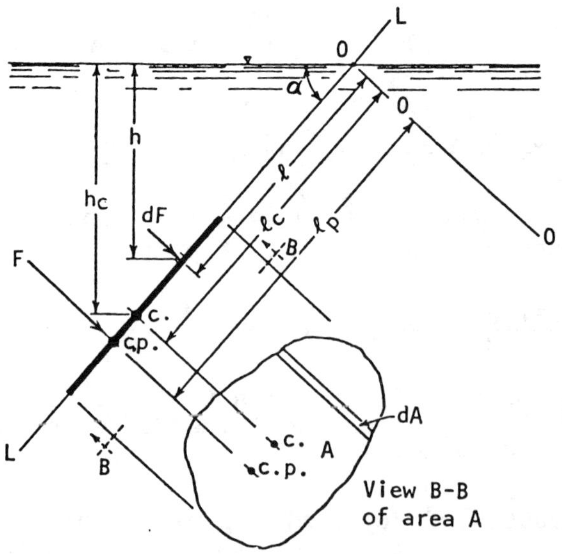

Fig. 2

Vertical location of resultant force:

$$\ell_p - \ell_c = I_c/\ell_c A \text{ for triangle} = \frac{1.2 \times (1.8)^3}{36 \times 0.9 \times 1.08} = 0.2 \text{ m}$$

$$\ell_p - \ell_c = I_c/\ell_c A \text{ for quadrant} = \frac{0.576}{1.064 \times 2.54} = 0.213 \text{ m}$$

Taking moments about line AC,

$$9.53 \times 0.8 + 26.53 \times 0.977 = 36.06(\ell_p - 0.3);$$

$$\ell_p =$$

Lateral location of resultant force: Since the center of pressure of the triangle is on the median line x_{p_1} is given (from similar triangles) by

$$\frac{(0.4 - x_{p_1})}{0.6} = \frac{0.2}{1.8} \qquad x_{p_1} = 0.333 \text{ m}$$

Dividing the quadrant into horizontal strips of differential height, dy, the moment about BD of the force on any one of them is

$$dM = (x\, dy)\gamma h\left(\frac{x}{2}\right)$$

in which $h = y + 0.3$ and $x^2 + y^2 = 3.24$. Substituting and integrating gives the moment about BD of the total force on the quadrant,

$$M = \frac{9.8 \times 10^3}{2} \int_0^{1.8} (3.24 - y^2)(y + 0.3)\, dy = 18\ 575 \text{ N·m}$$

and thus

$$x_{p_2} = \frac{18\ 575}{26.53 \times 10^3} = 0.7 \text{ m}$$

Finally, taking moments about line BD,

$$0.7 \times 26.53 - 9.53 \times 0.333 = 36.06 x_p \ ;$$

$$x_p = 0.427 \text{ m}$$

Thus, the center of pressure of the composite figure is 0.427 m to the right of BD and 1.23 m below the water surface.

FORCES ON A CURVED SURFACE

• PROBLEM 2-21

The end of a reservoir has the shape of quarter circle of radius R_0. It is hinged at the bottom and restrained by a horizontal strap at the top. The reservoir is filled with fluid of specific weight ρg. Determine the strap force T per unit width of reservoir.

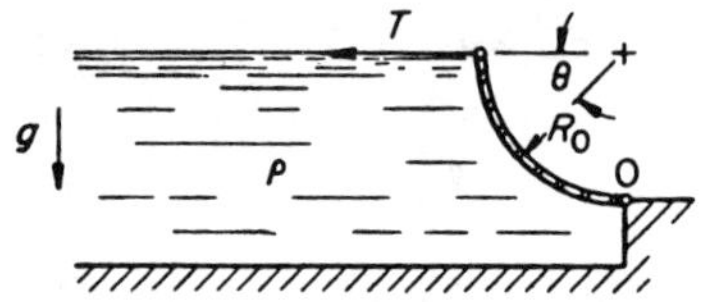

Solution: The pressure at an arbitrary point on the circle is $p = \rho g R_0 \sin\theta$ and the moment of the elemental force $pR_0 d\theta$ which is normal to the cylindrical surface about point 0 is $pR_0^2 \cos\theta d\theta$. Equating this to the restraining moment TR_0 about 0 we have

$$T = \frac{1}{R_0}\int_0^{\pi/2} \rho g R_0^3 \sin\theta\cos\theta d\theta = \frac{1}{2}\rho g R_0^2.$$

Note that the vertical reaction at point 0 is just the weight of the fluid displaced by the cylinder but pointing downwards and that the horizontal reaction there is zero.

• PROBLEM 2-22

Find the magnitude and direction of the force acting on the curved portion of the water tank shown in the figure.

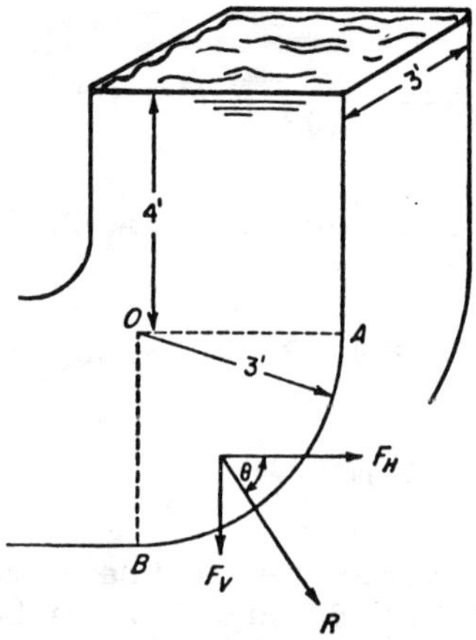

Solution: For the horizontal force F_H,

$$F_H = \text{pressure at centroid of OB} \times (\text{OB} \times 3)$$

$$= 5.5 \times 62.4 \times 3 \times 3$$

$$= 3{,}090 \text{ lb}$$

And for the vertical force F_V,

$$F_V = \text{weight of water above AB}$$

$$= \left(4 \times 3 = \frac{\pi 3^2}{4}\right) \times 3 \times 62.4$$

$$= 3{,}560 \text{ lb}$$

Hence the net force $R = \sqrt{3{,}090^2 + 3{,}560^2} = 4{,}700$ lb.

Since AB is circular, the resultant, being normal to the surface, must pass through O at an angle θ to the horizontal given by

$$\tan\theta = \frac{F_V}{F_H} = \frac{3{,}560}{3{,}090}$$

$$\theta = 49°$$

• PROBLEM 2-23

Consider the cross-section shown below of the hull of a 330,000 tonne (1 tonne = 10^3 kg) oil tanker. Calculate the magnitude, direction, and location of the resultant force per meter exerted by the sea water ($\gamma = 10$ kN/m^3) on the curved surface AB (which is a quarter cylinder) at the corner of the hull.

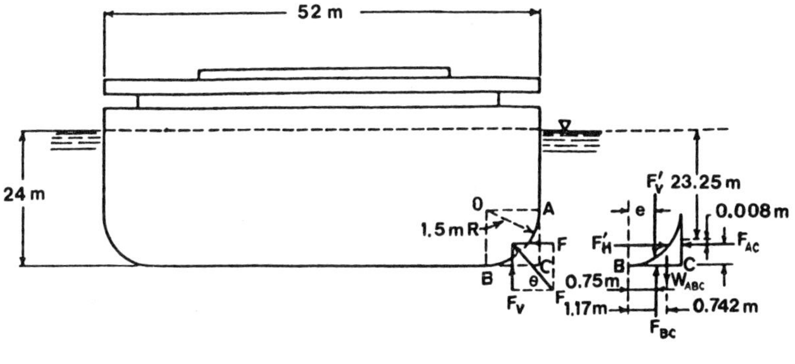

Solution: Isolate the free body of water ABC.

Horizontal component: By inspection, this force (on the free body) has direction to the right:

$$F_H' = F_{AC} = 1.5 \times 10^4 \times 23.25 = 348{,}8 \text{ kN/m}$$

$$(\ell_p - \ell_c)_{AC} = \frac{I_c}{\ell_c A} = \frac{0.28}{23.25 \times 1.5} = 0.008 \text{ m}$$

Vertical component: By inspection, this force (on the free body) has a downward direction:

$$\sum F_z = 24 \times 10^4 \times 1.5 - F_V' - 10^4\ 2.25 - \frac{2.25\pi}{4} = 0;$$

$$F_V' = 355.2 \text{ kN/m}$$

From statics the center of gravity of ABC is found to be 1.17 m to the right of B, and, taking moments of the forces (on the free body) about O,

$$355.2 \times e + 4.8 \times 1.17 - 360 \times 0.75 = 0;$$

$$e = 0.74 \text{ m}$$

Resultant force of water on AB:

Direction: upward to the left, $\theta = \arctan 355.2/348.8$

$= 45.5°$

Magnitude:

$$F = \sqrt{(348.8)^2 + (355.2)^2} = 497.8 \text{ kN/m}$$

Location: through a point 0.742 m above and 0.74 m to the right of B. Because the pressure forces on the elements of the cylinder, although different in magnitude, all pass through O, and thus form a concurrent force system, their resultant will also be expected to pass through O. Since (1.5 - 0.742)/0.74 = 355.2/348.8, this expectation is confirmed. Note that these tankers are of the order of 350 m long!

A sluice gate is in the form of a circular arc of radius 6 m as shown in the figure. Calculate the magnitude and direction of the resultant force on the gate, and the location with respect to O of a point on its line of action.

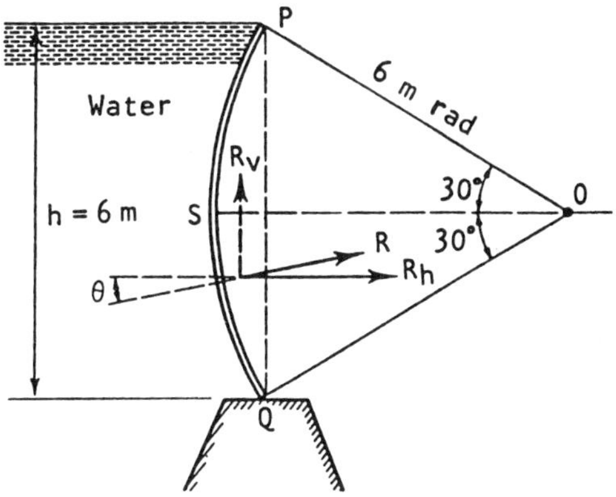

Sector gate.

Solution: Since the water reaches the top of the gate,

Depth of water, $h = 2 \times 6 \sin 30° = 6$ m,

Horizontal component of force on gate = R_h per unit length

= Resultant pressure on PQ per unit length

$= \rho g \times h \times h/2 = \rho g h^2/2$

$= (10^3 \times 9.81 \times 36)/2 \text{ Nm}^{-1} = 176.58 \text{ kN m}^{-1}$,

Vertical component of force on gate = R_V per unit length

= Weight of water displaced by segment PSQ

= (Sector OPSQ - ΔOPQ)ρg

$= ((60/360) \times \pi \times 6^2 - 6 \sin 30° \times 6 \cos 30°) \times$

$10^3 \times 9.81 \text{ N m}^{-1}$

$= 32.00 \text{ kN m}^{-1}$,

Resultant force on gate, $R = \sqrt{(R_h^2 + R_V^2)}$

$= \sqrt{(176.58^2 + 32.00^2)} = 179.46 \text{ kN m}^{-1}$.

If R is inclined at an angle θ to the horizontal,

$$\tan \theta = R_V/R_h = 32.00/176.58 = 0.181\ 22$$

$$\theta = 10.27° \text{ to the horizontal.}$$

Since the surface of the gate is cylindrical, the resultant force R must pass through O.

• PROBLEM 2-25

The water level is 4 ft from the top of the cylindrical gate shown in Fig. (1). Calculate the resultant water pressure force acting on the gate per unit length. (See Fig. 2)

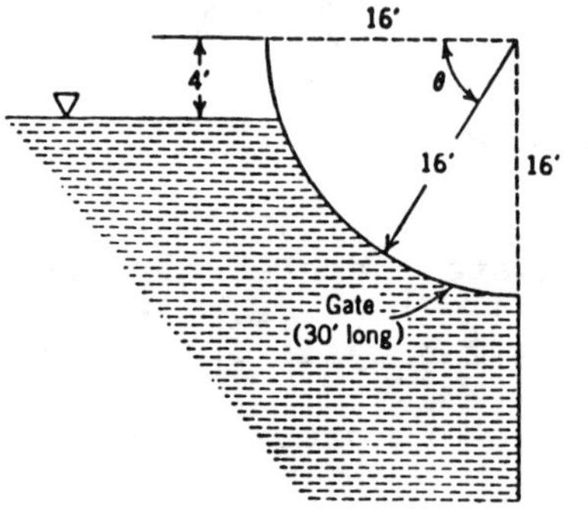

Fig. 1

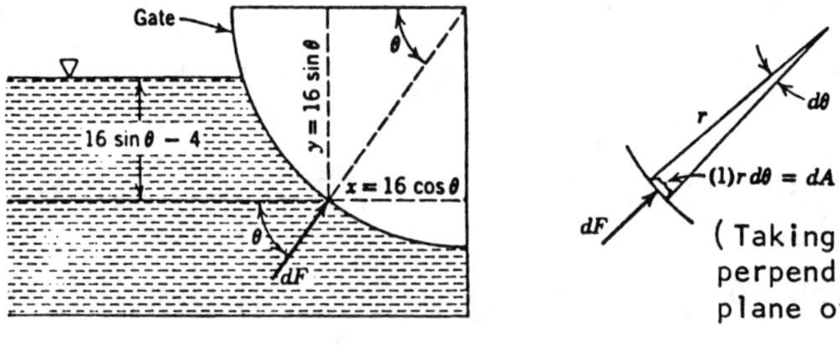

Fig. 2

(Taking unit length perpendicular to the plane of the paper.)

Solution: $dF = pdA = \gamma(16 \sin \theta - 4)(1)rd\theta$

$$= 4\gamma(4 \sin \theta - 1)(16)d\theta$$

$$= 64\gamma(4 \sin \theta - 1)d\theta$$

Because integration is an algebraic sum, you must add all the forces in the same direction. Sum the horizontal components (x direction) and then the vertical components (y direction).

$$\overset{+}{\rightarrow} dF_x = \cos \theta\, dF = \cos \theta[64\gamma(4\sin \theta - 1)]d\theta$$

$$F_x = 64\gamma \int_{\theta = \arcsin 1/4}^{\theta = \pi/2} (4 \sin\theta - 1)\cos\theta \, d\theta$$

$$F_x = 72\gamma \text{ lb/ft of length}$$

$$+\uparrow \quad dF_y = \sin\theta \, dF = \sin\theta[64\gamma(4\sin\theta - 1)]d\theta$$

$$F_y = 64\gamma \int_{\arcsin 1/4}^{\pi/2} (4\sin^2\theta - \sin\theta)d\theta$$

$$F_y = 137.8\gamma \text{ lb/ft of length}$$

To find the location of the resultant, take moments about center of cylinder at 0.

$$F_x y_F = \int y \, dF_x = \int (16\sin\theta)dF_x$$

$$= \int_{\arcsin 1/4}^{\pi/2} (16\sin\theta)64\gamma(4\sin\theta - 1)\cos\theta \, d\theta$$

$$= 864\gamma \text{ ft-lb/ft of length}$$

$$y_F = \frac{864\gamma}{F_x} = \frac{864}{72} = 12 \text{ ft below 0}$$

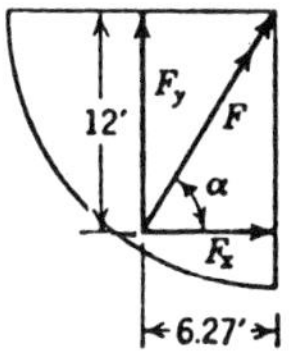

Fig. 3

$$F_y x_F = \int x \, dF_y =$$

$$= \int_{\arcsin 1/4}^{\pi/2} 16\cos\theta\sin\theta[64\gamma(4\sin\theta - 1)]d\theta$$

$$= 864\gamma \text{ ft-lb/ft of length}$$

$$x_F = \frac{864}{137.8} = 6.27 \text{ ft to left of 0}$$

$$\tan\alpha = \frac{137.8\gamma}{72\gamma}$$

$$\tan\alpha = 1.915$$

For the magnitude of resultant,

$$F = [F_x^2 + F_y^2]^{1/2}$$

$$= \gamma[(137.8)^2 + (72)^2]^{1/2}$$

$$= 155.5\gamma = 9700 \text{ lb/ft length}$$

Note that from the location of the point of application and the angle, the resultant force acts through center

0, as you could have predicted had you remembered that the pressure force must be normal to the surface. This implies each differential force dF is normal to the surface and, therefore, through the center of the cylinder.

• PROBLEM 2-26

A cross-section of a water channel is shown below. Find the magnitude of the hydrostatic force acting on the flat inclined surface and the curved surface given the equation of the two surfaces as $y = -z$ and $z = y^2$, respectively. Also find the line-of-action of the two forces.

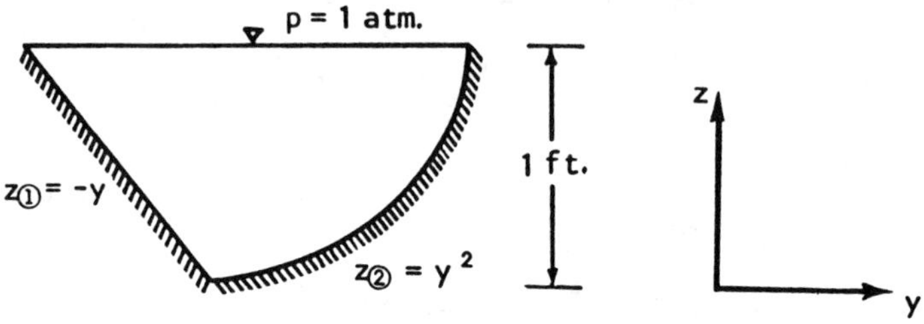

Solution: The pressure force per unit width on the inclined planar surface is

$$F_{p_{①}} = \gamma A \bar{z} \sin\theta = \gamma \bar{h} A$$

$$= \gamma\left(\frac{\sqrt{2}}{2} \sin 45°\right)\sqrt{2}$$

$$= \frac{\sqrt{2}}{2}\gamma = 44.1 \text{ lbs./unit width,}$$
(195.56 N/width)

The horizontal component of the hydrostatic force $F_{p_{y_②}}$ on the second face, i.e. the curved face, is

$$F_{p_{y_②}} = \gamma \int z dA_y$$

$$= \gamma \int_1^0 z dz = \frac{\gamma}{2}$$

Note, this is exactly the value of the y-component of the hydrostatic normal force $F_{①}$ on the first face, and thus

$$F_{p_{y_①}} = F_{p_{y_②}}$$

The magnitude of the vertical component of the hydrostatic force on the second face $F_{p_{z_{②}}}$ is

$$F_{p_{z_{②}}} = \gamma \int_{A_z} z_{②} dA_z$$

$$= \gamma \int_0^1 z_{②} dy$$

$$= \gamma \int_0^1 (y^2 - 1) dy$$

$$= -\frac{2\gamma}{3} = 41.5 \text{ lbs./unit width, } (184.34 \text{ N/unit width})$$

The magnitude of the normal force on the curved face is

$$F_{p_{②}} = \sqrt{(F_{p_{y_{②}}})^2 + (F_{p_{z_{②}}})^2}$$

$$= \gamma\sqrt{\frac{1}{4} + \frac{4}{9}}$$

$$= 52 \text{ lbs./unit width}$$

The location of the center-of-pressure is found by integrating equations (a) and (b) for the y and z plane.

$$z_{c.p.} = \frac{\int_0^1 z^2 dz}{\int_0^1 z dz} = \frac{2}{3} \text{ ft } (0.2032 \text{ m}) \qquad \text{(a)}$$

for both surfaces (the inclined plane and the curved surface), as measured from the free-surface. The coordinate $y_{c.p.}$ is found setting $dA_z = -dy$:

$$y_{c.p._{①}} = \frac{\int_0^{-1} yz dy}{\int_0^1 z dy}$$

$$= \frac{\int_0^{-1} y(1+y) dy}{\int_0^{-1} (1+y) dy} = -\frac{1}{3} \text{ft } (-0.1016 \text{ m}) \qquad \text{(b)}$$

for the y-location of the center-of-pressure on surface ①. For the curved surface ②.

$$y_{c.p._{②}} = \frac{\int_0^1 y(1-y^2)\,dy}{\int_0^1 (1-y^2)\,dy} = \frac{3}{8} \text{ ft } (0.1143 \text{ m})$$

Thus, the center-of-pressure is not even on the surface of the curved wall.

• PROBLEM 2-27

In Fig. (1) the surface AB is a circular arc with a radius of 2 ft. The distance DB is 4 ft. If water is the liquid supported by the surface and if atmospheric pressure prevails on the other side of AB, determine the magnitude and line of action of the resultant hydrostatic force on AB per unit length.

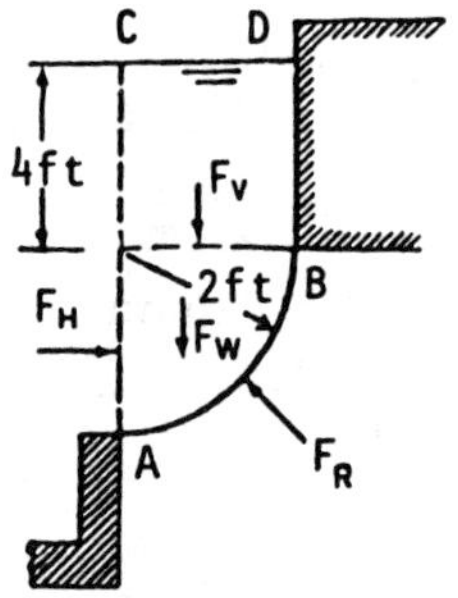

Anlysis of hydrostatic force on a curved surface. Fig. 1

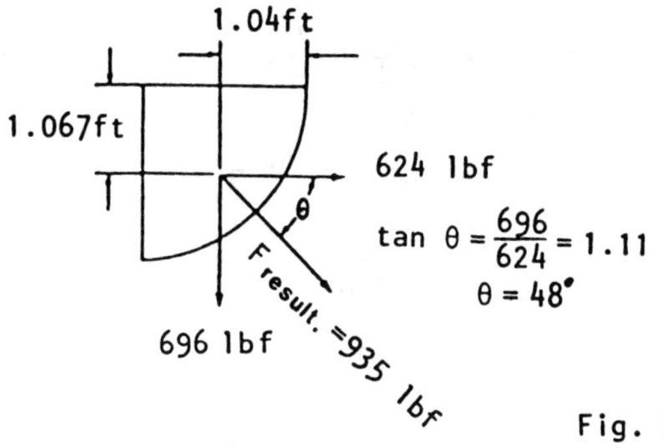

Fig. 2

Solution: This type of problem is analyzed by constructing horizontal and vertical projections of the surface (Fig. 1). Then the equations of equilibrium are applied to the fluid enclosed by these projections and by the surface in question. Consider the two-dimensional curved surface AB shown in Fig. 1. Assume that this surface has unit length normal to the paper. If we consider the mass of fluid OAB as a freebody and analyze the forces which act on it, we can identify the components that we are looking for. Because this is a hydrostatic condition, only normal forces act on the hypothetical surfaces OA and OB, and these forces are identified as F_H and F_V, respectively. The only other forces to act on this body of fluid are then the weight of the fluid itself, F_W, and the reaction from the curved surface, F_R. Therefore, all the forces acting on the body have been identified and the freebody is depicted as in Fig. 3. We now apply the equations of equilibrium to the freebody. In the horizontal direction $\Sigma F_x = 0$. The $F_H - F_{Rx} = 0$, from which we get $F_{Rx} = F_H$. In the vertical direction $\Sigma F_y = 0$. Hence, we have

$$-F_V - F_W + F_{Ry} = 0$$

$$F_{Ry} = F_V + F_W$$

Above we have shown that the magnitude of the horizontal reaction of the curved surface is equal to the hydrostatic force which acts on a vertical projection of the curved surface, and the magnitude of the vertical reaction is equal to the sum of the vertical forces acting above the curved surface--in this case, the weight of the fluid above.

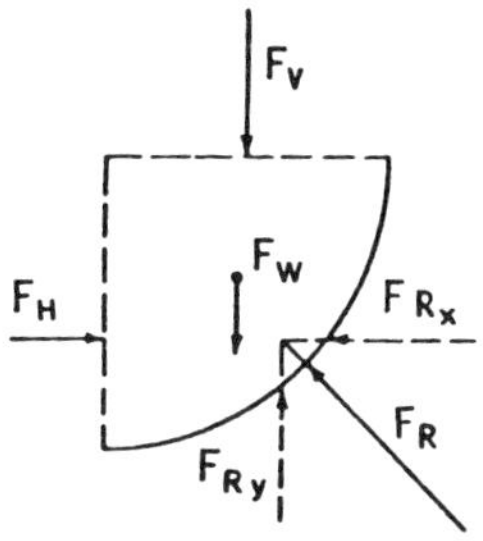

Freebody diagram. Fig. 3

Since we have found which forces make up the reaction of the curved surface, we need only reverse the signs on the components of the reaction to arrive at the components of the resultant hydrostatic force acting on the surface. By vectorally adding the component forces, we can determine both the magnitude and line of action of the resultant.

In summary, the horizontal component of force acting on a curved surface is equal to the force acting on a vertical projection of that surface--which includes both magnitude and line of action. Also, the magnitude of the vertical force acting on a curved surface is equal to the sum of all vertical forces acting on it. The lines of action of these vertical components are used to determine the line of action of the resultant vertical force.

The vertical component is equal to the weight of the water in volume AOCDB:

$$W_{OCDB} = \gamma \forall_{OCDB} = 62.4 \times 4 \times 2 \times 1 = 500 \text{ lbf}$$

$$W_{AOB} = \gamma \forall_{AOB} = \tfrac{1}{4}\pi r^2 = 196 \text{ lbf}$$

Therefore, the vertical component is $F_{Ry} = 696$ lbf. The line of action of the vertical force component acts through the centroid of the volume of water considered above, and this is calculated by taking moments about a horizontal axis through D and normal to plane ODB.

$$\bar{x}F_{Ry} = 1 \times 500 + r(1 - \frac{4}{3\pi})(196) = 725.5 \text{ ft lbf}$$

$$\bar{x} = \frac{725.5}{696} = 1.04 \text{ ft}$$

The magnitude of the horizontal component is given by the force on OA:

$$F_H = \bar{p}A$$

$$= 5 \times 62.4 \times 2 \times 1 = 624 \text{ lbf}$$

Location of the line of action of the horizontal component is

$$y_{cp} - \bar{y} = \frac{\bar{I}}{\bar{y}A}$$

$$y_{cp} - \bar{y} = \frac{1 \times 2^3/12}{5 \times 2} = 0.0667 \text{ ft}$$

$$y_{cp} = 5.067 \text{ ft } (1.544\text{m})$$ ⇐ answer

The resultant force is then as shown in figure (2).

• PROBLEM 2-28

(a) Find the X- and Z-components of the resultant force acting on the submerged gate consisting of a ¼ circular cylinder, as shown in Fig. 1. (b) Find the coordinates for the center of pressure of the component of the resultant force, $\vec{F}_{R,1}$. The cylinder is 4 ft long and has a radius of 3 ft.

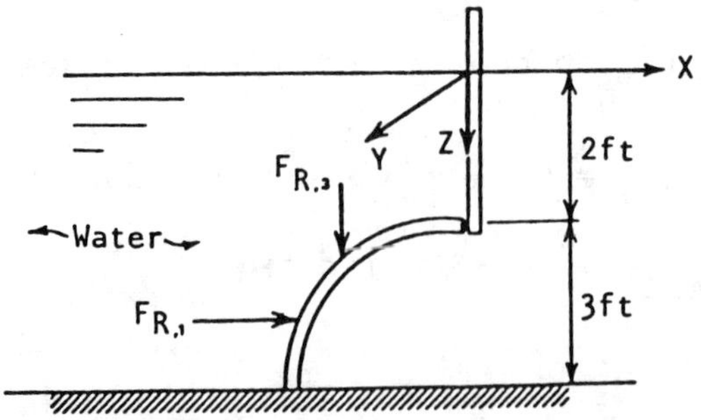

Fig. 1

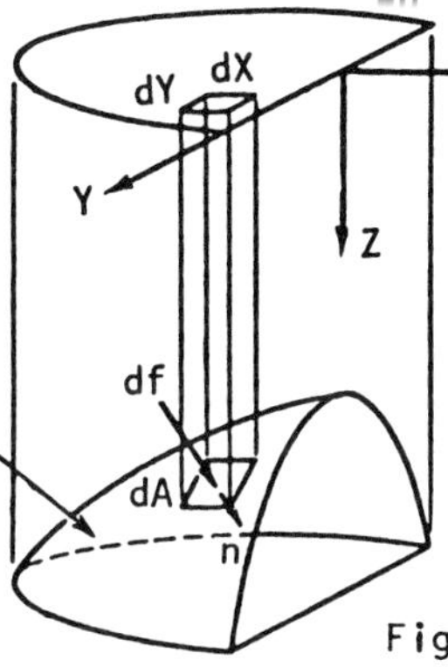

Fig. 2

Solution: a) Consider the forces on the submerged curved surface shown in Fig. 2. To obtain the Z-component of the resultant force, $\vec{F}_R$, first construct a grid on the surface by intersecting the surface with planes parallel to the (X,Z) and (Y,Z) planes. Then the force, $\vec{df}$, acting on area dA, is given by

$$\vec{df} = p\, dA\, \vec{n} = p\, \vec{dA} = df_1\vec{i} + df_2\vec{j} + df_3\vec{k}$$

The component df_3 can be obtained by taking the dot product of $\vec{df}$ with the unit vector $\vec{k}$; that is,

$$df_3 = \vec{df} \cdot \vec{k} = p\, \vec{dA} \cdot \vec{k}$$

$\vec{dA} \cdot \vec{k}$ = the projection of dA on the (X,Y) plane

$= dX\, dY$

Then,

$df_3 = p \cdot$ [projection of dA on the (X,Y) plane]

Since $p = \gamma Z$,

$$df_3 = \gamma Z\, dX\, dY$$

$(Z\, dX\, dY)$ is the volume in the column above the surface element, dA, and thus df_3 equals the weight of liquid in that column. The Z-component of the resultant force, $F_{R,3}$, is the sum of the forces on all the infinitesimal areas, which of course is an integral; that is,

$$F_{R,3} = \oiint_{A_3^*} \gamma Z\, dX\, dY \qquad (1)$$

= the weight of liquid directly above the surface A

where A_3^* is the projection of surface A on the (X,Y) plane. Similarly, the X-component of the resultant force $F_{R,1}$ is given by

$$F_{R,1} = \oiint_{A_1^*} \gamma Z \, dY \, dZ \tag{2}$$

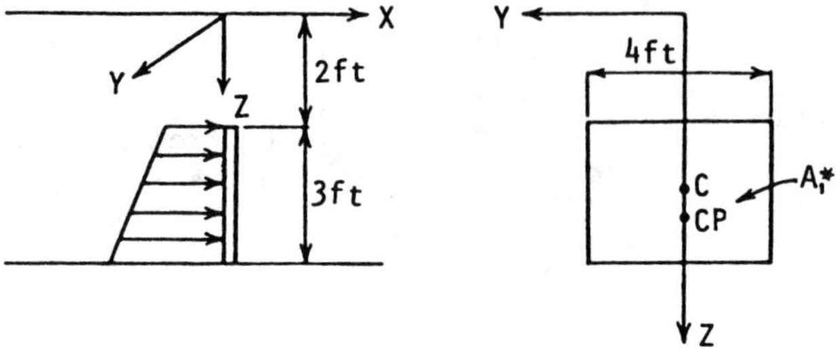

Fig. 3

$F_{R,1}$ is the same force that would exist on the vertical plane surface shown in Fig. 3. Its magnitude is given by Eq. 2; that is,

$$F_{R,1} = \oiint_{A_1^*} \gamma Z \, dY \, dZ = 4\gamma \int_2^5 Z \, dZ = \frac{4(62.4)}{2}[Z^2]_2^5 = 2621 \text{ lb}_f$$

The Z-component of the resultant force $F_{R,3}$, equals the weight of liquid directly above the surface, as given by Eq. 1. Therefore,

$$F_{R,3} = \gamma \cdot \text{(volume of water directly above the gate)}$$

$$= 62.4(4)\left[5(3) - \frac{\pi}{4}(3)^2\right] = 1980 \text{ lb}_f$$

(b) If (Y^*, Z^*) are the coordinates of the center of pressure of $F_{R,1}$, then

$$Z^* F_{R,1} = \iint_{A_1^*} \gamma Z^2 \, dY \, dX = 4\gamma \int_2^5 Z^2 \, dZ = \frac{4(62.4)}{3}[Z^3]_2^5$$

$$= 9734 \text{ lb}_f\text{-ft}$$

or

$$Z^* = \frac{9737}{2621} = 3.71 \text{ ft}$$

By symmetry, $Y^* = 0$.

FLUID UNDER ACCELERATION

• PROBLEM 2-29

At a particular instant an airplane is traveling downward at a velocity of 180 m/s in a direction that makes an angle of 40° with the horizontal. At this instant the airplane is gaining speed at the rate of 4 m/s² (see Fig. 1). Also it is moving on a concave upward circular path having a radius of 2,600 m. Determine for the given conditions the position of the free liquid surface in the fuel tank of this vehicle.

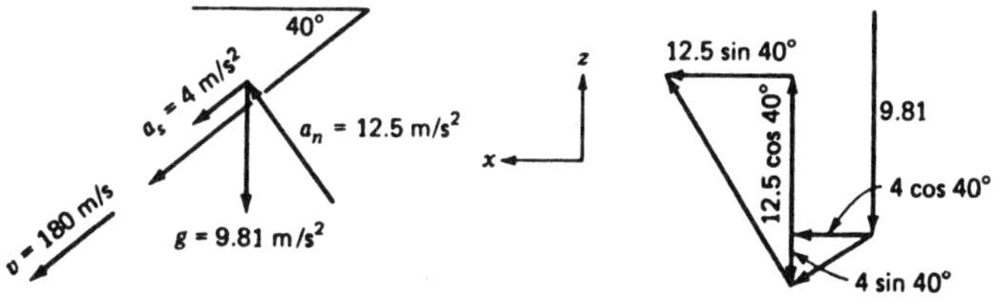

Fig. 1

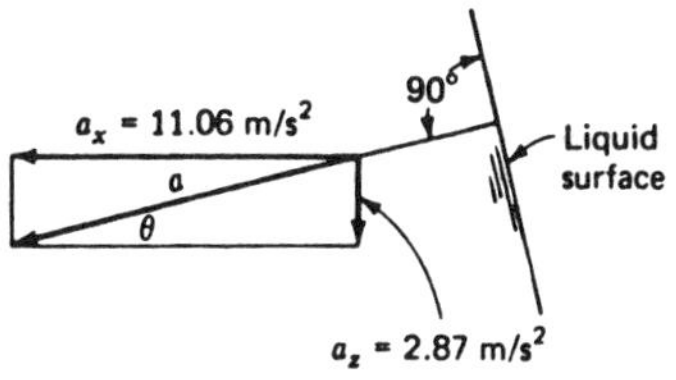

Fig. 2

Solution:

$$a_n = \frac{v^2}{r} = \frac{180^2}{2,600} = 12.5 \text{ m/s}^2$$

$$a_x = 4 \cos 40° + 12.5 \sin 40° = 11.06 \text{ m/s}^2$$

$$a_x = -9.81 - 4 \sin 40° + 12.5 \cos 40° = -2.87 \text{ m/s}^2$$

$$\theta = \tan^{-1}\frac{2.87}{11.06} = 14° \ 30' \quad \text{(see Fig. 2)}$$

Liquid surface makes an angle of 14° 30' with the vertical.

• PROBLEM 2-30

A U-tube filled with water is mounted on a drag racer. The racer accelerates at constant rate from start, and 5 sec later the water in the U-tube has the position shown in Fig. 1. What are the acceleration and velocity of the racer at that instant (neglect transient viscous effects of the water in the tube)?

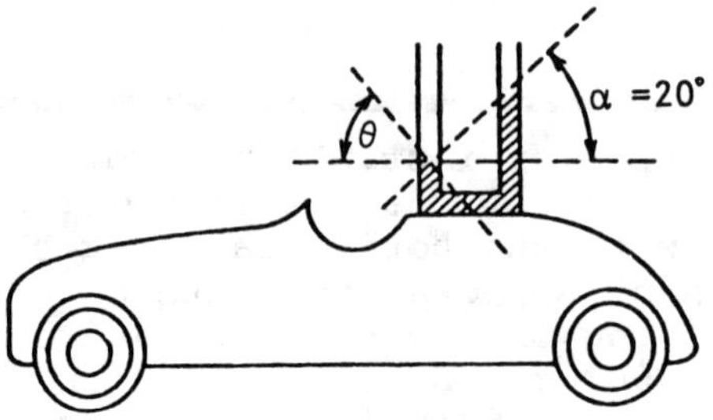

Fig. 1

Solution: From Fig. (2) we note the "affective gravitational acceleration" g_{eff} is normal to the line drawn through the ends of the water column. The ends are at atmospheric pressure and lie on a line of constant pressure. The magnitude of a_0 is found readily (see Fig. 2):

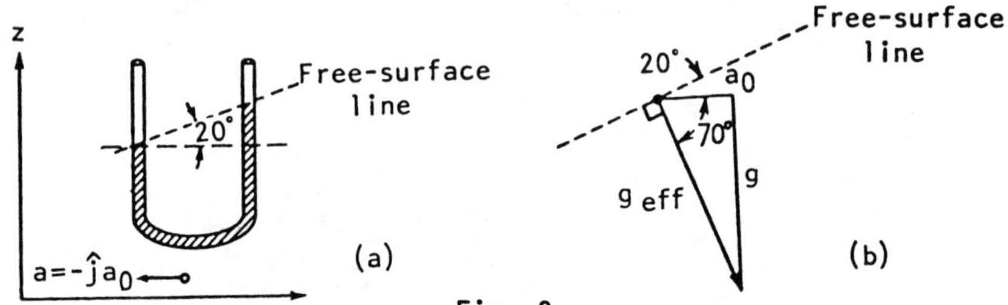

Fig. 2

$$a_0 = g \cot 70° = 32.2 \text{ ft/sec}^2 \times 0.364 = 11.72 \text{ ft/sec}^2$$

The magnitude of the racer velocity is found from

$$\frac{dV}{dt} = a_0 = 11.72 \text{ ft/sec}^2$$

By integrating and setting t = 5 sec,

$$V = 11.72t = 11.72 \times 5 \text{ ft/sec} = 58.60 \text{ ft/sec} \simeq 40 \text{ mph}$$

• PROBLEM 2-31

(a) An open tank filled with water rolls down an inclined plane which makes an angle α with the horizontal (see Fig. 1). The tank has a mass M, and the force produced by air resistance and rolling friction is F_{fric}. What angle will the surface of the water make with the bottom of the tank?

(b) A tank is in free fall (see Fig. 2). Find the pressure difference between two points separated by a vertical distance h.

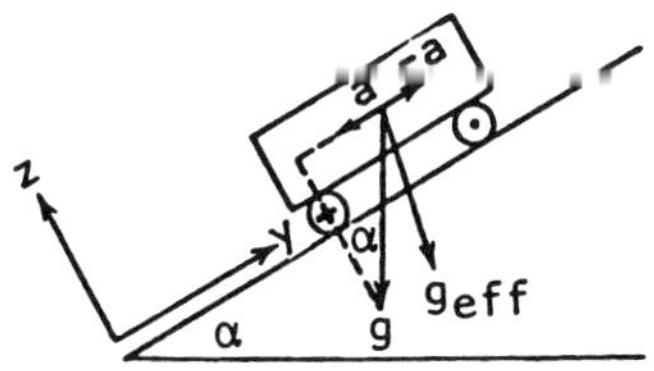

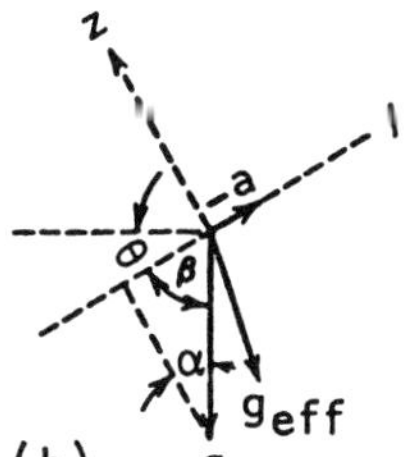

Fig. 1

Solution: The solution hinges on finding the acceleration $\vec{a}$ down the plane. The force from the gravitational field in the direction parallel to the plane surface and down the plane is g sin α (see Fig. 1). The retarding force is F_{fric}. Consequently the net force is

$$F_{net} = Mg \sin \alpha - F_{fric}$$

This force gives rise to the acceleration

$$\vec{a} = -j(g \sin \alpha - \frac{F_{fric}}{M}) \tag{1}$$

The effective body force $\vec{g}_{eff}$ can now be sketched (see Fig. 1(b)), and the angle β which $\vec{g}_{eff}$ makes with the negative y-axis is found from

$$\tan \beta = \frac{g \cos \alpha}{g \sin \alpha - F_{fric}/M}$$

(b) We can consider the tank in free fall to be a special case of the tank moving down a plane with α = 90°.

Using Eq. 1

$$\vec{a} = -j(g \sin \alpha - \frac{F_{fric}}{M})$$

and knowing that

$$P_A - P_B = e \cdot \vec{g}_{eff} \cdot h$$

and

$$\vec{g}_{eff} = \vec{g} - \vec{a}$$

with

$$\alpha = 90°,$$

$$\vec{a} = j\left[\frac{F_{fric}}{M} - g\right]$$

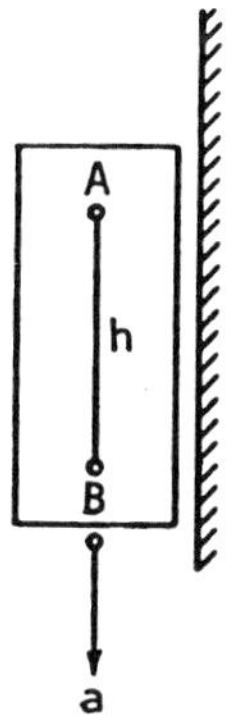

Fig. 2

$$\vec{g}_{eff} = -jg + jg - j\frac{F_{fric}}{M} = -j\frac{F_{fric}}{M}$$

Therefore

$$P_A - P_B = \rho\frac{F_{fric}}{M}h$$

If $F_{fric} = 0$ (for example, no air resistance), then $P_A = P_B$ throughout the tank. In any event, we would expect pressure differences to be small.

• PROBLEM 2-32

An open horizontal tank 2 ft high, 2 ft wide, and 4 ft long is full of water. How much water is spilled when the tank is accelerated horizontally at 8.05 ft/sec^2 in a direction parallel with its longest side? What are the forces on the ends under these conditions?

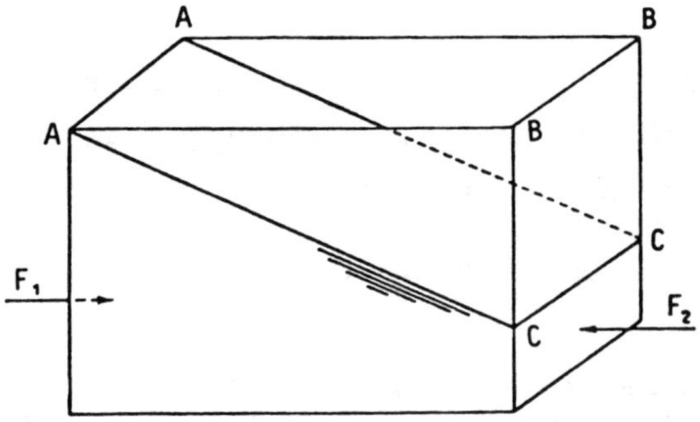

Solution: The surface will incline at an angle θ to the horizontal so that

$$\tan\theta = \frac{a}{g} = \frac{1}{4}$$

The water will spill until the level at the rear of the tank just reaches the top of the tank and the surface is inclined at an angle of arctan ¼ to the horizontal. The quantity spilled is the volume of the wedge ABCA'B'C', as shown in the figure. Now

$$\frac{BC}{4} = \tan\theta = \tfrac{1}{4}$$

and therefore

$$BC = 1 \text{ ft}$$

The volume spilled, then, is 4 ft^3.

The pressure force on the rear face is given by the area multiplied by the pressure at the centroid. Therefore

$$F_1 = 2 \times 2 \times 1 \times 62.4 = 250 \text{ lb}$$

and similarly

$$F_2 = 2 \times 1 \times \tfrac{1}{2} \times 62.4 = 62.4 \text{ lb}$$

[As a check, the difference between the end forces F_1 and F_2 equals the accelerating force:

$$F_1 - F_2 = Ma \text{ , hence}$$

$F_1 - F_2 = 187.6$ lb; the required accelerating force is

$$(16 - 4)\frac{62.4}{32.2} \times 8.05 = 187.4 \text{ lb]}$$

• PROBLEM 2-33

The tank of Fig. 1 contains oil with sp. gr. = 0.88. If the tank is 10 m long, the initial depth of oil is 2 m, and the tank accelerates to the right at 2.45 m/s^2, determine the slope of the surface and the minimum and maximum pressures at the bottom of the tank. Assume that the tank walls are sufficiently high so that there is no spillage.

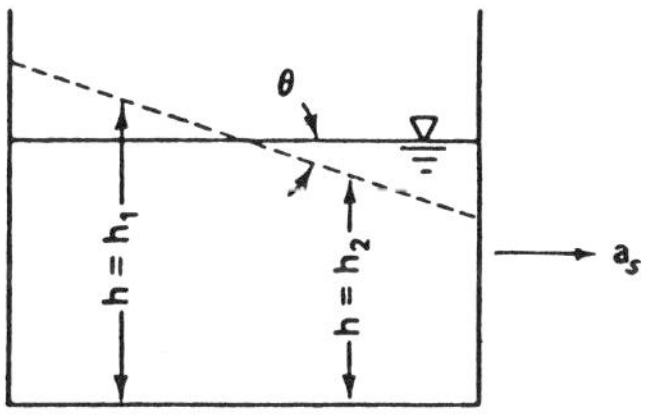

Fig. 1

Solution: Consider the open tank of fluid accelerated to the right with acceleration a_s in Fig. 1. If the acceleration a_s is a constant, the surface will, after an initial period of adjustment, align itself as shown by the dashed line. The value θ must now be determined. Bear in mind that after the period of adjustment there is no relative motion and that all of the fluid is exposed to the identical acceleration. The lines of constant piezometric head will be vertical lines decreasing in value in the direction of the acceleration, which is toward the right end of the tank. Consider two such lines shown on Fig. 1. If the datum is taken at the bottom of the tank, then, at the free surface where the pressure is zero, the piezometric head must be the elevation head or simply the depth. Thus, the piezometric heads h_1 and h_2 are the depths at the respective sections. Now, apply the Euler equation, so as to give

$$\frac{\partial h}{\partial s} = -\frac{a_s}{g} \tag{1}$$

We have seen that h is the elevation of the free surface; thus, $\partial h/\partial s$ is the slope of that surface and further

$$\frac{\partial h}{\partial s} = \tan\theta = \frac{a_s}{g}$$

yielding

$$|\theta| = \tan^{-1}\left(\frac{a_s}{g}\right) \tag{2}$$

$$= \tan^{-1}\left(\frac{2.45\ \text{m/s}^2}{9.80\ \text{m/s}^2}\right) = 14.04°$$

Since there is no spillage, the oil surface will rotate about the midpoint and rise and drop equal amounts:

$$\Delta y = (5\ \text{m})(\tan 14.04°) = 1.25\ \text{m}$$

At the front end of the tank, the pressure is due to a depth of 2.0 - 1.25 = 0.75 m, while at the rear the depth is 3.25 m.

If a tank is accelerated vertically, lines of constant piezometric head must be horizontal. An open tank is considered in Fig. 2.

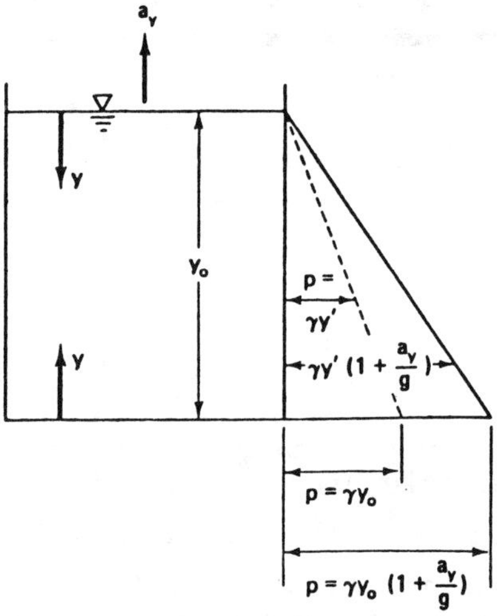

Fig. 2

Since lines of both constant piezometric head and constant elevation head are horizontal, lines of constant pressure, including the free surface, must also remain horizontal. The vertical pressure distribution, which is no longer hydrostatic, can be obtained by integrating Eq. 1. Using the usual vertical notation, Eq. 1 becomes

$$\frac{\partial}{\partial y}\left(\frac{p}{\gamma} + y\right) = \frac{-a_y}{g}$$

or

$$\frac{\partial}{\partial y}\left(\frac{p}{\gamma}\right) = -1 - \frac{a_y}{g}$$

Thus, the pressure decreases in the vertical direction at a rate given by

$$\Delta p = -\gamma \Delta y\left(1 + \frac{a_y}{g}\right)$$

Measuring vertically downward from the surface, the pressure increases with the depth y' according to

$$p = \gamma y'\left(1 + \frac{a_y}{g}\right) \qquad (3)$$

Since there is no vertical motion, $a_y = 0$, therefore Eq. 3 becomes

$$p = \gamma y'$$

The minimum and maximum pressures at the bottom of the tank are

$$p_{front} = \gamma y_{front}$$

$$= (9800 \times 0.88\ N/m^3)(0.75\ m) = 6470\ N/m^2$$

and

$$p_{rear} = \gamma y_{rear}$$

$$= (9800 \times 0.88\ N/m^3)(3.25\ m) = 28{,}000\ N/m^2$$

• PROBLEM 2-34

When the U-tube is not rotated, the water stands in the tube as shown. If the tube is rotated about the eccentric axis at a rate of 8 rad/sec, what is the new level of water in the tube? (Fig. 1)

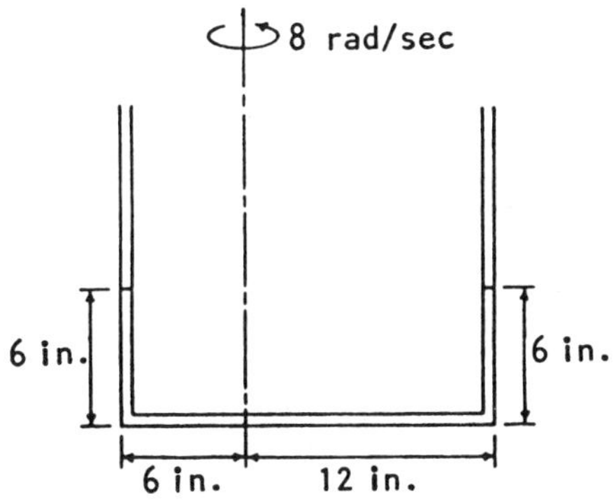

Fig. 1

Solution: Solution of this problem is based upon the equation for rotation of a tank of liquid

$$\frac{p}{\gamma} + z - \frac{V^2}{2g} = \text{const} \tag{1}$$

$$V = r\omega$$

and also upon the fact that the water occupies a given volume of the tube, which may be expressed in terms of a given length of tube. Let the elevation reference be at the level of the horizontal part of the tube; then, by considering a point at the water surface in the left tube where p = 0 and also a point at the surface in the right tube, Eq. (1) can be written as follows:

$$z_l - \frac{r_l^2\omega^2}{2g} = z_r - \frac{r_r^2\omega^2}{2g}$$

Another independent equation involving the volume of tube which the liquid occupies may be written as

$$z_l + z_r = 1.0$$

When $r_l = 1/2$ ft, $r_r = 1$ ft, and $\omega = 8$ rad/sec are substituted into the equation above and the equations are solved for z_l and z_r, one obtains

$$z_l = 0.12 \text{ ft} \qquad \Leftarrow \text{answer}$$

$$z_r = 0.88 \text{ ft} \qquad \Leftarrow \text{answer}$$

• PROBLEM 2-35

A thin, waferlike cylindrical tank, 1 ft in diameter, steadily rotates about its axis at 500 r/min, as shown in the figure. The tank is sealed and completely full of water. After some time, the water rotates with the tank as if it was frozen. What is the pressure difference between the outside edge and the center of the tank? Assume water density = 62.4 lb/ft³.

Solution: Since the fluid moves with the cylindrical tank, the streamlines are concentric circles. The fluid velocity at any radial location r is $r\Omega$. There are no pressure variations in the streamline direction because the magnitude of the velocity does not change in that direction or with time. Since the tank is thin, there is negligible pressure variation in the vertical direction. The only pressure variation in the radial direction, $p = p(r)$ and $\partial p/\partial r \equiv dp/dr$. Since the gravity body force is in the vertical direction, it has no component in the streamline direction. Then

$$\frac{dp}{dr} = \rho\frac{V^2}{r}$$

We substitute the velocity relation $V = r\Omega$, giving

$$\frac{dp}{dr} = \rho r\Omega^2$$

Integrating across the tank leads to

$$\int_{p(0)}^{p(R)} dp = \rho\Omega^2 \int_0^R r\ dr$$

$$p(R) - p(0) \equiv \Delta p = \tfrac{1}{2}\rho R^2\Omega^2$$

Then substituting numerical values, we have

$$\Delta p = \frac{(62.4\ \text{lb/ft}^3)(0.5\ \text{ft})^2(500\ \text{r/min})^2(2\pi/\text{r})^2}{2(60\ \text{s/min})^2 \dfrac{32.17\ \text{lb}}{1\ \text{lbf}\cdot\text{s}^2/\text{ft}}}$$

$$= 664.7\ \text{lbf/ft}^2 = 4.62\ \text{lbf/in}^2$$

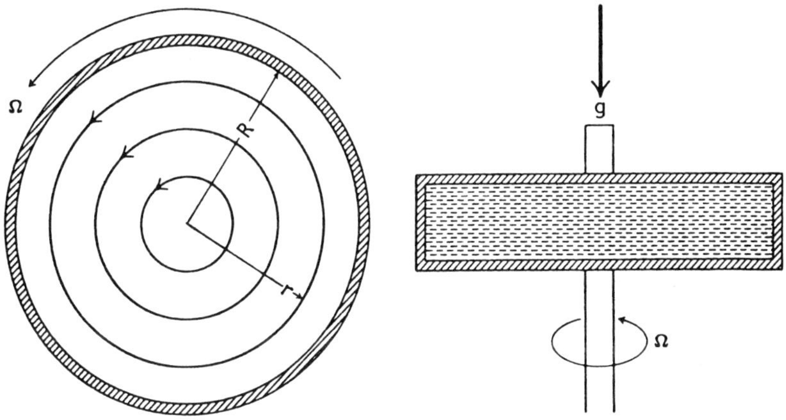

(a) Top view (b) Front view

• PROBLEM 2-36

A closed tank 2 ft in diameter and 3 ft high is filled with water and rotates about its axis at 100 rpm (Fig. 1). A pressure gage at the center of the top reads 4 psi during the rotation. Determine the pressure at the outer perimeter at the bottom of the tank.

Solution: A liquid rotating at a constant angular velocity is shown in Fig. 2. We know from solid body dynamics that the streamlines are concentric circles and that the velocity at any distance r from the axis is a constant equal to $r\omega$. Here ω (Greek lowercase omega) is the angular velocity in radians per second. The acceleration must be radially inward and of magnitude.

$$a_n = \frac{v^2}{r} = r\omega^2$$

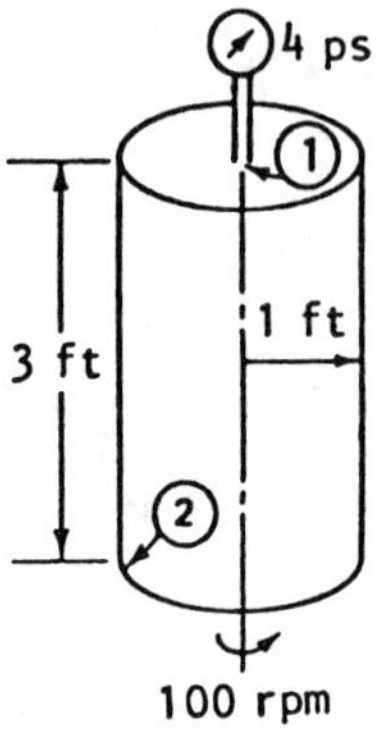

Fig. 1

Since the accéleration is horizontal and radially inward, lines of constant piezometric head will again be vertical lines (actually, vertical concentric cylinders) with decreasing magnitude toward the axis and with magnitude equal to the height of the free surface above the datum conveniently taken again at the tank bottom.

The Euler equation may be written as

$$\frac{\partial h}{\partial n} = -\frac{v^2}{gr} = -\frac{r\omega^2}{g}$$

Since the direction n is radially inward while r is measured positive outward, this may be changed to

$$\frac{\partial h}{\partial r} = \frac{r\omega^2}{g}$$

and subsequently integrated from r_1 to r_2 to give

$$h_2 - h_1 = \frac{\omega^2}{2g}(r_2^2 - r_1^2) \tag{1}$$

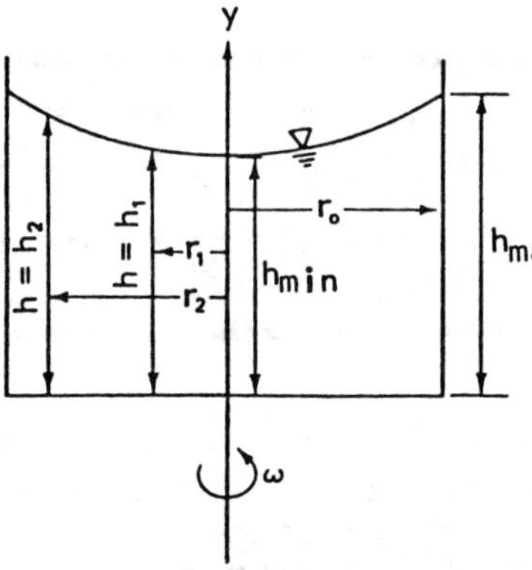

Fig. 2

By applying Eq. 1 between points 1 and 2 as shown in Fig. (2) and replacing the piezometric heads in equation (1) with

$$h = \frac{p}{\gamma} + y$$

$$\left(\frac{p_2}{\gamma} + y_2\right) - \left(\frac{p_1}{\gamma} + y_1\right) = \frac{\omega^2}{2g}(r_2^2 - r_1^2)$$

whence

$$\left(\frac{p_2 \text{lb/ft}^2}{62.4 \text{ lb/ft}^3} + 0 \text{ ft}\right) - \left[\frac{(4)(144)}{62.4} + 3\right]$$

$$= \frac{(100 \text{ rpm})^2 (2\pi \text{ rad/rev})^2 (1^2 - 0^2 \text{ ft}^2)}{(60 \text{ s/min})^2 (2)(32.2 \text{ ft/s}^2)}$$

Solving, we obtain

$$p_2 = 869.5 \text{ psf} = 6.04 \text{ psi}$$

• PROBLEM 2-37

A cylinder of radius 1.5 ft and height 4 ft is rotated at 10 rad/sec about its vertical axis. If the cylinder was originally full of water, how much is spilled, and what is the pressure intensity at the center of the base of the cylinder? (See Fig. 1)

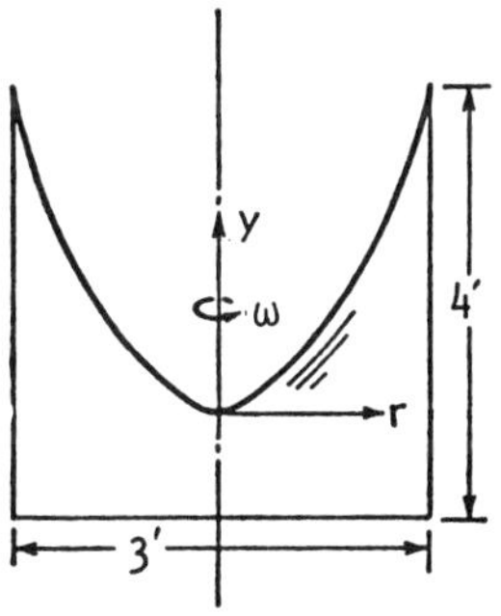

Fig. 1

Solution: Consider a small element of a fluid of unit depth, rotating at a constant angular velocity ω as shown in Fig. (2). In a horizontal sense the forces acting outward on the element are given by

$$pr\,d\theta - (p + dp)r\,d\theta$$

and these must sustain a radial acceleration of $r\omega^2$ toward the center of rotation. Hence

$$pr\,d\theta - (p + dp)r\,d\theta = -r\omega^2 r\,d\theta\,dr\,\frac{\gamma}{g}$$

which reduces to

$$\frac{dp}{dr} = \omega^2 r \frac{\gamma}{g} \tag{1}$$

Since all the terms on the right-hand side of the equation are positive, it is apparent that pressure is increasing with radius.

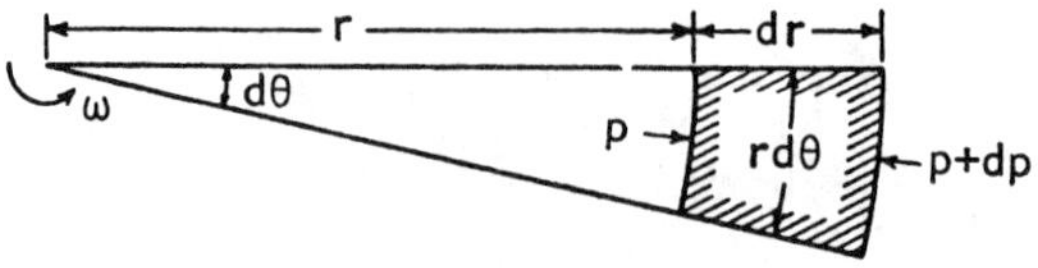

Fluid mass under radial acceleration. Fig. 2

Integrating this between the center of rotation (r = 0, $p = p_0$) and any general radius r, where the pressure is p,

$$\int_{p_0}^{p} dp = \gamma \int_0^r \frac{\omega^2 r}{g}\, dr$$

Therefore

$$p - p_0 = \frac{\omega^2 \gamma}{g}\left[\frac{r^2}{2}\right]_0^r$$

or

$$\frac{p - p_0}{\gamma} = \frac{\omega^2 r^2}{2g} \tag{2}$$

Now the left-hand side of Eq. (1) has units of length and represents a head of fluid. This term can be replaced by y, which is the height of the fluid above the central height, yielding

$$y = \frac{\omega^2 r^2}{2g} \quad \text{a parabola} \tag{3}$$

This is shown in Fig. 3. The pressure at any point can now be determined by using the relationship $p = \gamma h$.

From Eq. (3);

$$y_{max} = \frac{\omega^2 r_{max}^2}{2g}$$

$$= \frac{100 \times 1.5^2}{64.4}$$

$$= 3.5 \text{ ft}$$

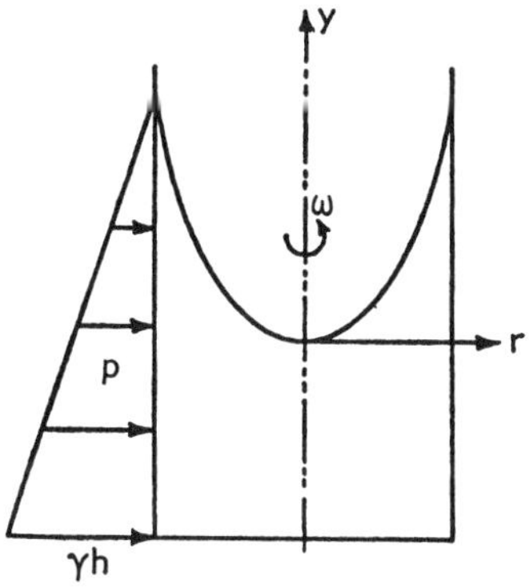

Rotating tank of fluid. Fig. 3

Volume spilled is the volume of the paraboloid (equal to half the volume of the surrounding cylinder), or

$$\frac{\pi}{2} \times 1.5^2 \times 3.5 = 12.4 \text{ ft}^3$$

The pressure intensity at the center of the base of the cylinder is due to a head of water of 0.5 ft. Therefore

$$p = 0.5 \times 62.4 = 31.2 \text{ psf}$$

• PROBLEM 2-38

> At what angular velocity must a U-tube 1 ft in width and filled with water to a depth of 1 ft be rotated about its axis to cause the water at point A to vaporize? (Fig. 1)

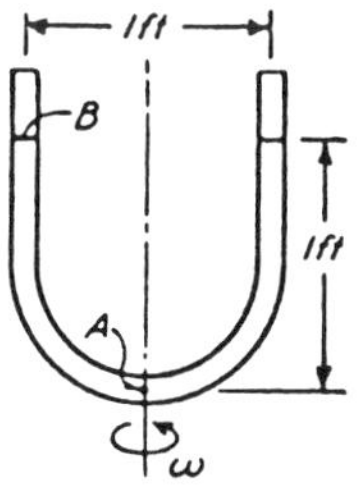

Fig. 1

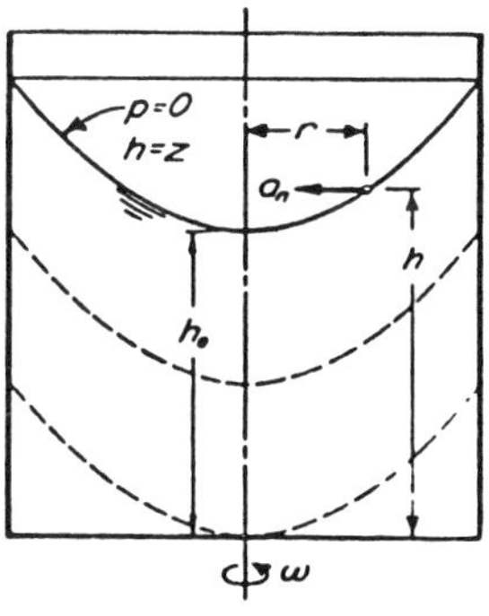

Fig. 2

Solution: Consider the rotation of an open cylindrical tank of liquid about its vertical axis at the constant angular velocity ω (Fig. 2). Because the acceleration is radially inward at every point, the pressure intensity will now be hydrostatically distributed over any coaxial cylindrical surface. The piezometric head on any such cylinder will be equal to the corresponding surface elevation, the form of the surface being found by integrating the acceleration equation with respect to radial distance from the axis. Thus, since $v = \omega r$, $a_n = v^2/r$, and $dr = -dn$,

$$-\frac{dh}{dn} = \frac{dh}{dr} = \frac{v^2}{gr} = \frac{\omega^2 r}{g}$$

$$\int_0^r \frac{dh}{dr} dr = \frac{\omega^2}{g}\int_0^r r\, dr$$

$$h - h_0 = \frac{\omega^2 r^2}{wg} = \frac{v^2}{2g}$$

$$\therefore h_B - h_A = \frac{\omega^2 r_B^2}{2g}$$

For the purpose of this problem assume vapor pressure of water to be absolute zero, or the required pressure head at A to be -34 ft,

$$\frac{p_B}{\gamma} + z_B - \frac{p_A}{\gamma} - z_A = 0 + 1 - (-34) - 0 = \frac{\omega^2 \times \overline{0.5}^2}{2 \times 32.2}$$

$$\omega = \sqrt{\frac{64.4 \times 35}{\overline{0.5}^2}} = 94.8 \text{ rad/sec} = 907 \text{ rpm}$$

• PROBLEM 2-39

Two immiscible liquids of densities $\rho_1 = 826$ kg/m^3 and $\rho_2 = 999$ kg/m^3 are contained in an open vessel as shown in Fig. 1. The vessel rotates about its vertical axis with a constant angular velocity ω. Calculate the ratio of the angular velocities $\frac{\omega_2}{\omega_1}$ of the two liquids so that $\frac{h_2}{h_1} = 4$.

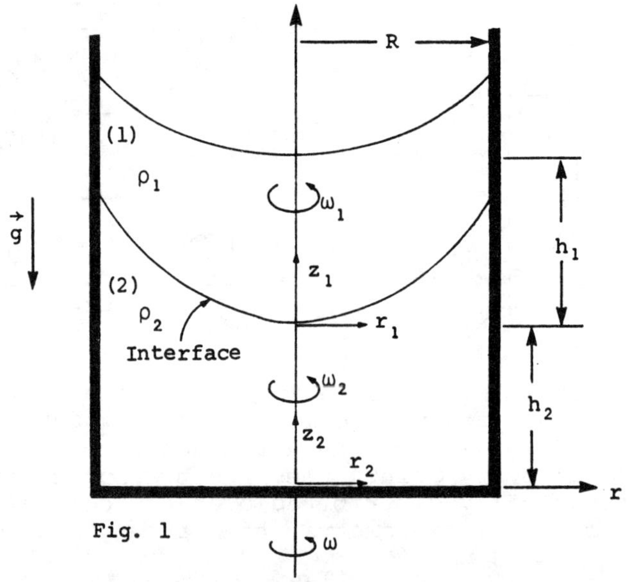

Fig. 1

Solution: Since the pressure is a function of r and z, the differential change in pressure,. dp, between two points is given by

$$dp = \left(\frac{\partial p}{\partial r}\right)_z + \left(\frac{\partial p}{\partial z}\right)_r dz$$

By writing Newton's second law of motion (see Problem 2-37) in the z and r directions, the above equation reduces to

$$dp = \rho\omega^2 r dr - \rho g dz$$

For the liquid (1), we can write the above equation as

$$dp = \rho_1\omega_1^2 r dr - \rho_1 g dz$$

Integration of this equation gives,

$$\int_{p=p_{atm}}^{p_{atm}} dp = \int_0^R \rho_1\omega_1^2 r dr - \int_0^h \rho_1 g dz$$

or

$$0 = \rho_1\omega_1^2 \frac{R^2}{2} - \rho_1 g h_1 \Rightarrow h_1 = \frac{\omega_1^2 R^2}{2g} \quad (1)$$

For the liquid (2), we have

$$dp_2 = \rho_2\omega_2^2 r dr - \rho_2 g dz$$

Along the interface $d(p_2 - p_1) = 0$

or

$$(\rho_2\omega_2^2 - \rho_1\omega_1^2)\gamma d\gamma = (\rho_2 - \rho_1) g dz$$

or

$$\int_0^R (\rho_2\omega_2^2 - \rho_1\omega_1^2) r dr = \int_0^{h_2} (\rho_2 - \rho_1) g dz$$

or

$$(\rho_2\omega_2^2 - \rho_1\omega_1^2)\frac{R^2}{2} = (\rho_2 - \rho_1)h_2 = h_2 = \frac{(\rho_2\omega_2^2 - \rho_1\omega_1^2)}{(\rho_2 - \rho_1)} \quad (2)$$

From equations (1) and (2) we have

$$\frac{\omega_2}{\omega_1} = \sqrt{\frac{h_2}{h_1}\left(1 - \frac{\rho_1}{\rho_2}\right) + \frac{\rho_1}{\rho_2}}$$

For $\rho_1 = 826 \frac{kg}{m^3}$, $\rho_2 = 999 \frac{kg}{m^3}$ and $\frac{h_2}{h_1} = 4$

$$\frac{\omega_2}{\omega_1} = \sqrt{(4)\left(1 - \frac{826}{999}\right) + \frac{826}{999}} \Rightarrow \frac{\omega_2}{\omega_1} = 1.237$$

• PROBLEM 2-40

A cylindrical vessel of diameter 100 mm and height = 0·3 m is arranged to rotate about its axis, which is vertical. When at rest, water is poured in to a depth of 225 mm. Calculate the speed of rotation (in radians per second) at which (a) the water will spill over the edge and (b) the axial depth is zero.

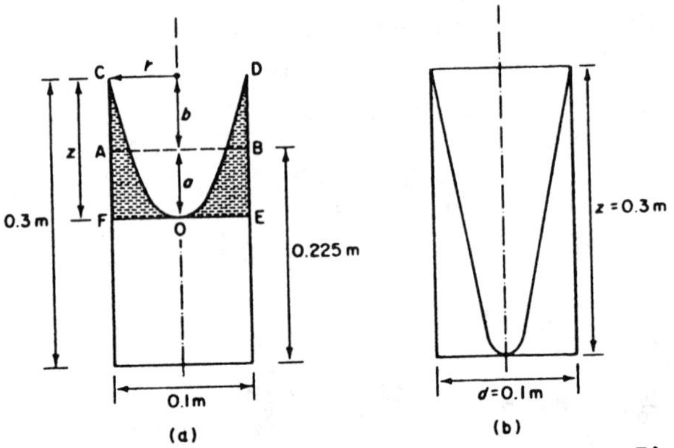

Rotating cylinder containing water

Fig. 1

Solution: (a) When the water is just about to spill over the edge, the conditions are as shown in Fig. 1(a), with the water surface forming a paraboloid COD. Since the volume of this paraboloid is half of that of the circumscribing cylinder,

$$\text{Volume of paraboloid COD} = \tfrac{1}{2}\text{ volume of cylinder CDEF}$$

$$= \tfrac{1}{2}\pi r^2 z,$$

therefore,

$$\text{Volume of shaded portion CFOED} = \tfrac{1}{2}\pi r^2 z = \tfrac{1}{2}\pi r^2 (a + b).$$

If AB is the original water level when stationary, since no liquid has been spilt,

$$\text{Volume CFOED} = \text{Volume of cylinder ABEF},$$

$$\tfrac{1}{2}\pi r^2 (a + b) = \pi r^2 b,$$

$$a = b,$$

and

$$z = (a + b) = 2a = 2(0{\cdot}3 - 0{\cdot}225) = 0{\cdot}15 \text{ m.}$$

At the rim

$$r = \tfrac{1}{2}d = 0{\cdot}05 \text{ m.}$$

If $\partial p/\partial x$, $\partial p/\partial y$ and $\partial p/\partial z$ are the rates of change of pressure p in the x, y and z directions (Fig. 2) and a_x, a_y and a_z the accelerations,

$$\text{Force in x direction, } F_x = p\delta y\delta z - (p + \frac{\partial p}{\partial x}\delta x)\,\delta y\delta z$$

$$= -\frac{\partial p}{\partial x}\delta x\delta y\delta z$$

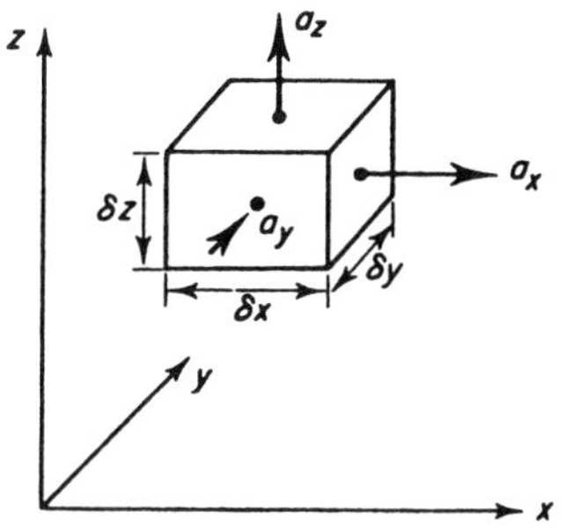

Fig. 2

By Newton's second law, $F_x = \rho\delta x\delta y\delta z \times a_x$, therefore,

$$-\frac{\partial p}{\partial y} = \rho a_y. \qquad (1)$$

Similarly, in the y direction

$$-\frac{\partial p}{\partial y} = \rho a_y. \qquad (2)$$

In the vertical z direction, the weight of the element $\rho g\delta x\delta y\delta z$ must be considered:

$$F_z = p\delta x\delta y - (p + \frac{\partial p}{\partial z}\delta z)\,\delta x\delta y - \rho g\delta x\delta y\delta z$$

$$= -\frac{\partial p}{\partial z}\delta x\delta y\delta z - \rho g\delta x\delta y\delta z.$$

By Newton's second law,

$$F_z = \rho\delta x\delta y\delta z \times a_z,$$

therefore,

$$-\frac{\partial p}{\partial z} = \rho(g + a_z). \qquad (3)$$

For an acceleration a_s in any direction in the x - z plane making an angle ϕ with the horizontal, the components of the acceleration are

$$a_x = a_s \cos \phi \quad \text{and} \quad a_z = a_s \sin \phi.$$

Now

$$\frac{dp}{ds} = \frac{\partial p}{\partial x}\frac{dx}{ds} + \frac{\partial p}{\partial z}\frac{dz}{ds}. \tag{4}$$

For the free surface and all other planes of constant pressure, $dp/ds = 0$. If θ is the inclination of the planes of constant pressure to the horizontal, $\tan \theta = dz/dx$. Putting $dp/ds = 0$ in equation (4)

$$\frac{\partial p}{\partial x}\frac{dx}{ds} + \frac{\partial p}{\partial z}\frac{dz}{ds} = 0$$

$$\frac{dz}{dx} = \tan \theta = -\frac{\partial p}{\partial x} \Big/ \frac{\partial p}{\partial z} .$$

Substituting from equations (1) and (3)

$$\tan \theta = -a_x/(g + a_z), \tag{5}$$

A body of fluid, contained in a vessel which is rotating about a vertical axis with uniform angular velocity, will eventually reach relative equilibrium and rotate with the same angular velocity ω as the vessel, forming a forced vortex. The acceleration of any particle of fluid at radius r due to rotation will be $-\omega^2 r$ perpendicular to the axis of rotation, taking the direction of r as positive outward from the axis. Thus, from equation (1),

$$\frac{dp}{dr} = -\rho\omega^2 r.$$

Referring to Fig. 1(a) at any point P on the free surface, the inclination θ of the free surface is given by equation (5),

$$\tan \theta = -\frac{a_x}{g + a_z} = \frac{\omega^2 r}{g} = \frac{dz}{dr}. \tag{9}$$

The inclination of the free surface varies with r and, if z is the height of P above O, the surface profile is given by integrating equation (9):

$$z = \int_0^x \frac{\omega^2 r}{g}\, dr = \frac{\omega^2 r^2}{2g} + \text{constant}. \tag{10}$$

From equation (10),

$$\omega^2 = 2gz/r^2$$

$$\omega = \sqrt{(2gz)}/r = \sqrt{(2 \times 9{\cdot}81 \times 0{\cdot}15)}/0{\cdot}05$$

$= 34{\cdot}31$ rad s^{-1}.

(b) When the axial depth is zero, conditions are as shown in Fig. 1(b) and at the rim, when $r = 0{\cdot}05$ m, $z = 0{\cdot}3$ m. From equation (10), as above,

$$\omega = \sqrt{(2gz)}/r = \sqrt{(2 \times 9{\cdot}81 \times 0{\cdot}3)}/0{\cdot}05$$

$$= 48{\cdot}52 \text{ rad s}^{-1}.$$

• PROBLEM 2-41

The impeller of centrifugal pump has a radius of 6 in. At the so-called shutoff condition no water flows through the pump. If the impeller is rotating at 1200 rpm and if the water in the impeller passages is moving with the same angular velocity as the impeller, calculate the pressure rise from the pump inlet where r = 0 to a point corresponding to the impeller tip (r = 6 in.).

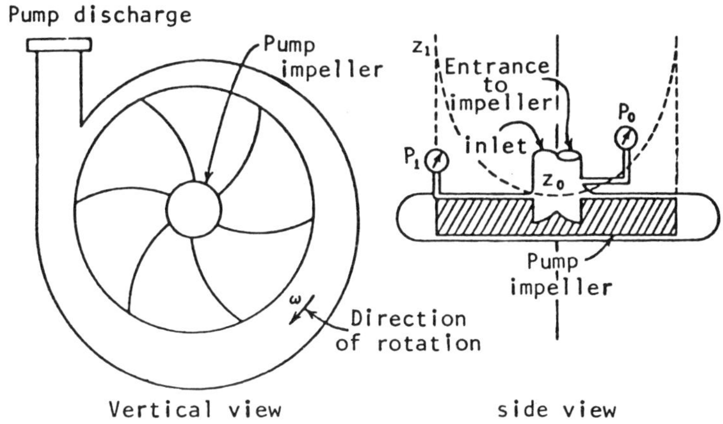

Fig. 1

Solution: Consider a particle at a distance r from the axis of rotation of the coordinate system and assume that the axis is aligned with gravitational field which points downward as shown in Fig. 2. The centrifugal body force per unit mass acts outward and is given by $\hat{i}_r\omega^2 r$. The gravitational force per unit mass is simply $-\hat{k}g$, as shown in Fig. 2. It follows that

$$g_{eff} = \hat{i}_r\omega^2 r - \hat{k}g \tag{1}$$

The slope of the g_{eff} body force vector in the z-r-plane is

$$(g_{eff} \text{ slope}) = \frac{dz}{dr} = -\frac{g}{\omega^2 r} \tag{2}$$

We know that the lines of constant pressure are perpendicular to the effective body force. The slope of a constant

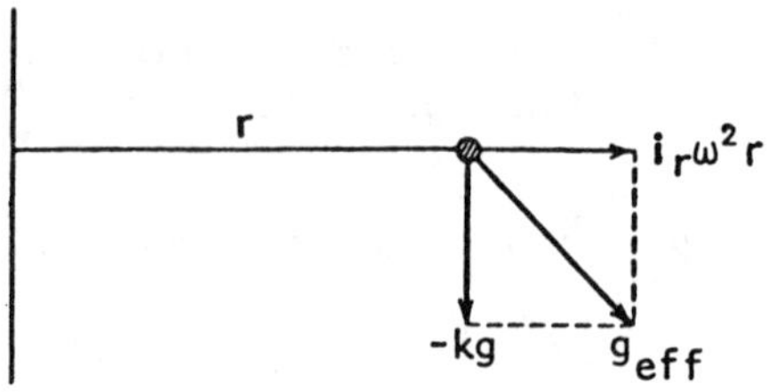

Fig. 2

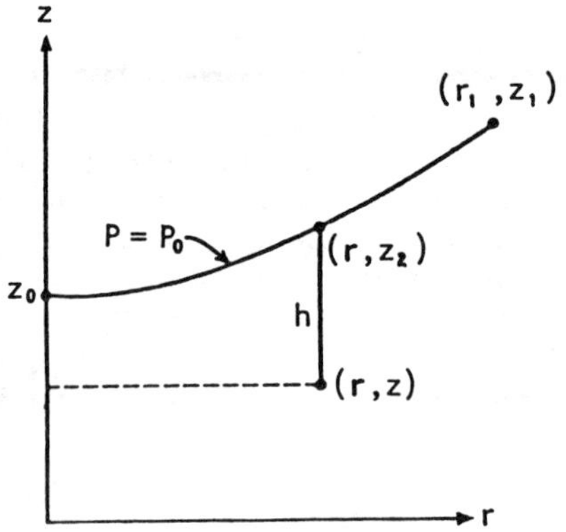

Fig. 3

pressure line in the z-r-plane is therefore the negative reciprocal of the slope of g_{eff}. From equation 2

$$\text{slope of constant-pressure curve} = \frac{dz}{dr} = \frac{\omega^2 r}{g} \tag{3}$$

The constant-pressure curve is found by integrating equation 3 with the result

$$z = \frac{\omega^2 r^2}{2g} + z_0 \tag{4}$$

where z_0 is the value of z at $r = 0$. Suppose that we know that pressure at a point (r_1, z_1) in the fluid (our reference pressure) is P_0, and we wish to find the pressure at some arbitrary point with coordinates (r,z). As shown in Fig. 3, a constant-pressure surface passes through (r_1, z_1) intersecting the z-axis at z_0. Let a vertical line pass through (r,z) and denote the distance from (r,z) to the $P = P_0$ curve by h. Along the vertical line r is constant, and because the centrifugal body force acts normal to a constant r-line, it will have no influence or pressure vibrations. Looking at the matter in a different light, in the absence of a gravitational field, g_{eff} would point radially outward at all points, and vertical position change would result in no pressure change. What would produce vertical pressure changes in the actual situation? Quite simply, the gravitational field and the relationship between the pressure P at (r,z) and P_0 is

$$P - P_0 = \rho g h$$

But $h = z_2 - z$ (Fig. 3) where (from equation 4) $z_2 = z_0 + \omega^2 r^2/2g$. It follows that

$$P - P_0 = \rho g \left[(z_1 - z) + \frac{\omega^2 r^2}{2g} \right] \tag{5}$$

In this problem, we take $z_0 = 0$ at the inlet. Then

$$\rho = 62.4 \frac{\text{lbm}}{\text{ft}^3}, \qquad \omega = 1200 \frac{\text{rev}}{\text{min}} \times 2\pi \frac{\text{radians}}{\text{rev}} \times \frac{1 \text{ min}}{60 \text{ sec}}$$

$$= 40\pi \frac{\text{radians}}{\text{sec}}$$

$$z_1 - z_0 = 0, \qquad r_1 = \tfrac{1}{2} \text{ ft}$$

Therefore

$$P - P_0 = 62.4 \frac{\text{lbm}}{\text{ft}^3} \times \frac{1 \text{ lbf}}{32.2 \text{ lbm-ft/sec}^2} \times \frac{(40\pi)^2}{2 \text{ sec}^2} \times \frac{1}{2} \text{ ft}^2$$

$$= 3820 \text{ psfa} = 26.6 \text{ psia}$$

BUOYANCY

• PROBLEM 2-42

What percent of the total volume of an iceberg floats above the water surface? Assume the density of ice to be 57.2 lb_m/ft^3, the density of water 62.4 lb_m/ft^3.

Solution: At equilibrium, the weight of the iceberg is balanced by the buoyant force of the displaced water. In other words,

$$\rho_{ice} V_{iceberg} = \rho_{water} V_{submerged}$$

Solving,

$$\frac{V_{submerged}}{V_{iceberg}} = \frac{57.2}{62.4} = 0.92$$

Therefore only 8 percent of the iceberg is above the surface.

• PROBLEM 2-43

A 100,000 deadweight ton sunken ship is to be raised from 200 ft to the water surface by pumping air into its sealed holds. How much water is to be expelled from the ship to make it float?

Solution: We know that 10^5 tons equal 2×10^8 lb, and that 1 ft^3 of water weighs 62.4 lb. Hence,

$$2 \times 10^8 \text{ lb of water}$$

equals

$$3.205 \times 10^6 \text{ ft}^3$$

which is the minimum amount of air space required.

The required air pressure to expel the water from the ship's holds

$$200 \text{ ft} \times 62.4 \text{ lb/ft}^3 = 12{,}480 \text{ lb/ft}^2$$

$$= 60{,}940 \text{ psi}$$

• PROBLEM 2-44

A 4-in.-dia. steel ball is suspended by a string in a liquid composed of two layers of different liquids. The bottom layer 6 in. deep is carbon tetrachloride, and the upper layer 8 in. thick is water. Calculate the tension in the string when (a) the ball is half submerged in water and half in carbon tetrachloride; (b) the ball is completely submerged in carbon tetrachloride; (c) the ball is submerged half in air and half in water.

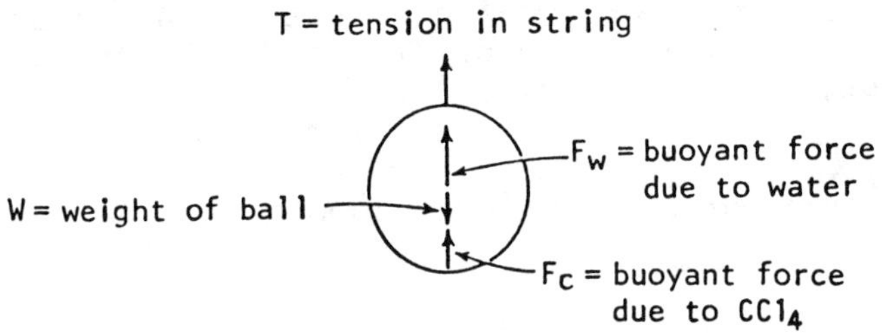

Fig. 1

Solution: (a) For the free body diagram of the ball shown in Fig. 1

γ_s = specific weight of steel = 487 lb/ft^3

γ_{CCl_4} = specific weight of carbon tetrachloride

$= 99.5\ lb/ft^3$

$$W = \gamma_s \frac{4}{3}\pi R^3 = (487)\frac{4}{3}(3.1416)\left(\frac{2}{12}\right)^3 = 9.45\ lb$$

$$F_w = \frac{1}{2}\gamma_{H_2O}\frac{4}{3}\pi R^3 = \frac{1}{2}(62.4)\frac{4}{3}\pi\left(\frac{2}{12}\right)^3 = 0.61\ lb$$

$$F_c = \frac{1}{2}\gamma_{CCL_4}\frac{4}{3}\pi R^3 = \frac{1}{2}(99.5)\frac{4}{3}\pi\left(\frac{2}{12}\right)^3 = 0.96\ lb$$

$$\Sigma F_V = 0 \quad +\uparrow$$

$$T + F_w + F_c - W = T + 0.61 + 0.96 - 9.45 = 0$$

$$T = 7.88\ lb$$

(b) For the free body diagram of the ball shown in Fig. 2

$W = 9.45\ lb$

$$F_c = \frac{4}{3}\pi R^3\gamma_c = \frac{4}{3}\pi\left(\frac{2}{12}\right)^3 99.5 = 1.93\ lb$$

$$\Sigma F_V = 0 \quad +\uparrow$$

$$T + F_c - W = 0$$

$$T = W - F_c = 9.45 - 1.93 = 7.52\ lb$$

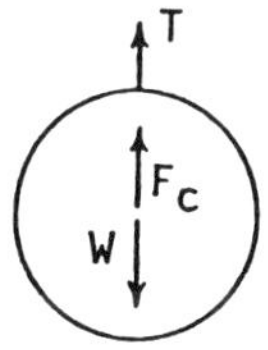

Fig. 2

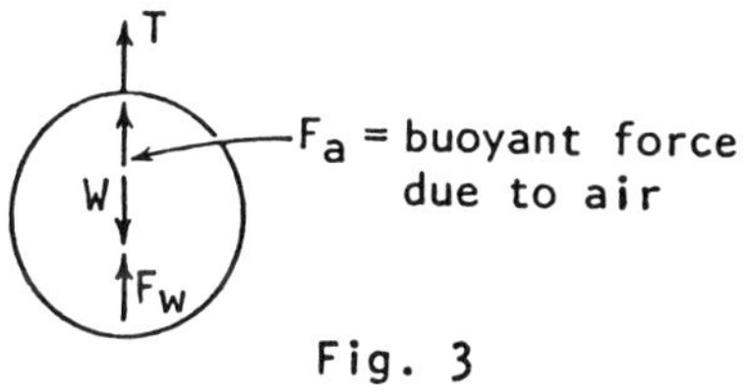

Fig. 3

(c) For the free body diagram of the ball shown in Fig. 3

$$\Sigma F_V = 0 \quad +\uparrow$$

$$T + F_a + F_w - W = 0$$

$$W = 9.45\ lb$$

$$F_w = 0.61\ lb$$

$$F_a = \frac{1}{2}\left(\frac{4}{3}\pi R^3\right)\gamma_a = \frac{1}{2}\,\frac{4}{3}\pi\left(\frac{2}{12}\right)^3 (0.0763) = 0.00074$$

$$T + 0.0074 + 0.61 - 9.45 = 0$$

$T = 8.83\ lb$

The buoyant force of air is so small when it is compared to liquid buoyant forces that you can neglect it. Hereafter in problems in which buoyant forces include liquids, neglect the buoyant forces of gases.

• PROBLEM 2-45

A 4-in diameter solid cylinder of height 3.75 in weighing 0.85 lb is immersed in liquid (γ = 52 lb/ft^3) contained in a tall, upright metal cylinder having a diameter of 5 in (see Figure). Before immersion the liquid was 3-in deep. At what level will the solid cylinder float?

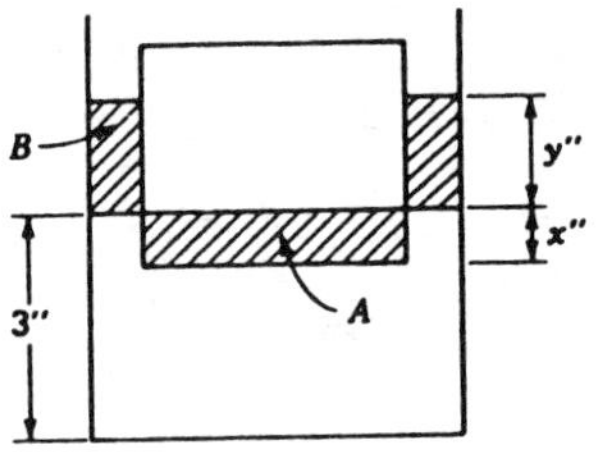

Solution: x = distance solid cylinder falls below original liquid surface

y = distance liquid rises above original liquid surface

x + y = depth of submergence

Volume A = Volume B $\quad \pi 2^2 x = \pi(2.5^2 - 2.0^2)y$

$4x = 2.25y \qquad x = 0.56y$

$F_B = \text{weight} = 0.85 = 52\pi\left(\frac{2}{12}\right)^2 \frac{x + y}{12}$

x + y = 2.24 in $\qquad$ x = 0.81 in $\qquad$ y = 1.43 in

Bottom of solid cylinder will be 3.0 - 0.81 = 2.19 in above bottom of hollow cylinder.

• PROBLEM 2-46

A container ship has a cross-sectional area of 3,000 m^2 at the water line when the draft is 9 m. What weight of containers can be added before the normal draft of 9.2 m is reached. Assume salt water of weight density 10 kN/m^3.

Solution: As the ship floats, the weight of water displaced by the containers equals the weight of the containers. Therefore,

$$\text{Weight of containers} = 3\,000 \times 0.2 \times 10$$

$$= 6\,000 \text{ kN}$$

An empty drum (Fig. 1) has a height of 3m (h=3m) and a radius of 0.80m (R=0.80m). The empty drum is submerged in water. The air originally present in the drum remains composed therein at the submerged depth. Determine the magnitude of the force required to maintain the drum at a depth of 16m (D=6m) below the free surface of the water.

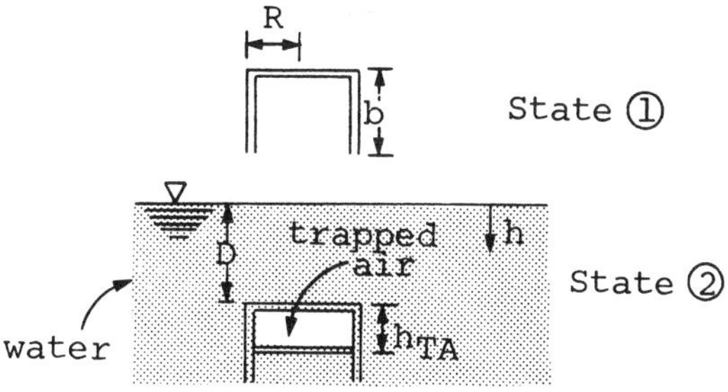

Fig. 1

Solution: Assumptions: (a) The wall thickness and the weight of the cylinder are considered negligible.

(b) The trapped air in the drum remains at constant temperature.

Equations:

Applying the ideal gas equation for states 1 and 2 respectively

$$P_{atm} V_B = nRT_1 \quad \text{and}$$

$$P_W V_{TA} = nRT_2$$

But $T_1 = T_2$, therefore

$$P_{atm} V_B = P V_{TA}$$

or

$$P_{atm} \cancel{\pi R^2} h = P_W \cancel{\pi R^2} h_{TA}$$

or

$$P_{atm} h = P_W h_{TA} \tag{1}$$

where, P_W is the pressure of the water at the free surface in the drum.

From the pressure height relation

$$\frac{dp}{dh} = \rho g \quad \text{or} \quad dp = \rho g dh$$

By assuming ρ = constant

$$\int_{P_{atm}}^{P_W} dp = \rho g \int_0^{h'} dh \quad \text{or} \quad p_w - p_{atm} = \rho g h'$$

or $$p_W = p_{atm} + \rho g\left[(D+h)-(h-h_{TA})\right]$$

or
$$p_W = p_{atm} + \rho g(D+h_{TA})$$

Substituting equation (1) into (2)

$$\frac{p_{atm}h}{h_{TA}} = p_{atm} + \rho g(D+h_{TA})$$

or $$\rho g h_{TA}^2 - h_{TA}(p_{atm} + \rho g D) - p_{atm}h = 0$$

With, $\rho_{H_2O} = 999 \frac{kg}{m^3}$, $g = 9.81 \text{ m/sec}^2$, $D = 6m$, $h = 3m$

and $p_{atm} = 1.01 \times 10^5 \text{ N/m}^2$

$$h_{TA}^2 + 13.33\, h_T - 31.017 = 0$$

and $$h_{TA} = 2.02 \text{ m}$$

The buoyancy of the drum is due to the trapped air. Therefore the buoyant force must be equal to the weight of the water whose volume is that of the entrapped air,

$$|\vec{F}_B| = \rho_{H_2O}\, g\, \mathcal{V}_{TA} = \rho_{H_2O} g \pi R^2 h_{TA}$$

$$= (999)(9.81)\pi(.80)^2 2.02$$

or $$|\vec{F}_B| = 39{,}803 \text{ N}$$

• PROBLEM 2-48

Determine the specific weight of a body consisting of a 30-cm-diameter hemisphere and a 50-cm-long by 30-cm-diameter cylinder which floats, as shown in the figure, at an oil (sp. gr. = 0.87)/water interface.

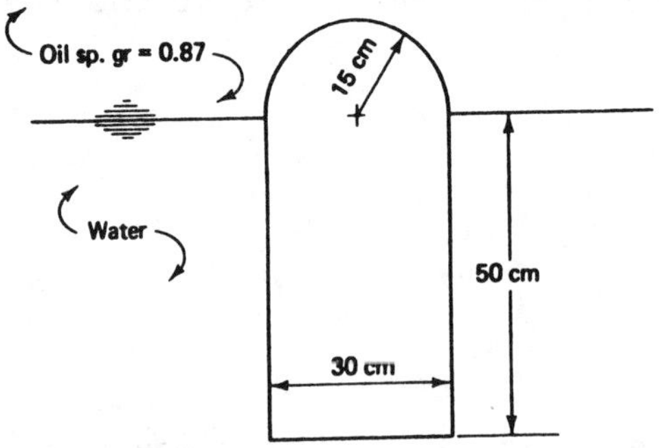

Solution: The weight of the body must exactly equal the weight of the displaced oil and water. Hence,

$$\left[\frac{\pi}{4}(0.30)^2(0.50) + \frac{\pi}{12}(0.30)^3\right]\gamma_{body}$$

$$\frac{\pi}{4}(0.30)(0.50)\gamma_{water} + \frac{\pi}{12}(0.30)^3\gamma_{oil}$$

With $\gamma_{water} = 9800\ N/m^3$ and $\gamma_{oil} = (0.87)(9800) = 8526\ N/m^3$, this expression reduces to

$$\gamma_{body} = 9588\ N/m^3$$

• PROBLEM 2-49

The hydrometer is a device used to measure the specific gravity of a liquid. As shown in the figure, the hydrometer consists of a weighted bulb and a stem of constant cross-sectional area. When floating in pure water, the hydrometer reaches the equilibrium position shown in Figure (a), with $\forall_0$ the total volume submerged. When floating in a liquid of different density than water, the hydrometer will reach a new equilibrium position (Figure (b). Obtain a correlation between h and specific gravity s. The cross-sectional area of the stem is A_s.

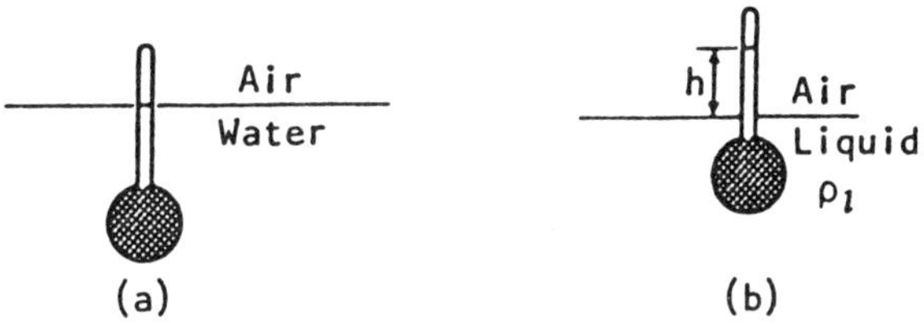

Solution: In water, the hydrometer reaches an equilibrium position at which its weight W is balanced by the buoyant force $\forall_0 \rho_{water}(g/g_c)$. In other words,

$$W = \rho_{water}\frac{g}{g_c}\forall_0$$

In the second liquid, the weight must again be balanced by the buoyant force.

$$W = \rho_L \frac{g}{g_c}(\forall_0 - Ah)$$

Equating the two expressions for W, we obtain

$$\rho_{water}\frac{g}{g_c}\forall_0 = \rho_L\frac{g}{g_c}(\forall_0 - Ah)$$

But, by definition,

$$s = \frac{\rho_L}{\rho_{water}}$$

Therefore,
$$s = \frac{\forall_0}{\forall_0 - A_s h}$$
$$= \frac{1}{1 - A_s h/\forall_0}$$

• PROBLEM 2-50

A hemispherical bowl whose mass is 100 grams is placed with its rim downwards on a horizontal plane which it fits closely. Water is poured into the bowl through a hole in the curved surface. Find the height in centimeters at which the water must be in the bowl in order that the bowl may be lifted and the water begin to escape between the plane and the bowl.

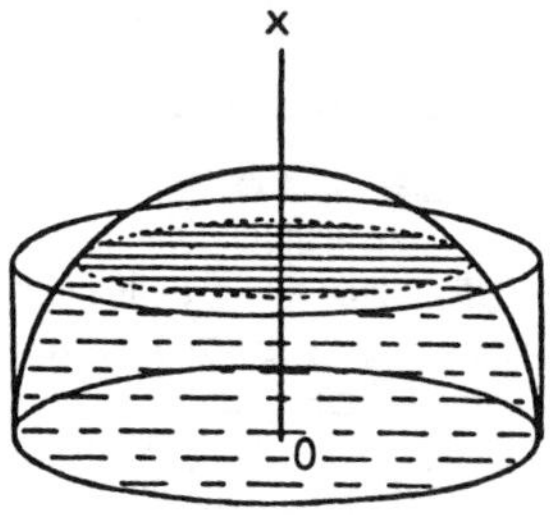

Solution: Let a be the internal radius of the bowl and h the required height of the water.

The volume of water required is

$$\int_0^A \pi y^2 dx = \int_0^A \pi(a^2 - x^2)dx = \pi(a^2h - \frac{1}{3}h^3).$$

But the volume of a cylinder of height h on the same base is $\pi a^2 h$, therefore the volume between the spherical water surface and this cylinder is

$$\frac{1}{3}\pi h^3 .$$

The upward thrust on the hemisphere is equal to the weight of this volume of water, i.e.,

$$\frac{1}{3}\pi h^3 \text{ c.c. of water.}$$

Therefore,

$$\frac{1}{3}\pi h^3 = 100$$

or

$$h = (300/\pi)^{1/3} .$$

By using logarithms we find that h = 4.57 cm.

• PROBLEM 2-51

The solid wooden sphere (γ = 8350 N/m^3) of diameter 0.4 m is held in the orifice (0.2 m diameter) by the water. Calculate the force exerted between sphere and orifice plate when the depth is 0.7 m. The sphere will float away if this force becomes zero; can this ever happen?

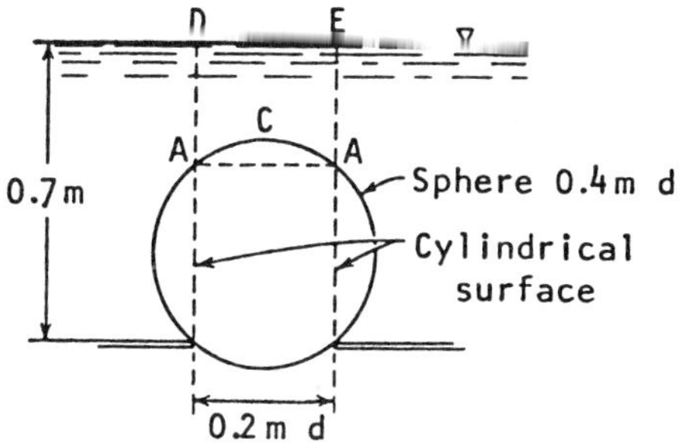

Solution:

Volume of sphere = $(\pi/6)(0.4)^3 = 0.0335 \text{ m}^3$

Weight of sphere = $0.0335 \times 8350 = 280$ N downward

Volume of section of sphere outside of cylindrical surface (from solid geometry) is 0.0218 m^3. From this volume the buoyant force F_B may be calculated

$$F_B = \gamma(\text{volume displaced}) = 0.0218 \times 9800 = 214 \text{ N upward}$$

The force downward on surface ACA is computed from the volume of ACADE, which (from solid geometry) is found to be 0.0098 m^3:

$$F_{ACA} = 0.0098 \times 9800 = 96 \text{ N downward}$$

For static equilibrium of the sphere,

Force between sphere and plate = 280 + 96 - 214 = 162 N

When the water surface coincides with the horizontal plane A-A, the force on surface ACA will be zero. Here the force between sphere and plate will be 280-214 = 66 N. Below this point the buoyant force will be less than 214 N; accordingly, the force between sphere and plate can never be zero.

• PROBLEM 2-52

Consider a spherical balloon of constant diameter of ten feet moving vertically in an isothermal atmosphere. Given the balloon's weight as 20 pounds, a ground pressure and temperature of 14.5 psia. and 70°F respectively, determine the altitude z at which the balloon will come to rest.

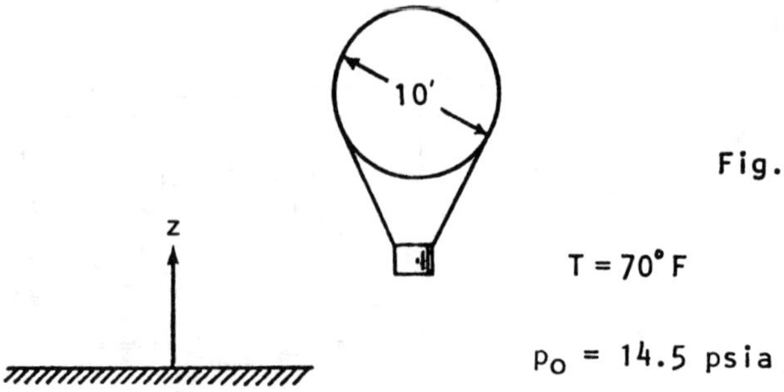

Fig. 1

Solution: The balloon will come to rest when the buoyant force equals 20 pounds. The volume of the balloon is

$$\forall = \frac{4}{3}\pi r^3 = \frac{4}{3}\pi(5)^3 = 523 \text{ ft.}^3 \quad (14.792 \text{ m}^3)$$

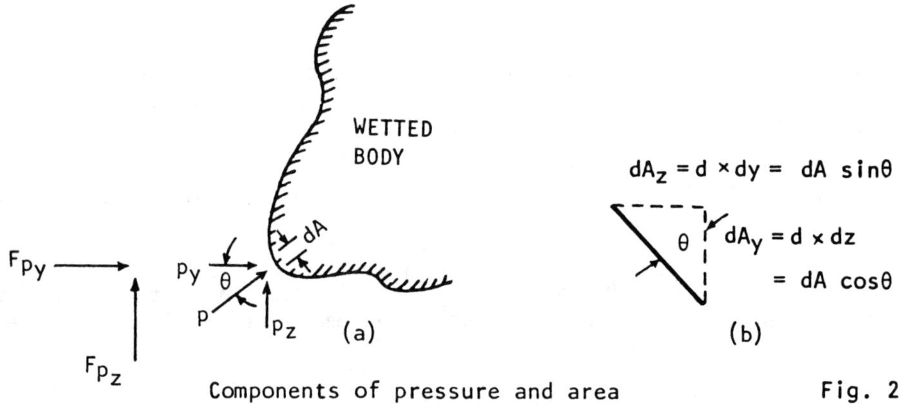

Components of pressure and area

Fig. 2

From the geometry of Figure 2(a) and 2(b)

$$p_z = p \sin\theta \tag{1}$$

$$dA_z = dA \sin\theta \tag{2}$$

From the definition of the pressure force $\vec{F}_p$, the z-component of the pressure force is

$$F_{p_z} = \int_A p_z \, dA \tag{3}$$

Substituting the pressure and area relationships of Equations (1) and (2) into the pressure force expression of Equation (3) one obtains

$$F_{pz} = \int_{A_z} p \, dA_z \tag{4}$$

for the vertical component of z pressure force.

The z-component of the pressure force F_{pz} is evaluated using Equations (4) and (2), such that

$$F_{pz} = \gamma \int_{A_z} z \, dA_z \qquad (5)$$

But as seen from Figure 3, the integrand $z\,dA_z$ is the volume of fluid between the free-surface and the differential area dA_z, so that the z-component of pressure force is

$$F_{pz} = \gamma \forall \qquad (6)$$

where $\forall$ is the volume of fluid above the wetted surface being considered.

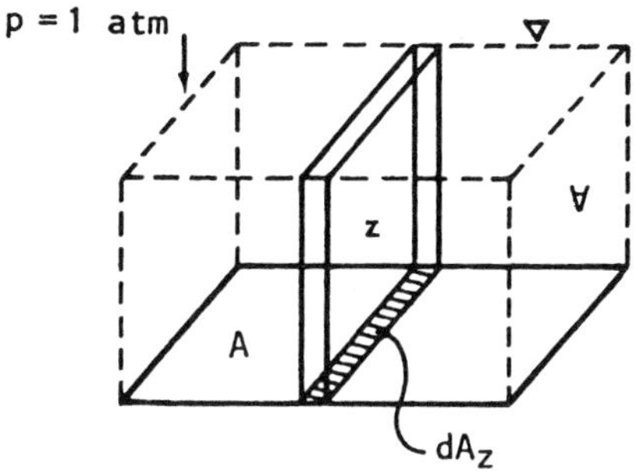

Volume of fluid above wetted surface A

Fig. 3

From Equation (6) the density of the terminal altitude assuming constant acceleration due to gravity will be

$$\rho = \frac{F_{p_z}}{g\forall} = \frac{20}{32.2 \times 523} = 0.00119 \text{ slug/ft.}^3$$

The pressure at this density is

$$p = \rho RT \qquad (7)$$

$$= 0.0019 \times 1715 \times 530$$

$$= 1080 \text{ psfa}$$

$$= 7.5 \text{ psia}$$

Assume the equation of state of air is governed by the perfect gas law relating density to pressure as

$$\rho = p/RT$$

where the absolute temperature field T is a prescribed function of altitude z. The density ρ of Equation (7) is substituted into the static equilibrium

$$\frac{\partial p}{\partial z} = - \rho g$$

to yield

$$\frac{dp}{dz} = - \frac{pg}{RT(z)} \tag{8}$$

Equation (8) is integrable if the temperature field is assumed isothermal to say temperature T_0. Integration of Equation (8) yields

$$p = p_0 \exp \frac{-gz}{RT_0} \tag{9}$$

where p_0 is the pressure at the earth's surface z equal 0. From Equation (9)

$$\ln\left(\frac{7.5}{14.5}\right) = - \frac{gz}{RT_0}$$

Solving for the altitude z gives

$$z = - \frac{1715(530)}{32.2} \ln\left(\frac{7.5}{14.5}\right)$$

or

$$z = 18{,}600 \text{ ft., } (5684 \text{ m})$$

is the altitude when the balloon will come to rest.

STABILITY

• PROBLEM 2-53

For a ship with a water-line cross-section as shown in Fig. 1 and a displacement of 600 tons, determine the maximum distance GB that the center of gravity may lie above the center of buoyancy if the ship is to remain stable.

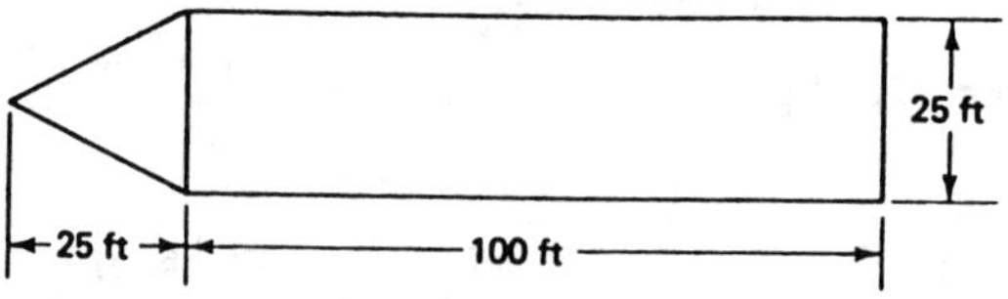

Fig. 1

Solution: With reference to Fig. 2 and 3, there exists the relationship

$$MG = \frac{\bar{I}}{V} - GB \tag{1}$$

where I is the moment of inertia of the area A about the longitudinal axis O.

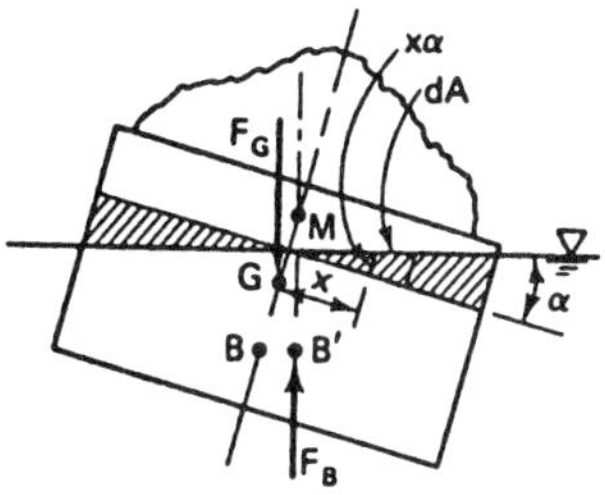

Fig. 2

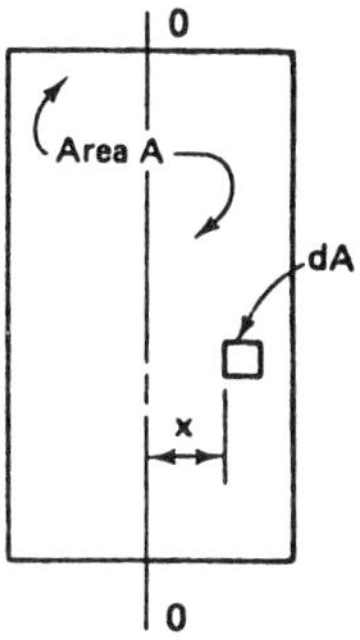

Fig. 3

Clearly, the greater the distance MG, the greater the stability. Since a barge or other floating body becomes unstable as M falls below G, Eq. 1 becomes a direct indicator of this condition. In particular,

$$\frac{\bar{I}}{V} > GB \qquad \text{stable body}$$

$$\frac{\bar{I}}{V} < GB \qquad \text{unstable body}$$

At the point of incipient instability, $GB = \bar{I}/V$, where

$$\bar{I} = \frac{(100)(25)^3}{12} + \frac{(2)(25)(12.5)^3}{12} = 1.383 \times 10^5 \text{ ft}^4$$

and

$$V = \frac{(600 \text{ tons})(2000 \text{ lb/ton})}{62.4 \text{ lb/ft}^3} = 1.923 \times 10^4 \text{ ft}^3$$

Finally,

$$GB = \frac{1.383 \times 10^5}{1.923 \times 10^4} = 7.19 \text{ ft}$$

• PROBLEM 2-54

A ship has a displacement of 6000 metric tons. A body of 30 metric tons mass is moved laterally on the deck 12 m, and the end of a 1.80-m plumb bob moves 92 mm. What is the transverse metacentric height?

Solution: Floating bodies may be in a condition of stable equilibrium even though the center of gravity G is above the center of buoyancy B. The position of B generally changes when a floating body tips about a horizontal axis because the shape of the volume of displaced liquid generally changes.

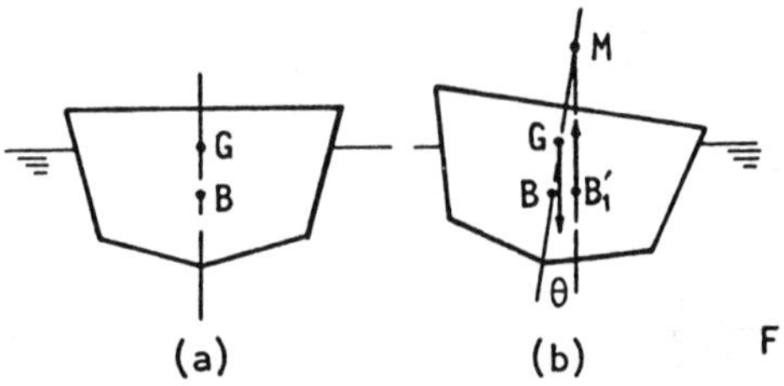

Fig. 1

In Fig. 1a the floating body is in a state of equilibrium; the net force on the body is zero and there are no net moments on the body. In Fig. 2b the body is tipped through a small angle θ and the new center of buoyancy is at B_1. Since the weight of the body acts downward through G and the buoyant force acts upward through B_1, there is a restoring couple which tends to turn the body back to its original position shown in Fig. 1a. A vertical line through B_1 intersects the line through BG (extended) at M, called the metacenter. When M is above G the body is stable, but when M is below G the system is unstable because an overturning couple would tend to overturn the floating body. The distance from G to M is called the metacentric height, and it must be positive (M above G) for stability.

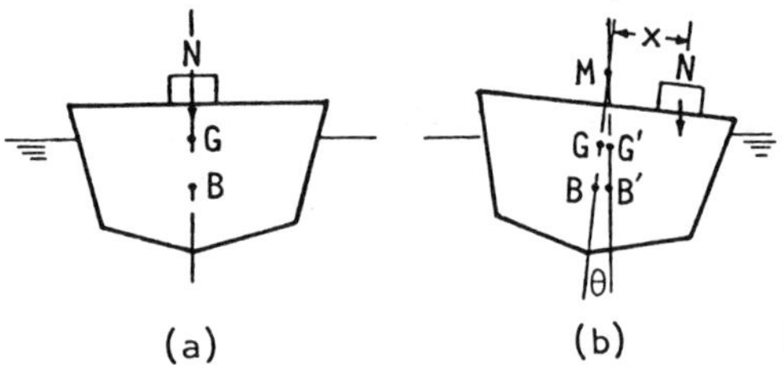

Fig. 2

The metacentric height $\overline{GM}$ for tilting about a horizontal axis may be determined experimentally. In Fig. 2 let a weight N be moved athwartship from the center of an initially horizontal deck by a distance x. The ship will tilt through a small angle θ which can be measured by a plumb bob or other suitable means. The new center of gravity G' and the new center of buoyancy B' are then in a new vertical line. The center of gravity shifts from G to G', and equating moments give

$$Nx = WGG' = W\overline{GM} \tan \theta$$

and thus

$$\overline{GM} = \frac{Nx}{W} \cot \theta \qquad (1)$$

Here W is the weight of the floating body (including N) and N is the weight of the movable body.

In Eq. (1) N = 30 tons, W = 6000 tons, x = 12 m, and $\cot\theta = \frac{1800}{92}$. Thus the metacentric height is

$$\overline{GM} = \frac{(30)(12)}{6000}\;\frac{1800}{92} = 1.17 \text{ m}$$

• PROBLEM 2-55

A barge such as that shown in Fig. 1 is 10 m wide by 30 m long. When loaded, the barge displaces 5 MN, and its center of gravity is 0.5 m above the waterline. Determine the metacenter height above the center of gravity and the righting moment for a roll angle of 10°.

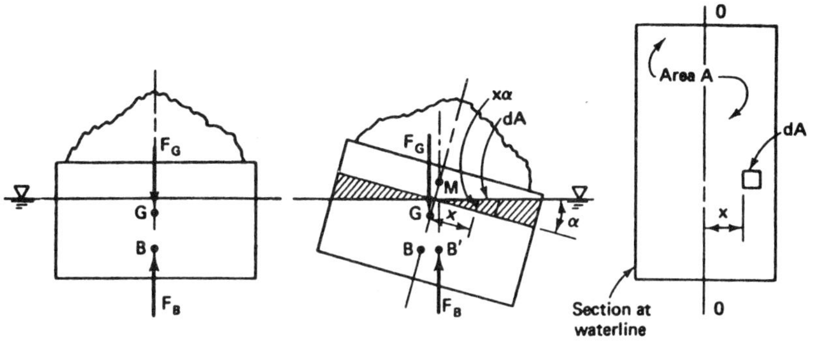

(a) Undisturbed barge with G above B (b) Barge in disturbed position (c) Plan view of barge

Fig. 1

Solution: For the barge of Fig. 1, which has a rectangular cross section, if the center of buoyancy lies above the center of gravity, as may be the case when the barge is unloaded, the vessel is stable and need not be considered further. As the barge is loaded, raising the center of gravity to a point above the center of buoyancy, the situation depicted in Fig. 1a may occur.

When the barge is displaced through an angle α known as the angle of heel, as shown in Fig. 1b, there is a shift in the center of buoyancy from B to B'. Thus, the forces give rise to a counterclockwise couple or righting moment. The shift in location of the buoyant force occurs because the volume of the left-hand shaded area no longer can contribute to the buoyancy, whereas an additional volume the right-hand shaded area, is now submerged, the effect of which moves the centroid of the submerged volume to the right. The intersection of the original line of action containing points B and G with the new line of action of the buoyancy force through B' locates the metacenter M. As long as the point M remains above point G, the body will be stable. If the angle α is further increased, the metacenter will eventually drop below point G and the body will heel over or capsize.

Defining the displaced volume (which remains constant regardless of α) as V and the horizontal area of the barge at the water line as A, the distance BB' may be determined for small angles of α as follows. The moment of the displaced volume V about point B must equal the moment of the shaded triangular portions. Thus, as determined from Fig. 1b and c,

$$(V)(BB') = \int_A x(x\alpha \, dA) = \alpha \int_A x^2 \, dA$$

The latter integral is the second moment, or moment of inertial $\bar{I}$, of the area A about the longitudinal axis 0 (which is the centroidal axis of the area). Therefore,

$$BB' = \frac{\bar{I}\alpha}{V} \tag{1}$$

Also, for small angles of heel,

$$BB' = (MB)(\alpha)$$

and therefore

$$MG = MB - GB = \frac{\bar{I}}{V} - GB \tag{2}$$

For a specific weight of 9800 N/m^3, the depth of water displaced, y, is calculated by

$$(y \text{ m})(10 \text{ m})(30 \text{ m})(9800 \text{ N/m}^3) = 5 \times 10^6 \text{ N}$$

or

$$y = 1.7 \text{ m}$$

and the center of buoyancy is normally 0.85 m below the waterline. Therefore, the center of gravity is 0.5 + 0.85 = 1.35 m above the center of buoyancy. Using the notation of Eq. 2

$$MG = \frac{\bar{I}}{V} - GB$$

$$= \frac{(30 \text{ m})(10 \text{ m})^3/12}{(1.7 \text{ m})(10 \text{ m})(30 \text{ m})} - 1.35 \text{ m}$$

$$= 4.90 - 1.35 = 3.55 \text{ m}$$

Hence, the metacenter is 3.55 m above the center of gravity. From Eq. 1 the distance BB' is

$$BB' = \frac{\bar{I}\alpha}{V}$$

where

$$\bar{I}/V = 4.90 \text{ m}$$

and

$\alpha = 10° = 0.175$ rad,

yielding

$$BB' = (4.90(0.175) = 0.858 \text{ m}$$

Therefore, the righting moment is

$$(F_G)(BB') = (5 \times 10^6 \text{ N})(0,858 \text{ m}) = 4.29 \times 10^6 \text{ N·m}$$

or 4.29 MN·m.

• PROBLEM 2-56

A cylindrical buoy (see Figure) 1·8 m in diameter, 1·2 m high and weighing 10 kN floats in salt water of density 1025 kg m^{-3}. Its center of gravity is 0·45 m from the bottom. If a load of 2 kN is placed on the top, find the maximum height of the center of gravity of this load above the bottom if the buoy is to remain in stable equilibrium.

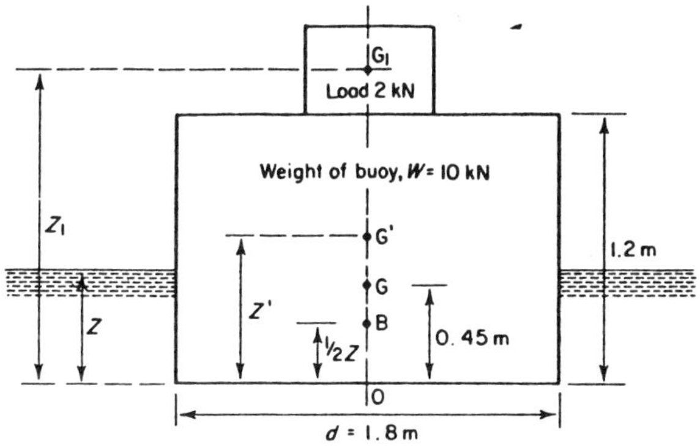

Stability of a cylindrical buoy

Solution: In the Figure, let G be the center of gravity of the buoy, G_1 the center of gravity of the load at a height Z_1 above the bottom, G' the combined center of gravity of the load and the buoy at a height Z' above the bottom.

When the load is in position, let V be the volume of salt water displaced and Z the depth of immersion of the buoy.

Buoyancy force = Weight of salt water displaced

$$= \rho g V = \rho g(\pi/4)d^2 Z.$$

For equilibrium, buoyancy force must equal the combined weight of the buoy and the load $(W + W_1)$, therefore,

$$W + W_1 = \rho g(\pi/4)d^2 Z,$$

Depth of immersion,

$$Z = 4(W + W_1)/\rho g \pi d^2$$

$$= 4(10 + 2) \times 10^3/1025 \times 9{\cdot}81 \times 1{\cdot}8^2 \times \pi = 0{\cdot}47 \text{ m.}$$

The center of buoyancy B will be at the center of gravity of the displaced water, so that OB = $\frac{1}{2}Z$ = 0·235 m.

If the buoy and the load are just in stable equilibrium, the metacenter M must coincide with the center of gravity G' of the buoy and load combined. The metacentric height G'M will then be zero and BG' = BM. With I = second moment of area about O and V = volume of liquid displaced, BG' may be calculated using the equation

$$BG' = BM = \frac{I}{V} = \frac{\pi d^4/64}{\pi d^2 y/4} = \frac{1{\cdot}8^2}{16 \times 0{\cdot}47} = 0{\cdot}431 \text{ m.}$$

Thus, the position of G' is given by

$$Z' = \tfrac{1}{2}Z + BG' = 0{\cdot}235 + 0{\cdot}431 = 0{\cdot}666 \text{ m.}$$

The value of Z_1 corresponding to this value of Z' is found by taking moments about O:

$$W_1 Z_1 + 0{\cdot}45\ W = (W + W_1)Z',$$

Maximum height of load above bottom,

$$Z_1 = \frac{(W + W_1)Z' - 0{\cdot}45\ W}{W_1}$$

$$= \frac{12 \times 10^3 \times 0{\cdot}666 - 0{\cdot}45 \times 10 \times 10^3}{2 \times 10^3} \text{m}$$

$$= 1{\cdot}746 \text{ m.}$$

• PROBLEM 2-57

A uniform cube of specific gravity s floats in water with two of its faces vertical and one specified edge above the water, the other three horizontal edges being immersed; show that if s lies between $\frac{23}{32}$ and $\frac{3}{4}$, there are three positions of equilibrium.

Prove that, if $s = \frac{93}{128}$, the cube can float with two of its faces inclined to the vertical at an angle $\tan^{-1} 1{\cdot}4$.

Solution: Let ODCE be the vertical central section cutting the water surface in AB. Let a denote an edge of the cube. Take OD, OE as axes of x and y. Then if G is the center of the cube, H the center of buoyancy and H' the centroid of the triangle OAB, the points H', G, H are in a vertical line.

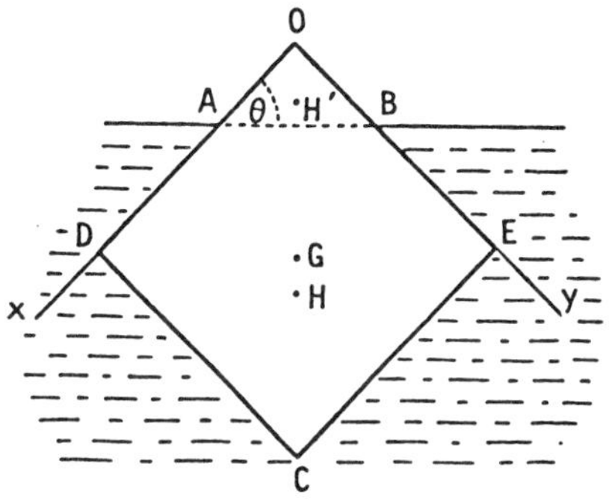

Also if the angle OAB is θ, and OA = c, the coordinates of H' are $\frac{1}{3}c$ and $\frac{1}{3}c \tan\theta$, and those of G are $\frac{1}{2}a$, $\frac{1}{2}a$, so that the gradient of H'G is $\dfrac{\frac{1}{2}a - \frac{1}{3}c\tan\theta}{\frac{1}{2}a - \frac{1}{3}c}$, and expressing the fact that this line is at right angles to AB gives

$$\frac{3a - 2c\tan\theta}{3a - 2c} = \cot\theta,$$

which, on reduction, gives

$$\tan\theta = 1 \quad \text{or} \quad (3a - 2c)/2c \qquad (1)$$

The first value corresponds to a position of equilibrium in which the diagonal OC is vertical.

The further condition for floating in the general position is

$$sa^2 = a^2 - \tfrac{1}{2}c^2\tan\theta,$$

and, substituting the second value for tan θ from (1), we get

$$sa^2 = a^2 - \frac{3ac - 2c^2}{4}$$

or

$$2c^2 - 3ac + 4a^2(1 - s) = 0 \qquad (2)$$

giving

$$c = \frac{3a \pm \sqrt{(32s - 23)}}{4}.$$

For real roots in c, we must have

$$s \geq \frac{23}{32} \qquad (3)$$

and for neither root of c/a to exceed unity, we require

$$\sqrt{(32s - 23)} \leq 1,$$

or

$$s \leq \frac{3}{4}.$$

Consequently when s lies between $\frac{23}{32}$ and $\frac{3}{4}$ there are three possible positions of equilibrium which satisfy the condition that one edge only is above the surface.

Again when $s = \frac{93}{128}$, the roots of (2) are $\frac{7}{8}a$ and $\frac{5}{8}a$, and substituting in (1), we get $\tan\theta = \frac{5}{7}$ or $\frac{7}{5}$, so that in either case two faces are inclined to the vertical at an angle $\tan^{-1} 1 \cdot 4$.

• PROBLEM 2-58

> Show that when a uniform hemisphere of density ρ and radius a floats with its plane base immersed in homogeneous liquid of density σ, the equilibrium is stable and the metacentric height is $\frac{3}{8}a(\sigma - \rho)/\rho$.

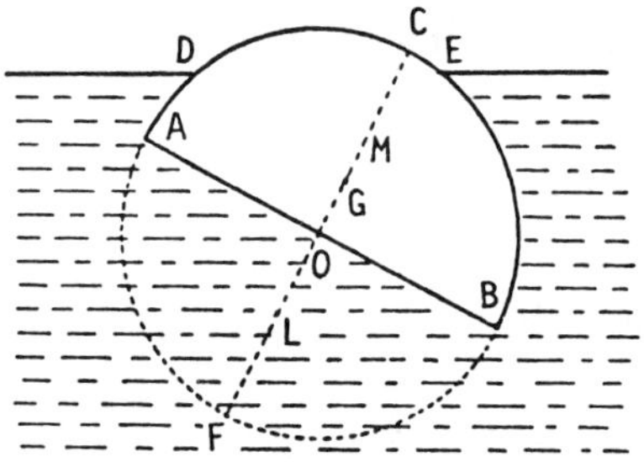

Solution: Let CAB be the solid hemisphere and DE its intersection with the surface of the liquid. Imagine the sphere to be completed below the surface of the liquid and let CF be the diameter at right angles to the base AB of the hemisphere.

Let V denote the volume immersed, viz. DABE. Then the force of buoyancy is an upward force gσV acting through the center of gravity of DABE. Now if we add to the liquid displaced the hemisphere AFB and also subtract it we do not alter the force of buoyancy. But the total upward force will now be the weight of liquid in DAFBE, viz.

$$g\sigma\left(V + \frac{2}{3}\pi a^3\right),$$

acting vertically through O since its components are all

normal to the sphere, together with a downward force

$$\frac{2}{3} g\sigma\pi a^3$$

(the added hemisphere) acting vertically through the center of gravity L of the hemisphere AFB, where

$$OL = \frac{3}{8} a .$$

These two parallel forces have an upward vertical resultant $g\sigma V$ which cuts FC in M, such that

$$g\sigma V \cdot OM = \frac{2}{3} g\sigma\pi a^3 \ . \ OL = \frac{1}{4} g\sigma\pi a^4 .$$

This point M is the metacenter, for its position is independent of the inclination of FC to the vertical so long as AB is immersed.

And from the condition for floating

$$g\sigma V = \frac{2}{3} g\rho\pi a^3 ,$$

therefore

$$OM = \frac{3}{8} \frac{\sigma}{\rho} a .$$

But if G is the center of gravity of the solid hemisphere $OG = \frac{3}{8} a$, and since σ is by hypothesis greater than ρ, therefore M is above G, the position in which GM is vertical is stable, and the metacentric height

$$GM = \frac{3}{8} a(\sigma - \rho)/\rho .$$

• PROBLEM 2-59

Discuss the stability of a uniform right circular cone of density σ floating in a liquid of density ρ with its axis vertical and vertex downwards.

<u>Solution:</u> Let h be the height of the cone, 2α its angle and h' the length of axis immersed.

In this case A is a circle of radius $h' \tan \alpha$ so that

$$Ak^2 = \frac{1}{4} \pi h'^4 \tan^4 \alpha^*$$

and $V = \frac{1}{3}\pi h'^3 \tan^2\alpha$,

therefore

$$HM = \frac{3}{4} h' \tan^2\alpha .$$

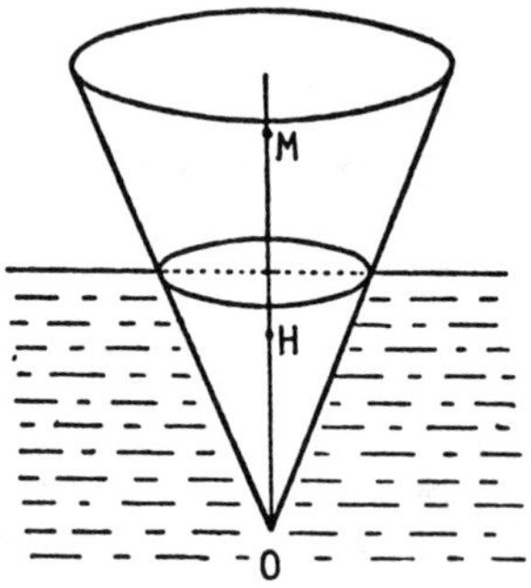

But, if O is the vertex, $OH = \frac{3}{4} h'$, so that $OM = \frac{3}{4} h' \sec^2\alpha$.

But $OG = \frac{3}{4} h$. Therefore the equilibrium is stable or unstable according as

$$h' \sec^2\alpha > \text{ or } < h.$$

But, since the cone floats, $\rho h'^3 = \sigma h^3$, so that the equilibrium is stable or unstable according as

$$\sigma/\rho > \text{ or } < \cos^6\alpha .$$

• PROBLEM 2-60

A thin metal circular cylinder contains water to a depth h and floats in water with its axis vertical immersed to a depth h'. Show that the vertical position is stable if the height of the center of gravity of the cylinder above its base is less than 1/2(h + h').

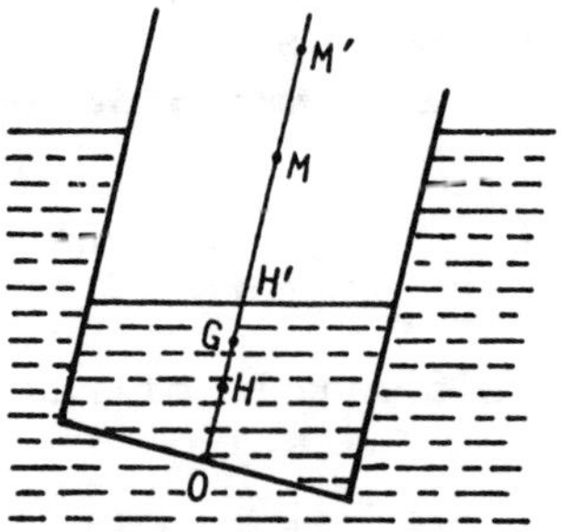

Solution: Let G be the center of gravity of the cylinder, O the center of the base, and H,H' the center of gravity of the water contained and displaced in the equilibrium position of the cylinder; M, M' the corresponding metacenters, a the radius of the cylinder and let OG = z.

Then

$$H'M' = Ak^2/V$$

$$= \frac{1}{4}\pi a^2/\pi a^2/h' = a^2/4h',$$

so that

$$GM' = OH' + H'M' - OG = \frac{h'}{2} + \frac{a^2}{4h'} - z$$

Similarly

$$GM = \frac{h}{2} + \frac{a^2}{4h} - z$$

The upward force of buoyancy acting through M' is $g\rho\pi a^2 h'$ and the weight of water contained acting through M is $g\rho\pi a^2 h$, so the equilibrium is stable if

$$g\rho\pi a^2 h'\left(\frac{h'}{2} + \frac{a^2}{4h'} - z\right) > g\rho\pi a^2 h\left(\frac{h}{2} + \frac{a^2}{4h} - z\right),$$

or if

$$\frac{1}{2}(h'^2 - h^2) > z(h' - h),$$

or

$$z < \frac{1}{2}(h + h').$$

CHAPTER 3

FLUID KINEMATICS

Basic Attacks and Strategies for Solving Problems in this Chapter. See pages 120 to 151 for step-by-step solutions to problems.

There are two primary viewpoints or descriptions of fluid flows. The first is called *Lagrangian* and involves tracking individual fluid particles in the flow. A complete Lagrangian description would require the vector position

$$\mathbf{r}(t) = x(t)\mathbf{i} + y(t)\mathbf{j} + z(t)\mathbf{k}$$

for each particle in the flow. The velocity and acceleration of each particle can be obtained by differentiation. While this is appropriate for solid mechanics, it is usually not feasible for fluid flows. Instead, the *Eulerian* description is more useful, since it is concerned with a *field* of fluid flow, where fluid particles pass in and out of the field. Velocity, pressure, etc., are described as field variables, i.e., as functions of space and time ($\mathbf{V}(x,y,z,t)$, $p(x,y,z,t)$, etc.).

Newton's second law of motion is

$$\mathbf{F} = m\mathbf{a}.$$

For fluid flows, this is written per unit volume as

$$\mathbf{f} = \rho\mathbf{a},$$

and the law applies to fluid particles. In the Eulerian description, acceleration vector **a** must contain both unsteady (local) terms and convective terms. To write Newton's law in the Eulerian viewpoint, acceleration is written as

$$\mathbf{a} = \frac{d\mathbf{V}}{dt} = \frac{\partial \mathbf{V}}{\partial t} + (\mathbf{V} \cdot \nabla)\mathbf{V}. \tag{1}$$

In Cartesian (x, y, z), (u, v, w) coordinates, (1) becomes

$$\mathbf{a} = \frac{d\mathbf{V}}{dt} = \frac{\partial \mathbf{V}}{\partial t} + u\frac{\partial \mathbf{V}}{\partial x} + v\frac{\partial \mathbf{V}}{\partial y} + w\frac{\partial \mathbf{V}}{\partial z}. \tag{2}$$

Note that the above is a vector equation which can be further split into three components in the **i, j,** and **k** directions.

The force per unit volume, **f**, is composed of body (gravitational) forces, pressure forces, and viscous (frictional) forces. For Newtonian fluids, Newton's second law reduces to the Navier-Stokes equation,

$$\rho\left[\frac{\partial \mathbf{V}}{\partial t}(\mathbf{V}\cdot\nabla)\mathbf{V}\right] = \rho\mathbf{g} - \nabla p + \mu\nabla^2\mathbf{V}. \tag{3}$$

Again, this is a vector equation. Fully expanded, the x-component is

$$\rho\left(\frac{\partial u}{\partial t} + u\frac{\partial u}{\partial x} + v\frac{\partial u}{\partial y} + w\frac{\partial u}{\partial z}\right) = \rho g_x - \frac{\partial p}{\partial x} + \mu\left(\frac{\partial^2 u}{\partial x^2} + \frac{\partial^2 u}{\partial y^2} + \frac{\partial^2 u}{\partial z^2}\right) \tag{4}$$

with similar expressions for the other two components.

The acceleration vector may be written in any coordinate system. A particularly useful choice is to decompose **a** into its component tangent to the direction of the flow (i.e., *along* a streamline), and its component *normal* to the streamline towards the center of curvature. If R represents the radius of curvature of a streamline, the normal component of acceleration, a_n, is given by

$$a_n = \frac{V^2}{R} \tag{5}$$

due to centripetal acceleration effects.

It is important to distinguish between streamlines, pathlines, and streaklines in a flow. *Streamlines* are everywhere tangent to the velocity vector at any instant in time and can be found by applying

$$\frac{dx}{u} = \frac{dy}{v} = \frac{dz}{w}. \tag{6}$$

A *pathline* is the actual path followed by a fluid particle. Pathlines are found from the relation between velocity and position, i.e.,

$$\frac{d\mathbf{r}}{dt} = \mathbf{V}(x,y,z,t). \tag{7}$$

For a known velocity field, the three components of Equation (7) can be integrated over time to yield the pathline. A *streakline* is defined as the locus of fluid particles which have earlier passed through a prescribed point. To obtain the streakline, integrate Equation (7) with respect to time as for the pathline, but apply the *family* of initial conditions $x = x_0$, $y = y_0$, $z = z_0$ at a continuous sequence of times prior to the time in question. Note that for a steady flow, streamlines, pathlines, and streaklines are all coincident.

For inviscid flow, the terms containing viscosity in Equation (3) drop out. The remaining equation is called Euler's equation. This is often a useful simplification of the full Navier-Stokes equation and is valid away from regions where viscosity is important (such as close to solid walls). The integration of Euler's equation along a streamline yields the Bernoulli equation

$$p + \tfrac{1}{2}\rho V^2 + \rho g z = \text{constant along a streamline.} \tag{8}$$

If the flow is furthermore irrotational, the same constant is applicable everywhere in the flow. For a known velocity field, the pressure field can easily be found by the integration of Equation (8).

Step-by-Step Solutions to Problems in this Chapter, "Fluid Kinematics"

COORDINATE TRANSFORMATION

• PROBLEM 3-1

We wish to express

$$\frac{\partial \phi}{\partial x} + \frac{\partial^2 \psi}{\partial x\ \partial y} + 4\ \frac{\partial \phi}{\partial y}$$

in terms of ξ and η, where

$$\xi = x \qquad \text{and} \qquad \eta = y/x$$

This represents a transformation of coordinates from (x,y) to (ξ, η).

Solution: If we wish to transform coordinates from (x,y,z) to (ξ, η, ζ) we use

$$\frac{\partial \phi}{\partial x} = \frac{\partial \phi}{\partial \xi}\frac{\partial \xi}{\partial x} + \frac{\partial \phi}{\partial \eta}\frac{\partial \eta}{\partial x} + \frac{\partial \phi}{\partial \zeta}\frac{\partial \zeta}{\partial x}$$

$$\frac{\partial \phi}{\partial y} = \frac{\partial \phi}{\partial \xi}\frac{\partial \xi}{\partial y} + \frac{\partial \phi}{\partial \eta}\frac{\partial \eta}{\partial y} + \frac{\partial \phi}{\partial \zeta}\frac{\partial \zeta}{\partial y}$$

$$\frac{\partial \phi}{\partial z} = \frac{\partial \phi}{\partial \xi}\frac{\partial \xi}{\partial z} + \frac{\partial \phi}{\partial \eta}\frac{\partial \eta}{\partial z} + \frac{\partial \phi}{\partial \zeta}\frac{\partial \zeta}{\partial z}$$

Higher-order derivatives follow from these relations: that is,

$$\frac{\partial^2 \psi}{\partial x^2} = \frac{\partial}{\partial x}\left(\frac{\partial \phi}{\partial x}\right)$$

and we simply substitute $\partial\phi/\partial x$ for ϕ in the first of the equations.

Then

$$\frac{\partial \phi}{\partial x} = \frac{\partial \phi}{\partial \xi}\frac{\partial \xi}{\partial x} + \frac{\partial \phi}{\partial \eta}\frac{\partial \eta}{\partial x}$$

$$= \frac{\partial \phi}{\partial \xi} - \frac{y}{x^2}\ \frac{\partial \phi}{\partial \eta}$$

$$= \frac{\partial \phi}{\partial \xi} - \frac{\eta}{\xi}\frac{\partial \phi}{\partial \eta}$$

where we have used $\eta/\xi = y/x^2$. We also have

$$\frac{\partial \phi}{\partial y} = \frac{\partial \phi}{\partial \xi}\overset{\nearrow 0}{\cancel{\frac{\partial \xi}{\partial y}}} + \frac{\partial \phi}{\partial \eta}\frac{\partial \eta}{\partial y}$$

$$= \frac{1}{x}\frac{\partial \phi}{\partial \eta}$$

$$= \frac{1}{\xi}\frac{\partial \phi}{\partial \eta}$$

Finally,

$$\frac{\partial^2 \phi}{\partial x\ \partial y} = \frac{\partial}{\partial x}\left(\frac{\partial \phi}{\partial y}\right) = \frac{\partial}{\partial \xi}\left(\frac{\partial \phi}{\partial y}\right)\frac{\partial \xi}{\partial x} + \frac{\partial}{\partial \eta}\left(\frac{\partial \phi}{\partial y}\right)\frac{\partial \eta}{\partial x}$$

$$= \frac{\partial}{\partial \xi}\left(\frac{1}{\xi}\frac{\partial \phi}{\partial \eta}\right) + \frac{\partial}{\partial \eta}\left(\frac{1}{\xi}\frac{\partial \phi}{\partial \eta}\right)\left(-\frac{y}{x^2}\right)$$

$$= \frac{1}{\xi}\frac{\partial^2 \phi}{\partial \xi\ \partial \eta} - \frac{1}{\xi^2}\frac{\partial \phi}{\partial \eta} - \frac{\eta}{\xi^2}\frac{\partial^2 \phi}{\partial \eta^2}$$

The expression is thus

$$\frac{\partial \phi}{\partial x} + \frac{\partial^2 \phi}{\partial x\ \partial y} + 4\frac{\partial \phi}{\partial y} = \frac{\partial \phi}{\partial \xi} + \frac{1}{\xi}\frac{\partial^2 \phi}{\partial \xi\ \partial \eta} + \left(\frac{4}{\xi} - \frac{\eta}{\xi} - \frac{1}{\xi^2}\right)\frac{\partial \phi}{\partial \eta} - \frac{\eta}{\xi^2}\frac{\partial^2 \phi}{\partial \eta^2}$$

VELOCITY AND ACCELERATION

• PROBLEM 3-2

If a very thin plate is placed in a fast-moving air stream, as shown in the figure, a boundary-layer flow will develop on the plate. That is, a steady flow is obtained in which the retarding effect of the plate is felt only near the plate. The path of a particle which starts in the free stream and a typical velocity profile are shown on the sketch. Describe the acceleration of the particle from both a Lagrangian viewpoint and an Eulerian viewpoint.

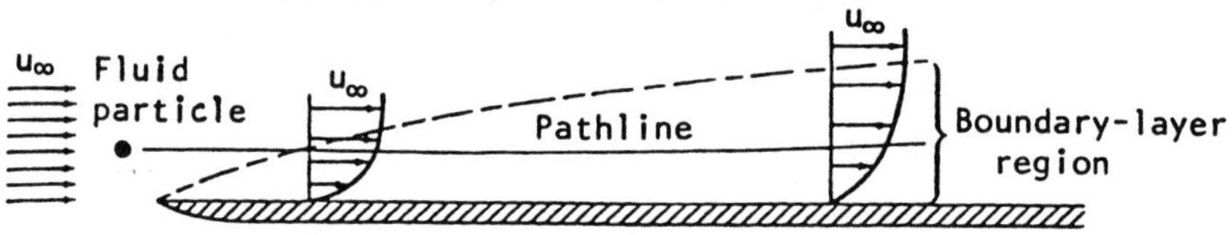

Solution: In the free stream outside the boundary layer, the fluid feels no effect from the plate (no retarding force) and consequently feels no acceleration. Thus in the free stream:

Lagrangian: $\frac{Du}{Dt} = 0$ Eulerian: $\frac{\partial u}{\partial t} = 0$

$$u \neq 0 \text{ but } \frac{\partial u}{\partial x} = 0$$

As the fluid enters the boundary-layer region, the viscous shear effects of the plate have caused the fluid in the outer layers to be retarded. Since the force on the particle is in the negative direction, we have

Lagrangian: $\frac{Du}{Dt} < 0$ Eulerian: $\frac{\partial u}{\partial t} = 0$,

$$u > 0 \; , \; \left.\frac{\partial u}{\partial x}\right|_y < 0$$

Because $\partial u/\partial x < 0$, the velocity u must decrease at a fixed y as x is increased. Hence the profiles appear as shown. The fluid velocity is zero at y = 0 because of the no-slip condition, and u_∞ outside the boundary layer.

Note that this example emphasizes that the two descriptions give the acceleration of a particle at some point at some instant of time. They provide identical numerical values at the instant in time when the particle being followed by the Lagrangian description occupied the Eulerian point of interest.

• PROBLEM 3-3

Along the straight streamline shown, the velocity is given by $v = 3\sqrt{x^2 + y^2}$. Calculate the velocity and acceleration at the point (8,6).

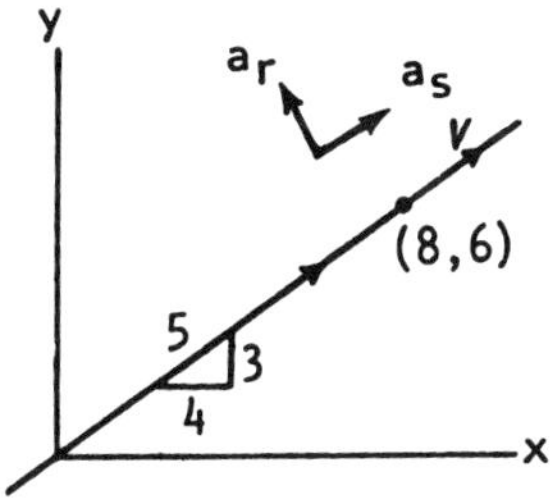

Solution: Since $s = \sqrt{x^2 + y^2}$, $v = 3s$.

At (8,6), s = 10; therefore v = 3 × 10 = 30 m/s.

$$a_s = \frac{d^2s}{dt^2} = \frac{d}{dt}\left(\frac{ds}{dt}\right) = \frac{dv}{dt} = \frac{ds}{dt}\frac{dv}{ds} = v\frac{dv}{ds} \qquad (1)$$

$$a_r = -\frac{v^2}{r} \qquad (2)$$

$$a_s = (3s)3 = 9s = 90 \text{ m/s}^2 \qquad (3)$$

a_r is obviously zero because the radius of curvature of the streamline is infinite.

• PROBLEM 3-4

Given $x = \frac{3x_0y_0t^2}{z_0}$, $y = \frac{5x_0z_0t}{y_0}$, $z = \frac{2y_0z_0t^3}{x_0}$

find the velocity of the fluid particle and acceleration at

$x_0 = 1$ cm, $y_0 = 2$ cm, $z_0 = 3$ cm at t = 2 .

Solution: The velocity is given by

$$\vec{V} = \frac{\partial x}{\partial t}\vec{i} + \frac{\partial y}{\partial t}\vec{j} + \frac{\partial z}{\partial t}\vec{k}$$

$$= \frac{6x_0y_0}{z_0}t\,\vec{i} + \frac{5x_0z_0}{y_0}\vec{j} + \frac{6y_0z_0}{x_0}t^2\,\vec{k}$$

$$\therefore \quad \vec{V}(1,2,3,2) = 8\,\vec{i} + 7.5\,\vec{j} + 144\,\vec{k}$$

The acceleration is given by

$$\vec{a} = \frac{\partial^2 x}{\partial t^2}\,\vec{i} + \frac{\partial^2 y}{\partial t^2}\,\vec{j} + \frac{\partial^2 z}{\partial t^2}\,\vec{k}$$

so that

$$\vec{a} = \frac{6x_0y_0}{z_0}\,\vec{i} + \frac{12y_0z_0}{x_0}\,t\,\vec{k}$$

Thus

$$\vec{a}(1,2,3,2) = 4\,\vec{i} + 144\,\vec{k}$$

• PROBLEM 3-5

The velocity of the fluid particle P relative to a fixed frame 0 is $3\vec{i} + 2\vec{j}$. The velocity of a moving frame is $4\vec{j}$. What is the angular speed $\vec{\omega}$ of the rotating frame when the position vector from the origin of the moving frame to the particle is $2\vec{k}$?

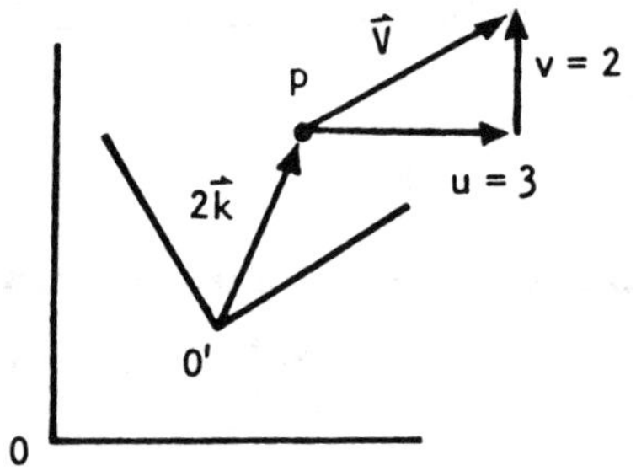

Solution: For a moving frame of reference and using the notation where $\vec{r}_{o'}$ is the absolute position vector of the moving origin 0' with respect to the fixed origin 0, $\vec{r}_p$ is the absolute position vector of the point P with respect to the fixed origin 0, and $\vec{r}_{p/o'}$ is the relative position vector of the point P with respect to the moving origin 0', an equation may be written

$$\vec{r}_p = \vec{r}_{o'} + \vec{r}_{p/o'} \qquad (1)$$

The relative velocity $\vec{V}_{p/o'}$, which is the velocity of the point P relative to the moving frame o', is, from Equation (1)

$$\vec{V}_{p/o'} = \vec{V}_p - \vec{V}_{o'} \quad (2)$$

From kinematics,

$$\vec{V}_{p/o'} = \dot{r}_{p/o'}\,\vec{e}_r + \vec{\omega} \times \vec{r}_{p/o'} \quad (3)$$

where $\vec{\omega}$ is the angular speed of the moving frame. If both points P and O' move together as a rigid body, then $\dot{r}_{p/o'}$ is zero, and one obtains

$$\vec{V}_p = \vec{V}_{o'} + \vec{\omega} \times \vec{r}_{p/o'} \quad (4)$$

$$3\vec{i} + 2\vec{j} = r\vec{j} + \vec{\omega} \times 2\vec{k}$$

Therefore

$$(\omega_x\vec{i} + \omega_y\vec{j} + \omega_z\vec{k}) \times 2\vec{k} = 3\vec{i} - 2\vec{j}$$

Simplifying the above results in

$$-2\omega_x\vec{j} + 2\omega_y\vec{i} = 3\vec{i} - 2\vec{j}$$

$$\therefore \quad \omega_y = 3/2 \; , \; \omega_x = 1$$

Thus, the angular speed $\vec{\omega}$ of the moving frame is

$$\vec{\omega} = \vec{i} + \frac{3}{2}\vec{j} \, .$$

• PROBLEM 3-6

Let the position of a fluid particle at any time t be given by $x = a$, $y = bt^2$, where a and b are constants. Find the velocity components u, v, v_r, v_θ at any time t, and magnitude of the velocity $|\vec{V}|$.

Solution: The velocity vector $\vec{V}$ has components u, v, w in rectangular coordinates in the x, y, z directions, respectively:

$$\vec{V} = u\vec{i} + v\vec{j} + w\vec{k}$$

where from the definition of velocity

$$\vec{V} = \frac{d\vec{r}}{dt}$$

and the position vector $\vec{r}$

$$\vec{r} = x\vec{i} + y\vec{j} + z\vec{k}$$

we obtain

$$u \equiv \frac{dx}{dt} \quad (1a)$$

$$v \equiv \frac{dy}{dt} \quad (1b)$$

$$w \equiv \frac{dz}{dt} \quad (1c)$$

since $\vec{i}$, $\vec{j}$ and $\vec{k}$ are constants.
from equations 1a and 1b

$$u = \frac{dx}{dt} = 0$$

$$v = \frac{dy}{dt} = 2bt$$

Using the equations to convert to cylindrical coordinates

$$r = \sqrt{x^2 + y^2} \quad (2)$$

$$\theta = \tan^{-1} y/x \quad (3)$$

We obtain

$$r = \sqrt{a^2 + b^2t^4}$$

$$\theta = \tan^{-1}\left(\frac{bt^2}{a}\right)$$

For a cylindrical coordinate system, the volocity vector is

$$\vec{V} = v_r\vec{e}_r + v_\theta\vec{e}_\theta + w\vec{k}$$

where the components are

$$v_r \equiv \frac{dr}{dt} \tag{4}$$

$$v_\theta \equiv r\frac{d\theta}{dt} \tag{5}$$

Then

$$v_r = \frac{dr}{dt} = \frac{d}{dt}\left(\sqrt{a^2 + b^2t^4}\right) = \frac{2b^2t^3}{\sqrt{a^2 + b^2t^4}}$$

$$v_\theta = r\frac{d\theta}{dt} = \frac{2abt}{\sqrt{a^2 + b^2t^4}}$$

The speed is given by

$$|\vec{V}| = V = \sqrt{u^2 + v^2} = \sqrt{v_r^2 + v_\theta^2} = \sqrt{(2bt)^2}$$

$$= \sqrt{\frac{(2bt)^2(b^2t^4 + a^2)}{(a^2 + b^2t^4)}}$$

$$= 2bt$$

• PROBLEM 3-7

Consider a very elementary flow problem. Let the flow be steady and inviscid. Calculate the velocity of the flow in terms of the pressure p given $u = u(x)$, $v = w = 0$.

Solution: The equations of motion can be written in the forms

$$a_x = \frac{\partial u}{\partial t} + u\frac{\partial u}{\partial x} + v\frac{\partial u}{\partial y} + w\frac{\partial u}{\partial z} = g_x + \frac{1}{\rho}\left(\frac{\partial p_{xx}}{\partial x} + \frac{\partial p_{yx}}{\partial y} + \frac{\partial p_{zx}}{\partial z}\right) \tag{1}$$

$$a_y = \frac{\partial v}{\partial t} + u\frac{\partial v}{\partial x} + v\frac{\partial v}{\partial y} + w\frac{\partial v}{\partial z} = g_y + \frac{1}{\rho}\left(\frac{\partial p_{xy}}{\partial x} + \frac{\partial p_{yy}}{\partial y} + \frac{\partial p_{zy}}{\partial z}\right) \tag{2}$$

$$a_z = \frac{\partial w}{\partial t} + u\frac{\partial w}{\partial x} + v\frac{\partial w}{\partial y} + w\frac{\partial w}{\partial z} =$$

$$g_z + \frac{1}{\rho}\left(\frac{\partial p_{xz}}{\partial x} + \frac{\partial p_{yz}}{\partial y} + \frac{\partial p_{zz}}{\partial z}\right) \tag{3}$$

For inviscid steady one-dimensional flow, Equation (1) becomes

$$u\frac{du}{dx} = -\frac{1}{\rho}\frac{\partial p}{\partial x} \tag{4}$$

and from Equations (2) and (3), $\frac{\partial p}{\partial y} = 0$ and $\frac{\partial p}{\partial z} = -g$

Integrating (4) gives

$$\frac{u^2}{2} = -\frac{1}{\rho}p + \text{const.}$$

where

$$\text{constant} = gz/\rho$$

Thus the velocity becomes

$$u = \sqrt{\frac{2gz-p}{\rho}}$$

• PROBLEM 3-8

Starting with the Navier-Stokes equations, obtain the differential equations which describe the incompressible flow over an impulsively accelerated infinitely long horizontal flat plate.

Solution: The Navier-Stokes equations in rectangular coordinates for laminar incompressible flow of a constant-viscosity newtonian fluid for the x-, y-, and z- components are given by the equations

$$\rho\left(\frac{\partial v_x}{\partial t} + v_x\frac{\partial v_x}{\partial x} + v_y\frac{\partial v_x}{\partial y} + v_z\frac{\partial v_x}{\partial z}\right) =$$

$$-\frac{\partial p}{\partial x} + \rho B_x + \mu\left(\frac{\partial^2 v_x}{\partial x^2} + \frac{\partial^2 v_x}{\partial y^2} + \frac{\partial^2 v_x}{\partial z^2}\right)$$

$$\rho\left(\frac{\partial v_y}{\partial t} + v_x\frac{\partial v_y}{\partial x} + v_y\frac{\partial v_y}{\partial y} + v_z\frac{\partial v_y}{\partial z}\right) =$$

$$- \frac{\partial p}{\partial y} + \rho B_y + \mu\left(\frac{\partial^2 v_y}{\partial x^2} + \frac{\partial^2 v_y}{\partial y^2} + \frac{\partial^2 v_y}{\partial z^2}\right)$$

$$\rho\left(\frac{\partial v_z}{\partial t} + v_x \frac{\partial v_z}{\partial x} + v_y \frac{\partial v_z}{\partial y} + v_z \frac{\partial v_z}{\partial z}\right) =$$

$$- \frac{\partial p}{\partial z} + \rho B_z + \mu\left(\frac{\partial^2 v_z}{\partial x^2} + \frac{\partial^2 v_z}{\partial y^2} + \frac{\partial^2 v_z}{\partial z^2}\right)$$

We shall need to use the following assumptions and physical arguments:

1. There is no flow or change of flow variables in the z direction; $v_z = 0$, $\partial/\partial z = 0$, and $\partial^2/\partial z^2 = 0$.

2. There can be no changes in any of the flow variables in the x direction because the plate is infinitely long in that direction; $\partial/\partial x = 0$ and $\partial^2/\partial x^2 = 0$.

3. The flow is one-directional in x; $v_y = 0$ and $v_z = 0$.

4. There are no body forces in the x direction; $B_x = 0$.

The x component of the Navier-Stokes equations is

$$\rho\left(\frac{\partial v_x}{\partial t} + v_x \overset{(2)}{\cancel{\frac{\partial v_x}{\partial x}}} + \overset{(3)}{\cancel{v_y}} \frac{\partial v_x}{\partial y} + \overset{(3)}{\cancel{v_z}} \overset{(1)}{\cancel{\frac{\partial v_x}{\partial z}}}\right) =$$

$$- \underset{(2)}{\cancel{\frac{\partial p}{\partial x}}} + \underset{(4)}{\cancel{\rho B_x}} + \mu\left(\underset{(2)}{\cancel{\frac{\partial^2 v_x}{\partial x^2}}} + \frac{\partial^2 v_x}{\partial y^2} + \underset{(1)}{\cancel{\frac{\partial^2 v_x}{\partial z^2}}}\right) \qquad +$$

The various terms which are zero have been canceled and the number of the reason given above. The result is

$$\rho \frac{\partial v_x}{\partial t} = \mu \frac{\partial^2 v_x}{\partial y^2}$$

Assuming the y direction to be that of gravity, $B_y = -g$, the y component of the Navier-Stokes equations becomes

$$\frac{\partial p}{\partial y} = - \rho g$$

which is the equation of a hydrostatic pressure distribution. Finally, the z component of the Navier-Stokes equations gives $B_z = 0$. There can be no body-force component in the z direction.

• PROBLEM 3-9

> Suppose a fluid particle moves according to the equations $x = at$, $y = bt - ct^2$, where a, b, c are constants. Find the normal and tangential accelerations.

Solution: We first calculate the speed V, where

$$V = \sqrt{u^2 + v^2} = \sqrt{\left(\frac{dx}{dt}\right)^2 + \left(\frac{dy}{dt}\right)^2}$$

$$= \sqrt{a^2 + (b - 2ct)^2} \tag{1}$$

Next, calculate the radius of curvature R. From calculus,

$$R = \frac{[1 + (dy/dx)^2]^{3/2}}{\left|\frac{d^2y}{dx^2}\right|} \tag{2}$$

From $x = at$, $y = bt - ct^2$, one obtains

$$y = \frac{bx}{a} - \frac{cx^2}{a^2}$$

Thus

$$\frac{dy}{dx} = \frac{b}{a} - \frac{2cx}{a^2} \tag{3}$$

$$\frac{d^2y}{dx^2} = -\frac{2c}{a^2} \tag{4}$$

Substituting Equations (3) and (4) into Equation (2) yields

$$R = \frac{\left[1 + \left(\frac{b}{a} - \frac{2cx}{a^2}\right)^2\right]^{3/2}}{2c/a^2} \tag{5}$$

Next, substitute Equations (1) and (5) into the equation for the centripetal acceleration toward the center of curvature,

$$a_n = \frac{V^2}{R}$$

which yields

$$a_n = \frac{\left[a^2 + (b - 2ct)^2\right] \frac{2c}{a^2}}{\left[1 + \left(\frac{b}{a} - \frac{2cx}{a^2}\right)^2\right]^{3/2}} \qquad (6)$$

Transform x in Equation (6) back to time t using x = at, which results in

$$a_n = \frac{2ca}{\sqrt{a^2 + (b - 2ct)^2}} \qquad (7)$$

for the normal acceleration.

The easiest way to find the tangential acceleration a_t is to first evaluate the total acceleration using the x and y components of accelerations.

$$a = \sqrt{a_x^2 + ay^2} = \sqrt{a_t^2 + a_n^2}$$

$$\therefore \quad a_t = \sqrt{a_x^2 + a_y^2 - a_n^2} \qquad (8)$$

where

$$a_x = \frac{du}{dt} = 0 \qquad (9)$$

$$a_y = \frac{dv}{dt} = - 2c \qquad (10)$$

Substitution of equations (7), (9) and (10) into equation (8) yields

$$a_t = \sqrt{\frac{- 4c^2a^2}{[a^2 + (b - 2ct)^2]} + 4c^2}$$

or

$$a_t = \frac{2c(b - 2ct)}{\sqrt{a^2 + (b - 2ct)^2}}$$

as the tangential acceleration.

• PROBLEM 3-10

For the nozzle shown, the equation describing incompressible flow, assuming the flow is steady and one-dimensional along the axis, is given by

$$\vec{V} = V_a\left(1 + \frac{x}{L}\right)$$

where V is the flow velocity at a section x; V_a is the velocity at inlet section, a, and L is the length of the nozzle.

(1) For a particle moving through the nozzle, calculate the x-component of its acceleration.

(2) For a particle located initially at the inlet section, a, find its position along the axis of the nozzle as a function of time, and determine its x-component of acceleration as a function of time.

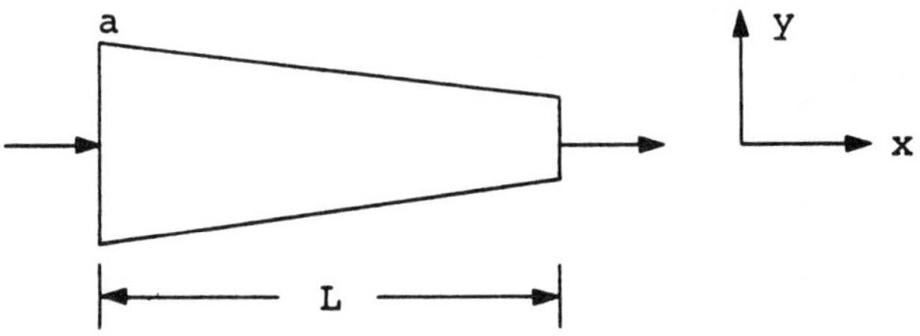

Solution: (1) The general equation of motion for the acceleration of a particle in a flow field is given by

$$\frac{D\vec{V}}{Dt} = u\frac{\partial \vec{V}}{\partial x} + v\frac{\partial \vec{V}}{\partial y} + w\frac{\partial \vec{V}}{\partial z} + \frac{\partial \vec{V}}{\partial t} \qquad (1)$$

For the x-component of acceleration,

$$\frac{Du}{Dt} = u\frac{\partial u}{\partial x} + v\frac{\partial u}{\partial y} + w\frac{\partial u}{\partial z} + \frac{\partial u}{\partial t} \qquad (2)$$

Restricting the flow to the conditions given,

$$v = w = 0 \quad \text{and} \quad u = V_a\left(1 + \frac{x}{L}\right)$$

Substituting in equation (2)

$$\frac{Du}{Dt} = u\frac{\partial u}{\partial x} = V_a\left(1 + \frac{x}{L}\right)\frac{V_a}{L} = \frac{V_a^{\,2}}{L}\left(1 + \frac{x}{L}\right) \tag{3}$$

which is the equation for determining the x-component of acceleration of the particle.

(2) To find the particle position, x, as a function of time f(t), we refer back to the given equation,

$$V = \frac{dx}{dt} = \frac{df}{dt} = V_a\left(1 + \frac{x}{L}\right)$$

since $x = f(t)$

$$\frac{df}{dt} = V_a\left(1 + \frac{f}{L}\right)$$

from which

$$\frac{df}{1 + \frac{f}{L}} = V_a dt$$

results from separation of variables.

Integrating both sides and letting $x = 0$ at inlet section, a, when $t = 0$,

$$\int_o^f \frac{df}{(1 + f/L)} = \int_o^t V_a dt \tag{4}$$

from which

$$L \ln\left(1 + \frac{f}{L}\right) = V_a t \tag{5}$$

Dividing both sides by L

$$\ln\left(1 + \frac{f}{L}\right) = \frac{V_a t}{L} \tag{6}$$

and solving for f

$$1 + \frac{f}{L} = e^{V_a t/L} \tag{7}$$

$$f = L\left[e^{V_a t/L} - 1\right] \tag{8}$$

Since x = f(t), equation (8) gives the position of the particle as a function of time.

It is also possible to derive the x-component of acceleration from the expression

$$\frac{d^2x}{dt^2} = \frac{d^2f}{dt^2} = \frac{V_a^2}{L}\, e^{V_a t/L} \tag{9}$$

It is interesting to investigate whether both methods for determing the x-component of acceleration of a particle yield the same results:

First Method using equation (3),

$$\frac{Du}{Dt} = \frac{V_a^2}{L}\left(1 + \frac{x}{L}\right)$$

Since x = 0 at t = 0,

$$\frac{Du}{Dt} = \frac{V_a^2}{L}(1 + 0) = \frac{V_a^2}{L}$$

For x = 0.5L,

$$\frac{Du}{Dt} = \frac{V_a^2}{L}(1 + 0.5) = 1.5\,\frac{V_a^2}{L}$$

For x = L,

$$\frac{Du}{Dt} = \frac{V_a^2}{L}(1 + 1) = \frac{2V_a^2}{L}$$

Second Method using equation (9),

$$\frac{d^2x}{dt^2} = \frac{V_a^2}{L}\, e^{V_a t/L}$$

Again, x = 0 at t = 0, and

$$\frac{d^2x}{dt^2} = \frac{V_a^2}{L} e^0 = \frac{V_a^2}{L}$$

For x = 0.5L,

$$x = L\left(e^{V_a t/L} - 1\right)$$

from which $e^{V_a t/L} = 1.5$ and

$$\frac{d^2x}{dt^2} = \frac{V_a^2}{L} e^{V_a t/L} = \frac{V_a^2}{L}(1.5)$$

For x = L,

$$\frac{d^2x}{dt^2} = L\left(e^{V_a t/L} - 1\right)$$

from which $e^{V_a t/L} = 2$ and substituting in

$$\frac{d^2x}{dt^2} = \frac{V_a^2}{L} e^{V_a t/L},$$

we obtain

$$\frac{d^2x}{dt^2} = \frac{2V_a^2}{L}$$

Accordingly, both methods yield the same results.

• PROBLEM 3-11

a) A fluid moves along the circular streamline with a constant tangential velocity component of 1.04 m/s. Calculate the tangential and radial components of acceleration at any point on the streamline (shown in Figure 1).

b) For the circular streamline described in part (a) along which the velocity is 1.04 m/s, calculate the horizontal, vertical, tangential, and normal components of the velocity and acceleration at the point P (2,60°). (Fig. 2)

c) When an incompressible, nonviscous fluid flows against a plate in a plane (two-dimensional) flow, an exact solution for the equations of motion for this flow is u = Ax, v = -Ay, with A > 0 for the sketch shown. The coordinate origin is located at the stagnation point 0, where the flow divides and the local velocity is zero. Find the velocities and accelerations in the flow. (Fig. 3)

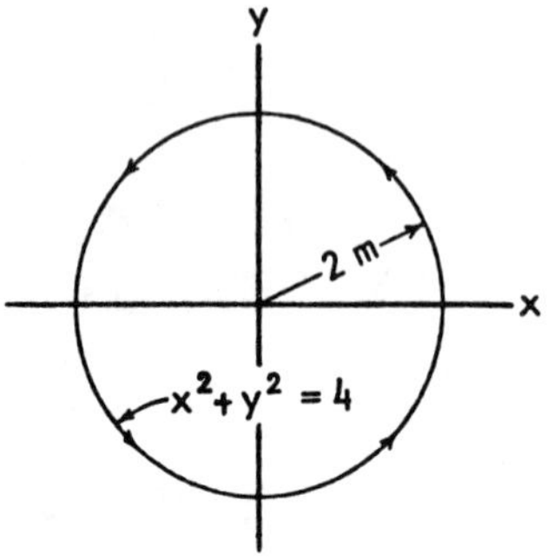

Fig. 1

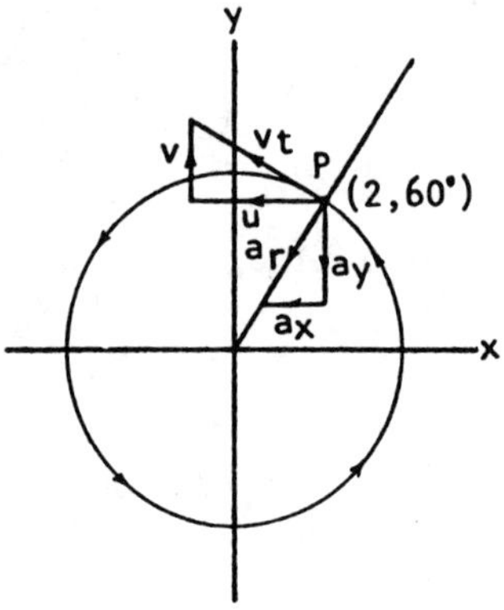

Fig. 2

Solution: a)

$$a_s = \frac{d^2s}{dt^2} = \frac{d}{dt}\left(\frac{ds}{dt}\right) = \frac{dv}{dt} \tag{1}$$

$$= \frac{ds}{dt}\frac{dv}{ds} = v\frac{dv}{ds}$$

$$a_r = -\frac{v^2}{r} \tag{2}$$

Because the velocity is constant $a_s = 0$. As the radius of the streamline is 2 m and the velocity along it is 1.04 m/s,

$$a_r = \frac{(1.04)^2}{2} = 0.541 \text{ m/s}^2$$

directed toward the center of the circle.

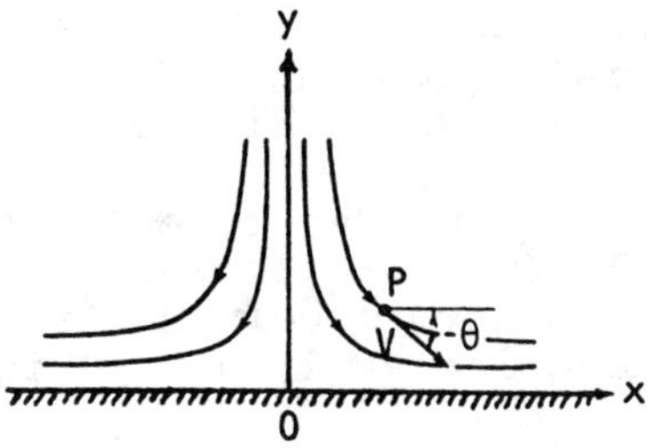

Fig. 3

b) Written mathematically in the Eulerian view,

$$u = u(x,y) \qquad \text{and} \qquad v = v(x,y)$$

In terms of displacement and time (the Lagrangian view), however,

$$u = \frac{dx}{dt} \qquad\qquad v = \frac{dy}{dt}$$

where x and y here are actually the coordinates of a fluid particle. Of course, the velocity at a point is the same in both the Eulerian and the Lagrangian view. Accelerations a_x and a_y are

$$a_x = \frac{du}{dt} \qquad \text{and} \qquad a_y = \frac{dv}{dt} \tag{3}$$

Writing the differentials du and dv in terms of partial derivatives,

$$du = \frac{\partial u}{\partial x}\,dx + \frac{\partial u}{\partial y}\,dy \qquad \text{and} \qquad dv = \frac{\partial v}{\partial x}\,dx + \frac{\partial v}{\partial y}\,dy$$

Substituting these relations in equations (3) and recognizing the velocities u and v as dx/dt and dy/dt, respectively,

$$a_x = u\,\frac{\partial u}{\partial x} + v\,\frac{\partial u}{\partial y} \qquad \text{and} \qquad a_y = u\,\frac{\partial v}{\partial x} + v\,\frac{\partial v}{\partial y} \tag{4}$$

A similar analysis for polar coordinates, in which v_r and v_t are both functions of r and θ (Fig. 4) leads to

$$v_r = \frac{dr}{dt} \qquad \text{and} \qquad v_t = r\,\frac{d\theta}{dt} \tag{5}$$

and, for the components of acceleration in a steady flow,

$$a_r = v_r\,\frac{\partial v_r}{\partial r} + v_t\,\frac{\partial v_r}{r\partial\theta} - \frac{v_t^2}{r} \tag{6}$$

$$a_t = v_r\,\frac{\partial v_t}{\partial r} + v_t\,\frac{\partial v_t}{r\partial\theta} + \frac{v_r v_t}{r} \tag{7}$$

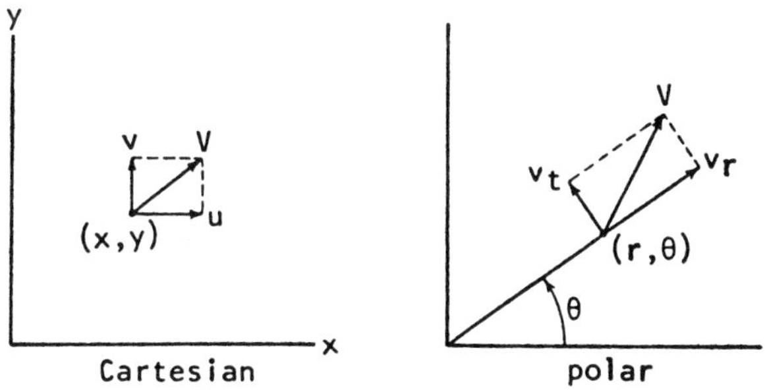

Fig. 4

From similar triangles,

$$u = \frac{-1.04y}{\sqrt{x^2 + y^2}} \qquad \text{and} \qquad v = \frac{1.04x}{\sqrt{x^2 + y^2}}$$

At P, $x = 1$, $y = \sqrt{3}$, so

$$u = -0.90 \text{ m/s}, \qquad v = 0.52 \text{ m/s}$$

Using equations (4),

$$a_x = -\frac{1.04y}{\sqrt{x^2 + y^2}} \frac{\partial}{\partial x}\left(\frac{-1.04y}{\sqrt{x^2 + y^2}}\right) + \frac{1.04x}{\sqrt{x^2 + y^2}} \frac{\partial}{\partial y}\left(\frac{-1.04y}{\sqrt{x^2 + y^2}}\right)$$

$$= -\frac{2.16x}{8}$$

$$a_y = -\frac{-1.04y}{\sqrt{x^2 + y^2}} \frac{\partial}{\partial x}\left(\frac{1.04x}{\sqrt{x^2 + y^2}}\right) + \frac{1.04x}{\sqrt{x^2 + y^2}} \frac{\partial}{\partial y}\left(\frac{1.04x}{\sqrt{x^2 + y^2}}\right)$$

$$= -\frac{2.16y}{8}$$

Substituting $x = 1$, $y = \sqrt{3}$,

$$a_x = -0.27 \text{ m/s}^2, \qquad a_y = -0.47 \text{ m/s}^2$$

By inspection,

$$v_t = 1.04 \text{ m/s}, \qquad v_r = 0 \text{ m/s}$$

Using equations (6) and (7),

$$a_r = 0 \frac{\partial}{\partial r}(0) + 1.04 \frac{\partial}{r\partial \theta}(0) - \frac{1.04 \times 1.04}{2}$$

$$= -0.54 \text{ m/s}^2$$

$$a_t = 0 \frac{\partial}{\partial r}(1.04) + 1.04 \frac{\partial}{r\partial \theta}(1.04) + \frac{0 \times 1.04}{2}$$

$$= 0 \text{ m/s}^2$$

Note that a_x and a_y might have been obtained more easily (in this problem) by calculating them as the horizontal and vertical components of a_r. Does $a_r^2 = a_x^2 + a_y^2$?

c) Along any streamline (say, at point P), the tangential component

$$|V| = v = (u^2 + v^2)^{1/2} = A(x^2 + y^2)^{1/2}$$

while the velocity direction (tangent to the streamline) is defined by

$$\theta = \tan^{-1} \frac{v}{u} = - \tan^{-1} \frac{y}{x}$$

The accelerations are

$$a_x = Ax \frac{\partial}{\partial x} (Ax) - Ay \frac{\partial}{\partial y} (Ax) = A^2 x$$

$$a_y = Ax \frac{\partial}{\partial x} (-Ay) - Ay \frac{\partial}{\partial y} (-Ay) = A^2 y$$

Consider the point P(8,6). There,

$$v = 10A, \qquad \theta = -\tan^{-1} \frac{6}{8} = -\tan^{-1} \frac{3}{4} \sim 37°$$

$$a_x = 8A^2, \qquad a_y = 6A^2$$

Along 0y, $a_x = 0$ but $a_y > 0$ and the flow decelerates as it moves down toward 0. Along 0x, $a_y = 0$ while $a_x > 0$ and the flow accelerates as it moves away horizontally from 0.

STREAMLINE, PATHLINE AND STREAKLINE

• PROBLEM 3-12

Sketch the streamlines in the first quadrant for the two-dimensional flow field specified by

$$u = x + 2y \quad \text{(m/s)}$$

$$v = 2x - y \quad \text{(m/s)}$$

Solution: Velocity components at representative points may be calculated:

These components, along with others, have been plotted to scale in the figure. The respective velocity vectors are shown by the dashed arrows and the streamlines sketched from them.

x (m)	y (m)	u (m/s)	v (m/s)
0	0	0	0
0	1	2	−1
0	2	4	−2
0	4	8	−4
1	0	1	2
2	0	2	4
4	0	4	8
1	1	3	1
1	2	5	0
1	4	9	−2
2	1	4	3
2	2	6	2
2	4	10	0
4	1	6	7
4	2	8	6
4	4	12	4

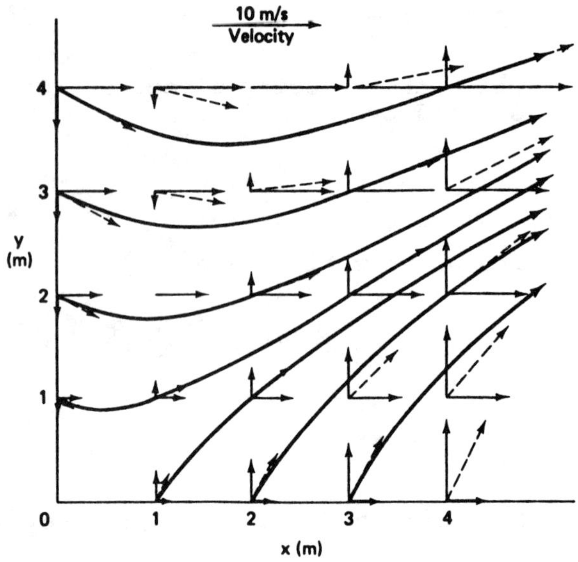

• PROBLEM 3-13

In the steady flow shown in the sketch below, find the equation of the streamline passing the point (1,2,3) if $u = 3ax$, $v = 4ay$ and $w = -7az$.

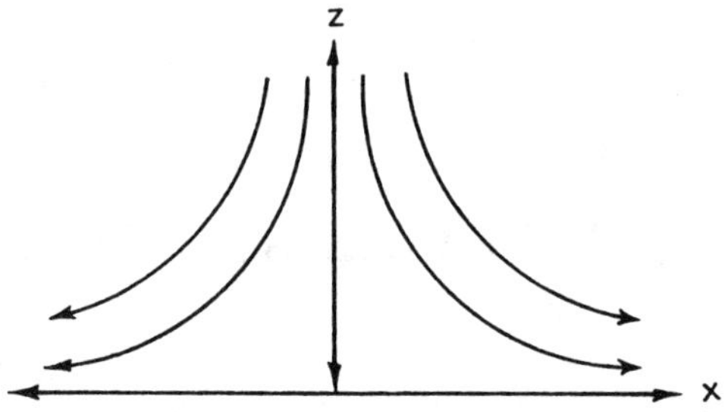

Fig. 1

Solution: Consider the two-dimensional (planar) flow pattern of Figure 2. At point P(x,y) one and only one streamline can pass. By definition the streamline is tangent to the velocity vector $\vec{V}$ at P. Using Cartesian coordinates, one obtains from the geometry of Figure 2

$$\frac{v}{u} = \tan\theta = dy/dx \tag{1}$$

From Equation (1) one obtains

$$udy - vdx = 0 \tag{2}$$

or in vector form,

$$\vec{V} \times d\vec{r} = 0 \tag{3}$$

which is known as the equation of a streamline.

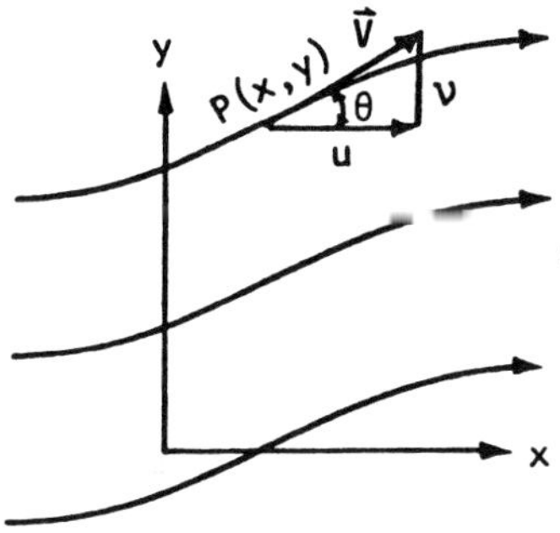

Fig. 2

From Equation (3), one obtains

$$\frac{dx}{u} = \frac{dy}{v} = \frac{dz}{w} \tag{4}$$

Substituting given values for the three velocity components into Equation (4) gives

$$\frac{dx}{3x} = \frac{dy}{4y} = -\frac{dz}{7z} \tag{5}$$

Considering x,y equation of Equation (5) gives after integration

$$\frac{1}{3}\ln x = \frac{1}{4}\ln y + \ln c \tag{6}$$

or taking antilogs

$$y = c_1\ x^{4/3} \tag{7}$$

Similarly, one considers the x, z equation of equation (5) and integrates to obtain

$$\frac{1}{3}\ln x = -\frac{1}{7}\ln z + \ln c \tag{8}$$

Taking antilogs of the terms in Equation (8) produces

$$z = \frac{c_2}{x^{7/3}} \tag{9}$$

These two equations, Equation (7) and (9), with different values c_1 and c_2, describe all the streamlines in the flow field. We next evaluate c_1 and c_2. At the point (1,2,3)

$$2 = c_1$$

from Equation (7), and

$$3 = c_2$$

from Equation (9). This streamline that passes through the space location (1,2,3) is then described by

$$y = 2x^{4/3}$$

$$z = 3x^{-7/3}$$

• **PROBLEM 3-14**

An idealized velocity distribution is given by

$$u = \frac{x}{1 + t} \qquad v = \frac{y}{1 + 2t} \qquad w = 0$$

Calculate and plot (a) the streamline; (b) the pathline; and (c) the streakline which pass through the point $(x_0, y_0, 0)$ at time t = 0.

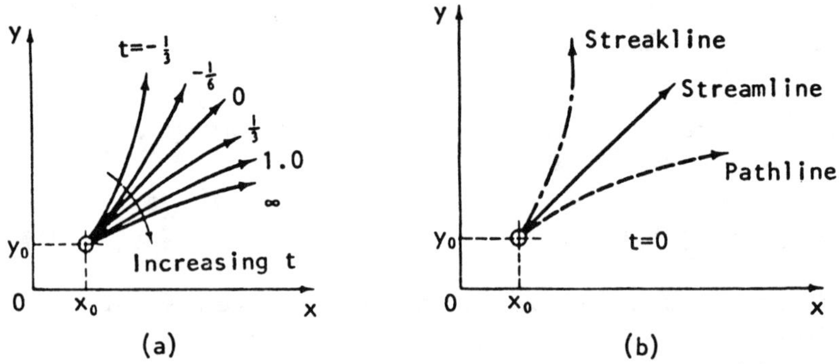

Fig. 1

(a) Streamlines through(x_0,y_0)as a function of time;(b) Streamline, Pathline and Streakline through (x_0,y_0) at time t=0.

Solution: (a) Since w = 0, there is no motion anywhere in the z direction and the flow pattern is two-dimensional; i.e., all streamlines, etc. are confined to planes parallel to the xy plane.

Streamline:

$$\frac{dx}{u} = \frac{dy}{v} = \frac{dz}{w} = \frac{dr}{V} \tag{1}$$

If the components u, v, and w are known functions of position and time, Eq. (1) can be integrated to find the streamline passing through (x_0, y_0, z_0, t_0). The integration may be rather laborious. One effective idea is to introduce a parameter ds equal to the ratios in Eq. (1). Thus

$$\frac{dx}{ds} = u \qquad \frac{dy}{ds} = v \qquad \frac{dz}{ds} = w \tag{2}$$

To find the streamlines, introduce (u,v) into the defining relation (2)

$$\frac{dx}{ds} = \frac{x}{1 + t} \qquad \frac{dy}{ds} = \frac{y}{1 + 2t}$$

Separate the variables and integrate each relation, with t held constant

$$x = C_1 \exp \frac{s}{1 + t} \qquad y = C_2 \exp \frac{s}{1 + 2t}$$

To evaluate C_1 and C_2, use the condition $x = x_0$, $y = y_0$ at $s = 0$, letting t remain unspecified for a moment. We find that $C_1 = x_0$, $C_2 = y_0$. Finally, eliminate the parameter s which is no longer needed

$$s = (1 + t) \ln \frac{x}{x_0} = (1 + 2t) \ln \frac{y}{y_0}$$

Rearrange this for a relation between y and x:

$$y = y_0 \left(\frac{x}{x_0}\right)^n$$

where

$$n = \frac{1 + t}{1 + 2t}$$

This is the equation of the streamlines which pass through (x_0, y_0) for all times t. They are plotted in Fig. 1a, where we see that the flow is moving down and to the right as time increases. The limiting streamline as $t \to \infty$ is given by $y/y_0 = (x/x_0)^{1/2}$. At $t = 0$ we have

$$\frac{y}{y_0} = \frac{x}{x_0} \qquad \text{Ans. (a)}$$

which is a 45° line through (x_0, y_0). This is plotted in Fig. 1b.

(b) The pathline is defined by integration of the relation between velocity and displacement

$$\frac{dx}{dt} = u(x,y,z,t)$$

$$\frac{dy}{dt} = v(x,y,z,t) \qquad (3)$$

$$\frac{dz}{dt} = w(x,y,z,t)$$

Integrate with respect to time using the condition (x_0, y_0, z_0, t_0). Then eliminate time to give the pathline function $f(x,y,z,t)$.

Integrating equation (3) yields:

$$\frac{dx}{dt} = \frac{x}{1 + t} \qquad \frac{dy}{dt} = \frac{y}{1 + 2t}$$

This time, of course, we integrate with t as a variable, with the result

$$x = C_1(1 + t) \qquad y = C_2(1 + 2t)^{1/2} \qquad (1)$$

Introduce the initial condition $x = x_0$, $y = y_0$ at $t = 0$, giving $C_1 = x_0$ and $C_2 = y_0$. To plot the pathline, eliminate t between the two relations

$$y = y_0\left[1 + 2\left(\frac{x}{x_0} - 1\right)\right]^{1/2} \qquad \text{Ans. (b)}$$

This is the pathline which passes through (x_0, y_0) at $t = 0$. It is plotted in Fig. 1b and does not coincide with the streamline at $t = 0$.

(c) To find the streakline, return to Eq. (1) and find the family of particles which passed through (x_0, y_0) at a continuous sequence of times ξ such that $\xi < t$. That is, leave t alone and use the family of initial conditions $x = x_0$, $y = y_0$ at $t = \xi$. The result is

$$C_1 = \frac{x_0}{1 + \xi} \qquad C_2 = \frac{y_0}{(1 + 2\xi)^{1/2}}$$

or

$$x = \frac{x_0(1 + t)}{1 + \xi} \qquad y = y_0\left(\frac{1 + 2t}{1 + 2\xi}\right)^{1/2}$$

These are the streaklines through (x_0, y_0) for any time t. They can be plotted by holding t constant and letting ξ take on all values $\leq t$. Or ξ can simply be eliminated from the two relations

$$\xi = (1 + t)\,\frac{x_0}{x} - 1 = \frac{1}{2}\left[(1 + 2t)\left(\frac{y_0}{y}\right)^2 - 1\right]$$

Rearrange and solve for y/y_0

$$\left(\frac{y}{y_0}\right)^2 = \frac{1 + 2t}{1 + 2[(1 + t)(x_0/x) - 1]}$$

At time $t = 0$, we obtain the particular streakline asked for in this problem:

$$\frac{y}{y_0} = \left[1 + 2\left(\frac{x_0}{x} - 1\right)\right]^{-1/2} \qquad \text{Ans. (c)}$$

This is also plotted in Fig. 1b and does not coincide with either the equivalent streamline or pathline. Physically, the streakline reflects the behavior of the streamlines before the specified time $t = 0$, while the pathline reflects the streamline behavior after $t = 0$.

Suppose a two-dimensional velocity field exists in the region $x \geq 0$ and $y \geq 0$. The velocity field is $v_1 = x$ and $v_2 = -y$. Find the equation of the streamline passing through the point (3,4).

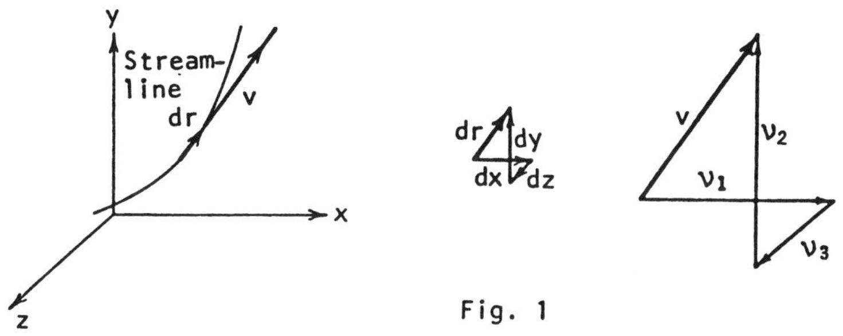

Fig. 1

Solution: A streamline is a curve in the flow field whose tangent points in the direction of the fluid-velocity vector at a given instant of time. If the flow is steady then a fluid particle will move along a streamline, since it moves in the direction of its velocity vector.

For a steady flow, the equation for a streamline in differential form can be obtained in terms of the velocity components. If $d\vec{r} = dx\vec{i} + dy\vec{j} + dz\vec{k}$ is a displacement vector along the streamline and $\vec{v} = v_1\vec{i} + v_2\vec{j} + v_3\vec{k}$ is the fluid velocity, then $d\vec{r}$ equals some scalar multiple of $\vec{v}$, since both vectors point in the same direction. Thus, $\vec{v} = \alpha d\vec{r}$, where α = constant and

$$\frac{dx}{v_1} = \frac{dy}{v_2} = \frac{dz}{v_3} = \frac{1}{\alpha} \tag{1}$$

See Fig. 1.

From Eq. 1, the differential equation for the streamline can be obtained; that is,

$$\frac{dy}{dx} = \frac{v_2}{v_1} = -\frac{y}{x}$$

The variables can be separated giving

$$\frac{dy}{y} = -\frac{dx}{x}$$

Integration gives

$$\ln y = -\ln x + \ln C$$

or

$$yx = C$$

By taking different numerical values for C, a family of curves is obtained, as shown in Fig. 2. To obtain the value of C for the curve passing through the point (3,4) substitute (3,4) for (x,y) in the above equation. This gives C = 12. Thus, the equation of the streamline is $y = 12/x$. The family of streamlines gives the flow pattern. The direction of the flow can be determined by noting that $v_1 \geq 0$ and $v_2 \leq 0$.

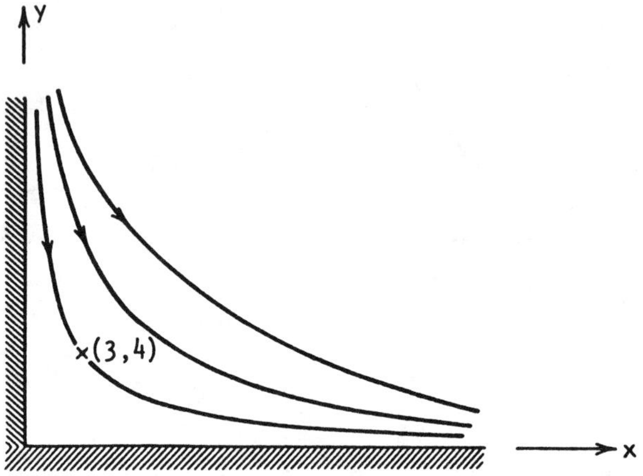

Fig. 2

• PROBLEM 3-16

> A two-dimensional steady flow has velocity components defined by
>
> $$\vec{V}_x = \frac{y}{x^2 + y^2}, \qquad \vec{V}_y = \frac{x}{x^2 + y^2}$$
>
> What are the equations of the streamlines?

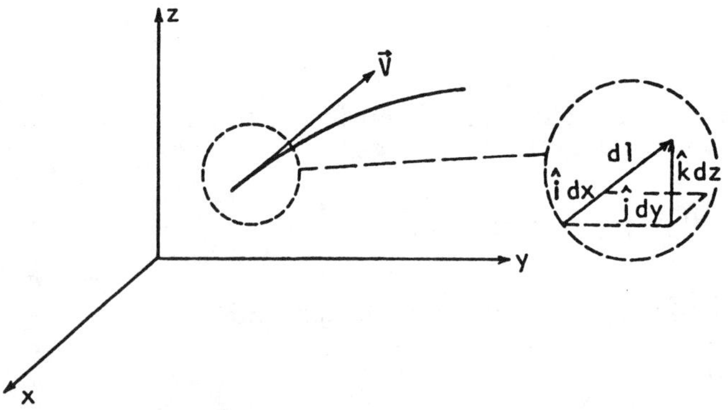

Fig. 1

Solution: A streamline is a line that is everywhere tangent to a velocity vector. Thus if $\vec{V}$ is the fluid velocity,

$$\frac{\vec{V}}{V} = \hat{1} = \frac{dl}{dt} \qquad (1)$$

where $dl = \hat{i}dx + \hat{j}dy + \hat{k}dz$ is the vector differential or arc length of a streamline, and $\hat{l}$ is the unit tangent to the streamline (see Figure 1). Equation 1 can also be written as

$$\hat{i}\,\frac{V_x}{V} + \hat{j}\,\frac{V_y}{V} + \hat{k}\,\frac{V_z}{V} = \hat{i}\,\frac{dx}{dl} + \hat{j}\,\frac{dy}{dl} + \hat{k}\,\frac{dz}{dl}$$

$$= \hat{i}\cos\theta_x + \hat{j}\cos\theta_y + \hat{k}\cos\theta_z \qquad (2)$$

where $\cos\theta_x$, $\cos\theta_y$, and $\cos\theta_z$ are the cosines of the angles between the tangent to the streamline and the x-, y-, and z- axes, respectively.

It follows from equation (2) that

$$\frac{V_x}{dx} = \frac{V_y}{dy} = \frac{V_z}{dz} \qquad (3)$$

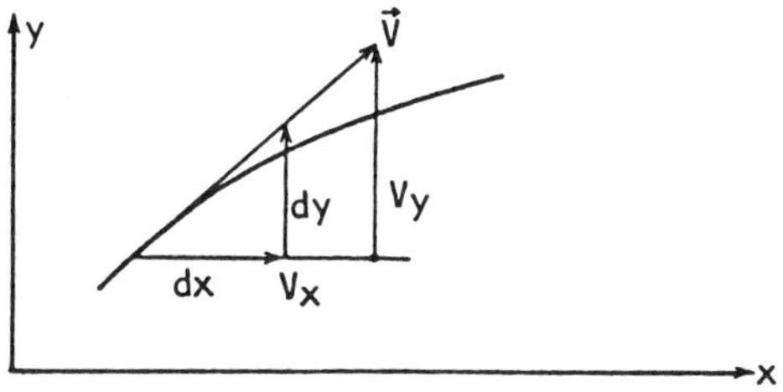

Fig. 2

If V_x, V_y, and V_z are known as functions of x, y, and z, equation 3 constitutes a system of differential equations for the streamline. A two-dimensional flow with $V_z = 0$ leads to the equation (Fig. 2)

$$\frac{dy}{dx} = \frac{V_y}{V_x} = \tan\alpha \qquad (4)$$

where, clearly, dy/dx is the slope of streamline and V_y/V_x is the angle of inclination of velocity vector.

Rearranging Eq. 4

$$\frac{dy}{V_y} = \frac{dx}{V_x}$$

and solving this equation we have

$$\frac{x^2 + y^2}{x}\,dy = -\,\frac{x^2 + y^2}{y}\,dx$$

or

$$ydy = -xdx$$

By integration,

$$x^2 + y^2 = C_1 = \text{const} \geq 0$$

• PROBLEM 3-17

The wind velocity 5 miles from the center of a tornado was measured as 30 mph, and the barometer was read as 29 in. of mercury. Calculate the wind velocity 1/2 mile from the tornado center and the barometric pressure. Assume

$$\rho_{air} = .00238 \text{ slug/ft}^3$$

and

$$1 \text{ ft Hg} = 13.55 \text{ ft water.}$$

Solution: The equation of motion of circular flow is given by

$$Vr = \text{constant}$$

At 5 miles the velocity is 30 mph. Therefore

$$Vr = 5 \times 30 = 150 \text{ miles}^2/\text{hr}$$

Thus at 1/2-mile radius, the velocity is given by

$$1/2\ V = 150$$

or

$$V = 300 \text{ mph}$$

At 5 miles the pressure is 29 in. mercury, or

$$29/12 \times 13.55 \times 62.4 = 2{,}040 \text{ psf}$$

Multiplying the Bernoulli equation through by γ gives

$$p + \frac{V^2\gamma}{2g} + z\gamma = \text{const} \tag{1}$$

The ratio γ/g is the density ρ, and for gases the quantity γz is negligible since the specific weight of gases is very small. Usually the change in potential z is also small in gas dynamic problems. Equation (1) then reduces to

$$p + 1/2\rho\ V^2 = \text{const} \tag{2}$$

which is the aerodynamic form of Bernoulli's equation for ideal incompressible flow of gases.

Applying the aerodynamic form of Bernoulli's equation between the two points,

$$2{,}040 + 1/2 \times 0.00238(30 \times 88/60)^2 =$$

$$p_2 + 1/2 \times 0.00238(300 \times 88/60)^2$$

Therefore

$$p_2 = 1{,}808 \text{ psf}$$

$$= 25.62 \text{ in. mercury}$$

• PROBLEM 3-18

A cylinder 85 mm in radius and 0.6 m in length rotates coaxially inside a fixed cylinder of the same length and 90 mm radius. Glycerin (μ = 1.48 Pa·s) fills the space between the cylinders. A torque of 0.7 N·m is applied to the inner cylinder. After constant velocity is attained, calculate the velocity gradients at the cylinder walls, the resulting r/min, and the power dissipated by fluid resistance. Ignore end effects.

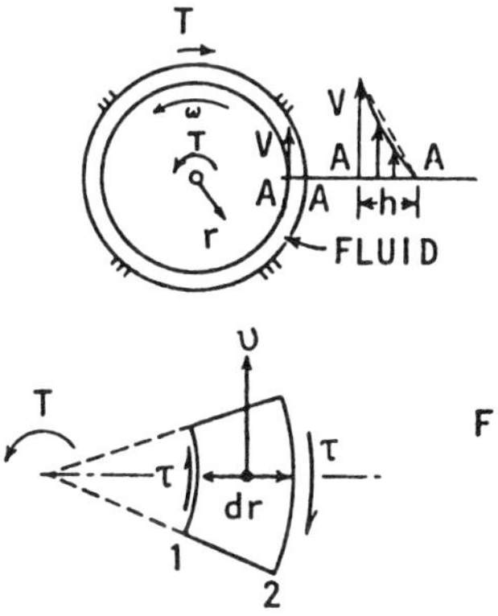

Fig. 1

Solution: Refer to Figure (1) for a definition sketch. The torque T is transmitted from inner cylinder to outer cylinder through the fluid layers; therefore (r is the radial distance to any fluid layer and ℓ is the length of the cylinders)

$$T = -\tau(2\pi r \times \ell)r;$$

$$\tau = -\frac{T}{2\pi r^2 \ell} = \frac{\mu r d(v/r)}{dr}$$

Consequently,

$$\frac{d(v/r)}{dr} = -\frac{T}{2\pi\mu\ell r^3}$$

and

$$\int_{V/r_i}^{0} d\left(\frac{v}{r}\right) = \frac{-T}{2\pi\mu\ell}\int_{r_i}^{r_0}\frac{dr}{r^3};$$

$$-\frac{V}{r_i} = \frac{T}{4\pi\mu\ell}\left[\frac{1}{r_0{}^2} - \frac{1}{r_i^2}\right]$$

$$V = \frac{-0.085 \times 0.7}{4\pi \times 1.48 \times 0.6}\left[\frac{1}{(0.090)^2} - \frac{1}{(0.085)^2}\right]$$

$$= 0.08 \text{ m/s} = 80 \text{ mm/s}$$

The power dissipated in fluid friction is

$$P = 2\pi r_i \tau_i V\ell = T\omega$$

where $\omega = V/r_i = 0.94\text{s}^{-1}$ is the rotational speed (rad/s).

Numerically

$$\left(\frac{dv}{dr}\right)_i = \frac{-T}{2\pi\mu\ell r_i^2} + \frac{V_i}{r_i} = -17.4 + 0.9 = -18.3 \text{ m/s·m}$$

$$\left(\frac{dv}{dr}\right)_0 = \frac{-T}{2\pi\mu\ell r_0{}^2} + \frac{V_0}{r_0} = -15.4 + 0 + -15.4 \text{ m/s·m}$$

$$\text{r/min} = \left(\frac{\omega}{2\pi}\right) \times 60 = 9.0$$

$$P = 0.7 \times 0.94 = 0.66 \text{ N·m/s} = 0.66\text{W}$$

This power will appear as heat, tending to raise the fluid temperature and decrease its viscosity; evidently a suitable heat exchanger would be needed to preserve the steady-state conditions given.

When $h = r_0 - r_i \to 0$, then h/r_i is small and $dv/dr \to -V/h$. Consequently, v/r is much less than dv/dr, and

$$\tau \approx \mu\frac{dv}{dr} = -\frac{\mu V}{h}$$

Assuming this linear profile case for an approximate calculation gives

$$h = 5 \text{ mm}; \qquad \frac{h}{r_1} \approx 0.06$$

$$V = \frac{T}{2\pi\mu\ell}\left(\frac{h}{r_i^2}\right) = 0.125\,\frac{0.005}{0.0072} = 0.087 \text{ m/s}$$

$$\text{r/min} = 9.8$$

Because these results differ from the former by almost 9%, the approximation is not satisfactory in this case.

CHAPTER 4

CONTINUITY EQUATION

Basic Attacks and Strategies for Solving Problems in this Chapter. See pages 152 to 175 for step-by-step solutions to problems.

The fundamental conservation laws (conservation of mass, conservation of linear momentum, conservation of angular momentum, and conservation of energy) can be applied directly to a *system* which is a quantity of mass (solid or fluid) of a fixed identity. While these laws are ideally suited for Lagrangian analysis, Eulerian analysis requires the transformation of these laws for application to a *control volume*, i.e., a specific region or field rather than a system of a fixed identity. The *Reynolds Transport Theorem* is used for this transformation and is applicable to all the fundamental conservation laws. For a fixed control volume,

$$\frac{DB}{Dt} = \frac{dB}{dt} = \frac{d}{dt}\int_m b\,dm = \iiint_{cv} \frac{\partial}{\partial t}(\rho b)dV + \iint_{cs} \rho b\mathbf{V} \cdot d\mathbf{A}, \qquad (1)$$

where B is any property of the fluid (mass, momentum, etc.), and

$$b = \frac{dB}{dm}$$

is the intensive value of B (i.e., B per unit mass). In the above, CV and CS represent integration over the control volume and control surface (closed boundary of the control volume) respectively. The surface area vector is A, defined as being positive *outward* from the control surface. An alternate notation is

$$dA = \hat{n}\,dA,$$

where $\hat{n}$ is the unit outward normal vector.

The control volume or integral conservation laws for fluid flows are thus obtainable simply by the substitution of mass, linear momentum, angular momentum, or energy for the dummy variable B in Equation (1), and application of the known fundamental conservation laws for a system on the left-hand side. For the conservation of mass, this is particularly simple since, for a system, mass must be conserved, i.e.,

$$\frac{Dm}{Dt} = \frac{dm}{dt} = 0$$

for a system. Letting $B = m$ and therefore $b = 1$, Equation (1) becomes

$$\iiint_{cv} \frac{\partial \rho}{\partial t}\, d\mathbf{V} + \iint_{cs} \rho \mathbf{V} \cdot \hat{\mathbf{n}} d\mathbf{A} = 0. \tag{2}$$

The above applies to a fixed control volume. If the control volume itself is deformable, the more general form for the conservation of mass is

$$\frac{d}{dt} \iiint_{cv} \rho\, d\mathbf{V} + \iint_{cs} \rho \mathbf{V}_r \cdot \hat{\mathbf{n}} d\mathbf{A} = 0, \tag{3}$$

where $\mathbf{V}_r$ is the fluid velocity vector relative to the control surface.

The first step in the solution of control volume problems is to choose a control volume. This basic step is most critical since the degree of difficulty of the problem can often be greatly reduced by a wise choice of control volume. Next, evaluate the net outflux of mass (the control surface integral) in Equation (2) or (3). Finally, equate this to the negative of the unsteady control volume integral. In many cases, if the flow is steady, this volume integral will vanish.

Often, a control volume will cut through a duct or pipe where the flow is predominantly in one direction. In such cases, an average velocity

$$V = \frac{Q}{A}$$

is defined where Q is the volume flow rate across the pipe's cross section and A is the cross-sectional area. In the control volume equation, then, the mass flux across this surface is simply

$$\dot{m} = \rho V A.$$

Finally, if the control volume is shrunk until is is infinitesimally small, the integral conservation of mass equation can be used to obtain the differential form of the conservation of mass, also called the continuity equation:

$$\frac{\partial \rho}{\partial t} + \nabla \cdot (\rho \mathbf{V}) = 0. \tag{4}$$

For incompressible flow (ρ = constant),

$$\nabla \cdot \mathbf{V} = 0. \tag{5}$$

In Cartesian coordinates, incompressible continuity is expressed as

$$\frac{\partial u}{\partial x} + \frac{\partial v}{\partial y} + \frac{\partial w}{\partial z} = 0. \tag{6}$$

This equation can be integrated to obtain one component of velocity when the other two components are known. Equation (6) can also serve as a useful check for a given velocity field. If the continuity equation is not satisfied, it *cannot* be a valid flowfield.

Step-by-Step Solutions to Problems in this Chapter, "Continuity Equation"

TRANSPORT THEOREM

• PROBLEM 4-1

Express the transport equation

$$\frac{D\Phi}{Dt} = \frac{\partial \Phi}{\partial t} + \int_A \phi \vec{V} \cdot d\vec{A} \qquad (1)$$

where the continuum function Φ is related to the continuum variable ϕ by

$$\Phi = \int_{\forall} \phi \, d\forall \qquad (2)$$

for a) density ρ, b) energy E, and c) enthalpy H.

Solution: (a) Density is related to the mass M by

$$\rho = \frac{dM}{d\forall} \qquad (3)$$

Substituting Equation (3) into Equations (1) and (2) yields

$$\Phi = \int \phi \, d\forall$$

$$= \int \rho \, d\forall$$

$$= \int dM = M$$

Thus, the transport equation for the continuum variable of density is

$$\frac{DM}{Dt} = \frac{\partial M}{\partial t} + \int \rho\vec{V} \cdot d\vec{A}$$

(b) Energy E is related to the energy e per unit mass M by

$$\frac{dE}{dM} = e \tag{4}$$

Substituting Equation (3) into Equation (4) results in

$$dE = \rho e d\forall$$

Thus,

$$\Phi = \int \phi \, d\forall$$

$$= \int \rho e \, d\forall$$

$$= \int dE = E$$

The transport equation for the continuum variables of energy is

$$\frac{DE}{Dt} = \frac{\partial E}{\partial t} + \int \rho e \vec{V} \cdot d\vec{A}$$

Frequently, the term $\frac{\partial E}{\partial t}$ is zero, which is a statement that the state of matter at each point in space is steady. This implies that the energy stored within the control volume is unchanging. Hence, to discuss how the energy of the system changes, one need only examine how the energy changes across the control surface of the control volume.

(c) The enthalpy H is related to the specific enthalpy h by

$$h = \frac{dH}{dM} \tag{5}$$

Substituting Equation (3) into Equation (5) results in

$$dH = \rho h \, d\forall$$

Thus,

$$\Phi = \int \phi \, d\forall$$

$$= \int \rho h \, d\forall$$

$$= \int dH = H$$

The transport equation for the continuum variable of enthalpy is

$$\frac{DH}{Dt} = \frac{\partial H}{\partial t} + \int \rho h \vec{V} \cdot d\vec{A}$$

• PROBLEM 4-2

Demonstrate the relationship given by

$$\frac{D}{Dt}(d\forall) = (\vec{\nabla} \cdot \vec{V})\ d\forall$$

Solution: From the definition of the symbol

$$\frac{D}{Dt} \equiv \left(\frac{\partial}{\partial t} + u\frac{\partial}{\partial x} + v\frac{\partial}{\partial y} + w\frac{\partial}{\partial z}\right)$$

$$\frac{D}{Dt}(d\forall) = \left(\frac{\partial}{\partial t} + u\frac{\partial}{\partial x} + v\frac{\partial}{\partial y} + w\frac{\partial}{\partial z}\right)(dxdydz)$$

$$= \frac{\partial(dxdydz)}{\partial t} + u\frac{\partial(dxdydz)}{\partial x} +$$

$$v\frac{\partial(dxdydz)}{\partial y} + w\frac{\partial(dxdydz)}{\partial z}$$

$$= dydz\ \partial(u) + dxdz\ \partial(v) + dxdy\ \partial(w) +$$

$$u\partial\ (dydz) + v\partial\ (dxdz) + w\partial\ (dxdy)$$

$$= \partial(udydz) + \partial(vdxdz) + \partial(wdxdy)$$

$$= \frac{\partial(udxdydz)}{\partial x} + \frac{\partial(vdxdydz)}{\partial y} +$$

$$\frac{\partial(wdxdydz)}{\partial z}$$

$$= \left(\frac{\partial u}{\partial x} + \frac{\partial v}{\partial y} + \frac{\partial w}{\partial z}\right)\ d\forall$$

$$= \vec{\nabla} \cdot \vec{V}\ d\forall$$

• PROBLEM 4-3

Internal energy U is known to be an extensive property in thermodynamics. (a) Express the time rate of change of internal energy of a system using the transport Eq. (1). (b) Give a word explanation of each of the integrals. (c) Express the time rate of change of internal energy of a system using the transport equation (1) for steady u and ρ fields.

Solution: a) Since U is a scalar quantity, B_i = U and, on a specific mass basis, b_i = u. That is, $U = \int_\forall u\rho\ d\forall$. The transport equation

$$\frac{DB_i}{Dt} = \frac{D}{Dt}\int_m b_i dm = \int_\forall \frac{\partial(b_i\rho)}{\partial t} d\forall + \oint_S b_i\rho\ \vec{V} \cdot d\vec{S}$$

can be expressed as

$$\frac{DU}{Dt} = \int_V \frac{\partial(u\rho)}{\partial t} d\forall + \oint_S \rho u \vec{V} \cdot d\vec{S}$$

b) The first integral represents the rate of change of internal energy of the matter within the control volume at a given instant of time. The second integral describes the net efflux of internal energy through the control surface at a given instant of time. The internal energy "flows" because it is associated with the mass flowing through the control surface.

c) Since internal energy and density fields are steady, i.e., not a function of time, the integrand of the volume integral vanishes, and so

$$\frac{DU}{Dt} = \oint_S \rho u \vec{V} \cdot d\vec{S}$$

• PROBLEM 4-4

(a) Consider the problem of a pneumatic cylinder as shown in the figure. For an arbitrary extensive property N, identify the meaning of the terms in the Reynolds transport theorem in the form

$$\frac{DN_{sys}}{Dt} = \frac{d}{dt} \int_{C.V.} \rho\eta \, d\forall + \int_{C.S.} \rho\eta \, V_r \cdot \hat{n} \, dA \qquad (1)$$

for the pneumatic cylinder and the control volume shown in these two cases: (1) zero leakage around the piston, and (2) leakage around the piston.

(b) Reevaluate the time-rate-of-change term and the flux term, using the Reynolds transport theorem in the forms

$$\frac{DN_{sys}}{Dt} = \int_{C.V.} \frac{\partial}{\partial t}(\rho\eta) d\forall + \int_{C.S.} \rho\eta \, V_b \cdot \hat{n} \, dA + \int_{C.S.} \rho\eta \, V_r \cdot \hat{n} \, dA \qquad (2)$$

where V_b is the velocity of the control surface and V_r is the velocity of the fluid with respect to the control surface, and

$$\frac{DN_{sys}}{Dt} = \int_{C.V.} \frac{\partial}{\partial t}(\eta\rho) d\forall + \int_{C.S.} \rho\eta \, V \cdot \hat{n} \, dA \qquad (3)$$

where $V = V_b + V_r$ is the total velocity of the fluid with respect to the chosen reference frame.

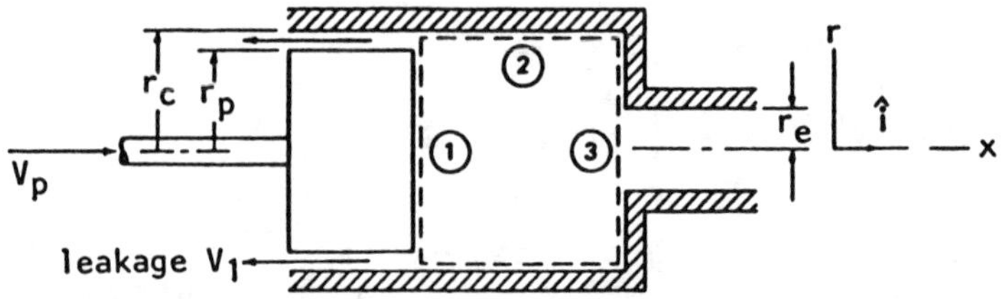

<u>Solution</u>: (a) D/Dt is on the left since we are following a particular group of fluid particles, and d/dt on the right since we are only looking at a volume in space and not particular particles.

We will refer to the first term on the right-hand side as the time-rate-of-change term and the second as the flux term; both expressions refer to the extensive property N. The time-rate-of-change term and the flux term will be separately considered; each will be examined for the "instructions" provided by its mathematical nature. The time-rate-of-change term $\frac{d}{dt}\int_{C.V.} \eta\rho \, d\forall$ states that the values of η and ρ are to be evaluated at the location of each differential volume element and that the product of $\eta\rho \, d\forall$ is to be summed over the control volume at a given instant. This provides N(t) for the control volume. The time-rate-of-change term is then the derivative of N(t).

(1) The flux term for the zero-leakage condition is

$$\int_{C.S.} \eta\rho \, V_r \cdot \hat{n} \, dA = \int_{A_3} \eta_3\rho_3 V_3 \cdot \hat{n}_3 \, dA$$

since, over the remainder of the control surface, $V_r \cdot \hat{n} = 0$. Because of the sharp-edged orifice at section ③, velocity V_3 will not be perpendicular to or constant over area A_3. However, we may write $V_3 \cdot \hat{n}_3 = (V_x)_3$ since $\hat{n}_3 = \hat{\imath}$. If η, V_x, and ρ are functions of r only, the flux integral may be written as

$$\int_{C.S.} \eta\rho \, V_r \cdot \hat{n} \, dA = 2\pi \int_0^{r_e} (\eta\rho V_x)_3 r \, dr$$

(2) If leakage occurs around the periphery of the piston, then such effects must be accounted for by the flux integral. Consequently, in addition to the foregoing, the flux term would include the term

$$\int_{A_1} \eta_1\rho_1 \, V_{r1} \cdot \hat{n}_1 \, dA,$$

where V_{r1} is the relative velocity at section ①. The unit vector $\hat{n}_1$ is given by $\hat{n}_1 = -\hat{\imath}$. Let the leakage velocity relative to the fixed $r - \theta - x$-coordinate system be $V_1(r)$;

This represents fluid motion in the negative x-direction. The relative component of velocity which is responsible for the flux of N across the boundary is given by $V_{rl} = -V_l\hat{i} + V_p\hat{i}$ since $V_r = V - V_b$. Hence $V_{rl} \cdot \hat{n} = V_l - V_p$

Again, for axisymmetric quantities,

$$\int_{C.S.} \eta\rho\, V_r \cdot \hat{n}\, dA = 2\pi \int_0^{r_e} (\eta\rho V_x)_3 r\, dr +$$

$$2\pi \int_{r_p}^{r_c} (\eta\rho)_1 (V_l - V_p) r\, dr$$

This formation recognizes that it is the velocity relative and normal to the control surface which accounts for the flux across the surface.

(b) The flux term of Eq. 2 is identical with that of Eq. 1, so it remains unchanged. The instructions given by the first two terms of Eq. 2 will now be carried out. The term

$$\int_{C.V.} \frac{\partial}{\partial t} (\rho\eta)\, d\forall$$

implies that one should examine the time rate of change of the $(\eta\rho)$ product at all the points inside the control volume at an instant. That is, at a particular point (x,y,z), $\eta(t)$ and $\rho(t)$ would provide the information from which we could find $\partial(\eta\rho)/\partial t$. With such values for all the points in the control volume the integral above could be evaluated at time t. An important feature is that only those points inside the control volume are considered. Consequently, in (a) a physical point near surface ① at time t will be excluded from the integral after the piston face passes the point at a later time $t + \Delta t$.

The operations described above are quite different from those described in (a). The results of the operations are also different. The difference is expressed by the term

$$\int_{C.S.} \eta\rho\, V_b \cdot \hat{n}\, dA.$$

This is nonzero only at the piston face, where

$$\hat{n} = -\hat{i}$$

$$V_b = V_p\hat{i}$$

$$\eta\rho = (\eta\rho)_1$$

For an axisymmetric condition the integral may be written as

$$\int_{C.S.} \eta\rho\, V_b \cdot \hat{n}\, dA = -2\pi \int_0^{r_p} (\eta\rho)_1 V_p r\, dr$$

It is important to note that this term, like the flux term, is evaluated at the control surface only. So we see that

$$\frac{d}{dt}\int_{C.V.} \rho\eta \, d\forall = \int_{C.V.} \frac{\partial}{\partial t}(\rho\eta) \, d\forall - 2\pi\int_0^{r_p} (\eta\rho)_1 V_p r \, dr$$

For Eq. 3, the time-derivative term is the same as that for Eq. 2. The term involving the area integral cannot properly be called a net flux with respect to the control volume; it is a flux of N from the region of space occupied by the control volume at the instant at which the term is evaluated. The reference frame for the evaluation of the velocity may be located on the piston or on the cylinder. For convenience, we will use a coordinate system attached to the cylinder. The area integral term can then be expressed as the sum of three terms:

$$\int_{C.S.} \eta\rho \, V \cdot \hat{n} \, dA = 2\pi\int_0^{r_e} (\eta\rho V_x)_3 r \, dr + 2\pi\int_{r_p}^{r_c} (\eta\rho)_1 V_l r \, dr$$

$$- 2\pi\int_0^{r_p} (\eta\rho)_1 \, V_p r \, dr$$

Note that one of these is identical with a part of the flux term in (a).

CONTINUITY EQUATION

• PROBLEM 4-5

In many important applications the flow is steady and the control volume is fixed. Express the conservation of mass if the fluid enters normal to the inlet area A_1 and exits normal to the exit area A_2.

Solution: The continuity equation for steady flow and a fixed control volume is

$$0 = \int_{C.S.} \rho \, V \cdot \hat{n} \, dA$$

For flow crossing the control surface at A_1 and A_2, this becomes

$$\int_{C.S.} \rho \, V \cdot \hat{n} \, dA = \int_{A_1} \rho \, V \cdot \hat{n} \, dA + \int_{A_2} \rho \, V \cdot \hat{n} \, dA = 0$$

or, for the velocity vector normal to the areas,

$$\int_{A_2} \rho_2 V_2 \, dA - \int_{A_1} \rho_1 V_1 \, dA = 0$$

since $V_1 \cdot \hat{n}_1 = -V_1$. Assume ρ_1 and ρ_2 are constant over their respective areas; then

$$\rho_2 \int_{A_2} V_2 \, dA = \rho_1 \int_{A_1} V_1 \, dA$$

If V_1 and V_2 are constant over A_1 and A_2, respectively, then

$$\rho_2 A_2 V_2 = \rho_1 A_1 V_1$$

If V_1 and V_2 are functions of the areas, then

$$\rho_2 A_2 \overline{V}_2 = \rho_1 A_1 \overline{V}_1$$

where $\overline{V}_2$ and $\overline{V}_1$ are the average velocities over the areas.

• PROBLEM 4-6

Apply the conservation of mass to the problem of an inflating balloon.

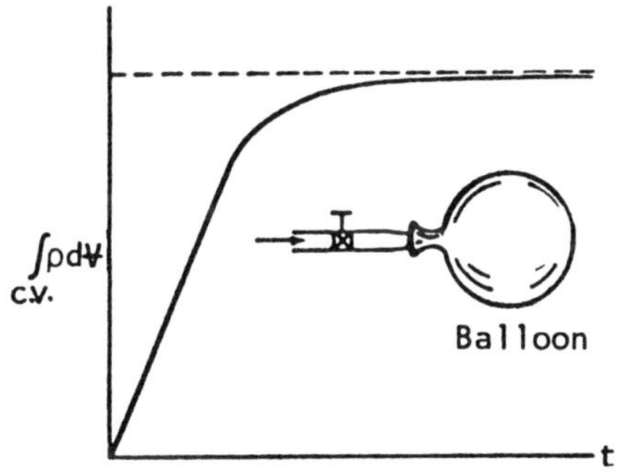

Solution: The integral form of the conservation of mass is given as

$$\frac{d}{dt}\int_{C.V.} \rho \, d\forall + \int_{C.S.} \rho \, V \cdot \hat{n} \, dA = 0 \tag{1}$$

Clearly, the physical problem involves an influx of mass to the balloon (represented by the last integral in Eq. 1), balanced by a change of mass inside the balloon. The integral over the volume in Eq. 1 represents the total mass inside the control volume (balloon). The air will stop flowing into the balloon when the pressure in the supply pipe equals the pressure in the balloon, as shown. The mass in the balloon changes as the volume changes and as the density changes, the density being dependent on the pressure and temperature of the air in the balloon. Equation 1 asserts that the slope of the curve of the figure shown is related to the entrance density, velocity, and area.

The flux term is nonzero only across the inlet, where it is reasonable to assert that the density is uniform over the inlet area; that is, the flux term may be expressed as

$$\int_{C.S.} \rho\, V \cdot \hat{n}\, dA = -\rho_i \bar{V}_i A_i \tag{2}$$

The full equation is then

$$\frac{d}{dt}\int_{C.V.} \rho\, d\forall - \rho_i A_i \bar{V}_i = 0 \tag{3}$$

It is also reasonable to consider the density to be uniform over the interior of the balloon at any time t. On this assumption, and approximating the balloon as a sphere of radius R, there results

$$\frac{d}{dt}\int_{C.V.} \rho d\forall = \frac{d}{dt}(\rho \frac{4}{3}\pi R^3)$$

$$= \frac{4}{3}\pi R^3 \frac{d\rho}{dt} + 4\pi\rho R^2 \frac{dR}{dt} \tag{4}$$

and the conservation of mass equation becomes

$$\frac{4}{3}\pi R^2 (R\frac{d\rho}{dt} + 3\rho \frac{dR}{dt}) = \rho_i A_i \bar{V}_i \tag{5}$$

• PROBLEM 4-7

Oil flows from a vertical pipe onto still water. The oil floats on the water and spreads as a circular slick. Develop the appropriate continuity equation to describe the fluid motion in the slick. Assume that the velocity at any radial location is uniform across the slick cross section, that the velocity direction is always horizontal, and that the oil is completely immiscible; i.e., none of the oil goes into solution with the water.

Solution: Take a ring-shaped control volume of infinitesimall thickness dr. The local thickness of the slick δ depends on r and t. Assume that the velocity varies only in the radial direction; i.e., the flow is axially symmetric. Now apply the control-volume continuity equation to the control volume shown in the Figure. We will use the continuity equation in the form

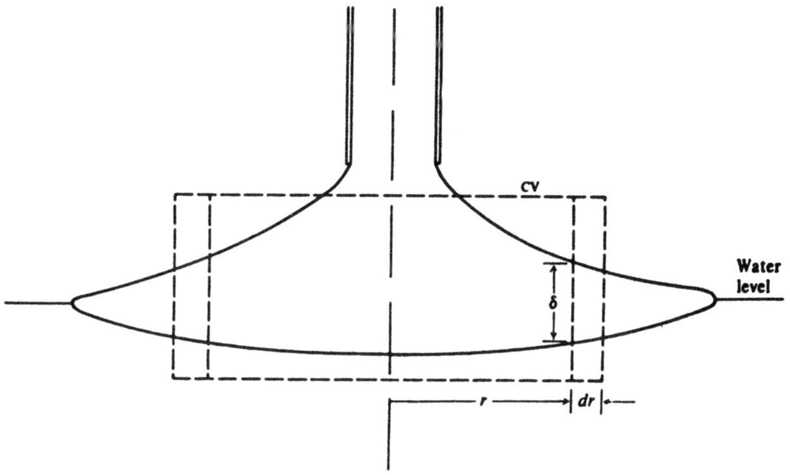

$$\frac{\partial}{\partial t}\iiint_{CV} \rho d\mathcal{V} + \oiint_{CS} \rho \mathbf{V} \cdot d\mathbf{A} = 0$$

The mass of fluid inside the control volume at any time is

$$\iiint_{CV} \rho \, d\mathcal{V} = \rho 2\pi r\delta \, dr$$

Then

$$\frac{\partial}{\partial t}\iiint_{CV} \rho \, d\mathcal{V} = \rho \, \frac{\partial\delta}{\partial t}2\pi r \, dr$$

Note that r is an independent variable and cannot depend on another independent variable t. When a Taylor series is used to express the leaving flow in terms of the entering flow, the net mass flow rate leaving the control volume is

$$\oiint_{CS} \rho \mathbf{V} \cdot d\mathbf{A} = \frac{\partial}{\partial r} \, (\rho V 2\pi r\delta) dr$$

Substituting the two terms back into the continuity equation results in the specific continuity equation for the oil-slick motion

$$\frac{\partial\delta}{\partial t} + \frac{1}{r}\frac{\partial}{\partial r} \, (r \, \delta V) = 0$$

δ, V, and certainly r depend on r and cannot be taken out from under the partial derivative operation $\partial/\partial r$.

a. Three kilonewtons of water per second flow through this pipeline reducer. Calculate the flowrate in cubic meters per second and the mean velocities in the 300 mm and 200 mm pipes. (Fig. 1)

b. Thirty newtons of air per second flow through the reducer of the preceding problem, the air in the 300 mm pipe having a weight density of 9.8 N/m³. In flowing through the reducer, the pressure and temperature will fall, causing the air to expand and producing a reduction of density; assuming that the weight density of the air in the 200 mm pipe is 7.85 N/m³, calculate the mass and volume flowrates, the mass velocities, and the velocities in the two pipes.

c. Taking Fig. 2 to represent an axisymmetric parabolic velocity distribution in a cylindrical pipe of radius R, calculate the mean velocity in terms of the maximum velocity, V_c.

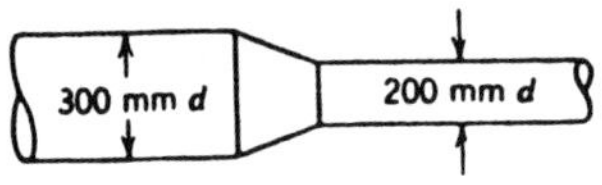

Fig. 1

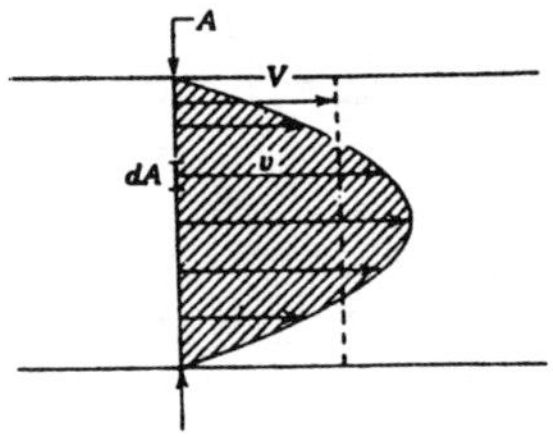

Fig. 2

Solution: a. Q: volumetric flowrate

$$Q = A_{cs}V$$

if density is essentially constant

$$Q_{300\text{ mm pipe}} = Q_{200\text{ mm pipe}}$$

if

$$\rho = \text{const}$$

else

$$\dot{m}_{300\text{ mm}} = \dot{m}_{200} = \rho A_{cs}V \qquad A_{cs} \text{ - cross sectional area}$$

$$Q = \frac{3 \times 10^3}{9.8 \times 10^3} = 0.306 \text{ m}^3/\text{s}$$

$$V_3 = \frac{0.306}{\frac{\pi}{4}(0.3)^2} = 4.33 \text{ m/s}$$

$$V_2 = \frac{0.306}{\frac{\pi}{4}(0.2)^2} = 9.74 \text{ m/s}$$

or

$$V_2 = 4.33\left(\frac{300}{200}\right)^2 = 9.74 \text{ m/s}$$

b. $Q_3 = \frac{30}{9.8} = 3.06 \text{ m}^3\text{/s}, \qquad Q_2 = \frac{30}{7.85} = 3.82 \text{ m}^3\text{/s}$

$$V_3 = \frac{3.06}{\frac{\pi}{4}(0.3)^2} = 43.3 \text{ m/s}, \quad V_2 = \frac{3.82}{\frac{\pi}{4}(0.2)^2} = 121.6 \text{ m/s}$$

$$\dot{m} = A\rho V = \text{Constant} = \frac{\dot{G}}{g_n} = \frac{30}{9.81} = 3.06 \text{ kg/s}$$

$\dot{m}$: mass flowrate

$\dot{G}$: weight flowrate

To check:

$$\frac{\gamma_3 Q_3}{g_n} = 9.8 \times \frac{3.06}{9.81} = 3.06 \text{ kg/s}$$

$$\frac{\gamma_2 Q_2}{g_n} = 7.85 \times \frac{3.82}{9.81} = 3.06 \text{ kg/s}$$

$$m_{v3} = \rho_3 V_3 = 1.0 \times 43.3 = 43.3 \text{ kg/m}^2\cdot\text{s}$$

$$m_{v2} = \rho_2 V_2 = 0.8 \times 121.6 = 97.3 \text{ kg/m}^2\cdot\text{s}$$

c. The velocity distribution is non uniform. But the flow is steady, 2 dimensional, and incompressible.

$$Q = AV$$

applies but V represents mean velocity which is obtained as follows,

$$V = \frac{Q}{A}$$

$$= \frac{1}{A}\int^A v\,dA$$

Taking r as radial distance to any local velocity, v, and element of area dA, $dA = 2\pi r\,dr$, the equation of the parabola is $v = v_c(1 - r^2/R^2)$.

$$V = \frac{1}{\pi R^2}\int_0^R v_c\left(1 - \frac{r^2}{R^2}\right)2\pi r\,dr$$

Performing the integration shows that for the parabolic axisymmetric profile, the mean velocity is half of the centeline velocity, that is,

$$V = \frac{V_c}{2}$$

• PROBLEM 4-9

A circular pipe of radius R_B is connected to a larger circular pipe of radius R_A (see Fig. 1). A pitot tube is used to obtain the velocity profile in the larger pipe at section A. The measurements indicate that at section A

$$\vec{v} = 2\beta\left[1 - \left(\frac{r}{R_A}\right)^2\right]\vec{i}$$

(a) Find the average velocity, V_A, at section A. (b) Find the maximum velocity at section A. (c) Determine the volume flow rate, Q. (d) Determine the average velocity, V_B, at section B. (e) Sketch the region swept out, in Δt, by the fluid located at section A at t = 0, and determine its volume.

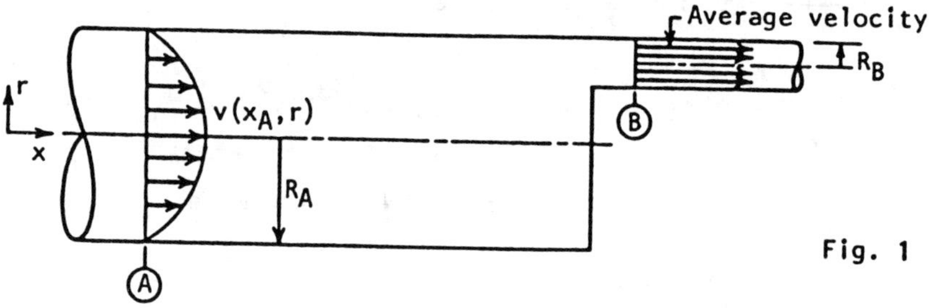

Fig. 1

Solution: (a) The average velocity, V_A, is given by

$$\pi R_A{}^2 V_A = \oiint_{S_1} \vec{v} \cdot d\vec{S} = \int_0^{R_A}\int_0^{2\pi} 2\beta\left[1 - \left(\frac{r}{R_A}\right)^2\right]\vec{i} \cdot \vec{i} r\, d\theta\, dr$$

$$= 2\beta(2\pi)\int_0^{R_A}\left(r - \frac{r^3}{R_A{}^2}\right)dr = 4\pi\beta\left[\frac{R_A{}^2}{2} - \frac{R_A{}^4}{4R_A{}^2}\right]$$

Eq. 1

Thus,

$$V_A = \beta$$

(b) By observation we see that the maximum velocity at section A occurs at r = 0. (For a nonobvious case, we would search for a relative extremum by setting $dv/dr = 0$). Thus,

$$v_{max} = 2\beta$$

(c) The volume flow rate, Q, is

$$Q = A_A V_A = \pi R_A^{\,2} \beta$$

(d) The average velocity at section B is given by

$$Q = A_B V_B \quad \text{or} \quad V_B = \frac{\pi R_A^{\,2} \beta}{\pi R_B^{\,2}} = \beta \left(\frac{R_A}{R_B}\right)^2$$

(e) A sketch of the region swept out in Δt by the fluid located at t = 0 at section A is shown in Fig. 2. The volume, $\mathcal{V}$, of this region is

$$\mathcal{V} = \oiint_A v(x_A, r)\,\Delta t\; dA = \Delta t \int_0^{R_A} \int_0^{2\pi} 2\beta \left[1 - \left(\frac{r}{R_A}\right)^2\right] r\, d\theta\, dr$$

Eq. 2

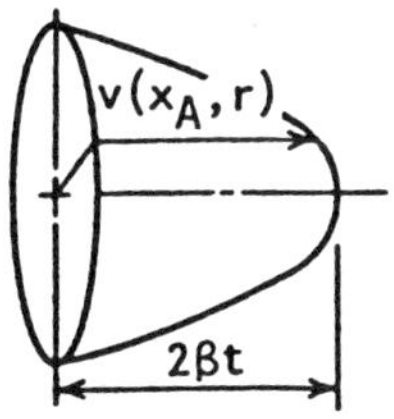

Fig. 2

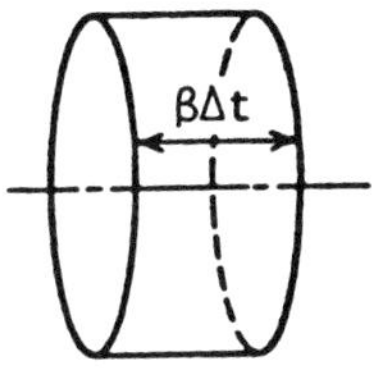

Fig. 3

The integral on the right-hand side of Eq. 2 is the same as the integral on the right-hand side of Eq. 1 and therefore

$$\mathcal{V} = \pi \beta R_A^{\,2}\, \Delta t = A_A V_A\, \Delta t = Q\, \Delta t$$

It should be obvious that this volume is the same as that which would be swept out by the fluid located at t = 0 at section A, having a uniform velocity, V_A (see Fig. 3).

• PROBLEM 4-10

The density of gas flowing through a section of pipe of constant cross section A and length L_0 varies according to the law

$$\rho = \rho_0 \left(1 - \frac{x}{2L_0}\right) \sin \frac{V_0 t}{L_0} \qquad \left(\frac{L_0}{V_0}\frac{\pi}{2} > t \geq 0,\; 0 \leq x \leq L_0\right)$$

where x is measured along the pipe axis and V_0 is a reference flow velocity. Find the difference in mass flux entering and leaving the pipe at any time.

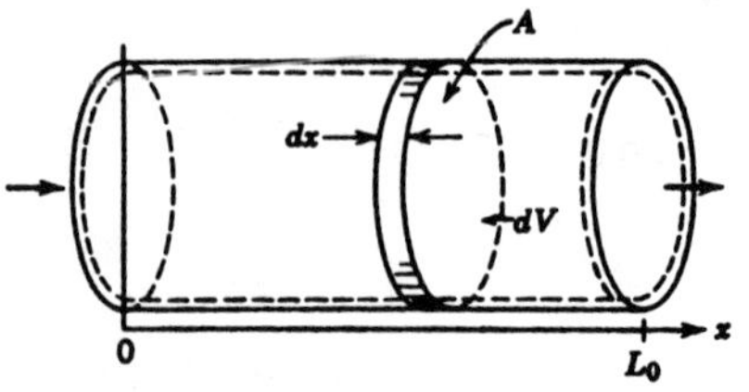

Solution: A non-deformable control volume is chosen inside of the pipe as shown in the Figure. We must use the continuity equation in the form

$$\frac{d}{dt}\int_{V_{C.V.}} \rho\, dV = -\int_{S_{C.V.}} \rho V_n dA$$

$$\left\{\begin{matrix}\text{net mass flux} \\ \text{out}\end{matrix}\right\} = -\frac{d}{dt}\int \rho dV = -\frac{d}{dt}\int \rho_0 (1 - \frac{x}{2L_0}) \sin \frac{V_0 t}{L_0} dV$$

Choose $dV = A\, dx$. Then

$$\frac{d}{dt}\int_{V_{C.S.}} \rho_0 (1 - \frac{x}{2L_0}) \sin \frac{V_0 t}{L_0} dV = \frac{d}{dt}\int_0^{L_0} \rho_0 (1 - \frac{x}{2L_0}) \sin \frac{V_0 t}{L_0} A\, dx$$

$$= \frac{d}{dt}\rho_0 \sin \frac{V_0 t}{L_0} A \int_0^{L_0} (1 - \frac{x}{2L_0}) dx = \frac{d}{dt}\rho_0 \sin \frac{V_0 t}{L_0} A \left[x - \frac{x^2}{4L_0}\right]_0^{L_0}$$

$$= \frac{d}{dt}(\rho_0 \sin \frac{V_0 t}{L_0} A \cdot \frac{3}{4}L_0) = \frac{3}{4}\rho_0 L_0 A \frac{d}{dt}\sin\frac{V_0 t}{L_0}$$

$$= \frac{3}{4}\rho_0 V_0 A \cos \frac{V_0 t}{L_0} .$$

Using

$$\int \frac{\partial \rho}{\partial t} dV$$

we have

$$\int_{V_{C.V.}} \frac{\partial \rho}{\partial t} dV = \int_{V_{C.V.}} \frac{\partial}{\partial t}\left[\rho_0 (1 - \frac{x}{2L_0}) \sin \frac{V_0 t}{L_0}\right] dV$$

$$= \int_{V_{C.V.}} \rho_0 (1 - \frac{x}{2L_0}) \cos \frac{V_0 t}{L_0} \frac{V_0}{L_0} dV$$

$$= \frac{V_0 \rho_0}{L_0} \cos \frac{V_0 t}{L_0} \int_0^{L_0} (1 - \frac{x}{2L_0}) A\, dx$$

$$= \frac{V_0 \rho_0}{L_0} \cos \frac{V_0 t}{L_0} \left[x - \frac{x^2}{4L_0}\right]_0^{L_0}$$

$$= \frac{3}{4}\rho_0 AV_0 \cos \frac{V_0 t}{L_0}$$

This equation is identical to the one just obtained. The analysis indicates a pulsating flow which at t = 0 shows the inflow at x = 0 to be greater than the outflow at x = L.

• PROBLEM 4-11

Water flows out of a tank through a pipe at the tank base. The velocity of flow across the pipe exit area varies according to

$$V = V_{max}\left[1 - \left(\frac{r}{R}\right)^2\right] \text{ ft/sec}$$

The velocity V_{max} is the maximum velocity of the flow at a given instant of time, and R is the pipe radius. Find the rate at which the mass of the tank and pipe combination is decreasing at any given time t_0.

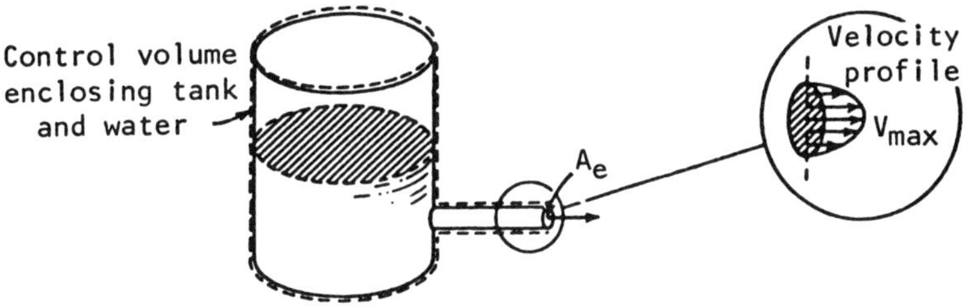

Solution: The control volume for this problem may be taken as either a non-deformable control volume around the entire tank or a deformable control volume with an upper surface that coincides with the falling water surface. We shall choose the latter, as shown in the Figure. We will use the continuity equation in the form

$$\frac{d}{dt}\int_{V_{C.V.(t)}} \rho \, dV = -\int_{S_{C.V.}} \rho V_{rn} \, dA$$

The only place where fluid crosses the boundary of the control volume is at the exit. At all the other points on the control volume surface $V_{rn} = 0$. Therefore, choosing $dA = 2\pi r \, dr$,

$$\int_{S_{C.V.}} \rho V_{rn} \, dA = \int_{A_r} \rho V_{rn} \, dA = \rho\int_0^R V_{max}\left[1 - \left(\frac{r}{R}\right)^2\right] 2\pi r \, dr$$

$$= \rho\frac{2\pi V_{max} R^2}{4} = \frac{\rho V_{max}}{2} A_e$$

We can now conclude that

$$\frac{d}{dt}\int_{V_{C.V.}} \rho\, dV = -\frac{\rho V_{max}}{2} A_e$$

is the rate at which mass is changing in the control volume. The negative value indicates that mass is decreasing.

• PROBLEM 4-12

The velocity distribution for a two-dimensional incompressible flow is given by

$$u = -\frac{x}{x^2 + y^2} \qquad v = -\frac{y}{x^2 + y^2}$$

Show that it satisfies continuity.

Solution: In two dimensions the continuity equation is,

$$\frac{\partial u}{\partial x} + \frac{\partial v}{\partial y} = 0$$

Then

$$\frac{\partial u}{\partial x} = -\frac{1}{x^2 + y^2} + \frac{2x^2}{(x^2 + y^2)^2}$$

$$\frac{\partial v}{\partial y} = -\frac{1}{x^2 + y^2} + \frac{2y^2}{(x^2 + y^2)^2}$$

and their sum does equal zero, satisfying continuity.

• PROBLEM 4-13

The velocity components of a steady two-dimensional, incompressible flow may be represented by

$$u = ky + 1$$

$$v = kx + 4$$

Determine if this is a possible flow.

Solution: You must satisfy the equation of continuity for a steady incompressible flow for the flow to be possible. Because this is two-dimensional flow, use

$$\frac{\partial u}{\partial x} + \frac{\partial v}{\partial y} = 0 \qquad (1)$$

$$\frac{\partial u}{\partial x} = \frac{\partial(ky + 1)}{\partial x} = 0, \quad \frac{\partial v}{\partial y} = \frac{\partial(kx + 4)}{\partial y} = 0$$

Substituting these values into Equation (1) gives

$$\frac{\partial u}{\partial x} + \frac{\partial v}{\partial y} = 0 + 0 = 0$$

which satisfies continuity.

• PROBLEM 4-14

Consider the three-dimensional incompressible vortex flow given by an axial velocity w = 2az, and a circumferential flow $v_\theta = \frac{A}{r}[1 - \exp(-ar)^2]$. Calculate the radial velocity v_r.

Solution: The three-dimensional incompressible continuity equation in cylindrical form is

$$\frac{1}{r}\frac{\partial(rv_r)}{\partial r} + \frac{1}{r}\frac{\partial v_\theta}{\partial \theta} + \frac{\partial w}{\partial z} = 0 \qquad (1)$$

Substitute the given expressions for the axial and circumferential velocity into Equation (1) yields

$$\frac{1}{r}\frac{\partial(rv_r)}{\partial r} = -2a \qquad (2)$$

Integrating Equation (2) yields

$$rv_r = -ar^2 + f(\theta,z)$$

or

$$v_r = -ar + \frac{f(\theta,z)}{r}$$

At the centerline of rotation, the radial velocity cannot be infinite, so $f(\theta,z) = 0$. Thus the radial velocity becomes

$$v_r = -ar$$

which says the flow is spiralling radially inward as it moves along the axis of rotation.

• PROBLEM 4-15

(a) A two-dimensional incompressible fluid has no velocity component in the z-direction. The velocity component V_x at any point is given by

$$V_x = x^2 - y^2$$

Find the V_y-component of velocity.

(b) Can the following velocity components characterize an incompressible flow?

$$V_x = x^2y,\ V_y = x + y + z,\ V_z = z^2 + x^2$$

Solution: For a homogenous, incompressible fluid with constant density, the continuity equation may be given in the form

$$\frac{\partial V_x}{\partial x} + \frac{\partial V_y}{\partial y} + \frac{\partial V_z}{\partial z} = 0$$

(a) Substituting the values of V_x and V_z into the continuity equation yields ($V_z = 0$)

$$2x + \frac{\partial V_y}{\partial y} = 0 \qquad (1)$$

Equation 1 is an expression for the derivative of V_y with respect to y with the variables x and z held constant. Thus equation 1 may be integrated by treating x as a parameter:

$$V_y = \int - 2x\ dy + f(x) \qquad (2)$$

Therefore

$$V_y = -2xy + f(x)$$

The function f(x) replaces the usual constant of integration in an ordinary differential equation and is completely arbitrary. This latter fact is readily confirmed by differentiating equation 2 with respect to y and noting that equation 1 is obtained.

(b) If the velocity components characterize an incompressible flow, the continuity equation must be satisfied.

Substitution of the various expressions into the continuity equation gives

$$\frac{\partial V_x}{\partial x} + \frac{\partial V_y}{\partial y} + \frac{\partial V_z}{\partial z} = 2xy + 1 + 2z$$

This expression is not identically zero, and therefore the velocity component expressions do not characterize an incompressible flow.

• PROBLEM 4-16

Determine which of the following functions of u and v are possible steady incompressible flow.

(a) $u = kxy + y$

$v = kxy + x$

(b) $u = x^2 + y^2$

$v = -2xy$

(c) $u = xy^2 + x + y^2$

$v = x(x - y) + 3y^3$

Solution: You must satisfy Equation of continuity to enable the foregoing to be possible steady flows of an incompressible fluid.

(a) $$\frac{\partial u}{\partial x} = \frac{\partial(kxy + y)}{\partial x} = ky$$

$$\frac{\partial v}{\partial y} = \frac{\partial(kxy + x)}{\partial y} = kx$$

$$\frac{\partial u}{\partial x} + \frac{\partial v}{\partial y} = ky + kx \neq 0$$

Therefore, this is not a possible flow.

(b) $$\frac{\partial u}{\partial x} = \frac{\partial(x^2 + y^2)}{\partial x} = 2x$$

$$\frac{\partial v}{\partial y} = \frac{\partial(-2xy)}{\partial y} = -2x$$

$$\frac{\partial u}{\partial x} = \frac{\partial v}{\partial y} = 2x - 2x = 0$$

which satisfies continuity. Therefore, this is a possible flow.

(c) $$\frac{\partial u}{\partial x} = \frac{\partial(xy^2 + x + y^2)}{\partial x} = y^2 + 1$$

$$\frac{\partial v}{\partial y} = \frac{\partial[x(x - y) + 3y^3]}{\partial y} = -x + 9y^2$$

$$\frac{\partial u}{\partial x} + \frac{\partial v}{\partial y} = y^2 + 1 - x + 9y^2$$

$$= 10y^2 - x + 1 \rightarrow \text{Not zero}$$

Therefore, this is not a possible flow.

• PROBLEM 4-17

The velocity distribution for the flow of an incompressible fluid is given by $v_x = 3 - x$, $v_y = 4 + 2y$, $v_z = 2 - z$. Show that this satisfies the requirements of the continuity equation.

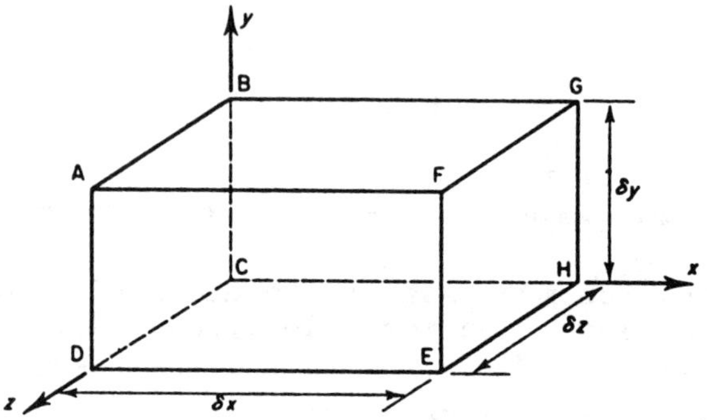

Continuity in three dimensions

Solution: The control volume ABCDEFGH in the Figure is taken in the form of a small rectangular prism with sides δx, δy and δz in the x, y and z directions, respectively. The mean values of the component velocities in these directions are v_x, v_y and v_z. Considering flow in the x direction,

Mass inflow through ABCD in unit time = $\rho v_x \delta y \delta z$.

In the general case, both mass density ρ and velocity v_x will change in the x direction and so

Mass outflow through EFGH in unit time =

$$\left\{\rho v_x + \frac{\partial}{\partial x}(\rho v_x)\,\delta x\right\}\delta y\,\delta z.$$

Thus,

Net outflow in unit time in x direction =

$$\frac{\partial}{\partial x}(\rho v_x)\,\delta x\,\delta y\,\delta z.$$

Similarly,

Net outflow in unit time in y direction =

$$\frac{\partial}{\partial y}(\rho v_y)\,\delta x \delta y \delta z,$$

Net outflow in unit time in z direction =

$$\frac{\partial}{\partial z}(\rho v_z)\,\delta x \delta y \delta z.$$

Therefore,

Total net outflow in unit time =

$$\left\{\frac{\partial}{\partial x}(\rho v_x) + \frac{\partial}{\partial y}(\rho v_y) + \frac{\partial}{\partial z}(\rho v_z)\right\}\delta x \delta y \delta z.$$

Also, since $\partial\rho/\partial t$ is the change in mass density per unit time.

Change of mass in control volume in unit time =

$$-\frac{\partial \rho}{\partial t}\delta x \delta y \delta z$$

(the negative sign indicating that a net outflow has been assumed). Then,

Total net outflow in unit time = change of mass in control volume in unit time

$$\left\{\frac{\partial}{\partial x}(\rho v_x) + \frac{\partial}{\partial y}(\rho v_y) + \frac{\partial}{\partial z}(\rho v_z)\right\}\delta x \delta y \delta z = -\frac{\partial \rho}{\partial t}\delta x \delta y \delta z$$

or

$$\frac{\partial}{\partial x}(\rho v_x) + \frac{\partial}{\partial y}(\rho v_y) + \frac{\partial}{\partial z}(\rho v_z) = -\frac{\partial \rho}{\partial t}. \tag{1}$$

Equation (1) holds for every point in a fluid flow whether steady or unsteady, compressible or incompressible. However, for incompressible flow, the density ρ is constant and the equation simplifies to

$$\frac{\partial v_x}{\partial x} + \frac{\partial v_y}{\partial y} + \frac{\partial v_z}{\partial z} = 0. \tag{2}$$

$$\frac{\partial v_x}{\partial x} = -1,\quad \frac{\partial v_y}{\partial y} = +2,\quad \frac{\partial v_z}{\partial z} = -1,$$

and, hence,

$$\frac{\partial v_x}{\partial x} + \frac{\partial v_y}{\partial y} + \frac{\partial v_z}{\partial z} = -1 + 2 - 1 = 0,$$

which satisfies the requirement for continuity.

Moist air flows steadily along a duct in an air-conditioning plant at the rate of 20 lbm/sec; the specific humidity is 0.004. Water is sprayed into the stream at the rate of 0.120 lbm/sec, and mixes with it but not uniformly. At one of the two outlets supplied by the duct the mass flow rates of air and water are 12.20 lbm/sec and 0.130 lbm/sec respectively. What is the discharge rate and the specific humidity at the other outlet? (Specific humidity is defined as the ratio of the masses of water and dry air in a quantity of mixture.)

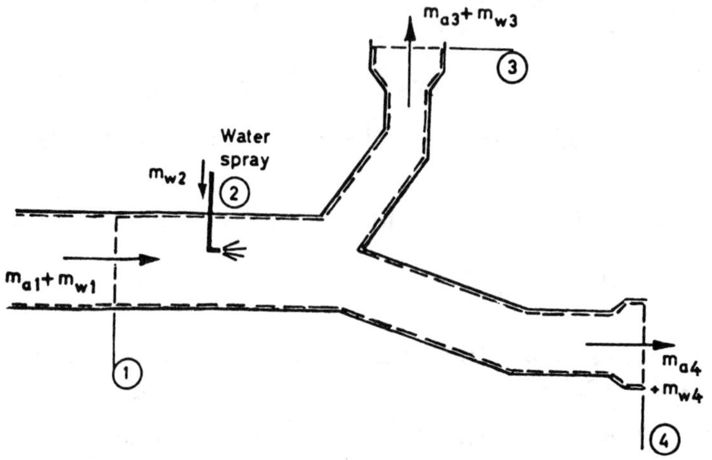

Solution: Let suffices a and w refer to air and water respectively and let the two inlet stations be indicated by suffices 1 and 2, and the two outlet stations by 3 and 4, as in the figure. In this notation the data are:

$$\dot{m}_{a_1} + \dot{m}_{\omega_1} = 20 \text{ lbm/sec},$$

$$\frac{\dot{m}_{\omega_1}}{\dot{m}_{a_1}} = 0.004,$$

$$\dot{m}_{\omega_2} = 0.120 \text{ lbm/sec},$$

$$\dot{m}_{a_3} = 12.20 \text{ lbm/sec},$$

$$\dot{m}_{\omega_3} = 0.130 \text{ lbm/sec} .$$

From the first two items we have

$$\dot{m}_{\omega_1} = \frac{20}{(1/0.004) + 1} \text{ lbm/sec} = 0.0797 \text{ lbm/sec},$$

$$\dot{m}_{a_1} = (20 - 0.0797) \text{ lbm/sec} = 19.92 \text{ lbm/sec} .$$

The mass continuity equation for the dry air stream is

$$\dot{m}_{a_1} = \dot{m}_{a_3} + \dot{m}_{a_4}$$

$$\therefore \quad \dot{m}_{a_4} = \dot{m}_{a_1} - \dot{m}_{a_3} = (19.92 - 12.20) \text{ lbm/sec}$$

$$= 7.72 \text{ lbm/sec} .$$

For the water

$$\dot{m}_{\omega_1} + \dot{m}_{\omega_2} = \dot{m}_{\omega_3} + \dot{m}_{\omega_4}$$

$$\therefore \quad \dot{m}_{\omega_4} = \dot{m}_{\omega_1} + \dot{m}_{\omega_2} - \dot{m}_{\omega_3}$$

$$= (0.0797 + 0.120 - 0.130) \text{ lbm/sec} = 0.0697 \text{ lbm/sec.}$$

Then the total discharge rate at the second outlet is

$$\dot{m}_{a_4} + \dot{m}_{\omega_4} = 7.79 \text{ lbm/sec,}$$

and the specific humidity is

$$\frac{\dot{m}_{\omega_4}}{\dot{m}_{a_4}} = \frac{0.0697}{7.72} = 0.00903$$

CHAPTER 5

ENERGY EQUATION

Basic Attacks and Strategies for Solving Problems in this Chapter. **See pages 176 to 237 for step-by-step solutions to problems.**

There are many forms of the Bernoulli equation, some of which may be obtained by the integration of the momentum equation; the most general form requires the integral conservation of energy equation. The following discussion will work up from the simplest to the most general.

First, define the Bernoulli constant for an incompressible flow as

$$\text{B.C.} = \frac{p}{\rho} + \frac{V^2}{2} + gz. \tag{1}$$

For steady incompressible frictionless flow along a streamline between points 1 and 2, the Bernoulli constant does not change:

$$\frac{p_1}{\rho} + \frac{V_1^2}{2} + gz_1 = \frac{p_2}{\rho} + \frac{V_2^2}{2} + gz_2. \tag{2}$$

Equation (2) is useful for duct or pipe flows, particularly when there are changes in cross-sectional areas and/or elevations. If the elevation increases ($z_2 > z_1$) or if the area decreases ($A_2 < A_1$ and thus $V_2 > V_1$ by the conservation of mass), it can be seen that the pressure p_2 must decrease accordingly to satisfy Equation (2). Often, it is more convenient to write the Bernoulli constant in terms of the equivalent column height of fluid, or *head*. Dividing by g, Equation (2) can be written

$$H = \frac{p_1}{\gamma} + \frac{V_1^2}{2g} + z_1 = \frac{p_2}{\gamma} + \frac{V_2^2}{2g} + z_2, \tag{3}$$

where $\gamma = \rho g$ and H is called the total head or total Bernoulli head. H is also equivalent to the height of the Energy Grade Line (EGL).

An unsteady form of Equation (3) can be obtained by the integration of the conservation of momentum equation, combined with the conservation of mass. The result is the unsteady Bernoulli equation,

$$\frac{p_1}{\rho} + \frac{V_1^2}{2} + gz_1 = \frac{p_2}{\rho} + \frac{V_2^2}{2} + gz_2 + \int_1^2 \frac{\partial V}{\partial t}\, ds. \tag{4}$$

Again, this is valid along a streamline from point 1 to point 2 in the flow and is restricted to frictionless incompressible flow in regions where no heat or work transfer occurs in the fluid. V is the velocity along the streamline and ds is the incremental distance along the streamline.

In most practical problems, friction cannot be neglected, and there may be work added to the flow (by a pump), work extracted from the flow (by a turbine), or heat transferred to or from the fluid. In this case, a much more general Bernoulli equation must be developed. Specifically, the total head H in Equation (3) does not remain constant, but rather changes whenever friction, work, or heat transfers are present. From the energy equation,

$$\frac{p_1}{\rho g} + \frac{V_1^2}{2g} + z_1 = \frac{p_2}{\rho g} + \frac{V_2^2}{2g} + z_2 + h_s + h_{\text{total}}. \tag{5}$$

Here, 1 and 2 are locations at an inlet and outlet of a control volume. The shaft work head,

$$h_s = \frac{\dot{W}_s}{(\dot{m}g)},$$

is the head associated with work done by the fluid,

where $\dot{W}_s$ is the shaft power (work per unit time) done *by* the fluid, and

h_{total} is the total amount of head loss due to fiction, heat transfer, etc.

For a pump, let

$$h_s = -E_p,$$

where E_p is the energy added per unit weight of the fluid (dimensions of head, i.e., height of fluid).

For a turbine, let E_T be the energy extracted from the fluid per unit weight. In this case,

$$h_s = E_T$$

in Equation (5). If the frictional losses are neglected, we will underestimate the work required to move the fluid from one location to another. Techniques used to find h_{total} are discussed in Chapter 7.

Finally, for flowfields with non-uniform inlets and outlets, compressibility effects, or heat transfers (through a heat exchanger, for example), it may be necessary to use the integral control volume form of the energy equation. From the Reynolds Transport Theorem (Chapter 4) and the First Law of Thermodynamics, one can obtain

$$\dot{Q} - \dot{W}_s = \frac{\partial}{\partial t}\iiint_{cv} \rho e \, dV + \iint_{cs} (\hat{h} + \tfrac{1}{2}V^2 + gz)\rho \mathbf{V} \cdot \mathbf{n} \, dA, \qquad (6)$$

where $\dot{Q}$ is the rate of heat transfer to the control volume,

$\dot{W}_s$ is the shaft work done by the control volume,

e is the energy per unit mass,

$e = \hat{u} + {}^1/_2 V^2 + gz$,

$\hat{u}$ is the internal energy per unit mass, and

$\hat{h}$ is the enthalpy per unit mass.

As with problems that utilize the integral conservation of mass, the approach to these problems is to first choose an appropriate control volume, evaluate the flux terms through all surfaces, and then apply Equation (6) to determine the unknown (typically $\dot{W}_s$ or $\dot{Q}$ for most pump or turbine problems).

Step-by-Step Solutions to Problems in this Chapter, "Energy Equation"

ENERGY EQUATION

• PROBLEM 5-1

Water is flowing in an open channel (Figure) at a depth of 2 m and a velocity of 3 m/s. It then flows down a chute into another channel where the depth is 1 m and the velocity is 10 m/s. Assuming frictionless flow, determine the difference in elevation of the channel floors.

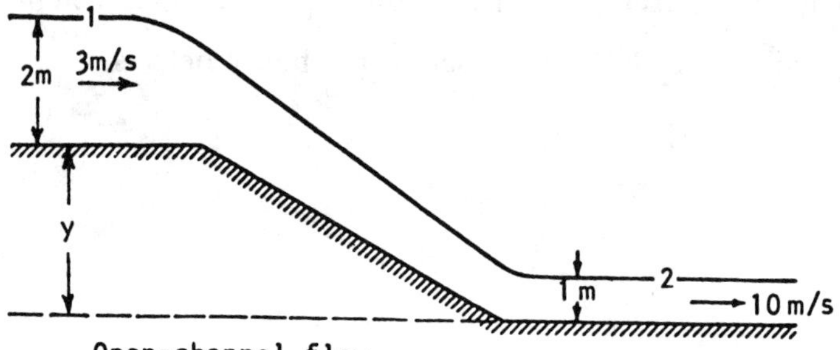

Open-channel flow.

Solution: The velocities are assumed to be uniform over the cross-sections, and the pressures hydrostatic. The points 1 and 2 may be selected on the free surface, as shown, or they could be selected at other depths. If the difference in elevation of floors is y, Bernoulli's equation is

$$\frac{V_1^2}{2g} + \frac{p_1}{\gamma} + z_1 = \frac{V_2^2}{2g} + \frac{p_2}{\gamma} + z_2$$

Then

$z_1 = y + 2,$

$z_2 = 1,$

$V_1 = 3 \text{ m/s}, V_2 = 10 \text{ m/s},$

and

$$p_1 = p_2 = 0,$$

$$\frac{3^2}{2 \times 9.806} + 0 + y + 2 = \frac{10^2}{2 \times 9.806} + 0 + 1$$

and

$$y = 3.64 \text{ m}.$$

• PROBLEM 5-2

Ideal fluid of weight density 7.9 kN/m³ flows down this pipe and out into the atmosphere through the "end cap" orifice. Gage B reads 41.0 kPa and gage A, 14.0 kPa. Calculate the mean velocity in the pipe.

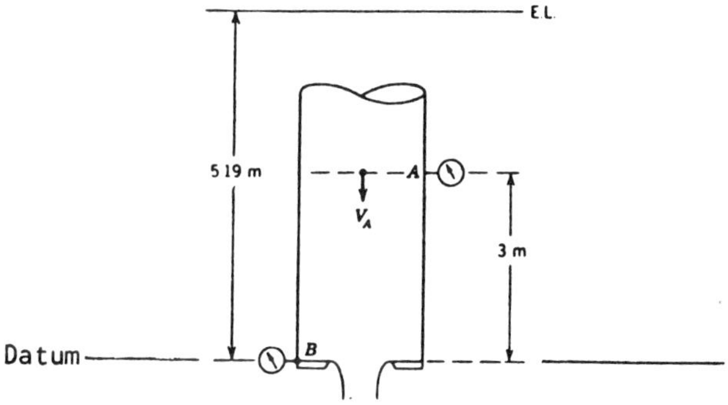

Solution: Identify point B as a stagnation point. The energy line will then be established $41.0 \times 10^3/7.9 \times 10^3 = 5.19$ m above B. Gage A measures the static pressure in the flow; p_A/γ is thus $14.0 \times 10^3/7.9 \times 10^3 = 1.77$ m. Therefore

$$\frac{V_A^2}{2g_n} = 5.19 - 3 - 1.77 = 0.42 \text{ m},$$

$$V_A = 2.87 \text{ m/s}$$

• **PROBLEM 5-3**

An open tank containing water has an orifice which is located near the bottom of the tank as shown in the figure. Show that for ideal flow, the discharge velocity at the orifice is $\sqrt{2gh}$. Assume steady flow.

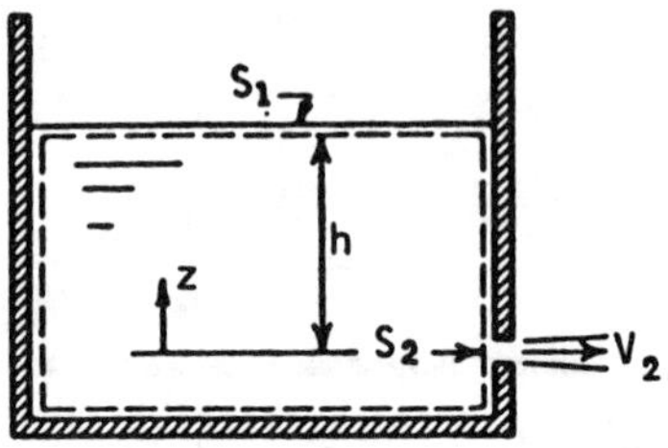

Solution: Consider the control volume shown in the figure. Apply the Bernoulli equation to sections S_1 and S_2.

$$\frac{p_1}{\gamma} + \frac{V_1^2}{2g_c} + \frac{gZ_1}{g_c} = \frac{p_2}{\gamma} + \frac{V_2^2}{2g_c} + \frac{gZ_2}{g_c} \tag{1}$$

Experiments indicate that the pressure at any section open to the atmosphere can be taken to be at atmospheric pressure. Therefore,

$$p_1 = p_2 = p_{atm}$$

Also, using continuity equation,

$$V_1 A_1 = V_2 A_2 \tag{2}$$

and by the fact that $A_1 \gg A_2$, we can take $V_1 \ll V_2$. Thus V_1 can be neglected. Under these circumstances, Eq. 1 reduces to

$$\frac{V_2{}^2}{2g} = z_1 - z_2 = h$$

or

$$V_2 = \sqrt{2gh}$$

• PROBLEM 5-4

(a) Determine the velocity of efflux from the nozzle in the wall of the reservoir of the figure. (b) Find the discharge through the nozzle.

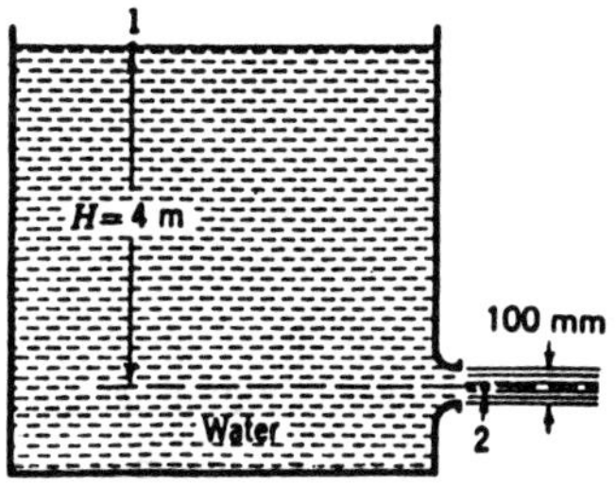

Flow through nozzle from reservoir.

Solution: (a) The jet issues as a cylinder with atmospheric pressure around its periphery. The pressure along its centerline is at atmospheric pressure for all practical purposes. Bernoulli's equation is applied between a point on the water surface and a point downstream from the nozzle,

$$\frac{V_1^2}{2g} + \frac{p_1}{\gamma} + z_1 = \frac{V_2^2}{2g} + \frac{p_2}{\gamma} + z_2$$

With the pressure datum as local atmospheric pressure, $p_1 = p_2 = 0$; with the elevation datum through point 2, $z_2 = 0$, $z_1 = H$. The velocity on the surface of the reservoir is zero (practically); hence,

$$0 + 0 + H = \frac{V_2^2}{2g} + 0 + 0$$

and

$$V_2 = \sqrt{2gH} = \sqrt{2 \times 9.806 \times 4} = 8.86 \text{ m/s}$$

which states that the velocity of efflux is equal to the velocity of free fall from the surface of the reservoir. This is known as Torricelli's theorem.

(b) The discharge Q is the product of velocity of efflux and area of stream,

$$Q = A_2V_2 = \pi(0.05 \text{ m})^2(8.86 \text{ m/s}) = 0.07 \text{ m}^3/\text{s} = 70 \text{ l/s}$$

• PROBLEM 5-5

What is the maximum pressure exerted on an aircraft flying at 200 mph at sea level? ($\rho = 0.00238$ slug/ft^3; $p_0 = 14.7$ psi.)

Solution: Multiplying the Bernoulli equation through by γ gives

$$p + \frac{V^2\gamma}{2g} + z\gamma = \text{const} \quad (1)$$

The ratio γ/g is the density ρ, and for gases the quantity γz is negligible since the specific weight of gases is very small. Usually the change in potential z is also small in gas dynamic problems. Equation (1) then reduces to

$$p + 1/2\rho V^2 = \text{const} \quad (2)$$

which is the aerodynamic form of Bernoulli's equation for ideal incompressible flow of gases.

In this equation p is referred to as the static pressure and $\rho V^2/2$ as the dynamic pressure. The units of both are pressure units, usually psf.

If at any point in a gas flow the velocity is reduced to zero, the point is referred to as a stagnation point, and the static pressure at this point is called the stagnation pressure p_s. This is the maximum pressure that may be recorded in a flow.

$$p_s = p_0 + 1/2\rho \quad V_0^2 \quad (3)$$

where

p_0 = free stream static pressure

V_0 = free stream velocity

$$p_s = 14.7 \times 144 + 1/2 \times 0.00238(200 \times 88/60)^2$$

$$= 2{,}220 \text{ psfa}$$

$$= 15.42 \text{ psia}$$

• PROBLEM 5-6

Determine the pressure drop in the entrance region of the channel flow (see figure). The entrance-velocity profile is uniform. The Reynolds number of the flow is 1500.

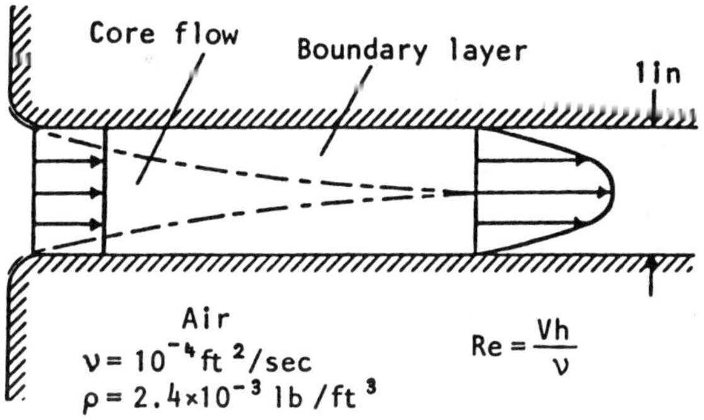

Solution: The average velocity is found from the definition of the Reynolds number.

$$V = \frac{Re\ \nu}{h}$$

$$= \frac{1500 \times 10^{-4}}{1/12}$$

$$= 1.8 \text{ fps}$$

In a channel the maximum velocity at the centerline at the end of the entrance length where the velocity profile is parabolic is given by

$$u_{max} = \frac{3}{2} V$$

$$= 2.7 \text{ fps}$$

Now, the flow is inviscid in the core region, so that Bernoulli's equation is applicable along the various streamlines. The air that flows down the streamline at the center of the channel has a velocity of 2.7 fps just at the end of the inviscid core. Hence, Bernoulli's equation allows us to write

$$\frac{V_1^2}{2} + \frac{p_1}{\rho} = \frac{V_2^2}{2} + \frac{p_2}{\rho}$$

or

$$\Delta p = \frac{V_1^2 - V_2^2}{2} \rho$$

$$= \frac{1.8^2 - 2.7^2}{2} \times 0.0024$$

$$= -0.0049 \text{ psf}$$

a very small pressure difference to measure with any type of irstrument.

Calculate the flowrate through this two-dimensional nozzle discharging to the atmosphere and designate any stagnation points in the flow.

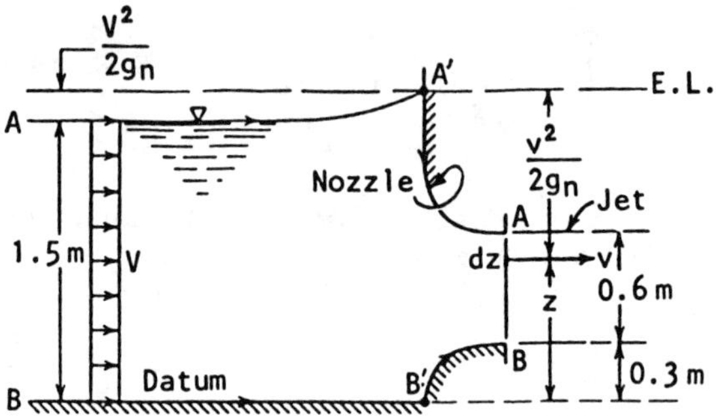

Solution: From the energy line drawn a distance $V^2/2g_n$ above the horizontal liquid surface, and with the pressure assumed zero throughout the free jet at the nozzle exit,

$$1.5 + \frac{V^2}{2g_n} = z + \frac{v^2}{2g_n} ;$$

$$v = \sqrt{2g_n\left(1.5 + \frac{V^2}{2g_n} - z\right)}$$

The flowrate q may then be expressed

$$q = 1.5V = \int_{0.3}^{0.9} v \, dz$$

$$= \int_{0.3}^{0.9} \sqrt{2g_n \left(1.5 + \frac{V^2}{2g_n} - z\right)} \, dz$$

which may be solved by trial to yield $q = 2.81 \text{ m}^3/\text{s·m}$.

Stagnation points on the boundary streamlines AA and BB are to be expected. The one at B' needs no comment. At some point on the plane A'B' there must be a stagnation point on the top boundary streamline; assume that this is somewhere below A'. Such a point could not be a stagnation point since its distance below the energy line would indicate a velocity head and thus a velocity there. Accordingly it is concluded that the stagnation point must be at A' and the liquid surface must rise to this point.

• PROBLEM 5-8

A siphon of uniform cross-sectional area is used to drain water from a tank (see figure). At the time of interest, let h be the elevation difference between the liquid level in the tank and the end of the tube. If the atmosphere is at standard atmospheric pressure and h is 6 ft, (a) Determine the average fluid velocity through the tube; (b) determine the pressure within the tube at point B; and (c) show that if the end of the tube is above the liquid level in the tank, then the liquid will not flow.

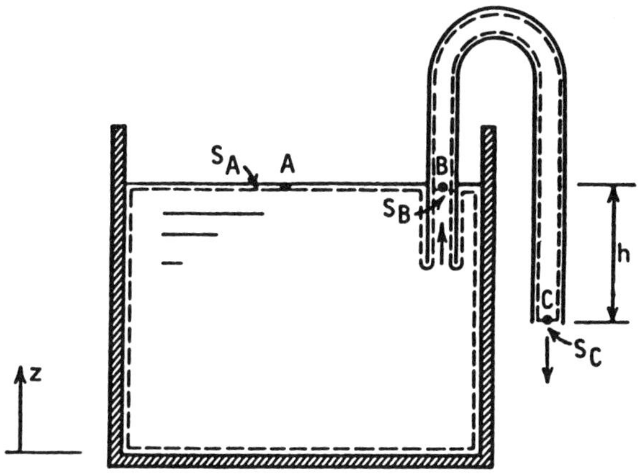

Solution: (a) Select a control volume shown by the dashed lines in the figure. First, write the energy equation

$$\frac{p_1}{\gamma} + \frac{V_1^2}{2g_c} + \frac{gz_1}{g_c} = \frac{p_2}{\gamma} + \frac{V_2^2}{2g_c} + \frac{gz_2}{g_c} \qquad (1)$$

between sections B and C. Since the tube is open to the atmosphere at section C, $p_C = p_{atm}$. By continuity and the given information that the cross-sectional area of the tube is uniform, we have that $V_B = V_C$. Then Eq. 1 reduces to

$$\frac{p_C - p_B}{\gamma} = \frac{p_{atm} - p_B}{\gamma} = z_B - z_C = h \qquad (2)$$

Now write the energy equation between sections A and B. Since the area at section A is much greater than the cross-sectional area of the tube, continuity gives that $V_A \ll V_B$ and may thus be neglected. Also, note that $z_A = z_B$ and that $p_A = p_{atm}$. Then Eq. 1 reduces to

$$\frac{p_A - p_B}{\gamma} = \frac{p_{atm} - p_B}{\gamma} = \frac{V_B^2}{2g}$$

Combining the last two equations, we obtain

$$V_B = \sqrt{2gh} \tag{3}$$

For h = 6 ft,

$$V_B = \sqrt{2(32.2)(6)} = 19.66 \text{ ft/s}$$

(b) From Eq. 2, we have

$$p_B = p_{atm} - \gamma h = 14.68 - \frac{62.4(6)}{144} = 12.08 \text{ psia}$$

(c) If the end of the tube is above the liquid level in the tank, then $(z_B - z_C)$ is negative. Then, by Eq. 3, no real solution exists for V_B.

THE BERNOULLI EQUATION

• PROBLEM 5-9

Assume frictionless flow for the water siphon shown, and that the water exits the siphon freely at atmospheric pressure. Find (1) the velocity of the water when exiting from the siphon, and (2) the water pressure in cross-section C in the base portion of the inverted U-shaped tube.

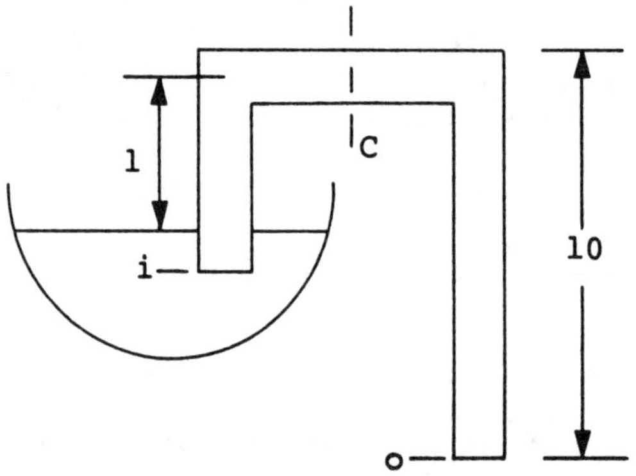

Solution: (1) Applying the flow (Bernoulli) equation between inlet, i, and outlet, o, of the siphon,

$$\frac{P_i}{\rho} + \frac{V_i^2}{2} + gz_i = \frac{P_o}{\rho} + \frac{V_o^2}{2} + gz_o \qquad (1)$$

The tank from which the water is to be siphoned may be assumed to have an area which exceeds by a large amount the area of the tube, and therefore V_i may be assumed to be substantially zero.

Since atmospheric pressure prevails substantially at inlet, i, and outlet, o,

$$P_i = P_o = \text{atmospheric pressure.}$$

Substituting into flow equation (1), we obtain

$$gz_i = \frac{V_o^2}{2} + gz_o$$

from which $\quad V_o^2 = 2g(z_i - z_o)$

and $\quad V_o = \sqrt{2g(z_i - z_o)}.$

Substituting the indicated quantities,

$$V_o = \sqrt{2 \times 9.81 \times (10 - 1)} = 13.3 \text{ meters/sec.}$$

(2) To find the pressure prevailing at cross-section, c, apply the flow (Bernoulli) equation between positions i and c.

$$\frac{P_i}{\rho} + \frac{V_i^2}{2} + gz_i = \frac{P_c}{\rho} + \frac{V_c^2}{2} + gz_c \qquad (2)$$

Since $V_i \approx 0$ for the reasons given previously, and $V_c = V_o$, where V_o has been calculated above, substituting in (2)

$$\frac{P_c}{\rho} = \frac{P_i}{\rho} + gz_i - gz_c - \frac{V_c^2}{2}$$

from which

$$P_c = P_i + \rho\{g(z_i - z_c) - \frac{V_c^2}{2}\}$$

Substituting numerical values,

$$P_c = 1.01 \times 10^5 \frac{N}{m^2} + 999 \frac{kg}{m^3}\{9.81(-1) - \frac{13.3^2}{2}\}$$

$$= 2840P_a \text{ - absolute}$$

• PROBLEM 5-10

Water at height, H, above a floor, flows beneath a gate. At the vena contracta below the gate, the depth is D where straight flow lines prevail. Find (a) the water velocity downstream from the gate; and (b) the volumetric flow per linear foot of width.

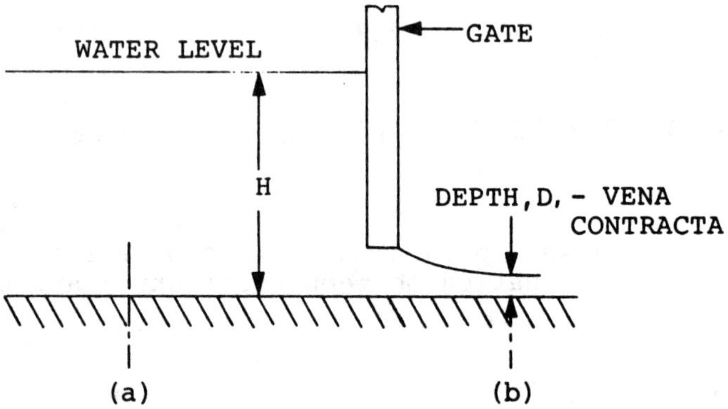

Solution: Applying the general flow (Bernoulli) equation between locations (a) and (b),

$$\frac{P_a}{\rho} + \frac{V_a^2}{2} + gz_a = \frac{P_b}{\rho} + \frac{V_b^2}{2} + gz_b \quad (1)$$

Assuming hydrostatic pressure distribution and $V_a \approx 0$, then

$$\frac{dP}{dz} = -q$$

and

$$P = P_{atm} + q(H - z)$$

from which

$$\frac{P}{\rho} = \frac{P_{atm}}{\rho} + g(H - z)$$

Substituting into equation (1),

$$\frac{P_{atm}}{\rho} + g(H - z_a) + \frac{V_a^2}{2} + gz_a$$

$$= \frac{P_{atm}}{\rho} + g(D - z_b) + \frac{V_b^2}{2} + gz_b$$

and

$$\frac{V_a^2}{2} + gH = \frac{V_b^2}{2} + gD$$

Solving for V_b,

$$V_b = \sqrt{V_a^2 + 2g(H - D)}$$

or

$$V_b = \sqrt{2g(H - D)} \quad \text{since } V_a \approx 0$$

(b) To find the volumetric flow, Q, per unit of width,

$$Q = VA = VDW,$$

assuming uniform flow, where W is the width.

Therefore

$$\frac{Q}{W} = V_bD$$

• PROBLEM 5-11

A large tank may be emptied by the pipe shown at its bottom. Find the velocity of water at the free end of the pipe as a function of time after a flow control valve at that free end is opened. The cross-sectional area of the tank is large compared to the diameter of the pipe.

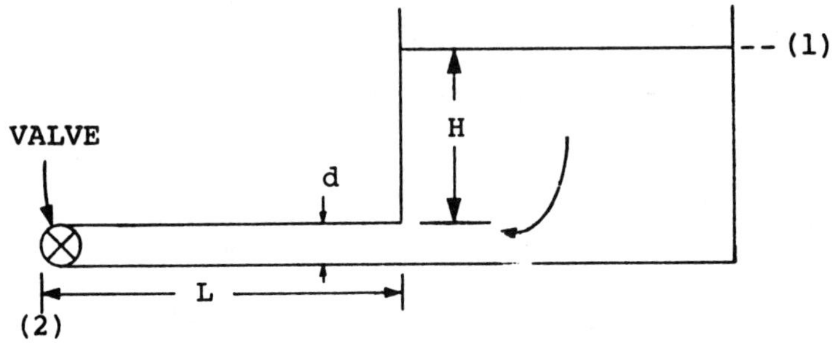

Solution: Applying the general flow (Bernoulli) equation between positions (1) and (2),

$$\frac{P_1}{\rho} + \frac{V_1^2}{2} + gh_1 = \frac{P_2}{\rho} + \frac{V_2^2}{2} + gh_2 + \int_1^2 \frac{\partial V_x}{\partial t}\, dx$$

Since the area of the tank is large compared to the pipe cross-section, $V_1 \approx 0$. If friction along the pipe is neglected, then

$$\int_1^2 \frac{\partial V_x}{\partial t}\, dx \approx \int_0^L \frac{\partial V_x}{\partial t}\, dx$$

and $$V_x = V_2$$

Consequently,

$$\int_0^L \frac{\partial V_x}{\partial t}\, dx = \int_0^L \frac{dV_2}{dt}\, dx = \frac{LdV_2}{dt}$$

Assuming locations (1) and (2) are both at atmospheric pressure, then $P_1 = P_2$. Since $V_1 \approx 0$, the general flow equation reduces to

$$gh_1 = \frac{V_2^2}{2} + gh_2 + \int_1^2 \frac{\partial V_x}{\partial t}\, dx$$

Assuming $h_2 - h_1 = H$ and substituting for $\int_1^2 \frac{\partial V_x}{\partial t}\, dx$,

$$gH = \frac{V_2^{\,2}}{2} + L\frac{dV_2}{dt}$$

To solve the equation by separating variables,

$$\frac{2gH - V_2^2}{2} = L\frac{dV_2}{dt}$$

from which

$$\frac{dt}{2L} = \frac{dV_2}{2gH - V_2^2}$$

Integrating with initial conditions $V = 0$ at $t = 0$ and $V = V_2$ at time t,

$$\int_0^{V_2} \frac{dV}{2gH - V^2} = \left[\frac{1}{\sqrt{2gH}} \tanh^{-1}\left(\frac{V}{\sqrt{2gH}}\right)\right]_0^{V_2}$$

but $\tanh^{-1}(0) = 0$.

Therefore

$$\frac{1}{\sqrt{2gH}} \tanh^{-1}\left(\frac{V_2}{\sqrt{2gH}}\right) = \frac{t}{2L}$$

from which

$$\tanh\left(\frac{t}{2L} \sqrt{2gH}\right) = \frac{V_2}{\sqrt{2gH}}$$

or $$V_2(t) = \sqrt{2gH}\tanh\left(\frac{t}{2L}\sqrt{2gH}\right)$$

If H = 10 meters and L = 5 meters, then

$$\sqrt{2gH} = \sqrt{2 \times 9.81 \times 10} = 14.0 \text{ m/sec.}$$

and

$$\frac{t}{2L}\sqrt{2gH} = \frac{t}{2(5)} \times 14 = 1.4t$$

Substituting in the equation for $V_2(t)$,

$$V_2(t) = 14 \tanh(1.4t)$$

• PROBLEM 5-12

How much power must be supplied for the pump to maintain readings of 250 mm of mercury vacuum and 275 kPa on gages 1 and 2, respectively, while delivering a flowrate of 0.15 m^3/s of water?

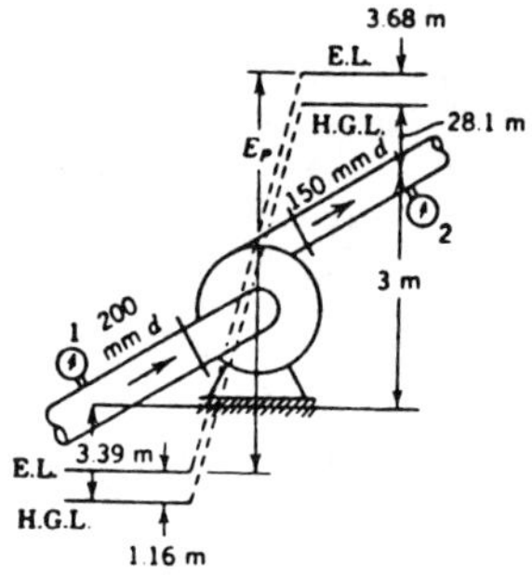

Solution: The addition of mechanical energy (E_P) to a fluid flow by a pump or its extraction (E_T) by a turbine will alter the basic Bernoulli equation to

$$\frac{p_1}{\gamma} + \frac{V_1^2}{2g_n} + z_1 + E_P = \frac{p_2}{\gamma} + \frac{V_2^2}{2g_n} + z_2 + E_T \quad (1)$$

in which the quantities E_p and E_r are in terms of energy added or subtracted per unit weight of fluid flowing and will appear as abrupt rises or falls of the energy line over the respective machines. Usually the engineer requires the total power of such machines, which may be computed from the product of weight flowrate (G) and (E_P or E_T), yielding total power. Thus

$$\text{Kilowatts of machine} = \frac{Q\gamma(E_p \text{ or } E_T)}{1000} \quad (2)$$

This equation can be used as a general equation for converting any unit energy, or head, to the corresponding total power. Converting the gage readings to meters of water, $p_1/\gamma = -3.39$ m and $p_2/\gamma = +28.1$ m. Thus the hydraulic grade lines are 3.39 m below and 28.1 m above points 1 and 2, respectively. The velocity heads at these points will be found to be 1.16 m and 3.68 m, respectively, so the energy lines will be these distances above the respective hydraulic grade lines. The rise in the energy line through the pump represents the energy supplied by the pump to each newton of fluid; from the sketch this is

$$E_p = 3.68 + 28.1 + 3 + 3.39 - 1.1.6 = 37.0 \text{ J/N}$$

$$\text{Pump power} = 0.15 \times 9.8 \times 10^3 \times 37.0 = 54.4 \text{ kW}$$

Note that the indicated H.G.L.'s are for the pipes only and do not include the pump passages, where the flow is not one-dimensional. The positions of these H.G.L.'s give no assurance that the pump will run cavitation-free, since local velocities in the pump passages may be considerably larger than the average velocities in the pipes.

• PROBLEM 5-13

Obtain:

(a) The differential equation of motion and the natural frequency of oscillation of the fluid in the U-tube shown in Fig. 1.

(b) If the tube is moving with a velocity $\vec{V}$ to the right, derive an expression for the acceleration, $\vec{a}$, of the tube in terms of the variation in liquid level h, tube geometry and fluid properties.

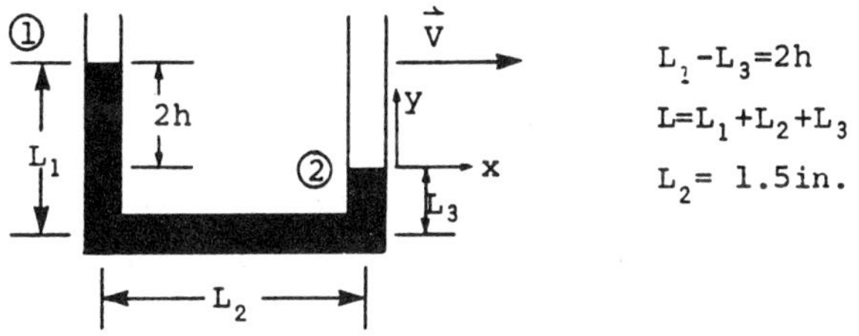

Fig. 1

Solution: (a) By applying the Bernoulli's equation to the unsteady flow along the streamline from point (1) to point (2), we have

$$\frac{p_1}{\rho} + \frac{v_1^2}{2} + gz_1 = \frac{p_2}{\rho} + \frac{v_2^2}{2} + gz_2 + \int_1^2 \frac{\partial V_s}{\partial t}\, ds \tag{1}$$

But

$$p_1 = p_2 = p_{atm}, \quad v_1 = v_2 = V_s$$

Therefore equation (1) becomes,

$$gL_1 = gL_3 + \int_1^2 \frac{\partial V_s}{\partial t}\, ds = gL_3 + L\frac{dV_s}{dt} \tag{2}$$

Realizing that $V_s = \frac{dh}{dt}$, equation (2) becomes,

$$gL_1 = gL_3 + L\frac{d^2h}{dt^2} \Rightarrow g(L_1 - L_3) = L\frac{d^2h}{dt^2} \Rightarrow 2gh = L\frac{d^2h}{dt^2}$$

or
$$\frac{d^2h}{dt^2} - \frac{2gh}{L} = 0 \qquad (3)$$

Equation (3) represents the differential equation of motion. The natural frequency of oscillation of the tube is

$$f = \frac{\omega}{2\pi} = \frac{1}{2\pi}\sqrt{\frac{2g}{L}} = \frac{1}{2\pi}\sqrt{\frac{2(32.2)}{0.41}} = 1.99\ \frac{\text{rev}}{\text{min}}$$

(b) By applying the hydrostatic equation, $-\text{grad } p + \rho\vec{g} = \rho\vec{a}$ in the x and y components, we have

$$\left\{\begin{aligned} -\frac{\partial p}{\partial x} + \rho g_x &= \rho a_x, \quad a_x = a \text{ and } g_x = 0 \\ -\frac{\partial p}{\partial y} + \rho g_y &= \rho a_y, \quad a_y = 0 \text{ and } g_y = -g \end{aligned}\right\} \Rightarrow \begin{aligned} \frac{\partial p}{\partial x} &= -\rho a \\ \frac{\partial p}{\partial y} &= -\rho g \end{aligned}$$

The difference in pressure at points (1) and (2) is,

$$dp = \frac{\partial p}{\partial x}dx + \frac{\partial p}{\partial y}dy = 0 = (-\rho a)(-L_2) + (-\rho g)(h)$$

or
$$-\rho a L_2 - \rho g h = 0 \Rightarrow a = -g\left(\frac{h}{L_2}\right)$$

• PROBLEM 5-14

A liquid (s = 0.86) with a vapor pressure of 3.8 psia flows through the horizontal constriction in the accompanying figure. Atmospheric pressure is 26.7 in Hg. Find the maximum theoretical flow rate (i.e., at what Q does cavitation occur?). Neglect head loss. $\gamma_{water} = 62.4\ \text{lb/ft}^3$.

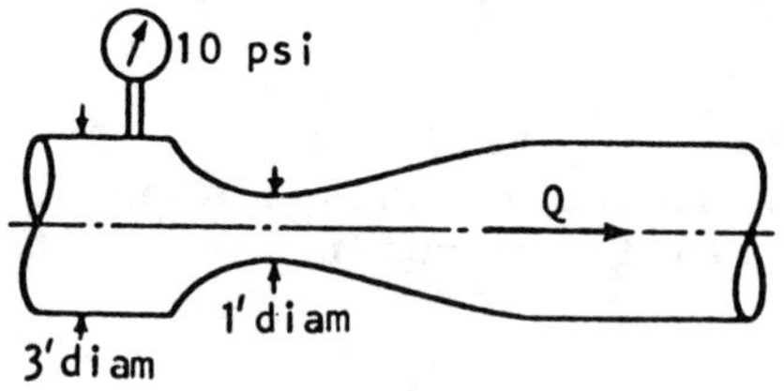

Solution: According to the Bernoulli theorem, if at any point the velocity head increases, there must be a corresponding decrease in the pressure head. For any liquid there is a minimum absolute pressure possible, namely, the vapor pressure of the liquid. The vapor pressure depends upon the liquid and its temperature. If the conditions are such that a calculation results in a lower absolute pressure than the vapor pressure, this simply means that the assumptions upon which the calculations are based no longer apply.

Expressed in equation form, the criterion with respect to cavitation is as follows:

$$\left(\frac{Pcrit}{\gamma}\right)_{abs} = \frac{Pv}{\gamma}$$

But

$$\left(\frac{Pcrit}{\gamma}\right)_{abs} = \frac{Patm}{\gamma} + \left(\frac{Pcrit}{\gamma}\right)_{gage}$$

Thus,

$$\left(\frac{Pcrit}{\gamma}\right)_{gage} = -\left(\frac{Patm}{\gamma} - \frac{Pv}{\gamma}\right)$$

where P_{atm}, P_v, and P_{crit} represent the atmospheric pressure, the vapor pressure, and the critical (or minimum) possible pressure, respectively, in liquid flow.

$$P_{atm} = 26.7 \text{ in. Hg} \left(\frac{14.7 \text{ psia}}{29.9 \text{ in. Hg}}\right) = 13.2 \text{ psia}$$

$$\left(\frac{Pcrit}{\gamma}\right)_{gage} = -\left[\frac{13.2 - 3.8}{0.86(62.4)}\right] 144 = -25.2 \text{ ft}$$

Apply the Bernoulli equation in the form

$$\frac{P_1}{\gamma} + z_1 + \frac{V_1^2}{2g} = \frac{P_2}{\gamma} + z_2 + \frac{V_2^2}{2g}$$

and with $V = Q/A$, where Q is the fluid flow rate and A is the cross sectional area of flow, and $z_1 = z_2$.

$$0 + \frac{10(144)}{0.86(62.4)} + \frac{(Q/2.25\pi)^2}{64.4} = -25.2 + \frac{(Q/0.25\pi)^2}{64.4}$$

$$Q = 45.8 \text{ cfs}$$

Water flows in a horizontal pipe of constant diameter (Figure). Assume that the water is frictionless and that the velocity is constant across and along the pipe. A tube connected to the side of the pipe bends into a U shape and returns to the inside of the pipe. The right end of the tube bends with the open end of the tube facing upstream. The U part of the tube is filled partially with mercury. The part of the tube connected to a hole in the side of the pipe at point B measures the pressure of the liquid at that point. The velocity at B is undisturbed and equals the uniform velocity of the water. The other end of the tube is a stagnation point and so it measures the stagnation pressure. The tube is filled with water except for the part filled with mercury. Compute the average velocity of the water.

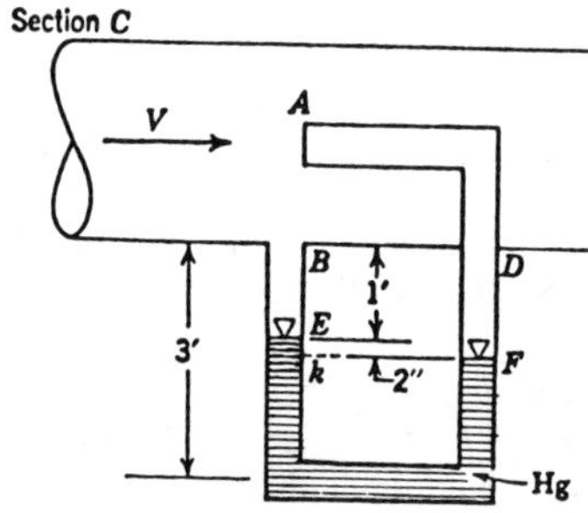

Solution: The Bernoulli equation can be applied to a pipe where the pipe is considered a stream tube. Since this pipe is a stream tube, the streamline that goes through point B and the streamline that goes through point A must come from streamlines that go past section C, which is far enough upstream to have uniform velocity and pressure. Write the Bernoulli equation between A and C along the streamline that goes through A.

$$\frac{p_A - p_C}{\rho} + \frac{V_A^2 - V_C^2}{2} = 0$$

or $$\frac{p_A}{\rho} + \frac{V_A^2}{2} = \frac{p_C}{\rho} + \frac{V_C^2}{2}$$

Do the same for the streamline through point B

$$\frac{p_B - p_C}{\rho} + \frac{V_B^2 - V_C^2}{2} = 0$$

or $$\frac{p_B}{\rho} + \frac{V_B^2}{2} = \frac{p_C}{\rho} + \frac{V_C^2}{2}$$

From this see that

$$\frac{p_A}{\rho} + \frac{V_A^2}{2} = \frac{p_C}{\rho} + \frac{V_C^2}{2} = \frac{p_B}{\rho} + \frac{V_B^2}{2}$$

$$\frac{p_A}{\rho} + \frac{V_A^2}{2} = \frac{p_B}{\rho} + \frac{V_B^2}{2}$$

Since point A is a stagnation point, $V_A = 0$ and from the above expression

$$V_B = \left[\frac{2(p_A - p_B)}{\rho}\right]^{\frac{1}{2}}$$

The velocity V_B equals the uniform velocity of the water in the pipe. Determine the difference in pressure between A and B to calculate V_B.

The pressure at point E can be computed from fluid statics because the water in the tube is stationary.

$$p_E = p_B + \gamma_{H_2O}(1)$$

The radius of the pipe is small; therefore assume that the pressure p_D at point D inside the tube equals p_A.

$$p_D = p_A$$

Compute the pressure p_F at the surface of the mercury from fluid statics.

$$p_F = p_D + \gamma_{H_2O}(1\ \frac{2}{12}) = p_A + \gamma_{H_2O}(1\ \frac{2}{12})\ .$$

The pressure in the other side of the tube at point k equals p_F, because points F and k are at the same elevation in a continuum. The pressure at k is

$$p_k = p_E + (\frac{2}{12})\gamma_{Hg}$$

but $p_k = p_F$

and so

$$p_B + (\frac{2}{12})\gamma_{Hg} = p_A + \gamma_{H_2O}(1\frac{2}{12})$$

but $p_E = p_B + \gamma_{H_2O}(1)$

Therefore,

$$p_B + \gamma_{H_2O}(1) + (\frac{2}{12})\gamma_{Hg} = p_A + \gamma_{H_2O}(1\frac{2}{12})$$

$$p_A - p_B = (1)\gamma_{H_2O} + \frac{1}{6}\gamma_{Hg} - 1\frac{1}{6}\gamma_{H_2O}$$

$$= \frac{1}{6}(\gamma_{Hg} - \gamma_{H_2O}) = 130.4 \text{ lb/ft}^2$$

Substitute this result into the equation for V_B

$$V_B = \left[\frac{2(p_A - p_B)}{\rho}\right]^{\frac{1}{2}} = \left[\frac{2(130.4)}{1.94}\right]^{\frac{1}{2}} = 11.5 \text{ fps}$$

• PROBLEM 5-16

A nozzle meter consists of a nozzle throat of circular area A_2 in a pipe of circular area A_1 and is used to measure the volumetric flow rate of a liquid of specific weight γ_f. A manometer connected as shown in the figure uses a heavier liquid of specific weight γ_m and indicates a deflection of h_m meters. What is the flow rate? Neglect viscous effects.

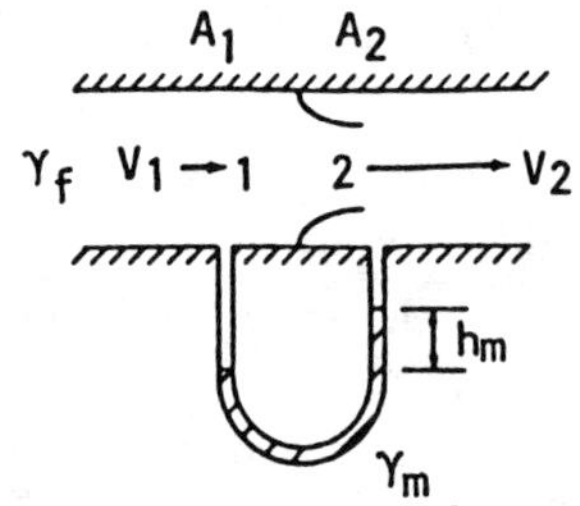

Flow through a nozzle meter.

Solution: The continuity equation and the Bernoulli equation written between sections 1 and 2 are solved simultaneously. These are

$$Q = V_1A_1 = V_2A_2$$

and

$$\frac{V_1^2}{2g} + \frac{p_1}{\gamma_f} = \frac{V_2^2}{2g} + \frac{p_2}{\gamma_f}$$

From the manometer deflection

$$\frac{p_1 - p_2}{\gamma_f} = \frac{h_m(\gamma_m - \gamma_f)}{\gamma_f}$$

Thus

$$Q = V_2 A_2 = \frac{A_2}{\sqrt{1 - (A_2/A_1)^2}} \sqrt{\frac{2gh_m(\gamma_m - \gamma_f)}{\gamma_f}}$$

The neglect of viscous effects for low-viscosity liquids gives quite accurate results in this instance.

• PROBLEM 5-17

An elbow of 30- x 30-cm cross section is made up of two circular arcs with an inner radius r_1 = 30 cm and an outer radius r_2 = 60 cm (Figure). Assume water flows through this elbow as an irrotational vortex. The pressure difference between the inner and outer walls is 20.0 kPa. What is the flow rate through the elbow?

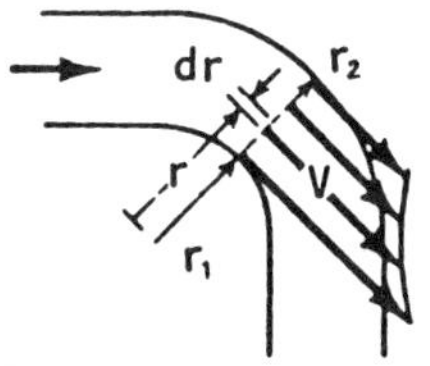

Irrotational vortex flow in a square elbow.

Solution: For an irrotational vortex, Vr = constant and the Bernoulli equation is used to find either the velocity at any radius or the value of the constant. Then we integrate across the section (since the velocity varies) to find the flow rate: $V_1 r_1 = V_2 r_2 = c$ so that $V_1 = 2V_2$:

$$\frac{\rho V_1^2}{2} + p_1 + \rho g z_1 = \frac{\rho V_2^2}{2} + p_2 + \rho g z_2 \qquad (1)$$

$$V_1 r_1 = V_2 r_2 = c \qquad (2)$$

$$p_2 - p_1 = \frac{\rho}{2}(V_1^2 - V_2^2)$$

$$= \frac{\rho}{2}\left[\left(\frac{V_2 r_2}{r_1}\right)^2 - \left(\frac{V_1 r_1}{r_2}\right)^2\right] \qquad \text{substitute (2)}$$

$$= \frac{\rho c^2}{2}\left[\frac{1}{r_1^2} - \frac{1}{r_2^2}\right]$$

$$p_2 - p_1 = \frac{\rho}{2}(V_1^2 - V_2^2) = \frac{c^2\rho}{2} \quad \frac{1}{r_1^2} - \frac{1}{r_2^2}$$

Thus,

$$c^2 = \frac{2(p_2 - p_1)}{\rho[1/r_1^2) - (1/r_2^2)]} = \frac{(2)(20{,}000)}{(1000)[1/(0.3)^2 - 1/(0.6)^2]} = 4.80$$

$$c = 2.191$$

$$Q = \int_{r_1}^{r_2} V(0.3)dr = \int_{r_1}^{r_2}\left(\frac{c}{r}\right)(0.3)dr = (2.191)(0.3)\ln\left(\frac{r_2}{r_1}\right)$$

$$= (2.191)(0.3)(0.693) = 0.456 \text{ m}^3/\text{s}$$

• PROBLEM 5-18

Air flows steadily and at a low speed through the horizontal nozzle shown below. At the nozzle inlet, the area is 0.1m^2. At the nozzle exit, the area is 0.02m^2,

(a) determine the pressure difference between points 1 and 2 if the inlet speed of the air is 10 m/sec.

(b) determine the gage pressure required at the nozzle inlet to produce an outlet speed of 50 m/sec.

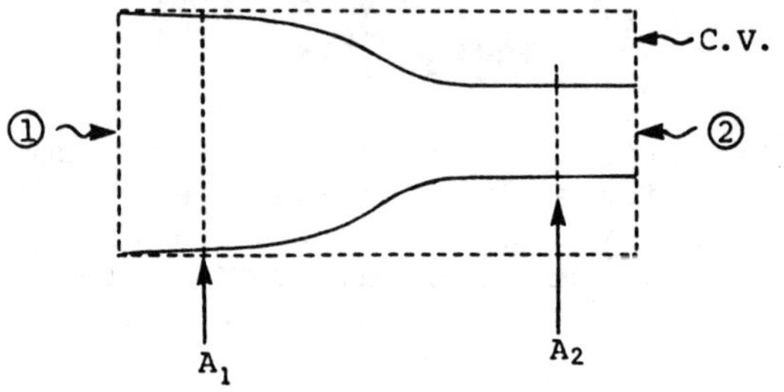

Solution: (a) Assuming a steady incompressible and frictionless flow, the continuity equation becomes,

$$0 = \frac{\partial}{\partial t} \int_{CV} \rho dV + \int_{CS} \rho \vec{V} d\vec{A} \Rightarrow \rho_1 V_1 A_1 = \rho_2 V_2 A_2$$

But $\rho_1 = \rho_2$, therefore

$$V_1 A_1 = V_2 A_2 = Q, \qquad (1)$$

where Q is the volume flow rate.

The Bernoulli's equation between points (1) and (2) is given by

$$\frac{p_1}{\rho} + \frac{V_1^2}{2} + gz_1 = \frac{p_2}{\rho} + \frac{V_2^2}{2} + gz_2$$

$$\Rightarrow \quad p_1 - p_2 = \frac{\rho}{2} (V_2^2 - V_1^2) \qquad (2)$$

By combining (1) and (2), we have

$$p_1 - p_2 = \rho \frac{V_1^2}{2} \left[1 - \left(\frac{A_2}{A_1} \right)^2 \right] \qquad (3)$$

Making numerical substitutions, we have

$$p_1 - p_2 = (1.23) \frac{(10)^2}{2} \left[1 - \left(\frac{0.02}{0.1} \right)^2 \right]$$

or

$$p_1 - p_2 = 59.04 \text{ kPa}$$

(b) From equation (1), we have

$$V_1 A_1 = V_2 A_2 \Rightarrow V_1 = V_2 \frac{A_2}{A_1} = 50 \times \frac{0.02}{0.1} \Rightarrow V_1 = 10 \text{ m/sec}$$

From equation (2), we have

$$p_1 - p_{atm} = \frac{\rho}{2} (V^2 - V_1^2) = \frac{(1.23)}{2} \left[50^2 - 10^2 \right] \Rightarrow$$

$$p_1 - p_{atm} = 1.48 \text{ kPa}$$

• PROBLEM 5-19

Calculate the flowrate of gasoline (r.d. = 0.82) through this pipeline, using first the gage readings and then the manometer reading.

Solution: With sizes of pipe and constriction given, the problem centers around the computation of the value of the quantity $(p_1/\gamma + z_1 - p_2/\gamma - z_2)$ of the Bernoulli equation.

Taking datum at the lower gage and using the gage readings,

$$\frac{p_1}{\gamma} + z_1 - \frac{p_2}{\gamma} - z_2 = \frac{138.0 \times 10^3}{0.82 \times 9.8 \times 10^3} - \frac{69.0 \times 10^3}{0.82 \times 9.8 \times 10^3} - 1.2$$

$$= 7.39 \text{ m of gasoline}$$

To use the manometer reading, construct the hydraulic grade line levels at 1 and 2. It is evident at once that the difference in these levels is the quantity ($p_1/\gamma + z_1 - p_2/\gamma - z_2$) and, visualizing p_1/γ and p_2/γ as liquid columns, it is also apparent that the difference in the hydraulic grade line levels is equivalent to the manometer reading. Therefore,

$$\left(\frac{p_1}{\gamma} + z_1 - \frac{p_2}{\gamma} - z_2\right) = 0.475\left(\frac{13.57 - 0.82}{0.82}\right) = 7.39 \text{ m of gasoline}$$

Insertion of 7.39 m into the Bernoulli equation followed by its simultaneous solution with the continuity equation yields a flow-rate Q of 0.22 m^3/s.

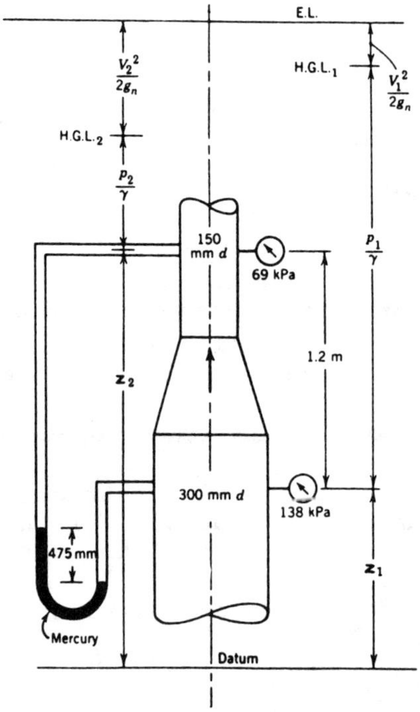

• PROBLEM 5-20

A liquid, of density 60 lb/ft^3, is conveyed up a conical tube at the rate of 3.14 ft^3/sec. The tube is 6 in. diameter at the lower end and 12 in. at the upper end, and the two ends are separated by a vertical distance of 10 ft. If the pressure at the lower end is 50 psia, what is the pressure at the upper end? Assume the liquid to behave like an ideal fluid.

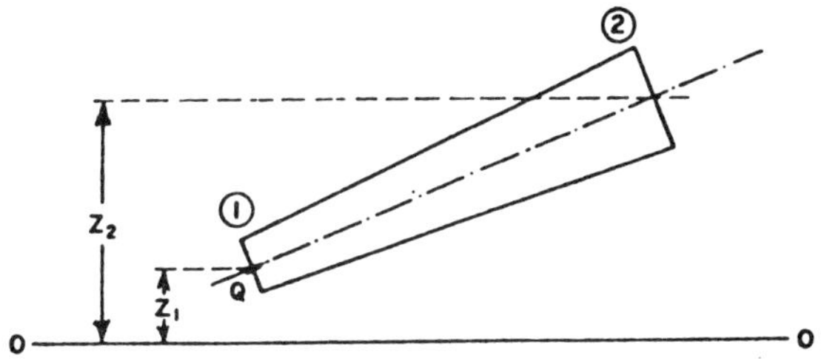

Solution: Applying the law of continuity to the end sections

$$Q = v_1A_1 = v_2A_2 = 3.14 \text{ cfs}$$

$$A_1 = (6/12)^2(\pi/4) = \pi/16$$

$$A_2 = (12/12)^2(\pi/4) = \pi/4$$

$$v_1 = \frac{(3.14)(16)}{\pi} = 16 \text{ fps}$$

$$v_2 = \frac{(3.14)(4)}{\pi} = 4 \text{ fps}$$

Bernoulli's theorem, when applied between the two sections leads to

$$\frac{p_1}{\gamma} + \frac{v_1^2}{2g} + Z_1 = \frac{p_2}{\gamma} + \frac{v_2^2}{2g} + Z_2$$

from which

$$\frac{p_2 - p_1}{\gamma} = \frac{v_1^2 - v_2^2}{2g} + (Z_1 - Z_2)$$

Substituting in this equation the data, and noting that

$$p_2 - p_1 = (p_2 - 50)144 \text{ psf}$$

$$\gamma = 60 \frac{\text{lb-f} = \text{lb-wt}}{\text{ft}^3}$$

$$2g = (2)(32.2) = 64.4 \text{ ft/sec}^2$$

$$Z_1 - Z_2 = -(Z_2 - Z_1) = -10 \text{ ft}$$

$$\frac{(p_2 - 50)144}{60} = \frac{16^2 - 4^2}{64.4} - 10 = -\frac{101}{16.1}$$

$$p_2 - 50 = -\frac{(101)(60)}{(16.1)(144)} = -2.614,$$

say -2.6

$$p_2 = 47.4 \text{ psia}$$

• **PROBLEM** 5-21

Calculate the flowrate through this pipeline and nozzle. Calculate the pressures at points 1, 2, 3, and 4 in the pipe and the elevation of the top of the jet's trajectory.

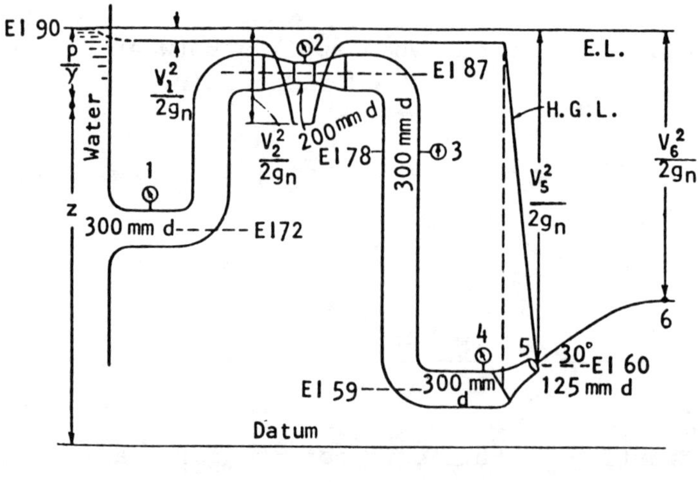

Solution: First sketch the energy line; it is evident that at all points in the reservoir where the velocity is negligible, $(z + p/\gamma)$ will be the same. Thus the energy line has the same elevation as the water surface. Next sketch the hydraulic grade line; this is coincident with the energy line in the reservoir where velocity is negligible but drops below the energy line over the pipe entrance where velocity is gained. The velocity in the 300 mm pipe is everywhere the same, so the hydraulic grade line must be horizontal until the flow encounters the constriction upstream from section 2. Here, as velocity increases, the hydraulic grade line must fall (possibly to a level below the constriction). Downstream from the constriction the hydraulic grade line must rise to the original level over the 300 mm pipe and continue at this level to a point over the base of the nozzle at section 4. Over the nozzle the hydraulic grade line must fall to the nozzle tip and after that follow the jet, because the pressure in the jet is everywhere zero.

Since the vertical distance between the energy line and the hydraulic grade line at any point is the velocity head at that point, it is evident that

$$\frac{V_5^2}{2g_n} = 30 \text{ and therefore } V_5 = 24.3 \text{ m/s}$$

Thus

$$Q = AV$$

$$Q = 24.3 \times \frac{\pi}{4}(0.125)^2 = 0.3 \text{ m}^3/\text{s}$$

From continuity considerations,

$$V_1 = \left(\frac{125}{300}\right)^2 V_5$$

and thus

$$\frac{V_1^2}{2g_n} = \left(\frac{125}{300}\right)^4 \frac{V_5^2}{2g_n} = \left(\frac{125}{300}\right)^4 30 = 0.9\text{m}$$

Similarly

$$\frac{V_2^2}{2g_n} = \left(\frac{125}{200}\right)^4 30 = 4.58 \text{ m}$$

Since the pressure heads are conventionally taken as the vertical distances between the pipe center line and hydraulic grade line, the pressures in pipe and constriction may be computed as follows:

$p_1/\gamma = 18 - 0.9 = 17.1 \text{ m} \quad p_1 = 167.6\text{kPa}$

$p_2/\gamma = 3 - 4.58 = -1.58\text{m} \quad p_2 = 116 \text{ mm Hg vacuum}$

$p_3/\gamma = 12 - 0.9 = 11.1 \text{ m} \quad p_3 = 108.8\text{kPa}$

$p_4/\gamma = 31 - 0.9 = 30.1 \text{ m} \quad p_4 = 295.0\text{kPa}$

The velocity of the jet at the top of its trajectory (where there is no vertical component of velocity) is given by

$V_6 = 24.3 \cos 30° = 21.0 \text{ m/s}$

and the elevation here is $90 - (21.0)^2/2g_n = 67.5$ m.

With increasing velocity or potential head, the pressure within a flowing fluid drops. However, this pressure does not drop below the absolute zero of pressure as it has been found experimentally that fluids will not sustain the tension implied by pressures less than absolute zero. Thus, a practical physical restriction is placed upon the Bernoulli equation. Such a restriction is, in fact, not appropriate for gases because they expand with reduction in pressure, but it frequently assumes great importance in the flow of liquids. Actually, in liquids the absolute pressure can drop only to the vapor pressure of the liquid, whereupon spontaneous vaporization (boiling) takes place. This vaporization or formation of vapor cavities is called cavitation. The formation, translation with the fluid motion, and subsequent rapid collapse of these cavities produces vibration, destructive pitting, and other deleterious effects on hydraulic machinery, hydrofoils, and ship propellers.

• **PROBLEM 5-22**

The barometric (absolute) pressure is 96.5 kPa. What diameter of constriction can be expected to produce incipient cavitation at the throat of the constriction?

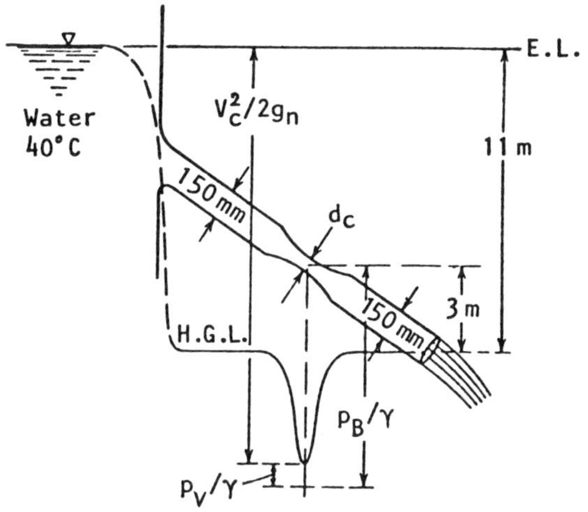

Solution: From a table of physical properties of water, find γ and p_v, so

$$\gamma = 9.73 \text{ kN/m}^3 \text{ and } p_v/\gamma = 0.76 \text{ m.}$$

$$\frac{p_B}{\gamma} = \frac{96.5 \times 10^3}{9.73 \times 10^3} = 9.92 \text{ m}$$

Construct E.L. and H.G.L. as indicated on the sketch. Evidently the velocity head at the constriction is given by

$$\frac{V_c^2}{2g_n} = 11.0 - 3.0 + 9.92 - 0.76 = 17.2 \text{ m}$$

However,

$$\frac{V_{150}^2}{2g_n} = 11.0 \text{ m}$$

Since, from continuity considerations,

$$\left(\frac{d_c}{150}\right)^2 = \frac{V_{150}}{V_c}, \quad \left(\frac{d_c}{150}\right)^4 = \left(\frac{V_{150}}{V_c}\right)^2 = \frac{11.0}{17.2}$$

yielding d_c = 134 mm. Incipient cavitation must be assumed in such ideal fluid flow problems for losses of head to be negligible; also with small cavitation there is more likelihood of the pipe flowing full at its exit. For a larger cavitation zone the same result will be obtained if losses are neglected and pipes assumed full at exit; however, the low point of the H.G.L. will be considerably flattened.

• PROBLEM 5-23

Both tanks of Fig. 1 have a liquid depth z. Tank (a) discharges a jet of diameter D through the rounded orifice from whence it falls freely, whereas tank (b) has a pipe with rounded entrance connected to it, and after a length L, it discharges a jet also of diameter D.

(A) Which of the systems has the greater discharge through it? In answering this question first compare the velocities and discharges at points 1 and 2 in system (a) with the corresponding points in system (b).

(B) If the depth z in the tank of Fig. 1b is 5 m, determine the maximum length L of connecting pipe which may be used without the occurrence of cavitation if the liquid is water at (a) 20°C; (b) 70°C. What is the discharge in each case if the pipe diameter is 20 cm?

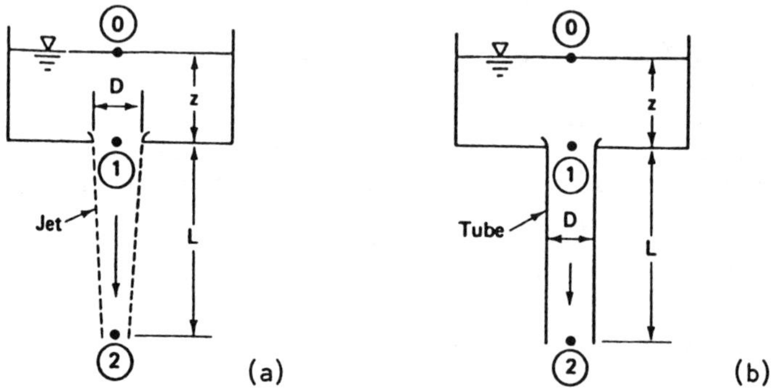

Fig. 1

Solution: (A) In solving problems involving Bernoulli's equation along a streamline, a suitable streamline must be justified. In this problem it is easy to accept that a streamline can be found in both tanks that passes from the free surface through points 1 and 2. The elevation of the points 2 is taken as the datum. We will label the surface points as 0, where, for a sufficiently large tank, it is possible to ignore the velocity and velocity head. The Bernoulli equation will be applied from 0 to 2 (i.e., from point 0 to point 2) in both tanks: In (a) the total head at the surface relative to the datum through point 2 is

$$H_0 = \frac{p_0}{\gamma} + y_0 + \frac{v_0^2}{2g} = 0 + (z + L) + 0$$

Temperature (°C)	Specific Weight, γ (N/m^3)	Density, ρ (kg/m^3)	Dynamic Viscosity, $\mu \times 10^3$ ($N \cdot s/m^2$)	Kinematic Viscosity, $\nu \times 10^6$ (m^2/s)	Surface Tension, $\sigma \times 10^2$ (N/m)	Vapor Pressure, p_v (kN/m^2)	Modulus of Compressibility, $E \times 10^{-9}$ (N/m^2)
0	9805	999.8	1.794	1.794	7.62	0.61	2.02
5	9806	1000.0	1.519	1.519	7.54	0.87	2.06
10	9802	999.7	1.308	1.308	7.48	1.23	2.11
15	9797	999.1	1.140	1.141	7.41	1.70	2.14
20	9786	998.2	1.005	1.007	7.36	2.34	2.20
25	9777	997.1	0.894	0.897	7.26	3.17	2.22
30	9762	995.7	0.801	0.804	7.18	4.24	2.23
35	9747	994.1	0.723	0.727	7.10	5.61	2.24
40	9730	992.2	0.656	0.661	7.01	7.38	2.27
45	9711	990.2	0.599	0.605	6.92	9.55	2.29
50	9689	988.1	0.549	0.556	6.82	12.33	2.30
55	9665	985.7	0.506	0.513	6.74	15.78	2.31
60	9642	983.2	0.469	0.477	6.68	19.92	2.28
65	9616	980.6	0.436	0.444	6.58	25.02	2.26
70	9588	977.8	0.406	0.415	6.50	31.16	2.25
75	9560	974.9	0.380	0.390	6.40	38.57	2.23
80	9528	971.8	0.357	0.367	6.30	47.34	2.21
85	9497	968.6	0.336	0.347	6.20	57.83	2.17
90	9473	965.3	0.317	0.328	6.12	70.10	2.16
95	9431	961.9	0.299	0.311	6.02	84.36	2.11
100	9398	958.4	0.284	0.296	5.94	101.33	2.07

This follows since the pressure at the free surface is atmospheric, the height of point 0 is (z + L) above the datum and, as previously mentioned, $v_0 = 0$. Similarly, the total head at point 2 is

$$H_2 = \frac{p_2}{\gamma} + y_2 + \frac{v_2^2}{2g} = 0 + 0 + \frac{v_2^2}{2g}$$

This time the pressure head is zero, since the jet is discharging into the atmosphere, which is at zero pressure, and the elevation head is zero since point 2 is on the datum. Thus, the Bernoulli equation gives

$$0 + (z + L) + 0 = 0 + 0 + \frac{v_2^2}{2g}$$

or

$$v_2 = \sqrt{2g(z + L)} \quad \text{for (a)}$$

Similarly, in tank (b), also from 0 to 2,

$$0 + (z + L) + 0 = 0 + 0 + \frac{v_2^2}{2g}$$

$$v_2 = \sqrt{2g(z + L)} \quad \text{for (b)}$$

The velocity v_2 is seen to be identical in each case. Now writing the equation between 0 and 1 for tank (a) only,

$$0 + (z + L) + 0 = 0 + L + \frac{v_1^2}{2g}$$

and

$$v_1 = \sqrt{2gz} \quad \text{for (a)}$$

The velocity in tank (b) at point 1 will not be the same, since the pressure will not equal zero as it does in tank (a). Rather, note by continuity that

$$v_1 = v_2 \quad \text{in (b)}$$

Summing up, the following information is obtained:

		Tank (a)	Tank (b)
Velocity	Point 1	$\sqrt{2gz}$	$\sqrt{2g(z + L)}$
	Point 2	$\sqrt{2g(z + L)}$	$\sqrt{2g(z + L)}$
Discharge (= AV)		$\sqrt{2gz}\frac{\pi}{4}D^2$	$\sqrt{2g(z + L)}\frac{\pi}{4}D^2$

In conclusion, it is apparent that the discharge will be greater through (b).

(B) It will be assumed that the inlet from the tank to the pipe is so designed that the local velocity at any point along the curvature does not exceed the average in the pipe. Further, an atmospheric pressure of 1.013×10^5 N/m^2 will be used. Then from the Table, $p_v = 2340$ N/m^2 (abs) at 20°C and 31,160 N/m^2 at 70°C. Since the velocities are identical at sections 1 and 2 of Fig. 1b, the Bernoulli equation for any streamline between the two sections is

$$\frac{p_1}{\gamma} + L = \frac{p_2}{\gamma} + 0$$

if the datum is taken at the outlet. Noting that $p_1 = p_v$, $p_2 = p_{atmos}$ and using $\gamma_{20°C} = 9790$ N/m^3 and $\gamma_{70°C} = 9590$ N/m^3,

$$L = \frac{p_2 - p_v}{\gamma}$$

$$= \frac{101,300 - 2340}{9790} = 10.11 \text{ m at } 20°\text{C}$$

while at 70°C,

$$L = \frac{101,300 - 31,160}{9590} = 7.31 \text{ m}$$

From (a), the discharge is

$$Q = \sqrt{2g(z + L)}\frac{\pi}{4}D^2$$

Therefore, the flow rates associated with the maximum non-cavitating pipe lengths are at 20°C,

$$Q = \sqrt{(2)(9.81)(5 + 10.11)}\frac{\pi}{4}(0.2)^2 = 0.541 \text{ m}^3/\text{s}$$

and at 70°C,

$$Q = \sqrt{(2)(9.81)(5 + 7.31)}\frac{\pi}{4}(0.2)^2 = 0.488 \text{ m}^3/\text{s}$$

• PROBLEM 5-24

A cylindrical vessel of diameter D = 50mm drains through an orifice of diameter D_o = 5mm at the bottom of the vessel. The pressure exerted on the liquid in the vessel is maintained at the atmospheric pressure p_a.

(a) Calculate the velocity of the efflux V_2.

(b) If the vessel is initially filled with water to a depth y_o = 0.4m, determine the water depth y, at time t = 12 sec.

(c) Determine the time required to drain the vessel completely.

Solution: (a) Assuming incompressible uniform flow and applying Bernoulli's equation for section (1) and (2), we have

$$\frac{V_1^2}{2} + \frac{p_{at}}{\rho_{H_2O}} + gy = \frac{V_2^2}{2} + \frac{p_{at}}{\rho_{H_2O}}$$

$$\Rightarrow \quad V_1^2 - V_2^2 = -2gy \quad \Rightarrow \quad V_2^2 - V_1^2 = 2gy$$

$$\Rightarrow \quad V_2^2 = V_1^2 + 2gy \tag{1}$$

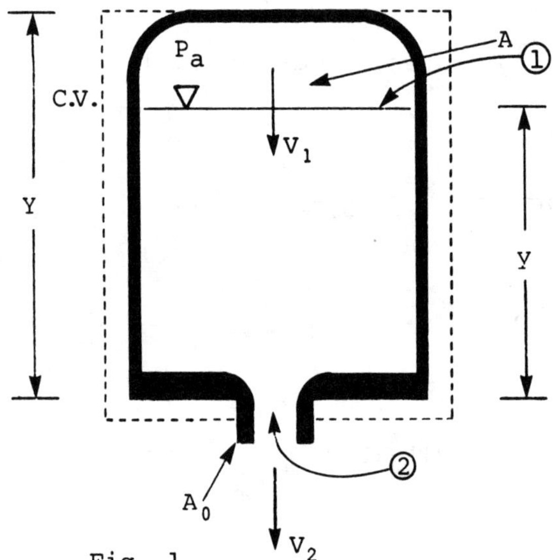

Fig. 1

Equation (1) may be approximated by,

$$V_2^2 = 2gy \quad \Rightarrow \quad V_2 = \sqrt{2gy} \tag{2}$$

(b) Using the C.V. shown in Fig. 1 and applying the law of conservation of mass, we have

$$\frac{\partial}{\partial t}\int_{CV} \rho_{H_2O} dV + \int_{CS} \rho_{H_2O} \overline{V} d\overline{A} = 0$$

$$\Rightarrow \quad \frac{\partial}{\partial t}\int_0^y \rho_{H_2O} A dy + \rho_{H_2O} V_2 A_o = 0$$

$$\Rightarrow \quad \rho A \frac{dy}{dt} + \rho A_o \sqrt{2gy} = 0$$

$$\Rightarrow \quad \frac{dy}{y^{\frac{1}{2}}} = -\sqrt{2g}\,\frac{A_o}{A}\int_0^t dt \qquad (3)$$

$$\Rightarrow \quad \int_{y_o}^{y} \frac{dy}{y^{\frac{1}{2}}} = 2\left[y^{\frac{1}{2}} - y_o^{\frac{1}{2}}\right] = -\sqrt{2g}\,\frac{A_o}{A}\,t$$

$$\Rightarrow \quad y = y_o\left[1 - \sqrt{\frac{g}{2y_o}}\,\frac{A_o}{A}\,t\right] \Rightarrow \quad y = y_o\left[1 - \sqrt{\frac{g}{2y_o}}\left(\frac{D_o}{D}\right)^2 t\right]$$

At t = 12 sec,

$$y = 0.4\left[1 - \sqrt{\frac{1}{2}\,\frac{(9.81)}{(0.4)}}\left(\frac{5}{50}\right)^2(12)\right]^2 \Rightarrow \quad y = 0.134\text{m}$$

(c) Integrating equation (3) from y_o at t = 0 to 0 at t, we have

$$\int_{y_o}^{0} \frac{dy}{y^{\frac{1}{2}}} = 2\left[0 - y_o^{\frac{1}{2}}\right] = -\sqrt{2g}\,\frac{A}{A_o}\,t = -\sqrt{2g}\,\frac{D^2}{D_o^2}\,t$$

$$\therefore \quad t = \frac{2}{\sqrt{2g}}\sqrt{y_o}\left(\frac{D_o}{D}\right)^2$$

$$= \sqrt{\frac{2y_o}{g}} \left(\frac{D_o}{D}\right)^2$$

or

$$t = \sqrt{\frac{2 \times 0.4}{9.81}} \left(\frac{50}{5}\right)^2 = 28.6 \text{ sec}$$

• PROBLEM 5-25

> What is the minimum efflux velocity required by a fire stream in order to reach a window at a horizontal distance of 40 ft and a vertical distance of 50 ft from the nozzle if the vertex of the trajectory coincides with the window?

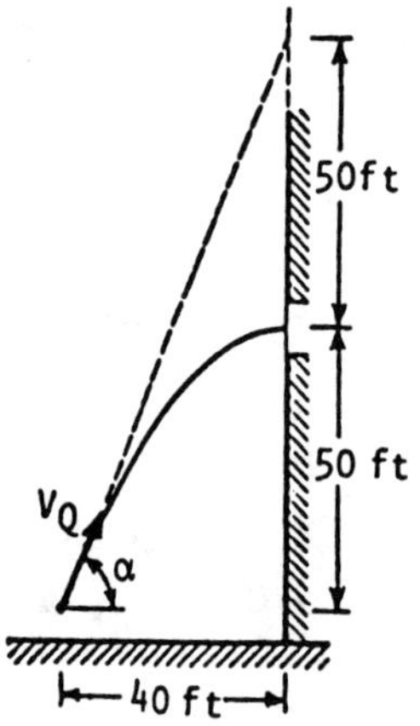

Fig. 1

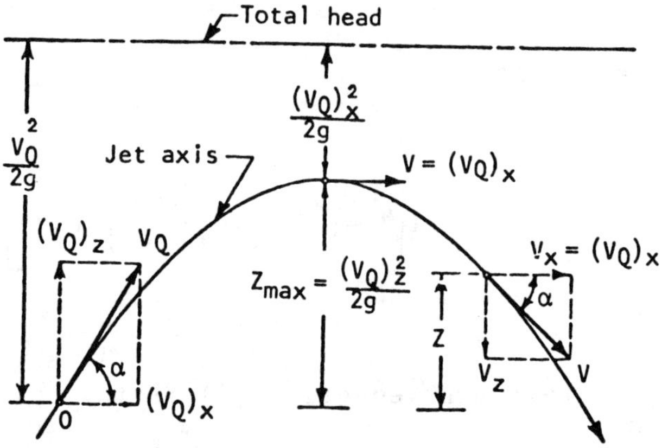

Kinematics of a liquid jet.

Fig. 2

Solution: From the characteristics of a parabola (Fig. 2) $-z = x^2$ and $dz/dx = -2x$; in other words, the nozzle must be aimed at a point twice as high as the window. Then

$$z_{max} = 50 - \frac{(V_0)_z^2}{2g}$$

or

$$(V_0)_z = \sqrt{64.4 \times 50} = 56.8 \text{ fps}$$

and

$$V_0 = (V_0)_z \csc\alpha = 56.8 \frac{\sqrt{100^2 + 40^2}}{100^2} = 61.1 \text{ fps}$$

• PROBLEM 5-26

A tank (figure) is partially filled with water which drains out a 1 in. diameter hole in the side. The x axis is horizontal through the center of the hole and perpendicular to the tank's centerline which is the y axis positive upward. The inside surface of the tank is a surface of revolution. The y axis is the axis of revolution. The radius r of the inside surface is a function of the distance y above the hole. This function is

$$y = r^2 - 1$$

How long does it take the water level in the tank to fall from 6 ft above the hole to 4 ft above the hole?

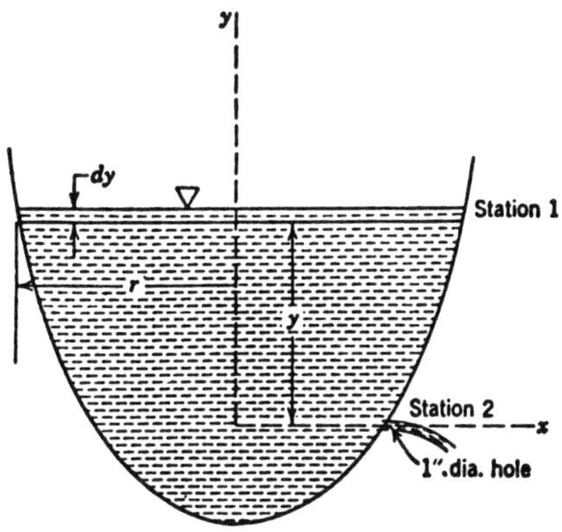

Solution: As the water drains out the hole in the tank, the water level in the tank drops, which changes the velocity of the water as it leaves the tank. This makes the flow unsteady, and the $\partial V/\partial t$ term is not zero. Assume that, although the $\partial V/\partial t$ term is not zero, you can neglect it. The accuracy of this assumption will be checked later. With this term neglected you can use the Bernoulli equation.

When the water surface (station 1) is a distance y above the hole, you can compute the velocity of the water at the vena contracta (station 2). Writing the Bernoulli equation between these two stations gives

$$\frac{V_2^2 - V_1^2}{2} + \frac{p_2 - p_1}{\rho} + g(z_2 - z_1) = 0$$

The pressure at the water surface and at the vena contracta is atmospheric pressure p_a. The velocity V_1 of the water is small compared with V_2 and may be neglected. Use the xz plane as the datum plane, which reduces the Bernoulli equation to

$$\frac{V_2^2}{2} + g(-y) = 0$$

$$V_2 = \sqrt{2gy}$$

In time dt the level of the water surface will fall a distance dy. Also in time dt an amount of water equal to Q dt will have flowed out of the hole. This amount of water Q dt must equal the change in volume in the tank. The change in volume of the tank is the cross-sectional area of the tank at the water surface times the distance the water drops.

$$Q\,dt = \pi r^2(-dy) = -\pi r^2\,dy \qquad (1)$$

The term dy is the negative because the surface drops.

The flow rate Q is the velocity multiplied by the cross-sectional area. The effective area is three-fifths of the area of the hole.

$$Q = V_2\,\frac{\pi\left(\frac{1}{12}\right)^2}{4}\;\frac{3}{5} = \left(\frac{3}{5}\right)\sqrt{2gy}\;\frac{\pi}{576}$$

Substitute this into Equation (1).

$$Q\,dt = \frac{3}{5}\sqrt{2gy}\,\frac{\pi}{576}\,dt = -\pi r^2\,dy$$

Given

$$y = r^2 - 1$$

or

$$r = \sqrt{y + 1}$$

Therefore,

$$\frac{\pi}{576}\left(\frac{3}{5}\right)\sqrt{2g}\,\sqrt{y}\,dt = -\pi\,(\sqrt{y + 1})^2\,dy$$

$$\int_0^t dt = -\frac{5}{3}\,\frac{576}{\sqrt{2g}}\int_6^4 \frac{y + 1}{\sqrt{y}}\,dy$$

$$t = 542 \text{ sec}$$

To check and get an idea of the error introduced by omitting the $\partial V/\partial t$ term, make some rough approximation. The true acceleration dV/dt is

$$\frac{dV}{dt} = V\,\frac{\partial V}{\partial s} + \frac{\partial V}{\partial t}$$

Change this to a finite form

$$\frac{\Delta V}{\Delta t} = V\left(\frac{\Delta V}{\Delta s}\right)_{t=\text{constant}} + \left(\frac{\Delta V}{\Delta t}\right)_{s=\text{constant}}$$

$$\left(\frac{\Delta V}{\Delta t}\right)_{s=\text{constant}} = \frac{V_{y=6} - V_{y=4}}{\Delta t}$$

$$= \frac{\sqrt{2g6} - \sqrt{2g4}}{542} = \frac{(8.08)\,(0.453)}{542} = \frac{1}{100}$$

From the water surface to the vena contracta there

is a change in velocity equal to the velocity of the water at the vena contracta. Assume the distance Δs traveled is 2y. The 2 is introduced as a factor to put the result on the conservative side. Thus,

$$V \frac{\Delta V}{\Delta s} = \sqrt{2gy}\ \frac{\sqrt{2gy} - 0}{2y} = g$$

Therefore,

$$\frac{\Delta V}{\Delta t} = g + \frac{1}{100} = 32.2 + \frac{1}{100}$$

The $\frac{1}{100}$ is obviously too small to be considered. This is certainly a rough estimate, but it does give some idea of the order of magnitude of the error. Even if this calculation of error is off by a factor of 100, the error is only 3 per cent and still can be neglected.

• PROBLEM 5-27

If air (ρ = 0.0030 slugs/ft^3) flows through the two-dimensional passage of Fig. 1, what is the difference in pressure between points P_5 and P_{10} when the flow rate is 300 cfs per foot of width? If water flows through the passage, what is the difference in pressure between the same two points? Assume the half-size of the passage at the reference section is 6 ft and the downstream size is 3 ft. The view shown in Fig. 1 is an elevation view.

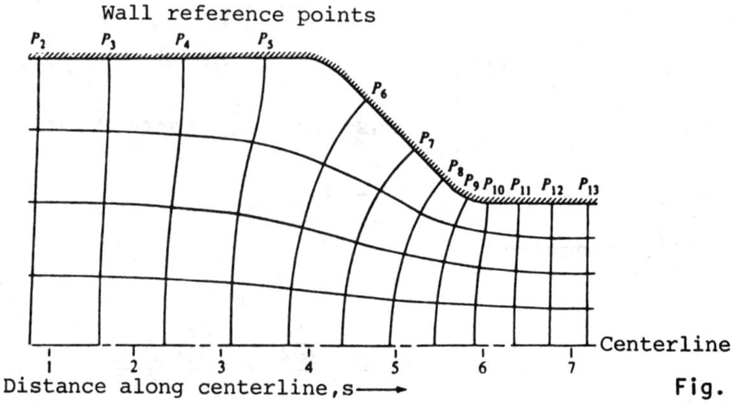

Fig. 1

Flow net for transition (half-section)

Solution: To solve for the pressure difference, we use Fig. 2 which shows that at point P_5

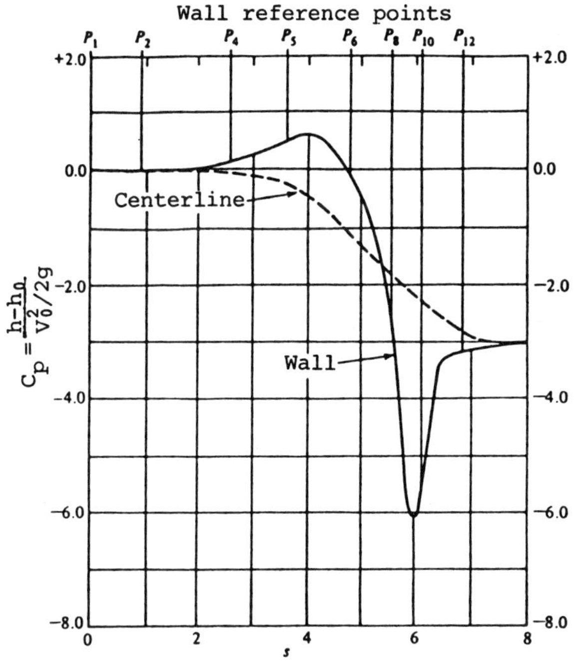

Fig. 2

Relative piezometric head along the wall and centerline of transition in figure 1

$$\frac{h_5 - h_0}{V_0{}^2/2g} = 0.5$$

or

$$h_5 - h_0 = 0.5 \frac{V_0{}^2}{2g}$$

while at point P_{10}

$$h_{10} - h_0 = -5.6 \frac{V_0{}^2}{2g}$$

Then

$$h_5 - h_{10} = \frac{V_0{}^2}{2g} [0.5 - (-5.6)]$$

$$= 6.1 \frac{V_0{}^2}{2g}$$

But

$$h = \frac{p}{\gamma} + z$$

so

$$\frac{p_5}{\gamma} + z_5 - \frac{p_{10}}{\gamma} - z_{10} = 6.1 \frac{V_0^2}{2g}$$

$$p_5 - p_{10} = \gamma \left[(z_{10} - z_5) + 6.1 \frac{V_0^2}{2g} \right]$$

For air flow.

$$p_5 - p_{10} = 0.0030g \left[(3-6) + 6.1 \frac{V_0^2}{64.4} \right]$$

But

$$V_0 = \frac{300}{12} = 25 \text{ ft/sec}$$

hence

$$p_5 - p_{10} = 0.097(-3.0 + 59.2) = 5.45 \text{ psf}$$

answer

For water flow,

$$p_5 - p_{10} = 62.4 \left[(+3 - 6) + \frac{6.1}{64.4} (25)^2 \right]$$

$$p_5 - p_{10} = 62.4(-3 + 59.2) = 3{,}507 \text{ psf}$$ answer

$$= 24.4 \text{ psi}$$ answer

• PROBLEM 5-28

A hydraulic turbine operates from a water supply with a 200-ft head above the turbine inlet, as shown in the figure below. It discharges the water to atmosphere through a 12-in.-diameter duct, with a velocity of 45 fps. Calculate the horsepower output of the turbine.

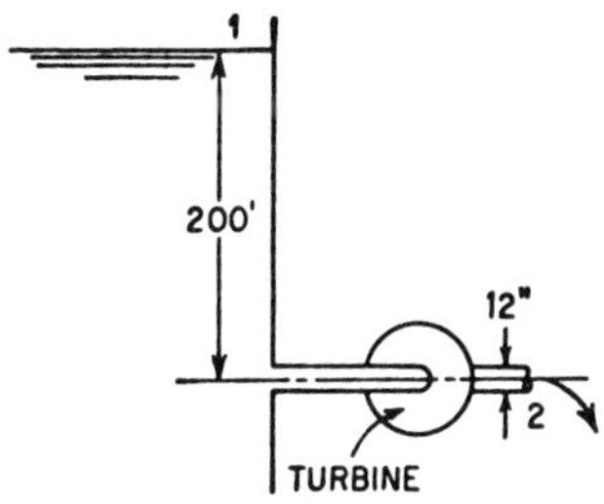

Solution: If E is the energy extracted per pound of fluid flowing, then the energy equation may be written as

$$\frac{p_1}{\gamma} + \frac{V_1^2}{2g} + z_1 = E + \frac{p_2}{\gamma} + \frac{V_2^2}{2g} + z_2$$

where suffix 1 refers to a point upstream of the turbine and suffix 2 to a point downstream of the turbine.

If the exit from the turbine is defined as the potential datum, then $z_2 = 0$ and

$$\frac{p_1}{\gamma} + \frac{V_1^2}{2g} + z_1 = 200 \text{ ft-lb/lb}$$

Therefore

$$200 = E + \frac{p_2}{\gamma} + \frac{V_2^2}{2g}$$

Now since the discharge is to atmosphere, $p_2 = 0$. Hence

$$200 = E + \frac{45^2}{2g}$$

Thus

$$E = 200 - 31.3 = 168.7 \text{ ft-lb/lb}$$

The rate of fluid flow is given by $Q = AV$ cfs, and the weight of fluid flowing by $Q\gamma$ lb/sec. Therefore the work done on the turbine is $EQ\gamma$ ft-lb/sec or

$$\frac{EQ\gamma}{550} \text{ hp} = \frac{168.7 \times \pi \times 45 \times 62.4}{4 \times 550}$$

$$= 675 \text{ hp}$$

• PROBLEM 5-29

Air enters a gas turbine with an average velocity of 450 ft/sec and a temperature of 1600°F, and leaves with a velocity of 750 ft/sec and a temperature of 1200°F. The mass flow rate through the turbine is 30 lbm/sec. Disregarding heat losses from the fluid and change in elevations, estimate the energy transfer from the air to the turbine rotor in horsepower. Assume C_p of air = .24 BTU/lbm°F.

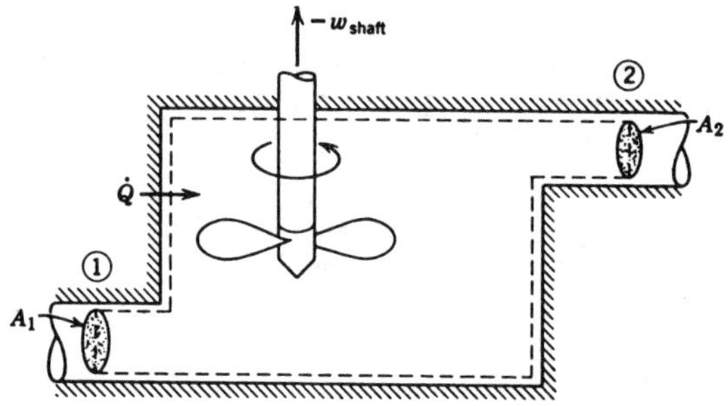

Solution: Suppose that a flow passes through a device having solid boundaries such as that shown in the figure. The flow is assumed to be uniform at points 1 and 2. For a flow and control volume defined in this manner, the energy equation is given by

$$\dot{Q} = \dot{W}_{shaft} + \left(h_2 + \frac{V_2^2}{2} + gz_2\right)\rho_2 V_2 A_2$$

$$-\left(h_1 + \frac{V_1^2}{2} + gz_1\right)\rho_1 V_1 A_1 \qquad (1)$$

The corresponding form for the continuity equation is given by

$$\rho_1 V_1 A_1 = \rho_2 V_2 A_2 = \dot{m}$$

Dividing equation 1 by $\dot{m}$ gives

$$q = w_{shaft} + (h_2 - h_1) + \frac{V_2^2 - V_1^2}{2} + g(z_2 - z_1) \qquad (2)$$

or, with $\dot{Q} = 0$ and $z_2 - z_1 = 0$, Eq. 1 may be rearranged to give the equation

$$\dot{W}_{shaft} = \left[(h_1 - h_2) + \frac{V_1^2 - V_2^2}{2}\right]\dot{m}$$

Employing the ideal gas relation, we have

$$C_{po}(T_1 - T_2) = h_1 - h_2$$

and taking $C_{po} = 0.24$ Btu/lbm°F, we have

$$\dot{W}_{shaft} = \left[\frac{0.24 \text{ Btu}}{\text{lbm°F}} \times (1600 - 1200)\text{°F} + \frac{(450)^2 - (750)^2}{2}\right]$$

$$\times 30 \frac{\text{lbm}}{\text{sec}}$$

$$= 0.24 \times 400 \frac{\text{Btu}}{\text{lbm}} \times 30 \frac{\text{lbm}}{\text{sec}} \times \frac{1.41 \text{ hp}}{1 \text{ Btu/sec}} -$$

$$5.4 \times 10^6 \frac{\text{ft}^2\text{-lbm}}{\text{sec}^3}$$

$$\times \frac{1 \text{ lbf-sec}^2/\text{ft}}{32.174 \text{ lbm}} \times \frac{1 \text{ hp}}{550 \text{ ft-lbf/sec}} = 3754 \text{ hp}$$

• PROBLEM 5-30

A centrifugal pump has been set up in the laboratory so that its characteristics may be determined. A sketch of the laboratory arrangement is shown in the Figure. During a particular run, the volume flow rate through the pump was measured to be 20.3 gal/min. The pressure gage on the discharge side read 15 ft of water, while the vacuum gage on the suction side read 4.3 in. of Hg. The pipe diameter on the discharge side of the pump is $\frac{1}{2}$ in., while the pipe diameter on the suction side is $\frac{3}{4}$ in. Determine (a) the head developed by the pump, and (b) the power transmitted to the water by the pump.

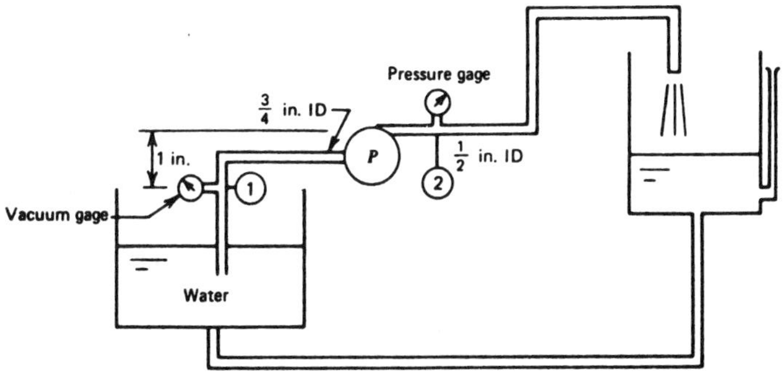

Solution: (a) Select a control volume containing the pump and the pipe lines between sections (1) and (2). Let us first determine pressures p_1 and p_2. Take section (1) to be at the vacuum gage and section (2) at the pressure gage; then

$$\frac{p_1}{\gamma} = \frac{p_{atm}}{\gamma} - \frac{4.3\gamma_{hg}}{12\gamma_w} = \frac{p_{atm}}{\gamma} - \frac{4.3(13.56)}{12}$$

$$= \frac{p_{atm}}{\gamma} - 4.86 \text{ ft of water}$$

and

$$\frac{p_2}{\gamma} = \frac{p_{atm}}{\gamma} + \frac{15\gamma}{\gamma} = \frac{p_{atm}}{\gamma} + 15 \text{ ft of water}$$

Now determine the velocities at sections (1) and (2).

$$Q = A_1V_1 = A_2V_2$$

Therefore,

$$V_1 = \frac{20.3 \text{ gal/min} \times (1/7.48)\text{ft}^3\text{/gal} \times (1/60)\text{min/s}}{\pi\left(\frac{3}{4}\right)^2 / (4 \times 144)}$$

$$= 14.74 \text{ ft/s}$$

Similarly,

$$V_2 = \frac{20.3 \times 4 \times 144 \times 4}{7.48\pi \times 60} = 33.17 \text{ ft/s}$$

Since work is done by the surroundings and is exerted on the system, then the energy equation becomes

$$\frac{p_2}{\gamma} + \frac{V_2^2}{2g} + z_2 = \frac{p_1}{\gamma} + \frac{V_1^2}{2g} + z_1 + \frac{1}{\dot{w}}\left(\frac{\delta W}{\delta t}\right)_{pump}$$

where $\dot{w} = \dot{m}g/g_c$ = weight flow rate and $(\delta w/\delta t)_{pump}$ is the rate of work done by the pump.

We see that the total head at section (2) is greater than the total head at section (1) by the term $(1/\dot{w})(\delta W/\delta t)_{pump}$. We designate this term as h_p. It represents the total head developed by the pump; that is,

$$h_p = \frac{1}{\dot{w}}\left(\frac{\delta W}{\delta t}\right)_p = \left(\frac{p_2}{\gamma} + \frac{V_2^2}{2g} + z_2\right) - \left(\frac{p_1}{\gamma} + \frac{V_1^2}{2g} + z_1\right) \qquad \text{Eq. 1}$$

or

$$h_p = \frac{p_2}{\gamma} - \frac{p_1}{\gamma} + \frac{V_2^2 - V_1^2}{2g} + (z_2 - z_1)$$

$$= (15 + 4.86) + \frac{(33.17)^2 - (14.74)^2}{64.4} + 1 = 34.57 \text{ ft}$$

(b) The power transmitted to the water by the pump = $(\delta W/\delta t)_p$. Applying Eq. 1

$$\left(\frac{\delta W}{\delta t}\right)_p = \dot{w}h_p = \gamma Q h_p = \frac{62.4 \times 20.3 \times 34.57}{7.48 \times 60}$$

$$= 97.57 \text{ ft-lb}_f/\text{s}$$

$$= 0.177 \text{ hp}$$

• PROBLEM 5-31

An incompressible, frictionless fluid flows steadily into a machine at section 1 and out at section 2. Heat transfers to the machine at the rate of 300 Btu per minute. The area at section 1 is $\frac{1}{10}$ ft^2 and at section 2 is $\frac{1}{5}$ ft^2. The fluid, water, flows in section 1 at the rate of 2 slugs/sec. The pressure at section 1 is 40 psia and at section 2 is 30 psia. Neglecting the change in elevation, compute the work per slug added to or taken from the machine.

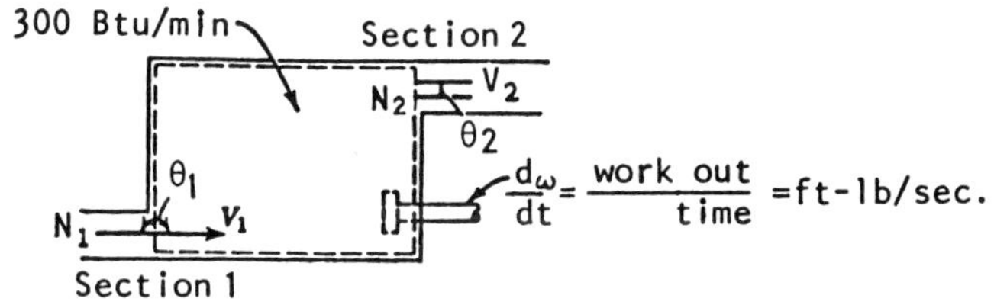

Solution: The dotted lines in the figure represent the control surfaces of the control volume. Select this control volume because the cos θ is different from zero only at sections 1 and 2. This selection of the control volume reduces the amount of work.

The general energy equation is

$$\frac{dQ}{dt} - \frac{dW}{dt} = \int_A \left(\frac{p}{\rho} + \frac{V^2}{2} + gz + u\right)\rho V \cos\theta \, dA$$

At section 1, $\theta_1 = \pi$ over the entire section; at section 2, $\theta_2 = 0$ over the entire section. The value of the integral is zero over the remaining surface of the control volume since $\theta = \pi/2$. From this you can get

$$\frac{dQ}{dt} - \frac{dW}{dt} = -\int_{A_1} \left(\frac{p_1}{\rho_1} + \frac{V_1^{\,2}}{2} + gz_1 + u_1\right)\rho_1 V_1 \, dA$$

$$+ \int_{A_2} \left(\frac{p_2}{\rho_2} + \frac{V_2^2}{2} + gz_2 + u_2\right)\rho_2 V_2 dA$$

All properties of the water at section 1 are constant and can be taken from inside the integral sign. The same is true at section 2.

$$\frac{dQ}{dt} - \frac{dW}{dt} = -\rho_1 V_1 \left(\frac{p_1}{\rho_1} + \frac{V_1^2}{2} + gz_1 + u_1\right) \int_{A_1} dA$$

$$+ \rho_2 V_2 \left(\frac{p_2}{\rho_2} + \frac{V_2^2}{2} + gz_2 + u_2\right) \int_{A_2} dA$$

By definition

$$\int_{A_1} dA = A_1$$

and

$$\int_{A_2} dA = A_2$$

The fluid is water and may be considered incompressible. Therefore, $\rho_1 = \rho_2 = \rho$, and

$$\frac{dQ}{dt} = +300 \text{ Btu/min} = \frac{(300)(778)}{60} = 3890 \text{ ft-lb/sec}$$

Substitute these values in the energy equation.

$$3890 - \frac{dW}{dt} = -\rho A_1 V_1 \left(\frac{p_1}{\rho_1} + \frac{V_1^2}{2} + gz_1 + u_1\right)$$

$$+ \rho A_2 V_2 \left(\frac{p_2}{\rho} + \frac{V_2^2}{2} + gz_2 + u_2\right)$$

From the equation of continuity for steady flow you get

$$\int_A \rho V \cos\theta \, dA = 0$$

and

$$-\rho A_1 V_1 + \rho A_2 V_2 = 0, \qquad \rho A_1 V_1 = \rho A_2 V_2 = \rho AV$$

Divide the energy equation by ρAV and rearrange.

$$\frac{3890}{\rho AV} - \frac{1}{\rho AV}\frac{dW}{dt}$$

$$= \left(\frac{p_2 - p_1}{\rho} + \frac{V_2^2 - V_1^2}{2} + g(z_2 - z_1)\right.$$

$$\left. + u_2 - u_1\right) \text{ ft-lb/slug}$$

For a frictionless incompressible fluid $u_2 - u_1 - q = 0$, and h_L is zero. The q in the above equation is $3890/\rho AV$. Therefore,

$$-\frac{1}{\rho AV}\frac{dW}{dt} = -\omega_s = \frac{p_2 - p_1}{\rho} + \frac{V_2^2 - V_1^2}{2} + g(z_2 - z_1)$$

Given in the problem that $p_2 = 30$ psia, $\rho = 1.94$ slugs/ft^3, $p_1 = 40$ psia, and $z_1 = z_2$. Therefore,

$$-\omega_s = \frac{(30 - 40)(144)}{1.94} + \frac{V_2^2 - V_1^2}{2} + g(z_2 - z_1) \rightarrow 0$$

$$-\omega_s = \frac{-1440}{1.94} + \frac{V_2^2 - V_1^2}{2} \qquad (1)$$

From continuity

$$\rho A_1 V_1 = \rho A_2 V_2 = 2 \text{ slugs/sec}$$

$$(1.94)\left(\frac{1}{10}\right)V_1 = (1.94)\left(\frac{1}{5}\right)V_2 = 2 \text{ slugs/sec}$$

Therefore,

$$V_1 = \frac{20}{1.94}, \quad V_2 = \frac{10}{1.94}$$

and

$$\frac{V_2^2 - V_1^2}{2} = \frac{1}{2}\left[\left(\frac{10}{1.94}\right)^2 - \left(\frac{20}{1.94}\right)^2\right] = \frac{1}{2}\left(\frac{1}{1.94}\right)^2[100 - 400]$$

$$= -39.86 \text{ ft-lb/slug}$$

Substitute this into the energy equation (1).

$$-\omega_s = -\frac{1440}{1.94} - 39.86 = -782.13$$

$$\omega_s = +782.13 \text{ ft-lb/slug}$$

The plus sign signifies work taken out of the machine.

(a) Consider a large reservoir of diameter D filled with a liquid to a height h (see Fig. 1). The reservoir is discharging through a pipe attached to its base. The pipe has a length L and diameter d. The pressure on the liquid surface of the reservoir is P_t and the pressure at the exit of the pipe is P_e. The exit flow velocity is V_e. Find an expression for the outlet velocity V_e at any instant of time as a function of $P_e - P_t$, the liquid height h at any instant, and the pipe dimensions. Assume uniform flow at liquid surface and at the pipe discharge.

(b) A 4-in.-diameter pipe, 1000 ft long, is attached to a reservoir 30 ft below the surface. The pipe has a sharp-edged inlet. Take an average value of the friction factor to be f = 0.008.

Assuming that the head is maintained relatively constant, find the steady-state value of the exit velocity.

(c) If the pipe has been closed off until t = 0, find the time after opening for the flow velocity to reach 90 per cent of the steady-state velocity. The reservoir head is again assumed to be constant.

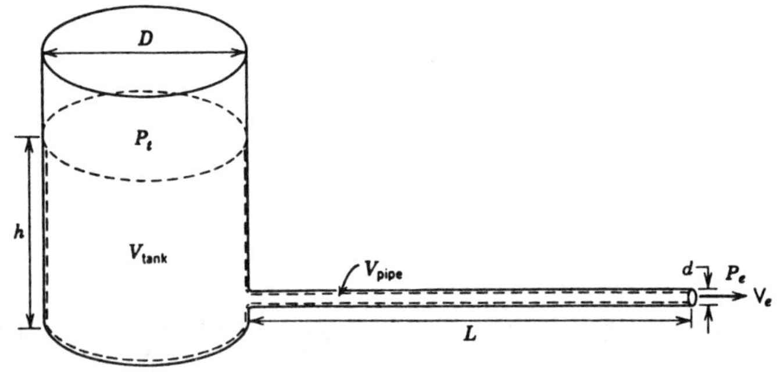

Fig. 1

Solution: (a) An appropriate control volume is selected, as shown in the Figure.

The continuity equation for uniform flow conditions can be written

$$\rho V_e A_e = \rho V_b A_{tank}$$

$$A_{tank} = \frac{\pi D^2}{4}$$

With reference to the energy equation, given in the form

$$\dot{Q} = \frac{d}{dt}\int_{V_{C.V.}} \rho\left(u + \frac{V^2}{2} + gz\right)dV + \int_{S_{C.V.}} PV_b \cdot \hat{n}\, dA$$

$$+ \int_{S_{C.V.}} \left(u + \frac{P}{\rho} + \frac{V^2}{2} + gz\right)\rho V_{rn}\, dA$$

$$+ \dot{W}_{shear} = 0$$

the volume integral in the energy equation can be written as the sum of two volume integrals, one over the tank and one over the pipe:

$$\frac{d}{dt}\int_{V_{C.V.}} \rho\left(u + \frac{V^2}{2} + gz\right)dV = \frac{d}{dt}\int_{V_{tank}} \rho\left(u + \frac{V^2}{2} + gz\right)dV$$

$$+ \frac{d}{dt}\int_{V_{pipe}} \rho\left(u + \frac{V^2}{2} + gz\right)dV$$

Assuming that the tank fluid velocity can be approximated by V_b, that the average tank height is $h/2$, and that the average velocity in the pipe is V_e, we may write

$$\frac{d}{dt}\int_{V_{tank}} \rho\left(u + \frac{V^2}{2} + gz\right) dV = \frac{d}{dt}\left(\frac{V_b^{\,2}}{2}\,\frac{\pi D^2}{4}h + \frac{h^2}{2}\,\frac{\pi D^2}{4}\right)\rho$$

$$+ \frac{d}{dt}\int_{V_{tank}} \rho u\, dV \qquad (1)$$

$$\frac{d}{dt}\int_{V_{pipe}} \rho\left(u + \frac{V^2}{2} + gz\right) dV = \frac{d}{dt}\left(\frac{V_e^{\,2}}{2}\,\frac{\pi d^2}{4}L\right)\rho$$

$$+ \frac{d}{dt}\int_{V_{pipe}} \rho u\, dV \qquad (2)$$

Other terms in the energy equation become

$$\dot{W}_{shear} = 0$$

$$\int_{S_{C.V.}} \left(u + \frac{P}{\rho} + \frac{V^2}{2} + gz\right)\rho V_{rn}\, dA = \left(\frac{P_e}{\rho} + \frac{V_e^{\,2}}{2}\right)\rho V_e \frac{\pi d^2}{4}$$

$$+ \int_{A_e} u_e \rho V_e\, dA \qquad (3)$$

From continuity, as $\partial\rho/\partial t = 0$,

$$\rho V_b \frac{\pi D^2}{4} = \rho V_e \frac{\pi d^2}{4} = \dot{m}$$

Therefore

$$-\frac{dh}{dt} = V_b = V_e \left(\frac{d}{D}\right)^2 \tag{4}$$

From equations 1 through 4 and with a reasonable amount of algebraic manipulation, we may transform the energy equation to obtain

$$\rho\frac{\pi D^2}{4} \cdot h\frac{d}{dt}\frac{V_b^2}{2} - \dot{m}gh + \frac{V_e^2}{2}\left[1 - \left(\frac{d}{D}\right)^4\right]\dot{m}$$

$$+ \dot{m}\frac{P_e - P_t}{\rho} + \rho\frac{\pi d^2}{\rho} \cdot L\frac{d}{dt}\frac{V_e^2}{2}$$

$$\left\{+ \frac{d}{dt}\int_{V_{C.V.}} u\rho\, dV + \int u_e \rho V_e\, dA - \dot{Q}\right\}$$

$$= 0 \tag{5}$$

We now let the term in braces in equation 5 be the energy "loss" and replace the expression by $K(V_e^2/2)\dot{m}$ where K is called a loss coefficient which may or may not vary with time and flow conditions. Substituting this term into equations 5 and dividing by $\dot{m} = \rho V_e \pi (d^2/4)$ gives

$$\left[L + h\left(\frac{d}{D}\right)^2\right]\frac{d}{dt}V_e - gh + \frac{V_e^2}{2}\left[1 - \left(\frac{d}{D}\right)^4\right]$$

$$+ \frac{P_e - P_t}{\rho} + K\frac{V_e^2}{2} = 0 \tag{6}$$

For a large tank and small-diameter exit pipe it would be reasonable to neglect $(d/D)^4$. Furthermore, for a large tank we would expect the rate of velocity change dV_e/dt also to be small. Neglecting these terms in equation 6, we have

$$\frac{P_e - P_t}{\rho} + \frac{V_e^2}{2}(1 + K) = gh \tag{7}$$

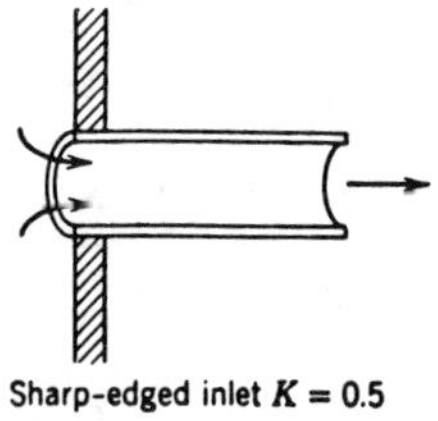

Sharp-edged inlet $K = 0.5$

Fig. 2

(b) The loss coefficient K is usually determined experimentally as a function of the Reynolds number. Nevertheless, in many problems the loss coefficient K can be assumed to be essentially constant. Flow entering a pipe from a reservoir undergoes an increase in internal energy or heat transfer which leads to a loss. The loss coefficient for various types of pipe inlets has been determined as shown in Fig. 2. The loss coefficient for flow in pipes is usually expressed as

$$K = \frac{4fL}{d}$$

where L is the pipe length, d is the pipe diameter, and f is the friction factor. Then K can be calculated

$$K = 0.5 + \frac{4fL}{d} = 0.5 + \frac{4 \times 0.008 \times 1000}{1/3} = 96.5$$

Using Eq. 7 and knowing that, $P_t = P_e$, gives

$$V_e = \sqrt{\frac{2gh}{1 + K}} = \sqrt{\frac{2 \times 32.2 \times 30}{97.4}}\ \frac{\text{ft}}{\text{sec}} = 4.45 \text{ ft/sec}$$

(c) Using Eq. 6, assuming that is $(\frac{d}{D})^4$ is negligible, and $P_e = P_t$ gives

$$\tfrac{1}{2}V_e^{\,2}(1 + K) + \left[L + h\left(\frac{d}{D}\right)^2\right]\frac{dV_e}{dt} = gh$$

Therefore

$$\frac{dV_e}{dt} = \frac{gh - \tfrac{1}{2}V_e^{\,2}(1 + K)}{\left[L + h\left(\frac{d}{D}\right)^2\right]}$$

$$\frac{2\left[L + h\left(\frac{d}{D}\right)^2\right] dV_e}{2gh - V_e^{\,2}(1 + K)} = dt$$

$$\frac{2\left[L + h\left(\frac{d}{D}\right)^2\right]}{1 + K}\int_0^{V_e}\frac{dV_e}{\left(\sqrt{\frac{2gh}{1 + K}}\right)^2 - V_e^{\,2}} = \int_0^t dt$$

$$\frac{\left[L + h\left(\frac{d}{D}\right)^2\right]}{1 + K}\ \frac{1}{\sqrt{\frac{2gh}{1 + K}}}\ln\frac{\sqrt{\frac{2gh}{1 + K}} + V_e}{\sqrt{\frac{2gh}{1 + K}} - V_e} = t$$

$$\frac{\left[L + h\left(\frac{d}{D}\right)^2\right]}{\sqrt{2gh(1 + K)}}\ln\left\{\frac{1 + V_e\Big/\sqrt{\frac{2gh}{1 + K}}}{1 - V_e\Big/\sqrt{\frac{2gh}{1 + K}}}\right\} = t$$

We choose $V_e = 0.9\sqrt{2gh/(1 + K)}$, where $\sqrt{2gh/(1 + K)}$ is the steady-state velocity. It follows that for $d/D \approx 0$, the value for t is

$$t = \frac{1000}{\sqrt{64.4 \times 30 \times 97.5}} \times \ln \frac{1.9}{0.1} = 6.76 \text{ sec}$$

• PROBLEM 5-33

A perfect gas flows steadily in a 1 ft dia. pipe that suddenly enlarges to 2 ft in diameter. The temperature is constant at 70°F. The engineering gas constant is $53\text{-lb}_f\text{-ft/lb}_m{}^\circ R$.

Compute the rate of heat transfer either to or from the pipe enlargement.

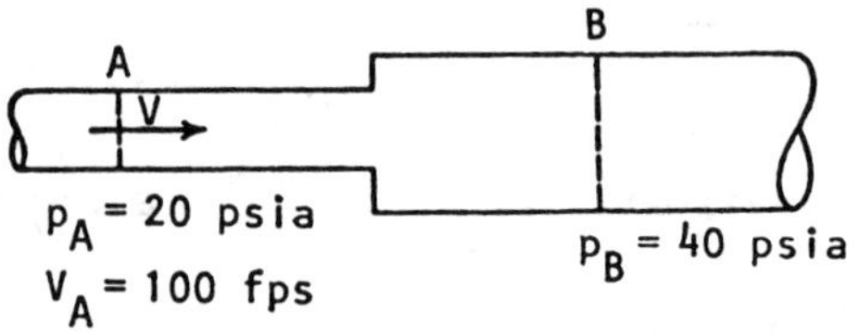

Solution: The equation of state for a perfect gas may be written as

$$\frac{p}{\rho} = g_c RT$$

where p = pressure, in lb/ft^2

ρ = mass density, in $slugs/ft^3$

g_c = Newton proportionality constant =

$= 32.2\ lb_m\text{-ft}/lb_f sec^2$

T = temperature, in °R

R = engineering gas constant, in $lb_f\text{-ft}/lb_m{}^\circ R$

Select the control volume as the walls of the pipe and as two surfaces normal to flow. The control volume then is

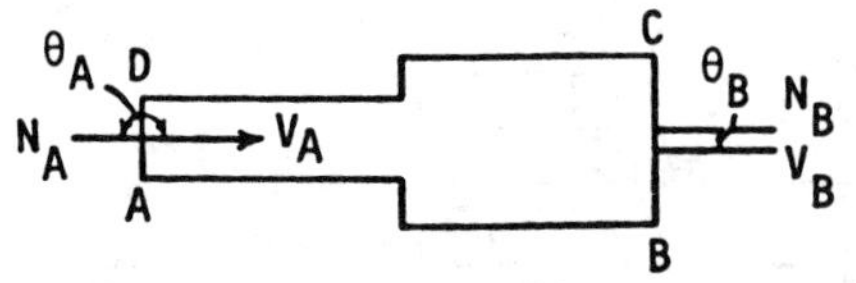

The angle θ is $\pi/2$ between the outward-drawn normal and the velocity vector at every surface of the control volume except AD and BC. Therefore, evaluate the integral which appears in the general energy equation for steady flow

$$\frac{dQ}{dt} - \frac{dW}{dt} = \int_A \left(\frac{p}{\rho} + \frac{V^2}{2} + gz + u\right) \rho V \cos\theta \, dA \tag{1}$$

as follows:

$$\int_A \left(\frac{p}{\rho} + \frac{V^2}{2} + gz + u\right) \rho V \cos\theta \, dA = \int_{AD+BC} \left(\frac{p}{\rho} + \frac{V^2}{2} + gz + u\right) \rho V \cos\theta \, dA$$

Here $\theta = \pi$ at surface AD, and $\theta = 0$ at surface BC.

$$\int_A \left(\frac{p}{\rho} + \frac{V^2}{2} + gz + u\right) \rho V \cos\theta \, dA - \int_{AD} \left(\frac{p}{\rho} + \frac{V^2}{2} + gz + u\right) \rho V \, dA + \int_{BC} \left(\frac{p}{\rho} + \frac{V^2}{2} + gz + u\right) \rho V \, dA \tag{2}$$

Since the properties p, ρ, V, z, and u are considered constant across the surface of integration, you may take them outside the integral sign. The integration over surface AD is

$$(-1)\left(\frac{p_A}{\rho_A} + \frac{V_A^{\;2}}{2} + gz_A + u_A\right)\rho_A V_A \int_{AD} dA = -\left(\frac{p_A}{\rho_A} + \frac{V_A^{\;2}}{2} + gz_A + u_A\right)\rho_A V_A A_A$$

Do the same across the face BC and substitute both into Equation (2) to get

$$\int_A \left(\frac{p}{\rho} + \frac{V^2}{2} + gz + u\right) \rho V \cos\theta \, dA = -\rho_A A_A V_A \left(\frac{p_A}{\rho_A} + \frac{V_A^{\;2}}{2} + gz_A + u_A\right) + \rho_B V_B A_B \left(\frac{p_B}{\rho_B} + \frac{V_B^{\;2}}{2} + gz_B + u_B\right)$$

With this substitution the energy equation becomes

$$\frac{dQ}{dt} - \frac{dW}{dt} = \rho_B A_B V_B \left(\frac{p_B}{\rho_B} + \frac{V_B^{\;2}}{B} + gz_B + u_B\right) - \rho_A A_A V_A \left(\frac{p_A}{\rho_A} + \frac{V_A^{\;2}}{2} + gz_A + u_A\right)$$

The intrinsic energy u in a perfect gas is a function of only the temperature. Therefore, $u_B = u_A$ since the temperature is constant and the gas may be considered a perfect gas. Measure the elevation z to the center of the pipe. Therefore, $z_B = z_A$.

Next consider the p/ρ terms. For a perfect gas

$$\frac{p}{\rho} = g_c RT$$

Since g_c, R, and T are constant, p/ρ is a constant; therefore,

$$\frac{p_B}{\rho_B} = \frac{p_A}{\rho_A}$$

No work transfers through the control volume and so $dW/dt = 0$. The energy equation becomes

$$\frac{dQ}{dt} = \rho_B A_B V_B \left(\frac{p_A}{\rho_A} + \frac{V_B^2}{2} + gz_A + u_A\right) - \rho_A A_A V_A \left(\frac{p_A}{\rho_A} + \frac{V_A^2}{2} + gz_A + u_A\right)$$

To simplify, it is necessary to apply the equation of continuity to the control volume.

$$\int_A \rho V \cos\theta \, dA = 0$$

$$\int_{AD} \rho V \cos\theta \, dA + \int_{BC} \rho V \cos\theta \, dA = 0$$

$$\rho_A V_A (-1) \int_{AD} dA + \rho_B V_B (+1) \int_{BC} dA = 0$$

$$-\rho_A V_A A_A + \rho_B V_B A_B = 0$$

$$\rho_A V_A A_A = \rho_B V_B A_B$$

When you make this substitution, the energy equation becomes

$$\frac{dQ}{dt} = \rho_A A_A V_A \left(\frac{p_A}{\rho_A} + \frac{V_B^2}{2} + gz_A + u_A\right) - \rho_A A_A V_A \left(\frac{p_A}{\rho_A} + \frac{V_A^2}{2} + gz_A + u_A\right)$$

$$\frac{dQ}{dt} = \rho_A A_A V_A \left[\left(\frac{p_A}{\rho_A} - \frac{p_A}{\rho_A}\right)^{0} + \frac{V_B^2 - V_A^2}{2} + g(z_A - z_A)^{0} + (u_A - u_A)^{0} \right]$$

$$\frac{dQ}{dt} = \rho_A A_A V_A \left(\frac{V_B^2 - V_A^2}{2}\right)$$

The mass flow rate is

$$\rho_A A_A V_A = \left(\frac{p_A}{g_c RT}\right) A_A V_A = \frac{(20)(144)}{(32.2)(53)(530)} \frac{\pi(1)^2}{4} (100)$$

$$= 0.25 \text{ slug/sec}$$

By continuity $\rho_A A_A V_A = \rho_B A_B V_B$.

Therefore,

$$(0.25) = \frac{(40)(144)}{(32.2)(53)(530)} \frac{\pi(4)}{(4)} V_B$$

$$V_B = 12.5 \text{ fps}$$

$$\frac{dQ}{dt} = \rho_A A_A V_A \left(\frac{V_B^2 - V_A^2}{2}\right)$$

$$= (0.25)\left(\frac{156.25}{2} - \frac{10,000}{2}\right) = -1230 \text{ ft-lb/sec}$$

The minus sign signifies that the heat transfers out of the control volume.

• PROBLEM 5-34

In a turbojet engine, the gaseous products of combustion enter the turbine with a velocity of 50 fps and temperature of 2000° R. At the turbine exit, the velocity and temperature of the gases are 200 fps and 1500° R (see Figure). If the mass flow rate through the turbine is 80 lb_m/sec, determine the horsepower output for steady flow. Neglect heat loss from the gases as they flow through the turbine, and assume the gases behave as a perfect gas with constant specific heat (c_p = 0.25 Btu-/lb_m-°R) so that the enthalpy difference $h_2 - h_1$ can be written as $h_2 - h_1 = c_p(T_2 - T_1)$.

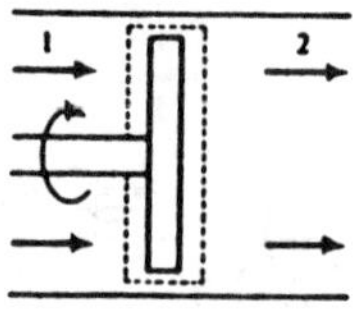

Solution: For a system of fluid particles, the law of conservation of energy states that the total energy E of the system increases, in going from state 1 to state 2, by an amount equal to the total heat energy $\tilde{Q}$ added to the system of particles less the work done W' by the system of fluid particles:

$$E_2 - E_1 = \tilde{Q} - W'$$

Here E represents the total energy possessed by the system in a given state and thereby includes the kinetic and potential energy of the entire system mass, the internal energy associated with the random motion of the molecules comprising the system, and other forms of storable energy, such as electrical energy (e.g., stored in a capacitor) and chemical energy. Heat and work are not properties of a system of particles but rather are forms of energy that are transferred across the system boundaries. Therefore, the heat and work transferred during a process are functions of the process itself, not just the end states.

The energy equation in differential form becomes

$$dE = d\tilde{Q} - dW'$$

We wish to apply the law of conservation of energy to a control volume in a fluid, in which mass is entering and leaving across the surface bounding the control volume. Let E equal the total energy of the fluid particles and e the energy per unit mass.

$$\left.\frac{dE}{dt}\right)_{\text{system}} = \left.\frac{\partial E}{\partial t}\right)_{\substack{\text{control}\\\text{volume}}} + \int_{\substack{\text{control}\\\text{surface}}} e\rho V\, dA$$

$$\left.\frac{dE}{dt}\right)_{\text{system}} = \frac{d}{dt}(\tilde{Q} - W') = \left.\frac{\partial E}{\partial t}\right)_{\substack{\text{control}\\\text{volume}}} + \int_{\substack{\text{control}\\\text{surface}}} e\rho V\, dA \qquad (1)$$

$$\frac{d}{dt}(\tilde{Q} - W') = \left.\frac{\partial E}{\partial t}\right)_{\substack{\text{control}\\\text{volume}}} + \int_{\substack{\text{control}\\\text{surface}}} e\rho\vec{V} \cdot d\vec{A}$$

The left-hand side represents the rate at which energy, in the form of heat and work, is transferred to the fluid in the control volume, with heat positive if it is added to the fluid and work positive if done by the fluid. The first term on the right represents the rate at which energy is stored in the control volume, the second term the net rate of efflux of energy from the control volume.

If the system possesses only internal, kinetic, and potential energies, then there results

$$E = U + KE + PE \quad \text{or} \quad e = u + \frac{1}{2}\frac{V^2}{g_c} + \frac{g}{g_c}z$$

with U equal to total internal energy and u equal to internal energy per unit mass. Substituting into (1),

$$\frac{d}{dt}(\tilde{Q} - W') = \left.\frac{\partial E}{\partial t}\right)_{\substack{\text{control}\\ \text{volume}}} + \int_{\substack{\text{control}\\ \text{surface}}} \left(u + \frac{1}{2}\frac{V^2}{g_c} + \frac{g}{g_c}z\right) \rho V \, dA$$

It is convenient to combine the expressions for internal energy per unit mass u and flow work per unit mass p/ρ into a property called enthalpy h:

$$h = u + \frac{p}{\rho}$$

Rewriting the energy equation for a control volume in terms of enthalpy, we obtain

$$\frac{d}{dt}(\tilde{Q} - W) = \left.\frac{\partial E}{\partial t}\right)_{\substack{\text{control}\\ \text{volume}}} + \int_{\substack{\text{control}\\ \text{surface}}} \left(h + \frac{V^2}{2g_c} + \frac{g}{g_c}z\right) \rho V \, dA \qquad (2)$$

Applying (2) to the control volume shown, we obtain for steady, adiabatic flow

$$\frac{-dW}{dt} = \int_{\substack{\text{control}\\ \text{surface}}} \left(h + \frac{V^2}{2g_c} + \frac{g}{g_c}z\right) \rho V \, dA$$

Any possible change in potential energy can be neglected in comparison with either the enthalpy change or kinetic energy change, so that after integrating

$$\frac{-dW}{dt} = \left[\left(h_2 + \frac{V_2^2}{2g_c}\right) - \left(h_1 + \frac{V_1^2}{2g_c}\right)\right] 80$$

$$= \left[c_p(T_2 - T_1) + \frac{V_2^2 - V_1^2}{2g_c}\right] 80$$

$$= \left[0.25(-500)\frac{\text{Btu}}{\text{lb}_m} + \frac{200^2 - 50^2}{2(32.2)} \frac{\text{ft-lb}_f}{\text{lb}_m} \times \frac{1}{778}\frac{\text{Btu}}{\text{ft-lb}_f}\right] 80 \text{ lt}_m/\text{sec}$$

$$= (-125 + 0.75)80$$

$$= -9940 \text{ Btu/sec}$$

$$\frac{dW}{dt} = \underline{14{,}100 \text{ hp}}$$

• PROBLEM 5-35

A pump is to be used to provide one cfs of water at atmospheric pressure through a 3-in. diameter pipe to a building 500 ft above sea level, the water coming from a reservoir at sea level (see Figure). Determine the pump horsepower required. The density of water can be taken as constant at 62.4 lb_m/ft^3. Moreover, neglect heat transfer and assume negligible change in the internal energy of the water as it flows through the pipe.

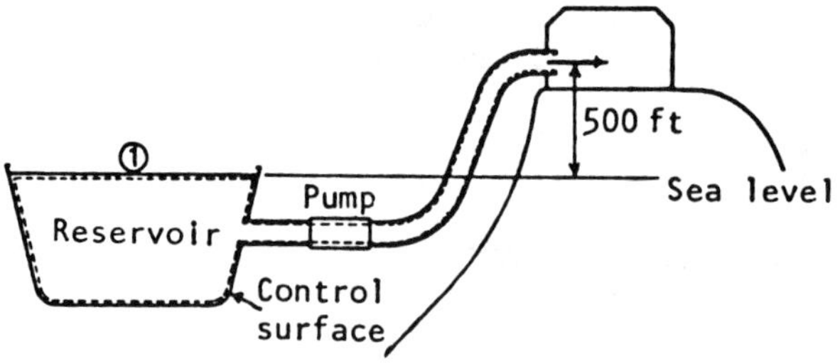

Solution: Select a control volume with a surface that includes those cross sections at which properties are known. For example, as shown in the figure, the inlet to the control volume has been taken at the reservoir surface, where the pressure is atmospheric and the velocity negligible. The outlet from the control volume has been taken at the building, where again flow and pressure are known.

The energy equation for a control volume in terms of enthalpy h, is

$$\frac{d}{dt}(\tilde{Q} - W) = \left.\frac{\partial E}{\partial t}\right)_{\substack{\text{control}\\ \text{volume}}} + \int_{\substack{\text{control}\\ \text{surface}}} \left(h + \frac{V^2}{2g_c} + \frac{g}{g_c}z\right)\rho V\, dA$$

or, in vector form,

$$\frac{d}{dt}(\tilde{Q} - W) = \left.\frac{\partial E}{\partial t}\right)_{\substack{\text{control}\\ \text{volume}}} + \int_{\substack{\text{control}\\ \text{surface}}} \left(h + \frac{V_2^2}{2g_c} + \frac{g}{g_c}z\right)(\rho\vec{V} \cdot d\vec{A}) \tag{1}$$

Where,

$\tilde{Q}$: total heat energy added to the system of particles,
W: work done by the system of fluid particles,
E: total energy possessed by the system in a given state,
h: enthalpy.

For steady flow, the first term on the right-hand side of (1) is equal to zero. Furthermore, if the inlet and outlet flows to a control volume are one dimensional with velocity vectors normal to the control surface the energy equation reduces to

$$\frac{d}{dt}(\tilde{Q} - W) = \left[\left(h_2 + \frac{V_2^2}{2g_c} + \frac{g}{g_c}z_2\right) - \left(h_1 + \frac{V_1^2}{2g_c} + \frac{g}{g_c}z_1\right)\right]\rho AV \tag{2}$$

For this problem, equation (2) is applicable, with $\tilde{Q} = 0$ and $h_2 - h_1 = (p_2 - p_1)/\rho$ since $u_2 - u_1 = 0$ and ρ = constant. Substituting we obtain

$$\frac{-dW}{dt} = \rho AV\left[\left(\frac{p_2}{\rho} + \frac{V_2^2}{2g_c} + \frac{g}{g_c}z_2\right) - \left(\frac{p_1}{\rho} + \frac{V_1^2}{2g_c} + \frac{g}{g_c}z_1\right)\right]$$

The mass flow rate of water is equal to 62.4 lb_m/ft^3 x 1 ft^3/sec = 62.4 lb_m/sec, and from continuity

$$V_2 = \frac{\rho AV}{A_2\rho} = \frac{62.4}{\frac{\pi}{4}\frac{9}{144}\,62.4} = 20.4 \text{ ft/sec}$$

Therefore,

$$\frac{-dW}{dt} = 62.4\left(\frac{(20.4)^2}{2(32.2)} + 500\right)$$

since $p_2 = p_1$ = atmospheric pressure.

Solving

$$\frac{-dW}{dt} = 62.4(6.46 + 500) = 31600 \text{ ft-lb}_f\text{/sec}$$

$$= \frac{31600}{550}\ \frac{\text{ft-lb}_f\text{/sec}}{\text{ft-lb}_f\text{/sec/hp}}$$

or

$$\frac{dW}{dt} = -57.5 \text{ hp}$$

where the negative sign denotes work done on the fluid.

CHAPTER 6

MOMENTUM EQUATION

Basic Attacks and Strategies for Solving Problems in this Chapter. See pages 238 to 302 for step-by-step solutions to problems.

In many fluid flow problems, the total force on a solid object or wall is desired, as, for example, in determining the required bolt strength on flanges in a piping system, or the total thrust produced by a jet engine. The control volume or integral technique may be applied here by utilizing the law of conservation of linear momentum. The Reynolds Transport Theorem in Chapter 4 can be applied directly, with

$$B = m\mathbf{V} \quad \text{and} \quad b = \mathbf{V},$$

and by utilizing Newton's second law for a system,

$$\frac{d(m\mathbf{V})}{dt} = \Sigma\mathbf{F}.$$

The result is the integral conservation of momentum law for a control volume,

$$\frac{d}{dt}\iiint_{cv} \rho\mathbf{V}dV + \iint_{cs} \rho\mathbf{V}(\mathbf{V}_r \cdot \mathbf{n})dA = \Sigma\mathbf{F}, \tag{1}$$

where $\mathbf{V}_r$ is the velocity of the fluid relative to the control surface, and $\Sigma\mathbf{F}$ is the total force acting on the control volume when the control volume is considered as a free body. Note that Equation (1) is a *vector* equation, and thus represents in general three components which may have to be evaluated separately.

In most cases, the control volume is fixed, $\mathbf{V}_r = \mathbf{V}$, and Equation (1) reduces to

$$\frac{\partial}{\partial t}\iiint_{cv} \rho\mathbf{V}dV + \iint_{cs} \rho\mathbf{V}(\mathbf{V} \cdot \mathbf{n})dA = \Sigma\mathbf{F}. \tag{2}$$

As with the integral conservation of mass and energy equations discussed in previous chapters, the technique for solving problems with the integral conservation of momentum equation involves first choosing an appropriate control volume, then determining the flux terms and (if non-zero) the unsteady control volume term. Typically, the unknown is some force on the right-hand side of Equation (2), which can then be obtained.

In general, the force term consists of surface forces due to pressure and friction, body forces such as gravity (and possibly electromagnetic forces), and other forces acting on the control surface such as the tension force in a bolt through which the control surface is sliced. It is important to keep in mind two things here:

1) $\Sigma\mathbf{F}$ is the *vector* sum of all forces acting on the control volume, and

2) $\Sigma\mathbf{F}$ include(s) *all* forces acting on both solid and fluid material in the control volume.

It is usually best to draw a free-body diagram of the control volume, showing all forces acting on it. If the desired result is the force acting on a solid wall *by* the fluid flow, remember to change the sign since any force in Equation (2) must be applied *on* the control volume or control surface.

For flowfields involving sections of pipe flow as inlets or exits, the surface integral in Equation (2) reduces to

$$\iint_{cs} \rho\mathbf{V}(\mathbf{V}\cdot\mathbf{n})dA = \sum_{\text{outlets}} \dot{m}\mathbf{V} - \sum_{\text{inlets}} \dot{m}\mathbf{V}. \tag{3}$$

Also, for such sections, the pressure is typically constant over the cross-sectional area A; thus, the pressure force on the area is simply pA acting in the direction opposite to the unit outward normal vector $\mathbf{n}$. Although absolute pressure is implied in the above, gage pressure is often more convenient since atmospheric pressure may be subtracted uniformly over the entire control surface without changing the problem.

Step-by-Step Solutions to Problems in this Chapter, "Momentum Equation"

MOMENTUM EQUATION

• PROBLEM 6-1

Consider the steady flow of water (ρ = 1.94 slug/ft^3) through the device shown in the diagram. The areas are: A_1 = 0.3 ft^2, A_2 = 0.5 ft^2, and A_3 = A_4 = 0.4 ft^2. Mass flow out through section 3 is given as 3.88 slug/sec. The volumetric flow rate in through section 4 is given as 1 ft^3/sec, and $\vec{V}_1 = 10\hat{i}$ ft/sec. If properties are assumed uniform across all inlet and outlet flow sections, determine the flow velocity at section 2.

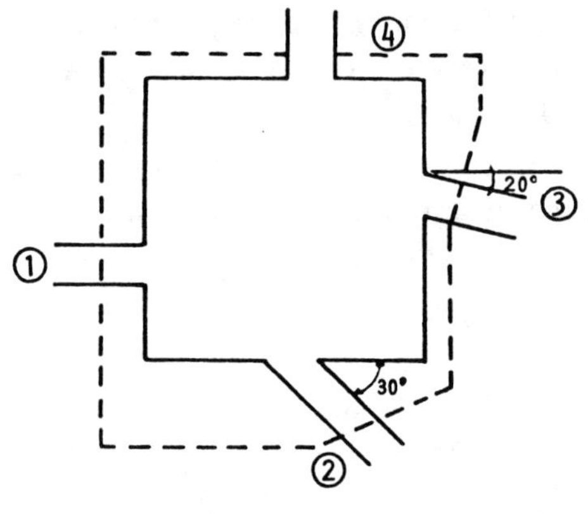

Solution: The dashed lines in the Figure represent a control volume. Equation (1) represents the control volume formulation of the conservation of mass.

$$0 = \frac{\partial}{\partial t} \int_{CV} \rho d\mathbf{V} + \int_{CS} \rho \bar{V} \cdot d\bar{A} \qquad (1)$$

Since the flow is steady (not time dependent), Equation (1) becomes

$$0 = \int_{CS} \rho\bar{V} \cdot d\bar{A}$$

In looking at the control volume, there are four sections where mass flows across the control surface; yielding:

$$\int_{CS} \rho\bar{V}\cdot d\bar{A} = \int_{A_1} \rho\bar{V}\cdot dA + \int_{A_2} \rho\bar{V}\cdot dA + \int_{A_3} \rho\bar{V}\cdot dA + \int_{A_4} \rho\bar{V}\cdot dA$$

$$= 0 \qquad (2)$$

Looking at each integral:

Since the flow is in at ①, $\bar{V}\cdot d\bar{A}$ is negative.

$$\int_{A_1} \rho\bar{V}\cdot dA = -\int_{A_1} |\rho V dA| = -|\rho V_1 A_1|$$

Section ③ is flowing out. Making $\bar{V}\cdot dA$ positive,

$$\int_{A_3} \rho\bar{V}\cdot d\bar{A} = \int_{A_3} |\rho V dA| = |\rho V_3 A_3| = \dot{m}_3$$

The flow is in at ④, making the product $\bar{V}\cdot d\bar{A}$ negative.

$$\int_{A_4} \rho\bar{V}\cdot d\bar{A} = -\int_{A_4} |\rho V dA| = -\rho|V_4 A_4| = -\rho|\dot{q}_4|$$

where $\dot{q}$ is the volumetric flow rate.

Substitute back into Eq. (2), and solving for section ② yields:

$$\int_{A_2} \rho\bar{V}_2\cdot d\bar{A}_2 = +|\rho V_1 A_1| - \dot{m}_3 + \rho|\dot{q}_4|$$

$$= \left|1.94\ \frac{\text{slug}}{\text{ft}^3} \times 10\ \frac{\text{ft}}{\text{sec}} \times 0.3\ \text{ft}^2\right| - 3.88\ \frac{\text{slug}}{\text{sec}}$$

$$+ 1.94\ \frac{\text{slug}}{\text{ft}^3}\left|1.0\ \frac{\text{ft}^3}{\text{sec}}\right|$$

$$\int_{A_2} \rho \bar{V} \cdot d\bar{A} = 3.88 \frac{\text{slug}}{\text{sec}} \tag{3}$$

Since $\bar{V} \cdot d\bar{A}$ is positive at section ②, the flow is out. Evaluating the integral at section ② V_2 can be found:

$$\rho V_2 A_2 = 3.88 \frac{\text{slug}}{\text{sec}}$$

$$|V_2| = 3.88 \frac{\text{slug}}{\text{sec}} \times \frac{\text{ft}^3}{1.94 \text{ slug}} \times \frac{1}{0.6 \text{ ft}^2} = 3.33 \frac{\text{ft}}{\text{sec}}$$

$$\bar{V}_2 = |V_2| (\sin\theta \hat{\imath} - \cos\theta \text{j})$$

$$\bar{V}_2 = (1.66\text{i} - 2.88\text{j}) \frac{\text{ft}}{\text{sec}}$$

• PROBLEM 6-2

A stream of fluid flows across a flat stationary surface. As a result of the boundary layer in which the fluid that is in contact with the stationary surface also becomes stationary, a flow pattern develops along the surface where the flow varies substantially from zero velocity at the surface to its maximum value, V_o, at a distance away from the surface.

Given the fluid density, ρ ; thickness of the boundary layer, h; and surface width, W, find the mass flow rate across the surface from the boundary layer where the fluid flow is zero, to the location where the fluid flow has velocity V_o.

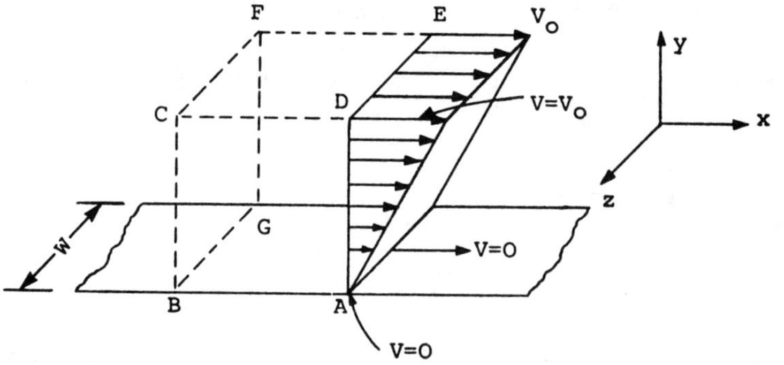

Solution: To facilitate analyzing the flow, we establish a control volume A-G shown in dashed lines. The general flow equation across the surface is given by

$$0 = \frac{\partial}{\partial t} \int_{CV} \rho d\forall + \int_{CS} \rho \overline{V} \cdot d\overline{A} \qquad (1)$$

Assuming steady flow

$$\frac{\partial}{\partial t} \int_{CV} \rho d\forall = 0 \qquad (2)$$

Consequently,

$$\int_{CS} \rho \overline{V} \cdot d\overline{A} = 0 \qquad (3)$$

Assuming that the flow velocity has only one component in the x-direction (there are no vector components along the y and z directions),

$$\int_{CS} \rho \overline{V} \cdot d\overline{A} = 0 = \int_{BC} \rho \overline{V} \cdot d\overline{A} + \int_{CD} \rho \overline{V} \cdot d\overline{A} + \int_{DA} \rho \overline{V} \cdot d\overline{A} + \int_{AB} \rho \overline{V} \cdot d\overline{A} \qquad (4)$$

Since no flow prevails across AB,

$$\int_{AB} \rho \overline{V} \cdot d\overline{A} = 0$$

For the remaining terms in equation (4),

$$\int_{CD} \rho \overline{V} \cdot d\overline{A} = - \int_{BC} \rho \overline{V} \cdot d\overline{A} - \int_{DA} \rho \overline{V} \cdot d\overline{A} \qquad (5)$$

Now $$\int_{BC} \rho \overline{V} \cdot d\overline{A} = - \int_{BC} |\rho V dA| = - \int_{Y_B}^{Y_C} |\rho V W dy|$$

$$= - \int_{0}^{h} |\rho V W dy| = - \left| \int_{0}^{h} \rho V_o W dy \right| = - |\rho V_o W h|$$

To evaluate $\int_{DA} \rho \overline{V} \cdot dA$, the term $\overline{V} \cdot dA$ is positive because the flow of part DA is outward. Therefore,

$$\int_{DA} \rho \overline{V} \cdot d\overline{A} = \int_{0}^{h} |\rho V W dy|$$

Expressing V as a function of y across the surface DA,

$$V = V_o \frac{y}{h}$$

from which

$$\int_o^h |\rho VWdy| = \left| \int_o^h \frac{\rho WV_o ydy}{h} \right| = \left| \frac{\rho WV_o h}{2} \right|$$

Substituting into equation (5),

$$-\int_{CD} \rho\bar{V} \cdot d\bar{A} = \rho V_o Wh - \frac{\rho V_o Wh}{2}$$

$$= \frac{\rho V_o Wh}{2} = \text{mass flow rate}$$

• **PROBLEM 6-3**

A jet plane is being refueled in mid-air by another plane. The refueling boom enters the aircraft at an angle of 30 degrees from its flight path. The fuel flow rate through the boom is 1.5 slugs/sec. at a velocity of 100 ft./sec. What additional lift force is necessary to overcome the increased force on the airplane due to momentum transfer during refueling. (lbf.)

Solution: Using the one dimensional momentum equation for steady flow, the vertical force due to momentum flux is given by

$$\Sigma F_z = \rho Q (V_2 - V_1)$$

$$= \dot{M} V \sin 30°$$

$$= 1.5 \times 100 \times 1/2$$

Therefore the additional lift force is 75 lbf., (333.6 N)

A 1-m-diameter pipe has a 30° horizontal bend in it, as shown, and carries crude oil (sp.gr. = 0.94) at a rate of 2 m^3/s. If the pressure in the bend is assumed to be constant at 75 kPa gage, if the volume of the bend is 1.2 m^3, and if the metal in the bend weighs 4 kN, what forces must be applied to the bend to hold it in place?

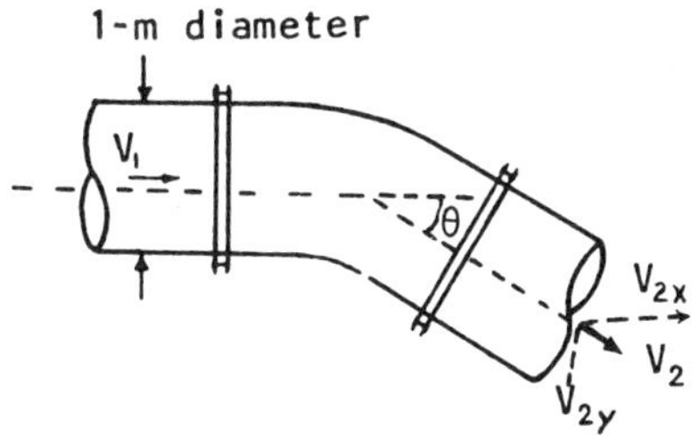

Fig. 1

Solution: For the bend in the pipe of Fig. 1, it is assumed that the flow is steady and that the velocity is uniform across sections 1 and 2. Also, assume that the bend is oriented so that y is vertically upward. In the design of such a bend, one of the primary design considerations is the force required to hold the bend in place. For small- or medium-sized pipes, such a force is supplied by the flange bolts or by welds if it happens to be a welded connection; however, if the pipe is large, specially designed anchor blocks may be required to hold the bend. In the latter case, we are primarily interested in the reactions R_x and R_y needed to hold the bend in place.

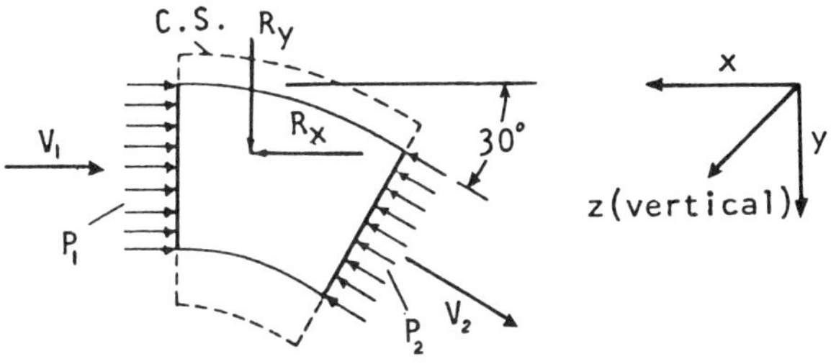

Fig. 2

The analysis is carried out by isolating the bend by means of a control surface as shown in Fig. 2, and the momentum equation is applied for the x, y, and z directions with due care to be certain that all relevant forces such as pressure and gravity are included in the solution.

Isolate the bend by a control surface as shown in Fig. 2 and solve for the unknown forces R_x, R_y, and R_z using the momentum equation. First solve for the forces in the x direction. The flow is steady; therefore, the relevant equation is

$$\sum F_x = \sum \rho u \vec{V} \cdot \vec{A}$$

which becomes

$$R_x - p_1 A_1 + \cos 30° p_2 A_2 = V_{1x} \rho(-V_1 A_1) + V_{2x} \rho(V_2 A_2)$$

$$V_{1x} = \frac{-Q}{A_1} = -2.54 \text{ m/s} \qquad V_{2x} = \frac{-Q}{A_1} \cos 30° = -2.20 \text{ m/s}$$

$$R_x = +p_1 A_1 - 0.866 p_2 A_2 + \rho Q(-V_{1x} + V_{2x})$$

$$= 59 \times 10^3 - 51 \times 10^3 + 0.94 \times 10^3 \times 2(2.54 - 2.20)$$

$$= 8 \times 10^3 + 0.64 \times 10^3 \text{N}$$

$$= 8.64 \text{ kN}$$

For the y direction we have

$$\sum F_y = \sum v \rho \vec{V} \cdot \vec{A}$$

$$R_y - p_2 A_2 \sin 30° = V_{1y} \rho(-V_1 A_1) + V_{2y} \rho(V_2 A_2)$$

But $V_{1y} = 0$ and $V_{2y} = +\frac{Q}{A_2} \sin 30° = +1.27 \text{ m/s}$

$$R_y = +p_2 A_2 \sin 30° + V_{2y} \rho Q$$

$$= 29.5 \text{ kN} + 1.27 \times 0.94 \times 10^3 \times 2\text{N}$$

$$= 31.9 \text{ kN}$$

There are no components of velocity in the z direction (normal to the xy plane); hence, the momentum equation reduces to

$$\sum F_z = 0$$

$$R_z - W_B - W_f = 0$$

$$R_z = W_B + W_f$$

where W_B = weight of bend

W_f = weight of the fluid in bend

$$W_B = 4 \text{ kN} \quad \text{and} \quad W_f = \rho g \forall$$

where $\forall$ = volume of fluid in bend

so $W_f = 0.94 \times 1{,}000 \times 9.81 \times 1.2 = 11.1$ kN

Thus $R_z = 4$ kN + 11.1 kN

$R_z = 15.1$ kN

$\vec{R} = 8.64\vec{i} + 31.9\vec{j} + 15.1\vec{k}$ kN

• PROBLEM 6-5

Consider a jet that is deflected by a stationary vane, such as is given in the figure. If the jet speed and diameter are 100 ft/sec. and 1 in., respectively, and the jet is deflected 60°, what force is exerted on the vane by the jet?

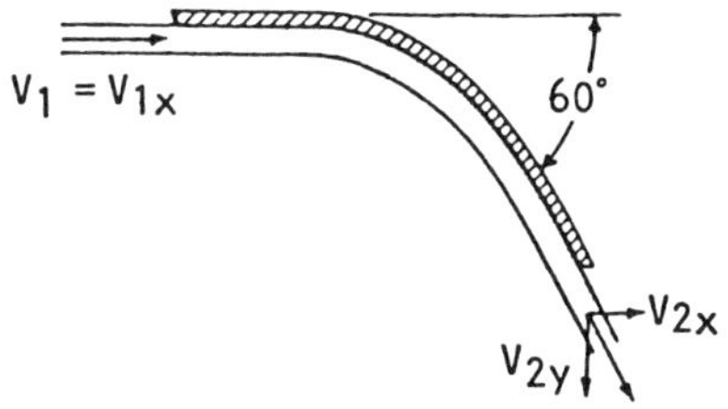

Solution:
$$F_x = \sum_{cs} u\rho\vec{V}\cdot\vec{A} + \frac{d}{dt}\int_{cv} u\rho d\forall \quad (1)$$

If the flow is steady, the second term on the right of Eq. (1) will be zero, which leaves the following:

$$F_x = \sum_{cs} u\rho\vec{V}\cdot\vec{A} \quad (2)$$

At section 1 the velocity is constant over the section and the area vector $\vec{A}_1$ is in the reverse direction of the velocity vector $\vec{V}$: therefore, for this part of the control surface we have

$$\sum_{cs} u\rho\vec{V}\cdot\vec{A} = V_{1x}\rho(-V_1A_1)$$

By a similar analysis for section 2 (in this case, the velocity $\vec{V}_2$ and area $\vec{A}_2$ have the same sense), we get

$$\sum_{cs} u\rho\vec{V}\cdot\vec{A} = V_{2x}\rho V_2A_2$$

However, $V_1A_1 = V_2A_2 = Q$, so when these substitutions are made, Eq. (2) becomes

$$F_x = \rho Q(V_{2x} - V_{1x}) \qquad (3)$$

In a similar manner the force in the y direction, F_y, will be

$$F_y = \rho Q(V_{2y} - V_{1y}) \qquad (4)$$

Since the forces given by Eqs. (3) and (4) are the forces exerted by the vane on the jet, we obtain the forces of the jet on the vane by simply reversing the sign on F_x and F_y.

First solve for F_x, the x component of force of the vane on the jet, by using Eq. (3)

$$F_x = \rho Q(V_{2x} - V_{1x}) \qquad (3)$$

Here, the final velocity in the x direction is given as

$$V_{2x} = 100 \cos 60°$$

Hence, $V_{2x} = 100 \times 0.500 = 50$ ft/sec

also, $V_{1x} = 100$ ft/sec

and $Q = V_1A_1 = 100 \dfrac{0.785}{144} = 0.545$ cfs

Therefore,

$$F_x = 1.94 \text{ lbf-sec}^2/\text{ft}^4 \times 0.545 \text{ ft}^3/\text{sec} \times (50-100) \text{ ft/sec}$$

$$= -52.9 \text{ lbf}$$

Similarly determined, the y component of force on the jet is

$$F_y = 1.94 \text{ lbf-sec}^2/\text{ft}^4 \times 0.545 \text{ ft}^3/\text{sec}$$

$$\times(-86.6 \text{ ft/sec} - 0) = -91.6 \text{ lbf}$$

Then the force on the vane will be the reactions to the forces of the vane on the jet, or

$$F_x = +52.9 \text{ lbf}$$

$$F_y = +91.6 \text{ lbf}$$

Air enters a right-angled elbow in a vertical direction at a mass flow rate of 2 slugs/sec. The air at a density of 0.002 slug/ft³ discharges to atmospheric pressure horizontally through a nozzle having an exit of 5 ft². The attachment of the elbow to the vertical inlet pipe is flexible. Find the horizontal force F_x required to keep the elbow stationary assuming one-dimensional flow conditions.

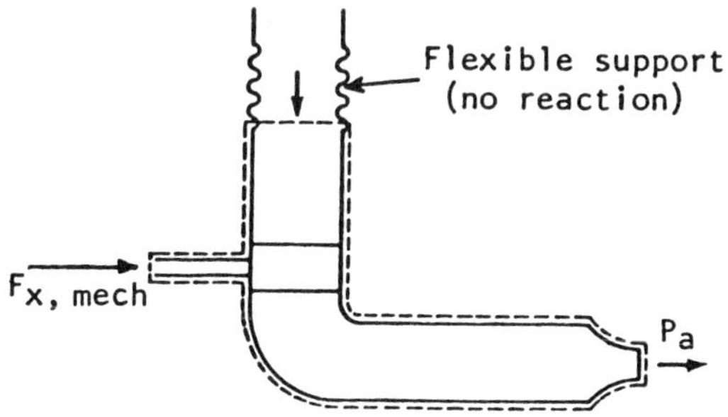

Solution: The control volume is chosen exterior to the elbow as shown in the figure.

For a uniform, steady-state flow, the momentum equation may be expressed by

$$F_{x,mech} = (P_2-P_a)A_2\cos\theta_2 - (P_1-P_a)A_1\cos\theta_1 + \dot{m}(V_{x2} - V_{x1})$$

where θ_1 and θ_2 are the angles between the fluid velocity and its x component at the inlet and outlet, respectively.

Pressure forces cancel out completely as $\cos\theta_1 = 0$ and $P_2 = P_a$.

The upper support is flexible; therefore there is no reaction force from this member and $F_{x,mech}$ is only in the horizontal support.

$$V_{x1} = 0 \qquad V_{x2} = \frac{2 \text{ slugs/sec}}{0.002 \text{ slug/ft}^3 \times 5 \text{ ft}^2} = 200 \text{ ft/sec}$$

$$F_{x,mech} = \dot{m}V_{x2} = 2\,\frac{\text{slugs}}{\text{sec}} \times \frac{1 \text{ lbf}}{\text{slug-ft/sec}^2} \times 200\,\frac{\text{ft}}{\text{sec}} = 400 \text{ lbf}$$

Water flows through the contraction shown at a rate of 7.85 cfs. The head loss due to this particular contraction is given by the empirical equation

$$h_L = 0.2 \frac{V_2^2}{2g}$$

Here V_2 is the velocity in the 8-in. pipe. What horizontal force is required to hold the transition in place if $p_1 = 10$ psi?

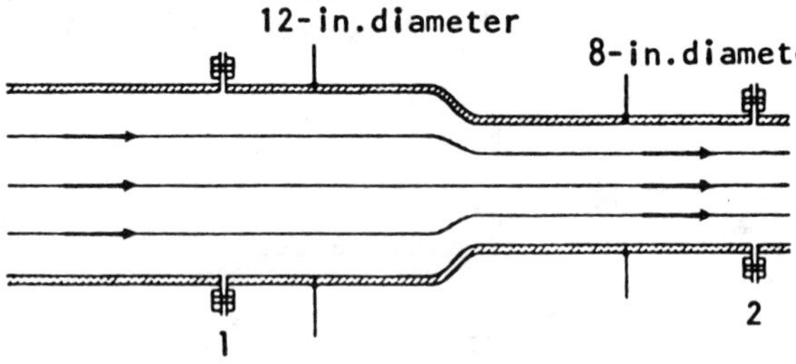

Fig. 1

Solution: Write the momentum equation for the transition between sections 1 and 2. The control surface is drawn so that it encloses the fluid and transition itself; consequently, the control volume is as shown.

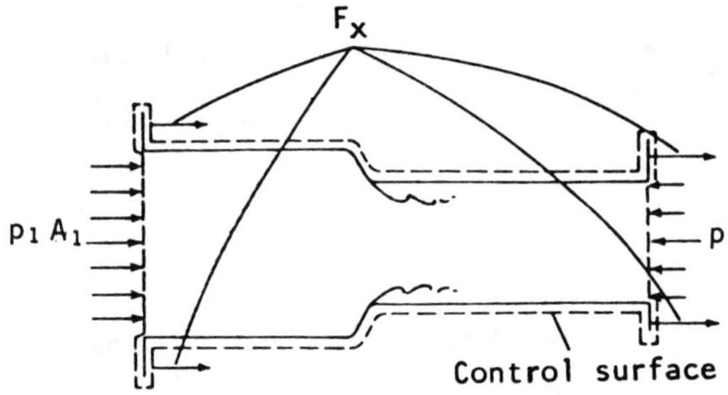

Fig. 2

Also, the pressures are gage pressures, so that the pressure on the exterior of the transition is zero; hence, the force on the exterior is zero. Then, for this control volume the momentum equation is

$$p_1A_1 - p_2A_2 + F_x = \sum_{cs} V_x \vec{V} \cdot \vec{A}$$

$$p_1A_1 - p_2A_2 + F_x = \rho V_2^2 A_2 - \rho V_1^2 A_1$$

$$F_x = \rho Q(V_2 - V_1) + p_2A_2 - p_1A_1$$

Here, Q and the velocities are known and F_x is the unknown. Also, p_2 is still unknown. Obtain p_2 by applying the energy equation between sections 1 and 2:

$$\frac{p_1}{\gamma} + \frac{V_1^2}{2g} + z_1 = \frac{p_2}{\gamma} + \frac{V_2^2}{2g} + z_2 + h_L$$

Assuming, $\alpha_1 = \alpha_2 = 1$. Also, for our problem, $z_1 = z_2$; hence, the foregoing equation reduces to

$$\frac{p_1}{\gamma} + \frac{V_1^2}{2g} = \frac{p_2}{\gamma} + \frac{V_2^2}{2g} + h_L$$

$$p_2 = p_1 - \gamma\left(\frac{V_2^2}{2g} - \frac{V_1^2}{2g} + h_L\right)$$

Substituting this expression for p_2 into the equation for the force on the transition yields

$$F_x = \rho Q(V_2 - V_1) + A_2\left[p_1 - \gamma\left(\frac{V_2^2}{2g} - \frac{V_1^2}{2g} + h_L\right)\right] - p_1 A_1$$

Obtain the velocities in this equation by use of the continuity equation:

$$V_1 = \frac{Q}{A_1} = \frac{7.85}{0.785 \times 1^2} = 10 \text{ ft/sec}$$

$$V_2 = \frac{Q}{A_2} = \frac{7.85}{0.785 \times (2/3)^2} = 22.5 \text{ ft/sec}$$

also $$h_L = \frac{0.2V_2^2}{2g} = \frac{0.2 \times 22.5^2}{64.4} = 1.58 \text{ ft}$$

Then $$F_x = 1.94 \times (7.85)(22.5 - 10)$$

$$+ 0.785 \times 2/3^2\left[10 \times 144 - 62.4\left(\frac{22.5^2}{64.4} - \frac{10^2}{64.4}\right.\right.$$

$$\left.\left. + 1.58\right)\right] - 10 \times 144 \times 0.785 \times 1^2$$

$$= -609 \text{ lbf}$$

Thus it is seen that a force of 609 lbf must be applied in the negative x direction to hold the transition in place for the given conditions.

The reducing bend of the figure is in a vertical plane. Water is flowing, $D_1 = 6$ ft, $D_2 = 4$ ft, $Q = 300$ cfs, $W = 18{,}000$ lb, $z = 10$ ft, $\theta = 120°$, $p_1 = 40$ psi, $x = 6$ ft, and losses through the bend are $0.5\ V_2^2/2g$ ft-lb/lb. Determine F_x, F_y, and the line of action of the resultant force. $\beta_1 = \beta_2 = 1$.

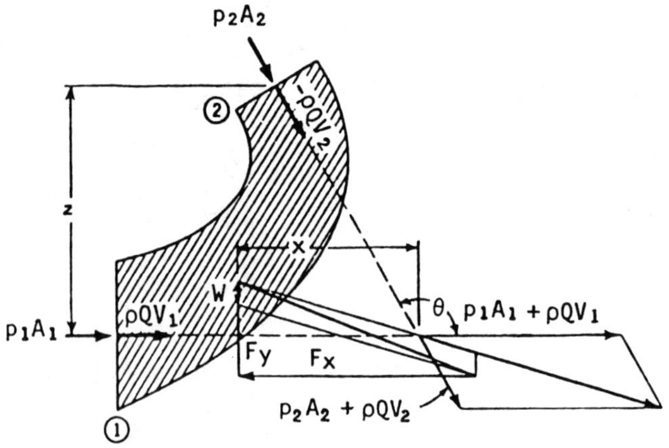

Forces on a reducing elbow, including the vector solution.

Solution: The inside surface of the reducing bend comprises the control-volume surface for the portion of the surface with no flow across it. The normal sections 1 and 2 complete the control surface.

$$V_1 = \frac{Q}{A_1} = \frac{300}{\pi(6^2)/4} = 10.61 \text{ ft/s}$$

$$V_2 = \frac{Q}{A_2} = \frac{300}{\pi(4^2)/4} = 23.87 \text{ ft/s}$$

By application of the energy equation,

$$\frac{p_1}{\gamma} + \frac{V_1^2}{2g} + z_1 = \frac{p_2}{\gamma} + \frac{V_2^2}{2g} + z_2 + \text{losses}_{1-2}$$

$$\frac{40\times144}{62.4} + \frac{10.61^2}{64.4} + 0 = \frac{p_2}{62.4} + \frac{23.87^2}{64.4} + 10 + 0.5 \times \frac{23.87^2}{64.4}$$

from which $p_2 = 4420$ lb/ft^2 = 30.7 psi.

To determine F_x, the momentum equation is:

$$\Sigma\vec{F} = \frac{\partial}{\partial t} \int_{cv} \rho\vec{v}dV + \int_{cs} \rho\vec{v}\vec{v}\cdot d\vec{A} \tag{1}$$

This vector relation may be applied for any component, say the x direction, reducing to

$$\Sigma F_x = \frac{\partial}{\partial t} \int_{cv} \rho v_x dV + \int_{cs} \rho v_x \vec{v}\cdot d\vec{A} \tag{2}$$

In selecting the arbitrary control volume, it is generally advantageous to take the surface normal to the velocity wherever it cuts across the flow. In addition, if the velocity is constant over the surface, the surface integral can be dispensed with. In the figure shown with steady flow Eq. (2) becomes

$$p_1A_1 - p_2A_2\cos\theta - F_x = \rho Q(V_2\cos\theta - V_1)$$

$$40 \times 144\pi(3^2) - 4420\pi(2^2)\cos 120° - F_x$$

$$= 1.935 \times 300(23.87 \cos 120° - 10.61)$$

Since $\cos 120° = -0.5$,

$$162{,}900 + 27{,}750 - F_x = 580.5(-11.94 - 10.61)$$

$$F_x = 203{,}740 \text{ lb}$$

For the y direction

$$\Sigma F_y = \rho Q(V_{y2} - V_{y1})$$

$$F_y - W - p_2A_2\sin\theta = \rho QV_2\sin\theta$$

$$F_y - 18{,}000 - 4420\pi(2^2)\sin 120°$$

$$= 1.935 \times 300 \times 23.87 \sin 120°$$

$$F_y = 78{,}100 \text{ lb}$$

To find the line of action of the resultant force, using the momentum flux vectors (Figure), $\rho QV_1 = 6160$ lb, $\rho QV_2 = 13{,}860$ lb, $p_1A_1 = 162{,}900$ lb, $p_2A_2 = 55{,}560$ lb. Combining these vectors and the weight W in the figure yields the final force, 218,000 lb, which must be opposed by F_x and F_y.

A 12-in.-diameter horizontal pipe terminates in a nozzle with an exit diameter of 3 in., as shown in the figure below. If water flows through the pipe at a rate of 5 cfs, calculate the force exerted on the nozzle.

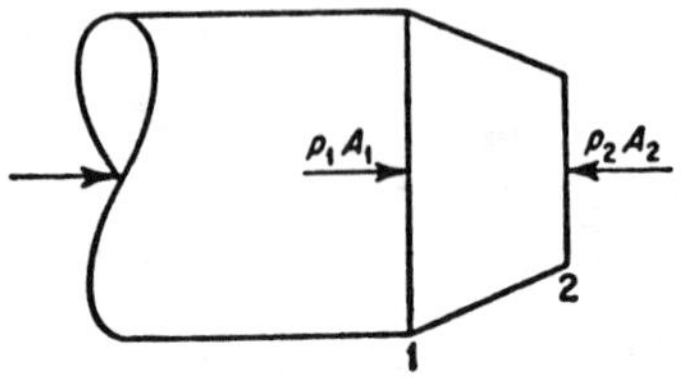

Solution: The fluid at section 1 reaches section 2 in a finite time Δt. If the flow rate through the streamtube is Q cfs, then the mass flowing between sections 1 and 2 is

$$Q\ \Delta t\rho$$

and the equation of motion becomes

$$F = ma = Q\ \Delta t\rho \times \text{acceleration}$$

(Note that F is considered a vector force.)

Now the acceleration is given by

$$\frac{\Delta V}{\Delta t}$$

and therefore

$$F = Q\Delta t\rho\ \frac{\Delta V}{\Delta t} = Q\rho\Delta V \qquad (1)$$

since Δt is a finite time interval.

Also, ΔV is the vector change in velocity and equals

$$V_2 - V_1$$

Hence Eq. (1) can be written in its final vector form,

$$F = Q\rho(V_2 - V_1) \qquad (2)$$

This is called the general impulse-momentum equation.

From symmetry, there is no F_y force. The velocities at stations 1 and 2 are given by

$$A_1V_1 = A_2V_2 = Q$$

Therefore $\qquad V_1 = \dfrac{5}{\pi/4} = 6.3 \text{ fps}$

and $$V_2 = 16 \times 6.3 = 102 \text{ fps}$$

Since the nozzle is discharging into atmosphere, $p_2 = 0$. Applying Bernoulli's equation between stations 1 and 2 to find p_1,

$$\frac{p_1}{62.4} + \frac{6.3^2}{2g} + 0 = \frac{102^2}{2g} + 0 + 0$$

since $z_1 = z_2 = 0$. Therefore

$$p_1 = \frac{102^2 - 6.3^2}{2g} 62.4 = 9{,}850 \text{ psfg}$$

Now applying Eq. (2) in the direction of flow gives

$$p_1A_1 - p_2A_2 - F_x = Q\rho(V_2 - V_1)$$

or $$9{,}850 \times \frac{\pi}{4} - F_x = 5 \times \frac{62.4}{32.2}(102 - 6.3)$$

Hence $$F_x = 9{,}850 \times \frac{\pi}{4} - \frac{5 \times 62.4 \times 95.7}{32.2}$$

$$= 6{,}820 \text{ lb}$$

Since F_x is positive, the assumed direction of F_x was correct. This is the force exerted by the nozzle on the fluid, and therefore the force exerted on the nozzle is 6,820 lb in the downstream direction.

• PROBLEM 6-10

In Fig. 1, the plate is parallel to the flow. The stream is a broad river, or free stream, of uniform velocity $V = U_0 i$. The pressure is assumed uniform, and so it has no net force on the plate. The no-slip condition at the wall brings the fluid there to a halt, and these slowly moving particles retard their neighbors above, so that at the end of the plate there is a significant retarded shear layer, or boundary layer, of thickness $y = \delta$. The viscous stresses along the wall can sum to a finite drag force on the plate. These effects are illustrated in the figure. The problem is to make an integral analysis and find the drag force D in terms of the flow properties ρ, U_0, and δ and the plate dimensions L and b^2.

Solution: This problem requires a combined mass and momentum balance. A proper selection of control volume is essential, and we select the four-sided region from 0 to h to δ to L and back to the origin 0, as shown in Fig. 1.

Had we chosen to cut across horizontally from left to right along the height y = h, we would have cut through the shear layer and exposed unknown shear stresses. Instead we follow the streamline which passes through (x,y) = (0,h), which is outside the shear layer and also has no mass flux across it. The four control-volume sides are thus

1. From (0,0) to (0,h): a one-dimensional inlet, $\vec{V}\cdot\vec{n} = -U_0$
2. From (0,h) to (L,δ): a streamline, no shear, $\vec{V}\cdot\vec{n} \equiv 0$
3. From (L,δ) to (L,0): a two-dimensional outlet, $\vec{V}\cdot\vec{n} = +u(y)$
4. From (L,0) to (0,0): a streamline just above the plate surface, $\vec{V}\cdot\vec{n} = 0$, shear forces summing to the drag force -Di acting from the plate onto the retarded fluid

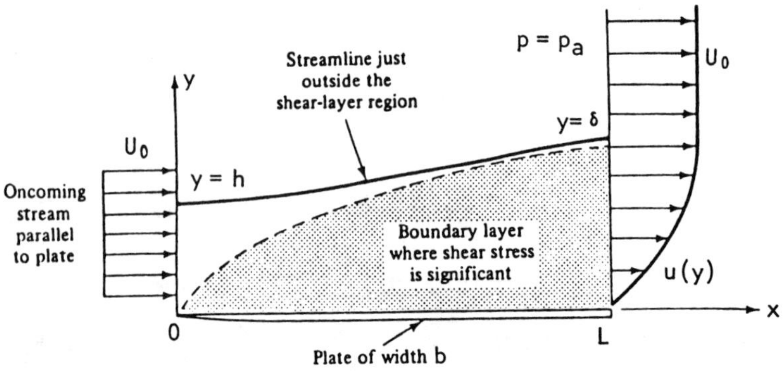

Control-volume analysis of drag force on a flat plate due to boundary shear.

The pressure is uniform, and so there is no net pressure force. Since the flow is assumed incompressible and steady,

$$\Sigma F = \frac{\partial}{\partial t}\left(\iiint_{CV} u\rho dV\right) + \iint_{CS} u\rho (V_r\cdot n)\, dA$$

applies with no unsteady term and fluxes only across sections 1 and 3:

$$\Sigma F_x = -D = \rho \iint_1 u(\vec{V}\cdot\vec{n})\, dA + \rho \iint_3 u(\vec{V}\cdot\vec{n})\, dA$$

$$= \rho \int_0^h U_0(-U_0) b\, dy + \rho \int_0^\delta u(+u) b\, dy$$

Evaluating the first integral and rearranging gives

$$D = \rho U_0^2 bh - \rho b \int_0^\delta u^2 dy \qquad (1)$$

This could be considered the answer to the problem, but it is not useful because the height h is not known with respect to the shear layer thickness δ. This is found by applying mass conservation, since the control volume forms a streamtube

$$\rho \iint_{cs} (\vec{V}\cdot\vec{n})\,dA = 0 = \rho \int_0^h (-U_0)b\,dy + \rho \int_0^\delta ub\,dy$$

or

$$U_0 h = \int_0^\delta u\,dy \qquad (2)$$

after canceling b and ρ and evaluating the first integral. Introduce this value of h into Eq. (1) for a much cleaner result

$$D = \rho b \int_0^\delta u(U_0 - u)\,dy \qquad (3)$$

This result relates the friction drag on one side of a flat plate to the integral of the momentum defect $u(U_0 - u)$ across the trailing cross section of the flow past the plate. Since $U_0 - u$ vanishes as y increases, the integral has a finite value. Equation (3) is an example of momentum-integral theory for boundary layers. To illustrate the magnitude of this drag force, we can use a simple approximation for the outlet-velocity profile u(y) which simulates low-speed, or laminar, shear flow

$$u \approx U_0\left(\frac{2y}{\delta} - \frac{y^2}{\delta^2}\right) \quad \text{for } 0 \le y \le \delta \qquad (4)$$

Substituting into Eq. (3) and letting $\eta = y/\delta$ for convenience, we obtain

$$D = \rho b U_0^2 \delta \int_0^1 (2\eta - \eta^2)(1 - 2\eta + \eta^2)\,d\eta = \frac{2}{15}\rho U_0^2 b\delta \qquad (5)$$

This is within 1 percent of the accepted result from laminar boundary-layer theory in spite of the crudeness of the Eq. (4) approximation. This is a happy situation and has led to the wide use of Kármán's integral theory in the analysis of viscous flows. Note that D increases with the shear-layer thickness δ, which itself increases with plate length and the viscosity of the fluid.

• **PROBLEM 6-11**

Water from a large reservoir is gated as shown. Find the force vector applied to unit width of the gate by the flowing water.

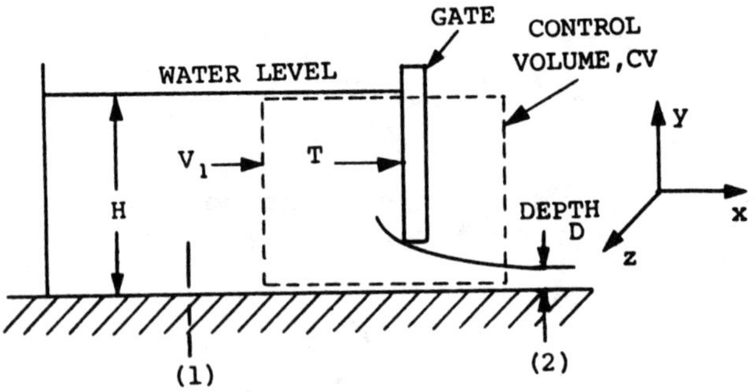

Solution: Using the control volume (CV) method of solution, and the momentum equation for the force component directed along the x-direction,

$$F = \frac{\partial}{\partial t}\int_{CV} q\rho d\forall + \int_{CS} q\rho\vec{V}\cdot d\vec{A} \tag{1}$$

Assuming steady flow,

$$\frac{\partial}{\partial t}\int_{CV} q\rho d\forall = 0$$

and $$F = q_1\{-|\rho V_1 WH|\} + q_2\{|\rho V_2 WD|\} \tag{2}$$

where W = width of the gate.

Assuming hydrostatic pressure distribution,

$$\frac{dP}{dy} = -\rho g,$$

the solution of which is

$$P = P_o + \rho g(y_o - y)$$

where P_o = atmospheric pressure.

Now,

$$F = \int_0^H P_1 dA_1 - \int_0^D P_2 dA_2 - P_o(H - D)W + T$$

$$= \int_0^H \left[P_o + \rho g(H - y)\right] W dy - \int_0^D \left[P_o + \rho g(D - y)\right] W dy$$

$$- P_o(H - D)W + T$$

$$= \frac{\rho g H^2}{2} W - \frac{\rho g D^2}{2} W + T$$

from which

$$F = \frac{\rho g W}{2}(H^2 - D^2) + T \qquad (3)$$

Equating equation (3) for F to equation (2) for F, and setting $q_1 = V_1$ and $q_2 = V_2$,

$$-V_1|\rho V_1 WH| + V_2|\rho V_2 WD| = \frac{\rho g W}{2}(H^2 - D^2) + T$$

Solving for $\frac{T}{W}$, the force on the gate per unit width,

$$\frac{T}{W} = \rho\{(V_2^2 D - V_1^2 H) - \frac{g}{2}(H^2 - D^2)\} \tag{4}$$

• PROBLEM 6-12

a) Water flows through a 180° vertical reducing bend, as shown. The discharge is 7.85 cfs and the pressure at the center of the inlet of the bend is 10 psig. If the bend volume is 3 ft^3 and irrotational flow is assumed, what reaction is required to hold the bend in place? Assume that the metal in the bend weighs 100 lbf.

b) In part (a), if the bend is to be supported from above by an external system of supports, what moment about an axis which is level with the top of the pipe and 18 in. to the right of the flange must the support system be designed for? Assume that the center of mass of the bend and the water taken together is 1 ft below the top of the pipe and 2 ft to the right of the flanges of the bend.

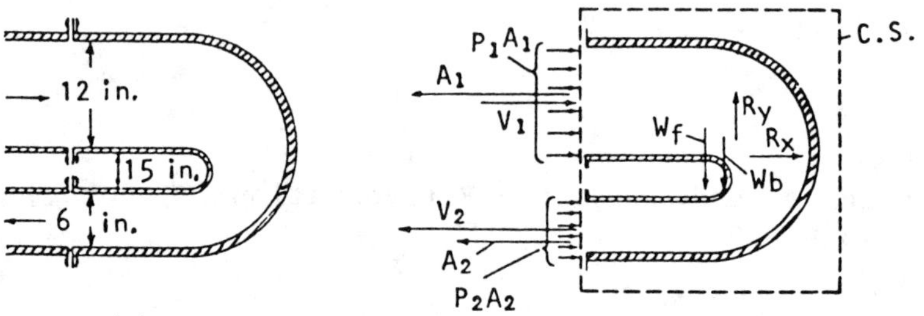

(a) (b)

Fig. 1

Solution: a) The momentum equation assuming steady state is applied with the control volume shown. For the x direction we have

$$\sum F_x = \sum_{cs} V_x \rho \vec{V} \cdot \vec{A}$$

which becomes the following for this example:

$$R_x + p_1 A_1 + p_2 A_2 = u_1 \rho (-V_1 A_1) - u_2 \rho (V_2 A_2)$$

Note that in the x direction, the forces acting on the system are the pressure forces p_1A_1 and p_2A_2, both acting in the positive x direction for this example, and R_x. Here R_x is the net force acting across the metal at the flange. Note also at section 1 that the velocity is assumed to be constant across the section and the area vector $\vec{A}_1$ is pointing in the opposite direction to $\vec{V}_1$; therefore, $u_1 \rho \vec{V} \cdot \vec{A} = V_1 \rho (-V_1 A_1)$. At section 2 the area and velocity vectors point in the same direction, so the sign on $\vec{V} \cdot \vec{A}$ is positive. However $u_2 = -V_2$; consequently, there is also a negative sign on the last term on the right. The reaction in the x direction is then given as

$$R_x = -p_1 A_1 - p_2 A_2 - \rho V_1^2 A_1 - \rho V_2^2 A_2$$

However, $$V_1 A_1 = V_2 A_2 = Q$$

so this reduces to

$$R_x = -p_1 A_1 - p_2 A_2 - \rho Q (V_1 + V_2)$$

The velocities in the foregoing equation are determined by the continuity equation:

$$V_1 = \frac{Q}{A_1} = \frac{7.85}{(\pi/4) \times 1^2} = 10 \text{ ft/sec}$$

$$V_2 = \frac{Q}{A_2} = \frac{7.85}{(\pi/4) \times (1/2)^2} = 40 \text{ ft/sec}$$

The initial pressure p_1 is given; however, p_2 will have to be obtained by the Bernoulli equation written between the center of section 1 and the center of section 2.

$$\frac{p_1}{\gamma} + \frac{V_1^2}{2g} + z_1 = \frac{p_2}{\gamma} + \frac{V_2^2}{2g} + z_2$$

Here $z_2 - z_1 = -2$ ft, $V_1 = 10$ ft/sec, and $V_2 = 40$ ft/sec, so we obtain

$$\frac{p_2}{\gamma} = \frac{10 \times 144}{62.4} + \frac{100}{64.4} + 2 - \frac{40^2}{64.4}$$

$p_2 = 111$ psf

Now we can solve for R_x as follows:

$$R_x = -10 \times 144 \times 0.785 - 111 \times 0.785 \times 1/2^2$$
$$- 1.94 \times 7.85 \times (10 + 40)$$
$$= -1{,}130 - 21.8 - 761$$
$$= -1{,}914 \text{ lbf}$$

Because there are no pressure forces or velocities in the y direction where the flow passes across the control surface, R_y is obtained by static equilibrium:

$$R_y - 100 - 3 \times 62.4 = 0$$
$$= 100 + 187$$
$$= 287 \text{ lbf}$$

The determination of the sign on the unknown reaction R is determined from one basic rule. In writing out the momentum equation, the direction of R is designated and the appropriate sign is given to it. If the assumption is correct, the sign that is obtained when R is solved for will be the same as initially assigned. However, if the assumption is incorrect, the sign that will be obtained will be reversed. To yield the proper sign automatically, most engineers prefer always to write the initial momentum equation with a positive sense on the unknown; then the resulting sign will be the correct one. This latter procedure is the one that was followed in this problem.

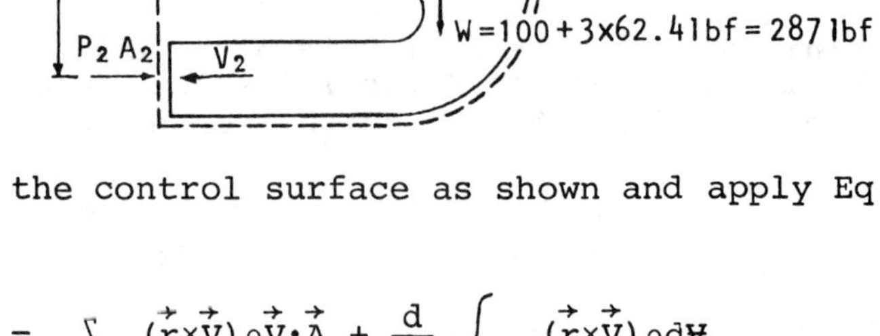

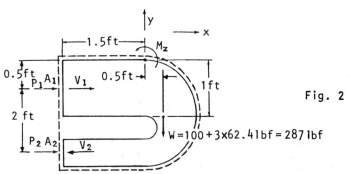

Fig. 2

b) We draw the control surface as shown and apply Eq. (1a).

$$\sum \vec{M} = \sum_{cs} (\vec{r}\times\vec{V})\rho\vec{V}\cdot\vec{A} + \frac{d}{dt}\int_{cv} (\vec{r}\times\vec{V})\rho d\forall \qquad (1a)$$

Equation (1a) is the basic moment-of-momentum equation for cases where the velocity distribution is uniform across the flow section. When the velocity is variable across the flow section, we then use the more general form:

$$\sum \vec{M} = \int_{cs} (\vec{r}\times\vec{V})\rho\vec{V}\cdot d\vec{A} + \frac{d}{dt}\int_{cv} (\vec{r}\times\vec{V})\rho d\forall \qquad (1b)$$

We have steady flow so the last term of Eq. (1a) drops out. Then we have

$$\sum M_{z\ axis} = \sum_{cs} rV\rho\vec{V}\cdot\vec{A}$$

Here the moments will be from the pressure forces, weight of the bend, weight of the water, and external moment M_z applied at the axis in question. When these moments, the velocities, and their moment arms, are inserted in the equation above we get

$$M_z + 0.5p_1A_1 + 2.5p_2A_2 - 0.5 \times 287 = (0.5 \times 10)(-1.94 \times 10$$

$$\times 0.785) + (-2.5 \times 40)\left(1.94 \times 40 \times \frac{0.785}{4}\right)\text{ft-lbf}$$

Here p_1 = 10 psi = 1,440 psf and p_2 = 111 psf, so we solve for M_z:

$$M_z = -0.5 \times 1{,}440 \times 0.785 - 2.5 \times 111 \times \frac{0.785}{4} + 0.5 \times 287$$

$$- 76 - 1{,}523$$

$$= -565 - 54 + 143 - 76 - 1{,}523$$

$$= 2075 \quad \text{ft-lbf}$$

Thus, a moment of 2,075 ft-lbf applied in the clockwise sense about the axis in question is needed. Stated differently, the support system must be designed to withstand a counterclockwise moment of 2,075 ft-lbf.

• PROBLEM 6-13

A jet of water from a nozzle is deflected through an angle θ = 60° from its original direction by a curved vane which it enters tangentially (see Fig. 1) without shock with a mean velocity $\bar{v}_1$, of 30 ms^{-1} and leaves with a mean velocity $\bar{v}_2$ of 25 ms^{-1}. If the discharge $\dot{m}$ from the nozzle is 0·8 kg s^{-1}, calculate the magnitude and direction of the resultant force on the vane if the vane is stationary.

Solution: Assuming one-dimensional flow in a straight line and that the incoming and outgoing velocities v_1 and

v_2 are in the same direction, the momentum equation can be written in the form

$$F = \dot{m}(v_2 - v_1). \quad (1)$$

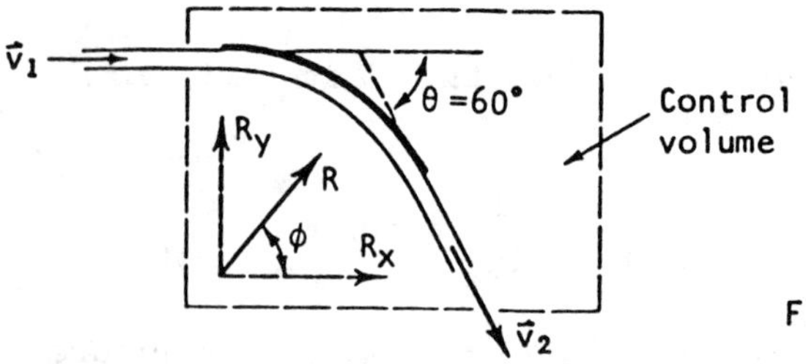

Fig. 1

Force exerted on a curved vane

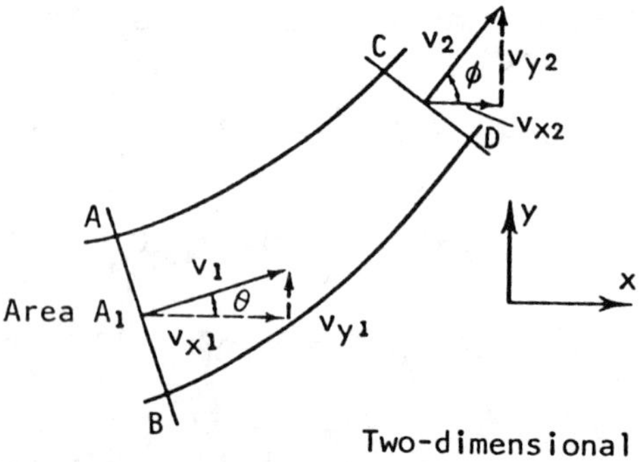

Fig. 2

Two-dimensional

Figure 2 shows a two-dimensional problem in which v_1 makes an angle θ with the x axis while v_2 makes a corresponding angle φ. Since both momentum and force are vector quantities, they can be resolved into components in the x and y directions and equation (1) applied. Thus, if F_x and F_y are the components of the resultant force on the element of fluid ABCD,

F_x = Rate of change of momentum of fluid in x direction

= Mass per unit time × Change of velocity in x direction

$$= \dot{m}(v_2\cos\phi - v_1\cos\theta) = \dot{m}(v_{x2} - v_{x1}).$$

Similarly,

$$F_y = \dot{m}(v_2\sin\phi + v_1\sin\theta) = \dot{m}(v_{y2} - v_{y1}).$$

These components can be combined to give the resultant force,

$$F = \sqrt{(F_x^2 + F_y^2)}.$$

Again, the force exerted by the fluid on the surroundings will be equal and opposite.

For three-dimensional flow, the same method can be used, but the fluid will also have component velocities v_{z1} and v_{z2} in the z direction and the corresponding rate of change of momentum in this direction will require a force

$$F_z = \dot{m}(v_{z2} - v_{z1}).$$

To summarize the position, we can say, in general that

Total force exerted on the fluid in a control volume in a given direction	=	Rate of change of momentum in the given direction of the fluid passing through the control volume,

$$F = \dot{m}(v_{out} - v_{in}).$$

The value of F being positive in the direction in which v is assumed to be positive.

For any control volume, the total force F which acts upon it in a given direction will be made up of three component forces:

F_1 = Force exerted in the given direction on the fluid in the control volume by any solid body within the control volume or coinciding with the boundaries of the control volume;

F_2 = force exerted in the given direction on the fluid in the control volume by body forces such as gravity;

F_3 = force exerted in the given direction on the fluid in the control volume by the fluid outside the control volume.

Thus,

$$F = F_1 + F_2 + F_3 = \dot{m}(v_{out} - v_{in}). \quad (2)$$

The force R exerted by the fluid on the solid body inside or coinciding with the control volume in the given direction will be equal and opposite to F_1 so that $R = -F_1$.

In this problem, the control volume will be as shown in Fig. 1. The resultant force R exerted by the fluid on the vane is found by determining the component forces R_x and R_y in the x and y directions, as shown. Using equation (2),

$$R_x = -F_1 = F_2 + F_3 - \dot{m}(v_{out} - v_{in})_x.$$

Neglecting force F_2 due to gravity and assuming that for a free jet the pressure is constant everywhere, so that $F_3 = 0$,

$$R_x = \dot{m}(v_{in} - v_{out})_x, \quad (3)$$

and, similarly,

$$R_y = \dot{m}(v_{in} - v_{out})_y. \quad (4)$$

Since the nozzle and vane are fixed relative to each other,

$$\text{Mass per unit time entering control volume} = \dot{m} = \text{Mass per unit time leaving nozzle.}$$

In the x direction,

$$v_{in} = \text{Component of } \bar{v}_1 \text{ in x direction} = \bar{v}_1,$$

$$v_{out} = \text{Component of } \bar{v}_2 \text{ in x direction} = \bar{v}_2\cos\theta.$$

Substituting in (3),

$$R_x = \dot{m}(\bar{v}_1 - \bar{v}_2\cos\theta). \quad (5)$$

Putting $\dot{m} = 0{\cdot}8$ kg s^{-1}, $\bar{v}_1 = 30$ m s^{-1}, $\bar{v}_2 = 25$ m s^{-1}, $\theta = 60°$,

$$R_x = 0{\cdot}8\ (30 - 25\cos 60°) = 14\ \text{N}.$$

In the y direction,

$$v_{in} = \text{Component of } \bar{v}_1 \text{ in y direction} = 0,$$

$$v_{out} = \text{Component of } \bar{v}_2 \text{ in y direction} = \bar{v}_2\sin\theta.$$

Thus, from (4),

$$R_y = \dot{m}\ \bar{v}_2 \sin\theta. \quad (6)$$

Putting in the numerical values,

$$R_y = 0{\cdot}8 \times 25 \sin 60° = 17{\cdot}32\ \text{N}.$$

Combining tne rectangular components R_x and R_y,

$$\text{Resultant force exerted by fluid on vane, } R = \sqrt{(R_x^2 - R_y^2)}$$

$$= \sqrt{(14^2 + 17{\cdot}32^2)}$$

$$= 22{\cdot}27\ \text{N}.$$

This resultant force R will be inclined to the x direction at an angle $\phi = \tan^{-1}(R_y/R_x) = \tan^{-1}(17{\cdot}32/14) = 51°$.

A steady laminar flow exists in a conduit, with all streamlines parallel to the walls. Determine the simplified differential equations which govern the flow.

Solution: The equation of continuity in differential form is given by

$$\frac{\partial u}{\partial x} + \frac{\partial v}{\partial y} + \frac{\partial w}{\partial z} = 0$$

If all streamlines are parallel to the wall (and in the x-direction) then $u \neq 0$, and $v = w = 0$. The equation of continuity

$$\frac{\partial u}{\partial x} + \overbrace{\frac{\partial v}{\partial y}}^{0} + \overbrace{\frac{\partial w}{\partial z}}^{0} = 0$$

shows that $u \neq u(x)$; that is, $u = u(y,z)$ where y and z are normal to the walls. The velocity profile is independent of x. The momentum equations

$$\rho\left[\frac{\partial u}{\partial t} + u\frac{\partial u}{\partial x} + v\frac{\partial u}{\partial y} + w\frac{\partial u}{\partial z}\right]$$
$$= -\frac{\partial p}{\partial x} + \rho g_x + \mu\left[\frac{\partial^2 u}{\partial x^2} + \frac{\partial^2 u}{\partial y^2} + \frac{\partial^2 u}{\partial z^2}\right]$$

$$\rho\left[\frac{\partial v}{\partial t} + u\frac{\partial v}{\partial x} + v\frac{\partial v}{\partial y} + w\frac{\partial v}{\partial z}\right]$$
$$= -\frac{\partial p}{\partial y} + \rho g_y + \mu\left[\frac{\partial^2 v}{\partial x^2} + \frac{\partial^2 v}{\partial y^2} + \frac{\partial^2 v}{\partial z^2}\right]$$

$$\rho\left[\frac{\partial w}{\partial t} + u\frac{\partial w}{\partial x} + v\frac{\partial w}{\partial y} + w\frac{\partial w}{\partial z}\right]$$
$$= -\frac{\partial p}{\partial z} + \rho g_z + \mu\left[\frac{\partial^2 w}{\partial x^2} + \frac{\partial^2 w}{\partial y^2} + \frac{\partial^2 w}{\partial z^2}\right]$$

take the forms

$$0 = -\frac{\partial p}{\partial x} + \rho g_x + \mu\left(\frac{\partial^2 u}{\partial y^2} + \frac{\partial^2 u}{\partial z^2}\right)$$

$$0 = -\frac{\partial p}{\partial y} + \rho g_y$$

$$0 = -\frac{\partial p}{\partial z} + \rho g_z$$

Consider the conduit to be horizontal so that $g_x = 0$, and neglect the pressure variation normal to the conduit (this is permissible if the conduit is small); then we would have

$$\frac{dp}{dx} = \mu\left(\frac{\partial^2 u}{\partial y^2} + \frac{\partial^2 u}{\partial z^2}\right)$$

With the appropriate boundary conditions this can be solved.

• PROBLEM 6-15

A common type of viscometer employs a cylinder rotating inside of another cylinder as shown in Fig. 1. Assuming that (1) flow between the cylinders is laminar as the inner cylinder rotates, (2) fluid properties are constant, (3) radial and vertical velocity components are zero, (4) the flow does not vary in the z-direction, estimate the torque required to turn the cylinder in terms of the cylinder geometry, rotational speed, and fluid viscosity. Neglect the influence of the base of the viscometer and consider only the cylinder walls.

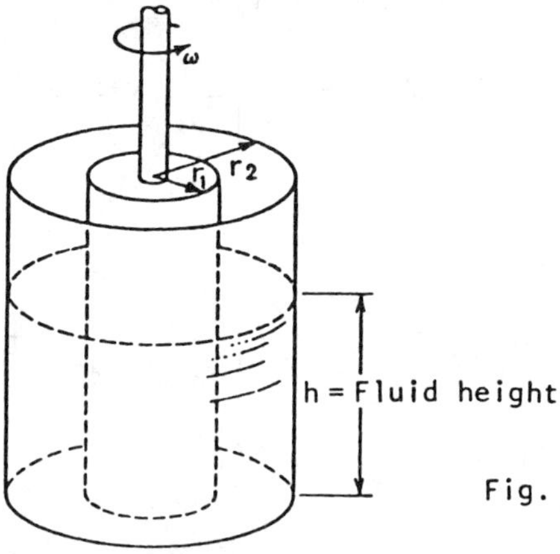

Fig. 1

Solution: Choose a polar cylindrical system with the z-coordinate vertically upward (see Fig. 2). The continuity equation, in cylindrical coordinates, is given by

$$\frac{\partial r V_r}{\partial r} + \frac{\partial V_\theta}{\partial \theta} + \frac{\partial r V_z}{\partial z} = 0$$

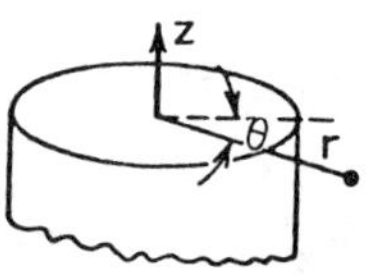

Fig. 2

Under the basic assumptions, the continuity equation reduces to

$$\frac{\partial V_\theta}{\partial \theta} = 0$$

Because V_θ is not dependent on z, it follows that

$$V_\theta = f(r)$$

The momentum equation, in cylindrical coordinates, is given by

$$\frac{\partial V_\theta}{\partial t} + V_r \frac{\partial V_\theta}{\partial r} + \frac{V_\theta}{r}\frac{\partial V_\theta}{\partial \theta} + V_z \frac{\partial V_\theta}{\partial z} + \frac{V_r V_\theta}{r}$$

$$= F_\theta - \frac{1}{\rho r}\frac{\partial P}{\partial \theta}$$

$$+ v\left(\frac{\partial^2 V_\theta}{\partial r^2} + \frac{1}{r}\frac{\partial V_\theta}{\partial r} + \frac{1}{r^2}\frac{\partial^2 V_\theta}{\partial \theta^2} + \frac{\partial^2 V_\theta}{\partial z^2}\right.$$

$$\left. + \frac{2}{r^2}\frac{\partial V_r}{\partial \theta} - \frac{V_\theta}{r^2}\right)$$

The momentum equation becomes ($F_\theta = 0$; $\partial P/\partial\theta = 0$, because quantities are assumed not to vary in the θ-direction):

$$v\left(\frac{\partial^2 V_\theta}{\partial r^2} + \frac{1}{r}\frac{\partial V_\theta}{\partial r} - \frac{V_\theta}{r^2}\right) = 0$$

This equation can be rewritten as

$$\frac{d}{dr}\left(\frac{dV_\theta}{dr} + \frac{V_\theta}{r}\right) = 0$$

Therefore

$$\frac{dV_\theta}{dr} + \frac{V_\theta}{r} = C_1$$

or

$$\frac{1}{r}\frac{drV_\theta}{dr} = C_1$$

It follows from direct integration that

$$rV_\theta = \frac{C_1 r^2}{2} + C_2$$

Therefore

$$V_\theta = \frac{C_1 r}{2} + \frac{C_2}{r} \tag{1}$$

The boundary conditions at the inner and outer cylinder walls are

$$V_\theta = r_1\omega \qquad \text{at } r = r_1$$

$$V_\theta = 0 \qquad \text{at } r = r_2$$

Substituting these values into Eq. (1) gives

$$r_1\omega = \frac{C_1 r_1}{2} + \frac{C_2}{r_1}$$

$$0 = \frac{C_1 r_2}{2} + \frac{C_2}{r_2}$$

Solving for C_1 and C_2 gives

$$C_1 = -\frac{2r_1^2\omega}{r_2^2 - r_1^2}, \qquad C_2 = \frac{r_2^2 r_1^2}{r_2^2 - r_1^2}\,\omega$$

It follows, after substituting C_1 and C_2 into Eq. (1), that

$$V_\theta = \frac{\omega \dfrac{r_1^2}{r}\left[\left(\dfrac{r_2}{r_1}\right)^2 - \left(\dfrac{r}{r_1}\right)^2\right]}{\left(\dfrac{r_2}{r_1}\right)^2 - 1}$$

The torque on the fluid exerted by the surroundings at $r = r_1$ is found from the shear stress relation as given by

$$\tau_{r-\theta} = -\mu r \frac{d}{dr}\left(\frac{V_\theta}{r}\right) = -\mu r\left[\frac{r_2^2 r_1^2\omega}{r_2^2 - r_1^2}\left(-\frac{2}{r^3}\right)\right]$$

At $r = r_1$

$$\tau_{r-\theta} = \frac{2\mu\omega r_2^2}{r_2^2 - r_1^2}$$

The torque exerted by the cylinder is

$$\tau_{r-\theta} 2\pi r_1 h r_1 = \frac{4\pi\omega\mu r_1^2 r_2^2 h}{r_2^2 - r_1^2}$$

• PROBLEM 6-16

Find the torque developed on the runner of a Francis turbine (Fig. 1) which has the following known data:

(a) Outlet velocity polygon (see Fig. 2)

(b) $Q, \rho, \dot{m}$ are constant

(c) The inlet velocity V_1, is tangent to the stationary vanes

(d) Tip and hub diameters

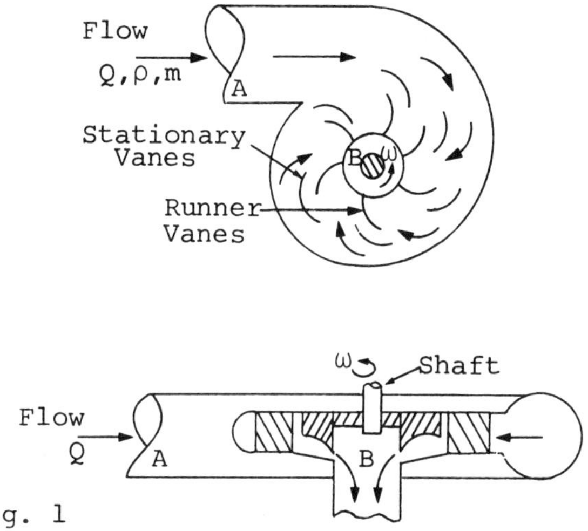

Fig. 1

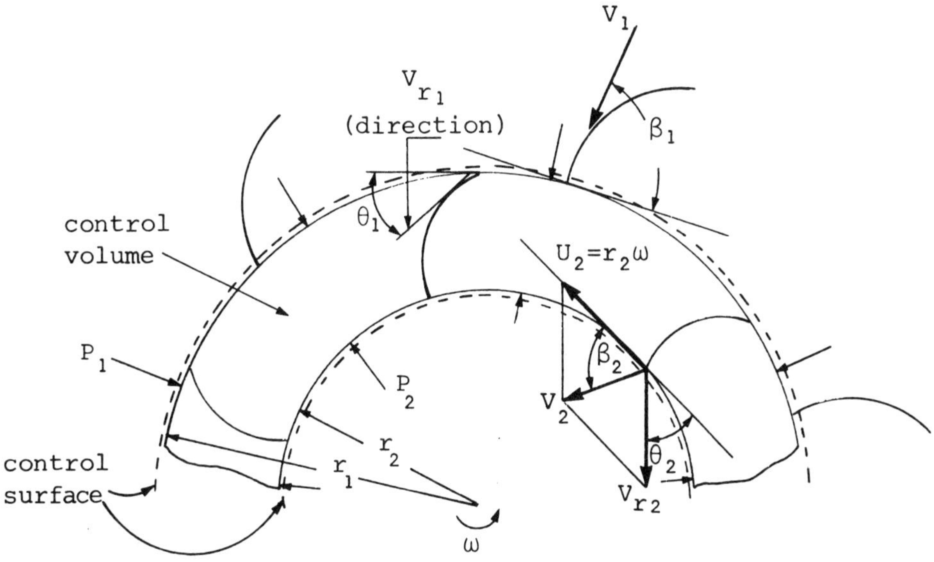

Fig. 2

Solution: A stationary control volume enclosing the runner is chosen, as shown in Fig. 2. The basic moment of momentum equation for a stationary control volume is

$$\vec{r} \times \vec{F} + \int_{cv} (\vec{r} \times \vec{g})\rho d\vec{V} + \vec{T}_{shaft}$$

$$= \int_{cs} (\vec{r}\times\vec{V})\rho\vec{V} \cdot d\vec{A} + \frac{\partial}{\partial t} \int_{cv} (\vec{r} \times \vec{V})\rho d\vec{V} \quad (1)$$

Assumptions: (1) Torque due to body forces equal to zero
(2) Steady flow
(3) Constant angular velocity
(4) Incompressible flow
(5) No friction

Under these assumptions, equation (1) reduces to:

$$T_{shaft} = \int_{cs} (\vec{r} \times \vec{V})\rho\vec{V} \cdot d\vec{A} \quad (2)$$

There are only normal stresses from the fluid on the control surface. Consequently, only the surfaces parallel to the axis of rotation will be considered. Here, we have surfaces of concentric cylinders about the axis where the torque contribution due to the normal stresses (P_1, P_2) is zero. It can be seen that on the surface of the runner a torque can be developed. This torque has to be the one transmitted by the shaft to the runner, T_{shaft}. Thus, equation (2) reduces to:

$$T_{shaft} = \vec{r}\vec{V}\rho Q \quad (3)$$

$$V_{in} = V_1 \cos \beta_1$$

$$V_{out} = wr_2 - V_2 \cos\theta_2$$

Introducing these values in eq.(3) we get,

$$T_{shaft} = - r_1(V_1 \cos\beta_1)\rho Q + r_2(wr_2 - Vr_2 \cos\theta_2)\rho Q$$

or

$$T_{shaft} = -\rho Q\left[r_1 V_1 \cos\beta_1 + r_2(V_{r_2}\cos\theta_2 - wr_2)\right]$$

Considering the control volume as a free body, it can be seen that the torque due to the fluid acting on the runner vanes is equal and opposite to the torque on the shaft. Thus, the torque developed on the runner is

$$T = \rho Q\left[r_1 V_1 \cos\beta_1 + r_2(V_{r_2} \cos\theta_2 - wr_2)\right]$$

• PROBLEM 6-17

A. A water jet of velocity V_j impinges normal to a flat plate which moves to the right at velocity V_c, as shown in Fig. 1a. Find the force required to keep the plate moving at constant velocity if the jet density is 1000 kg/m^3, the jet area is 3 cm^2, and V_j and V_c are 20 and 15 m/s, respectively. Neglect the weight of the jet and plate and assume steady flow with respect to the moving plate with the jet splitting into an equal upward and downward half-jet.

B. Repeat part (A) with the plate and its cart unrestrained horizontally and thus allowed to accelerate to the right. Derive (a) the equation of motion for cart velocity $V_c(t)$ and (b) the time required for the cart to accelerate from rest to 95 percent of the jet velocity and (c) compute actual numerical values from (b) for the conditions of part (A) and a cart mass of 3 kg. Neglect cartwheel friction.

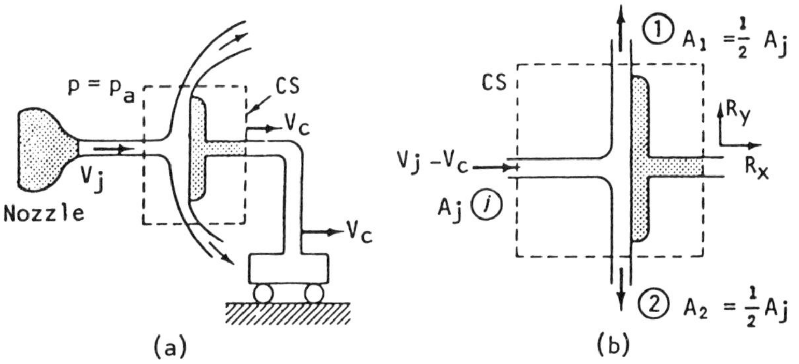

Force on a plate moving at constant velocity: (*a*) jet striking a moving plate normally; (*b*) control volume fixed relative to the plate. Fig. 1

Solution: A. The suggested control volume in Fig. 1a cuts through the plate support to expose the desired forces R_x and R_y. This control volume moves at speed V_c and thus is fixed relative to the plate, as in Fig. 1b. We must satisfy both mass and momentum conservation for the assumed steady-flow pattern in Fig. 1b. There are two outlets and one exit, and mass conservation applies,

$$\dot{m}_{out} = \dot{m}_{in}$$

or $$\rho_1 A_1 V_1 + \rho_2 A_2 V_2 = \rho_j A_j (V_j - V_c) \quad (1)$$

We assume that the water is incompressible, $\rho_1 = \rho_2 = \rho_j$, and we are given that $A_1 = A_2 = \frac{1}{2}A_j$. Therefore Eq. (1) reduces to

$$V_1 + V_2 = 2(V_j - V_c) \quad (2)$$

Strictly speaking, this is all that mass conservation tells us. However, from the symmetry of the jet deflection and the neglect of fluid weight, we conclude that the two velocities V_1 and V_2 must be equal, and hence (2) becomes

$$V_1 = V_2 = V_j - V_c \quad (3)$$

For the given numerical values, we have

$$V_1 = V_2 = 20 - 15 = 5 \text{ m/s}$$

Now we can compute R_x and R_y from the two components of momentum conservation. Equation applies with the unsteady term zero

$$\sum F_x = R_x = \dot{m}_1 u_1 + \dot{m}_2 u_2 - \dot{m}_j u_j \quad (4)$$

where from the mass analysis, $\dot{m}_1 = \dot{m}_2 = \frac{1}{2}\dot{m}_j = \frac{1}{2}\rho_j A_j (V_j - V_c)$. Now check the flow directions at each section: $u_1 = u_2 = 0$, and $u_j = V_j - V_c = 5$ m/s. Thus Eq. (4) becomes

$$R_x = -\dot{m}_j u_j = -[\rho_j A_j (V_j - V_c)](V_j - V_c) \quad (5)$$

For the given numerical values, we have

$$R_x = -(1000\ \mathrm{kg/m^3})(0.0003\ \mathrm{m^2})(5\ \mathrm{m/s})^2 = -7.5(\mathrm{kg\cdot m})/\mathrm{s}^2$$

$$= -7.5\ \mathrm{N} \qquad \text{Ans. (a)}$$

This acts to the left; i.e., it requires a restraining force to keep the plate from accelerating to the right due to the continuous impact of the jet. The vertical force is

$$F_y = R_y = \dot{m}_1 v_1 + \dot{m}_2 v_2 - \dot{m}_j v_j$$

Check directions again: $v_1 = V_1, v_2 = -V_2, v_j = 0$. Thus

$$R_y = \dot{m}_1(V_1) + \dot{m}_2(-V_2) = \frac{1}{2}\dot{m}_j(V_1 - V_2) \tag{6}$$

But, since we found earlier that $V_1 = V_2$, this means that $R_y = 0$, as we could expect from the symmetry of the jet deflection. Two other results are of interest. First, the relative velocity at section 1 was found to be 5 m/s up, from Eq. (3). If we convert this to absolute motion by adding on the control-volume speed $V_c = 15$ m/s to the right, we find that the absolute velocity $\vec{V}_1 = 15\vec{i} + 5\vec{j}$ m/s, or 15.8 m/s at an angle of 18.4° upward, as indicated in Fig. 1a. Thus the absolute jet speed changes after hitting the plate. Second, the computed force R_x does not change if we assume the jet deflects in all radial directions along the plate surface rather than just up and down. Since the plate is normal to the x-axis, there would still be zero outlet x-momentum flux when Eq. (4) was rewritten for a radial-deflection condition.

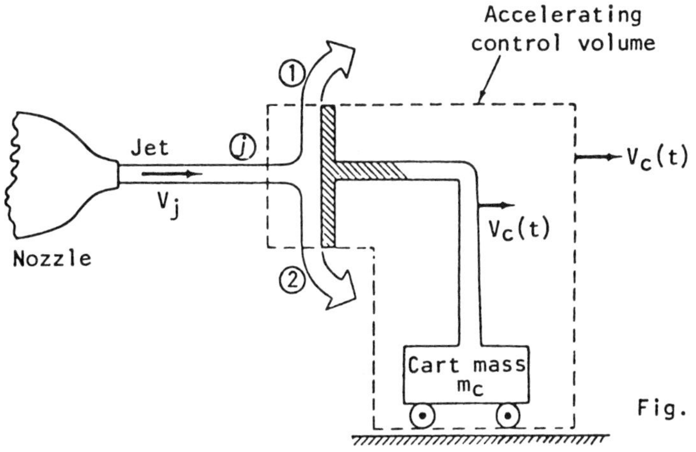

Fig. 2

B. (a) The control volume is shown in Fig. 2. to include the cart and its wheel. The control volume moves at speed $V_c(t)$. The flow within the control volume is unsteady because the inlet velocity $V_j - V_c$ varies with time. The mass-conservation relation is

$$\frac{\partial}{\partial t}\left(\iiint_{CV} \rho dV\right) + \rho_1 A_1 V_1 + \rho_2 A_2 V_2 - \rho_j A_j (V_j - V_c) = 0 \tag{7}$$

Again assume incompressible flow so that the density cancels out. The triple integral is the mass within the control volume, which remains constant if the jet cross section is constant. Therefore we conclude that the time derivative of the triple integral is zero, and the previous result holds

$$V_1 = V_2 = V_j - V_c \tag{8}$$

These speeds V_1 and V_2 are relative to the moving cart. The momentum relation is now applied to the x direction with the coordinates xyz fixed to the accelerating cart

$$\sum F_x - \iiint_{CV} a_{x,rel}\, dm = \frac{\partial}{\partial t}\left(\iiint_{CV} u\rho dV\right) + \iint_{CS} u\rho(\vec{V}\cdot\vec{n})\,dA$$

Now the cart is unrestrained, and the pressure is constant; hence $\sum F_x = 0$. The relative acceleration of the control volume is none other than the rate of change of cart speed since the cart is not rotating relative to inertial coordinates

$$a_{x,rel} = \frac{d}{dt}(V_c) \tag{9}$$

The area integrals in the right-hand side are the same as found in part (A)

$$\iint_{CS} u\rho(\vec{V}\cdot\vec{n})\,dA = -\dot{m}_j u_j = -\rho_j A_j (V_j - V_c)^2 \tag{10}$$

with only the jet inlet flux having nonzero u velocity. Finally, evaluation of the two remaining (triple) integrals requires an assumption about the mass and x momentum of the fluid within the control volume. Since these are probably small with respect to the cart mass and momentum, we apply a little art: we neglect the contribution of fluid mass to the volume integrals, so that the integrals are, approximately,

$$\iiint_{CV} a_{x,rel}\, dm \approx m_c \frac{dV_c}{dt} \qquad \frac{\partial}{\partial t}\left(\iiint_{CV} u\rho dV\right) \approx 0$$

where m_c is the mass of the cart and plate. Substituting back gives

$$0 - m_c \frac{dV_c}{dt} = 0 - \rho A_j (V_j - V_c)^2$$

or

$$\frac{dV_c}{dt} = K(V_j - V_c)^2 \quad K = \frac{\rho A_j}{m_c} \qquad (11)$$

Ans. (a)

This is the desired differential equation for the cart speed.

(b) We can find the time to reach 95 percent of the jet speed by integrating Eq.(11), assuming that K and V_j are constant. Separate the variables and use a lower limit starting from rest, that is, $V_c = 0$ at $t = 0$

$$\int_0^{V_c} (V_j - V_c)^{-2} dV_c = \int_0^t K\, dt$$

Evaluate the integrals

$$\frac{1}{V_j - V_c} - \frac{1}{V_j} = Kt$$

or

$$\frac{V_c}{V_j} = \frac{V_j Kt}{1 + V_j Kt} \qquad (12)$$

This is the general solution for the cart speed $V_c(t)$ when starting from rest. The time t* when $V_c/V_j = 0.95$ is thus given by

$$0.95 = \frac{V_j Kt^*}{1 + V_j Kt^*}$$

or

$$t^* = \frac{19}{KV_j} = \frac{19 m_c}{\rho A_j V_j}$$

Ans. (b)

(c) The constant $1/KV_j$, is an indication of the time response of the cart when struck by the jet. For the given conditions,

$$t^* = \frac{19(3\ \text{kg})}{(1000\ \text{kg/m}^3)(0.0003\ \text{m}^2)(20\ \text{m/s})} = 9.5\ \text{s}$$

Ans. (c)

MOMENT OF MOMENTUM EQUATION

• PROBLEM 6-18

a.) The sprinkler shown in Fig. 3 discharges water upward and outward from the horizontal plane so that it makes an angle of θ° with the t axis when the sprinkler arm is at rest. It has a constant cross-sectional flow area of A_0 and discharges q cfs starting with $\omega = 0$ and $t = 0$. The resisting torque due to bearings and seals is the constant T_0, and the moment of inertia of the rotating empty sprinkler head is I_s. Determine the equation for ω as a function of time.

b.) A turbine discharging 10 m^3/s is to be so designed that a torque of 10,000 N·m is to be exerted on an impeller turning at 200 rpm that takes all the moment of momentum out of the fluid. At the outer periphery of the impeller, r = 1 m. What must the tangential component of velocity be at this location?

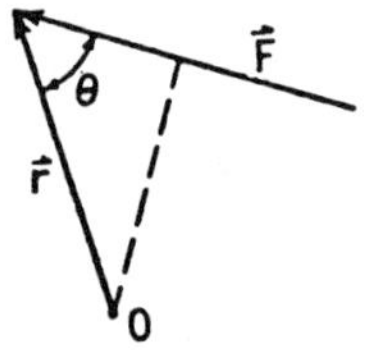

Fig. 1

Notation for moment of a vector

Solution: The general unsteady linear-momentum equation applied to a control volume is

$$\vec{F} = \frac{\partial}{\partial t}\int_{cv} \rho\vec{v}dV + \int_{cs} \rho\vec{v}\vec{v}\cdot d\vec{A} \qquad (1)$$

The moment of a force $\vec{F}$ about a point O (Fig. 1) is given by

$$r \times \vec{F}$$

or

$$Fr \sin\theta$$

By taking $\vec{r} \times \vec{F}$, using Eq. (1),

$$\vec{r} \times \vec{F} = \frac{\partial}{\partial t}\int_{cv} \rho\vec{r} \times \vec{v}dV + \int_{cs} (\rho\vec{r} \times \vec{v})(\vec{v}\cdot d\vec{A}) \qquad (2)$$

The left-hand side of the equation is the torque exerted by any forces on the control volume, and terms on the right-hand side represent the rate of change of moment of momentum within the control volume plus the net efflux of moment of momentum from the control volume. This is the general moment-of-momentum equation for a control volume. It has great value in the analysis of certain flow problems, e.g., in turbomachinery, where torques are more significant than forces.

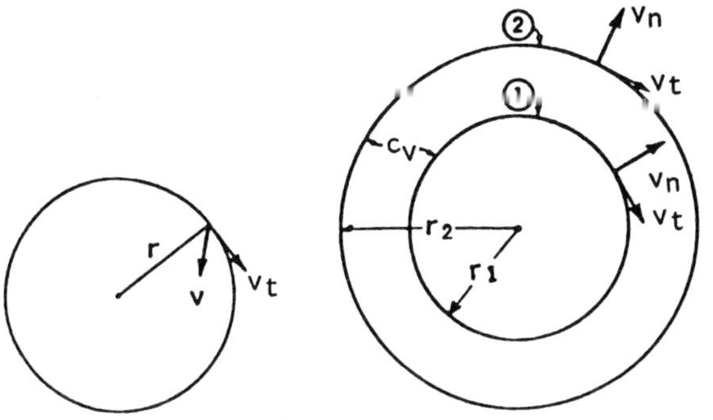

Fig. (2a) Fig. (2b)

Two-dimensional flow in a centrifugal pump impeller.

When Eq. (2) is applied to a case of flow in the xy plane, with r the shortest distance to the tangential component of the velocity v_t, as in Fig. 2a, and v_n is the normal component of velocity,

$$F_t r = T_z = \int_{cs} \rho r v_t v_n dA + \frac{\partial}{\partial t} \int_{cv} \rho r v_t dV \qquad (3)$$

in which T_z is the torque. A useful form of Eq. (3) applied to an annular control volume, in steady flow (Fig. 2b), is

$$T_z = \int_{A2} \rho_2 r_2 v_{t2} v_{n2} dA_2 - \int_{A1} \rho_1 r_1 v_{t1} v_{n1} dA_1 \qquad (4)$$

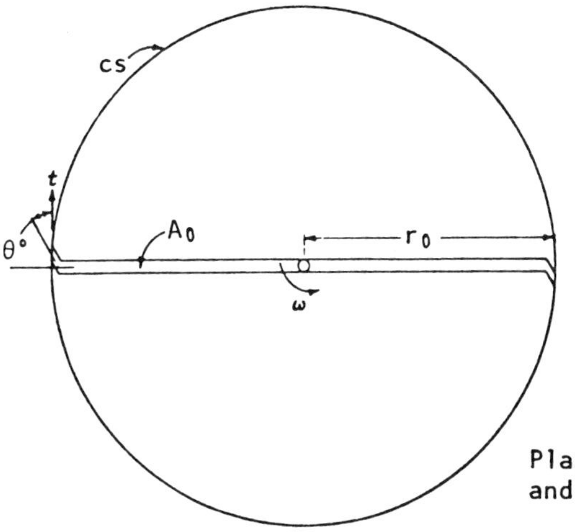

Fig. 3

Plan view of sprinkler and control surface.

For complete circular symmetry, where r, ρ, v_t, and v_n are constant over the inlet and outlet control surfaces, it takes the simple form

$$T_z = \rho Q[(r v_t)_2 - (r v_t)_1] \qquad (5)$$

since $\int \rho v_n dA = \rho Q$, the same at inlet or outlet.

a.) Equation (2) may be applied. The control volume is the cylindrical area enclosing the rotating sprinkler head. The inflow is along the axis, so that it has no moment of momentum; hence, the torque $-T_0$ due to friction is equal to the time rate of change of moment of momentum of sprinkler head and fluid within the sprinkler head plus the net efflux of moment of momentum from the control volume. Let $V_r = q/2A_0$

$$-T_0 - 2\frac{d}{dt}\int_0^{r_0} A_0\rho\omega r^2 dr + I_s\frac{d\omega}{dt} - \frac{2\rho q r_0}{2}(V_r\cos\theta - \omega r_0)$$

The total derivative may be used. Simplifying gives

$$\frac{d\omega}{dt}\left(I_s + \frac{2}{3}\rho A_0 r_0^3\right) = \rho q r_0 (V_r\cos\theta - \omega r_0) - T_0$$

For rotation to start, $\rho q r_0 V_r \cos\theta$ must be greater than T_0. The equation is easily integrated to find ω as a function of t. The final value of ω is obtained by setting $d\omega/dt = 0$ in the equation.

b.) Equation (5) is

$$T = \rho Q(rv_t)_{in}$$

in this case, since the outflow has $v_t = 0$. By solving for $v_{t_{in}}$,

$$v_{t_{in}} = \frac{T}{\rho Q r} = \frac{10,000 \text{ N}\cdot\text{m}}{(1000 \text{ kg/m}^3)(10 \text{ m}^3/\text{s})(1 \text{ m})} = 1,000 \text{ m/s}$$

• PROBLEM 6-19

A nozzle discharges water with a speed of 30 m/sec at an angle of 45° to the horizontal (Fig. 1). The 50mm diameter nozzle is 2m above the ground. Calculate the moment that tends to overturn the nozzle.

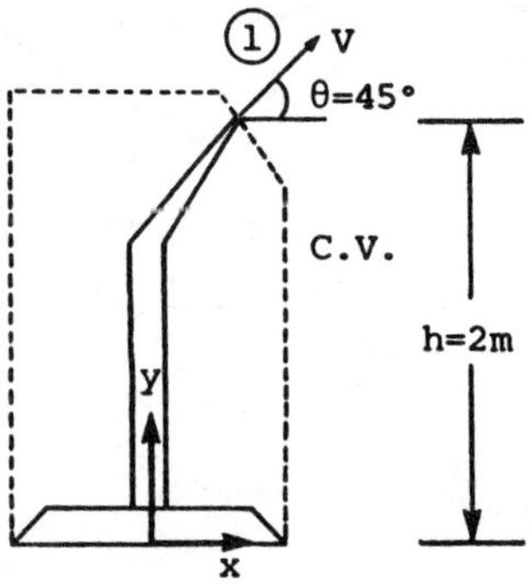

Solution: By assuming steady uniform flow, we will apply the moment of momentum equation to the control volume shown in Fig. 1.

Therefore,

$$\vec{T} + \int_{CV} \vec{r} \times \cancelto{0}{\vec{g}}\rho dV = \cancelto{0}{\frac{\partial}{\partial t}} \int_{CV} \vec{r} \times \vec{V}\rho dV + \int_{CS} \vec{r} \times \vec{V}\rho \vec{V}dA$$

$$\vec{T} = \vec{r}_1 \times \vec{V}_1 \cancel{\{-|\rho V_1 A_1|\}} + \vec{r}_2 \times \vec{V}_2\{|\rho V_2 A_2|\} \qquad (1)$$

But

$$\vec{r}_2 = h\hat{j}, \quad \vec{V}_2 = V\cos\theta\hat{i} + V\sin\theta\hat{j}$$

Therefore,

$$\vec{r}_2 \times \vec{V}_2 = hV\cos\theta(-\hat{k}) + hV\sin\theta(0)$$

$$= -hV\cos\theta\hat{k}$$

and equation (1) can be written as

$$\vec{T} = -hV\cos\theta(\hat{k})\rho VA = -h\rho V^2 A\cos\theta(\hat{k}) = -h\rho \frac{\pi}{4} D^2 V^2 \cos\theta(\hat{k})$$

Making numerical substitutions,

$$\vec{T} = -h\rho \frac{\pi}{4} D^2 V^2 \cos\theta\hat{k} = -(2)(999)\frac{\pi}{4}(0.040)^2(40)^2\cos45°(\hat{k})$$

$$\Rightarrow \vec{T} = -2.84 \text{ kN.m}$$

The moment tending to overturn the nozzle is

$$\vec{M} = -\vec{T} = 2.84 \text{ kN.m}$$

● **PROBLEM** 6-20

Consider the motion of a lawn sprinkler that discharges water from two nozzles as shown in the figure. Let α denote the angle of the jet velocity with the normal to the radius r. Calculate a) the torque required to keep the sprinkler from rotating and b) the resultant circular speed for zero torque. Let the fluid enter the sprinkler rotary arms in the z direction.

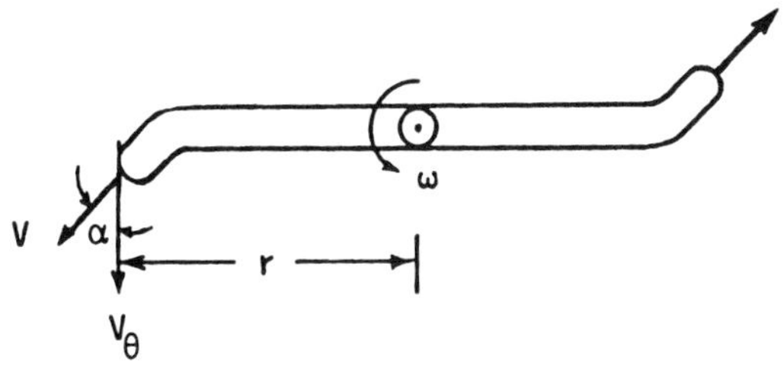

Solution: The control volume is the sprinkler in the horizontal x-y plane, so that the torque is about the z-

279

axis through the pivot of the sprinkler. Since the fluid enters the system in the z-direction, there is no entering momentum. Thus the torque is due solely to the efflux of flow out of the two nozzles. Using the moment of momentum equation, given by

$$T_z = \rho\, Q(r_{②}V_{②} \cos\alpha_{②} - r_{①}V_{①} \cos\alpha_{①})$$

gives

$$\Sigma T_z = \rho\, Qr\, V \cos\alpha \tag{1}$$

where Q is the volume rate out of both nozzles, and V is the absolute velocity of the fluid. We next express the absolute velocity V in terms of the relative velocity V_r of the flow (relative to the moving nozzle) and rotational velocity of the nozzle. The tangential velocity component V_θ of the absolute velocity V is

$$V_\theta = V_r \cos\alpha - \omega r \tag{2}$$

where $V_r \cos\alpha$ is the tangential component of the relative velocity. The tangential component of the absolute velocity V_θ is related to the absolute velocity V by

$$V_\theta = V \cos\alpha \tag{3}$$

Substituting Equations (2) and (3) into Equation (1) yields

$$\Sigma T_z = \rho Qr\,(V_r \cos\alpha - \omega r) \tag{4}$$

The volume rate of flow Q is found using the relative velocity V_r and diameter D of the nozzle:

$$Q = 2\,\frac{\pi}{4}\,D^2\,V_r \tag{5}$$

so that one can express the relative velocity in terms of volume rate Q:

$$V_r = \frac{2Q}{\pi D^2} \tag{6}$$

Substituting Equation (6) into Equation (4) gives

$$\Sigma T_z = \rho\, Qr\left(\frac{2Q}{\pi D^2}\cos\alpha - \omega r\right) \tag{7}$$

Thus, for a given density ρ, nozzle diameter D, nozzle setting α, sprinkler arm r, volume rate of flow Q and speed of rotation ω the torque can be calculated.

(a) The torque required to keep the sprinkler from rotating is found by setting the angular speed ω equal to zero in Equation (7) so that

$$T_z = \frac{2 \rho Q^2 r \cos \alpha}{\pi D^2}$$

(b) The rotational speed of the sprinkler for no torque (the case or maximum speed) is found by setting ΣT_z equal to zero in Equation (7) and solving for ω:

$$\omega = \frac{2Q}{\pi D^2 r} \cos \alpha$$

Thus, the maximum speed is obtained for a nozzle setting of zero degrees, large volume flow rate, very small nozzle diameter and sprinkler arm.

• PROBLEM 6-21

An axial air blowing fan has the following operating characteristics:

Angular speed = 1000 RPM

Inlet angle = 30^0

Exit angle = 60^0

Hub radius = 0.5m

Blade tip radius = 0.7m

Find

(1) volumetric rate of flow;

(2) shape of the rotor blade;

(3) torque about the rotor; and

(4) power used to operate the fan.

Assume the air is at standard conditions, and is not compressed substantially in passage through the fan.

Solution: To carry out the calculations we assume the following additional conditions:

(a) The flow velocity has an axial component that remains unchanged across the rotor; and

(b) the relative flow enters and exits the rotor at geometric blade angles.

Additional assumptions will be indicated as they pertain to the calculations.

Applying the moment of momentum to the flow across the rotor,

$$\bar{r} \times \bar{F}_S + \int_{CV} \bar{r} \times g\rho d\forall + \bar{T} = \frac{\partial}{\partial t} \int_{CV} \bar{r} \times \bar{V}\rho dV + \int_{CS} \bar{r} \times \bar{V}\rho\bar{V} \cdot d\bar{A} \qquad (1)$$

where T = shaft torque

r = blade radius.

In analyzing torque reactions in turbomachinery, it is useful to enclose the rotating member such as the rotor in a fixed control volume as represented below:

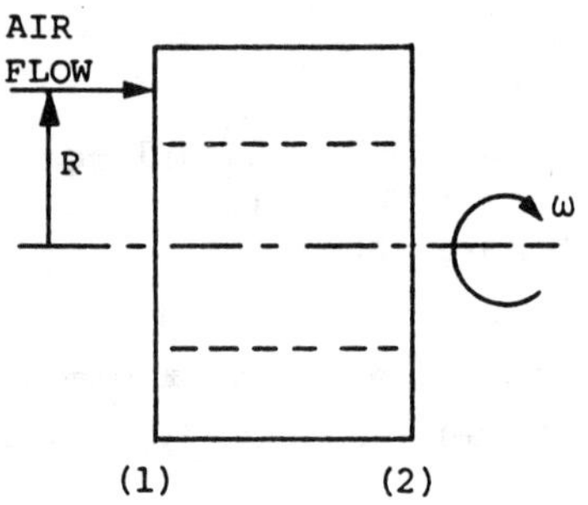

Fig. 1

Assuming torques arising from surface forces are negligible, and that the body force contribution is negligible by symmetry, then for steady flow equation (1) reduces to

$$\bar{T} = \int_{CS} \bar{r} \times V\rho V_{xyz} \cdot d\bar{A} \qquad (2)$$

From the basic equations of flow,

$$\bar{T} = \int_{CS} \bar{r} \times \bar{V}\rho\bar{V} \cdot d\bar{A} \qquad (3)$$

$$\frac{\partial}{\partial t} \int_{CV} \rho dV + \int_{CS} \rho\bar{V} \cdot d\bar{A} = 0 \qquad (4)$$

To determine the shape of the blades, we draw the vector diagram of the inlet velocity:

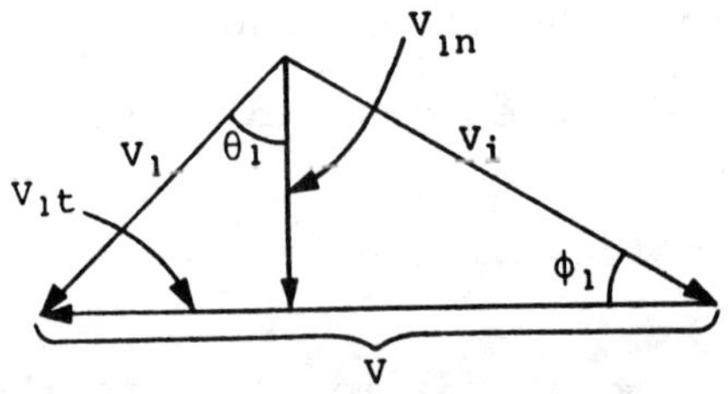

Fig.2

where R = mean blade radius.

To obtain the geometry of the blades

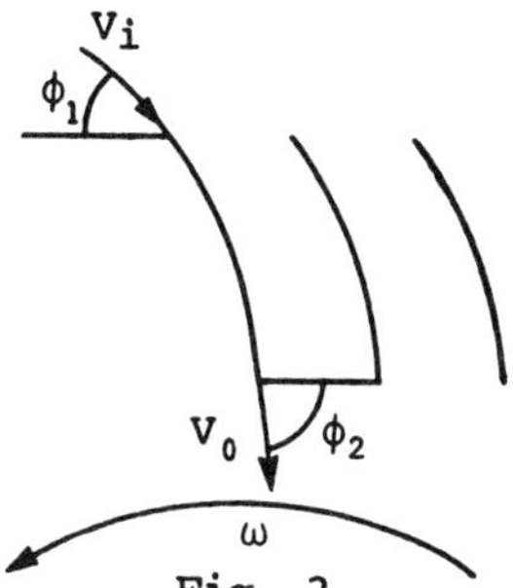

Fig. 3

The vector diagram of the outlet velocity is obtained from

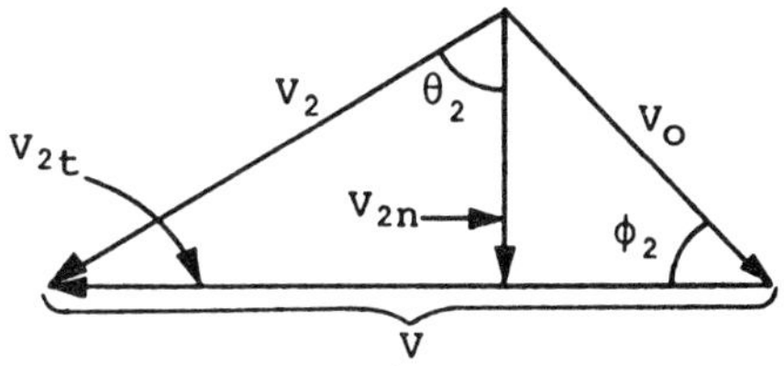

Fig.4

Using the continuity equation,

$$V_{1n} A_1 = V_{2n} A_2 \tag{5}$$

Assuming the axial flow area remains constant,

$$A_1 = A_2$$

and therefore $V_{1n} = V_{2n}$

Referring to the vector diagram of the inlet velocity,

$$V_{1n} \tan\theta_1 + \frac{V_{1n}}{\tan\phi_1} = V \tag{6}$$

where V = rotor surface velocity at the mean blade radius, R. For angular rotor velocity ω,

$$V = \omega R \tag{7}$$

To solve for V_1, V_i, and V_{1t}, we first solve for V_{1n} from equation (6),

$$V_{1n} = \frac{V}{\tan\theta_1 + 1/\tan\phi_1} \tag{8}$$

from which we can calculate

$$V_1 = \frac{V_{1n}}{\cos\theta_1} \tag{9}$$

$$V_i = \frac{V_{1n}}{\sin\phi_1} \tag{10}$$

and

$$V_{1t} = V_{1n}\tan\theta_1 \tag{11}$$

From the diagram for the outlet velocity,

$$V_2 = \frac{V_{2n}}{\cos\theta_2} = \frac{V_{1n}}{\cos\theta_2} \quad \text{since } V_{2n} = V_{1n} \tag{12}$$

$$V_o = \frac{V_{2n}}{\sin\phi_2} = \frac{V_{1n}}{\sin\phi_2} \tag{13}$$

$$V_{2t} = V_{2n}\tan\theta_2 = V_{1n}\tan\theta_2 \tag{14}$$

To calculate the torque from the moment of momentum equation (3)

$$\overline{T} = \int_{CS} \overline{R} \times \overline{V}\rho\overline{V} \cdot d\overline{A}$$

from which we obtain

$$T = RV_{1t}(-|\rho V_{1n}A_1|) + RV_{2t}(|\rho V_{2n}A_2|) \tag{15}$$

assuming uniform flow.

Since $V_{1n}A_1 = V_{2n}A_2 = Q$ from continuity

$$T = QR\rho(V_{2t} - V_{1t}) \tag{16}$$

Substituting now numerical values to evaluate Q, first calculate V_{1n} from (8). Thus, $V = \omega R$, where

$$R = \frac{0.7 + 0.5}{2} = 0.6$$

and

$$V = 1000\ \frac{\text{Rev}}{\text{Min}} \times \frac{\text{Min}}{60\ \text{sec}} \times 2\pi\ \frac{\text{rad}}{\text{rev}} \times 0.6 = 62.8\ \frac{\text{m}}{\text{sec}}$$

Note that R is computed as the average of the blade tip radius and the hub radius. Therefore,

$$V_{1n} = \frac{62.8}{\tan 30 + 1/\tan 30} = 27.3\ \frac{\text{m}}{\text{sec}}$$

and

$$A_1 = A_2 = \pi(0.7^2 - 0.5^2) = .752m^2$$

from which

$$Q = 27.3 \times .752 = 20.5 \frac{m^3}{sec}$$

To calculate T from (16), find V_{1t} and V_{2t}.

From (11),

$$V_{1t} = 27.3 \times \tan 30 = 15.8 \frac{m}{sec}$$

and from (14),

$$V_{2t} = 27.3 \times \tan 60 = 47.2 \frac{m}{sec}$$

Therefore

$$T = 20.5 \times 0.6 \times 1.23 \frac{kg}{m^3}(47.2 - 15.8)$$

$$= 475Nm$$

Since the power, P, applied to the rotor is $T\omega$,

$$P = 475 \times 1000 \times \frac{2\pi}{60} = 49.7KW$$

• PROBLEM 6-22

Determine the torque and power developed by the water on the pelton water wheel shown in Fig. 1a. The shape of the bucket is shown in Fig. 1b. Neglect the effects of friction and gravity. The velocity, V_i, of the water leaving the nozzle is constant and the runner wheel rotates at constant angular speed ω under load.

Solve the problem using:

(a) The linear momentum method
(b) The moment of momentum method.

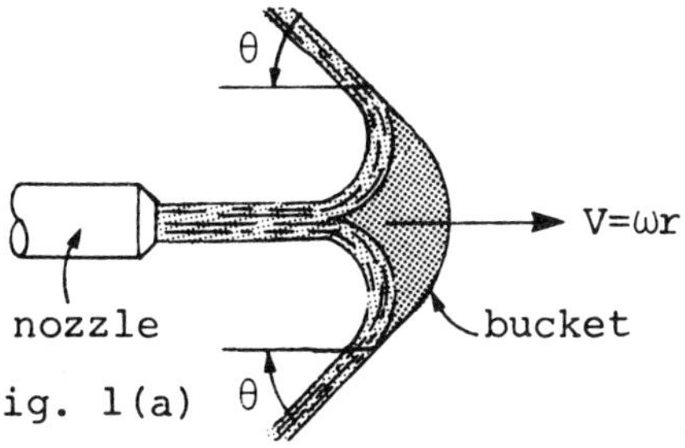

Fig. 1(a)

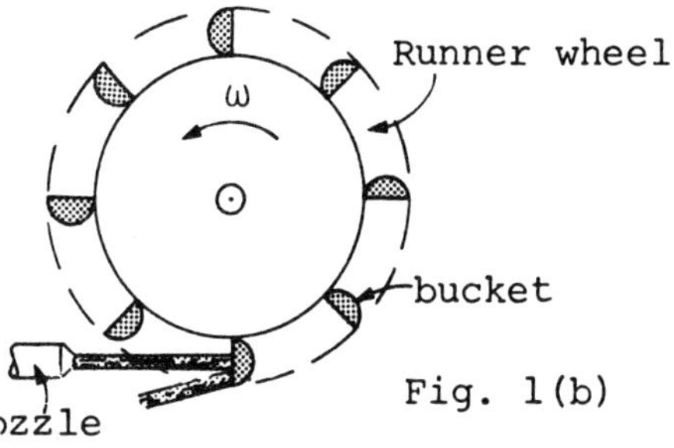

Fig. 1(b)

<u>Solution</u>: Method (a) Linear Momentum Equation

A stationary control volume (shown in Fig. 2) moving with the bucket at angular velocity, $V_R = \omega r$, is selected.

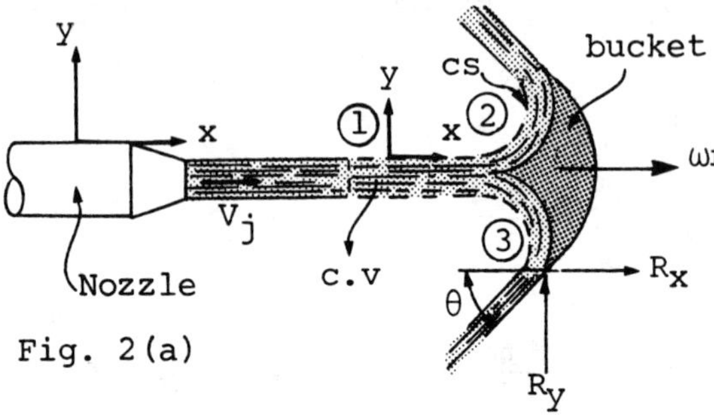

Fig. 2(a)

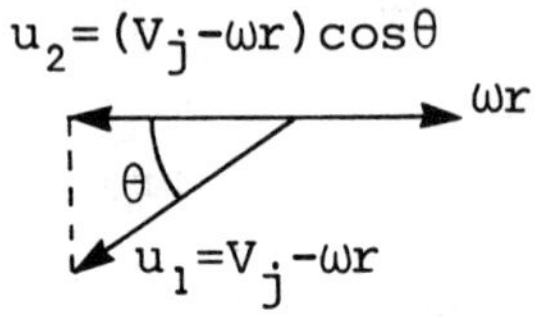

Fig. 2(b)

Assumptions

(1) Flow is steady relative to the wheel.
(2) Properties are uniform at sections 1, 2 and 3.
(3) Magnitude of the relative velocities along the control volume are constant.
(4) The body forces, F_{Bx}, of the control volume are assumed to be zero.
(5) Incompressible flow

Equations

The x component of the linear momentum equation is

$$F_{sx} + \cancelto{0}{F_{Bx}} = \frac{\partial}{\partial t}\int_{cv} \cancelto{0}{u\rho dV} + \int_{cs} u\rho\vec{V}\cdot d\vec{A}$$

There is no net pressure force since p_{atm} acts on all sides of the cv. Thus

$$Rx = \int_{A_1} u\{-|\rho VdA|\} + \int_{A_2} u\{|\rho VdA|\} + \int_{A_3} u\{|\rho VdA|\}$$

$$= \int_{A_1} u\{-|\rho VdA|\} + 2\int_{A_2,A_3} u\{|\rho VdA|\} \qquad (1)$$

$$= -u_1|\rho V_1 A_1| + 2u_2|\rho V_2 A_2|$$

From the continuity equation $0 = \frac{\partial}{\partial t}\int_{cv} \cancelto{0}{\rho dV} + \int_{cs} \rho\vec{V}d\vec{A}$

or

$$0 = \int_{A_1} \{-|\rho V dA|\} + 2\int_{A_2 . A_3} \{|\rho V dA|\} = -\rho V_1 A_1 + 2\rho V_2 A_2$$

or $$V_1 A_1 = 2V_2 A_2$$

Therefore,

$$R_x = (u_2 - u_1)\rho V_1 A_1$$

The velocities measured relative to the cv are

$$u_1 = V_j - \omega r$$

$$u_2 = -(V_j - \omega r)\cos\theta$$

$$V_1 = V_2 = V_j - \omega r$$

Substituting,

$$R_x = \left[-(V_j - \omega r)\cos\theta - (V_j - \omega r)\right]\left[\rho(V_j - \omega r)A_1\right]$$

$$= -(V_j - \omega r)\left[\cos\theta + 1\right]\left[\rho(V_j - \omega r)A_1\right]$$

$$= -Q\rho(V_j - r)\left[1 + \cos\theta\right]$$

Thus the total average force is $(K_T)_x = -R_x$ or

$$(K_T)_x = Q\rho(V_j - \omega r)\left[1 + \cos\theta\right]$$

Then, the average torque on the water wheel is

$$T = (K_T)_x \cdot r$$

$$= Q\rho r(V_j - \omega r)\left[1 + \cos\theta\right]$$

and the power developed is

$$\text{Power} = T \cdot (\omega) = Q\rho r\omega(V_j - \omega r)(1 + \cos\theta).$$

Method b)

Moment of Momentum Equation

A stationary control volume (shown in Fig. 3) including the water wheel is selected.

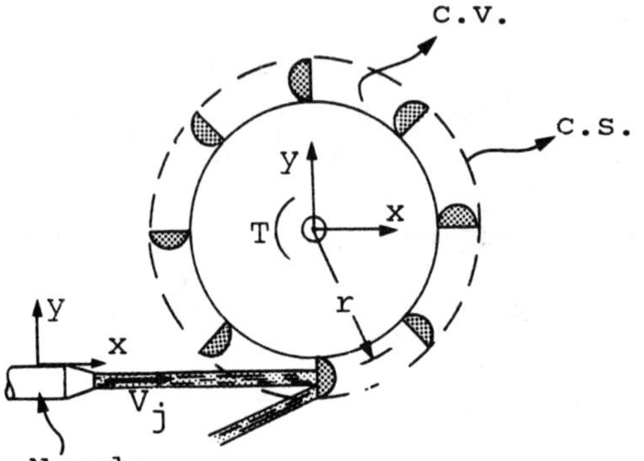

Fig. 3

Assumptions :

(1) Steady flow
(2) Uniform flow at each section
(3) $\bar{\omega}$ = constant
(4) Neglect torque due to body forces
(5) Incompressible flow

Equations:

The moment of momentum equation is;

$$\vec{r} \times \vec{F}_s^{\,0} + \int_{cv} \vec{r} \times g\rho d\Psi^{\,0} + \vec{T}_{shaft} = \frac{\partial}{\partial t}\int_{cv} \vec{r} \times \vec{V}\rho d\Psi + \int_{cs} \vec{r} \times \vec{V}_\phi \rho \vec{V} dA$$

or
$$\vec{T}_{shaft} = \int_{cs} \vec{r} \times V_\phi \rho \vec{V}_d A$$

$$\vec{T}_{shaft} = \int_{in} \vec{r} \times \vec{V}_j \rho V_j dA + \int_{out} \vec{r} \times V_\phi \rho \vec{V}_j dA =$$

$$= -rV_j\rho V_j A + r\{\omega r-(V_j-\omega r)\cos\theta\}\rho V_j A =$$

$$= -rV_j\rho Q + r\{\omega r-(V_j-\omega r)\cos\theta\}\rho Q$$

or
$$T_{shaft} = -r(V_j-\omega r)(1+\cos\theta)\rho Q$$

The above value is equal and opposite to the torque developed by the action of the jet on the water wheel.

Therefore, the desired torque from the jet is

$$T = r(V_j-\omega r)(1+\cos\theta)\rho Q$$

Water flows through the capillary tube shown in the Figure. The tube has a constant diameter with steady flow through it and uniform velocity. From temperature measurements, the increase in internal energy $(u_2 - u_1)$ between the two sections has been determined. Evaluate the frictional force exerted on the capillary tube by the fluid in terms of the internal energy change.

Solution: Since the tube is of constant diameter, the continuity equation with $\rho_1 = \rho_2$ and $A_1 = A_2$ requires that $V_2 = V_1$. We can then find p_2 from the energy equation. Use of the momentum equation allows evaluation of the force exerted on the tube by the fluid.

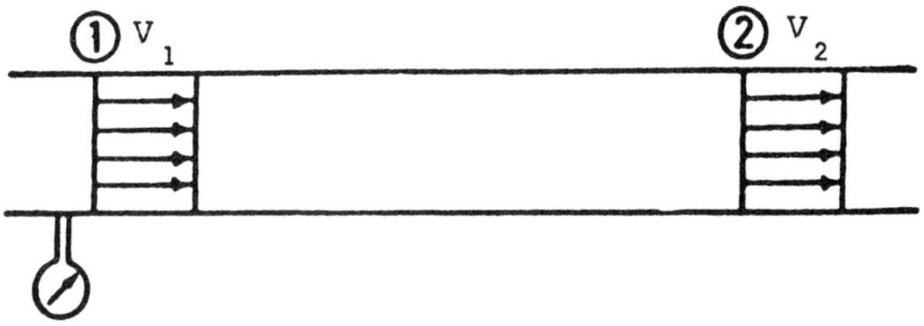

In the absence of external work and heat addition, the energy equation becomes

$$\frac{p_1}{\rho} + \frac{V_1^2}{2g_c} + \frac{g}{g_c}z_1 = \frac{p_2}{\rho} + \frac{V_2^2}{2g_c} + \frac{g}{g_c}z_2 + u_2 - u_1$$

Rewriting by using $V_2 = V_1$ and $z_2 - z_1 = 0$, we obtain

$$\frac{p_2}{\rho} = \frac{p_1}{\rho} - (u_2 - u_1)$$

The momentum equation becomes

$$\sum F_x = f_x + p_1 A - p_2 A = 0$$

with f_x the force exerted by the tube on the fluid due to friction. Therefore,

$$f_x = -p_1 A + [p_1 - \rho(u_2 - u_1)]A$$

$$f_x = -\rho(u_2 - u_1)A$$

Hence the frictional force exerted on the tube is $-f_x$, that is, a force in the direction of flow, since there is an increase in internal energy in the direction of flow.

• PROBLEM 6-24

> **Water flows from a dam to run the turbine (T) as shown in Fig. 1. Assuming no frictional losses and no heat transfer, determine the power developed by the turbine due to the flow of water through it.**

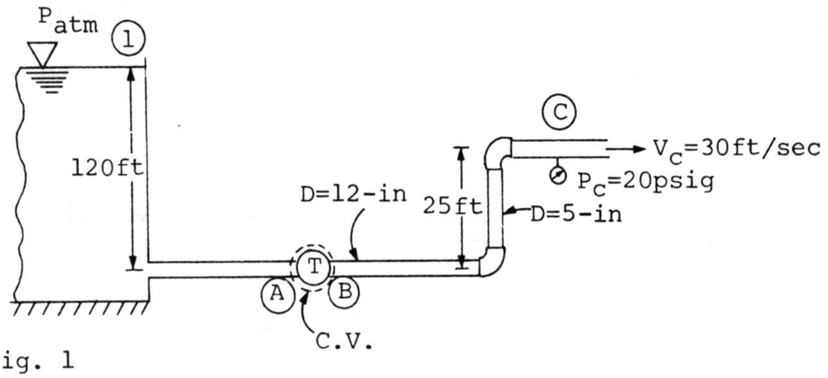

Fig. 1

Solution: Assumptions:
(1) No frictional losses and no heat transfer through the system.
(2) One-dimensional flow
(3) Incompressible flow
(4) Free surface velocity equal to zero
(5) Steady flow

A control volume is chosen around the turbine. From the first law of thermodynamics,

$$\frac{V_A^2}{2g} + \frac{P_1}{\gamma} = \frac{V_B^2}{2g} + \frac{P_B}{\gamma} + W_{shaft} \qquad (1)$$

where

$$W_{shaft} = \frac{P}{\gamma Q_1}$$

$$P = \text{Power (hp)}$$

Applying Bernoulli's equation between the free surface 1 and point A

$$\frac{V_1^2}{2g} + \frac{P_1}{\gamma} + Z_1 = \frac{V_A^2}{2g} + \frac{P_A}{\gamma} + Z_2$$

$$0 + 0 + 120 = \frac{V_A^2}{2g} + \frac{P_A}{\gamma} + 0 \qquad (2)$$

$$P_1 = \text{Patm} = 0 \text{ psig}$$

Bernoulli's equation between B and exit C,

$$\frac{V_B^2}{2g} + \frac{P_B}{\gamma} + 0 = \frac{V_C^2}{2g} + \frac{P_C}{\gamma} + 25 \qquad (3)$$

Applying the continuity equation between B and C, we have

$$A_B V_B = A_C V_C$$

$$\pi\left(\frac{12}{2}\right)^2 V_B = \pi\left(\frac{5}{2}\right)^2 V_C$$

$$V_B = \left(\frac{5}{12}\right)^2 V_C = 0.1736 \, V_C$$

$$V_B = 5.208 \text{ ft/sec} \qquad (4)$$

and around the turbine, we get

$$V_A = V_B = 5.208 \text{ ft/sec} \qquad (5)$$

Introducing eq.(5) in eq.(2) and (3), we have

$$\frac{P_A}{\gamma} = 120 \text{ ft} - \frac{(5.208)^2}{2(32.2)} = 119.578 \text{ ft}$$

$$\frac{P_B}{\gamma} = \frac{30^2 \text{ ft}}{2(32.2)} + \frac{20 \text{ lb/in}^2 \times \frac{2116 \text{ lb/ft}^2}{14.7 \text{ lb/in}^2}}{62.4 \text{ lb/ft}^3} + 25 \text{ ft} - \frac{(5.208)^2}{2(32.2)} \text{ ft}$$

$$\frac{P_B}{\gamma} = 96.87 \text{ ft}$$

Substituting these results in equation (1),

$$W_{shaft} = \frac{P}{\gamma Q_1} = \frac{P_A}{\gamma} - \frac{P_B}{\gamma}$$

Then,

$$P = \frac{119.578(\gamma)Q}{550} = \frac{119.578\ \text{ft}(62.4\ \text{lb/ft}^3)(1.30\ \text{ft}^3/\text{sec})}{550\ \frac{\text{ft-lb/sec}}{\text{hp}}}$$

$$P = 17.66\ \text{hp}$$

• PROBLEM 6-25

A turbine installed in a conduit as shown in Fig. 1 develops 15 hp. Determine the reaction force in the x-direction on the conduit due to the water and the atmospheric pressure.

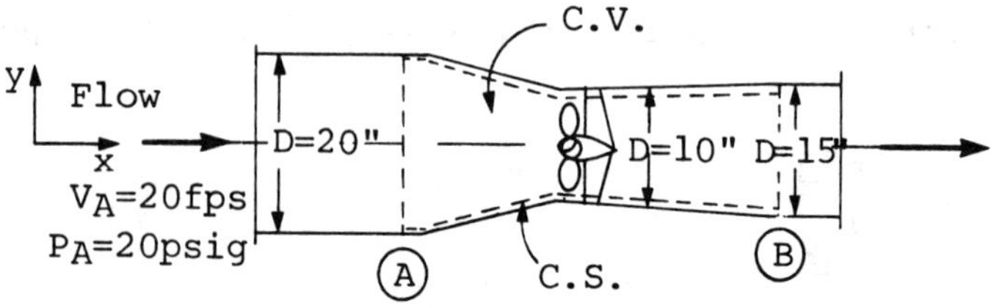

Fig. 1

Solution: Assumptions: (1) Incompressible flow
(2) Isothermal flow
(3) No heat transfer
(4) Steady flow

A control volume that includes the support of the turbine is chosen so that the reaction force F_{RX} can be introduced in the momentum equation. The momentum equation for steady incompressible flow is given by

$$\Sigma F = \iint_{CS} \vec{V}(\rho\vec{V}\cdot d\vec{A}) \tag{1}$$

For the chosen control volume in the x-direction, we have

$$P_A A_A - P_B A_B + F_{RX} = \rho(V_B^2 A_B - V_A^2 A_A) \tag{2}$$

$$A_A = \frac{\pi}{4}\left(\frac{20}{12}\right)^2 = 2.18\ \text{ft}^2$$

$$A_B = \frac{\pi}{4}\left(\frac{15}{12}\right)^2 = 1.23\ \text{ft}^2$$

Substituting the values of all variables in eq.(2),

$$\left[20(2116)/14.7\right](2.18) - 1.23\ P_B + F_{RX} = 1.96(1.23V_B^2 - 2.18\times 20)$$

$$F_{RX} - 1.23\ P_B = 2.41\ V_B^2 - 6{,}361.47 \qquad (3)$$

From the continuity equation, V_B can be found

$$Q_A = Q_B$$

$$\not\rho V_A A_A = \not\rho V_B A_B$$

$$V_B = \frac{A_A}{A_B} V_A = \frac{2.18}{1.23}(20)$$

$$V_B = 35.45 \text{ ft/sec.}$$

Using the first law of thermodynamics between A and B, P_B can be found. Bernoulli's equation does not apply due to frictional effects and unsteady flow around the turbine.

$$\frac{V_A^2}{2g} + \frac{P_A}{\gamma} = \frac{V_B^2}{2g} + \frac{PB}{\gamma} + W_{shaft}$$

$$\frac{(20)^2}{2g} + \frac{20 \times 2116/14.7}{62.4} = \frac{(35.45)^2}{2g} + \frac{P_B}{62.4} + \frac{(15)(550)}{(62.4)(43.6)}$$

$$P_B = 1851.26 \text{ lb/ft}^2$$

Introducing this value and the V_B value in eq.(3), we get

$$F_{RX} = -1{,}055.8 \text{ lbs.}$$

This is considered as the reaction force developed on the turbine due to the water. The outside force will be,

$$F_{out} = -(14.7)(144)\left[\frac{\pi}{4}\left(\frac{30}{12}\right)^2 - \frac{\pi}{4}(1.25)^2\right]$$

$$F_{out} = -7{,}793.11 \text{ lbs}$$

The total force on the system from air and water is

$$(F_{total})_X = 1{,}055.8 - 7{,}793.11$$
$$= -6{,}737.31 \text{ lbs.}$$

$$(F_{total})_X = -6{,}737.31 \text{ lbs.}$$

A nozzle attached to the end of a vertical pipe discharges water against a small flat surface, ℓ = 5 m, below the nozzle exit. The diameter of the pipe is D = 10 cm, and the nozzle exit is d = 4 cm. The pressure gage shows a gage pressure of 50,000 N/m^2. If frictional resistance of the atmosphere on the jet is neglected, what fluid force is exerted on the surface?

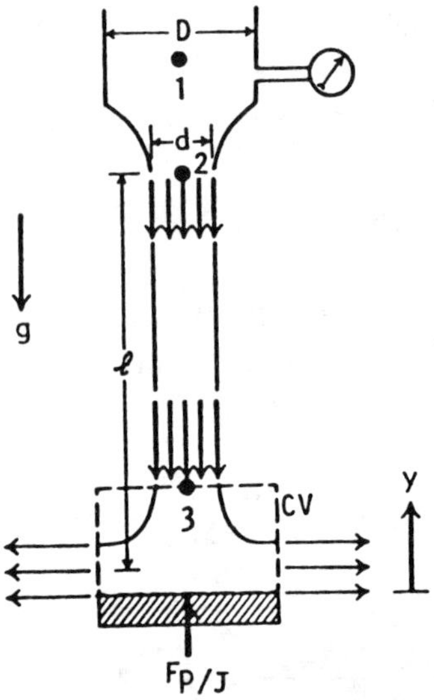

Solution: We select the control volume shown in the figure and apply the vertical component of the momentum equation to the control volume. We assume that the control volume is so small that it contains a negligible weight of fluid and that it is essentially a distance ℓ below the nozzle exit. The only force acting on the control volume is the reaction force of the plate to the jet $F_{P/J}$. The equal and opposite force, that of the jet on the plate, is the force to be found. The vertical component of the momentum equation is then

$$F_{P/J} = \rho(-V_3)(-V_3A_3)$$

V_3 is found by first applying the Bernoulli equation between 1 and 2. p_{1_g} is given by the pressure gage, and point 2 is at atmospheric pressure. We assume that the length of the nozzle is negligible, so that $y_1 \approx y_2$. V_2 and V_1 are related through the steady incompressible continuity equation

$$V_1 = \left(\frac{d}{D}\right)^2 V_2$$

Then

$$V_2 = \sqrt{\frac{2}{\rho}\,\frac{p_1 - p_{atm}}{1-(d/D)^4}} = \sqrt{\frac{2}{\rho}\,\frac{p_{1_g}}{1-(d/D)^4}}$$

Now applying the Bernoulli equation between 2 and 3, which are both at atmospheric pressure, gives

$$V_3 = \sqrt{V_2^2 + 2g\ell}$$

The area A_3 is obtained by applying the steady continuity equation between 2 and 3,

$$A_3 = \frac{V_2}{V_3} A_2$$

Finally, substituting numerical values, we get

$$V_2 = \sqrt{\frac{2(50{,}000\ \text{N/m}^2)(1\ \text{kg}\cdot\text{m/N}\cdot\text{s}^2)}{(1000\ \text{kg/m}^3)[1 - (4/10)^4]}} = 10.13\ \text{m/s}$$

$$V_3 = \sqrt{(10.13\ \text{m/s})^2 + 2(9.8\ \text{m/s}^2)(5\ \text{m})} = 14.16\ \text{m/s}$$

$$A_3 = \frac{10.13\ \text{m/s}}{14.16\ \text{m/s}} A_2 = 0.715 A_2$$

$$F_{P/J} = (1000\ \text{kg/m}^3)(14.16\ \text{m/s})^2(0.715)\frac{\pi}{4}(0.04\ \text{m})^2(1\ \text{N}\cdot\text{s}^2/\text{kg}\cdot\text{m})$$

$$= 180.2\ \text{N}$$

$$F_{J/P} = -F_{P/J} = -180.2\ \text{N}$$

• PROBLEM 6-27

A steady jet of water comes from a hydrant and hits the ground some distance away. If the water outlet is 1 m above the ground and the hydrant water pressure is 125 lbf/in^2 absolute, what distance from the hydrant does the jet hit the ground?

Solution: The distance from the hydrant 1 where the water jet hits the ground is

$$\ell = \int_0^T v_x\, dt$$

where v_x is the x, or horizontal, component of the velocity of a fluid particle and T is the time required for the particle to hit the ground. Both v_x and T are unknown.

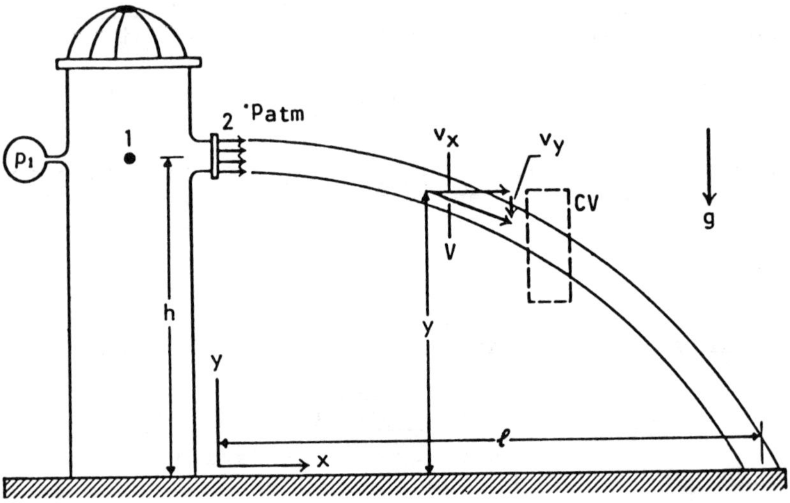

To determine v_x, we draw a control volume enclosing some small portion of the jet (see Figure). Applying the steady continuity equation to the control volume gives

$$\oiint_{CS} \rho \vec{V}\cdot d\vec{A} = \oiint_{CS} d\dot{m} = -\dot{m}_{CV_{in}} + \dot{m}_{CV_{out}} = 0$$

The mass flow rate entering and leaving the control volume is the same. Now apply the x component of the momentum equation to the control volume. There are no forces acting on the control volume in the x direction, and hence there can be no change in x-linear momentum.

$$\oiint_{CS} \rho v_x (\vec{V}\cdot d\vec{A}) = \oiint_{CS} v_x d\dot{m} = -\dot{m}_{CV_{in}} v_{x_{in}} + \dot{m}_{CV_{out}} v_{x_{out}} = 0$$

The x component of the velocity entering and leaving the control volume is the same. Since the control volume is arbitrary and could have been drawn anywhere along the jet, the important result is that the x component of the fluid velocity is constant everywhere.

The magnitude of v_x can be obtained by noting that at the hydrant outlet the flow is entirely in the x direction, $v_x = V_2$. Applying the Bernoulli equation between the interior of the hydrant and the outlet gives

$$p_1 + \rho g y_1 + \frac{1}{2}\rho V_1^2 = p_2 + \rho g y_2 + \frac{1}{2}\rho V_2^2$$

The pressure in the hydrant p is given, and the outlet is open to the atmosphere, $p_2 = p_{atm}$. The elevation of points 1 and 2 is the same, $y_1 = y_2$. We assume that the outlet area is small enough compared with the hydrant cross-sectional area for the hydrant to be essentially a reservoir, $V_1^2 << V_2^2$. Neglecting V_1, we get

$$v_x = \sqrt{\frac{2}{\rho}(p_1 - p_{atm})}$$

Since v_x is constant, it can be brought outside the integral for ℓ, giving $\ell = v_x T$.

To find the time T required for a fluid particle to hit the ground, we apply the Bernoulli equation between point 1 and some arbitrary point on the jet having elevation y and velocity V:

$$p_1 + \rho g h = p_{atm} + \rho g y + \frac{1}{2}\rho V^2$$

Now $V^2 = v_x^2 + v_y^2$. When we use the previously determined value of v_x and note that $v_y = dy/dt$, the Bernoulli equation becomes

$$\left(\frac{dy}{dt}\right)^2 = 2g(h - y)$$

We take the square root (the negative root is the appropriate one since dy/dt must be negative)

$$\frac{dy}{dt} = -\sqrt{2g(h - y)}$$

Then we separate variables and integrate between the limits of $y = h$, $t = 0$ and $y = 0$, $t = T$

$$\int_h^0 \frac{dy}{\sqrt{h - y}} = -\sqrt{2g}\int_0^T dt$$

Integrating and solving for T gives

$$T = \sqrt{2h/g}$$

The y component of the fluid motion is that of a body freely falling under the influence of gravity.

Finally, we substitute numerical values to get

$$v_x = \left[\frac{2(125 - 14.7)\,\text{lbf/in}^2}{1000\ \text{kg/m}^3}\ \frac{1\ \text{N}}{0.224\ \text{lbf}}\ (12\ \text{in/ft})^2 \ (3.281\ \text{ft/m})^2\ \frac{1\ \text{kg m/s}^2}{1\ \text{N}}\right]^{1/2}$$

$$= 39.1\ \text{m/s}$$

$$T = \sqrt{\frac{2(1\ \text{m})}{9.8\ \text{m/s}^2}} = 0.452\ \text{s} \quad \text{and} \quad \ell = (39.1\ \text{m/s})(0.452\ \text{s})$$

$$= 17.7\ \text{m}$$

• PROBLEM 6-28

A liquid jet is issuing upward against a flat board of weight W and supporting it as indicated in the figure. Determine the equilibrium height of the board above the nozzle exit as a function of nozzle area and nozzle exit velocity.

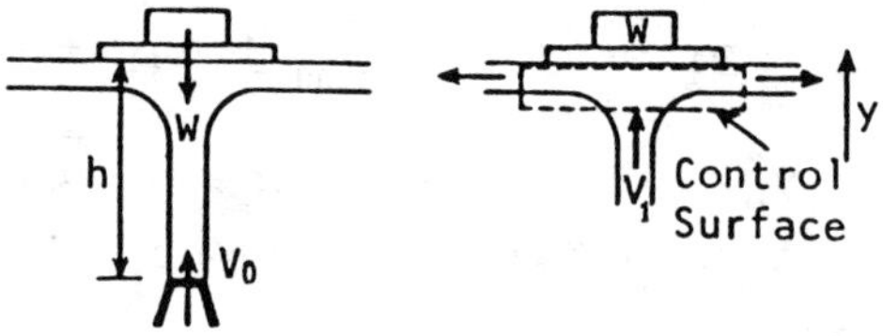

Solution: Since the jet is issuing into the atmosphere, the pressure acting on the surface of the jet is constant. Assuming one-dimensional flow, the pressure within the jet is also constant and equal to atmospheric pressure. Bernoulli's equation, given by

$$\frac{p_2 - p_1}{\rho} + \frac{V_2^2 - V_1^2}{2g_c} + \frac{g}{g_c}(z_2 - z_1) = 0$$

thus gives us a relation between local jet velocity and distance above nozzle:

$$\frac{p_1 - p_0}{\rho} + \frac{V_1^2 - V_0^2}{2g_c} + \frac{g}{g_c} h = 0$$

where

$$p_0 = p_1 = p_a$$

Therefore

$$V_1^2 = -2gh + V_0^2$$

To determine the equilibrium height (h) at which the force exerted on the board by the jet having a local velocity V_1 is exactly balanced by the weight placed on the board, use the momentum equation given by

$$\Sigma F_y = -\frac{\rho_1 V_1^2 A_1}{g_c} + \frac{\rho_2 V_2^2 A_2}{g_c}$$

but in this problem,

$$\frac{\rho_2 V_2^2 A_2}{g_c} = W \quad \text{and} \quad \Sigma F_y = 0,\ \text{giving}$$

$$F_y - W = -\frac{\rho V_1^2 A_1}{g_c}$$

where F_y is the force exerted on the fluid within the control volume in the positive y direction. Substituting the expression for V_1 obtained above into the momentum equation, we get

$$\frac{g_c W}{\rho A_1} = (V_0^2 - 2gh)$$

or

$$h = \frac{1}{2g}\left(V_0^2 - \frac{g_c W}{\rho A_1}\right)$$

where A_1, the cross-sectional area of the jet at height h, is still to be determined. To determine A_1, use the continuity equation

$$\rho_1 V_1 A_1 = \rho_2 V_2 A_2$$

but $\rho_1 = \rho_2$ so that

$$V_0 A_0 = V_1 A_1$$

where A_0 is the cross-sectional area of the jet at the nozzle exit. Substituting the expression for V_1 obtained previously, we have

$$V_0 A_0 = \sqrt{V_0^2 - 2gh}\, A_1$$

and

$$A_1 = \frac{A_0}{\sqrt{1 - 2gh/V_0^2}}$$

Combining with the expression for h obtained above, there results an equation for h

$$h + \frac{g_c}{2g}\frac{W}{\rho A_0}\sqrt{1 - \frac{2gh}{V_0^2}} - \frac{V_0^2}{2g} = 0$$

Solving for h, we obtain

$$h = \frac{V_0^2}{2g} - \frac{1}{2g}\left(\frac{g_c W}{V_0 \rho A_0}\right)^2$$

• PROBLEM 6-29

Two pipe lengths are connected by a tapered 70° bend which lies in a horizontal plane. The pressure head (gauge) upstream of the bend is 25 m where the flow is 0·6 m^3/s, and the pipe diameter is 0·5 m. Downstream of the bend the pipe diameter is 0·25 m. Calculate the force acting upon the fluid specifying its magnitude and direction.

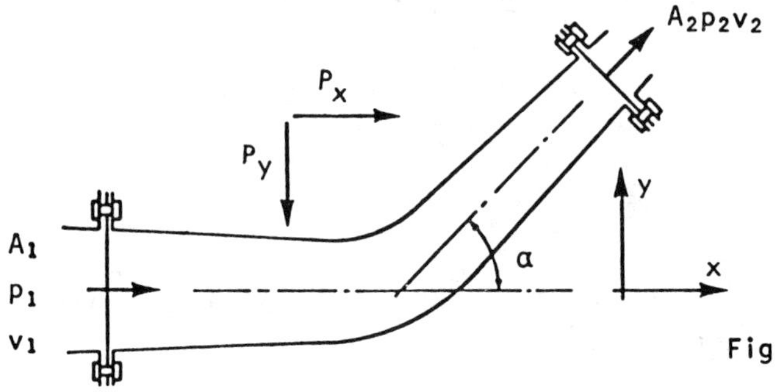

Fig. 1

Solution: A tapered bend is illustrated in Fig. 1. Applying the energy equation and the continuity equation, and assuming no energy loss occurs between points 1 and 2 (Fig. 1)

$$h_1 + \frac{v_1^2}{2g} = h_2 + \frac{v_2^2}{2g} \qquad \text{and}$$

$$\frac{\pi}{4} \times 0\cdot5^2 \times v_1 = \frac{\pi}{4} \times 0\cdot25^2 \times v_2$$

$\therefore$

$$h_2 = h_1 + \frac{v_1^2 - v_2^2}{2g}$$

but

$$v_2 = 4v_1$$

so

$$h_2 = h_1 - 15(v_1^2/2g)$$

but

$$v_1 = \frac{0\cdot6}{\pi/4 \times 0\cdot5^2} = 3\cdot06 \text{ m/s}$$

$$h_2 = 17\cdot863$$

The tapered bend force equations are

$$\left(\frac{p_1}{\gamma} + \frac{v_1^2}{g}\right)a_{1x} - \left(\frac{p_2}{\gamma} + \frac{v_2^2}{g}\right)a_{2x} = -\frac{F_x}{\gamma}$$

and

$$\left(\frac{p_1}{\gamma} + \frac{v_1^2}{g}\right)a_{1y} - \left(\frac{p_2}{\gamma} + \frac{v_2^2}{g}\right)a_{2y} = \frac{F_y}{\gamma}$$

In the x direction

$$a_{1x} = A_1 \qquad a_{2x} = A_2 \cos\alpha$$

$$\therefore \qquad \left(\frac{p_1}{w} + \frac{v_1^2}{g}\right)A_1 - \left(\frac{p_2}{w} + \frac{v_2^2}{g}\right)A_2 \cos\alpha = -\frac{F_x}{w}$$

In the y direction

$$a_1 = 0, \qquad a_{2y} = A_2 \sin\alpha,$$

$$\therefore \qquad -\left(\frac{p_2}{w} + \frac{v_2^2}{g}\right)A_2 \sin\alpha = -\frac{F_y}{w}$$

Now F_x and F_y are the forces that the bend exerts upon the fluid in the x and y direction so as to change its direction. The forces that the fluid exerts upon the bend, P_x and P_y, are equal in magnitude but opposite in direction by Newton's third law.

In the x direction

$$\left(\frac{p_1}{\gamma} + \frac{v_1^2}{g}\right)A_1 - \left(\frac{p_2}{\gamma} + \frac{v_2^2}{g}\right)A_2 + \frac{F_x}{\gamma} = 0$$

$$A_1 = \frac{\pi}{4} \times 0{\cdot}5^2 = 0{\cdot}196; \qquad A_2 = 0{\cdot}0491 \times \cos 70° = 0{\cdot}0168$$

$$v_2 = 4 \times v_1 = 12{\cdot}2$$

$$\therefore F_x = -9810\left[\left(25 + \frac{3{\cdot}056^2}{9{\cdot}81}\right) \times 0{\cdot}19635 - \left(17{\cdot}863 + \frac{12{\cdot}224^2}{9{\cdot}81}\right) \times 0{\cdot}01679\right]$$

$$= -44{\cdot}54 \text{ kN}$$

In the y direction

$$A_1 = 0, \quad A_2 = 0{\cdot}0491 \times \sin 70° = 0{\cdot}0461$$

$$\therefore \qquad F_y = -9810 - \left[\left(17{\cdot}863 + \frac{12{\cdot}224^2}{9{\cdot}81}\right) \times 0{\cdot}0461\right]$$

$$= +14{\cdot}967 \text{ kN}$$

P_x, the force acting on the bend in the x direction is the reaction to F_x so

$$P_x = 44{\cdot}54 \text{ kN}$$

Similarly

$$P_y = -14{\cdot}967 \text{ kN}$$

The resultant force $R = \sqrt{(P_x^2 + P_y^2)} = 46{\cdot}99$ kN. The angle this resultant makes with the extended upstream pipe centre line α is given by

$$\alpha = \tan^{-1} \frac{14{\cdot}967}{44{\cdot}54} = 18{\cdot}57°$$

CHAPTER 7

PIPE FLOW

Basic Attacks and Strategies for Solving Problems in this Chapter. See pages 303 to 432 for step-by-step solutions to problems.

Fluid flow can be categorized into two fundamental types: internal (or bounded) flow and external (or unbounded) flow. This chapter considers the former, where the flow is surrounded by walls, as in the flow of water through a pipe. An important non-dimensional parameter in pipe flows is the Reynolds number, defined as

$$\mathrm{Re} = \frac{\rho V d}{\mu} = \frac{Vd}{\nu},$$

where V is the average velocity through a cross section of the pipe,

d is the pipe diameter,

μ is the viscosity, and

ν is the kinematic viscosity ($\nu = \mu/\rho$).

Average velocity is simply the volume flow rate Q divided by cross-sectional area $\pi d^2/4$. For pipe flows where Re is less than about 2,300, the flow is *laminar*, i.e., smooth and steady. For Re ≥ 2,300, the pipe flow becomes *turbulent*, i.e., unsteady, three-dimensional, and irregular, with lots of fluctuating eddies or vortices mixing and churning up the flow. Laminar pipe flows can be predicted analytically, while experiments must be used to guide any attempts at analyzing turbulent flow. Most practical pipe flow problems are turbulent.

For flow along a pipe, the steady one-dimensional energy equation may be applied:

$$\frac{p_1}{\rho g} + \alpha_1 \frac{V_1^2}{2g} + z_1 = \frac{p_2}{\rho g} + \alpha_2 \frac{V_2^2}{2g} + z_2 + \frac{\dot{W}_s}{\dot{m} g} + h_{\text{total}}.$$

Here, $\dot{W}_s$ is the shaft power (work per unit time) done by the fluid,

z is the vertical height,

$\dot{m}$ is the mass flow rate through the pipe,

α is the kinetic energy correction factor, and

h_{total} is the total head loss (dimensions of length) from inlet 1 to outlet 2.

$\dot{W}_s$ is positive for a turbine, which draws power *from* the fluid, but negative for a pump, which supplies power *to* the fluid. The total head loss is typically split into two parts,

$$h_{\text{total}} = \Sigma h_f + \Sigma h_m.$$

The first term on the right is the sum of all the Moody-type frictional losses. These losses are associated with frictional losses along the inner wall of long, straight sections of pipe. The Moody Chart is a collection of semi-empirically obtained values of this frictional loss as a function of Reynolds number Re and pipe roughness factor ε/d. These losses are plotted in terms of the non-dimensional Darcy friction factor f, i.e.,

$$f = \frac{h_f}{\left(\frac{L}{d}\right)\left(\frac{V^2}{2g}\right)},$$

where L is the total length of the pipe.

The second component of h_{total} comes from the so-called "minor" losses. These are losses associated with parts of the piping system other than long, straight pipe sections, such as valves, bends or elbows, sudden changes in pipe diameter, inlets, exits, etc. Again, empirical values of these losses can be obtained from tables or charts. The non-dimensional minor loss coefficient K is typically the listed value, defined as

$$K = \frac{h_m}{V^2/(2g)}.$$

For a constant diameter section of the pipe system, Σh_m is simply $V^2/(2g)$ multiplied by ΣK, the sum of all the minor loss coefficients along the pipe section.

For pipes of different diameters in *series*, volume flow rate Q must be the same along each section (for steady flow in the mean), and h_f and h_m must be summed independently in each section, then added to obtain the total head loss h_{total}. For pipes in *parallel*, however, the volume flow rate may be different in each parallel section. If the parallel pipes branch off at one point A and later rejoin at a point B as sketched below,

$$Q_A = Q_B = Q_1 + Q_2 + Q_3,$$

whereas the total head loss along any of the three pipes from A to B must be identical since p_A and p_B are the same regardless of which pipe (1, 2, or 3) is under consideration. Another general rule is that the net volume flow rate into any junction must be zero, analogous to the statement that the net current into any junction in an electrical circuit must be zero. Also, the net head loss around any closed loop must be zero, just as in electrical circuits the net voltage drop around any closed loop must be zero.

Finally, for pipes with a cross section other than circular, the *hydraulic diameter* is defined as

$$D_h = \frac{4A}{P},$$

where A is the cross-sectional area and

P is the wetted perimeter,

which is to say the portion of the perimeter in contact with the fluid. Once D_h has been calculated, the Moody Chart can be used to obtain friction factors based on D_h, which is basically the diameter of an equivalent round pipe that would give the same losses as the actual non-round pipe.

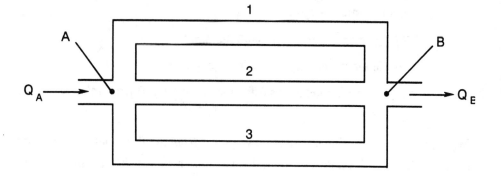

Step-by-Step Solutions to Problems in this Chapter, "Pipe Flow"

LAMINAR/TURBULENT FLOW

• PROBLEM 7-1

The velocity distribution in turbulent flow in a pipe is given approximately by Prandtl's one-seventh-power law,

$$\frac{\upsilon}{\upsilon_{max}} = \left(\frac{y}{r_0}\right)^{1/7}$$

with y the distance from the pipe wall and r_0 the pipe radius. Find the kinetic-energy correction factor.

Solution: The average velocity V is expressed by

$$\pi r_0^2 V = 2\pi \int_0^{r_0} r\upsilon \, dr$$

in which $r = r_0 - y$. By substituting for r and υ.

$$\pi r_0^2 V = 2\pi\upsilon_{max} \int_0^{r_0} (r_0 - y) \left(\frac{y}{r_0}\right)^{1/7} dy = \pi r_0^2 \upsilon_{max} \frac{98}{120}$$

or $$V = \frac{98}{120}\upsilon_{max} \qquad \frac{\upsilon}{V} = \frac{120}{98} \left(\frac{y}{r_0}\right)^{1/7}$$

The kinetic energy correction factor is

$$\alpha = \frac{1}{A} \int_A \left(\frac{\upsilon}{V}\right)^3 dA$$

a) The velocity profile for laminar flow between two parallel plates is given by

$$u = u_m \left(1 - \frac{b^2}{B^2}\right)$$

where u_m is the centerplane velocity, B is the half-spacing between plates, and b is the normal distance from the center plane. (1) What is the average velocity in terms of u_m? (2) What is the momentum correction factor β?

b) The velocity profile for turbulent flow in a circular tube may be approximated by

$$u = u_m \left(\frac{y}{R}\right)^{1/7} = u_m \left(1 - \frac{r}{R}\right)^{1/7}$$

where u_m is the centerline velocity, R is the tube radius, and y is the radial distance from the tube wall $(r = R - y)$. (1) What is the average velocity in terms of u_m? (2) What is the momentum flux factor β?

c) Velocities measured at the center of equal increments of $(r/R)^2$, representing equal increments of area, at the downstream end of a diffuser in a cavitation-testing water tunnel are as follows: 18.2, 16.8, 14.9, 12.75, 10.9, 9.4, 7.9, 6.5, 5.6, and 4.5 m/s. (a) What is the average flow velocity? (b) What is the momentum flux factor β?

Solution: We have used the average velocity across a section in determining the momentum flux $V^2 A\rho$ through a section in a one-dimensional analysis. If the velocity varies across the section, the true momentum flux is greater than that based on the average velocity. The true momentum flux is found by integrating the product of the mass flow rate and velocity over the area of flow. This is

$$\int_A (u\rho \, dA) u$$

while that based on the average velocity is

$$(V \rho A) V = V^2 A\rho$$

The ratio of the true momentum flux to that based on the average velocity is called the momentum-flux correction factor β, where

$$\beta = \frac{\int_A u^2 \, dA}{V^2 A} \geq 1 \qquad (1a)$$

$$\approx \frac{1}{V^2 n} \sum_{1}^{i=n} u_i^2 \qquad (1b)$$

where n is the number of equal incremental areas making up the total area A.

a) $\vec{V}$ represents the average velocity vector, and V the average scalar velocity in a direction parallel to the mean streamline direction at a cross section normal to the streamline direction. When the velocity u varies across this section, the average velocity $\vec{V}$ is

$$V = \frac{1}{A} \int u \, dA = \frac{1}{n} \sum_{1}^{i=n} u_i \qquad (2)$$

where u_i are point velocities measured at the centroids of n equal areas throughout a flow cross section.

$$\bar{u} = V = \frac{2}{2B} \int_0^B u \, db = \frac{u_m}{B} \int_0^B \left(1 - \frac{b^2}{B^2}\right) db = \frac{2}{3} u_m$$

From Eq. (1b),

$$\beta = \frac{2\int_0^B u^2 \, db}{(2u_m/3)^2 (2B)} = \frac{9}{4B} \int_0^B \left(1 - \frac{b^2}{B^2}\right)^2 db = 1.20$$

b)

$$V(\pi R^2) = \int_0^R u(2\pi r) dr$$

$$V = \frac{2}{R^2} \int_0^R \frac{u_m y^{1/7}}{R^{1/7}} r \, dr = \frac{2u_m}{R^{15/7}} \int_0^R y^{1/7} (R-y) dy = \frac{49}{60} u_m$$

The momentum flux factor is, from Eq. (1a),

$$\beta = \frac{\int_0^R u_m^2 (y^{2/7}/R^{2/7})\, 2\pi r\, dr}{(49/60)^2 (u_m^2)\ R^2}$$

$$= \frac{7200}{2401\ R^{16/7}} \int_0^R y^{2/7}(R-y)\, dy = \frac{50}{49} = 1.020$$

c) From Eq. (2), the average velocity is $\bar{u} = V = 107.45/10 = 10.75$ m/s. From Eq. (1b), the momentum flux factor is

$$\beta = \frac{(18.2)^2 + (16.8)^2 + \ldots + (4.5)^2}{10(10.75)^2} = \frac{1361.4}{(10)(115.56)} = 1.178$$

For one-dimensional flow, which has a flat velocity profile, $\beta = 1$, and this is an assumption often made. Part (b) indicates a value of 1.020; measurements for fully developed turbulent flow in a circular pipe are about 1.03. Thus from a momentum flux point of view, turbulent flow often may be assumed to be one dimensional. Part (a) showed $\beta = 1.20$ for laminar flow with a parabolic velocity profile between parallel plates. For fully developed laminar flow in a circular pipe with a parabolic velocity profile, $\beta = 1.333$.

• PROBLEM 7-3

The velocity distribution for laminar flow in a pipe is given by the equation

$$V = V_{max}\left[1 - \left(\frac{r}{r_0}\right)^2\right]$$

Here r_0 is the radius of the pipe and r is the radial distance from the center. Determine $\bar{V}$ in terms of V_{max} and evaluate the kinetic energy correction factor α.

Solution: A sketch for the velocity distribution is shown in the figure. The discharge is given by $Q = \int V\, dA$, or

$$Q = \bar{V}A = \int_0^{r_0} V_{max}\left(1 - \frac{r^2}{r_0^2}\right) 2\pi r\, dr$$

or

$$\bar{V} = \frac{-\pi}{A} r_0^2 V_{max} \left. \frac{(1 - r^2/r_0^2)}{2} \right|_0^{r_0}$$

$$= \frac{1}{2} V_{max}$$

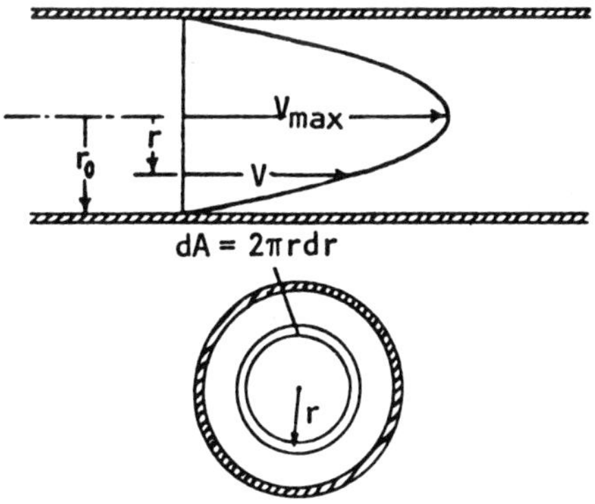

The mean velocity is one-half the maximum velocity. To evaluate α we apply the equation

$$\alpha = \frac{1}{A} \int_A \left(\frac{V}{\bar{V}}\right)^3 dA$$

$$\alpha = \frac{1}{r_0^2} \int_0^{r_0} \frac{V_{max}(1 - r^2/r_0^2)^3}{\left(\frac{1}{2}\right)^3 V_{max}^3} 2\pi r\,dr$$

$$= 2.$$

• PROBLEM 7-4

Assuming the following velocity distribution in a circular pipe:

$$u = u_{max}(1 - r/R)^{1/7},$$

where u_{max} is the maximum velocity, calculate (a) the ratio between the mean velocity and the maximum velocity, (b) the radius at which the actual velocity is equal to the mean velocity.

Solution: (a) The elementary discharge through an annulus dr is given by

$$dQ = 2\pi ru\, dr$$

$$= 2\pi ru_m(1 - r/R)^{1/7}\, dr$$

and discharge through the pipe by

$$Q = 2\pi u_m \int_0^R r(1 - r/R)^{1/7}\, dr$$

Let 1 - r/R = x, then

$$\frac{dx}{dr} = -\frac{1}{R} \quad \text{and} \quad dr = -R\, dx$$

so that

$$R-r = xR, \qquad \text{when } r = 0, x = 1,$$

$$= R-xR = R(1-x), \qquad \text{when } r = R, x = 0 .$$

Therefore, substituting

$$Q = 2\pi u_{max} \int_1^0 R(1-x)x^{1/7}(-R\, dx) =$$

$$2\pi R^2 u_{max} \int_0^1 (1-x)x^{1/7}\, dx$$

$$= 2\pi R^2 u_{max} \left[\frac{7}{8} x^{8/7} - \frac{7}{15} x^{15/7}\right]_0^1$$

$$= 2\pi R^2 u_{max} \left(\frac{7}{8} - \frac{7}{15}\right) = 2\pi R^2 u_{max} \left(\frac{105-56}{120}\right)$$

$$= \pi R^2 u_{max} \frac{49}{60} .$$

and

$$\bar{u} = Q/\pi R^2 = \pi R^2 u_{max} \frac{49}{60} \Big/ \pi R^2 = \frac{49}{60} u_{max} .$$

With the result that

$$\bar{u}/u_{max} = 49/60 \ .$$

(b)

$$u = \bar{u} = 49/60\ u_{max} = u_{max}(1 - r/R)^{1/7} \ .$$

Therefore,

$$(49/60)^7 = 1 - R/r$$

and

$$r/R = 1 - (49/60)^7 = 1 - 0.242 = 0.758.$$

Hence,

$$r = 0.758R \ .$$

• PROBLEM 7-5

a) What is the pressure drop in 15 m of 6-mm tubing for a flow of Univis J-43 hydraulic fluid (μ = 0.014 kg/ms and s = 0.848 at 50°C at a velocity of 2.00 m/s? What is the wall shear stress?

b) The same oil as in part (a) flows in an equilateral triangle duct 2 cm on each side. What is the pressure drop in 15 m of this duct for a flow velocity of 2 m/s?

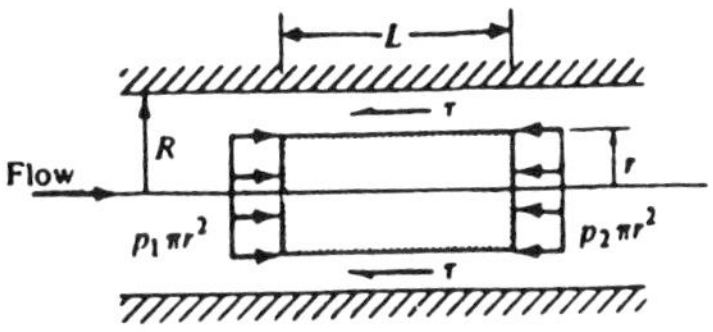

Fig. 1

Cylindrical element in a round pipe.

Solution: a) The density is ρ = (0.848)(1000) = 848 kg/m^3. Then

$$Re_D = \frac{VD\rho}{\mu} = \frac{(2.00)(0.006)(848)}{0.014} = 727$$

The friction factor f for laminar flow is

$$f = \frac{64}{VD\rho/\mu} = \frac{64}{Re_D} \qquad (1)$$

This expression has been verified experimentally and is valid for engineering calculations of both smooth and rough circular pipes for Reynolds numbers up to about 2000. It is possible to achieve laminar flow for Reynolds numbers as high as 40,000 or more in carefully controlled laboratory experiments, but instabilities are usually present in piping systems, limiting laminar flow to Reynolds numbers no greater than about 2000.

$$f = \frac{64}{Re_D} = \frac{64}{727} = 0.0880 \qquad (1)$$

$$\frac{\Delta p}{L} = \frac{f}{D}\frac{\rho V^2}{2} \qquad (2)$$

where f is the friction factor, $\rho V^2/2$ is the dynamic pressure of the mean flow, and D is the pipe diameter. An alternative form in terms of the head loss due to friction h_f is

$$h_f = \frac{\Delta \rho}{\gamma} = f\frac{L}{D}\frac{V^2}{2g}$$

$$\Delta p = \frac{f(L/D)\rho V^2}{2} = \frac{(0.0880)(15/0.006)(848)(2.00)^2}{2} = 373 \text{ kPa} \qquad (2)$$

For fully developed incompressible flow in a round pipe, the momentum theorem applied to a cylindrical element of length L and radius r (Fig. 1) gives

$$p_1\pi r^2 - p_2\pi r^2 - \tau 2\pi r L = 0$$

or

$$\tau = \frac{p_1 - p_2}{L}\frac{r}{2} \qquad (3)$$

which is valid for both laminar and turbulent flow. Equation (3) shows that the shear stress varies linearly with the pipe radius, being zero at the center and a maximum at the pipe walls, where the wall shear stress τ_0 becomes

$$\tau_0 = \frac{p_1 - p_2}{L}\frac{R}{2} = \frac{\Delta p}{L}\frac{D}{4} \qquad (4)$$

$$= \frac{\Delta p D}{4L} = \frac{(373{,}000)(0.006)}{(4)(15)} = 37.3 \text{ Pa}$$

Table 1

LAMINAR FLOW FRICTION FACTORS

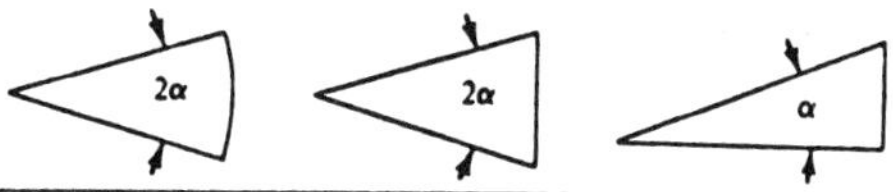

α	CIRCULAR SECTOR $f\,Re$	ISOSCELES TRIANGLE $f\,Re$	RIGHT TRIANGLE $f\,Re$
0	48.0	48.0	48.0
10	51.8	51.6	49.9
20	54.5	52.9	51.2
30	56.7	53.3	52.0
40	58.4	52.9	52.4
50	59.7	52.0	52.4
60	60.8	51.1	52.0
70	61.7	49.5	51.2
80	62.5	48.3	49.9
90	63.1	48.0	48.0

b) From Table 1, f = 53.3/Re. The duct area is A = $\sqrt{3}$ cm^2, the Perimeter is P = 6 cm, and the hydraulic diameter is $D_h = 4A/P = (4)(\sqrt{3})/6 = 1.155$ cm = 0.01155 m. Then

$$Re = \frac{VD_h\rho}{\mu} = \frac{(2)(0.01155)(848)}{0.014} = 1399$$

$$f = \frac{53.3}{1399} = 0.0381$$

$$\Delta p = \frac{fL}{D_h}\frac{\rho V^2}{2} = \frac{(0.0381)(15/0.01155)(848)(2)^2}{2}$$

$$= 83.9 \text{ kPa.}$$

• PROBLEM 7-6

In fig. (1) one plate moves relative to the other as shown. μ = 0.80 P; ρ = 850 kg/m^3. Determine the velocity distribution, the discharge, and the shear stress exerted on the upper plate.

Solution: The general case of steady flow between parallel inclined plates is first developed for laminar flow, with the upper plate having a constant velocity U (fig. 2). Flow between fixed plates is a special case obtaied by setting U = 0. In Fig. 2 the upper plate moves parallel to the flow direction, and there is a pressure variation in the ℓ direction. The flow is analyzed by taking a thin lamina of unit width as a free body. In steady flow the lamina moves at constant velocity u. The equation of motion yields

$$p\ \delta y - \left(p\ \delta y + \frac{dp}{dl}\ \delta \ell\ \delta y\right) - \tau\ \delta \ell + \left(\tau\ \delta \ell + \frac{d\tau}{dy}\ \delta y\ \delta \ell\right)$$

$$+ \gamma\ \delta \ell\ \delta y \sin\theta = 0$$

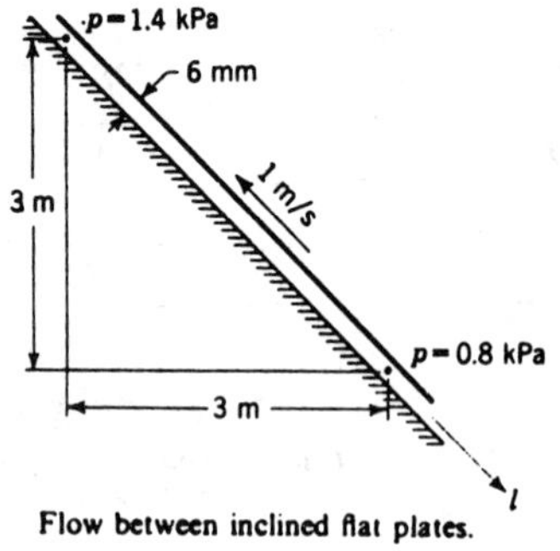

Fig. 1

Flow between inclined flat plates.

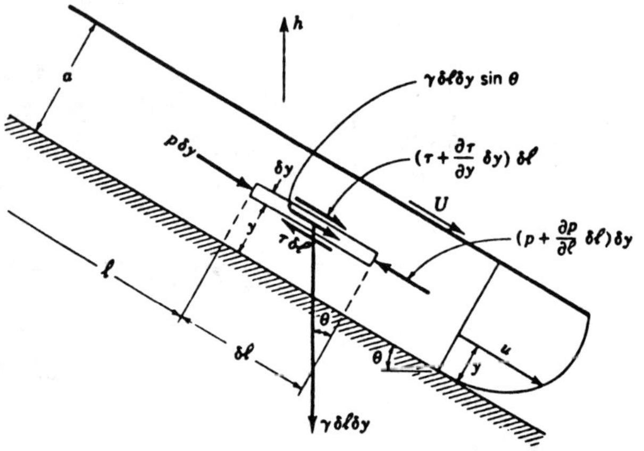

Fig. 2

Flow between inclined parallel plates with the upper plate in motion.

Dividing through by the volume of the element, using $\sin\theta = -\ \partial h/\partial y$, and simplifying yields

$$\frac{\partial\tau}{\partial y} = \frac{\partial}{\partial \ell}\ (p + \gamma h)$$

Since u is a function of y only, $\partial\tau/\partial y = d\tau/dy$; and since $p + \gamma h$ does not change value in the y direction (no acceleration), $p + \gamma h$ is a function of ℓ only. Hence,

$$\partial(p + \gamma h)/\partial \ell = d(p + \gamma h)/d\ell\ ,$$

and

$$\frac{d\tau}{dy} = \mu \frac{d^2u}{dy^2} = \frac{d}{d\ell}(p + \gamma h) \qquad (1)$$

This equation can be determined from the Navier-Stokes equations.

Integrating Eq. (1) with respect to y yields

$$\mu \frac{du}{dy} = y \frac{d}{d\ell}(p + \gamma h) + A$$

Integrating again with respect to y leads to

$$u = \frac{1}{2\mu} \frac{d}{dl}(p + \gamma h)y^2 + \frac{A}{\mu} y + B$$

in which A and B are constants of integration. To evaluate them, take y = 0, u = 0 and y = a, u = U and obtain

$$B = 0 \qquad U = \frac{1}{2\mu} \frac{d}{d\ell}(p + \gamma h)a^2 + \frac{Aa}{\mu} + B$$

Eliminating A and B results in

$$u = \frac{Uy}{a} - \frac{1}{2\mu} \frac{d}{d\ell}(p + \quad h)(ay - y^2) \qquad (2)$$

For horizontal plates, h = C; for no gradient due to pressure or elevation, i.e., hydrostatic pressure distribution, $p + \gamma h = C$ and the velocity has a straight-line distribution. For fixed plates, U = 0, and the velocity distribution is parabolic.

The dischrage past a fixed cross section is obtained by integration of Eq. (2) with respect to y:

$$Q = \int_0^a u\, dy = \frac{Ua}{2} - \frac{1}{12\mu} \frac{d}{d\ell}(p + \gamma h)a^3 \qquad (3)$$

In general, the maximum velocity is not at the midplane.

Now, considering Fig. 1

At the upper plate

$$p + \gamma h = 1400 \text{ Pa} + (850 \text{ kg/m}^3)(9.806 \text{ m/s}^2)(3\text{m})$$

$$= 26,405 \text{ Pa}$$

and at the lower point

$$p + \gamma h = 800 \text{ Pa}$$

to the same datum. Hence,

$$\frac{d}{dl}(p + \gamma h) = \frac{800 \text{ Pa} - 26{,}405 \text{ Pa}}{3\sqrt{2} \text{ m}} = -6035 \text{ N/m}^3$$

From fig. (1), a = 0.006 m, U = -1 m/s; and from Eq. (2)

$$u = \frac{(-1 \text{ m/s})(y \text{ m})}{0.006 \text{ m}} + \frac{6035 \text{ N/m}^3}{2(0.08 \text{ N} \cdot \text{s/m}^2)}(0.006y - y^2 \text{m}^2)$$

$$= 59.646y - 37{,}718y^2 \text{ m/s}$$

The maximum velocity occurs where du/dy = 0, or y = 0.00079 m, and it is u_{max} = 0.0236 m/s. The discharge per meter of width is

$$Q = \int_0^{0.006} u\,dy = 29.823y^2 - 12{,}573y^3 \Big]_0^{0.006} = -0.00164 \text{ m}^2/\text{s}$$

which is upward. To find the shear stress on the upper plate,

$$\left.\frac{du}{dy}\right|_{y=0.006} = 59.646 - 75{,}436y \Big|_{y=0.006}$$

$$= -392.97 \text{ s}^{-1}$$

and

$$\tau = \mu\frac{du}{dy} = 0.08(-392.97) = -31.44 \text{ Pa}$$

This is the fluid shear at the upper plate; hence, the shear force on the plate is 31.44 Pa resisting the motion of the plate.

• PROBLEM 7-7

a) A liquid of specific weight $\rho g = 58$ ft^3/h flows by gravity through a 1-ft tank and a 1-ft capillary tube at a rate of 0.15 ft^3/h, as shown. Sections 1 and 2 are at atmospheric pressure. Neglecting entrance effects, compute the viscosity of the liquid in slugs per foot-second.

b) See whether there is any possible turbulent-flow solution for a smooth-walled pipe.

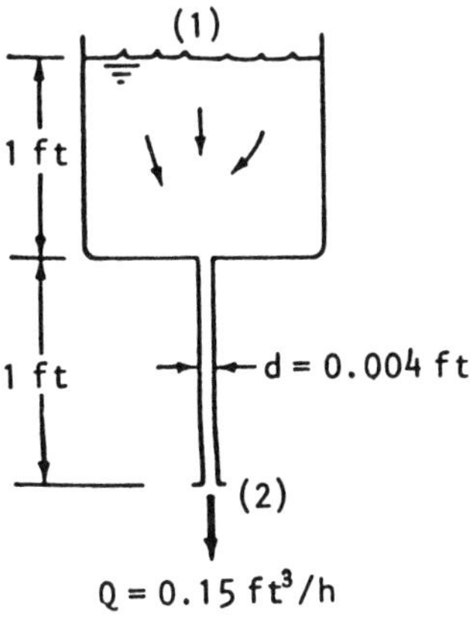

Solution: a) Apply the steady-flow energy equation with no heat transfer or shaft work

$$\frac{p_1}{\rho g} + \frac{V_1^2}{2g} + z_1 = \frac{p_2}{\rho g} + \frac{V_2^2}{2g} + z_2 + h_f$$

But $p_1 = p_2 = p_a$, and V_1 is negligible. Therefore, approximately,

$$h_f = z_1 - z_2 - \frac{V_2^2}{2g} = 2 \text{ ft} - \frac{V_2^2}{2g} \qquad (1)$$

But V_2 can be computed from the known volume flux and pipe diameter

$$V_2 = \frac{Q}{2R^2} = \frac{0.15/3600 \text{ ft}^3/\text{s}}{\pi(0.002 \text{ ft})^2} = 3.32 \text{ ft/s}$$

Substitution into Eq. (1) gives the net head loss

$$h_f = 2.0 - \frac{(3.32)^2}{2(32.2)} = 1.83 \text{ ft} \qquad (2)$$

Note that h_f includes the entire 2-ft drop through the system and not just the 1 ft of capillary pipe length.

Up to this point we have not specified laminar or turbulent flow. For laminar flow with negligible entrance loss, the head loss is given by

$$h_f = 1.83 \text{ ft} = \frac{32\mu LV}{\rho g d^2} = \frac{32\mu(1.0 \text{ ft})(3.32 \text{ ft/s})}{(58 \text{ lb/ft}^3)(0.004 \text{ ft})^2} = 114{,}500\mu$$

or

$$\mu = \frac{1.83}{114{,}500} = 1.60 \times 10^{-5} \text{ slug/(ft} \cdot \text{s)} \qquad \text{Ans.}$$

Note that L in this formula is the pipe length of 1 ft. Check the Reynolds number to see whether it is really laminar flow

$$\rho = \frac{\rho g}{g} = \frac{58.0}{32.2} = 1.80 \text{ slugs/ft}^3$$

$$Re_d = \frac{\rho V d}{\mu} = \frac{(1.80)(3.32)(0.004)}{1.60 \times 10^{-5}} = 1500 \qquad \text{laminar}$$

Since this is less than 2300, we seem to verify that the flow is laminar. Actually, we may be quite wrong, as part (b) will show.

b) As mentioned in part (a), we determined the head loss $h_f = 1.83$ ft independent of any assumption of laminar or turbulent flow. Then the friction factor is

$$f = h_f \frac{d}{L}\frac{2g}{V^2} = (1.83 \text{ ft}) \frac{0.004 \text{ ft}}{1.0 \text{ ft}} \frac{2(32.2 \text{ ft/s}^2)}{(3.32 \text{ ft/s})^2} = 0.0428$$

Assuming laminar flow $f = 64/Re_d$, we could then predict $Re_d = 64/0.0428 = 1500$, which is what we did in part (a). However, the Moody chart shows that a turbulent flow can also have $f = 0.0428$. For a smooth wall, we read $f = 0.0428$ at about $Re_d = 3400$, very near the shaded transition area. We can also compute Re from other formulas given in most text, such as

$$\frac{1}{f^{1/2}} = -2.0 \log\left(\frac{\varepsilon/d}{3.7} + \frac{2.51}{Re_d f^{1/2}}\right) \qquad (3)$$

This is the accepted design formula for turbulent friction. It was plotted in 1944 by L. F. Moody into what is now called the Moody chart for pipe friction

$$f \approx \begin{cases} 0.316\, Re_d^{-1/4} \quad 4000 < Re_d < 10^5 & \text{H. Blasius (1911)} \\ 1.02(\log Re_d)^{-2.5} & \text{F. White (1974)} \end{cases} \tag{4}$$

$$Re_d = \begin{cases} 3600 & (4b) \\ 3200 & (3) \end{cases}$$

So the flow might have been turbulent, in which case the viscosity of the fluid would have been

$$\mu = \frac{\rho V d}{Re_d} = \frac{1.80(3.32)(0.004)}{3500} = 7.0 \times 10^{-6} \text{ slug/(ft} \cdot \text{s)}$$

Ans.

This is about 60 percent less than our laminar estimate in part (a).

• PROBLEM 7-8

The mixing-length concept is used to discuss turbulent velocity distributions for the flat plate and the pipe. For turbulent flow over a smooth plane surface (such as the wind blowing over smooth ground) the shear stress in the fluid is constant, say τ_0. An equation similar to Newton's law of viscosity may be written for turbulent flow:

$$\tau = \eta \frac{du}{dy} \tag{1}$$

The factor η, however, is not a fluid property alone; it depends upon the fluid motion and the density. It is called the eddy viscosity.

In many practical flow situations, both viscosity and turbulence contribute to the shear stress:

$$\tau = (\mu + \eta) \frac{du}{dy} \tag{2}$$

Equation (2) is applicable, but η approaches zero at the surface and μ becomes unimportant away from the surface. If η is negligible for the film thickness $y = \delta$, in which μ predominates, Eq. (2) becomes

$$\frac{\tau_0}{\rho} = \frac{\mu u}{\rho y} = v\,\frac{u}{y} \qquad y \le \delta \tag{3}$$

The term $\sqrt{\tau_0/\rho}$ has the dimensions of a velocity and is called the shear-stress velocity u_* . Hence,

$$\frac{u}{u_*} = \frac{u_* y}{v} \qquad y \le \delta \tag{4}$$

shows a linear relation between u and y in the laminar film. For $y > \delta$, μ is neglected, and Eq. (2) produces

$$\tau_0 = \rho l^2 \left(\frac{du}{dy}\right)^2 \tag{5}$$

Since l has the dimensions of a length and from dimensional consideration would be proportional to y (the only significant linear dimension), assume $l = ky$. Substituting into Eq. (5) and rearranging gives

$$\frac{du}{u_*} = \frac{1}{k}\frac{dy}{y} \tag{6}$$

and integration leads to

$$\frac{u}{u_*} = \frac{1}{k}\ln y + \text{const} \tag{7}$$

Von Kármán suggested, after considering similitude relations in a turbulent fluid, that

$$l = k\,\frac{du/dy}{d^2 u/dy^2} \tag{8}$$

in which k is a universal constant in turbulent flow, regardless of the boundary configuration or value of the Reynolds number. This value of u substituted in Eq. (8) also determines l proportional to y ($d^2 u/dy^2$ is negative, since the velocity gradient decreases as y increases). Equation (7) agrees well with experiment and, in fact, is also useful when τ is a function of y, because most of the velocity change occurs near the wall, where τ is substantially constant. It is quite satisfactory to apply the equation to turbulent flow in pipes.

In evaluating the constant in Eq. (7) following the methods of Bakhmeteff, $u = u_w$, the wall velocity, when

$y = \delta$. According to Eq. (4),

$$\frac{u_w}{u_*} = \frac{u_* \delta}{v} = N \tag{9}$$

from which it is reasoned that $u_* \delta / v$ should have a critical value N at which flow changes from laminar to turbulent, since it is a Reynolds number in form. Substituting $u = u_w$ when $y = \delta$ into Eq. (7) and using Eq. (9) yields

$$\frac{u_w}{u_*} = N = \frac{1}{k} \ln \delta + \text{const} = \frac{1}{k} \ln \frac{Nv}{u_*} + \text{const}$$

Eliminating the constant gives

$$\frac{u}{u_*} = \frac{1}{k} \ln \frac{yu_*}{v} + N - \frac{1}{k} \ln N$$

or

$$\frac{u}{u_*} = \frac{1}{k} \ln \frac{yu_*}{v} + A \tag{10}$$

in which $A = N - (1/k) \ln N$ has been found experimentally by plotting u/u_* against $\ln yu_*/v$. For flat plates $k = 0.417$, $A = 5.84$, but for smooth-wall pipes Nikuradse's experiments yield $k = 0.40$ and $A = 5.5$.

Prandtl has developed a convenient exponential velocity-distribution formula for turbulent pipe flow,

$$\frac{u}{u_m} = \left(\frac{y}{r_0}\right)^n \tag{11}$$

in which n varies with the Reynolds number. This empirical equation is valid only at some distance from the wall. For R less than 100,000, $n = 1/7$, and for greater values of R, n decreases. The velocity-distribution equations, Eqs. (10) and (11), both have the fault of a nonzero value of du/dy at the center of the pipe.

a) By integration of Eq. (7) find the relation between the average velocity V and the maximum velocity u_m in turbulent flow in a pipe.

$$\frac{u}{u_*} = \frac{u_m}{u_*} + \frac{1}{k} \ln \frac{y}{r_0}$$

b) Find an approximate expression for mixing length distribution in turbulent flow in a pipe from Prandtl's one-seventh-power law.

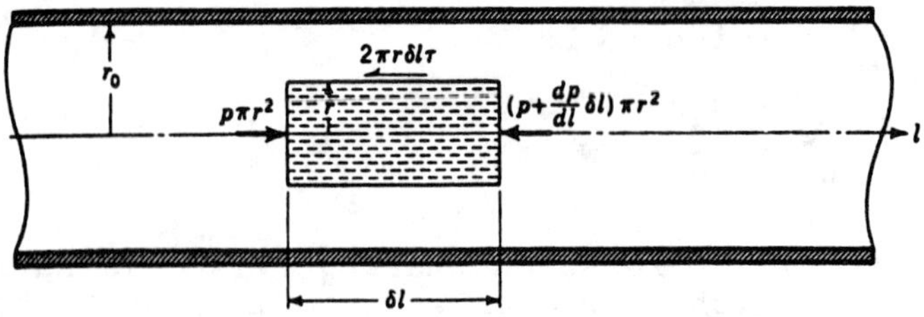

Free-body diagram for steady flow through a round tube.

Solution: The discharge $V\pi r_0^2$ is obtained by integrating the velocity over the area,

$$V\pi r_0^2 = 2\pi \int_0^{r_0-\delta} ur\,dr = 2\pi \int_\delta^{r_0} \left(u_m + \frac{u_*}{k} \ln \frac{y}{r_0}\right)(r_0 - y)\,dy$$

The integration cannot be carried out to y = 0, since the equation holds in the turbulent zone only. The volume per second flowing in the laminar zone is so small that it may be neglected. Then

$$V = 2 \int_{\delta/r_0}^{1} \left(u_m + \frac{u_*}{k} \ln \frac{y}{r_0}\right)\left(1 - \frac{y}{r_0}\right) d\,\frac{y}{r_0}$$

in which the variable of integration is y/r_0. Integrating gives

$$V = 2\left\{ u_m \left[\frac{y}{r_0} - \frac{1}{2}\left(\frac{y}{r_0}\right)^2\right] + \frac{u_*}{k}\left[\frac{y}{r_0} \ln \frac{y}{r_0} - \frac{y}{r_0} - \frac{1}{2}\left(\frac{y}{r_0}\right)^2 \ln \frac{y}{r_0} + \frac{1}{4}\left(\frac{y}{r_0}\right)^2\right]\right\}_{\delta/r_0}^{1}$$

Since δ/r_0 is very small, such terms as δ/r_0 and $\delta/r_0 \ln (\delta/r_0)$ become negligible ($\lim_{x\to 0} x \ln x = 0$); thus

$$V = u_m - \frac{3u_*}{2k}$$

or

$$\frac{u_m - v}{u_*} = \frac{3}{2k}$$

b) Writing a force balance for steady flow in a round tube (Figure) gives

$$\tau = -\frac{dp}{dl}\frac{r}{2}$$

At the wall

$$\tau_0 = -\frac{dp}{dl}\frac{r_0}{2}$$

hence,

$$\tau = \tau_0\frac{r}{r_0} = \tau_0\left(1 - \frac{y}{r_0}\right) = \rho l^2\left(\frac{du}{dy}\right)^2$$

Solving for l gives

$$l = \frac{u_*\sqrt{1 - y/r_0}}{du/dy} .$$

From Eq. (11)

$$\frac{u}{u_m} = \left(\frac{y}{r}\right)^{1/7}$$

the approximate velocity gradient is obtained,

$$\frac{du}{dy} = \frac{u_m}{r_0}\,\frac{1}{7}\left(\frac{y}{r_0}\right)^{-6/7}$$

and

$$\frac{l}{r_0} = \frac{u_*}{u_m}\,7\left(\frac{y}{r_0}\right)^{6/7}\sqrt{1 - \frac{y}{r_0}} .$$

The dimensionless velocity deficiency, $(u_m - u)/u_*$ is a function of y/r_0 only for large Reynolds numbers (part a) whether the pipe surface is smooth or not.

• PROBLEM 7-9

Water is discharged from a large reservoir through a straight pipe of 3 in. diameter and 1200 ft long at a rate of 12 cfm. The discharge end is open to the atmosphere. If the open end is 40 ft below the surface level in the reservoir, what is the Darcy friction factor? Losses other than pipe friction may be ignored.

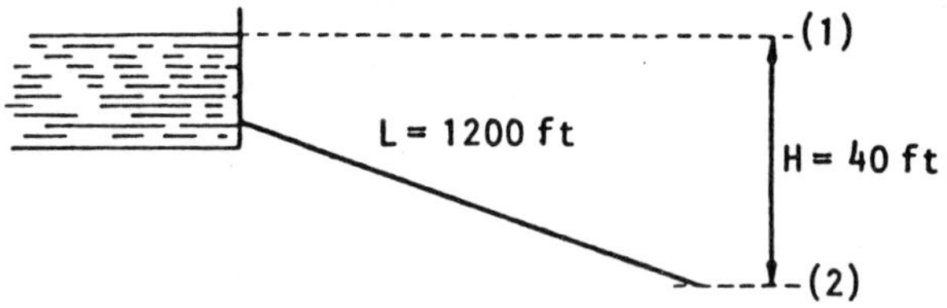

Solution: Applying the flow equation between the levels 1 and 2, as shown in the figure

$$Z_1 + \frac{p_1}{\gamma} + \frac{v_1^2}{2g} = Z_2 + \frac{p_2}{\gamma} + \frac{v_2^2}{2g} + H_f$$

Both ends of the system are open to the atmosphere, then

$$p_1 = p_2$$

Ignoring losses other than pipe friction, and using the Darcy equation

$$H_f = h_f = 4f\left(\frac{L}{D}\right)\frac{v^2}{2g}$$

where v is the velocity in the pipe. It follows that $v_2 = v$, and since $Z_1 - Z_2 = H$, and $v_1 = 0$, the flow equation assumes the form

$$H = \frac{v^2}{2g} + 4f\left(\frac{L}{D}\right)\frac{v^2}{2g}$$

$$H = \frac{v^2}{2g}\left(1 + 4f\frac{L}{D}\right)$$

$$Q = \frac{12}{60} = 0.2 \text{ cfs}$$

$$A = D^2\pi/4 = (3/12)^2\pi/4 = \pi/64$$

$$v = \frac{Q}{A} = \frac{0.2}{\pi/64} = \frac{12.8}{\pi} \text{ fps}$$

Substituting

$$40 = \frac{(12.8)^2}{(64.4)(\pi^2)} \left(1 + 4f \frac{1200}{3/12}\right)$$

from which

$$f = 0.008$$

• PROBLEM 7-10

A liquid of specific gravity 0.9 and viscosity 20 cp flows through the annulus formed by two concentric pipes at a rate of 10 ft/sec. If the inner diameter of the larger pipe is 2 ft and the outer diameter of the smaller pipe is 1.5 ft, what is the Reynolds number corresponding to the flow?

Solution: Let D_o and D_i be the diameters of the larger and smaller pipes, respectively, then the equivalent diameter is given by

$$D_e = 4R_H = 4 \frac{(D_o^2 - D_i^2)(\pi/4)}{(D_o + D_i)(\pi)}$$

$$D_e = D_0 - D_i = 2.0 - 1.5 = 0.5 \text{ ft}$$

The symbol G represents the mass velocity, and is given by

$$G = vQ = 10(\text{ft/sec})(0.9)(62.4)(\text{lb/ft}^3)$$

$$G = 561.6 \text{ lb/(ft}^2\text{)(sec)}$$

The Reynolds number, Re, is given by

$$Re = \frac{D_e G}{\mu} = \frac{(0.5)(561.6)}{(20)(0.000672)} \; \frac{(\text{ft})(\text{lb/ft}^2 \times \text{sec})}{\text{lb/ft} \times \text{sec}}$$

$$Re = 20{,}890$$

A liquid flows in the rectangular duct shown in the figure below to a depth of 2 ft. If the duct is 6 ft wide and 3 ft deep, compute the hydraulic radius R_h and the equivalent diameter D_e .

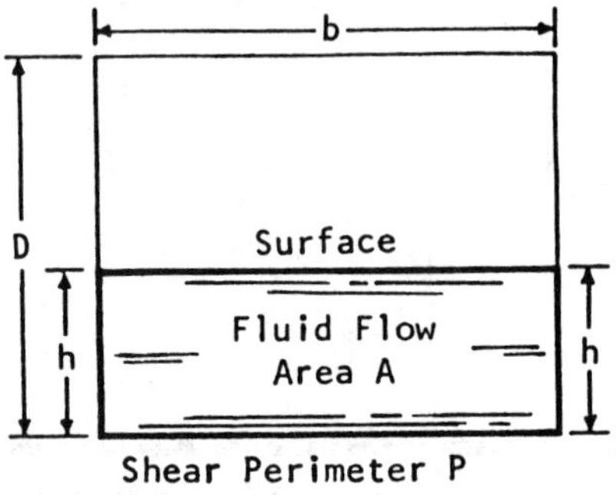

Solution: The ratio of the fluid flow area (dA) to shear area dP dL is a function of streamtube flow shape and is a constant for a given geometry. The hydraulic radius R_h may be used to define this ratio per unit of length as follows:

$$\text{Hydraulic radius} = R_h = \frac{dA}{dP} = \frac{\text{fluid flow area}}{\text{shear perimeter}} = \frac{A}{P}$$

The hydraulic radius is used to compute flow losses in noncircular flow passages and circular conduits flowing partly full of liquids. For this reason, it is important to relate the hydraulic radius to the diameter D of a circular pipe:

$$R_h = \frac{\pi D^2/4}{\pi D} = \frac{D}{4}$$

The equivalent diameter D_e is

$$D_e = 4R_h$$

Fluid flow area, A = bh

Shear perimeter, P = h + b + h = 2h + b

$$R_h = \frac{A}{P} = \frac{bh}{2h + b}$$

Compute R_h and D_e

$$R_h = \frac{6 \times 2}{2 \times 2 + 6} = 1.2 \text{ ft}$$

$$D_e = 4R_h = 4 \times 1.2 = 4.8 \text{ ft}$$

• PROBLEM 7-12

Water flows through a 1-in. diameter pipe. The kinematic viscosity of water is 0.1×10^{-4} ft^2/s. Calculate the largest flow rate for which the flow will definitely be laminar. For a flow through a pipe $(Re)_c \approx 2100$.

Solution: The Reynolds number, Re, is defined by the following equation

$$Re = \frac{Vd\rho}{\mu} = \frac{Vd}{\nu}$$

where d is the diameter of the pipe or some other physical dimension, such as size of body in a flow field; V is the mean velocity or free-stream velocity; ρ is the mass density; μ is the absolute viscosity; and ν is the kinematic viscosity. The Reynolds number also indicates when a flow will turn turbulent. The critical Reynolds number $(Re)_c$ is defined to be the maximum Reynolds number at which a flow will remain laminar.

$$(Re)_c = 2100 = \frac{V_c d}{\nu}$$

$$V_c = \frac{2100\nu}{d} = \frac{2100(0.1 \times 10^{-4})}{\frac{1}{12}} = 0.252 \text{ ft/s}$$

$$Q_c = AV_c = \frac{\pi}{4}\left(\frac{1}{12}\right)^2 \times 0.252 = 1.38 \times 10^{-3} \text{ ft}^3/\text{s}$$

• PROBLEM 7-13

A 24-in. pipe line is to carry 30 cfs of oil ($\nu = 10^{-3}$ ft^2/sec, s = 0.85) across country. Estimate the maximum heights of surface imperfections which will have no effect upon the resistance.

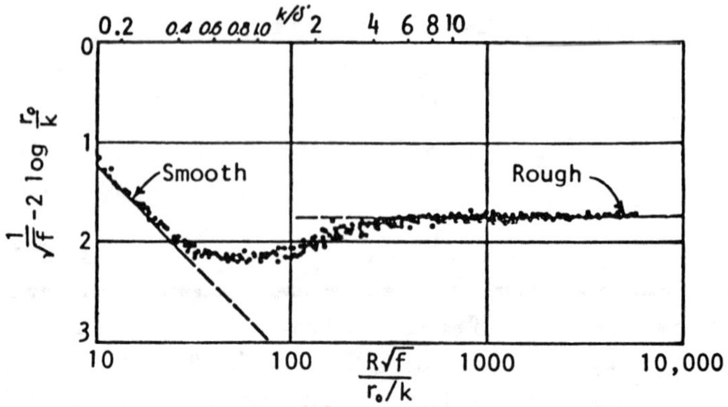

Superposition of transition curves of generalized plot of resistance coefficients for smooth and artificially roughened pipes.

Solution: The Reynolds number of the flow is,

$$R = \frac{VD}{\nu} = \frac{4Q}{\pi D \nu} = \frac{4 \times 30 \times 1000}{\pi \times 2} = 19{,}100$$

The thickness δ' of the laminar sublayer in its ratio to the pipe radius r_0 varies inversely with the Reynolds number of the flow according to the equation,

$$\delta' = \frac{116 r_0}{R^{7/8}} = \frac{116 \times 1}{(19{,}100)^{7/8}} = 0.021 \text{ ft}$$

From the figure it is seen that roughness will have no effect upon the resistance if $k/\delta' < 0.3$. Hence the maximum permissible surface irregularity is about

$$k = 0.3\delta' = 0.3 \times 0.021 = 0.006 \text{ ft.}$$

• PROBLEM 7-14

An oil (density = 59.3 lb/ft^3, viscosity = 50 cp) flows through a horizontal pipe of 4 in. diameter. What is the drop in pressure, and the theoretical horse-power required in a mile length (5280 ft) of the pipe, if the oil flows at a rate of (a) 10 lb/sec, (b) 50 lb/sec?

It may be assumed that in turbulent flow, the Darcy friction factor is 10 per cent higher than calculated for a smooth pipe.

Solution: a) $$Q = \frac{10 \text{ lb/sec}}{59.3 \text{ lb/ft}^3} = 0.1686 \text{ ft}^3\text{/sec}$$

$$A = D^2\pi/4 = (4/12)^2\pi/4 = \pi/36 \text{ ft}^2$$

$$v = \frac{Q}{A} = \frac{0.1686}{\pi/36} = 1.932 \text{ fps}$$

$$Re = \frac{Dv\rho}{\mu} = \frac{(4/12)(1.932)(59.3)}{(50)(0.000672)}$$

$$Re = 1145$$

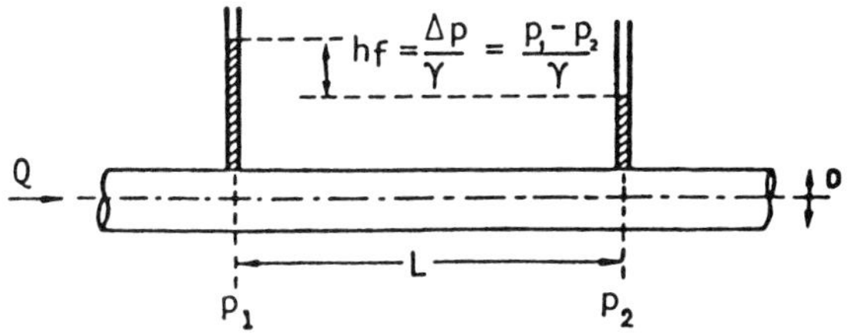

The Reynolds number is less than 2000, hence the flow is laminar and the Darcy friction factor is given by

$$f = \frac{16}{Re} = \frac{16}{1145}$$

Consider in the figure, a fully developed flow in a straight pipe of diameter D. The cross-sectional area of the pipe is then $D^2\pi/4$, and the surface area in a length L is πDL. Let $\Delta p = p_1 - p_2$ be the drop in pressure experienced in this length of the pipe, then by a balance of forces

$$R'\pi DL = (\Delta p)g_c(D^2\pi/4)$$

from which

$$R' = (\Delta p)g_c \frac{D}{4L} \qquad (1)$$

where g_c is the gravitational conversion factor to allow for the fact that, unlike the resistance R, the pressure drop Δp is commonly expressed in the gravitational system of units.

Substituting for R' in eq. (1)

$$C_D \rho v^2 = (\Delta p) g_c \frac{D}{4L}$$

from which

$$\frac{(\Delta p) g_c}{\rho} = \frac{4C_D v^2 L}{D}$$

Substituting for $\rho = \gamma (g_c/g)$

$$\frac{(\Delta p) g_c}{\gamma (g_c/g)} = \frac{4C_D v^2 L}{D}$$

Simplifying and rearranging the terms

$$\frac{\Delta p}{\gamma} = \frac{4v^2 L C_D}{gD}$$

Let

$$f = 2C_D = 2 \frac{R'}{\rho v^2}$$

and

$$h_f = \frac{\Delta p}{\gamma} \tag{2}$$

then

$$h_f = 4f \frac{L}{D} \frac{v^2}{2g} \tag{3}$$

$$h_f = 4f \frac{L}{D} \frac{v^2}{2g} = (4) \frac{16}{1145} \frac{5280}{4/12} \frac{1.932^2}{64.4}$$

$$h_f = 51.7 \text{ ft}$$

From eq. (2)

$$\Delta p = \gamma h_f = 59.3 \left(\frac{\text{lb-f}}{\text{ft}^3}\right) (51.7 \text{ ft})$$

$$\Delta p = 3065.8 \text{ psf}$$

or

$$\frac{3065.8}{144} = 21.3 \text{ psi}$$

The theoretical horsepower is

$$\frac{(51.7)(10)}{550} = 0.93 \text{ hp}$$

b)

$$v = 5(1.932) = 9.660$$

$$Re = 5(1145) = 5725$$

This is more than 3000 and the flow is turbulent. Adding 10 per cent to the Darcy friction factor calculated for smooth pipes from

$$f = .079/Re^{.25} \quad ,$$

the Blasius relationship for turbulent flow,

$$f = \frac{(1.1)(0.079)}{Re^{0.25}} = \frac{(1.1)(0.079)}{5725^{0.25}}$$

$$f = 0.01$$

Substituting this value and the other data in eq. (3)

$$h_f = 4(0.01)\ \frac{5280}{4/12}\ \frac{9.66^2}{64.4}$$

$$h_f = 918.3 \text{ ft}$$

$$\Delta p = (918.3)(59.3) = 54{,}460 \text{ psf}$$

or

$$\frac{54{,}460}{144} = 378.2 \text{ psi}$$

The theoretical power required is

$$\frac{(918.3)(50)}{550} = 83.5 \text{ hp}$$

• PROBLEM 7-15

The siphon in the figure is filled with water and is discharging at 150 l/s. Find the losses from point 1 to point 3 in terms of velocity head $V^2/2g$. Find the pressure at point 2 if two-thirds of the losses occur between points 1 and 2.

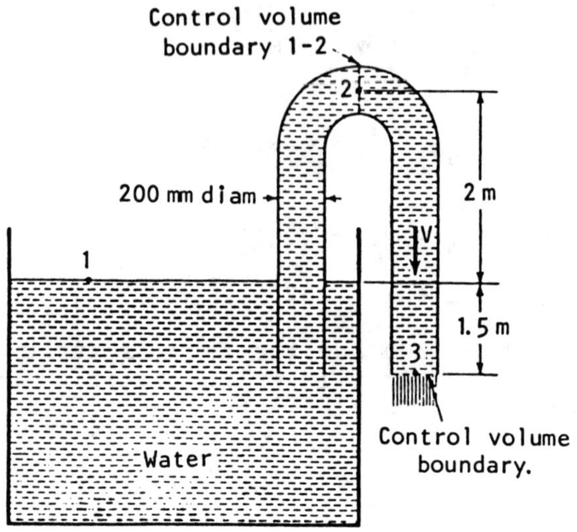

Solution: The energy equation is first applied to the control volume consisting of all the water in the system upstream from point 3, with elevation datum at point 3 and gage pressure zero for pressure datum:

$$\frac{V_1^2}{2g} + \frac{p_1}{\gamma} + z_1 = \frac{V_3^2}{2g} + \frac{p_3}{\gamma} + z_3 + \text{losses}_{1-3}$$

or

$$0 + 0 + 1.5 = \frac{V_3^2}{2g} + 0 + 0 + K\frac{V_3^2}{2g} \qquad (1)$$

in which the losses from 1 to 3 are expressed as $KV_3^2/2g$. From the discharge

$$V_3 = \frac{Q}{A} = \frac{150\ \text{l/s}\ 1\ \text{m}^3\text{/s}}{\pi(0.1^2)1000\ \text{l/s}} = 4.77\ \text{m/s}$$

and $V_3^2/2g = 1.16$ m. Hence from equation (1) $K = 0.29$, and the losses are $0.29\ V_3^2/2g = 0.34$ m·N/N.

The energy equation applied to the control volume between points 1 and 2, with losses $\frac{2}{3}KV_3^2/2g = 0.23$ m, is

$$0 + 0 + 0 = 1.16 + \frac{p_2}{\gamma} + 2 + 0.23$$

The pressure head at 2 is -3.39 m H_2O, or $p_2 = -33.2$ kPa.

Water at 80 F flows through a horizontal, rough pipe at an average velocity of 9.30 ft/s. If the pipe diameter is 1.20 in. and the roughness ratio of the pipe is 0.02, find the pressure loss in pounds per square inch through 10 ft of pipe.

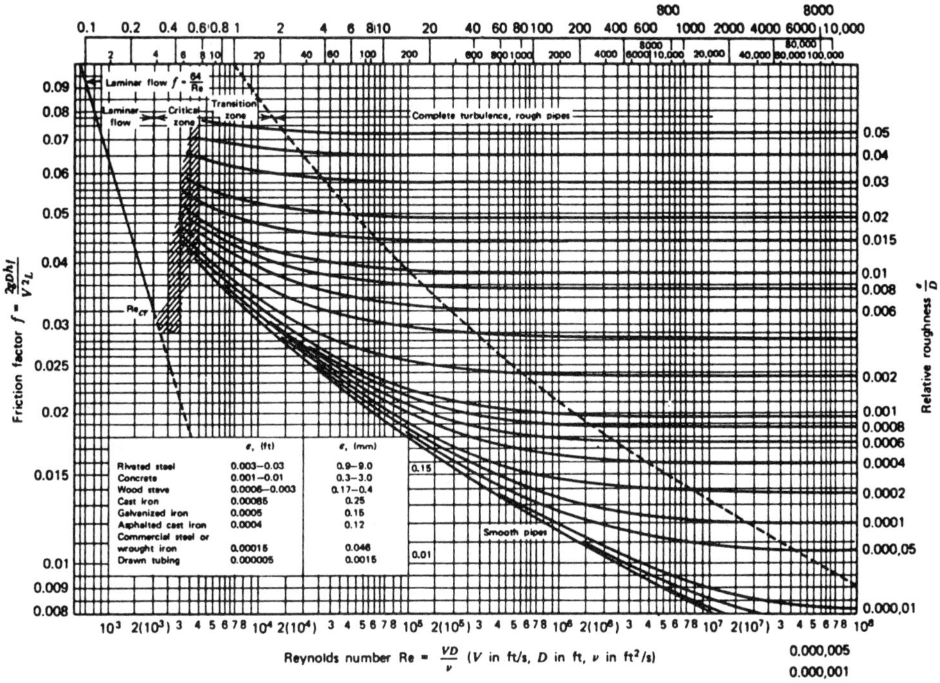

Solution: The pressure loss, which is a result of viscous effects, is given by the Darcy equation as

$$H_f = f \frac{L}{D} \frac{\rho V^2}{2g_c}$$

In order to find the friction factor, f, from the figure, it is necessary to know the Reynolds number, Re. To calculate the Reynolds number, we need the kinematic viscosity, v. From the table, for water at 80 F,

$$v = 0.930 \times 10^{-5} \text{ ft}^2/s.$$

Thus,

$$Re = \frac{VD}{v} = \frac{(9.3)(0.10)}{0.930 \times 10^{-5}} = 10^5$$

From the figure, f = 0.048 for a roughness ratio of 0.02 and a Reynolds number of 10^5. Therefore,

$$H_f = (0.048)\left(\frac{10}{0.10}\right)\frac{62.2(9.3)^2}{64.4} = 402\ lb_f/ft^2$$

$$= 2.78\ lb_f/in.^2$$

TABLE

Temperature (F)	Density ρ (lb_m/ft^3)	Kinematic Viscosity $\nu \times 10^5$ (ft^2/s)	Surface Tension $\sigma \times 10^2$ (lb_f/ft)	Vapor-Pressure Head p_v/γ (ft)	Bulk Modulus of Elasticity $K \times 10^{-3}$ (lb_f/in^2)
32	62.42	1.931	0.518	0.20	293
40	62.43	1.664	0.514	0.28	294
50	62.41	1.410	0.509	0.41	305
60	62.37	1.217	0.504	0.59	311
70	62.30	1.059	0.500	0.84	320
80	62.22	0.930	0.492	1.17	322
90	62.11	0.826	0.486	1.61	323
100	62.00	0.739	0.480	2.19	327
110	61.86	0.667	0.473	2.95	331
120	61.71	0.609	0.465	3.91	333
130	61.55	0.558	0.460	5.13	334
140	61.38	0.514	0.454	6.67	330
150	61.20	0.476	0.447	8.58	328
160	61.00	0.442	0.441	10.95	326
170	60.80	0.413	0.433	13.83	322
180	60.58	0.385	0.426	17.33	313
190	60.36	0.362	0.419	21.55	313
200	60.12	0.341	0.412	26.59	308
212	59.83	0.319	0.404	33.90	300

• PROBLEM 7-17

Water (density) = 62.4 lb/ft^3, viscosity = 0.000672 lb/(ft)(sec) flows in a tube of 0.25 in. bore at a velocity of 1 fps. What is the drop in pressure in a length of 10 ft?

Solution: In laminar flow, the friction factor is given by

$$f = \frac{16}{Re} = \frac{16\mu}{Dv\rho}$$

Substituting from this equation for f in the Darcy equation

$$h_f = 4\ \frac{16\mu}{Dv}\left(\frac{L}{D}\right)\frac{v^2}{2g}$$

Simplifying and taking

$$h_f = \frac{\Delta p}{\gamma}$$

$$\frac{\Delta p}{\gamma} = \frac{32Lv\mu}{gD^2\rho}$$

Substituting in this equation for

$$\gamma = \rho(g/g_c)$$

$$\frac{\Delta p}{\rho(g/g_c)} = \frac{32Lv\mu}{gD^2\rho}$$

from which

$$(\Delta p)g_c = \frac{32Lv\mu}{D^2} \qquad (1)$$

$$Re = \frac{Dv\rho}{\mu} = \frac{(0.25\ /12)(1.0)(62.4)}{0.000672}$$

$$Re = 1935$$

This is less than the critical Reynolds number 2000, and eq. (1) applies.

$$(\Delta p)g_c = \frac{32Lv\mu}{D^2} = \frac{(32)(10)(1.0)(0.000672)}{(0.25/12)^2}$$

$$(\Delta p)g_c = 495.5 \text{ poundals/ft}^2$$

$$\Delta p = \frac{495.5}{32.2} = 15.38 \text{ psf}$$

• PROBLEM 7-18

A water transmission pipe having the diameter shown, conducts water with flow rate of $0.5 m^3/sec$. The relative roughness of the pipe, e/D, is 3×10^{-4}. Find the pressure loss over unit length of the pipe.

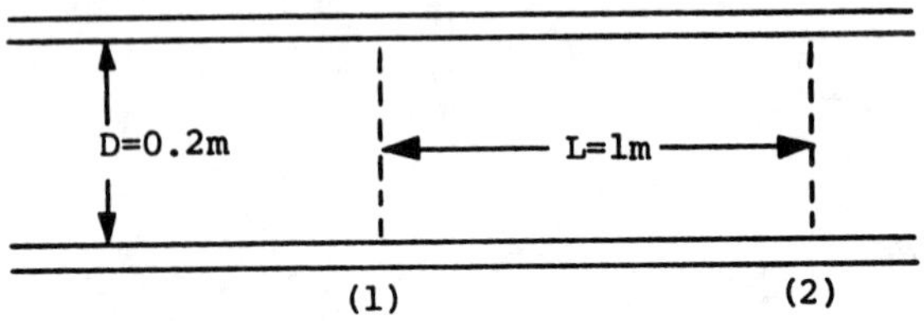

Solution: Applying the general flow equation to compute the total head loss,

$$h = \frac{P_1}{\rho} + K_1 \frac{V_1^2}{2} + gh_1 - \frac{P_2}{\rho} - K_2 \frac{V_2^2}{2} - gh_2 \tag{1}$$

Assuming that the pipe is horizontal, then $h_1 = h_2$. With uniform internal pipe cross-section, $V_1 = V_2$ since $A_1 = A_2$, and flow rate Q is constant throughout the pipe length. Substituting these conditions into (1), we obtain

$$h = \frac{1}{\rho}(P_1 - P_2) \tag{2}$$

Now,

$$h = f \frac{L}{D} \frac{V^2}{2} \tag{3}$$

and therefore

$$P_1 - P_2 = \rho h = f \rho \frac{L}{D} \frac{V^2}{2} \tag{4}$$

The friction factor, f, is a function of the Reynold's number and the relative roughness, e/d. The Reynold's number, Re, is obtained from

$$Re = \frac{\rho VD}{\mu} \tag{5}$$

The velocity of flow may be calculated from the flowrate equation

$$V = \frac{Q}{A} = \frac{0.5}{\pi(.1)^2} = 15.95 \text{ m/sec}$$

The parameters $\rho = 9.99 \times 10^2$ Kg/m^3 and $\mu = 10 \times 10^{-4}$ Kg/m-sec can be assumed for water at 20°C. The Reynold's number therefore becomes

$$Re = \frac{0.99 \times 10^2 \times 0.2 \times 15.95}{10 \times 10^{-4}}$$

$$= 3.18 \times 10^6$$

Referring to the Moody diagram to obtain the friction factor when knowing the Reynold's number and the relative roughness,

$$f = .014$$

Substituting values into equation (4),

$$P_1 - P_2 = .014 \times 9.99(10^2) \times \frac{1}{0.2} \times \frac{15.95^2}{2}$$

$$= 8.89 \times 10^3 Pa$$

• PROBLEM 7-19

Water flows steadily at the rate of 3 cfs in a 6 in. dia. straight smooth pipe. The flow is fully developed between section 1 and section 2, 1000 ft down the pipe. Calculate the change in pressure between the two sections when (a) the pipe is horizontal and (b) the elevation of the pipe drops 15 ft.

Solution: Apply the modified Euler Equation for incompressible flow between sections 1 and 2 for both parts a and b.

$$\frac{p_2 - p_1}{\rho} + \frac{V_2^2 - V_1^2}{2} + g(z_2 - z_1) + gh_L = 0 \qquad (1)$$

(a) In part a elevation and velocity remain unchanged, and Equation (1) becomes

$$\frac{p_2 - p_1}{\rho} + gh_L = 0$$

or

$$p_1 - p_2 = \rho g h_L$$

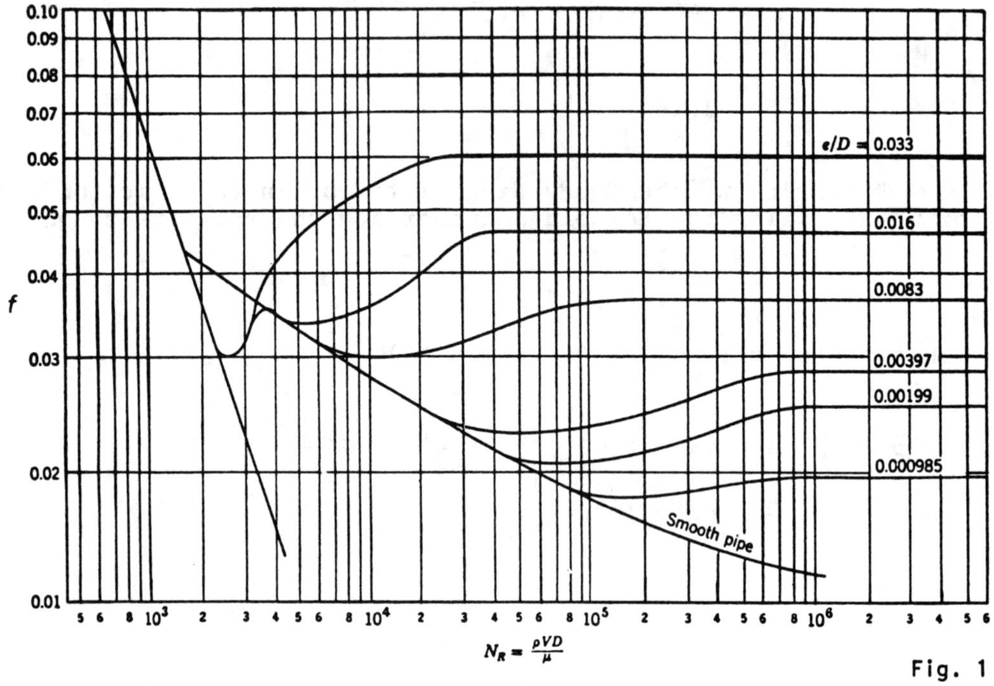

Fig. 1

From the Darcy equation evaluate the head loss as follows:

$$h_L = f \frac{L}{D} \frac{V^2}{2g}$$

The length L is 1000 ft, the diameter D is ½ ft, and g is 32.2 ft/sec^2. Determine the value of the velocity V and the friction factor f. The flow rate equals the product of the velocity and the cross-sectional area.

$$Q = VA = \frac{V\pi D^2}{4}$$

$$3 = V \frac{\pi(\frac{1}{2})^2}{4} = \frac{V\pi}{16}$$

Therefore

$$V = \frac{48}{\pi} = 15.2 \text{ fps}$$

Determine the friction factor f from the Stanton diagram in Fig. (1). (You can compute the Reynolds number $\rho VD/\mu$.)

$$\frac{\rho VD}{\mu} = \frac{(1.94)(15.2)(\frac{1}{2})}{(2.36)(10)^{-5}} = 625{,}000$$

From the Stanton diagram (Fig. 1) the value of f for a smooth pipe is

$$f = 0.011$$

Substitute these values in the Darcy equation to compute the head loss.

$$h_L = (0.011) \frac{1000}{\frac{1}{2}} \frac{(15.2)^2}{(2)(32.2)} = 79 \text{ ft}$$

Therefore,

$$p_1 - p_2 = \rho g h_L = (1.94)(32.2)(79)$$

$$p_1 - p_2 = 4940 \text{ lb/ft}^2$$

(b) In part b the velocity remains unchanged between sections 1 and 2, but $z_2 - z_1 = -15$ ft. Therefore,

$$\frac{p_2 - p_1}{\rho} + g(-15) + gh_L = 0$$

$$p_1 - p_2 = \rho(gh_L - 15g) \qquad (2)$$

There is no difference in the head loss between part a and part b. Therefore, substitute the head loss computed in part a into Equation (2).

$$p_1 - p_2 = (1.94)(32.2)(79-15) = 1.94(32.2)(64) = 4000 \text{ lb/ft}^2$$

• PROBLEM 7-20

A solution, of sp.gr. 1.2, is siphoned out of an open tank by means of a bent tube of 1 in. bore, as shown in the figure.

The rising end of the tube is vertical and 30 ft long, while the remaining length measures 100 ft. The discharging end is open to the atmosphere and the two ends are separated by a vertical distance of 6 ft. What must be the minimum height h of the solution, measured above the dipping end of the tube, if the siphon is to be operational, and what will be the rate of flow under this limiting condition?

The Darcy friction factor is f = 0.01, and it may be assumed that the siphon ceases to operate when the pressure head at the highest point of the tube C falls below 6 ft of water.

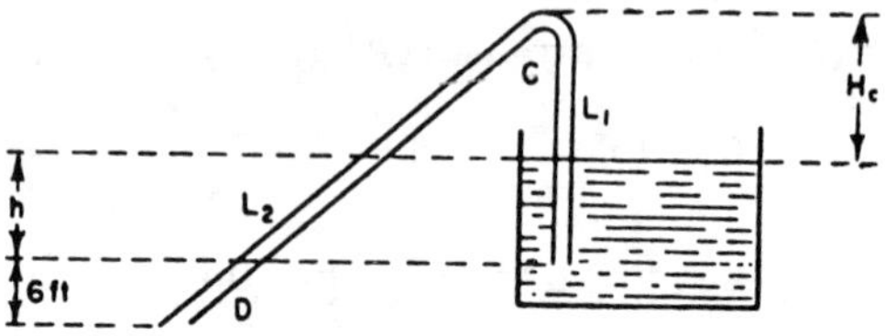

Solution: Let p be the atmospheric pressure, p_c the pressure at the highest point of the U-tube, and H_c the height of this point above the surface level in the tank, then ignoring the correcting factor (α) the following flow equation can be written.

$$\frac{p}{\gamma} = H_c + \frac{Pc}{\gamma} + \frac{v^2}{2g} + H'_f \qquad (1)$$

in which H'_f is the loss of head due to the contraction at the entrance to the tube (h_c), and due to friction in the rising end of the tube (h'_f).

The head loss due to a sudden contraction is given by the equation

$$h_c = 0.5 \frac{v^2}{2g}$$

Using the Darcy equation

$$h'_f = 4f \frac{L_1}{D} \frac{v^2}{2g} = (4)(0.01)\left(\frac{30}{1/12}\right) \frac{v^2}{2g}$$

$$h'_f = 14.4 \frac{v^2}{2g}$$

$$H'_f = h_c + h'_f = 0.5 \frac{v^2}{2g} + 14.4 \frac{v^2}{2g}$$

$$H'_f = 14.9 \frac{v^2}{2g}$$

The critical pressure head at C is 6 ft of water, thus

$$\frac{Pc}{\gamma} = \frac{6}{1.2} = 5 \text{ ft solution}$$

Taking 34 ft water for the atmospheric pressure head

$$\frac{p}{\gamma} = \frac{34}{1.2} \text{ ft solution}$$

With reference to the diagram (Fig.)

$$H_c = 30-h$$

Substituting the above in eq. (1)

$$\frac{34}{1.2} = (30-h) + 5 + \frac{v^2}{2g} + 14.9\frac{v^2}{2g}$$

from which

$$h = 6.6 + 15.9\frac{v^2}{2g} \qquad (2)$$

Writing a flow equation between the surface level in the tank and the open end of the tube

$$\frac{p}{\gamma} + (h+6) = \frac{p}{\gamma} + H_f$$

This equation reduces to

$$h = H_f - 6 \qquad (3)$$

$$H_f = h_c + h_f + h_e$$

$$h_f = 4f\frac{(L_1 + L_2)}{D}\frac{v^2}{2g} = (4)(0.01)\left(\frac{130}{1/12}\right)\frac{v^2}{2g}$$

$$h_f = 62.4\frac{v^2}{2g}$$

$$h_c + h_e = 0.5\frac{v^2}{2g} + \frac{v^2}{2g} = 1.5\frac{v^2}{2g}$$

$$H_f = 1.5\frac{v^2}{2g} + 62.4\frac{v^2}{2g} = 63.9\frac{v^2}{2g}$$

Substituting for H_f in eq. (3)

$$h = 63.9\frac{v^2}{2g} - 6 \qquad (4)$$

Solving the eqns. (2) and (4) simultaneously

$$h = 10.8 \text{ ft}$$

$$v = 4.11 \text{ fps}$$

For 1-in. bore, the flow area is

$$A = (1/12)^2\pi/4$$

$$Q = vA = \frac{(4.11)(\pi)}{(4)(144)} \text{ cfs}$$

or

$$\frac{(4.11)(\pi)(1.2)(62.4)}{(4)(144)} = 1.68 \text{ lb/sec}$$

• PROBLEM 7-21

Sulphuric acid (sp.gr. = 1.8) is to be pumped from an open tank to a process column at the rate of 18 lb/sec. The column operates at 19.65 psia, and the acid is sprayed into it from a nozzle situated 60 ft above the acid surface level in the tank, with a velocity of 8 fps. If the energy losses are estimated to be equivalent to 9 ft water head, what power will be required to run the pump with an overall efficiency of 60 per cent? The barometer reading is 750 mm.

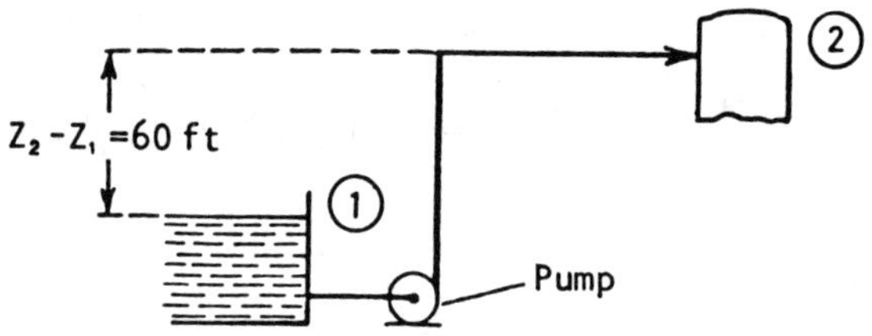

Solution: The Bernoulli equation can be modified to cover all possible situations in which the law of conservation of energy is applicable. A modified form which takes into account heat effects and external work, and also energy losses in the flow of real fluids is called the flow equation.

$$\frac{p}{\gamma} + \frac{v^2}{2\alpha g} + Z + q + W + H_f = \text{constant} \qquad (1)$$

Applying eq. 1 between the two sections shown in the figure

$$\frac{p_1}{\gamma} + \frac{v_1^2}{2g} + Z_1 + W + q = \frac{p_2}{\gamma} + \frac{v_2^2}{2\alpha g} + Z_2 + H_f \qquad (2)$$

In the absence of heat effects, q = 0, and for a large surface in the tank, $v_1 = 0$, and also taking $\alpha = 1$, eq. (2) reduces to

$$W = (Z_2 - Z_1) + \frac{p_2 - p_1}{\gamma} + \frac{v_2^2}{2g} + H_f$$

where W represents the net work to be done by the pump on each pound of the acid conveyed. The right-hand-side terms of this equation have the following values in terms of the acid head:

$$Z_2 - Z_1 = 60.0 \text{ ft}$$

$$v_2^2/2g = 8^2/(2)(32.2) = 1.0 \text{ ft}$$

$$H_f = \frac{9.0(\text{ft water})}{1.8} = 5.0 \text{ ft}$$

$$\text{Taking } p_1 = \frac{750}{760}(14.7) = 14.65 \text{ psia}$$

$$\frac{p_2 - p_1}{\gamma} = \frac{(19.65-14.65)(144)}{(1.8)(62.4)} = 6.4 \text{ ft}$$

$$W = 60 + 6.4 + 1.0 + 5.0$$

$$W = 72.4 \text{ ft}$$

Making use of the concept of weight, 1 lb-f = 1 lb-wt, this work can be expressed as

$$W = 72.4 \frac{(\text{lb-f})(\text{ft})}{(\text{lb-wt})}$$

The work to be done in pumping 18 lb of the acid per second is

$$72.4 \frac{(\text{lb-f})(\text{ft})}{(\text{lb-wt})} \times 18 \frac{(\text{lb-wt})}{\text{sec}} = (72.4)(18)(\text{lb-f})(\text{ft})/\text{sec}$$

The power required to run the pump with an overall efficiency of 60 per cent is then

$$\frac{(72.4)(18)(\text{lb-f})(\text{ft})/\text{sec}}{(0.6)(550)(\text{lb-f})(\text{ft})/\text{sec}} = 3.95 \text{ hp}$$

Water is discharged from a reservoir through a pipe, 300 ft long, branching into two pipes of 200 ft length each. All the pipes are of 2 in. diameter and horizontal. The branched pipes are uniformly perforated, one of them being blocked at the end, while in the other one-half the water entering it is drawn off through its end.

If the water surface in the reservoir is maintained at 15 ft above the center lines of the pipes, what is the discharge through each of the branched pipes? The Darcy friction factor is 0.006.

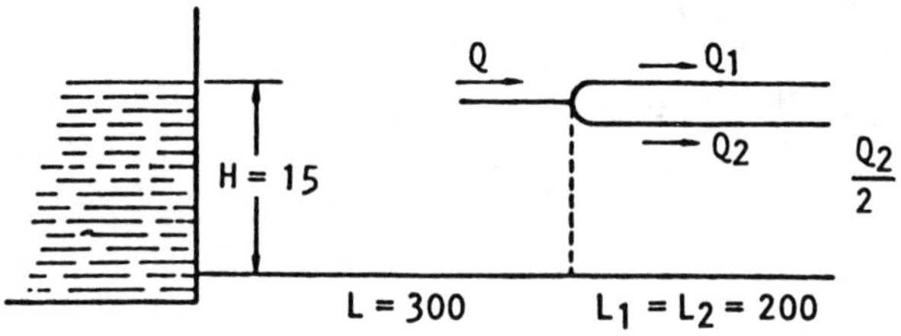

Fig. 1

Solution: Consider a uniformly perforated pipe discharging under a constant head, as shown in Fig. 2. Let v_x be the velocity in the pipe at a distance x from the inflow end, then the loss of head in a differential length dx is

$$dh_f = 4f\left(\frac{dx}{D}\right)\frac{v_x^2}{2g} \qquad (1)$$

where D is the pipe diameter, f is the Darcy friction factor, and v_x is some function of x.

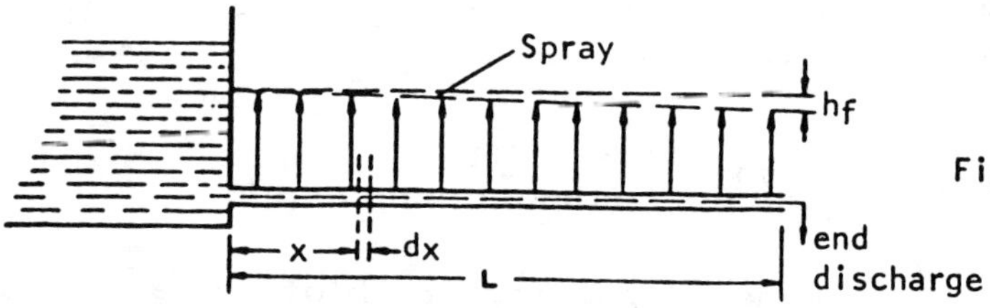

Fig. 2

Referring to Fig. 1, let Q, Q_1, and Q_2 be the rates of flow through the main, the blocked and partly open pipes respectively, then

$$Q = Q_1 + Q_2 \qquad (2)$$

Also

$$H = h_f + h_{f_1} = h_f + h_{f_2} \qquad (3)$$

where

$$h_f = \frac{fLQ^2}{10D^5} = \frac{(0.006)(300)Q^2}{10(2/12)^5}$$

$$h_f = 1400\ Q^2$$

Let v_x be the velocity at a distance x from the inlet to the blocked pipe then from the diagram in Fig. 3,

$$v_x = v_1 \frac{L_1 - x}{L_1}$$

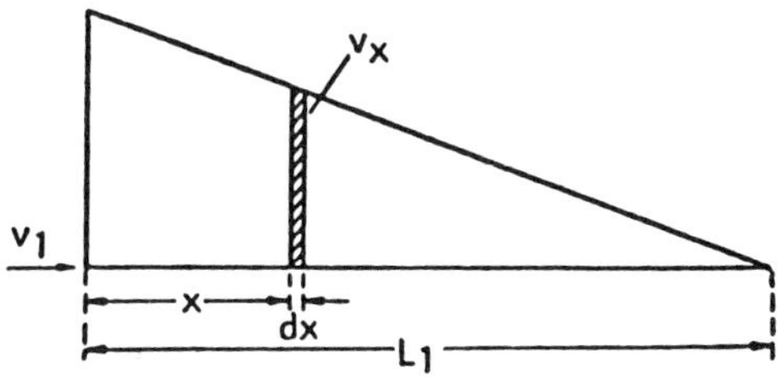

Fig. 3

Substituting for v_x in eq. (1),

$$dh_{f_1} = \frac{4fv_1^2(L_1 - x)^2(dx)}{2gDL_1^2}$$

$$h_{f_1} = \frac{4fv_1^2}{2gDL_1^2}\int_{x=0}^{x=L_1}(L_1 - x)^2(dx)$$

$$h_{f_1} = \frac{4fv_1^2}{2gDL_1^2}\left[L_1^2x - 2L_1\frac{x^2}{2} + \frac{x^3}{3}\right]_0^{L_1}$$

$$h_{f_1} = \frac{1}{3}\ \frac{4fL_1v_1^2}{2gD}$$

or

$$hf_1 = \frac{1}{3}\ \frac{fL_1Q_1^2}{10D^5} = \frac{1}{3}\ \frac{(0.006)(200)Q_1^2}{10(2/12)^5}$$

$$hf_1 = 311.1\ Q_1^2$$

Let v_2 be the velocity at the inlet to the partly open branched pipe, then the outlet velocity is $v_2/2$, and with reference to the diagram in Fig. 4,

$$v_2 = \frac{v_2}{2}\left(\frac{L_2 + z}{z}\right)$$

from which

$$\frac{L_2 + z}{z} = 2$$

Also

$$\frac{v_x}{v_2} = \frac{(L_2 - x) + z}{L_2 + z}$$

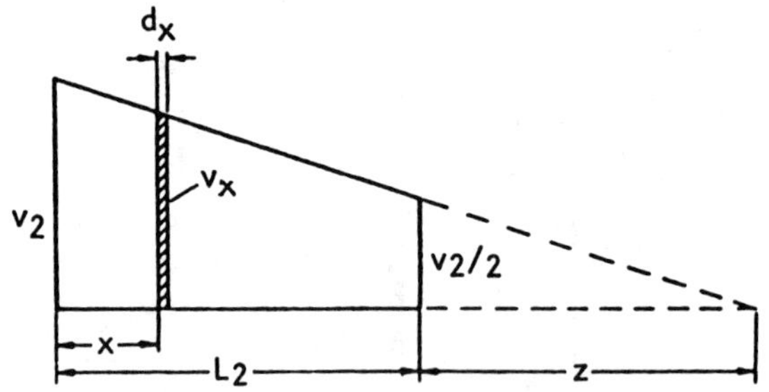

Fig. 4

The velocity at a distance x from the branching point is then

$$v_x = v_2 \left(1 - \frac{x}{2L_2}\right)$$

Again, substituting for v_x in eq. (1)

$$dh_{f_2} = \frac{4fv_2^2 \left(1 - \frac{x}{2L_2}\right)^2 (dx)}{2gD}$$

$$h_{f_2} = \frac{4fv_2^2}{2gD} \int_{x=0}^{x=L_2} \left(1 - \frac{x}{L_2} + \frac{x^2}{4L_2^2}\right) (dx)$$

$$h_{f_2} = \frac{4fv_2^2}{2gD} \left[x - \frac{x^2}{2L_2} + \frac{x^3}{12L_2^2}\right]_0^{L_2}$$

$$h_{f_2} = \frac{7}{12} \frac{4fL_2v_2^2}{2gD}$$

or

$$h_{f_2} = \frac{7}{12} \frac{fL_2Q_2^2}{10D^5} = \frac{7}{12} \frac{(0.006)(200)Q_2^2}{10(2/12)^5}$$

$$h_{f_2} = 554.3Q_2^2$$

Since $h_{f_2} = h_{f_1}$, then

$$554.3Q_2^2 = 311.1Q_1^2$$

$$Q_1 = 1.323Q_2$$

From eq. (2)

$$Q = Q_1 + Q_2 = 1.323Q_2 + Q_2$$

$$Q = 2.323Q_2$$

From eq. (3)

$$H = h_{f_1} + h_{f_2}$$

$$15 = 1400Q^2 + 554.3Q_2^2$$

$$15 = 1400(2.323)^2Q_2^2 + 554.3Q_2^2$$

from which

$$Q_2 = 0.043 \text{ cfs}$$

Also

$$Q_1 = 1.323Q_2 = 1.323\ (0.043)$$

$$Q_1 = 0.057 \text{ cfs}$$

• PROBLEM 7-23

a. A reservoir discharges water through a 200 ft long, horizontal, smooth brass pipeline open to the atmosphere at one end, as shown in Fig. 1. The flow rate is controlled by an angle valve near the exit of the pipe. If the pipe entrance is square edged and the water temperature is 50 F, determine the elevation of the water in the open reservoir if the inside diameter (ID) of the pipe is ½ in. and the flow rate is 0.01 ft^3/s when the angle valve is partly open. Assume that the loss coefficient, C, is 8.75.

b. Suppose z_1 is known to be 50 ft, but the flow rate, Q, is unknown. The problem is to find Q.

c. Suppose that z_1 is known to be 50 ft. The problem is to determine the diameter of the smooth, brass pipe that will result in a flow rate of 0.01 ft^3/s.

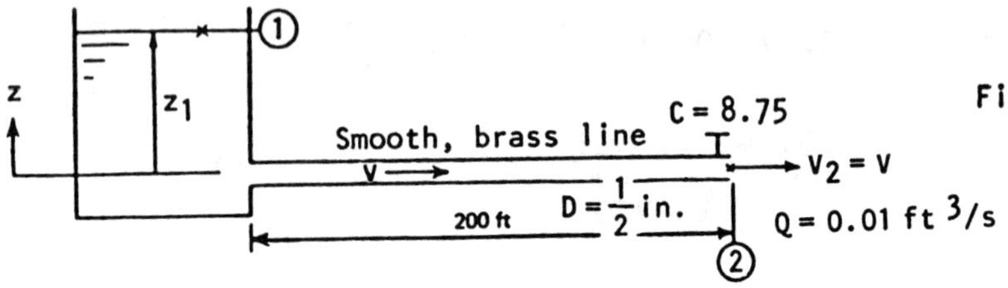

Fig. 1

Solution: In single-path pipeline problems there exist several different types of head losses between the end sections of the pipe system being analyzed. A useful equation which can readily be demonstrated is as follows. If the flow takes place from section α to section β, then the modified energy equation becomes

> The total head at section α equals the total head at section β plus the sum of the head losses which occur in the path between sections α and β.

This statement expressed mathematically is

$$\left(\frac{p}{\gamma} + z + \frac{V^2}{2g}\right)_A = \left(\frac{p}{\gamma} + z + \frac{V^2}{2g}\right)_\beta + \sum_i h_{L,i} \qquad \text{Eq. 1}$$

where i is an index enumerating the number of different head losses existing between sections α and β.

Applying Eq. 1 to the problem, we obtain

$$\frac{p_1}{\gamma} + z_1 + \frac{V_1^2}{2g} = \frac{p_2}{\gamma} + z_2 + \frac{V_2^2}{2g} + h_f + h_{L,V} + h_{L,\text{ entrance}} \qquad \text{Eq. 2}$$

From the specified conditions of the problem, we have

$$z_2 = 0, \; V_2 = V, \; \text{and } A_2 = A = \frac{\pi}{4}\,\frac{(0.5)^2}{144} = 0.1364 \times 10^{-2}\,\text{ft}^2$$

Since the cross-sectional area of the reservoir is very large compared to the cross-sectional area of the pipe, we may take $V_1 = 0$. Finally, since sections 1 and 2 are open to the atmosphere, $p_1 = p_2 = p_{atm}$. Then Eq. 2 reduces to

$$z_1 = \frac{V^2}{2g} + h_f + h_{L,V} + h_{L,\text{entrance}}$$

By the Darcy equation,

$$h_f = \frac{V^2}{2g} \frac{L}{D} f$$

the equation describing the head loss due to a pipe fitting,

$$h_{L,fitting} = C \frac{V^2}{2g}$$

the specified value for C, and Fig. 2

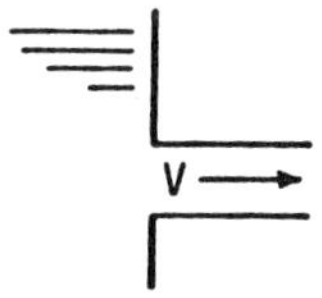

Sharp entrance
C = 0.5

Fig. 2

we have

$$z_1 = \frac{V^2}{2g}\left(1 + \frac{L}{D} f + 8.75 + 0.5\right) = \frac{V^2}{2g}\left(\frac{L}{D} f + 10.25\right) \qquad \text{Eq. 3}$$

Knowing the volume flow rate, the average fluid velocity in the pipe, V, can be determined; that is,

$$V = \frac{Q}{A} = \frac{0.01}{0.1364 \times 10^{-2}} = 7.33 \text{ ft/s}$$

TABLE 1

Temperature (F)	Density ρ (lb_m/ft^3)	Kinematic Viscosity $\nu \times 10^5$ (ft^2/s)	Surface Tension $\sigma \times 10^2$ (lb_f/ft)	Vapor-Pressure Head p_v/γ (ft)	Bulk Modulus of Elasticity $K \times 10^{-3}$ (lb_f/in^2)
32	62.42	1.931	0.518	0.20	293
40	62.43	1.664	0.514	0.28	294
50	62.41	1.410	0.509	0.41	305
60	62.37	1.217	0.504	0.59	311
70	62.30	1.059	0.500	0.84	320
80	62.22	0.930	0.492	1.17	322
90	62.11	0.826	0.486	1.61	323
100	62.00	0.739	0.480	2.19	327
110	61.86	0.667	0.473	2.95	331
120	61.71	0.609	0.465	3.91	333
130	61.55	0.558	0.460	5.13	334
140	61.38	0.514	0.454	6.67	330
150	61.20	0.476	0.447	8.58	328
160	61.00	0.442	0.441	10.95	326
170	60.80	0.413	0.433	13.83	322
180	60.58	0.385	0.426	17.33	313
190	60.36	0.362	0.419	21.55	313
200	60.12	0.341	0.412	26.59	308
212	59.83	0.319	0.404	33.90	300

The friction factor, f, can be determined from the Moody diagram. To enter the Moody diagram, we need the Reynolds number. To calculate Re, we need the kinematic viscosity, v. From Table 1, $v = 1.4 \times 10^{-5}$ ft²/s. Then

$$Re = \frac{VD}{v} = \frac{7.33(0.5)}{12(1.4 \times 10^{-5})} = 2.2 \times 10^4$$

From the Moody diagram, f = 0.0255. Then

$$z_1 = \frac{(7.33)^2}{64.4}\left[\frac{200(12)(0.0255)}{0.5} + 10.25\right]$$

$$= 0.8343[122.4 + 10.25] = 110.7 \text{ ft}$$

b. The solution to this problem is obtained by an iterative technique. Again, Eq. 3 is valid. Substituting for z_1 and multiplying both sides of Eq. 3 by 2g, we obtain

$$50(64.4) = V^2\left[\frac{200(12)}{0.5} f + 10.25\right] \qquad \text{Eq. 4}$$

Since Q is unknown, we cannot immediately determine either V or f. Thus, we must solve the problem by iteration. Here are the steps we may follow.

1. Assume a value for f. (For the case of a smooth pipe, a good starting value for f is 0.03. For the case of a rough pipe, a good starting value is the value determined for f in the completely turbulent region of the Moody diagram).

2. Solve Eq. 4 for V.

3. Solve for Re.

4. Use the Moody diagram, knowing e/D and Re from step 3, and determine f.

5. Compare the value of f obtained in step 4 with the assumed f. If different, take the value of f obtained in step 4 as the assumed value and repeat steps 2 to 5 until agreement is reached. This iterative process is usually a fast converging process. We can now carry out this procedure for this example.

We assume f = 0.03 and determine V from Eq. 4, giving V = 4.57 ft/s. We can then determine that $Re = 1.36 \times 10^4$. Since in this case the pipe is smooth, we may use the smooth curve on the Moody diagram. From the Moody diagram, f = 0.028. We repeat the process assuming f = 0.028; then V = 4.72 ft/s and $Re = 1.4 \times 10^4$. From the Moody diagram, f = 0.028. Thus, the assumed f agrees with the f obtained from the Moody diagram. Therefore, the correct V is 4.72 ft/s, giving

$$Q = AV = \frac{\pi(0.5)^2(4.72)}{4(144)} = 0.00643\ ft^3/s$$

c. This problem may be considered as a pipe design problem. This is so because, in many instances, the engineer will be faced with designing a piping system that will provide a specific flow rate to a certain part of a system at a specific pressure. The problem is to determine the pipe sizes that will satisfy the specified conditions.

Although Eq. 3 is still valid in this problem, the velocity, V, cannot immediately be determined, since the cross-sectional area, A, is unknown. Without knowing V, we can neither determine Re nor the friction factor, f. The solution is obtained by either an iterative process or by plotting two curves and obtaining the point of intersection. We will use the point-of-intersection method.

First let us solve for V in terms of Q and D; that is,

$$V = \frac{Q}{A} = \frac{4(0.01)}{\pi D^2} = \frac{0.01273}{D^2} \qquad \text{Eq. 5}$$

Substituting the values of z_1 and L and Eq. 5 into Eq. 3, we obtain

$$50 = \frac{(0.01273)^2}{64.4D^4}\left(10.25 + \frac{200f}{D}\right)$$

Solving for f in terms of D gives

$$f = 9.935 \times 10^4 D^5 - 0.05125D \qquad \text{Eq. 6}$$

We can also express Re in terms of D; that is,

$$Re = \frac{VD}{\nu} = \frac{0.01273}{1.4 \times 10^{-5}\ D} = \frac{9.093 \times 10^2}{D} \qquad \text{Eq. 7}$$

A method of solution is as follows.

1. Assume a value of D. (A good starting value can be obtained by neglecting the second term on the right-hand side of Eq. 6. We then assume that f = 0.03 and solve for D. This gives the assumed D.)

2. Determine f from Eq. 6.

3. Determine Re from Eq. 7 and e/D for those problems where the pipe is not smooth.

4. Determine f from the Moody diagram.

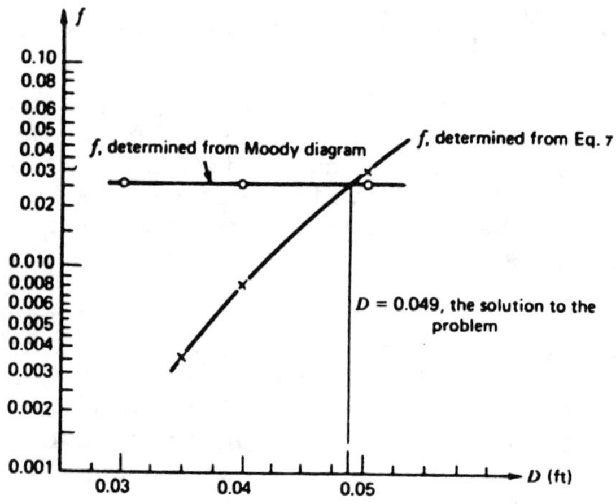

Fig. 3

By assuming other values of D, two curves may be plotted: one being f versus D, where f is determined from Eq. 6, the other being f versus D, where f is determined from the Moody diagram. The solution is the value of D at the intersection of the two curves.

TABLE 2

D (ft)	f (Determined from Eq. 7)	Re	f (Determined from Moody Diagram)
0.035	0.00342	2.60×10^4	0.0240
0.040	0.00812	2.30×10^4	0.0250
0.050	0.02848	1.80×10^4	0.0265
0.055	0.04718	1.65×10^4	0.0270

Table 2 gives points on each curve. The first assumed D, obtained according to step 1, was $D \simeq 0.05$ ft. It should be observed that a small change in D causes: (1) a large change in f when it is determined from Eq. 6 and (2) only a small change in f when it is determined from the Moody diagram. This fact can be employed in deciding whether to use a smaller or larger D for the next trial. That is, if in some particular trial f as determined from Eq. 6 is smaller than f as determined from the Moody diagram, then use a larger D in the next trial. A plot of the two curves is shown in Fig. 3. The solution is D = 0.049 ft = 0.588 in. In most design problems, it is sufficient to select the nearest standard pipe size.

Consider the following three situations for the single path system:

1. Water is to be delivered from reservoir R to tank T (Fig. 1) by a pump that delivers 185 hp to the system in order to maintain a flow of 8.6 ft^3/sec. Determine the pressure at the tank T. For this situation, the length (L) and the inside diameter (D) of the pipe are known, as is the rate of flow (Q) (see given data). The pressure (P) at a given point of the system is to be determined.

2. The system shown in Fig. 2 discharges water as a free jet. If the inside diameter of the pipe is 8 inches, determine the rate of flow leaving the system. In this situation, the length (L) and the inside diameter (D) of the pipe are known as is the pressure drop (ΔP) of the system. The flowrate is to be determined.

3. Oil flows at a rate of 1 ft^3/sec from tank T_1 to tank T_2 through a pipe system, as shown in Fig. 3. The pressure at T_1 is 100 psig and at T_2 is 50 psig. Determine the inside diameter. In this situation, the length (L) of the pipe, the pressure drop (ΔP) of the system, and the rate of flow (Q) are known. The diameter (D) is to be determined.

Solution: Problem 1

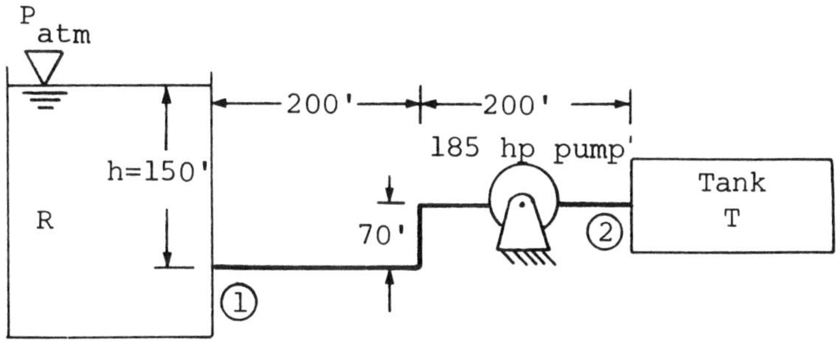

Fig. 1

Given data:

Water at 60°F

$\nu = 0.1217 \times 10^{-4}$ ft^2/sec (From water tables)

D = 8" steel commercial pipe (roughness e = 0.00015)

2-90° elbows (K=0.90)

Rounded entrance (K=0.05)

Q = 10 ft^3/sec

Control Volume: The interior of the pipe including the pump.

Applying the first law of thermodynamics between 1 and 2 we get:

$$\left(\frac{V_1^2}{2} + \frac{P_1}{\rho} + gZ_1\right) - \left(\frac{V_2^2}{2} + \frac{P_2}{\rho} + gZ_2\right) = h_T + \frac{dw_{pump}}{dm} \quad (1)$$

where $$h_T = (h_\ell)_{pipe} + (h_\ell)_{elbows} \quad (2)$$

Assumptions: (1) Frictionless and irrotational flow
(2) Uniform pressure along the system
(3) $\overline{V} \approx Q/A$ (average velocity)

To find P_1 we employ Bernoulli's equation between 1 and the free surface

$$\frac{P_1}{\rho} + \frac{V_1^2}{2} = \frac{Patm}{\rho} + gh \quad (3)$$

now, $V_1 \approx \overline{V} = \frac{Q}{A} = \dfrac{10 \ ft^3/sec}{\pi\left(\frac{0.667}{2}\right)^2} = 28.62$ ft/sec

introducing this value in eq.(3) and arranging we get

$$\frac{P_{1_{gage}}}{\rho} = gh - \frac{V_1^2}{2} = (32.2)(150) - \frac{(28.62)^2}{2} = 4{,}420.47 \ \frac{ft\text{-}lb}{slug}$$

We use the Moody diagram to find the friction factor

$$Re = \frac{\overline{V}D}{\nu} = \frac{(28.62)(0.667)}{0.1217 \times 10^{-4}} = 1.568 \times 10^6$$

$$e/D = \frac{0.00015}{0.667} = 0.00025$$

Thus, f = 0.0147.

The total minor loss h_T is evaluated.

$$(h_\ell)_{pipe} = f\,\frac{V_1^2}{2}\,\frac{L}{D} = 0.0147\,\frac{(28.62)^2}{2}\,\frac{470}{0.667} = 4{,}242.27 \ ft\text{-}lb/slug$$

$$(h_\ell)_{elbows} = (\Sigma k_i)\,\frac{V_1^2}{2} = (0.05+0.9+0.9)\,\frac{(28.62)^2}{2} = 757.67 \ \frac{ft\text{-}lb}{slug}$$

$h_T = 4{,}242.27 + 757.67 = 4999.94$ ft-lb/slug

Now,

$$\frac{dw_{pump}}{dm} = -\frac{P}{\rho Q} = \frac{-(185)(550)}{(8.6)(1.94)} = -6098.66 \text{ ft-lb/slug}$$

Finally inserting these results in eq.(1) and solving for P_2, we get

$$P_2 = 40.58 \text{ psig}$$

Problem 2

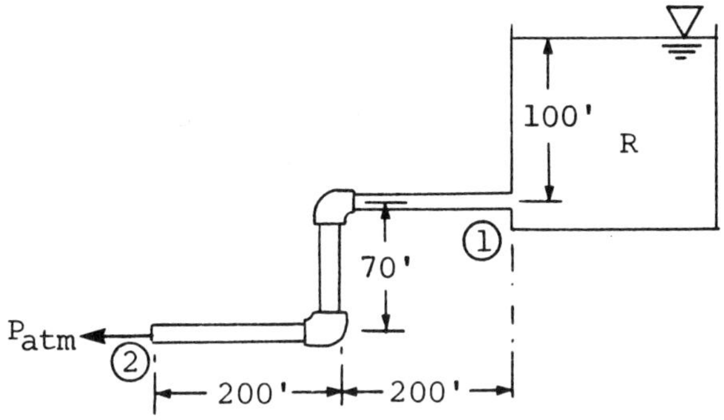

Fig. 2

Given data:

Water at 60^0F, $\nu = 0.1217 \times 10^{-4}$ ft^2/sec

$K_{elbows} = 0.90$, $K_{entrance} = 0.05$

Assumptions: the same as in problem 1.
Control Volume: The entire interior of the pipe.

Applying the first law between 1 and 2, and using Bernoulli's equation as in problem 1 we get

$$\frac{V_1^2}{2} + \left(\frac{P_{atm}}{\rho} - \frac{V_1^2}{2} + 100g\right) + gz_1 - \left(\frac{V_2^2}{2} + \frac{P_2}{\rho} + gz_2\right) \qquad (1)$$

$$= (h_\ell)_{pipe} + (h_\ell)_{elbows}$$

$P_2 = P_{atm}$ (subsonic free jet)

Introducing the elevation and using the head loss formulas in eq.(1), we get

$$\frac{V_1^2}{2} - \frac{P_{atm}}{\rho} - \frac{V_1^2}{2} + 100g + 70g - \frac{V_2^2}{2} - \frac{P_{atm}}{\rho} = f\,\frac{V_2^2}{2}\,(705)$$

$$+ (0.05+0.9+0.9)\,\frac{V_2^2}{2} \tag{2}$$

or

$$\frac{V_2^2}{2} + 1.85\,\frac{V_2^2}{2} + 705\,f\,\frac{V_2^2}{2} = 170(g)$$

or

$$V_2 = \sqrt{\frac{10948}{705f+2.85}} \tag{3}$$

the velocity is a function of the friction factor. f, cannot be found because the reynolds number cannot be calculated. Then, we have to use an iterative method. Let us guess a friction factor value (in the fully-developed region, with e/D = 0.00025), f = 0.0015. Thus,

$$(V_2)_1 = 28.55 \text{ ft/sec}$$

and $$Re = \frac{VD}{\nu} = \frac{28.55 \times 0.667}{0.1217 \times 10^{-4}} = 1.5647 \times 10^6$$

From the Moody diagram we find f for $Re = 1.5647 \times 10^6$ and e/D = 0.00025, f = 0.0147. With this value we compute V_2 once again.

$$(V_2)_2 = 28.78 \text{ ft/sec}$$

then, $Re = 1.577 \times 10^6$, and from the Moody diagram we can see that the friction factor is about the same. In that case V_2 = 28.78 ft/sec and,

$$Q = V.A = 28.78 \text{ ft/sec}\left[\pi\left(\frac{0.667}{2}\right)^2\right] \text{ft}^2 = 9.91 \text{ ft}^3\text{/sec}$$

<u>Problem 3</u>

Given data

Fluid: oil

$\mu = 50 \times 10^{-5}$ lb-sec/ft^2
$\gamma = 50$ lb/ft^3
$k_{elbows} = 0.90$, $k_{entrance} = 0.05$

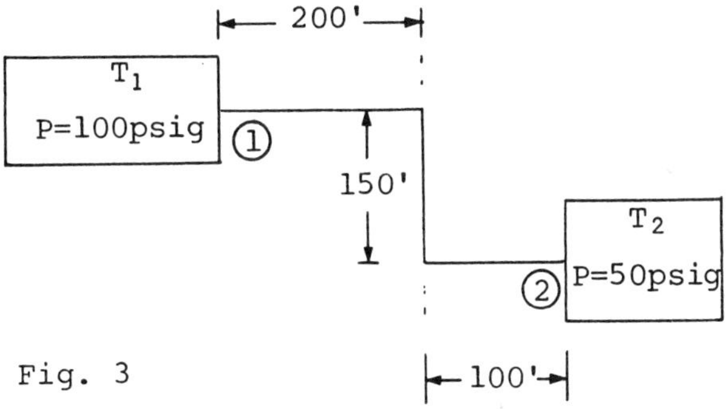

Fig. 3

Assumptions: 1 - 3 the same as in problem 1 plus
Control Volume: the entire interior of the pipe.

Applying the first law of thermodynamics between 1 and 2, we have

$$\left(\frac{V_1^2}{2} + \frac{P_1}{\rho} + gZ_1\right) - \left(\frac{V_2^2}{2} + \frac{P_2}{\rho} + gZ_2\right) = h_T \tag{1}$$

Introducing the values and the head loss formulas, we have

$$\frac{V_1^2}{2} + \frac{100(144)}{1.552} + 4,830 - \frac{V_1^2}{2} - \frac{50(144)}{1.552} = f\,\frac{450}{D}\,\frac{V_1^2}{2} + 1.85\,\frac{V_1^2}{2} \tag{2}$$

$$V_1 \simeq \overline{V} = \frac{Q}{A} = \frac{1}{\frac{\pi D^2}{4}} = \frac{4}{\pi D^2} \tag{3}$$

Introducing this value in eq.(2), multiplying by D^5 and arranging terms, we have

$$D^5 - 0.0002D - 0.0385f = 0 \tag{4}$$

we will solve this equation by iteration. First, we neglect the second term because it comes from minor losses in order to simplify the calculations, later it will be introduced, thus we have

$$D^5 - 0.0385f = 0 \tag{5}$$

second, a first guess of f, may be 0.015 for steel commercial pipe

$$D^5 - 5.775 \times 10^{-4} = 0 \tag{6}$$

from where D = 0.225 ft.

$$V = \frac{4}{\pi D^2} = \frac{4}{\pi(0.225)^2} = 25.13 \text{ ft/sec}$$

$$Re = 1.754 \times 10^4$$

Now, for a diameter of 0.225 ft (2.7 in) a relative roughness of 0.00062 for steel commercial pipe is given (p. 357). Then from the Moody chart, the friction factor is found

$$f = 0.0277$$

With this friction factor value we repeat the above calculations once again. We find

$$D = 0.254 \text{ ft}$$

$$V = 19.66 \text{ ft/sec}$$

$$Re = 1.547 \times 10^4$$

$$f = 0.028$$

Further iteration will result in very small changes in the friction value. Introducing the diameter and function factor in eq.(4), we can see that they satisfy the equation. The trial and error method should be used in the case that the values do not satisfy the equation.

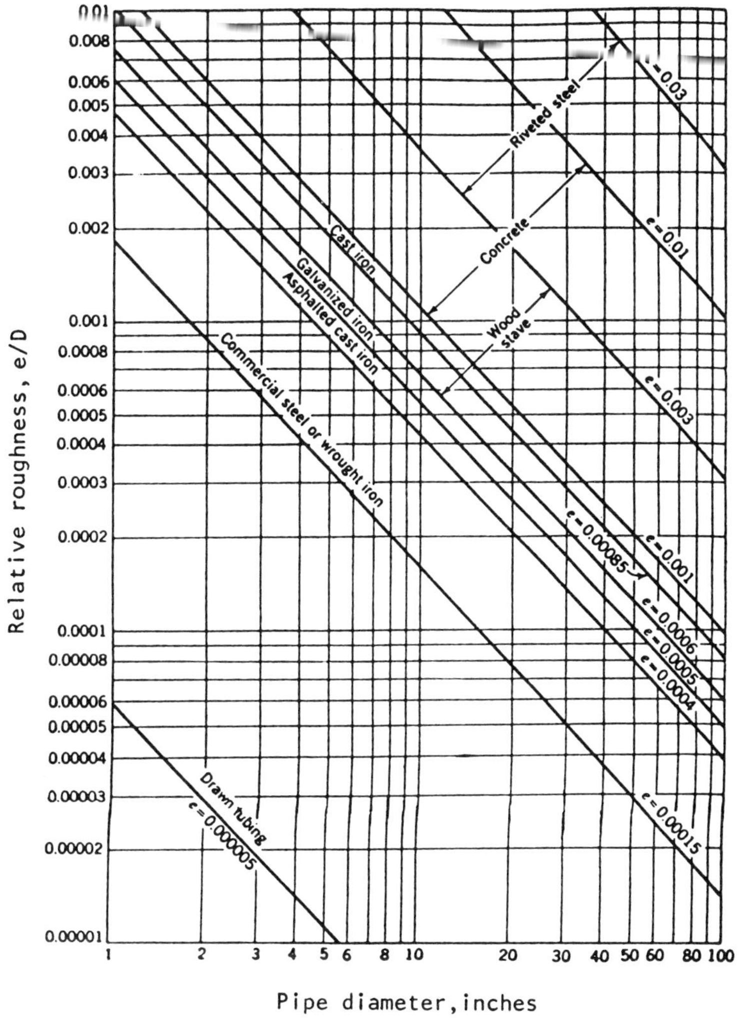

Fig. 4

• PROBLEM 7-25

A 12 km long pipe carries $0.2\text{m}^3/\text{s}$ of water. The pipe diameter is 30 cm; its relative roughness is 0.004. Compute the change in head loss if the temperature of the water changes from 65°C to 30°C.

Solution: The average velocity of the flow =

$$\frac{Q}{A} = \frac{0.2}{\frac{\pi}{4}(0.3)^2} = 2.829 \ \frac{\text{m}}{\text{s}}$$

The Darcy equation for head loss is given by $h_L = f\frac{L}{D}\frac{V^2}{2g}$

Table 1
Viscosity of Water

Temperature in °C	Absolute Viscosity Centipoise	Absolute Viscosity lb sec/ft²
0	1.792	0.374×10^{-4}
4	1.567	0.327×10^{-4}
10	1.308	0.272×10^{-4}
20	1.005	0.209×10^{-4}
20.2	1.000	0.208×10^{-4}
30	0.8807	0.183×10^{-4}
50	0.5494	0.114×10^{-4}
70	0.4061	0.084×10^{-4}
100	0.2838	0.059×10^{-4}
150	0.184	0.038×10^{-4}

From Table 1, at 30°C

$$\mu \cong 0.8807 \text{ centipoise}$$

$$= 0.8807 \times 10^{-3} \text{ poise}$$

$$= 8.807 \times 10^{-5} \frac{N \cdot s}{m^2}$$

At 65°C

$$\mu \cong (0.4276 \times 10^{-3}) \text{ poise}$$

$$= 4.284 \times 10^{-5} \frac{N \cdot s}{m^2}$$

Table 2
Density of Water

Temperature °C	Density g/cm³	Density slugs/ft³
0 ice	0.917	1.779
0 water	0.9998	1.9406
3.98	1.0000	1.941
10	0.9997	1.940
25	0.9971	1.935
100	0.9584	1.860

From Table 2, at 30°C

$$\rho \simeq 0.9970 \frac{g}{cm^3} = 997 \frac{kg}{m^3}$$

$$= \frac{997}{9.81} = 101.63 \frac{N \cdot s^2}{m^4}$$

At 65°C

$$\rho \cong 0.980 \text{ g/cm}^3 = 980 \text{ kg/m}^3$$

$$= \frac{980}{9.81} = 99.897 \frac{\text{N}\cdot\text{s}^2}{\text{m}^4}$$

The Reynolds number is given by

$$Re = \frac{\rho VD}{\mu}$$

Then

$$\text{Re at } 30°\text{C} = \frac{101.63 \times 2.829 \times 0.3}{8.8237 \times 10^{-5}} = 9.775 \times 10^5$$

$$\text{Re at } 65°\text{C} = \frac{99.897 \times 2.829 \times 0.3}{4.284 \times 10^{-5}} = 19.791 \times 10^5$$

The relative roughness $e/D = 0.004$.

Entering Moody's diagram we find that because of the large Reynolds numbers

$$f \text{ at } 65°\text{C} \cong f \text{ at } 30°\text{C} = 0.0282$$

The head loss may be computed from the Darcy equation

$$h_L = f\frac{L}{D}\frac{V^2}{2g} = 0.0282 \frac{(12{,}000)}{0.3} \frac{(2.829)^2}{2 \times 9.81}$$

$$= 460.0 \text{ m}$$

over the entire length of the 12 km pipeline.

• PROBLEM 7-26

Derive the equation for the velocity profile in the annulus shown in the accompanying figure. The steady flow is fully developed and $v = w = 0$.

Solution: The velocity components in the r and θ-directions are v and w, respectively. For the fully developed flow $\partial u/\partial x = 0$. The describing equations are identical to those which describe pipe flow. The x-momentum equation

$$\frac{\partial u}{\partial t} + u\frac{\partial u}{\partial x} + v\frac{\partial u}{\partial r} + \frac{w}{r}\frac{\partial u}{\partial \theta} =$$

$$-\frac{1}{\rho}\frac{\partial p_k}{\partial x} + \nu\left[\frac{1}{r}\frac{\partial}{\partial r} r\left(\frac{\partial u}{\partial r}\right) + \frac{1}{r^2}\frac{\partial^2 u}{\partial \theta^2} + \frac{\partial^2 u}{\partial x^2}\right]$$

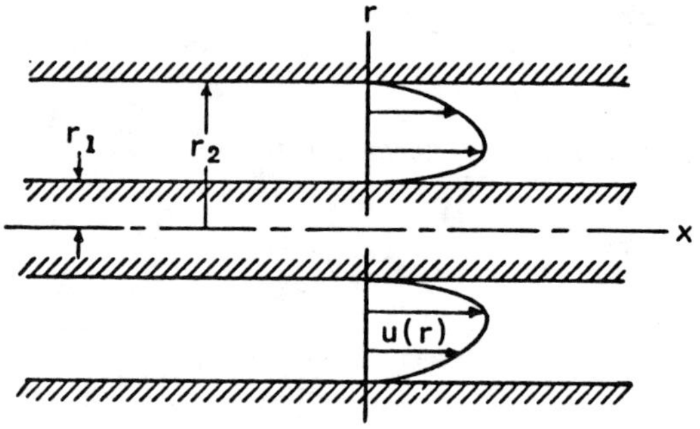

with $v = w = 0$, becomes

$$0 = -\frac{dp}{dx} + \frac{\mu}{r}\frac{d}{dr}\, r\frac{du}{dr}$$

The general solution is

$$u = \frac{\lambda r^2}{4} + C_1 \ln r + C_2$$

where $\lambda = (1/\mu)(dp/dx)$. The boundary conditions are $u = 0$ at $r = r_1$ and r_2. This results in

$$0 = \frac{\lambda r_1^2}{4} + C_1 \ln r_1 + C_2$$

$$0 = \frac{\lambda r_2^2}{4} + C_1 \ln r_2 + C_2$$

Solving for C_1 and C_2 yields

$$C_1 = \frac{\lambda}{4}\,\frac{r_1^2 - r_2^2}{\ln(r_2/r_1)}$$

$$C_2 = \frac{\lambda}{4}\,\frac{(r_1^2 - r_2^2)\ln r_2}{\ln(r_2/r_1)} - \frac{\lambda r_2^2}{4}$$

The velocity distribution is then

$$u = \frac{1}{4\mu}\frac{dp}{dx}\left[(r^2 - r_2^2) + \frac{r_2^2 - r_1^2}{\ln(r_1/r_2)}\ln(r/r_2)\right]$$

As we let r_1 approach 0, $\ln r_1/r_2$ approaches $-\infty$ and the last term in the brackets drops out, resulting in the pipe flow parabolic velocity profile.

• PROBLEM 7-27

A pipeline of inside diameter d = 100mm and length l = 200m is attached to a water reservoir as shown in Fig. 1. The loss coefficient, k, for the inlet is k = 0.5. Determine the height, 'h', that has to be maintained in the reservoir to produce a volume flow rate of 0.03 m^3/sec of water.

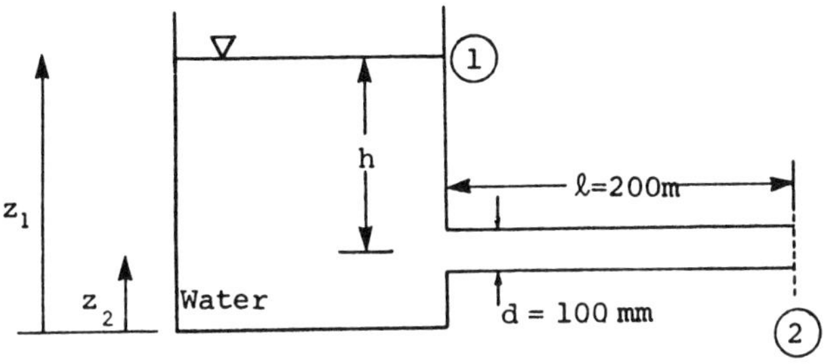

Fig. 1

Solution: Using the energy equation, and taking into account the losses due to friction between points 1 and 2, we have

$$\left(\alpha_1 \frac{V_1^2}{2} + gz_1 + \frac{p_1}{\rho}\right) - \left(\frac{p_2}{\rho} + \alpha_2 \frac{V_2^2}{2} + gz_2\right) = h_l + h_{lm} \qquad (1)$$

where $\quad h_l = f \frac{l}{d} \frac{V^2}{2} \qquad h_{lm} = k \frac{V^2}{2}$

h_l = head loss due to friction

h_{lm} = minor head loss due to the shape of the pipe

$\alpha_2 = 1,$

$p_1 = p_2 = p_{atm}, \quad V_1 \simeq 0, \; V_2 = V, \; z_1 - z_2 = h$

Simplifying equation (1) gives,

$$gh - \frac{V^2}{2} = f \frac{l}{d} \frac{V^2}{2} + k \frac{V^2}{2}$$

or

$$h = \frac{V^2}{2g}\left[f \frac{l}{d} + k + 1\right]$$

Since

$$V = \frac{Q}{A} = \frac{4Q}{\pi D^2}, \quad \text{then}$$

$$h = \frac{8Q^2}{\pi^2 D^4 g}\left[f\,\frac{1}{d} + k + 1\right]$$

The Reynolds number, Re, is

$$Re = \frac{\rho \bar{V} D}{\mu} = \frac{4\rho Q}{\pi \mu D} = \frac{(4)}{\pi} \cdot \frac{(999)(0.03)}{(1 \times 10^{-3})(0.10)} = 3.82 \times 10^5$$

For smooth pipe from Fig. 2 (see Problem 7-33), f = 0.0141. Then,

$$h = \frac{8Q^2}{\pi^2 D^4 g}\left[f\,\frac{1}{d} + k + 1\right] = \frac{(8)(0.03)^2}{(\pi^2)(0.100)^4(9.81)} \times$$

$$\left[(0.0141)\,\frac{200}{0.100} + 0.5 + 1\right]$$

or

$$h = 22.08\text{m}$$

• PROBLEM 7-28

a. What is the flow rate for water at 15°C in a 25-cm cast iron pipe when the head loss is 5.0 m in 300 m of pipe?

b. What is the flow rate for water at 15°C in a commercial steel pipe, 250-mm diameter, when the head loss in 300 m of pipe is 5.00 m?

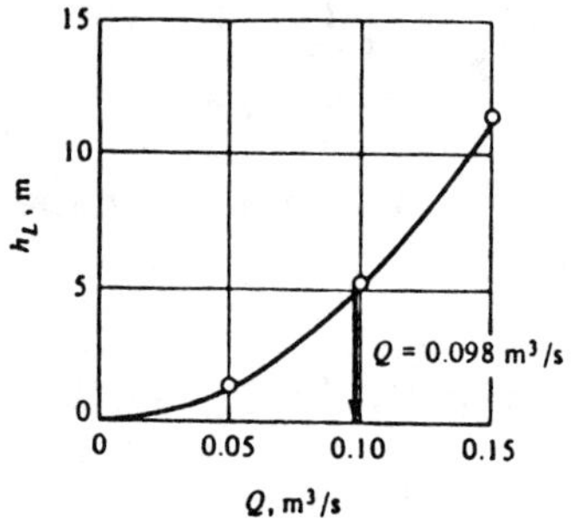

Fig. 1

Solution: a. Method 1. k/D = 0.00026/0.25 = 0.00104 and, for a high Re_D, f = 0.020. Then from the Darcy-Weisbach equation, which was also developed by a dimensional analysis:

$$\frac{\Delta p}{L} = \frac{f}{D} \frac{\rho V^2}{2} \tag{1}$$

where f is the friction factor, $\rho V^2/2$ is the dynamic pressure of the mean flow, and D is the pipe diameter. An alternative form in terms of the head loss due to friction h_f is

$$h_f = \frac{\Delta p}{\gamma} = f \frac{L}{D} \frac{V^2}{2g} \tag{2}$$

$$\frac{V^2}{2g} = \frac{h_f D}{fL} = \frac{(5)(0.25)}{(0.020)(300)} = 0.208 \text{ and } V = 2.02 \text{ m/s}$$

For this first estimate of velocity, $Re_D = VD/\nu = (2.02)(0.25)/1.14 \times 10^{-6} = 4.4 \times 10^5$, and for this Re_D and given relative roughness, f = 0.021. Then

$$\frac{V^2}{2g} = \frac{(5)(0.25)}{(0.021)(300)} = 0.198 \text{ and } V = 1.97 \text{ m/s}$$

The Reynolds number for this velocity is essentially the same as above, and thus f = 0.021. The flow rate is

$$Q = VA = (1.97) \frac{\pi}{4} (0.25)^2 = 0.097 \text{ m}^3/\text{s}$$

Method 2. If flow rates of 0.05, 0.10, and 0.15 m^3/s are assumed, the corresponding head losses are calculated to be 1.33, 5.13, and 11.42 m, respectively. A graphic interpolation gives Q = 0.098 m^3/s (Fig. 2).

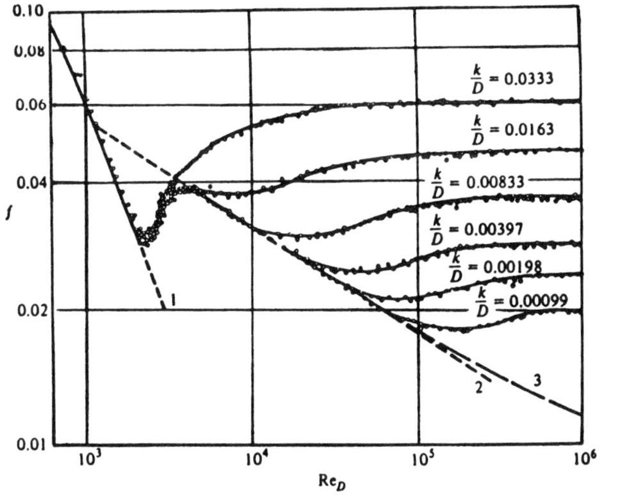

Fig. 2

Results of Nikuradse's measurements on pipes artificially roughened with sand grains. Curve 1: $f = 64/Re_D$. Curve 2: $f = 0.316/Re_D^{0.25}$. Curve 3: $1/\sqrt{f} = 2\log(Re_D\sqrt{f}) - 0.8$.

b. k = 0.000045, so the relative roughness is k/D = 0.000045/0.25 = 0.00018, and for high Re, f = 0.0133. Then

$$V^2/2g = h_f D/fL = (5.00)(0.25)/(0.0133)(300) = 0.313 \text{ m}$$

and

$$V = \sqrt{(2)(9.807)(0.313)} = 2.48 \text{ m/s}$$

For this velocity, Re = VD/ν = $(2.48)(0.25)/1.140 \times 10^{-6}$ = 5.4×10^5. (The viscosity is found in Table 1. For this Re, f = 0.0152.

Table 1

VISCOSITY OF WATER

TEMPERATURE (°C)	DYNAMIC μ (kg/m s)	KINEMATIC ν (m^2/s)
0	1.787×10^{-3}	1.787×10^{-6}
5	1.519	1.519
10	1.307	1.307
15	1.139	1.140
20	1.002	1.004
25	0.890	0.893
30	0.798	0.801
35	0.719	0.723
40	0.653	0.658
45	0.596	0.602
50	0.547	0.553
55	0,504	0.511
60	0.466	0.474
65	0.433	0.441
70	0.404	0.409

Then $V^2/2g = h_f D/fL$ = 0.274 and V = 2.32 m/s. For this velocity, Re = 5.1×10^5 and f is again 0.0152. Thus the flow rate is Q = VA = $(2.32)(\pi/64)$ = 0.114 m^3/s = 114 L/s.

• PROBLEM 7-29

A solution (viscosity = 2.4 cp, sp. gr. = 1.2) is to be pumped from a storage tank to an overhead tank, both open to the atmosphere. The pipe line, 2.067 in. diameter, is 160 ft long, and the fittings include three 90° standard elbows, one tee-piece, and one conventional globe valve. The vertical distance between the surface levels in the tanks is 80 ft. If the solution is to flow through the pipe at 6 fps, what horse-power will be consumed by the pump at 50 per cent efficiency? The roughness of the pipe may be assumed to be 0.0018 in.

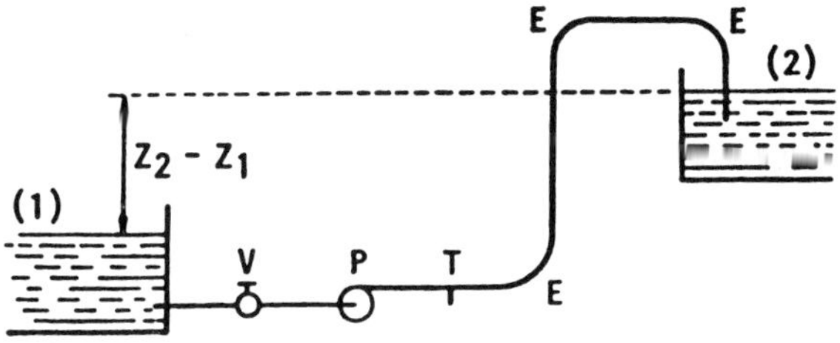

Fig. 1

Table 1

Description	Equivalent number of pipe diameters (*N*)
90° standard elbow	30
90° long radius elbow	20
45° standard elbow	16
180° close return bend	50
Standard tee	
(a) with flow through run	20
(b) with flow through branch	60
Conventional globe valve, fully open	340
Conventional angle valve, fully open	145
Conventional gate valve, fully open	13
Butterfly valve, fully open	20
Straight-through cock	18

Solution: With reference to the Table 1, the equivalent length of the fittings is obtained as follows.

		N
1 tee-piece	=	20
1 globe valve	=	340
3 x 90° elbows = 3 x 30	=	90
Total	=	450

$$ND = (450)\ \frac{2.067}{12} = 77.5 \text{ ft}$$

$$L = 160 + 77.5 = 237.5 \text{ ft}$$

$$Re = \frac{Dv\rho}{\mu} = \frac{(2.067)/12)(6)(1.2)(62.4)}{(2.4)(0.000672)}$$

$$Re = 48{,}000$$

For this Reynolds number and the relative roughness

$$e/D = 0.0018/2.067 \cong 0.0009$$

the Moody friction factor, from the Moody diagram, is

$$f' = 0.024$$

Using the Darcy equation

$$h_f = f' \frac{L}{D} \frac{v^2}{2g} = (0.024) \frac{237.5}{2.067/12} \frac{v^2}{2g}$$

$$h_f = 33.1 \frac{v^2}{2g}$$

where v is the velocity in the pipe.

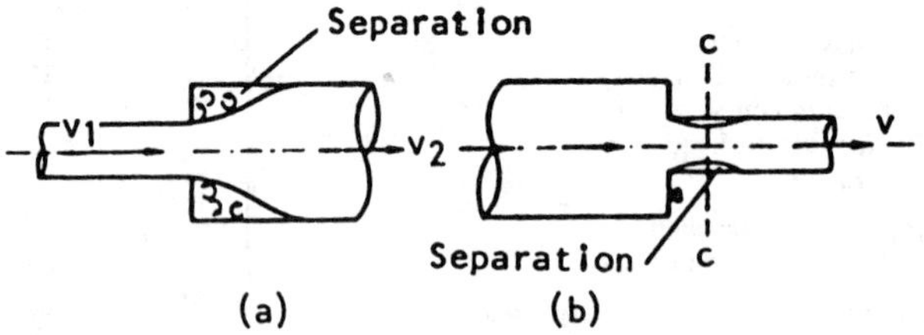

Fig. 2

Suppose that a pipe changes its cross-sectional area suddenly, as shown in Fig. 2 , and consider at first a sudden enlargement. The loss of energy experienced in this case is due to the eddies in the portion of the fluid trapped at the corner of the larger section of the pipe. These eddies draw appreciable energy from the stream with the result that there is a loss of head. Let this loss be h_e, then with reference to Fig. 2(a)

$$h_e = \frac{(v_1 - v_2)^2}{2g} \tag{1}$$

where v_1 and v_2 are the velocities in the smaller and larger sections of the pipe.

Now consider the flow through a pipe which suddenly reduces its cross-section, and let h_c be the loss of head due to the sudden contraction of the stream. As shown in Fig. 2(b), it contracts in a similar way to the stream passing through an orifice, with the effect that vena contracta forms just beyond the change in section. The loss is expressed by the equation

$$h_c = \frac{(v_c - v)^2}{2g} \tag{2}$$

in which v is the average velocity in the smaller pipe, and v_c is that at the vena contracta.

A generally accepted approximation of Eq. 2 is given by

$$h_c = 0.5 \frac{v^2}{2g} \tag{3}$$

The contraction and enlargement losses, at the respective ends of the pipe line, are

$$h_c + h_e = 0.5 \frac{v^2}{2g} + \frac{v^2}{2g} = 1.5 \frac{v^2}{2g}$$

For v = 6 fps, the total loss of head is

$$H_f = h_f + h_c + h_e = (33.1 + 1.5)\frac{v^2}{2g} = 34.6 \frac{6^2}{64.4}$$

$$h_f = 19.34, \text{ say } 19.4 \text{ ft}$$

Writing the flow equation between the levels (1) and (2), in Fig. 1

$$Z_1 + \frac{v_1^2}{2g} + \frac{p_1}{\gamma} + W = Z_2 + \frac{v_2^2}{2g} + \frac{p_2}{\gamma} + H_f$$

Since $v_1 = v_2 = 0$, and $p_1 = p_2$, this equation reduces to

$$W = (Z_2 - Z_1) + H_f$$

where W is the net work to be done by the pump on 1 lb of the solution.

Since

$$Z_2 - Z_1 = 80 \text{ ft}$$

$$W = 80 + 19.4 = 99.4 \text{ ft}$$

or

99.4(ft)(lb-f)/lb-wt

The cross-sectional area of the pipe is $(2.067/12^2\pi/4 = 0.0233 \text{ ft}^2$, and the rate of flow is

$$(6)(0.0233)(1.2)(62.4) = 10.47 \text{ lb/sec}$$

At 50 per cent efficiency of the pump, the power consumption is

$$\frac{(99.4)(10.47)}{(550)(0.5)} = 3.78 \text{ hp}$$

A 3000 ft long pipe of 10 in. diameter carries 5 cfs of water. The C coefficient is assumed to be 100. Calculate the loss of energy due to friction using the Hazen-Williams nomograph shown in the Figure.

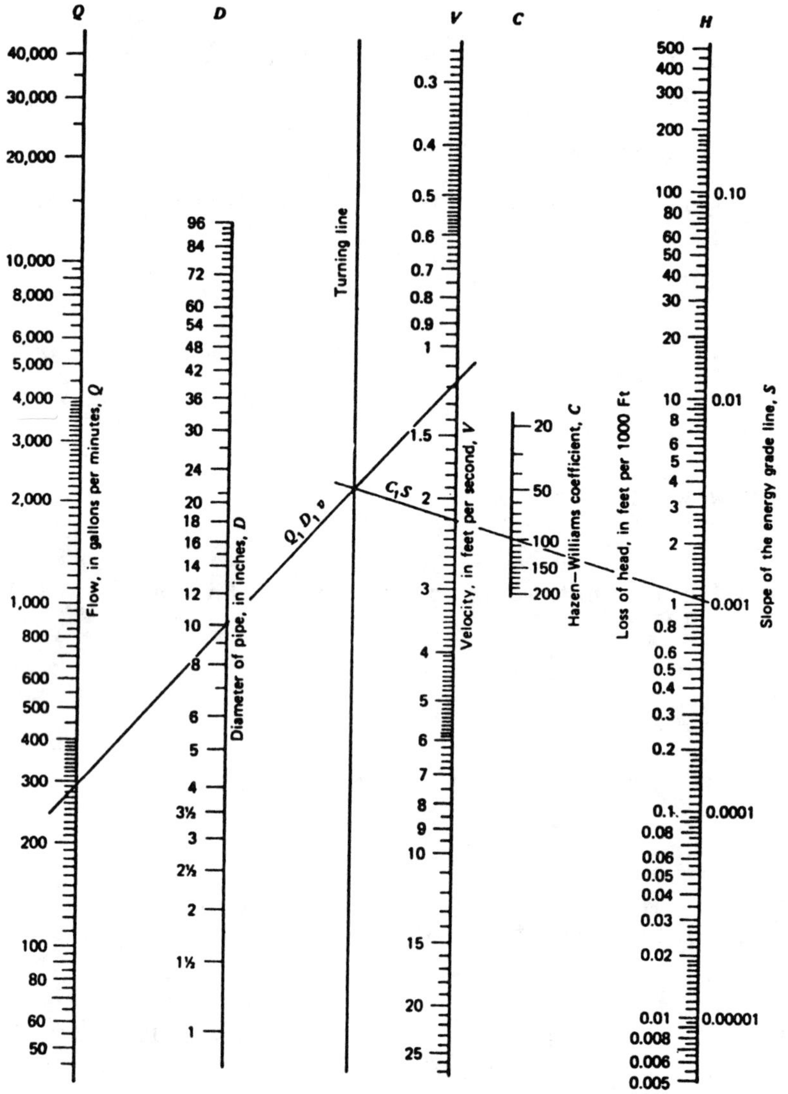

The Hazen-Williams nomograph.

Solution: The variables are:

$Q = 5 \text{ cfs} = 300 \text{ ft}^3/\text{min}$

$D = 10 \text{ in.}$

$C = 100$

In the nomograph connect Q and D with a straight line and mark off the intersection on the turning line. Connect the marked point with C and extend the line to S. The resulting S is 1/1000 ft per ft. For 3000 ft the head loss is 3000 x 0.001 = 3 ft.

• PROBLEM 7-31

Two large open-topped water tanks are connected by a galvanised iron pipe 4 in. in diameter, 400 ft long, containing a gate valve; the bends in the pipe are of such large radius that the extra losses in them are negligible. (a) What difference in height between the levels in the tanks is required to obtain a flow rate Q of $0.5\,ft^3/sec$, with the valve wide open? Take the density ρ and viscosity μ of the water to be 62.4 lbm/ft^3 and 2.74×10^{-5} lbf sec/ft^2 respectively, and gravitational acceleration to be 32.1 ft/sec^2.

(b) What would be the water flow rate through the pipe specified in (a) if the difference in level were 20 ft? Take the same values of density and viscosity as before.

(c) It is desired to replace the pipe specified in (a) by a larger one so that a flow rate of 0.75 ft^3/sec could be obtained with a difference in level of 20 ft. Calculate the diameter required, for the same values of ρ, μ, and e as before.

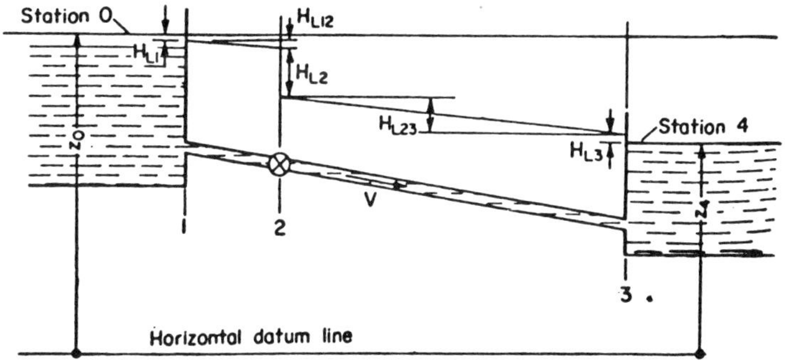

Fig. 1

Solution: Figure 1 illustrates the layout. Using values of loss coefficients from Table 1 and Fig. 2 and 3, the total loss of head. H_{L04} is made up of the following components (V ≡ flow velocity in pipe, d ≡ pipe diameter):

The head lost at the pipe entrance,

$$H_{L1} = 0.5\,\frac{V^2}{2g}.$$

The head lost in pipe friction between station 1 and 2,

Table 1

Fitting	Loss coefficient, C
Globe valve, fully open	10·0
Gate valve, fully open	0·2
Standard elbow, 90°	0·9
Long sweep elbow, 90°	0·6
Standard *T* (equal diameters)	1·8

$$H_{L12} = f \frac{l_{12}}{d} \frac{V^2}{2g}.$$

The head lost at the valve,

$$H_{L2} = 0.2 \frac{V^2}{2g}.$$

The head lost in pipe friction between stations 2 and 3,

$$H_{L23} = f \frac{l_{23}}{d} \frac{V^2}{2g}.$$

The head lost at the pipe exist,

$$H_{L3} = \frac{V^2}{2g}.$$

We know that

$$H_0 = H_4 + H_{L04},$$

or,

$$H_0 - H_4 = H_{L04} = H_{L1} + H_{L12} + H_{L2} + H_{L23} + H_{L3}$$

$$= 0.5 \frac{V^2}{2g} + f \frac{l_{12}}{d} \frac{V^2}{2g} + 0.2 \frac{V^2}{2g} + f \frac{l_{23}}{d} \frac{V^2}{2g} + \frac{V^2}{2g}$$

$$= \frac{V^2}{2g} \left[1.7 + \frac{f}{d} (l_{12} + l_{23}) \right] \qquad (1)$$

From equation 1 it is clear that, at least to the degree of accuracy afforded by the simple one-dimensional pipe-flow analysis, the position of the valve in the line does not affect the result; we simply need to know the sum of l_{12} and l_{23}, which is 400 ft. If the tanks are large the velocity at each surface (stations 0 and 4) is negligible, and since the pressures at these surfaces are practically equal, then

$$H_0 - H_4 = z_0 - z_4,$$

and equation 1 can be used directly once the value of the friction factor f has been found.

The mean flow velocity is

$$V \equiv \frac{4Q}{\pi d^2} \quad \text{(d = internal diameter)}$$

$$= \frac{4 \times 0.5}{\pi \times \left(\frac{1}{3}\right)^2} \text{ ft/sec} = 5.73 \text{ ft/sec.}$$

Thus the Reynolds number, $Re \equiv \dfrac{\rho V d}{\mu g_0}$

$$= \frac{62.4 \times 5.73 \times \frac{1}{3}}{2.74 \times 10^{-5} \times 32.2} = 1.35 \times 10^5$$

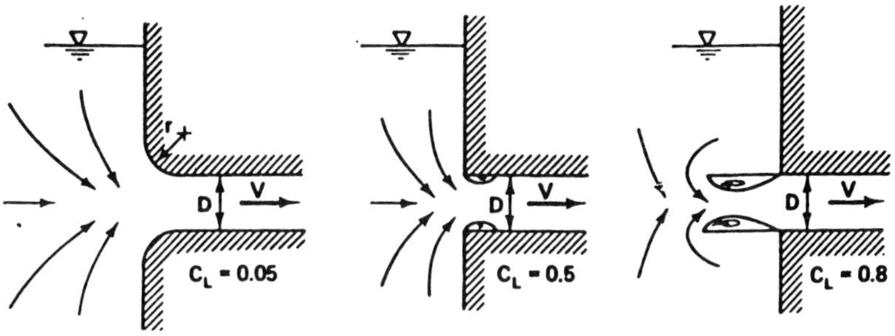

Fig. 2

(a) Rounded entrance (b) Square entrance (c) Re-entrant entrance

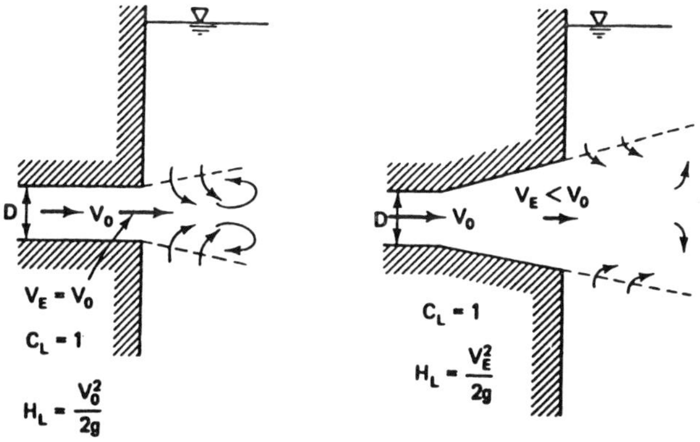

Fig. 3

(a) Standard exit. (b) Expanded exit section.

For galvanised iron the surface roughness is given in Table 2 as 0.0005 ft, so that the relative roughness is

$$\frac{e}{d} = \frac{0.0005}{\frac{1}{3}} = 0.0015.$$

For these values of Re and e/d the friction factor is found from the Moody Diagram to be 0.0235; then inserting known values into equation 1,

$$z_0 - z_4 = H_0 - H_4 = \frac{V^2}{2g}\left[1.7 + \frac{f}{d}(l_{12} + l_{23})\right]$$

$$= \frac{(5.73)^2}{2 \times 32.1}\left[1.7 + \frac{0.0235}{\frac{1}{3}} \times 400\right] \text{ ft} = 15.3 \text{ ft.}$$

Table 2

Material	Representative roughness dimension, *e* (ft)
Glass, Drawn brass, Drawn copper	Negligible
Steel, Wrought iron	0·00015
Asphalted cast iron	0·0004
Galvanised iron	0·0005
Cast iron	0·00085
Concrete	0·001–0·01

(b) A trial-and-error solution is required; this can be carried out in several ways, and the most direct method is shown in Table 3. It consists simply of a series of calculations of the difference of level required for a number of values of mean flow velocity V. The first two values of V are chosen more or less arbitrarily, but succeeding ones are selected on the basis of the results of the previous calculations; the object is to reduce the error (that is, the difference between the required $(z_0 - z_4)$ and the given value of 20 ft) to an acceptably small figure.

Table 3

V (ft/sec)	*Re*	*f*	1200*f* + 1·7	$\frac{V^2}{2g}$ (ft)	$(z_0-z_4) = (H_0-H_4)$ (ft)	Error (ft)
6·0	$1{\cdot}42 \times 10^5$	0·0233	29·75	0·561	16·6	−3·4
6·5	$1{\cdot}53 \times 10^5$	0·0230	29·3	0·658	19·3	−0·75
6·6	$1{\cdot}56 \times 10^5$	0·0229	29·2	0·678	19·8	−0·2
6·64	$1{\cdot}565 \times 10^5$	0·0229	29·2	0·687	20·02	+0·02

Table 4

d (ft)	Re	e/d	f	$\frac{l}{d}f+1.7$	$\frac{8Q^2}{\pi^2 d^4 g}$ (ft)	$(z_0 - z_4) = (H_0 - H_4)$ (ft)	Error (ft)
0·36	1.88×10^5	0·00139	0·0216	25·7	0·846	21·7	+1·7
0·38	1.78×10^5	0·00132	0·0217	24·3	0·681	16·7	−3·5
0·368	1.84×10^5	0·00136	0·0216	25·2	0·773	19·5	−0·5
0·366	1.85×10^5	0·00137	0·0216	25·3	0·791	20·0	0·0

The value of friction factor f is obtained from the Moody diagram, in each case, using the value of 0.0015 for e/d: the corresponding values of $(z_0 - z_4)$ are calculated in the same way as in (a).

It will be appreciated that the third significant figure is unlikely to be accurate because of the difficulty of reading values of f from diagram. From the final value of V the flow rate Q is quickly found:

$$Q = \frac{\pi}{4} d^2 V$$

$$= \frac{\pi}{4} \times \left(\frac{1}{3}\right)^2 \times 6.64 \text{ ft}^3/\text{sec} = 0.579 \text{ ft}^3/\text{sec}.$$

(c) A trial-and-error solution is required in this case also. For each of a series of values of diameter d the Reynolds number $(Re = 4Q/\pi d \mu g_0)$ and of e/d are calculated; the corresponding value of friction factor f is found from the Moody Diagram, and the required difference between the water levels is calculated and compared with the given value, 20 ft. The loss coefficients for the valve, pipe entry and pipe exit are taken as 0.2, 0.5 and 1.0 as before.

Required pipe diameter = 0.366 ft = 4.39 in.

PARALLEL/SERIES PIPES

• PROBLEM 7-32

Two pipes, 6-in. diameter, are 100 ft. and 10,000 ft. long respectively and have each a total loss of head of 10 ft.: pipes not bell-mouthed. Each contains a gate valve, the pipe being of square section, 6 in. × 6 in. at the valve. Find the ratio of the discharge to the maximum discharge in each pipe when the gate is 4, 2, 1 and ½ in. open. Take f = 0.005 and c_c for the valve as .60.

Solution: When the gate is x in. open, the area of the contracted section is $.60 \times \frac{x}{12} \times \frac{1}{2}$ ft.^{2}, the whole area being, $\frac{1}{2} \times \frac{1}{2}$. If v is the velocity in the pipe, the velocity v_c at the contraction is

$$\frac{.5 \times 12}{.6x} v = \frac{10v}{x} .$$

The loss of head at the valve is, therefore,

$$\left(\frac{10}{x} - 1\right)^2 \frac{v^2}{2g} .$$

The loss at entrance is

$$\frac{.5v^2}{2g}$$

and the loss in friction is $\dfrac{.020 \times 100}{\frac{1}{2}} \dfrac{v^2}{2g} = 4 \dfrac{v^2}{2g}$

in the short pipe and

$$\frac{.020 \times 10000}{\frac{1}{2}} \cdot \frac{v^2}{2g} = 400 \frac{v^2}{2g}$$

We therefore prepare the following table:

x.	$\frac{10}{x}$.	$\frac{10}{x}-1$.	$\left(\frac{10}{x}-1\right)^2$.	(Vel. Head + Entry) × $\frac{v^2}{2g}$.	Local Losses × $\frac{v^2}{2g}$.	Short Pipe. Friction × $\frac{2}{2g}$.	Short Pipe. Total × $\frac{v^2}{2g}$.	Short Pipe. v.	Short Pipe. $\frac{v}{v_0}$.	Long Pipe. Friction × $\frac{v^2}{2g}$.	Long Pipe. Total × $\frac{v^2}{2g}$.	Long Pipe. v.	Long Pipe. $\frac{v}{v_0}$.
6	--	--	--	1.5	1.5	4	5.5	10.82	1.00	400	401.5	1.266	1.00
4	2.5	1.5	2.25	1.5	3.75	4	7.75	9.11	.84	400	403.75	1.263	.997
2	5.0	4.0	16.00	1.5	17.50	4	21.50	5.47	.51	400	417.5	1.241	.980
1	10.0	9.0	81.00	1.5	82.50	4	86.50	2.73	.25	400	482.5	1.156	.913
$\frac{1}{2}$	20.0	19.0	361.00	1.5	362.50	4	366.50	1.32	.12	400	762.5	0.919	.726

We thus see that, while in the short pipe the flow is reduced roughly proportionately to the opening of the valve, in the long pipe the valve only begins to reduce the flow appreciably when the opening is reduced to about a third, and when the opening is reduced to $\frac{1}{12}$ the flow is still

.726 of the maximum. There is, therefore, much more risk of producing "waterhammer" in a long main by the sudden closing of a valve, a sound wave of high pressure travelling to and fro along the pipe when this occurs.

• PROBLEM 7-33

In figure 1 the open reservoirs are 20 m apart in elevation. Pipe 1 is 60 cm in diameter and 200 m long. Pipe 2 is 30 cm in diameter and 100 m long. Both are made of commercial steel. The entrance to pipe 1 is square (k_L = 0.5) and the junction between pipes 1 and 2 is an abrupt contraction. Assume the water viscosity to be 10^{-6} m^2/s. What is the flow rate?

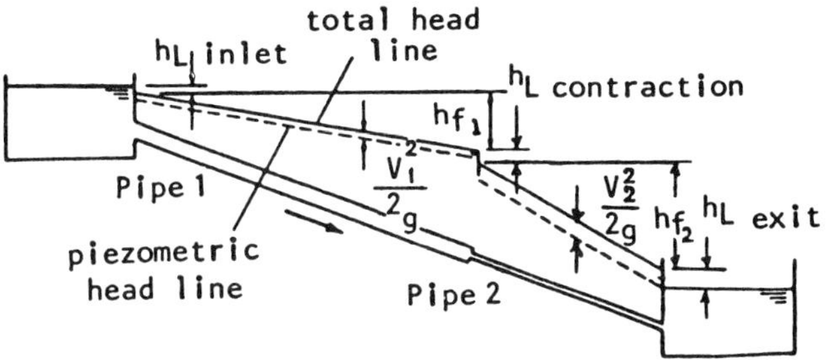

Fig. 1

Total head line and piezometric head line for flow between two open reservoirs.

Solution:

k_1/D_1 = 0.000045/0.6 = 0.000075 and f_1 = 0.0115 if Re_1 is large;

k_2/D_2 = 0.000045/0.3 = 0.00015 and f_2 = 0.0128 if Re_2 is large.

The energy equation written between the free surfaces of the reservoirs, with all losses shown in figure 1 included, is

$$0 + 0 + 20 = 0 + 0 + 0 + h_{L\ inlet} + h_{f-1} + h_{L\ contraction}$$

$$+ h_{f-2} + h_{L\ exit}$$

$$20 = \frac{(0.5)V_1^2}{2g} + \frac{(f_1L_1/D_1)V_1^2}{2g} + \frac{k_LV_2^2}{2g} + \frac{(f_2L_2/D_2)V}{2g} + \frac{V_2^2}{2g}$$

From continuity $V_2 = 4V_1$, and thus the energy equation may be written as

$$20 = \frac{V_1^2}{2g}\left\{0.5 + \frac{(0.0115)(200)}{0.6} + (0.35)(16) + \frac{(0.0128)(100)}{0.3}\quad 16 + 16\right\}$$

so that $V_1 = 2.04$ m/s and $V_2 = 8.16$ m/s. Better values of the friction factors are obtained after calculating the Reynolds numbers for the first approximations to the velocities.

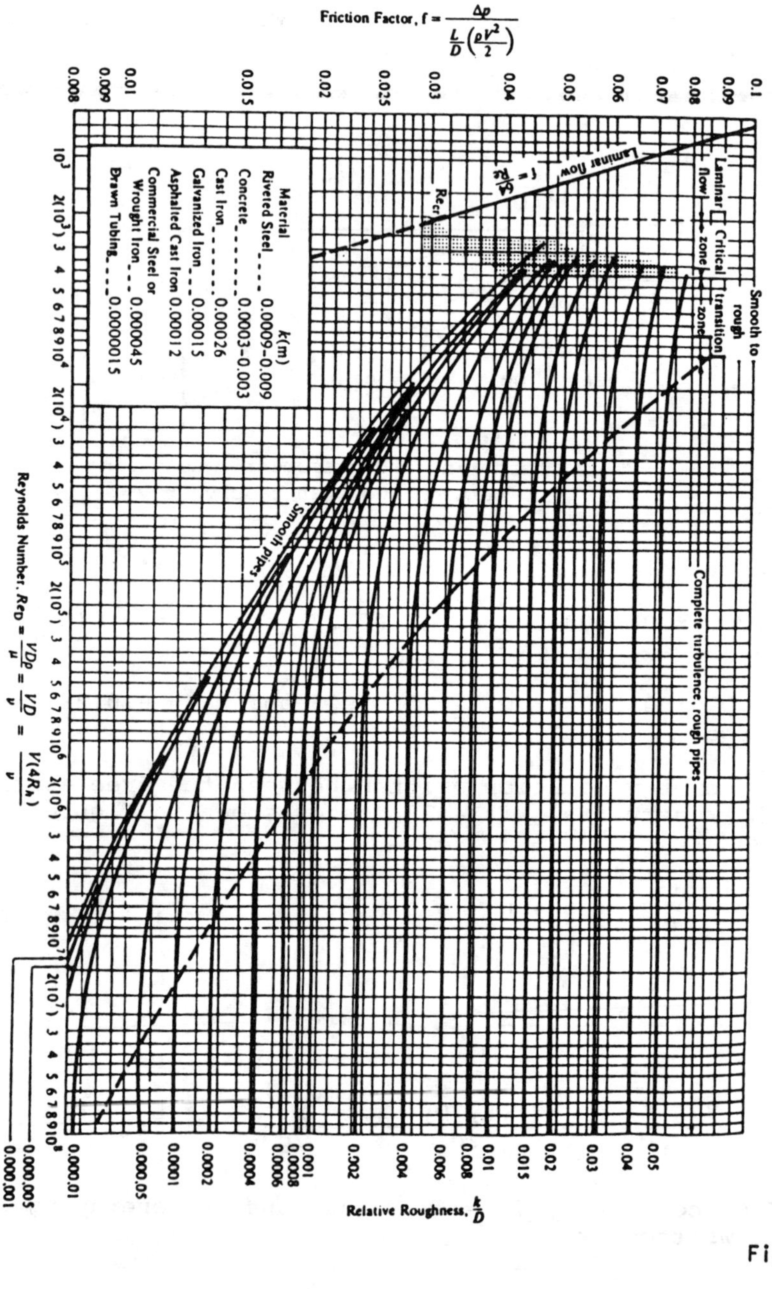

Friction factors for commercial pipe.

Fig. 2

$$Re_1 = \frac{(2.04)(0.60)}{10^{-6}} = 1.22 \times 10^6 \qquad \text{and then} \qquad f_1 = 0.013$$

$$Re_2 = \frac{(8.16)(0.30)}{10^{-6}} = 2.45 \times 10^6 \qquad \text{and then} \qquad f_2 = 0.0135$$

A recalculation of the value of V_1 with these new f values is

$$20 = \frac{V^2}{2g}\left[0.5 + \frac{(0.013)(200)}{0.6} + (0.35)(16) + \frac{(0.0135)(100)(16)}{0.3} + 16\right]$$

so that $V_1 = 2.00$ m/s and $V_2 = 8.00$ m/s.

Both new Reynolds numbers are essentially the same as above. Then the flow rate is:

$$Q = V_1 A_1 = (2.00)\frac{\pi}{4}(0.60)^2 = 0.565 \text{ m}^3/\text{s}$$

• PROBLEM 7-34

Two water pipes, 12-in. and 6-in. diameter, of the same length are coupled in parallel and together are to deliver 3 ft.^{3} per sec. Find the loss of head per mile if f = .0075.

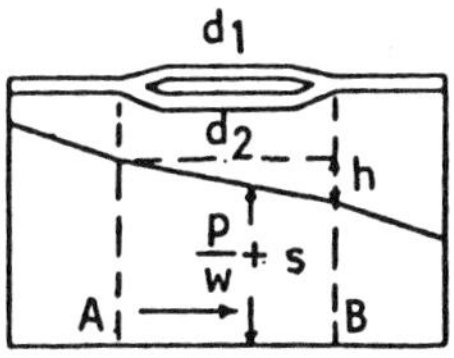

Fig. 1

Solution: Two Pipes in Parallel (Fig. 1) -- If the discharge Q is divided between two pipes of diameter d_1 and d_2 coupled together at the ends of lengths l_1 and l_2 respectively, the head h, lost between A and B, is the same for each pipe;

$$\therefore \quad h = \frac{4f_1 l_1 v_1^2}{2gd_1} = \frac{4f_2 l_2 v_2^2}{2gd_2};$$

$$\therefore \quad \frac{v_1}{v_2} = \left(\frac{d_1}{f_1 l_1} \cdot \frac{f_2 l_2}{d_2}\right)^{\frac{1}{2}}.$$

Also

$$Q = Q_1 + Q_2 = \frac{\pi}{4}(v_1 d_1^2 + v_2 d_2^2);$$

$$\therefore \quad \frac{Q_1}{Q_2} = \frac{v_1 d_1^2}{v_2 d_2^2} = \left(\frac{d_1^5}{f_1 l_1} \cdot \frac{f_2 l_2}{d_2^5}\right)^{\frac{1}{2}}.$$

Then,

$$Q = Q_2\left(\frac{Q_1}{Q_2} + 1\right) = \frac{\pi}{4} v_2 d_2^2 \left\{\left(\frac{d_1^5}{f_1 l_1} \cdot \frac{f_2 l_2}{d_2^5}\right)^{\frac{1}{2}} + 1\right\}.$$

But

$$v_2 = \left(\frac{2gd_2 h}{4f_2 l_2}\right)^{\frac{1}{2}};$$

$$\therefore \quad Q = \frac{\pi}{4}\left(\frac{2gh}{4f_2 l_2}\right)^{\frac{1}{2}} \cdot d_2^{5/2} \left\{\left(\frac{d_1^5}{f_1 l_1} \cdot \frac{f_2 l_2}{d_2^5}\right)^{\frac{1}{2}} + 1\right\}$$

$$= \frac{\pi}{4}\sqrt{\frac{2gh}{4}} \left\{\left(\frac{d_1^5}{f_1 l_1}\right)^{\frac{1}{2}} + \left(\frac{d_2^5}{f_2 l_2}\right)^{\frac{1}{2}}\right\}.$$

Here there is a difficulty in selecting f_1 and f_2 as we do not know the velocities until we have found Q_1 and Q_2, but as $\frac{Q_1}{Q_2}$ depends on $\frac{d_1^5 l_2}{d_2^5 l_1}$ and on $\frac{f_2}{f_1}$ it will be nearly correct to assume $f_1 = f_2$ at this stage and write:

$$\frac{Q_1}{Q_2} = \left(\frac{d_1^5 l_2}{d_2^5 \cdot l_1}\right)^{\frac{1}{2}},$$

and as the lengths are equal, take

$$\frac{Q_1}{Q_2} = \left(\frac{d_1}{d_2}\right)^{5/2} = 2^{5/2} = 5.656;$$

$$\therefore \quad Q_2 + 5.656\, Q_2 = 3;$$

$$\therefore \quad Q_2 = \frac{3}{6.656} = .4507 \text{ ft.}^3 \text{ per sec.},$$

$$v_2 = \frac{.4507}{\pi/4 \times \frac{1}{4}} = 2.296 \text{ ft. per sec.};$$

$$\therefore \quad h = \frac{.030 \times 5280 \times (2.296)^2}{64.4 \times \frac{1}{2}} = 25.92 \text{ ft.}$$

= loss of head per mile.

For a more exact result, after finding v_2, we should now find $Q_1 = 3 - .4507 = 2.5493$ ft.^{3} per sec. and

$$v_1 = \frac{4 \times 2.5493}{\pi \times 1} = 3.245 \text{ ft. per sec.}$$

Then from our curves for f and R_e, we should find f_1 and f_2 from the values of

$$\frac{v_1 d_1}{v}$$

and

$$\frac{v_2 d_2}{v},$$

then redetermine

$$\frac{Q_1}{Q_2} \text{ as } \left(\frac{d_1{}^5 f_2}{f_1 d_2{}^5}\right)^{\frac{1}{2}}$$

and proceed as before, using, of course, the revised value of f_2 in the equation for h.

• PROBLEM 7-35

Calculate the horsepower that can be delivered to a factory 6.5 km distant from a hydraulic power house through three horizontal pipes each 150 mm in diameter laid in parallel, if the inlet pressure is maintained constant at 540 N/cm^2 and the efficiency of transmission is 94%. If one of the pipes becomes unusable, what increase of pressure at the power station would be required to transmit the same delivery pressure as before. What would be the efficiency of transmission under these conditions? f = 0.0075 in both cases.

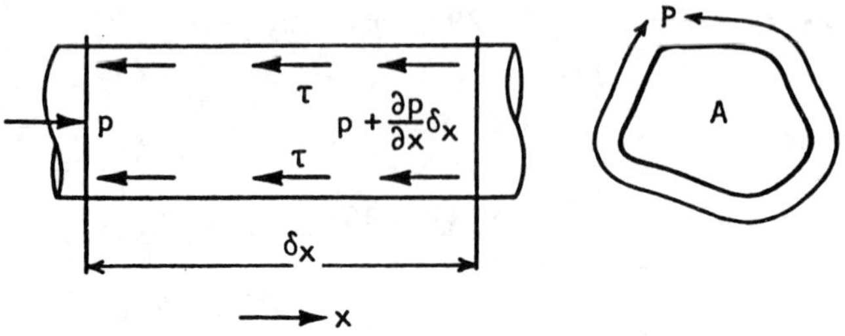

Fig. 1

Solution:

$$\text{head} = \frac{p}{w}$$

$$p = 540 \ \frac{N}{cm^2} = 5400\ 000 \ \frac{N}{m^2}$$

$$w_w \text{ (specific weight of water)} = 9810 \ \frac{N}{m^3}$$

$$\therefore \quad \text{Head at power station} = \frac{5\ 400\ 000}{9810} \text{ metres}$$

$$= 550.46 \text{ m}$$

$$\text{Head loss due to friction} = 6\% \text{ of } 550.5$$

$$= 33.03 \text{ m}$$

The force acting upon the fluid contained within a length of pipe L of cross sectional area A and wetted perimeter P is caused by a pressure drop which in turn is balanced by the previously mentioned viscous shear acting over the surface of contact of the fluid with the pipe wall (Fig. 1).

The force causing flow is $-(\partial p/\partial x)\delta xA$ and the force opposing flow is $P\tau\delta x$. (The pressure gradient must be negative if flow is to occur in the direction of x increasing.) Equating these forces as the flow is to be steady

$$-\frac{\partial p}{\partial x}\delta x\ A = P\tau\delta x$$

and substituting for τ

$$\frac{\partial p}{\partial x} = \frac{kv^n}{m}$$

where $m = A/P$ that is m is the mean thickness of flow.

To understand this last statement imagine the wetted perimeter laid out straight and construct a rectangle of area A as illustrated in Fig. 2. Its height will then be m that is A/P. This term is called the hydraulic mean radius.

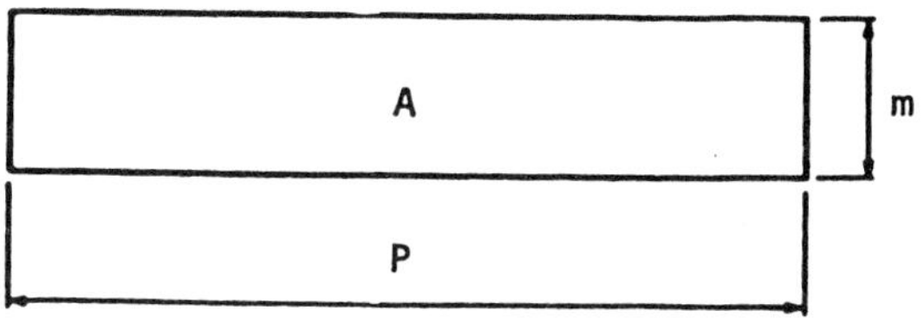

Fig. 2

$$\therefore \quad p = -kv^n x/m + C$$

When $x = 0$, $p = p_1$ and when $x = L$, $p = p_2$,

so

$$p_1 = p_2 = \Delta p = kv^n L/m$$

Dividing through by w to obtain the head equivalent to the Δp value

$$h_f = \frac{\Delta p}{w} = \frac{kLv^n}{\rho g m}$$

Resulting from experiments performed by Reynolds, the pressure drop in a pipe was linked to the velocity raised to a power of approximately 1.85.

As n is nearly 2 we may introduce a 2 into the denominator as this form of the equation facilitates the addition of energy loss and kinetic energy terms:

$$h_f = \frac{k'l}{\rho m}\frac{v^2}{2g}$$

k'/ρ is usually denoted by f. So the equation becomes

$$h_f = \frac{fL}{m}\frac{v^2}{2g}$$

This result is ascribed to Darcy and Weisbach. For circular pipes

$$m = \frac{A}{P} = \frac{\pi}{4} d^2/(\pi d) = \frac{d}{4}$$

so

$$h_f = \frac{4fL}{d}\frac{v^2}{2g}$$

Flow/pipe: $33.03 = \frac{4 \times 0.0075 \times 6500}{19.62 \quad 0.15} v^2$

$$v = 0.706 \text{ m/s}$$

$$Q/\text{pipe} = \frac{\pi}{4} \times 0.15^2 \times 0.706$$

$$= 0.01248 \text{ m}^3/\text{s}$$

Delivery head = 517.43 m

$$\text{Power delivered/pipe} = \frac{WQH}{1000} \text{ kW}$$

$$= \frac{9810 \times 0.01284 \times 517.43}{1000}$$

$$= 63.33 \text{ kW/pipe}$$

$$\text{Total power delivered} = 3 \times 63.33$$

$$= 189.99 \text{ kW}$$

When one pipe becomes unusable and the delivery pressure is to remain unaltered,

$$h_d = 517.43 \text{ m}$$

The power to be delivered must remain the same as before of course, i.e., 190 kW, so power/pipe = 95.0 kW

$$95.00 = \frac{WQh_d}{1000}$$

$$Qh_d = \frac{95\ 000}{9810} = 9.6845$$

Also if h_n is the new head required at the pumping station, then

$$h_n - h_d = \frac{4fL}{2gd} v_n^2$$

$$= \frac{4 \times 0.0075 \times 6500}{19.62 \times 0.15} v_n^2$$

$$v_n = 0.12285(h_n - h_d)^{\frac{1}{2}}$$

and

$$Q_n/\text{pipe} = \frac{\pi}{4} d^2 v_n = 2.171(h_n - h_d)^{\frac{1}{2}} \times 10^{-3}$$

$$2.171 \times 10^{-3}(h_n - h_d)^{\frac{1}{2}} h_d = 9.6835$$

$$(h_n - h_d)^{\frac{1}{2}} = \frac{9.6835}{2.171 \times 10^{-3} \times 517.43}$$

$$h_n = 74.309 + 517.43$$

$$= 591.739 \text{ m}$$

$$\text{Pumping pressure} = 591.74 \times 9810 \text{ N/m}$$

$$= 580.496 \text{ N/cm}^2$$

$$\text{Transmission efficiency} = \frac{517.43}{591.739} \times 100$$

$$= 87.4\%$$

• PROBLEM 7-36

A horizontal pipe of 7.5 cm diameter is joined by a sudden enlargement to a pipe of 15 cm diameter. Water is flowing through it at a rate of 0.0141 m^3/s. If the pressure just before the junction is 6 m head, what will be the pressure head in the 15 cm pipe downstream of the junction? What will be the power loss at the junction?

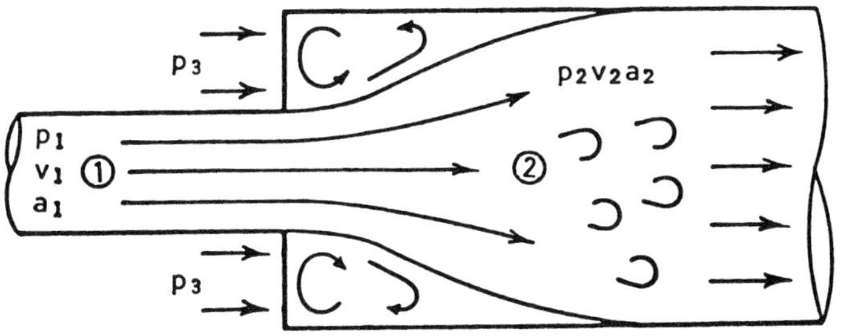

Fig. 1

Solution: In the sudden increase in pipe diameter (see Fig. 1), the annulus that joins the two pipes exerts a pressure on the fluid p_3. The magnitude of this pressure must be close to p_1 because it is the pressure in the fluid just outside the parallel sided jet in which the pressure is p_1. For the jet to be parallel sided

p_3 must equal p_1. This is one of the main assumptions underlying this analysis. The second assumption is that the pressure p_3 acts uniformly over the annulus. If this pressure distribution is significantly non-uniform the flow is curvilinear at high speeds and in the eddies in the corners of the section these conditions are not present. The assumption of uniform pressure distribution on the annulus is effectively true in practice.

Applying Bernoulii's equation from 1 to 2

$$\frac{p_1}{w} + \frac{v_1^2}{2g} = z_1 = \frac{p_2}{w} + \frac{v_2^2}{2g} + z_2 + h_L$$

where h_L is the energy loss/unit weight of fluid

$$\therefore \quad \frac{p_1 - p_2}{w} = \frac{v_2^2 - v_1^2}{2g} + h_L \qquad (1)$$

assuming that changes in z values are negligible.

Applying the total force equation

$$\left(\frac{p_1}{w} + \frac{v_1^2}{g}\right) a_1 - \left(\frac{p_2}{w} + \frac{v_2^2}{g}\right) a_2 + \frac{F_x}{w} = 0$$

Now

$$F_x = p_3(a_2 - a_1) = p_1(a_2 - a_1)$$

$$\therefore \quad \frac{p_1a_1 - p_2a_2 + p_1a_2 - p_1a_1}{w} = \frac{a_2v_2^2 - a_1v_1^2}{g}$$

$$\therefore \quad \frac{(p_1 - p_2)a_2}{w} = \frac{a_2v_2^2 - a_1v_1^2}{g}$$

$$\therefore \quad \frac{v_2^2 - v_1^2}{2g} + h_L = \frac{v_2^2 - a_1v_1^2/a_2}{g}$$

but

$$a_1v_1 = a_2v_2$$

$$\therefore \quad h_L = \frac{v_2^2 - v_2v_1}{g} - \frac{v_2^2 - v_1^2}{2g}$$

$$= \frac{v_1^2 - 2v_1v_2 + v_2^2}{2g}$$

$$= \frac{(v_1 - v_2)^2}{2g}$$

$$v_1 = \frac{0.0141}{(0.075)^2 \pi/4} = 3.192 \text{ m/s}$$

$$v_2 = \frac{d_1^2 \pi/4}{d_2^2 \pi/4} v_1 = \frac{1}{4} v_1 = 0.7979 \text{ m/s}$$

$$h_L = \frac{(3.192 - 0.7979)^2}{19.62} = 0.292 \text{ m}$$

Rearranging equation (1),

$$\frac{p_1}{w} + \frac{v_1^2}{2g} = \frac{p_2}{w} + \frac{v_2^2}{2g} + h_L$$

$$6 + 0.519 = \frac{p_2}{w} + 0.0324 + 0.292$$

$$\frac{p_2}{w} = 6.195 \text{ m}$$

Power loss - wQh_L = 9810 × 0.0141 × 0.292

= 40.39 watts (newton metres)

• PROBLEM 7-37

When two pipes of different sizes or roughnesses are so connected that fluid flows through one pipe and then through the other, they are said to be connected in series. A typical series-pipe problem, in which the head H may be desired for a given discharge or the discharge wanted for a given H, is illustrated in Fig. 1.

a) In Fig. 1, K_e = 0.5, L_1 = 300 m, D_1 = 600 mm, ε_1 = 2 mm, L_2 = 240 m, D_2 = 1 m, ε_2 = 0.3 mm, $v = 3 \times 10^{-6}$ m^2/s, and H = 6 m. Determine the discharge through the system.

b) Solve by means of equivalent pipes.

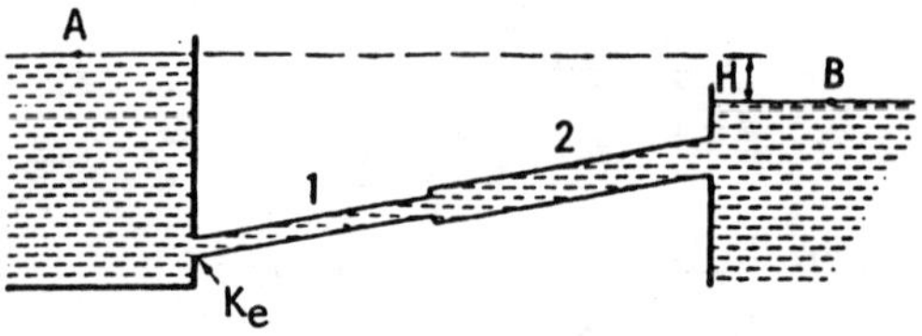

Pipes connected in series. Fig. 1

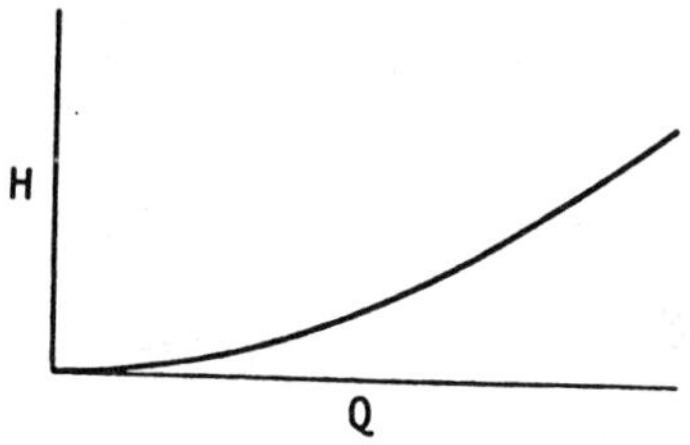

Plot of calculated H for selected values of Q. Fig. 2

Solution: a) Applying the energy equation from A to B, including all losses, gives

$$H + 0 + 0 = 0 + 0 + 0 + K_e \frac{V_1^2}{2g} + f_1 \frac{L_1}{D_1}\frac{V_1^2}{2g} + \frac{(V_1 - V_2)^2}{2g}$$

$$+ f_2 \frac{L_2}{D_2}\frac{V_2^2}{2g} + \frac{V_2^2}{2g}$$

in which the subscripts refer to the two pipes. The last item is the head loss at exit from pipe 2. With the continuity equation

$$V_1 D_1^2 = V_2 D_2^2$$

V_2 is eliminated from the equations, so that

$$H = \frac{V_1^2}{2g}\left\{K_e + \frac{f_1 L_1}{D_1} + \left[1 - \left(\frac{D_1}{D_2}\right)^2\right]^2 + \frac{f_2 L_2}{D_2}\left(\frac{D_1}{D_2}\right)^4 + \left(\frac{D_1}{D_2}\right)^4\right\}$$

For known lengths and sizes of pipes this reduces to

$$H = \frac{V_1^2}{2g}(C_1 + C_2 f_1 + C_3 f_2) \tag{1}$$

in which C_1, C_2, C_3 are known. With the discharge given, the Reynolds number is readily computed, and the f's may be looked up in the Moody diagram. Then H is found by direct substitution. With H given, V_1, f_1, f_2 are unknowns in Eq. (1). By assuming values of f_1 and f_2 (they may be assumed equal), a trial V_1 is found from which trial Reynolds numbers are determined and values of f_1, f_2 looked up. With these new values, a better V_1 is computed from Eq. (1). Since f varies so slightly with the Reynolds number, the trial solution converges very rapidly. The same procedures apply for more than two pipes in series.

In place of the assumption of f_1 and f_2 when H is given, a graphical solution may be utilized in which several values of Q are assumed in turn, and the corresponding values of H are calculated and plotted against Q, as in Fig. 2. By connecting the points with a smooth curve it is easy to read off the proper Q for the given value of H.
From the energy equation:

$$6 = \frac{V_1^2}{2g}\left[0.5 + f_1 \frac{300}{0.6} + (1 - 0.6^2)^2 + f_2 \frac{240}{1.0} 0.6^4 + 0.6^4\right]$$

After simplifying,

$$6 = \frac{V_1^2}{2g}(1.0392 + 500f_1 + 31.104f_2)$$

From $\varepsilon_1/D_1 = 0.0033$, $\varepsilon_2/D_2 = 0.003$, and the Moody diagram values of f are assumed for the fully turbulent range:

$$f_1 = 0.026 \qquad f_2 = 0.015$$

By solving for V_1 with these values, $V_1 = 2.848$ m/s,

$V_2 = 1.025$ m/s,

$$R_1 = \frac{2.848 \times 0.6}{3 \times 10^{-6}} = 569{,}600$$

$$R_2 = \frac{1.025 \times 1.0}{3 \times 10^{-6}} = 341{,}667$$

From the Moody diagram, $f_1 = 0.0265$, $f_2 = 0.0168$.
By solving again for V_1,

$$V_1 = 2.819 \text{ m/s},$$

and

$$Q = 0.797 \text{ m}^3\text{/s} .$$

b) Equivalent Pipes:

Series pipes can be solved by the method of equivalent lengths. Two pipe systems are said to be equivalent when the same head loss produces the same discharge in both systems. From the Darcy-Weisbach equation

$$h_{f_1} = f_1 \frac{L_1}{D_1} \frac{Q_1^2}{(D_1^2 \pi/4)^2 2g} = \frac{f_1 L_1}{D_1^5} \frac{8Q_1^2}{\pi^2 g}$$

and for a second pipe

$$h_{f_2} = \frac{f_2 L_2}{D_1^5} \frac{8Q_2^2}{\pi^2 g}$$

For the two pipes to be equivalent,

$$h_{f_1} = h_{f_2} \qquad Q_1 = Q_2$$

After equating $h_{f_1} = h_{f_2}$ and simplifying,

$$\frac{f_1 L_1}{D_1^5} = \frac{f_2 L_2}{D_2^5}$$

Solving for L_2 gives

$$L = L \frac{f_1}{f_2} \left(\frac{D_2}{D_1}\right)^5 \qquad (2)$$

which determines the length of a second pipe to be equivalent to that of the first pipe. For example, to replace 300 m of 250-mm pipe with an equivalent length of 150-mm pipe, the values of f_1 and f_2 must be approximated by selecting a discharge within the range intended for the pipes. Say $f_1 = 0.020$, $f_2 = 0.018$, then

$$L_2 = 300 \frac{0.020}{0.018} \left(\frac{150}{250}\right)^5 = 25.9 \text{ m}$$

For these assumed conditions 25.9 m of 150-mm pipe is equivalent to 300 m of 250-mm pipe.

Hypothetically, two or more pipes composing a system may also be replaced by a pipe which has the same discharge for the same overall head loss.

First, by expressing the minor losses in terms of

equivalent lengths, for pipe 1

$$K_1 = 0.5 + (1 - 016^2)^2 - 0.91$$

$$L_{e_1} = \frac{K_1 D_1}{f_1} = \frac{0.91 \times 0.6}{0.026} = 21 \text{ m}$$

and for pipe 2

$$K_2 = 1$$

$$L_{e_2} = \frac{K_2 D_2}{f_2} = \frac{1 \times 1}{0.015} = 66.7 \text{ m}$$

The values of f_1, f_2 are selected for the fully turbulent range as an approximation. The problem is now reduced to 321 m of 600-mm pipe and 306.7 m of 1-m pipe. By expressing the 1-m pipe in terms of an equivalent length of 600-mm pipe, by Eq. (2)

$$L_e = \frac{f_2}{f_1} L_2 \left(\frac{D_1}{D_2}\right)^5 = 306.7 \frac{0.015}{0.026} \left(\frac{0.6}{1.0}\right)^5 = 13.76 \text{ m}$$

By adding to the 600-mm pipe, the problem is reduced to finding the discharge through 334.76 m of 600-mm pipe, $\varepsilon_1 = 2$ mm, H = 6 m,

$$6 = f \frac{334.76}{0.6} \frac{V^2}{2g}$$

with f = 0.026, V = 2.848 m/s,

and

$$R = 2.848 \times 0.6/(3 \times 10^{-6}) = 569{,}600.$$

For $\varepsilon/D = 0.0033$, f = 0.0265, V = 2.821,

and

$$Q = \pi(0.3^2)(2.821) = 0.798 \text{ m}^3/\text{s}.$$

By use of

$$\frac{1}{\sqrt{f}} = -0.86 \ln \left(\frac{\varepsilon/D}{3.7} + \frac{2.51}{R\sqrt{f}}\right) \qquad (3)$$

which is the basis for the Moody diagram.

Solving for $1/\sqrt{f}$

$$\frac{1}{\sqrt{f}} \cong \frac{\sqrt{8}\,Q}{\pi\sqrt{gh_f D^5/L}}$$

By substitution of 1/ f into Eq. (3) and simplifying

$$Q = -0.955D^2 \sqrt{gDh_f/L}\,\ln\left(\frac{\varepsilon}{3.7D} + \frac{1.775v}{D\sqrt{gDh_f/L}}\right) \qquad (4)$$

This equation, first derived by Swamee and Jain, is as accurate as the Colebrook equation and holds for the same range of values of ε/D and R. Substitution of the variables from part (a) gives Q = 0.781.

• PROBLEM 7-38

In rough pipes the Darcy f is given by:

$$1/\sqrt{f} = 4.0\,\log_{10}(r/k) + 3.48 \qquad (1)$$

In a certain district, the flow through a pipe of 10 cm diameter was found to be 0.007 m^3/s when the hydraulic gradient was 0.01. Fifteen years later this flow rate was shown to have decreased by 20%. A new 22.5 cm diameter pipe is to be installed to carry a flow of at least 0.056 m^3/s when operating under a hydraulic gradient of 0.015. For how long can this pipe be expected to carry this minimum flow. Assume that the roughness of the new pipe is the same as the initial roughness of the 10 cm pipe and that the roughness will increase linearly with time at the same rate as did that of the 10 cm diameter pipe.

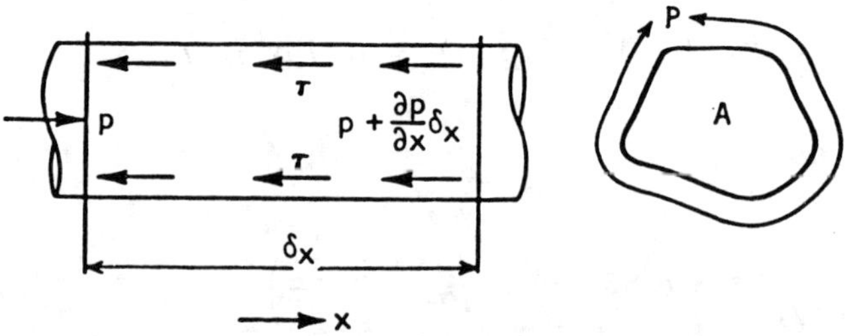

Fig. 1

Solution: The force acting upon the fluid contained within a length of pipe L of cross sectional area A and wetted perimeter P is caused by a pressure drop which in turn is balanced by the previously mentioned viscous shear acting over the surface of contact of the fluid with the pipe wall (Fig. 1).

The force causing flow is $-(\partial p/\partial x)\delta xA$ and the force opposing flow is $P\tau\,\delta x$. (The pressure gradient must be negative if flow is to occur in the direction of x increasing.) Equating these forces as the flow is to be steady

$$-\frac{\partial p}{\partial x}\delta x\, A = P\tau\delta x$$

and substituting for τ

$$-\frac{\partial p}{\partial x} = \frac{kv^n}{m}$$

where m = A/P that is m is the mean thickness of flow.

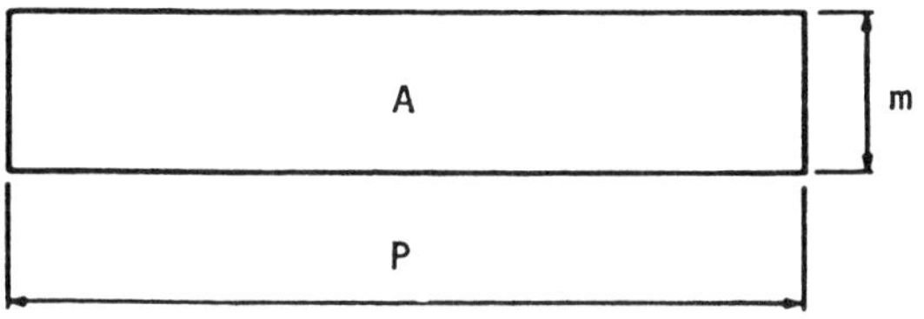

Fig. 2

To understand this last statement imagine the wetted perimeter laid out straight and construct a rectangle of area A as illustrated in Fig. 2. Its height will then be m that is A/P. This term is called the hydraulic mean radius. In American practice it is often denoted by R.

$$\therefore \qquad p = -\,kv^n x/m + C$$

When $x = 0$, $p = p_1$ and when $x = L$, $p = p_2$,

so

$$p_1 - p_2 = \Delta p = kv^n L/m$$

Dividing through by w to obtain the head equivalent to the Δp value

$$h_f = \frac{\Delta p}{w} = \frac{kLv^n}{\rho g m}$$

Resulting from experiments performed by Reynolds, the

pressure drop in a pipe was linked to the velocity raised to a power of approximately 1.85.

As n is nearly 2 we may introduce a 2 into the denominator as this form of the equation facilitates the addition of energy loss and kinetic energy terms:

$$h_f = \frac{k'L}{\rho m} \frac{v^2}{2g}$$

For reasons that emerge from the dimensional considerations mentioned above k'/ρ is usually denoted by f. So the equation becomes

$$h_f = \frac{fL}{m} \frac{v^2}{2g}$$

This result is ascribed to Darcy and Weisbach. For circular pipes

$$m = \frac{A}{P} = \frac{\pi}{4} d^2/(\pi d) = \frac{d}{4}$$

so

$$h_f = \frac{4fL}{d} \frac{v^2}{2g}$$

Initially for the 10 cm dia. pipe

$$\frac{h_f}{L} = \frac{4f}{d} \frac{v^2}{2g} = 0.01$$

$$v = \frac{0.007}{(\pi/4) \times 0.1^2} = 0.8913 \text{ m/s}$$

$$f = 6.175 \times 10^{-3}$$

$$1/\sqrt{f} = 12.73$$

After 15 years and with the same hydraulic gradient,

$$v = 0.8 \times 0.8913 = 0.713 \text{ m/s}$$

$$1/\sqrt{f} = 10.18$$

After rearranging equation (1)

Initially

$$\log_{10}(r/k_0) = \frac{12.73 - 3.48}{4}$$

$$= 2.312$$

$$r/k_0 = 204.9$$

$$k_0 = 0.05/204.9 = 2.44 \times 10^{-4} \text{ m}$$

After 15 years

$$\log_{10}(r/k_{15}) = \frac{10.18 - 3.48}{4} = 1.675$$

$$k_{15} = 0.05/47.34 = 1.056 \times 10^{-3}$$

$\therefore$ Rate of increase of k per year =

$$\frac{0.001056 - 0.0002441}{15} = 0.000054 \text{ m/year}$$

In the case of the 0.225 m pipe the flow will reduce to its minimum acceptable value of 0.056 m^3/s under a hydraulic gradient of 0.015 eventually.

$$\frac{h_f}{L} = 0.015 = \frac{4f}{0.225}\left.\frac{0.056}{(\pi/4)0.225^2}\right/19.62$$

$$\therefore \quad f = 0.008346$$

$$\therefore \quad 1/\sqrt{f} = 10.95$$

$$\therefore \quad \log_{10}\frac{r}{k} = \frac{10.95 - 3.48}{4} = 1.867$$

$$\therefore \quad r/k = 73.5$$

so

$$k = 0.1125/73.54$$

$$\therefore \quad k = 0.00153.$$

$$\therefore \qquad t = \frac{0.00153 - 0.000244}{0.000054}$$

$$= 24 \text{ years}$$

So the 22.5 cm diameter pipe will transport 0.056 m^3 for at least 24 years.

• PROBLEM 7-39

a. A new commercial steel pipe 200 mm in diameter and 1000 m long is in parallel with a similar pipe 300 mm in diameter and 3000 m long. The total flow in both pipes is 0.20 m^3/s. What is the head loss through the system? Use water at 20°C (kinematic viscosity is 10^{-6} m^2/s) and consider only frictional head loss.

b. Determine the flows in each of the two parallel pipes in part (a) by using the Hazen-Williams equation.

Solution: a. The relative roughness of the pipes

$$\frac{k}{d}$$

are 0.000225 and 0.00015, respectively. At large Reynolds number the corresponding friction factors are 0.014 and 0.013 (from the Moody Chart). These are estimates, and a trial solution for the velocity in each pipe is made from these data. Then Reynolds numbers and more accurate friction factors are obtained iteratively. Using subscripts 1 and 2 for the smaller and larger pipes, respectively,

$$\frac{V_2}{V_1} = \sqrt{\frac{f_1}{f_2}\frac{L_1}{L_2}\frac{D_2}{D_1}} = \sqrt{\frac{0.014}{0.013}\left(\frac{1000}{3000}\right)\frac{300}{200}} = 0.734$$

The pipe areas are 0.0314 and 0.0707 m^2. Then, from continuity,

$$Q = V_1A_1 + V_2A_2$$

or

$$0.20 = 0.0314V_1 + (0.734V_1)(0.0707)$$

and

$$V_1 = 2.40 \text{ m/s}$$

and

$V_2 = 1.76$ m/s.

Corresponding Reynolds numbers are

$$R_{e_1} = \frac{(2.40)(0.20)}{10^{-6}} = 4.8 \times 10^5 \quad \text{and} \quad f_1 = 0.0156$$

$$R_{e_2} = \frac{(1.76)(0.30)}{10^{-6}} = 5.3 \times 10^5 \quad \text{and} \quad f_2 = 0.0150$$

Then a recalculation of V_2/V_1 gives a value of 0.721, from which $V_1 = 2.43$ m/s. The head loss is the same for each pipe, and for pipe 1

$$h_f = \frac{f_1 L_1}{D_1}\frac{V_1^2}{2g} = \frac{(0.0156)(1000/0.20)(2.43)^2}{2g}$$

$$= 23.5 \text{ m}$$

b. A number of empirical pipe-flow equations for water in pipes have been used, and the equation of Hazen and Williams is perhaps the most widely used. This equation is

$$V = 1.318C(R_h)^{0.63}S^{0.54} \quad \text{ft/s}$$

$$Q = 1.318C(R_h)^{0.63}S^{0.54}A \quad \text{ft}^3\text{/s}$$

where

R_h = hydraulic radius of the pipe, A/P ($R_h = D/4$ for a round pipe), in feet;

S = slope of the total head line h_f/L;

A = pipe cross-sectional area;

C = roughness coefficient.

In SI units the Hazen-Williams equations are

$$V = 0.850CR_h^{0.63}S^{0.54}A \quad \text{m}^3\text{/s} \tag{3a}$$

and

$$Q = 0.850CR_h^{0.63}S^{0.54}A \quad \text{m}^3\text{/s} \tag{3b}$$

where the hydraulic radius R_h is in meters. The roughness cooefficient C is the same whether the equations in the technical English system or in the SI system are used. The different coefficients (1.318 and 0.850) indicate that the Hazen-Williams equations are not dimensionally consistent.

TABLE

HAZEN-WILLIAMS ROUGHNESS VALUE

Type of Pipe	C
Extremely smooth pipes	140
New steel or cast iron	130
Wood, average concrete	120
New riveted steel, clay	110
Old cast iron, brick	100
Old riveted steel	95
Badly corroded cast iron	80
Very badly corroded iron or steel	60

Roughness values C for the Hazen-Williams equation are given in a table. The Hazen-Williams equation is based on the premise that the Reynolds numbers are large and the pipes are reasonably rough so that the flow regime is in the range labeled complete turbulence, rough pipes in the figure. In this range the friction factors or roughness coefficients are independent of the Reynolds number.

Flow in parallel pipe systems may readily be solved. Since $R_h = D/4$ for a round pipe, we may write Eq. (3b) as

$$Q = \frac{0.850\pi CD^{2.63}}{4^{1.63}} \left(\frac{h_L}{L}\right)^{0.54}$$

Thus from Eq. (1)

$$Q_0 = h_L^{0.54} (C_1' + C_2' + C_3' + \dots + C_n')$$

where

$$C' = \frac{0.850\pi CD^{2.63}}{4^{1.63} L^{0.54}}$$

which has a fixed value for each pipe. Therefore, any assumed head loss h_L through the parallel system will give flows in each pipe in the correct proportion, though the total may not be correct. The flow in each branch may be corrected by the same factor needed to correct the total flow to the given Q_0, and the head loss may be determined directly from Eq. (3a).

C = 130 from the table. Assume a head loss of h_L = 20 m. Then for the 200-mm pipe, h_L/L = 20/1000 and

$$Q_{200} = (0.850)(130)\left(\frac{0.200}{4}\right)^{0.63}\left(\frac{20}{1000}\right)^{0.54}\frac{\pi}{4}(0.200)^2$$

$$= 0.0636\ \mathrm{m^3/s}$$

For the 300-mm pipe h_L/L = 20/3000 and

$$Q_{300} = (0.850)(130)\left(\frac{0.300}{4}\right.^{0.63}\left.\frac{20}{3000}\right)^{0.54}\frac{\pi}{4}(0.300)^2$$

$$= 0.1021\ \mathrm{m^3/s}$$

The total flow for the assumed 20-m head loss would be 0.1657, whereas the actual flow is 0.200 m^3/s. Thus a factor of 1.207 applied to each branch will result in a flow of 0.200 m^3/s total:

$$Q_{200} = (0.0636)(1.207) = 0.0768\ \mathrm{m^3/s}$$

$$Q_{300} = (0.1021)(1.207) = 0.1232\ \mathrm{m^3/s}$$

for a total of 0.200 m^3/s. These results compare with 0.0763 and 0.1237m^3/s, respectively, from part (a).

BRANCHED/NETWORK PIPES

• PROBLEM 7-40

A simple branching-pipe system is shown in the figure. In this situation the flow through each pipe is wanted when the reservoir elevations are given. The sizes and types of pipes and fluid properties are assumed known. The Darcy-Weisbach equation must be satisfied for each pipe, and the continuity equation must be satisfied. It takes the form that the flow into the junction J must equal the flow out of the junction. Flow must be out of the highest reservoir and into the lowest; hence, the continuity equation may be either

$$Q_1 = Q_2 + Q_3 \qquad \text{or} \qquad Q_1 + Q_2 = Q_3$$

If the elevation of hydraulic grade line at the junction is above the elevation of the intermediate reservoir, flow is into it; but if the elevation of hydraulic grade line at J is below the intermediate reservoir, the flow is out of it. Minor losses may be expressed as equivalent lengths and added to the actual lengths of pipe.

The solution is effected by first assuming an elevation of hydraulic grade line at the junction and then computing Q_1, Q_2, Q_3 and substituting into the continuity equation. If the flow into the junction is too great, a higher grade-line elevation, which will reduce the inflow and increase the outflow, is assumed.

In the figure find the discharges for water at 20°C with the following pipe data and reservoir elevations:

$L_1 = 3000$ m, $D_1 = 1$ m, $\varepsilon_1/D_1 = 0.0002$; $L_2 = 600$ m,

$D_2 = 0.45$ m, $\varepsilon_2/D_2 = 0.002$; $L_3 = 1000$ m, $D_3 = 0.6$ m,

$\varepsilon_3/D_3 = 0.001$; $z_1 = 30$ m, $z_2 = 18$ m, $z_3 = 9$ m.

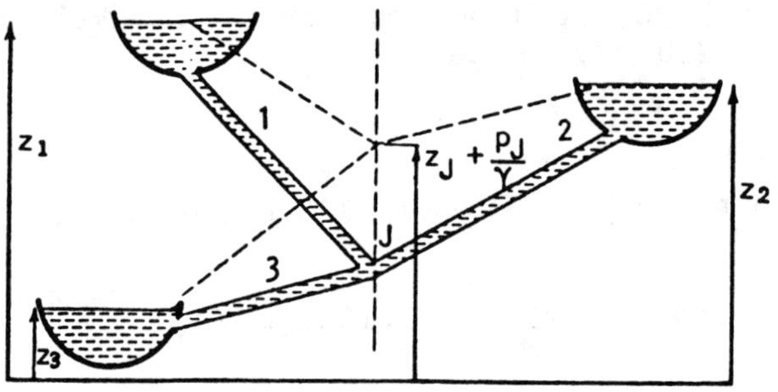

Three interconnected reservoirs.

Solution: Assume $z_J + p_J/\gamma = 23$ m. Then

$$7 = f_1 \frac{3000}{1} \frac{V_1^2}{2g} \qquad f_1 = 0.0014 \qquad V_1 = 1.75 \text{ m/s}$$

$$Q_1 = 1.380 \text{ m}^3/\text{s}$$

$$5 = f_2 \frac{600}{0.45} \frac{V_2^2}{2g} \qquad f_2 = 0.024 \qquad V_2 = 1.75 \text{ m/s}$$

$$Q_2 = 0.278 \text{ m}^3/\text{s}$$

$$14 = f_3 \frac{1000}{0.60} \frac{V_3^2}{2g} \qquad f_3 = 0.020 \qquad V_3 = 2.87 \text{ m/s}$$

$$Q_3 = 0.811 \text{ m}^3/\text{s}$$

so that the inflow is greater than the outflow by

$$1.380 - 0.278 - 0.811 = 0.291 \text{ m}^3/\text{s}$$

Assume $z_J + p_J/\gamma = 24.6$ m. Then

$$5.4 = f_1 \frac{3000}{1} \frac{V_1^2}{2g} \qquad f_1 = 0.015 \qquad V_1 = 1.534 \text{ m/s}$$

$$Q_1 = 1.205 \text{ m}^3/\text{s}$$

$$6.6 = f_2 \frac{600}{0.45} \frac{V_2^2}{2g} \qquad f_2 = 0.204 \qquad V_2 = 2.001 \text{ m/s}$$

$$Q_2 = 0.320 \text{ m}^3/\text{s}$$

$$15.6 = f_3 \frac{1000}{0.60} \frac{V_3^2}{2g} \qquad f_3 = 0.020 \qquad V_3 = 3.029 \text{ m/s}$$

$$Q_3 = 0.856 \text{ m}^3/\text{s}$$

The inflow is still greater by 0.029 m^3/s. By extrapolating linearly, $z_J + p_J/\gamma = 24.8$ m, $Q_1 = 1.183$, $Q_2 = 0.325$, $Q_3 = 0.862 \text{ m}^3/\text{s}$.

• PROBLEM 7-41

Branched Pipes: The figure shows a pipe, dividing at D to supply two other reservoirs. Let the diameters of the three pipes AD, DB and DC be d_1, d_2, d_3 and their lengths L_1, L_2 and L_3 respectively. Let H_1, H_2, H_3 be the water levels of the reservoirs at A, B and C. If the branch pipe to B be closed, ADC forms two pipes, d_1 and d_3 in series; similarly if the branch pipe to C be closed, ADB forms two pipes d_1 and d_2 in series, the hydraulic gradients in each case consisting of two straights, intersecting at D, the steeper over the smaller pipe. But opening both branches to B and C will depress the hydraulic gradient at D to some level H. The head lost at entry and the velocity head in each pipe and other local losses will be neglected: they must be corrected for afterwards if they are appreciable compared with the friction losses, as they will be if the lengths are short.

In the figure, $H_1 = 200$ ft., $H_2 = 151$ ft., $H_3 = 182$ ft. above datum; $d_1 = 15$ in., $d_2 = 9$ in., $d_3 = 12$ in. $L_1 = 2000$ ft., $L_2 = 3000$ ft., $L_3 = 3600$ ft. Taking f = .005, find the discharges in the three pipes, neglecting all small losses.

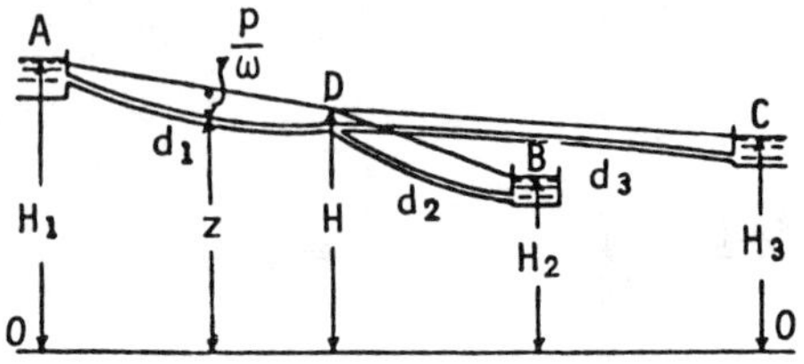

Solution: If v_1, v_2, v_3 are the velocities in the three pipes AD, DB, DC respectively,

$$H_1 - H = \frac{4f\ L_1 v_1^2}{2gd_1},$$

$$H - H_2 = \frac{4f L_2 v_2^2}{2gd_2},$$

$$H - H_3 = \frac{4f L_3 v_3^2}{2gd_3}$$

and

$$Q_1 = Q_2 + Q_3 ;$$

$$\therefore\ v_1 d_1^2 = v_2 d_2^2 + v_3 d_3^2 ,$$

viz 4 equations for the 4 unknown quantities, H, v_3, v_1 and v_1. Adding the first and second, we eliminate H and can express v_2 in terms of v_1, adding the first and third we again eliminate H and can express v_0 in terms of v_1; we then find v_1 from the last equation, and subsequently v_2 and v_3, also H if desired. Having found the v's, the Q's follow.

$$H_1 - H = \frac{.020 \times 2000 \times v_1^2}{64.4 \times 1.25} = .497v_1^2;$$

$$H - H_2 = \frac{.020 \times 3000 \times v_2^2}{64.4 \times .75} = 1.243v_2^2;$$

$$H - H_3 = \frac{.020 \times 3600\ v_3^2}{64.4 \times 1} = 1.118v_3^2;$$

$$1.5625v_1 = .5625v_2 + v_3;$$

$$\therefore \quad H_1 - H_2 = 49 = .497v_1^2 + 1.243v_2^2;$$

$$\therefore \quad v_2 = \sqrt{\frac{49 - .497v_1^2}{1.243}} = \sqrt{39.43 - .400v_1^2};$$

$$H_1 - H_3 = 18 = .497v_1^2 + 1.118v_3^2;$$

$$\therefore \quad v_3 = \sqrt{\frac{18 - .497v_1^2}{1.118}} = \sqrt{16.11 - .445v_1^2};$$

$$\therefore \quad 1.5625v_1 = .5625\sqrt{39.43 - .400v_1^2}$$

$$+ \sqrt{16.11 - 445v_1^2}$$

or,

$$\phi(v_1) = 0.$$

Putting $v_1 = 5$, $\phi(v_1) = 7.812 - 3.051 - 2.232 = 2.529.$

Putting $v_1 = 4$, $\phi(v_1) = 6.250 - 3.232 - 2.999 = 0.019.$;

$\therefore \quad v_1 = 3.99$ ft. per sec.;

$v_2 = \sqrt{39.43 - 6.37} = 5.75$ ft. per sec.;

$v_3 = \sqrt{16.11 - 7.08} = 3.005$ ft. per sec.;

$\therefore \quad Q_1 = 4.90$ ft.^{3} per sec.;

$Q_2 = 2.54$ ft.^{3} per sec.;

$Q_3 = 2.36$ ft.^{3} per sec.;

and these check.

Also $H = 200 - .497(3.99)^2 = 200 - 7.9 = 192.1$ ft. = level of hydraulic gradient at D.

• PROBLEM 7-42

A reservoir, A, with its surface at an elevation of 26 m above datum supplies water to two other reservoirs, B and C. The surface level in B is 7.6 m above datum and that in C is at datum level. From A to a junction, J, a pipe of 15 cm dia., 244 m length is used. The branch, JB, is 7.5 cm dia. and 61 m length. The branch JC is 10 cm dia. and 91 m length. Taking f as 0.0075 throughout and neglecting local losses, calculate the flow through the net.

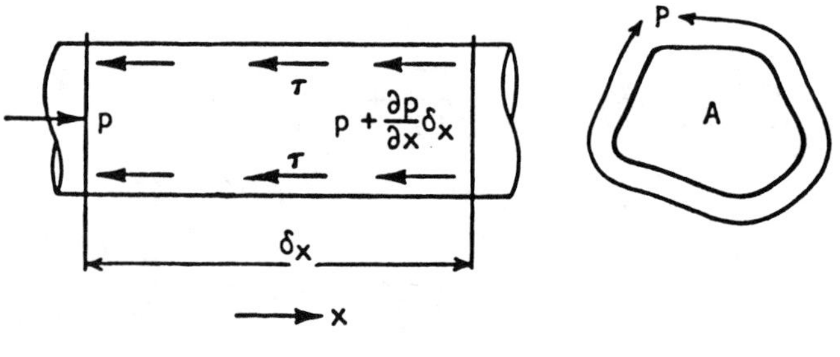

Fig. 1

Solution: The force acting upon the fluid contained within a length of pipe L of cross sectional area A and wetted perimeter P is caused by a pressure drop which in turn is balanced by the previously mentioned viscous shear acting over the surface of contact of the fluid with the pipe wall (Fig. 1).

The force causing flow is $-(\partial p/\partial x)\delta x A$ and the force opposing flow is $P\tau\delta x$. (The pressure gradient must be negative if flow is to occur in the direction of x increasing.) Equating these forces as the flow is to be steady

$$- \frac{\partial p}{\partial x} \delta x A = P\tau\delta x$$

and substituting for τ

$$- \frac{\partial p}{\partial x} = \frac{kv^n}{m}$$

where m = A/P that is m is the mean thickness of flow.

To understand this last statement imagine the wetted perimeter laid out straight and construct a rectangle of area A as illustrated in Fig. 2. Its height will then be m that is A/P. This term is called the hydraulic mean radius. In American practice it is often denoted by R.

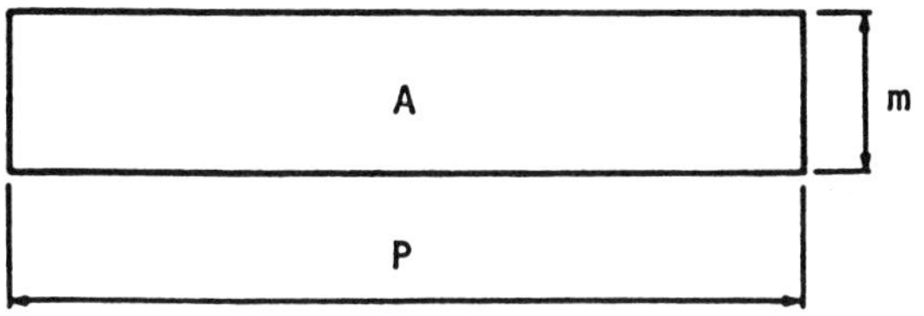

Fig. 2

$$\therefore \qquad p = -kv^n x/m + C$$

When x - 0, $p = p_1$ and when x = L, $p = p_2$,

so

$$p_1 - p_2 = \Delta p = kv^n L/m$$

Dividing through by w to obtain the head equivalent to the Δp value

$$h_f = \frac{\Delta p}{w} = \frac{kLv^n}{\rho gm}$$

Resulting from experiments performed by Reynolds, the pressure drop in a pipe was linked to the velocity raised to a power of approximately 1.85.

As n is nearly 2 we may introduce a 2 into the denominator as this form of the equation facilitates the addition of energy loss and kinetic energy terms:

$$h_f = \frac{k'L}{\rho m} \frac{v}{2g}$$

For reasons that emerge from the dimensional considerations mentioned above k'/ρ is usually denoted by f. So the equation becomes

$$h_f = \frac{fL}{m}\frac{v^2}{2g}$$

This result is ascribed to Darcy and Weisbach. For circular pipes

$$m = \frac{A}{P} = \frac{\pi}{4}d^2/(\pi d) = \frac{d}{4}$$

so

$$h_f = \frac{4fL}{d}\frac{v^2}{2g}$$

$$h_f = \frac{4fLv}{2gd} = \frac{4fL(4Q/\pi d^2)^2}{2gd} = \frac{64fLQ^2}{19.62\pi^2 d^5}$$

$$= \frac{fLQ^2}{3.026d^5}$$

$$h_A - h_B = \frac{fL_A Q_A{}^2}{3.026d_A^5} + \frac{fL_B Q_B^2}{3.026d_B^5} = 26 - 7.6 = 18.4$$

$$18.4 = \frac{0.0075 \times 244}{3.026 \times 0.15^5} Q_A^2 + \frac{0.0075 \times 61}{3.026 \times 0.075^5} Q_B^2$$

$$= 7963.9Q_A^2 + 63711.21Q_B^2$$

$$h_A - h_C = \frac{fL_A Q_A^2}{3.026d_A^5} + \frac{fL_C Q_C^2}{3.026d_C^5} = 26 - 0 = 26$$

$$\therefore \qquad 26 = 7963.9Q_A^2 + \frac{0.0075 \times 91}{3.026 \times 0.10^5} Q_C^2$$

$$7963.9Q_A^2 + 22554.5Q_C^2$$

Let

$Q_B = \alpha Q_A$

then $\quad Q_C = (1 - \alpha)Q_A$

$$\therefore \quad 18.4 = 7963.9Q_A^2 + 63711.21\alpha^2 Q_A^2 \quad (1)$$

$$26 = 30518.4Q_A^2 - 45109\alpha Q_A^2 + 22554.5\alpha^2 Q_A^2 \quad (2)$$

Dividing equation (1) by (2) and simplifying

$$\alpha^2 + 0.6686\alpha - 0.2855 = 0$$

$$\therefore \quad \alpha = 0.29598$$

$$Q_A = \sqrt{\left(\frac{18.4}{7963.9 + 63711.21 \times 0.29598^2}\right)} = 0.036853 \text{ m}^3/\text{s}$$

$$Q_B = 0.0369 \times 0.29598 = 0.01091 \text{ m}^3/\text{s}$$

$$Q_C = 0.0369(1 - \alpha) = 0.025943 \text{ m}^3/\text{s}$$

Check

$$h_j = 26.0 - 7963.9Q_A^2$$

$$= 15.18$$

$$h_j - 63711.21Q_B^2 = 7.596 \text{ (error = 0.004)}$$

$$h_j - 22554.5Q_C^2 = -0.000\ 063$$

Note that these calculations were performed on a desk top calculator. High accuracy is needed to obtain satisfactory results.

• PROBLEM 7-43

A combination of two or more pipes connected as in Fig. 1, so that the flow is divided among the pipes and then is joined again, is a parallel-pipe system. In series pipes the same fluid flows through all the pipes and the head losses are cumulative, but in parallel pipes the head losses are the same in any of the lines and the discharges are cumulative. In Figure 1 $L_1 = 3000$ ft, $D_1 = 1$ ft, $\varepsilon_1 = 0.001$ ft; $L_2 = 2000$ ft, $D_2 = 8$ in, $\varepsilon_2 = 0.0001$ ft; $L_3 = 4000$ ft, $D_3 = 16$ in, $\varepsilon_3 = 0.0008$ ft; $\rho = 2.00$ slugs/ft^2, $\nu = 0.00003$ ft^2/s, $p_A = 80$ psi, $Z_A = 100$ ft, $Z_B = 80$ ft. For a total flow of 12 cfs, determine flow through each pipe and the pressure at B. Assume $Q'_1 = 3$ cfs.

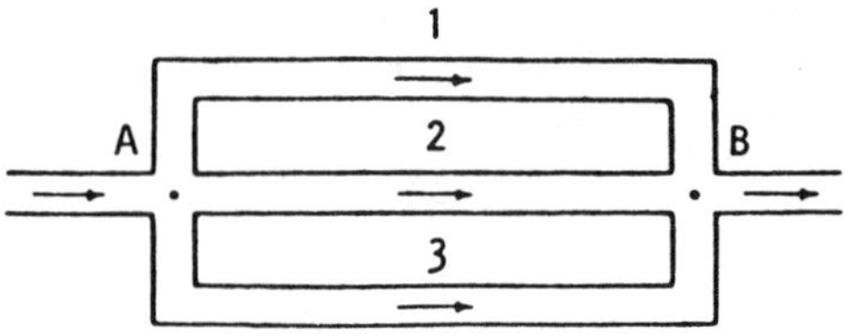

Parallel-pipe system Fig. 1

Solution: In analyzing parallel-pipe systems, it is assumed that the minor losses are added into the lengths of each pipe as equivalent lengths. From Fig. 1 the conditions to satisfied are

$$h_{f_1} = h_{f_2} = h_{f_3} = \frac{p_A}{\gamma} + Z_A - \left(\frac{p_B}{\gamma} + Z_B\right) \tag{1}$$

$$Q = Q_1 + Q_2 + Q_3$$

in which Z_A, Z_B are elevations of points A and B, and Q is the discharge through the approach pipe or the exit pipe.

Two types of problems occur: (1) with elevation of hydraulic grade line at A and B known, to find the discharge Q; (2) with Q known, to find the distribution of flow and the head loss. Sizes of pipe, fluid properties, and roughnesses are assumed to be known.

The first type is, in effect, the solution of simple pipe problems for discharge, since the head loss is the drop in hydraulic grade line. These discharges are added to determine the total discharge.

The second type of problem is more complex, as neither the head loss nor the discharge for any one pipe is known. The recommended procedure is as follows:

1. Assume a discharge Q'_1 through pipe 1.

2. Solve for h'_{f_1}, using the assumed discharge.

3. Using h'_{f_1}, find Q'_2, Q'_3.

4. With the three discharges for a common head loss, now assume that the given Q is split up among the pipes in the same proportion as Q'_1, Q'_2, Q'_3; thus

$$Q_1 = \frac{Q'_1}{\Sigma Q'} Q \qquad Q_2 = \frac{Q'_2}{\Sigma Q'} Q \qquad Q_3 = \frac{Q'_3}{\Sigma Q'} Q \qquad (2)$$

5. Check the correctness of these discharges by computing h_{f_1}, h_{f_2}, h_{f_3} for the computed Q_1, Q_2, Q_3.

This procedure works for any number of pipes. By judicious choice of Q'_1, obtained by estimating the percent of the total flow through the system that should pass through pipe 1 (based on diameter, length, and roughness), Eq. (2) produces values that check within a few percent, which is well within the range of accuracy of the friction factors.

Since $Q'_1 = 3$ cfs'

then

$V'_1 = 3.82$, $R'_1 = 3.82 \times 1/0.00003 = 127,000$,

$\varepsilon_1/D_1 = 0.001$, $f'_1 = 0.022$,

and

$$h'_{f_1} = 0.022 \frac{3000}{1.0} \frac{3.82^2}{64.4} = 14.97 \text{ ft.}$$

For pipe 2

$$14.97 = f'_2 \frac{2000}{0.667} \frac{V'^2_2}{2g}$$

Then $\varepsilon_2/D_2 = 0.00015$. Assume $f'_2 = 0.020$;

then

$V'_2 = 4.01$ ft/s, $R'_2 = 4.01 \times \frac{2}{3} \times 1/0.00003 = 89,000$,

$f'_2 = 0.019$, $V'_2 = 4.11$ ft/s, $Q'_2 = 1.44$ cfs.

For pipe 3

$$14.97 = f'_3 \frac{4000}{1.333} \frac{V'^2_3}{2g}$$

Then $\varepsilon_3/D_3 = 0.0006$ Assume $f'_3 = 0.020$;

then

$V'_3 = 4.01$ ft/s, $R'_3 = 4.01 \times 1.333/0.00003 = 178{,}000$,

$f'_3 = 0.020$, $Q'_3 = 5.60$ cfs.

The total discharge for the assumed conditions is

$$\Sigma Q' = 3.00 + 1.44 + 5.60 = 10.04 \text{ cfs}$$

Hence

$$Q_1 = \frac{3.00}{10.04} 12 = 3.58 \text{ cfs}$$

$$Q_2 = \frac{1.44}{10.04} 12 = 1.72 \text{ cfs}$$

$$Q_3 = \frac{5.60}{10.04} 12 = 6.70 \text{ cfs}$$

Check the values of h_1, h_2, h_3:

$V_1 = \frac{3.58}{\pi/4} = 4.56$ $R_1 = 152{,}000$ $f_1 = 0.021$

$$h_{f_1} = 20.4 \text{ ft}$$

$V_2 = \frac{1.72}{\pi/9} = 4.93$ $R_2 = 109{,}200$ $f_2 = 0.019$

$$h_{f_2} = 21.6 \text{ ft}$$

$V_3 = \frac{6.70}{4\pi/9} = 4.80$ $R_3 = 213{,}000$ $f_3 = 0.019$

$$h_{f_3} = 20.4 \text{ ft}$$

f_2 is about midway between 0.018 and 0.019. If 0.018 had been selected, h_2 would be 20.4 ft.

To find p_B,

$$\frac{p_A}{\gamma} + z_A = \frac{p_B}{\gamma} + z_B + h_f$$

or

$$\frac{p_B}{\gamma} = \frac{80 \times 144}{64.4} + 100 - 80 - 20.8 = 178.1$$

in which the average head loss was taken. Then

$$p_B = \frac{178.1 \times 64.4}{144} = 79.6 \text{ psi}$$

• PROBLEM 7-44

Consider the following two situations for the pipe network shown in Figure 1.

(1) Estimate the pressure at junction 2 if the volumetric flow rate (Q) passing through the system is 15 cfs and the pressure at junction 1 is 200 psig. The pipe roughness coefficients are known as is the entire cast iron pipe geometry.

(2) If the pressure at junction 2 were 180 psig, find the total volumetric flow rate (Q).

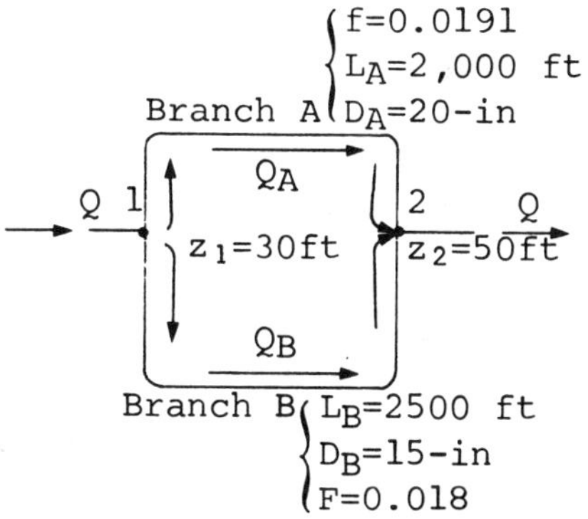

Fig. 1

Solution: Part (1) To solve this type of multiple-path pipe systems, various methods are available. An iterative method will be used in this case. The solution is presented step by step.

Step 1) Assume a flow passing through either branch and solve for the head loss. The cross-section area of each branch is constant, then the head loss is the same for all of them. Let us assume that a flow(Q'_A) of 9 cfs is passing through branch A. Then, the head loss is calculated (h'_ℓ)

$$h'_{\ell A} = f\frac{L_A}{D_A}\frac{V_A^2}{2} = f\frac{L_A}{D_A}\frac{(Q'_A/A_A)^2}{2}$$

$$A_A = \pi\frac{(20)^2}{4}\text{ in}^2 \times \frac{\text{ft}^2}{144\text{ in}^2} = 2.18\text{ ft}^2$$

$$h'_{\ell A} = (0.0191)\frac{2000\text{ ft}}{20\text{ in}\times\frac{1\text{ ft}}{12\text{ in}}} \times \frac{9\text{ ft}^3/2.18\text{ ft}^2}{2}$$

$$h'_{\ell A} = 195.02\text{ ft-lb/slug}$$

Step 2) With the help of this head loss, calculate the flow of branch B. (Remember $h_{\ell A} = h_{\ell B}$)

$$195.02 = f\frac{L_B}{D_B}\frac{V_B^2}{2}$$

$$V_B = \sqrt{2\times\frac{195.02}{0.018}\frac{(1.25)}{2,500}}$$

$$V_B = 3.29\text{ ft/sec}$$

Thus,

$$Q'_B = A_BV_B = \pi\frac{(15)^2}{4\times 144}\times 3.29$$

$$Q'_B = 4.039\text{ ft}^3/\text{sec}$$

Step 3) Calculate the actual flow assuming that it divides up in the same proportions as Q'_A and Q'_B.

This is evaluated in the following way:

$$Q_A = \frac{Q'_A}{Q'_A+Q'_B} \times Q = \frac{9\text{cfs}}{9\text{ cfs}+4.039\text{cfs}} \times 15\text{ cfs}$$

$$Q_A = 10.35\text{ ft}^3/\text{sec}$$

$$Q_B = \frac{Q'_B}{Q'_A+Q'_B} \times Q = \frac{4.030\text{ cfs}}{9\text{cfs}+4.039\text{cfs}} \times 15\text{ cfs}$$

$$Q_B = 4.65\text{ ft}^3/\text{sec}$$

Thus, $Q = Q_A + Q_B$

Step 4) Calculate the head losses using the actual flows from step 3.

$$h_{\ell A} = (0.0191)\left(\frac{2,000}{20}\right)(12)\,\frac{(10.35/2.18)^2}{2}$$

$$h_{\ell A} = 258.32\text{ ft-lb/slug}$$

$$h_{\ell B} = (0.018)\left(\frac{2500}{15}\right)(12)\frac{\left[4.65/\frac{\pi}{4}\left(\frac{15}{12}\right)^2\right]^2}{2}$$

$$h_{\ell B} = 258.44\text{ ft-lb/slug}$$

Here, the head losses are almost the same which means that the assumptions made so far are correct. If the head losses do not match with each other, then we consider Q_1 as the next trial flow and go back to step 1. This iteration is carried out until the head losses for all the branches are equal (two in this case).

Finally, we find the pressure at junction 2 applying the first law of thermodynamics for either branch. Using branch B, we have

$$\frac{V_B^2}{2g} + \frac{P_1}{\gamma} + Z_1 = \frac{V_B^2}{2g} + \frac{P_2}{\gamma} + Z_2 + \frac{258.4}{g}$$

$$\frac{P_2}{\gamma} = \frac{200}{\gamma} + (30-50) - 8.02$$

$$\frac{P_2}{\gamma} = 461.4\text{ ft} - 20\text{ ft} - 8.02\text{ ft} = 433.3\text{ ft}$$

$$P_2 = 433\text{ ft} \times 62.4\text{ lb/ft}^3 \times \frac{14.7\text{ psi}}{2116\text{ lb/ft}^2}$$

$$P_2 = 187.8\text{ psig}$$

Part (2)

This becomes a problem of a single-path system for each branch because the end pressures in the grid, have to be considered as the end pressures in each branch. All the flow information is known. Let us apply the first law of thermodynamics for branch B

$$\frac{V_B^2}{2g} + \frac{P_1}{\gamma} + Z_1 = \frac{V_B^2}{2g} + \frac{P_2}{\gamma} + Z_2 + \frac{h_{\ell B}}{g}$$

$$\frac{h_{\ell B}}{g} = \frac{P_1 - P_2}{\gamma} + (Z_1 - Z_2) = \frac{200-180}{\gamma} - 20$$

$$h_{\ell B} = g(46.13-20) = 841.6 \text{ ft-lb/slug}$$

$$h_{\ell B} = 841.6 = f \frac{L_B}{D_B} \frac{V_B^2}{2}$$

$$= (0.018) \frac{2500}{1.25} \frac{V_B^2}{2}$$

$$V_B^2 = 46.75$$

$$V_B = 6.83 \text{ ft /sec}$$

Now,
$$h_{\ell A} = 841.6 = (0.0191)\left(\frac{2000}{20}\right)(12) \frac{V_A^2}{2}$$

$$V_A = 8.57 \text{ ft}^2\text{/sec}$$

Then, $$Q_A = A_A V_A = 2.18 \times 8.57$$

$$Q_A = 18.68 \text{ ft}^3\text{/sec}$$

$$Q_B = A_B V_B = 1.23 \times 6.83$$

$$Q_B = 8.38 \text{ ft}^3\text{/sec}$$

Thus, the total volumetric flow rate is

$$Q = Q_A + Q_B = 18.68 + 8.38$$

$$Q = 27.06 \text{ cfs}$$

• **PROBLEM 7-45**

Water is to be supplied to a nozzle shown in Figure 1 at a pressure of 50 PSIG or somewhat higher from a pump that is 1,000 feet away through a pipe of, as yet, unknown internal diameter. The pump delivers 4.0 ft^3/sec at an output pressure of 100 PSIG. Assuming the pipe is substantially horizontal, (A) find the pipe size to be used. (B) In a water flow network, three arms are to be fed by a junction, J, connected to a source through a 6" pipe which supplies 3.0 ft^3/sec. Each arm conducts water through 2" pipe and two arms include couplings having fluid resistance equivalent to an additional length of pipe, L_c, such that $L_c/D = 20$. Thus, the frictional effects of pipe fittings and connections can be translated into pipe lengths which result in equivalent frictional effects. Each arm terminates in a 1" diameter nozzle outlet. Find the pressure at the junction, the volumetric flow rate in each arm, and the pressure prevailing at each nozzle inlet. The lengths of the arms and other dimensions are given in Figure 2.

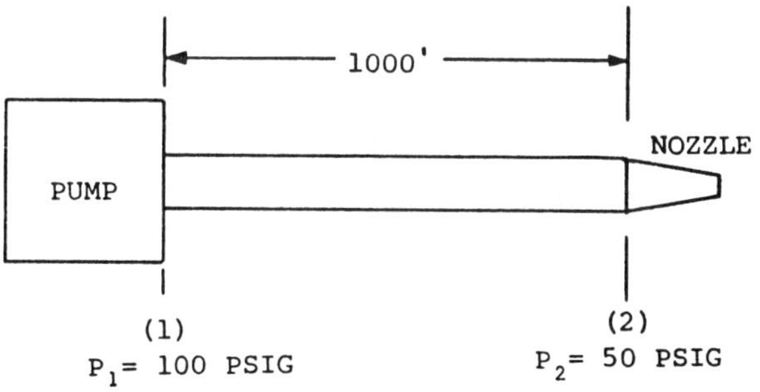

Fig. 1

Solution: (A) From the general equation of flow,

$$P_1 + K_1 \frac{V_1^2}{2} + gh_1 - P_2 - K_2 \frac{V_2^2}{2} - gh_2 = h \qquad (1)$$

where h is the head loss.

Assuming steady horizontal flow and uniform pipe cross-section, then

$$A_1 = A_2 \quad \text{and} \quad V_1 = V_2 = V, \quad \text{so that} \quad K_1 \simeq K_2.$$

Since $h_1 = h_2$, the head loss h becomes

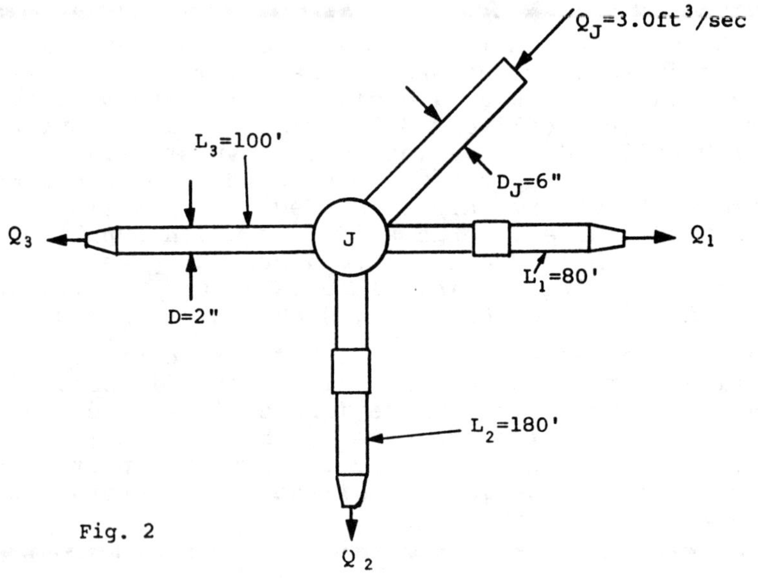

Fig. 2

$$h = P_1 - P_2 = f\rho \frac{L}{D}\frac{V^2}{2} \qquad (2)$$

where D is the pipe diameter which is to be determined. However D cannot be found because f is also unknown. To find f it is necessary to use the Reynold's number,

$$Re = \frac{VD\rho}{\mu} = \frac{VD}{\nu} \qquad (3)$$

where ν is the kinematic viscosity.

Since Re is a function of D also, it is not possible to calculate Re directly and a procedure of repeated substitution of trial values will have to be applied to find D.

To calculate V, we use

$$V = \frac{Q}{A} = \frac{4Q}{\pi D^2}$$

Substituting (4) into (3),

$$Re = \frac{4Q}{\nu\pi D}$$

Selecting a 4" pipe as a first guess, and given that $Q = 4.0$ ft^3/sec, then

$$Re = \frac{4 \times 4}{\pi \times 1.2 \times 10^{-5} \times \frac{4.03}{12}} = 1.27 \times 10^6$$

where the internal diameter of 4" pipe is approximately 4.03". For such pipe, the relative roughness is 1.4×10^{-5}, and referring to the Moody diagram, assuming commercial drawn tubing, the friction factor f is .0125. Returning to equation (2), we first express V in terms of Q,

$$V = Q/A = \frac{4Q}{\pi D^2}$$

and obtain

$$P_1 - P_2 = 8f \frac{\rho L Q^2}{\pi^2 D^5} \qquad (4)$$

so that

$$P_1 - P_2 = \frac{8 \times .0125 \times 1.94 \times 1000 \times 4^2 \times 12^3}{\pi^2 (4.03)^5}$$

$$= 511.8 \frac{\text{lbf}}{\text{in}^2}$$

Since we seek a pressure drop in the neighborhood of only 50 PSIG, we will try a 6" pipe which should provide less resistance to the flow and therefore less pressure drop.

The 6" pipe has an internal diameter of approximately 6.07". The friction factor is obtained from the previous method by noting that the relative roughness is 0.95×10^{-5} and that the Reynold's number is

$$Re = \frac{4Q}{\nu \pi D} = \frac{4 \times 4}{\pi \times 1.2 \times 10^{-5} \times \frac{6.07}{12}} = .84 \times 10^6$$

The friction factor for the 6" pipe is .011. Consequently

$$P_1 - P_2 = \frac{8 f \rho L Q^2}{\pi^2 D^5} = \frac{8 \times .011 \times 1.94 \times 1000 \times 4^2 \times 12^3}{\pi^2 (6.07)^5}$$

$$= 58.1 \frac{\text{lbf}}{\text{in}^2}$$

This value compares favorably to the allowable pressure drop of 50 $\frac{\text{lbf}}{\text{in}^2}$. Therefore a 6" standard pipe will provide the desired results closest. A 7" pipe will result in a pressure drop that is too small and not as close to the desired pressure drop obtained with the 6" pipe.

(B) Applying the general equation of flow to any one of the three branches,

$$\frac{P_J}{\rho} + K_J \frac{V_J^2}{2} + gz_J = \frac{P_O}{\rho} + K_O \frac{V_O^2}{2} + gz_O + h_t$$

where h_t is the head loss due to the pipe and the fitting. The subscripts J and O denote the locations at the junction and the nozzle outlet, respectively.

Now $\quad h_t = h + h_c$

where $\quad h$ = head loss in pipe

h_c = head loss due to fitting

$$h_t = f \frac{L}{D} \rho \frac{V^2}{2} + f \frac{L_c}{D} \rho \frac{V^2}{2}$$

where $\quad L_c$ = equivalent pipe length due to the loss in the coupling.

Assuming horizontal flow, $z_J = z_O$. Assuming the velocity at the nozzle outlet to be large compared to the velocity at the junction, $V_J \approx 0$ and $K_O \simeq 1.0$. Assuming further that P_O is atmospheric,

$$\frac{P_J}{\rho} = \frac{V_O^2}{2} + f\left(\frac{L}{D}\frac{V^2}{2} + \frac{L_c}{D}\frac{V^2}{2}\right) \tag{5}$$

But V_O can be expressed as a function V from

$$VA = V_O A_O \tag{6}$$

so that

$$V_O = \frac{VA}{A_O} = V\left(\frac{D}{D_O}\right)^2$$

Substituting into (5),

$$\frac{P_J}{\rho} = \frac{V^2}{2}\left(\frac{D}{D_O}\right)^4 + \frac{V^2}{2} f\left(\frac{L}{D} + \frac{L_c}{D}\right)$$

from which

$$\frac{2P_J}{\rho} = V^2\left[\left(\frac{D}{D_O}\right)^4 + f\left(\frac{L}{D} + \frac{L_c}{D}\right)\right]$$

and

$$Q = VA = \frac{A(2P_J/\rho)^{\frac{1}{2}}}{\left[\left(\frac{D}{D_O}\right)^4 + f\left(\frac{L}{D} + \frac{L_c}{D}\right)\right]^{\frac{1}{2}}} \tag{7}$$

We can now find the Reynold's number from

$$Re = \frac{VD}{\nu} = \frac{4Q}{\pi D\nu} \tag{8}$$

The distribution of volumetric water flow among the branches will depend on the resistance to flow in the pipes. For example, pipe branches with greater resistance to flow, will carry less water than the branches having lower resistance. Thus, the water tends to take the paths of least resistance. To find the actual flow distribution, we can start out by assuming equal volumetric flows in the branches, checking the results and recalculating as necessary. Progressive such calculations provide progressively more accurate results. Computing, then, the Reynold's number from (8),

$$Re = \frac{4(1.0)}{\pi(1.2)10^{-5}(2)/12} = 6.37 \times 10^5$$

Assuming the pipes are smooth, the friction factor is obtained from the Moody diagram as .0128. The frictional effect arising from the coupling can be translated into an additional length of pipe, L_c. Since the ratio L_c/D is given as 20, from (7)

$$Q_1 = \frac{A(2P_J/\rho)^{\frac{1}{2}}}{\left[\left(\frac{2.0}{1.0}\right)^4 + .0128\left(\frac{80}{2.0} \times 12 + 20\right)\right]}$$

$$Q_1 = .0446A(2P_J/\rho)^{\frac{1}{2}}$$

Similarly,

$$Q_2 = \frac{A(2P_J/\rho)^{\frac{1}{2}}}{\left[\left(\frac{2.0}{1.0}\right)^4 + .0128\left(\frac{180}{2.0} \times 12 + 20\right)\right]}$$

$$Q_2 = .033A(2P_J/\rho)^{\frac{1}{2}}$$

and

$$Q_3 = \frac{A(2P_J/\rho)^{\frac{1}{2}}}{\left[\left(\frac{2.0}{1.0}\right)^4 + .0128\left(\frac{100}{2.0} \times 12\right)\right]}$$

$$Q_3 = .0422A(2P_J/\rho)^{\frac{1}{2}}$$

Although we don't know P_J as yet, we can nevertheless calculate the flow rates in the separate branches by noting that

$$Q_J = 3.0 = Q_1 + Q_2 + Q_3$$

and that

$$\frac{Q_1}{Q_J} = \frac{Q_1}{Q_1 + Q_2 + Q_3}$$

Now, $$Q_1 + Q_2 + Q_3 = .12A(2P_J/\rho)^{\frac{1}{2}}$$

Therefore,

$$\frac{Q_1}{Q_J} = \frac{.0446}{.12} = .372$$

and

$$Q_1 = .372 \times 3.0 = 1.11 \text{ ft}^3/\text{sec}$$

Similarly,

$$\frac{Q_2}{Q_J} = \frac{.033}{.12} = .275$$

$$Q_2 = .275 \times 3.0 = .825 \text{ ft}^3/\text{sec}$$

$$\frac{Q_3}{Q_J} = \frac{.0422}{.12} = .352$$

$$Q_3 = .352 \times 3.0 = 1.05 \text{ ft}^3/\text{sec}$$

It is now possible to improve on the results for the flow rates by recomputing the Reynolds number for each branch based on the previously computed Q values, and then find new flow rates based on the new Reynolds numbers. Accordingly, from (8)

$$Re_1 = \frac{4Q_1}{\pi D\nu} = \frac{4(1.11)}{\pi(1.2)10^{-5}(2/12)} = 7.07 \times 10^5$$

$$Re_2 = \frac{4(.825)}{\pi(1.2)10^{-5}(2/12)} = 5.26 \times 10^5$$

$$Re_3 = \frac{4(1.05)}{\pi(1.2)10^{-5}(2/12)} = 6.69 \times 10^5$$

From the new values of the Reynolds numbers, we can obtain new friction factors from the Moody diagram, such as

$$f_1 = .0125$$

$$f_2 = .013$$

$$f_3 = .0127$$

Substituting the new friction values into equation (7),

$$Q_1 = \frac{A(2P_J/\rho)^{\frac{1}{2}}}{\left[\left(\frac{2.0}{1.0}\right)^4 + .0125\left(\frac{80}{2.0} \times 12 + 20\right)\right]}$$

$$Q_1 = .0449A(2P_J/\rho)^{\frac{1}{2}} \tag{9}$$

$$Q_2 = \frac{A(2P_J/\rho)^{\frac{1}{2}}}{\left[\left(\frac{2.0}{1.0}\right)^4 + .013\left(\frac{180}{2.0} \times 12 + 20\right)\right]}$$

$$Q_2 = .033A(2P_J/\rho)^{\frac{1}{2}} \tag{10}$$

$$Q_3 = \frac{A(2P_J/\rho)^{\frac{1}{2}}}{\left[\left(\frac{2.0}{1.0}\right)^4 + .0127\left(\frac{100}{2.0} \times 12\right)\right]}$$

$$Q_3 = .0423A(2P_J/\rho)^{\frac{1}{2}} \tag{11}$$

$$Q_1 + Q_2 + Q_3 = .1202A(2P_J/\rho)^{\frac{1}{2}}$$

To obtain numerical values for the new flow rates in the three branches,

$$Q_1 = \frac{Q_1 \times 3.0}{Q_1 + Q_2 + Q_3} = \frac{.0449}{.1202} \times 3.0 = 1.12 \text{ ft}^3/\text{sec}$$

$$Q_2 = \frac{Q_2 \times 3.0}{Q_1 + Q_2 + Q_3} = \frac{.033}{.1202} \times 3.0 = 0.824 \text{ ft}^3/\text{sec}$$

$$Q_3 = \frac{Q_3 \times 3.0}{Q_1 + Q_2 + Q_3} = \frac{.0423}{.1202} \times 3.0 = 1.056 \text{ ft}^3/\text{sec}$$

The newly calculated flow rates compare favorably with the values computed at first, and nothing will be gained by repeating the procedure with the aim of improving the accuracy, since it is improbable that more accurate values for f can be obtained.

The value for P_J can now be found from any one of equations (9), (10) or (11). Using first (9),

$$\sqrt{2P_J/\rho} = \frac{Q_1}{.0449A}$$

and

$$P_J = \frac{\rho}{2}\left(\frac{Q_1}{.0449A}\right)^2$$

$$P_J = \frac{1.94}{2}\left(\frac{1.12}{.0449\pi(1)^2}\right)^2 = 61.2 \text{ lbs/in}^2$$

Checking this value of P_J with equations (10) and (11),

$$P_J = \frac{\rho}{2}\left(\frac{Q_2}{.033A}\right)^2$$

$$P_J = \frac{1.94}{2}\left(\frac{.824}{.033\pi(1)^2}\right)^2 = 61.3$$

$$P_J = \frac{\rho}{2}\left(\frac{Q_3}{.0423A}\right)^2$$

$$P_J = \frac{1.94}{2}\left(\frac{1.056}{.0423\pi(1)^2}\right)^2 = 61.3$$

The small differences arising above from the calculations for P_J are due to rounding off values in the calculations.

After having found the value for P_J, it is now possible to derive the pressure at each nozzle inlet by applying the basic energy relationship between the distribution center and the nozzle inlet as follows:

$$\frac{P_J}{\rho} + \frac{V_J^2}{2} + gz_J = \frac{P_i}{\rho} + \frac{V^2}{2} + gz_i + h_t$$

where the subscript i denotes the nozzle inlet location, and

$$h_t = f\,\frac{L}{D}\,\rho\,\frac{V^2}{2} + f\,\frac{L_c}{D}\,\rho\,\frac{V^2}{2}$$

To express the velocities in terms of the known volumetric flow rates,

$$V_J = \frac{Q_J}{A_J}; \qquad V = \frac{Q}{A}$$

Since $z_J = z_i$,

$$P_i = P_J + \frac{\rho}{2}\left(\frac{Q_J}{A_J}\right)^2 - \frac{\rho}{2}\left(\frac{Q}{A}\right)^2 - \frac{f}{2}\left(\frac{Q}{A}\right)^2\left(\frac{L}{D} + \frac{L_c}{D}\right)\rho$$

$$= P_J + \frac{\rho}{2}\left\{\left(\frac{Q_J}{A_J}\right)^2 - \left(\frac{Q}{A}\right)^2\left[1 + f\left(\frac{L}{D} + \frac{L_c}{D}\right)\right]\right\}$$

Applying the relationship to branch arm 1,

$$P_{i_1} = 61.3 + \frac{1.94}{2}\left\{\left(\frac{3}{\pi 3^2}\right)^2 - \left(\frac{1.12}{\pi 1^2}\right)^2\left[1 + .0125\left(\frac{80}{2} \times 12 + 20\right)\right]\right\}$$

$$P_{i_1} = 60.4$$

Applying similar calculations for P_i to branch arms 2 and 3,

$$P_{i_2} = 61.3 + \frac{1.94}{2}\left\{\left(\frac{3}{\pi 3^2}\right)^2 - \left(\frac{.824}{\pi 1^2}\right)^2\left[1 + .013\left(\frac{180}{2} \times 12 + 20\right)\right]\right\}$$

$$P_{i_2} = 60.3$$

$$P_{i_3} = 61.3 + \frac{1.94}{2}\left\{\left(\frac{3}{\pi 3^2}\right)^2 - \left(\frac{1.056}{\pi 1^2}\right)^2\left[1 + .0127\left(\frac{100}{2} \times 12\right)\right]\right\}$$

$$P_{i_3} = 60.4$$

The calculated values for P_i in the three branch arms show that their differences from P_J is insignificant for the pipe dimensions and flow characteristics. Larger differences of the P_i values from P_J would be obtained by increasing the pipe length and frictional losses so that the pressure drops between the junction J and the nozzle inlets are increased.

Two open tanks, (2) and (3) in the figure, are to be supplied with water from a reservoir (1) through a pipe branching at (0), 50 ft below the surface level in the reservoir, the surface levels in the tanks 2 and 3 being 10 ft above and 10 ft below the junction point, respectively. The single pipe will be 200 ft long and 3.068 in. in diameter, while the branched pipes will have lengths of 100 ft and 150 ft for the higher and lower tanks, respectively. What should be the diameters of these branched pipes in order that each has the same capacity of 9 ft^3/min? The Darcy friction factor is 0.008.

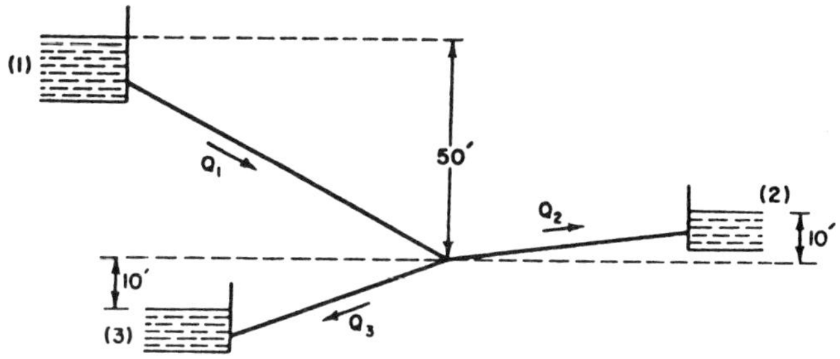

Solution: Ignoring the velocity head terms, the following flow equations can be written with reference to the figure:

Between (1) and (0)

$$Z_1 + \frac{p_1}{\gamma} = Z_0 + \frac{p_0}{\gamma} + h_{f_1}$$

But

$$Z_1 - Z_0 = 50,$$

then

$$\frac{p_0 - p_1}{\gamma} = 50 - h_{f_1} \tag{1}$$

Between (0) and (2)

$$Z_0 + \frac{p_0}{\gamma} = Z_2 + \frac{p_2}{\gamma} + h_{f_2}$$

$$Z_2 - Z_0 = 10,$$

and

$$\frac{p_0 - p_2}{\gamma} = 10 + h_{f_2} \tag{2}$$

Between (0) and (3)

$$Z_0 + \frac{p_0}{\gamma} = Z_3 + \frac{p_3}{\gamma} + h_{f_3}$$

$$Z_0 + Z_3 = 10$$

$$\frac{p_0 - p_3}{\gamma} = h_{f_3} - 10$$

$$Q_2 = Q_3 = \frac{9}{60} = 0.15 \text{ cfs}$$

$$Q_1 = 0.3 \text{ cfs}$$

In the Darcy equation substitute for

$$v = \frac{Q}{A} = \frac{4Q}{\pi D^2}$$

$$h_f = 4f \; \frac{L}{D} \; \frac{1}{2g} \; \frac{16Q^2}{\pi^2 D^4}$$

Rearranging and taking $2g = 64.4$

$$h_f = \frac{(4)(16)}{(\pi^2)(64.4)} \quad \frac{fLQ^2}{D^5}$$

But, since

$$\frac{(\pi^2)(64.4)}{(4)(16)} \cong 10$$

then

$$h_f \cong \frac{fLQ^2}{10D^5}$$

$$\therefore \qquad h_{f_1} = \frac{fL_1Q_1^2}{10D_1^5} - \frac{(0.008)(200)(0.3)^2}{(10)(3.068/12)^5}$$

$$h_{f_1} = 13.19 \text{ ft}$$

$$h_{f_2} = \frac{fL_2Q_2^2}{10D_2^5} = \frac{(0.008)(100)(0.15)^2}{10D_2^5}$$

$$h_{f_2} = \frac{0.0018}{D_2^5}$$

$$h_{f_3} = \frac{fL_3Q_3^2}{10D_3^5} = \frac{(0.008)(150)(0.15)^2}{10D_3^5}$$

$$h_{f_3} = \frac{0.0027}{D_3^5}$$

Substituting for the friction factors in the respective equations (1), (2), (3)

$$\frac{p_0 - p_1}{\gamma} = 50 - 13.19 = 36.81 \quad (4)$$

$$\frac{p_0 - p_2}{\gamma} = 10 + \frac{0.0018}{D_2^5} \quad (5)$$

$$\frac{p_0 - p_3}{\gamma} = \frac{0.0027}{D_3^5} - 10 \quad (6)$$

Since $p_1 = p_2 = p_3$ (these are all atmospheric pressures), then from Eq. 4 and Eq. 5

$$10 + \frac{0.0018}{D_2^5} = 36.81$$

from which

$$D_2^5 = \frac{0.0018}{26.81}$$

$$D_2 = 0.1463 \text{ ft } (1.756 \text{ in.})$$

From Eq. 4 and Eq. 6

$$\frac{0.0027}{D_3^5} - 10 = 36.81$$

$$D_3^5 = \frac{0.0027}{46.81}$$

$$D_3 = 0.1420 \text{ ft } (1.704 \text{ in.})$$

The nearest nominal sizes would be taken for the branched pipes.

Interconnected pipes through which the flow to a given outlet may come from several circuits are called a network of pipes, in many ways analogous to flow through electric networks. Problems on these in general are complicated and require trial solutions in which the elementary circuits are balanced in turn until all conditions for the flow are satisfied.

List these flow conditions and discuss their significance in the pipeline network shown in Fig. 1.

Since it is impractical to solve network problems analytically, methods of successive approximations are utilized. Possibly the most efficient of these methods is the Hardy-Cross method. Discuss the Hardy-Cross method for pipe networks and list the proper steps necessary to obtain a final solution.

The distribution of flow through the network of Fig. 2 is desired for the inflows and outflows as given. For simplicity n has been given the value 2.0.

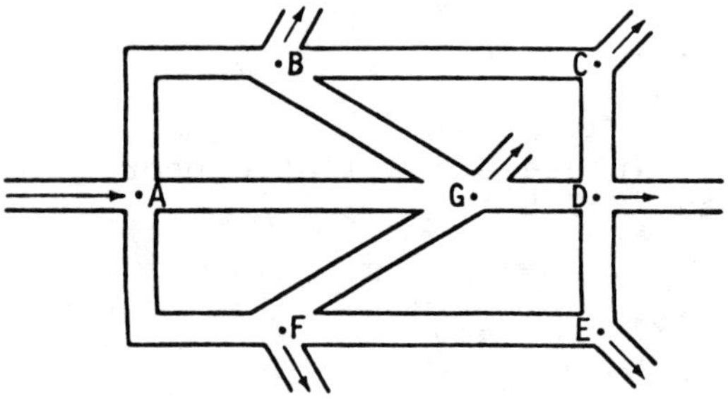

Pipe network

Fig. 1

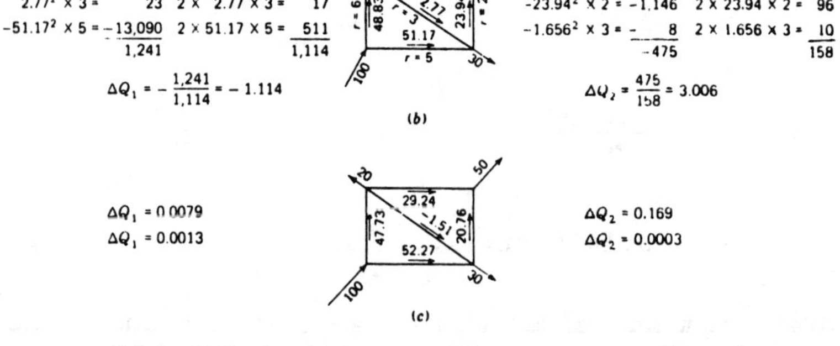

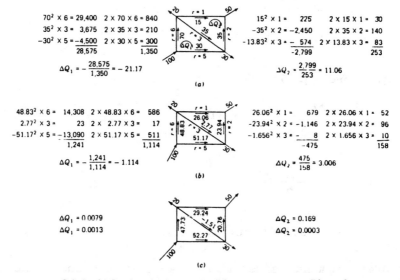

Solution for flow in a simple network.

Fig. 2

Solution: The following conditions must be satisfied in a network of pipes:

1. The algebraic sum of the pressure drops around each circuit must be zero.

2. Flow into each junction must equal flow out of the junction.

3. The Darcy-Weisbach equation, or equivalent exponential friction formula, must be satisfied for each pipe; i.e., the proper relation between head loss and discharge must be maintained for each pipe.

The first condition states that the pressure drop between any two points in the circuit, for example, A and G (Fig. 1), must be the same whether through the pipe AG or through AFEDG. The second condition is the continuity equation.

The Hardy Cross method is one in which flows are assumed for each pipe so that continuity is satisfied at every junction. A correction to the flow in each circuit is then computed in turn and applied to bring the circuits into closer balance.

Minor losses are included as equivalent lengths in each pipe. Exponential equations are commonly used, in the form $h_f = rQ^n$, where $r = RL/D^m$. The value of r is a constant in each pipeline (unless the Darcy-Weisbach equation is used) and is determined in advance of the loop-balancing procedure. The corrective term is obtained as follows.

For any pipe in which Q_0 is an assumed initial discharge

$$Q = Q_0 + \Delta Q \qquad (1)$$

where Q is the correct discharge and ΔQ is the correction. Then for each pipe,

$$h_f = rQ^n = r(Q_0 + \Delta Q)^n = r(Q_0^n + nQ_0^{n-1}\,\Delta Q + \ldots)$$

If ΔQ is small compared with Q_0, all terms of the series after the second may be dropped. Now for a circuit,

$$\Sigma h_f = \Sigma rQ|Q|^{n-1} = \Sigma rQ_0|Q|^{n-1} + \Delta Q\Sigma rn|Q_0|^{n-1} = 0$$

in which ΔQ has been taken out of the summation because it is the same for all pipes in the circuit and absolute-value signs have been added to account for the direction of summation around the circuit. The last equation is solved for ΔQ in each circuit in the network

$$\Delta Q = - \frac{\Sigma r Q_0 |Q_0|^{n-1}}{\Sigma r n |Q_0|^{n-1}} \tag{2}$$

When ΔQ is applied to each pipe in a circuit in accordance with Eq. (1), the directional sense is important; i.e., it adds to flows in the clockwise direction and subtracts from flows in the counterclockwise direction.

Steps in an arithmetic procedure may be itemized as follows:

1. Assume the best distribution of flows that satisfies continuity by careful examination of the network.

2. For each pipe in an elementary circuit, calculate and sum the net head loss $\Sigma h_f = \Sigma r Q^n$. Also calculate $\Sigma r n|Q|^{n-1}$ for the circuit. The negative ratio, by equation (2) yields the correction, which is then added algebraically to each flow in the circuit to correct it.

3. Proceed to another elementary circuit and repeat the correction process of 2. Continue for all elementary circuits.

4. Repeat 2 and 3 as many times as needed until the corrections (ΔQ's) are arbitrarily small.

The values of r occur in both numerator and denominator; hence, values proportional to the actual r may be used to find the distribution. Similarly, the apportionment of flows may be expressed as a percent of the actual flows. To find a particular head loss, the actual values of r and Q must be used after the distribution has been determined.

For the pipe network in Fig. 2, the assumed distribu- is shown in diagram a. At the upper left the term $\Sigma r Q_0 |Q_0|^{n-1}$ is computed for the lower circuit number 1. Next to the diagram on the left is the computation of $\Sigma n r |Q_0|^{n-1}$ for the same circuit. The same format is used for the second circuit in the upper right of the figure. The corrected flow after the first step for the top horizontal pipe is deterimined as 15 + 11.06 = 26.06 and for the diagonal as 35 + (-21.17) + (-11.06) = 2.77. Diagram (b) shows the flows after one correction, Diagram (c) shows the values after four corrections.

• PROBLEM 7-48

Given the pipe network shown in the figure, determine the flow rate through each pipe. The pipe sizes and lengths are shown alongside each pipe. Assume a C value of 100.

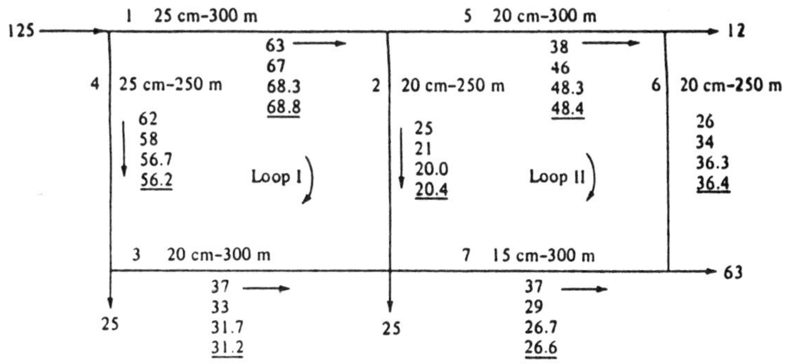

Flow distribution by Hardy–Cross method. Pipe numbers are labeled. Flow rates are in liters per second (L/s). See Table 1.

Solution: Complex pipe networks conveying water may be analyzed quite readily using the Hazen-Williams equation. The flow distribution for a given network may be desired, and this is generally an indeterminate problem which must be solved by successive trials, or iterations. In the design of a network, the flow and pressures at various points may be specified and the pipe sizes determined. This is also an indeterminate type of problem which must be solved in successive trials, or iterations.

A network consists of a finite number of loops containing any number of individual pipes, some of which may be common to two loops. In the figure the simple network has two loops, pipe 2 being common to both loops. Two conditions must be met for a balanced flow in the network:

1. The net flow into any junction must be zero. This means that the flow rate into the junction must equal the flow rate out of the junction.

2. The net head loss (or pressure drop) around a loop must be zero. If a loop is traversed in either direction, a balanced flow must result in a return to the original condition (head or pressure) at the starting point.

The procedure for determining the flow distribution in a given network involves assigning flows in each pipe so that continuity at each junction is satisfied (condition 1). Then the head loss around each loop is calculated and, if not zero, adjustments to the assumed flows are made either by pure estimate or by a method of iteration known as the Hardy-Cross method. The correction for each loop is given by

$$\Delta Q = -\frac{\text{net head loss for the assumed flows}}{1.85\Sigma h_L/Q_0 \text{ for the assumed flows}}$$

This equation is derived as follows:

In a given loop in a network let Q be the actual, or balanced, flow rate and Q_0 the assumed flow rate, so that $Q = Q_0 + \Delta Q$. Then, since the Hazen-Williams equation (as well as others) may be expressed as $h_L = nQ^x$, we may write

$$nQ^x = n(Q_0 + \Delta Q)^x$$

$$= n\left[Q_0^x + xQ_0^{x-1}\Delta Q + \frac{x(x-1)}{2}Q_0^{x-2}(\Delta Q)^2 + \ldots\right]$$

If ΔQ is truly small compared to Q_0, terms beyond the second may be neglected. For a balanced loop or network,

$$\Sigma h_L + \Sigma nQ^x = \Sigma nQ_0^x + \Delta Q\Sigma xnQ_0^{x-1} = 0$$

Solving for ΔQ we get

$$\Delta Q = \frac{\Sigma nQ_0^x}{\Sigma xnQ_0^{x-1}} = -\frac{\Sigma h_L}{1.85\Sigma h_L/Q_0} \qquad (1)$$

since x = 1.85 (the reciprocal of 0.54) in the Hazan-Williams equation.

The procedure is as follows:

1. Assume any reasonable flow distribution, in both magnitude and direction, in all pipes so that the total flow into each junction is algebraically zero. This should be indicated on a diagram of the pipe network.

2. Set up a table to analyze each closed loop in the network semi-independently.

3. Compute the head loss, h_L in each pipe.

4. For each loop, consider the flow rate Q_0 and the head loss h_L to be positive for clockwise flow in the loop and negative for counter-clockwise flow.

5. Compute the algebraic head loss Σh_L in each loop.

6. Compute the total head loss per unit discharge h_L/Q_0 for each pipe. Determine the sum of the quantities $\Sigma h_L/Q_0 = \Sigma n Q_0^{0.85}$ for each loop. From the definitions of head loss and flow direction, each term in this sum is necessarily positive.

7. Determine the flow correction for each loop from

$$Q = \frac{\Sigma h_L}{1.85 \Sigma h_L/Q_0} \qquad (2)$$

This correction is to be applied algebraically to each pipe in the loop. For a pipe which is in common with another loop, the flow correction for that pipe is the net effect of the corrections for both loops.

8. Indicate corrected flows on the diagram of the pipe network as in step 1. A check on the corrections of step 7 will be shown by a continuity check at each pipe junction.

9. Repeat steps 1 through 8 until either the head loss for a loop is balanced within desired limits or the flow corrections are made as small as desirable.

Flow corrections may be made by either of two methods:

1. Corrections for all loops may be made before any corrections are applied. The head loss and the value of h_L/Q_0 for a pipe in common with two loops need be calculated but once and the results used in both loops.

2. Corrections for a loop may be applied to each pipe in that loop before calculating correction in the next of successive loops.

Table 1

				FIRST TRIAL			SECOND TRIAL			THIRD TRIAL		
LOOP	PIPE	DIAMETER (cm)	L (m)	Q_0 (L/s)	h_L (m)	h_L/Q_0	Q_0 (L/s)	h_L (m)	h_L/Q_0	Q_0 (L/s)	h_L (m)	h_L/Q_0
I	1	25	300	+63	+3.23	0.0513	+67	+3.62	0.0540	+68.3	+3.75	0.0549
	2	20	250	+25	+1.44	0.0576	+21	+1.04	0.0495	+20.0	+0.95	0.0475
	3	20	300	−37	−3.58	0.0968	−33	−2.89	0.0875	−31.7	−2.69	0.0849
	4	25	250	−62	−2.61	0.0421	−58	−2.31	0.0398	−56.7	−2.22	0.0391
					−1.52	0.2478		−0.54	0.2308		−0.21	0.2264
				$\Delta Q = -\frac{(-1.52)}{(1.85)(0.2478)}$ = +3.3 L/s			$\Delta Q = -\frac{(-0.54)}{(1.85)(0.2308)}$ = +1.3 L/s			$\Delta Q = -\frac{(-0.21)}{(1.85)(0.2264)}$ = +0.5 L/s		
II	5	20	300	+38	+3.76	0.0989	+46	+5.35	0.1163	+48.3	+5.86	0.1213
	6	20	250	+26	+1.55	0.0596	+34	+2.55	0.0750	+36.3	+2.88	0.0793
	7	15	300	−37	−12.09	0.3268	−29	−9.25	0.3190	−26.7	−7.94	0.2973
	2	20	250	−25	−1.44	0.0576	−21	−1.04	0.0495	−20.0	−0.95	0.0477
					−8.22	0.5429		−2.39	0.5598		−0.15	0.5456
				$\Delta Q = -\frac{(-8.22)}{(1.85)(0.5429)}$ = +8.2 L/s			$\Delta Q = -\frac{(-2.39)}{(1.85)(0.5598)}$ = +2.3 L/s			$\Delta Q = -\frac{(-0.15)}{(1.85)(0.5456)}$ = +0.1 L/s		

Assumed flow and direction are indicated near each pipe, together with corrected flows for each trial as indicated in the table 1. Corrections were made to the nearest 1 L/s for the first and second trials, and to the nearest 0.1 L/s for the third trial. Final results are shown underlined in the figure. This type of problem may be solved readily on programmable computers.

CHAPTER 8

DIMENSIONAL ANALYSIS

Basic Attacks and Strategies for Solving Problems in this Chapter. See pages 433 to 485 for step-by-step solutions to problems.

It is generally desirable to express a given set of dimensional variables in a flowfield in terms of dimensionless parameters (or "Pi's"). These dimensionless parameters provide universal measures of the flow regimes and effectively reduce the number of independent variables. They are also necessary in order to obtain scaling laws to predict prototype performance based on measurements on a (typically smaller) model. Examples of dimensionless parameters are the Reynolds number, the Mach number, drag and lift coefficients, the Froude number, etc. The Buckingham Pi Technique provides a systematic way to determine dimensionless parameters from a given set of dimensional variables.

The following procedure is invoked in the Buckingham Pi Technique:

1) Count the total number of variables, n.

2) List the dimensions of each variable. This is typically done in terms of the four primary dimensions—mass, length, time, and temperature. Alternately, force can replace mass as a primary dimension.

3) Count the total number of primary dimensions (typically 3 or 4), and let j equal this number. Note: If the dimensional analysis fails in the steps below, return here, decrease j by one, and then repeat the analysis.

4) You now expect to find $k = n - j$ dimensionless parameters, $\prod_1, \prod_2, \ldots, \Pi_k$. To find these, you first need to select j variables as "repeating variables." Choose variables which do not by themselves form a dimensionless group, but which represent all of the primary dimensions involved in the problem. There are "preferred" choices, such as velocity and density, which will generate recognizable $\prod$s, such as the Reynolds number, etc.

5) Form a power product consisting of the j repeating variables and each of the remaining k variables. By forcing the exponent of each dimension to be zero, $\prod_1, \prod_2, \ldots, \prod_k$ are found.

6) Check that each Π is dimensionless.

Note that although your Πs may be dimensionless, they may not be of a form suitable for your particular application. It is perfectly valid to multiply or divide two or more Πs together to form a new Π set. In fact, since each Π by itself is dimensionless, any Π multiplied by another Π raised to any exponent must also be dimensionless. The new set of Πs formed in this manner is no more or less valid than the first set; the "correct" set is that which is most suitable to the problem at hand. Note also that such a rearrangement of parameters may also sometimes be used to obtain "standard" Πs such as the Reynolds number, the Froude number, etc., even if these do not result directly from your dimensional analysis.

The best way to learn this technique is to practice on many problems. True dynamic similarity between a model and a prototype can only exist if each Π for the model exactly matches the corresponding Π of the prototype. When such is the case, it is possible to scale up from the model to the prototype to predict its performance. These concepts of dimensional analysis and similitude are thus of paramount significance to designers who test small models before building a full-scale prototype

Step-by-Step Solutions to Problems in this Chapter, "Dimensional Analysis"

DIMENSIONAL ANALYSIS

• PROBLEM 8-1

A fluid-flow situation depends upon the velocity V, the density ρ, several linear dimensions l, l_1, l_2, pressure drop Δp, gravity g, viscosity μ, surface tension σ, and bulk modulus of elasticity K. Apply dimensional analysis to these variables to find a set of Π parameters.

Solution: $F(V, \rho, l, l_1, l_2, \Delta p, g, \mu, \sigma, K) = 0$

As three dimensions are involved, three repeating variables are selected. For complex situations, V, ρ, and l are generally helpful. There are seven Π parameters:

$$\Pi_1 = V^{x_1}\rho^{y_1} l^{z_1}\Delta p \qquad \Pi_2 = V^{x_2}\rho^{y_2} l^{z_2} g$$

$$\Pi_3 = V^{x_3}\rho^{y_3} l^{z_3}\mu \qquad \Pi_4 = V^{x_4}\rho^{y_4} l^{z_4}\sigma$$

$$\Pi_5 = V^{x_5}\rho^{y_5} l^{z_5}K \qquad \Pi_6 = \frac{l}{l_1}$$

$$\Pi_7 = \frac{l}{l_2}$$

By expanding the Π quantities into dimensions,

$$\Pi_1 = (LT^{-1})^{x_1}(ML^{-3})^{y_1}L^{z_1}ML^{-1}T^{-2}$$

$$L: \quad x_1 - 3y_1 + z_1 - 1 = 0$$

$$T: \quad -x_1 - 2 = 0$$

$$M: \quad y_1 + 1 = 0$$

from which $x_1 = -2$, $y_1 = -1$, $z_1 = 0$.

$$\Pi_2 = (LT^{-1})^{x_2}(ML^{-3})^{y_2}L^{z_2}LT^{-2}$$

$$\begin{aligned} L:&\quad x_2 - 3y_2 + z_2 + 1 = 0 \\ T:&\quad -x_2 - 2 = 0 \\ M:&\quad y_2 = 0 \end{aligned}$$

from which $x_2 = -2$, $y_2 = 0$, $z_2 = 1$.

$$\Pi_3 = (LT^{-1})^{x_3}(ML^{-3})^{y_3}L^{z_3}ML^{-1}T^{-1}$$

$$\begin{aligned} L:&\quad x_3 - 3y_3 + z_3 - 1 = 0 \\ T:&\quad -x_3 - 1 = 0 \\ M:&\quad y_3 + 1 = 0 \end{aligned}$$

from which $x_3 = -1$, $y_3 = -1$, $z_3 = -1$.

$$_4 = (LT^{-1})^{x_4}(ML^{-3})^{y_4}L^{z_4}MT^{-2}$$

$$\begin{aligned} L:&\quad x_4 - 3y_4 + z_4 = 0 \\ T:&\quad -x_4 - 2 = 0 \\ M:&\quad y_4 + 1 = 0 \end{aligned}$$

from which $x_4 = -2$, $y_4 = -1$, $z_4 = -1$.

$$\Pi_5 = (LT^{-1})^{x_5}(ML^{-3})^{y_5}L^{z_5}ML^{-1}T^{-2}$$

$$\begin{aligned} L:&\quad x_5 - 3y_5 + z_5 - 1 = 0 \\ T:&\quad -x_5 - 2 = 0 \\ M:&\quad y_5 + 1 = 0 \end{aligned}$$

from which $x_5 = -2$, $y_5 = -1$, $z_5 = 0$.

$$\Pi_1 = \frac{\Delta p}{\rho V^2} \quad \Pi_2 = \frac{gl}{V^2} \quad \Pi_3 = \frac{\mu}{Vl\rho} \quad \Pi_4 = \frac{\sigma}{V^2\rho l}$$

$$\Pi_5 = \frac{K}{\rho V^2} \quad \Pi_6 = \frac{l}{l_1} \quad \Pi_7 = \frac{l}{l_2}$$

and

$$f\left(\frac{\Delta p}{\rho V^2}, \frac{gl}{V^2}, \frac{\mu}{V_l\rho}, \frac{\sigma}{V^2\rho}, \frac{K}{\rho V^2}, \frac{l}{l_1}, \frac{l}{l_2}\right) = 0$$

It is convenient to invert some of he parameters and to take some square roots,

$$f_1\left(\frac{\Delta p}{\rho V^2}, \frac{V}{\sqrt{g_l}}, \frac{V l\rho}{\mu}, \frac{V^2 l\rho}{\sigma}, \frac{V}{\sqrt{K/\rho}}, \frac{l}{l_1}, \frac{l}{l_2}\right) = 0$$

The first parameter, usually written $\Delta p/(\rho V^2/2)$, is the pressure coefficient; the second parameter is the Froude number F; the third is the Reynolds number R; the fourth is the Weber number W; and the fifth is the Mach number M. Hence,

$$f_1\left(\frac{\Delta p}{\rho V^2}, F, R, W, M, \frac{l}{l_1}, \frac{l}{l_2}\right) = 0$$

After solving for pressure drop,

$$\Delta p = \rho V^2 f_2 (F, R, W, M, \frac{l}{l_1}, \frac{l}{l_2})$$

in which f_1, f_2 must be determined from analysis or experiment. By selecting other repeating variables, a different set of Π parameters could be obtained.

Table 1 Dimensions of physical quantities used in fluid mechanics

Quantity	Symbol	Dimensions (M, L, T)
Length	l	L
Time	t	T
Mass	m	M
Force	F	MLT^{-2}
Velocity	V	LT^{-1}
Acceleration	a	LT^{-2}
Area	A	L^2
Discharge	Q	L^3T^{-1}
Pressure	Δp	$ML^{-1}T^{-2}$
Gravity	g	LT^{-2}
Density	ρ	ML^{-3}
Specific weight	γ	$ML^{-2}T^{-2}$
Dynamic viscosity	μ	$ML^{-1}T^{-1}$
Kinematic viscosity	ν	L^2T^{-1}
Surface tension	σ	MT^{-2}
Bulk modulus of elasticity	K	$ML^{-1}T^{-2}$

• PROBLEM 8-2

A V-notch weir is a vertical plate with a notch of angle ϕ cut into the top of it and placed across an open channel. The liquid in the channel is backed up and forced to flow through the notch. The discharge Q is some function of the elevation H of upstream liquid surface above the bottom of the notch. In addition, the discharge depends upon gravity and upon the velocity of approach V_0 to the weir. Determine the form of discharge equation.

Solution: A functional relation

$$F(Q, H, g, V_0, \phi) = 0$$

is to be grouped into dimensionless parameters. ϕ is dimensionless; hence, it is one Π parameter. Only two dimensions are used, L and T. If g and H are the repeating variables,

$$\Pi_1 = H^{x_1} g^{y_1} Q = L^{x_1} (LT^{-2})^{y_1} L^3 T^{-1}$$

$$\Pi_2 = H^{x_2} g^{y_2} V_0 = L^{x_2} (LT^{-2})^{y_2} LT^{-1}$$

Then

$$L: \quad x_1 + y_1 + 3 = 0 \qquad x_2 + y_2 + 1 = 0$$

$$-2y_1 - 1 = 0 \qquad -2y_2 - 1 = 0$$

from which $x_1 = -\frac{5}{2}$, $y_1 = -\frac{1}{2}$, $x_2 = -\frac{1}{2}$, $y_2 = -\frac{1}{2}$, and

$$\Pi_1 = \frac{Q}{\sqrt{g}H^{5/2}} \qquad \Pi_2 = \frac{V_0}{\sqrt{gH}} \qquad \Pi_3 = \phi$$

or

$$f\left(\frac{Q}{\sqrt{g}H^{5/2}}, \frac{V_0}{\sqrt{gH}}, \phi\right) = 0$$

This may be written

$$\frac{Q}{\sqrt{g}H^{5/2}} = f_1\left(\frac{V_0}{\sqrt{gH}}, \phi\right)$$

in which both f, f_1 are unknown functions. After solving for Q,

$$Q = \sqrt{g}H^{5/2} f_1\left(\frac{V_0}{\sqrt{gH}}, \phi\right)$$

Either experiment or analysis is required to yield additional information about the function f_1. If H and V_0 were selected as repeating variables in place of g and H,

$$\Pi_1 = H^{x_1} V_0^{y_1} Q = L^{x_1} (LT^{-1})^y L^3 T^{-1}$$

$$\Pi_2 = H^{x_2} V_0^{y_2} g = L^{x_2} (LT^{-1})^y LT^{-2}$$

Then

$$x_1 + y_1 + 3 = 0 \qquad x_2 + y_2 + 1 = 0$$

$$-y_1 - 1 = 0 \qquad -y_2 - 2 = 0$$

from which $x_1 = -2$, $y_1 = -1$, $x_2 = 1$, $y_2 = -2$, and

$$\Pi_1 = \frac{Q}{H^2 V_0} \qquad \Pi_2 = \frac{gH}{V_0^2} \qquad \Pi_3 = \phi$$

or

$$f\left(\frac{Q}{H^2 V_0}, \frac{gH}{V_0^2}, \phi\right) = 0$$

Since any of the Π parameters may be inverted or raised to any power without affecting their dimesionless status,

$$Q = V_0 H^2 f_2\left(\frac{V_0}{\sqrt{gH}}, \phi\right)$$

The unknown function f_2 has the same parameters as f_1, but it could not be the same function. The last form is not very useful, in general, because frequently V_0 may be neglected with V-notch weirs. This shows that a term of minor importance should not be selected as a repeating variable.

Table 1 Dimensions of physical quantities used in fluid mechanics

Quantity	Symbol	Dimensions (M, L, T)
Length		L
Time	t	T
Mass	m	M
Force	F	MLT^{-2}
Velocity	V	LT^{-1}
Acceleration	a	LT^{-2}
Area	A	L^2
Discharge	Q	L^3T^{-1}
Pressure	Δp	$ML^{-1}T^{-2}$
Gravity	g	LT^{-2}
Density	ρ	ML^{-3}
Specific weight	γ	$ML^{-2}T^{-2}$
Dynamic viscosity	μ	$ML^{-1}T^{-1}$
Kinematic viscosity	ν	L^2T^{-1}
Surface tension	σ	MT^{-2}
Bulk modulus of elasticity	K	$ML^{-1}T^{-2}$

• PROBLEM 8-3

The variables controlling the motion of a floating vessel through water are the drag force F, the speed v, the length l, the density ρ and dynamic viscosity μ of the water and the gravitational acceleration g. Derive an expression for F by dimensional analysis.

<u>Solution</u>: The resistance to motion will be partly due to skin friction which depends on viscosity μ and partly due to to wave resistance which depends on the gravitational acceleration g. The relationship will be of the form

$$F = \phi\{v, l, \rho, \mu, g\}$$

The dimensions of the variables are

	F	v	l	ρ	μ	g
M	1	0	0	1	1	0
L	1	1	1	-3	-1	1
T	-2	-1	0	0	-1	-2

The dependent variable is F. The repeating variables could be v and l, both likely to be major factors. Since the dimensions of v and l do not include [M], a further repeating variable which contains M is needed, say ρ.

Total number of variables, n = 6,

Number of fundamental dimensions, m = 3,

Number of dimensionless groups to be formed = n - m = 3.

The required solution will be

$$\Pi_1 = \phi\{\Pi_2, \Pi_3\}.$$

To find Π_1. This group is formed so that it includes the dependent variable F. F contains all three fundamental dimensions M, L and T, all three of the repeating variables will be required to form a dimensionless group. We can write

$$\Pi_1 = Fv^a l^b \rho^c.$$

Replacing the variables by their dimensions and remembering that the dimensional formula of a dimensionless number is $M^0L^0T^0$,

$$[M^0L^0T^0] = [MLT^{-2}][LT^{-1}]^a[L]^b[ML^{-3}]^c.$$

Equating powers of M, L and T,

$$0 = 1 + c \text{ for } M,$$

$$0 = 1 + a + b - 3c \text{ for } L,$$

$$0 = -2 - a \text{ for } T.$$

From which $a = -2$, $c = -1$, $b = -2$ and $\Pi_1 = F/\rho v^2 l^2$.

To find Π_2. This group is formed by combining one of the non-repeating independent variables μ with the required number of repeating variables. Since the dimensional formula of μ is $ML^{-1}T^{-1}$ all three of the repeating variables will be needed to form a dimensionless group.

$$\Pi_2 = \mu v^d l^e \rho^f.$$

For dimensional homogeneity,

$$[M^0L^0T^0] = [ML^{-1}T^{-1}][LT^{-1}]^d[L]^e[ML^{-3}]^f.$$

Equating powers of M, L and T,

$$0 = 1 + f \text{ for } M,$$

$$0 = -1 + d + e - 3f \text{ for } L,$$

$$0 = -1 - d \text{ for } T.$$

From which $d = -1$, $f = -1$, $e = -1$ and $\Pi_2 = \mu/\rho v l = 1/Re$, where Re = Reynolds number.

To find Π_3. The remaining independent variable g contains the dimensions [L] and [T] only and so must be combined to form a dimensionless group with the repeating variables which do not contain M, namely v and l:

$$\Pi_3 = g v^p l^q.$$

For dimensional homogeneity,

$$[M^0L^0T^0] = [LT^{-2}][LT^{-1}]^p[L]^q.$$

Equating powers of L and T,

$$0 = 1 + p + q \text{ for } L,$$

$$0 = -2 - p \text{ for } T.$$

From which $p = -2$, $q = 1$ and $\Pi_3 = lg/v^2 = 1/Fr^2$, where Fr = Froude number.

The required relationship is therefore

$$F/\rho v^2 l^2 = \phi\{1/Re, 1/Fr^2\}$$

or, since ϕ is an unknown function, we can write

$$F/\rho v^2 l^2 = \phi^1(Re, Fr).$$

Alternative solution:

Instead of forming the dimensionless groups by applying the indicial method, they can be formed by expressing the fundamental dimensions M, L and T in terms of the repeating variables v, l and ρ:

$$[v] = [LT^{-1}], \quad [l] = [L], \quad [\rho] = [ML^{-3}],$$

giving

$$[L] = [l], \quad [M] = [\rho l^3], \quad [T] = [l/v].$$

Now, select each of the remaining variables in turn, starting with the dependent variable, and write down their dimensional formulae first in terms of M, L and T and then in terms of v, l and ρ.

To find Π_1. Selecting the dependent variable F,

$$[F] = [MLT^{-2}]$$

$$= [\rho l^3][l][v^2/l^2]$$

giving

$$F = \Pi_1 \rho l^2 v^2 \quad \text{and} \quad \Pi_1 = F/\rho v^2 l^2.$$

To find Π_2. Selected the independent variable μ,

$$[\mu] = [ML^{-1}T^{-1}]$$

$$= [\rho l^3][l^{-1}][v/l] = [\rho v l],$$

giving

$$\mu = \Pi_2 \rho v l \quad \text{and} \quad \Pi_2 = \mu/\rho v l.$$

To find Π_3. Select the remaining independent variable g

$$[g] = [LT^{-2}]$$

$$= [L][v^2/l^2] = [v^2/l],$$

giving

$$g = \Pi_3 v^2/l \quad \text{and} \quad \Pi_3 = lg/v^2.$$

In many cases the dimensionless groups can be determined by inspection from experience without formal calculation.

The flow through a sluice gate set into a dam is to be investigated by building a model of the dam and sluice at 1:20 scale. Calculate the head at which the model should work to give conditions corresponding to a prototype head of 20 metres. If the discharge from the model under this corresponding head is 0.5 m^3/s, estimate the discharge from the prototype dam.

A relationship between a set of physical variables can be stated as a relationship between a set of independent dimensionless groups made up from suitably chosen variables.

The variables involved in this problem are Q--the flow rate, d--a leading dimension of the sluice, h--the head over the sluice, ρ--the mass density of the water, g--the gravity field intensity and μ--the viscosity of water, so that there is some function,

$$f(Q, \rho, g, h, d, \mu) = 0$$

Solution: The number of dimensionless groups required to specify completely the relationship is the number of variables, n, minus the number of dimensions, m, involved in the variables. These dimensionless groups are called π groups.

A group can be formed by combining any four variables which between them contain the dimensions of mass (M), length (L), and time (T). If three of the variables chosen

There are 6 - 3 = 3π groups.

These are

$$\pi_1 = \rho^a g^b h^c Q$$

$$\pi_2 = \rho^a g^b h^c d$$

$$\pi_3 = \rho^a g^b h^c \mu$$

The Buckingham π theorems permit the equation

$$f(\pi_1, \pi_2, \pi_3) = 0$$

to be written

Each group must be dimensionless, thus for π_1

$$\pi_1 \rightarrow M^a L^{-3a} L^b T^{-2b} L^c L^3 T^{-1} = |0|$$

Equating indices of M, L and T respectively to zero gives

M: $a = 0$

T: $-2b - 1 = 0$, $\therefore b = -1/2$

L: $-3a + b + c + 3 = 0$, $\therefore c = -3 - b + 3a = -2\frac{1}{2}$

$$\therefore \pi_1 = \frac{Q}{g^{1/2}h^{5/2}}$$

$$\pi_2 \rightarrow M^a L^{-3a} L^b T^{-2b} L^c L = |0|$$

by inspection, π_2 can be seen to be d/h.

$$\pi_3 \rightarrow M^a L^{-3a} L^b T^{-2b} L^c ML^{-1} T^{-1} = |0|$$

M: $a + 1 = 0$, $\therefore a = -1$

T: $-2b - 1 = 0$, $\therefore b = -0.5$

L: $-3a + b + c - 1 = 0$

$\therefore c = 1 + 3a - b = 1 - 3 + \frac{1}{2} = -\frac{3}{2}$

$$\therefore \Pi_3 = \frac{\mu}{\rho g^{1/2}h^{3/2}}$$

The solution is thus:

$$\frac{Q}{g^{1/2}h^{5/2}} = \phi\left(\frac{\rho g^{1/2}h^{3/2}}{\mu}, \frac{d}{h}\right)$$

(Note that dimensionless groups can be inverted as they remain dimensionless).

As flow through a sluice resembles flow through a large orifice, this expression requires slight amendment.

Multiply $Q/g^{1/2}h^{5/2}$ by $(h/d)^2$

that is

$$\frac{Q}{g^{1/2}d^2h^{1/2}}$$

Then

$$Q = d^2\sqrt{(gh)}\,\phi\left(\frac{\rho g^{1/2}h^{3/2}}{\mu}, \frac{d}{h}\right)$$

This type of problem involves both the Reynolds and the Froude numbers.

$$\frac{Q^2}{d^2\sqrt{(gh)}} \propto \frac{Q}{A\sqrt{(gh)}}$$

as A is a function of d^2, but $Q/A = v$, so

$$\frac{Q}{d^2\sqrt{(gh)}} \propto \frac{v}{\sqrt{(gh)}} \propto \frac{v^2}{gh}$$

that is, the Froude number.

It is not possible to obtain Reynolds and Froude-number modelling simultaneously. If the Reynolds number for the model is large, the value for the prototype will be even larger and the effect of the Reynolds-number variation will be minimal. Therefore, it is only necessary to model for the Froude number and the d/h value.

$$d_m/h_m = d_p/h_p$$

$$h_p/h_m = d_p/d_m = 20$$

$$h_m = 20/20 = 1 \text{ m}$$

$$\frac{Q_p}{g^{1/2}h_p^{5/2}} = \frac{Q_M}{g^{1/2}h_m^{5/2}}$$

$$Q_p/Q_M = (h_p/h_m)^{5/2} = 20^{5/2}$$

$$Q_p = \frac{1}{2}\, 20^{5/2} = 894 \text{ m}^3/\text{s}$$

• PROBLEM 8-5

(a) Find a set of coordinates for the equation giving the frictional resistance in terms of pertinent variables of a shaft rotating in a well-lubricated bearing. Assume that the following variables are involved.

Variable	Symbol	Dimensions of variable
Tangential friction force	R	MLT^{-2}
Force normal to shaft	P	MLT^{-2}
Shaft revolutions per unit time	N	T^{-1}
Viscosity of lubricant	μ	$ML^{-1}T^{-1}$
Shaft diameter	D	L

(b) Obtain a set of convenient coordinates for the thrust of a screw propeller immersed fully in a fluid. In a study of variables that may be involved, the following has been found:

Variable	Symbol	Dimensions of variable
Thrust (axial force)	P	MLT^{-2}
Propeller diameter	D	L
Velocity of advance	V	LT^{-1}
Revolutions per unit time	N	T^{-1}
Gravitational acceleration	g	LT^{-2}
Density of fluid	ρ	ML^{-3}
Kinematic viscosity of fluid	ν	L^2T^{-1}

Solution: (a) Since there are five variables (n = 5) and three fundamental units (m = 3), the physical equation has two dimensionless ratios. Let π_1 and π_2 represent these ratios. Dimensional equations are written combining the three variables P, N, and D [selected as suggested in (1) of Article 52] with each of the remaining variables in turn. Thus

$$\pi_1 = P^{x_1}N^{y_1}D^{z_1}R, \qquad \pi_2 = P^{x_2}N^{y_2}D^{z_2}\mu$$

The six exponents are found from dimensional considerations. Substituting the dimensions for the symbols in the equation of π_1 gives

$$\left(\frac{ML}{T^2}\right)^{x_1}\left(\frac{1}{T}\right)^{y_1}L^{z_1}\,\frac{ML}{T^2} = L^0M^0T^0$$

where $L^0M^0T^0$ represents the fact that the π_1 ratio is dimensionless or has zero dimensions. Solving for each dimension separately gives

M	$x_1 + 1 - 0$	$x_1 = -1$
L	$x_1 + z_1 + 1 = 0$	$z_1 = 0$
T	$-2x_1 - y_1 - 2 = 0$	$y_1 = 0$

Therefore

$$\pi_1 = \frac{R}{P}$$

which is easily checked. R/P is commonly called a friction coefficient f.

Substituting the dimensions for the symbols in the equation for π_2 gives

$$\left(\frac{ML}{T^2}\right)^{x_2}\left(\frac{1}{T}\right)^{y_2}L^{z_2}\,\frac{M}{LT} = L^0M^0T^0$$

M	$x_2 + 1 = 0$	$x_2 = -1$
L	$x_2 + z_2 - 1 = 0$	$z_2 = 2$
T	$-2x_2 - y_2 - 1 = 0$	$y_2 = 1$

Thus

$$\pi_2 = \frac{N\mu D^2}{P}$$

The physical equation has the form

$$\pi_1 = \phi(\pi_2) \qquad \text{or} \qquad f = \phi\left(\frac{N\mu D^2}{P}\right)$$

where ϕ means "some function of"; that is, the friction coefficient f is some function of the dimensionless ratio $N\mu D^2/P$. The foregoing coordinates are used in lubrication studies. Experimental data are necessary in order to determine the exact nature of the functional relation; such data are available in current lubrication literature. Dimensional analysis provides a systematic and efficient approach. Practical engineering work has shown that the result is of aid in the design and prediction of performance of lubricated bearings.

R, N, and D were selected to give one solution. Another set of three variables could have been selected to give another solution. For example, the dimensionless ratio P/R, from a dimensional point of view, is just as satisfactory and sound as its reciprocal. Custom, however, has arbitrarily established the ratio R/P as a friction coefficient.

(b) Since there are seven variables and three fundamental units, there are four dimensionless ratios. Dimensional relations are written combining D, V, and ρ with each of the remaining variables in turn:

$$\pi_1 = D^{x_1}V^{y_1}\rho^{z_1}P$$

$$\pi_2 = D^{x_2}V^{y_2}\rho^{z_2}N$$

$$\pi_3 = D^{x_3}V^{y_3}\rho^{z_3}g$$

$$\pi_4 = D^{x_4}V^{y_4}\rho^{z_4}\nu$$

The twelve exponents in x, y, and z are determined such that each π function is dimensionless. The result is

$$\pi_1 = \frac{P}{D^2V^2\rho}, \quad \pi_2 = \frac{DN}{V}, \quad \pi_3 = \frac{Dg}{V^2}, \quad \pi_4 = \frac{\nu}{DV}$$

This dimensional analysis gives one solution to the problem. There are other solutions each involving different dimensionless groupings of the variables. The above indicates that the form of the physical equation involves the four ratios π_1, π_2, π_3, and π_4. One solution of a variety can be written as

$$\pi_1 = \phi(\pi_2, \pi_3, \pi_4)$$

meaning that π_1 is some function of the other three π ratios, or

$$\frac{P}{D^2V^2\rho} = \phi\left(\frac{DN}{V}, \frac{Dg}{V^2}, \frac{\nu}{DV}\right)$$

Proper experiments would give the functional relation. Sometimes the functional relation can be written as

$$P = KD^2V^2\rho\left(\frac{DN}{V}\right)^a\left(\frac{Dg}{V^2}\right)^b\left(\frac{\nu}{DV}\right)^c$$

where the coefficient K and the exponents a, b, and c are to be found from experimental data.

The last example includes one feature, in connection with the ratio $\pi_4 = \nu/DV$, which merits special attention. The reciprocal of this number is still a dimensionless ratio. The ratio DV/ν occurs frequently in fluid flow problems and is significant in establishing criteria of flow. Its significance will be discussed more fully throughout the remainder of this book. This particular number is called Reynolds number, in honor of Osborne Reynolds. Other dimensionless ratios occur frequently in fluid mechanics and heat transfer, and have been given special names.

The capillary rise h of a liquid in a tube varies with tube diameter d, gravity g, fluid density ρ, surface tension γ, and the contact angle θ. (a) Find a dimensionless statement of this relation. (b) If h = 3 cm in a given experiment, what will h be in a similar case if diameter and surface tension are half as much, density is twice as much, and the contact angle is the same?

Solution: (a) Step 1. Write down the function and count variables

$$h = f(d, g, \rho, \gamma, \theta) \qquad n = 6 \text{ variables}$$

Step 2. List the dimensions (FLT) from Table 1:

h	d	g	ρ	γ	θ
{L}	{L}	$\{LT^{-2}\}$	$\{FT^2L^{-4}\}$	$\{FL^{-1}\}$	None

Step 3. Find j. Several groups of three form no pi. γ, ρ, and g or ρ, g, and d. Therefore j = 3, and we expect n - j = 6 - 3 = 3 dimensionless groups. One of these is obviously θ, which is already dimensionless:

$$\Pi_3 = \theta \qquad \text{Ans. (a)}$$

If we chose carelessly to search for it using steps 4 and 5, we would still find $\Pi_3 = \theta$.

Step 4. Select j variables which do not form a pi group: ρ, g, d.

Step 5. Add one additional variable in sequence to form the pis:

Add h:

$$\Pi_1 = \rho^a g^b d^c h = (FT^2L^{-4})^a (LT^{-2})^b (L)^c (L) = F^0L^0T^0$$

Solve for

$$a = b = 0 \qquad c = -1 \qquad \text{Ans. (a)}$$

Therefore

$$\Pi_1 = \rho^0 g^0 d^{-1} h = \frac{h}{d} \qquad \text{Ans. (a)}$$

Finally add γ, again selecting its exponent to be 1

$$\Pi_2 = \rho^a g^b d^c \gamma = (FT^2L^{-4})^a (LT^{-2})^b (L)^c (FL^{-1}) = F^0L^0T^0$$

Solve for

$$a = b = -1 \qquad c = -2$$

Therefore

$$\Pi_2 = \rho^{-1}g^{-1}d^{-2}\gamma = \frac{1}{\rho g d^2} \qquad \text{Ans. (a)}$$

Step 6. The complete dimensionless relation for this problem is thus

$$\frac{h}{d} = F\left(\frac{\gamma}{\rho g d^2}, \theta\right) \qquad \text{Ans. (a)} \quad (1)$$

This is as far as dimensional analysis goes. Theory, however, establishes that h is proportional to γ. Since γ occurs only in the second parameter, we can slip it outside

$$\left(\frac{h}{d}\right)_{\text{actual}} = \frac{\gamma}{\rho g d^2}F_1(\theta) \qquad \text{or} \qquad \frac{h\rho g d}{\gamma} = F_1(\theta)$$

(b) We are given h_1 for certain conditions d_1, γ_1, ρ_1, and θ_1. If $h_1 = 3$ cm, what is h_2 for $d_2 = \frac{1}{2}d_1$, $\gamma_2 = \frac{1}{2}\gamma_1$, $\rho_2 = 2\rho_1$, and $\theta_2 = \theta_1$? We know the functional Eq. (1) must still hold at condition 2

$$\frac{h_2}{d_2} = F\left(\frac{\gamma_2}{\rho_2 g d_2^2}, \theta_2\right)$$

But

$$\frac{\gamma_2}{\rho_2 g d_2^2} = \frac{\frac{1}{2}\gamma_1}{2\rho_1 g(\frac{1}{2}d_1)^2} = \frac{\gamma_1}{\rho_1 g d_1^2}$$

Therefore, functionally,

$$\frac{h_2}{d_2} = F\left(\frac{\gamma_1}{\rho_1 g d_1^2}, \theta_1\right) = \frac{h_1}{d_1}$$

We are given a condition 2 which is exactly similar to condition 1, and therefore a scaling law holds

$$h_2 = h_1\frac{d_2}{d_1} = (3\text{ cm})\frac{\frac{1}{2}d_1}{d_1} = 1.5\text{ cm} \qquad \text{Ans. (b)}$$

If the pi groups had not been exactly the same for both conditions, we would have to know more about the functional relation F to calculate h_2.

Parameter	*Definition*	*Qualitative ratio of effects*	*Importance*
Reynolds number	$\mathrm{Re} = \dfrac{\rho U L}{\mu}$	$\dfrac{\text{Inertia}}{\text{Viscosity}}$	Always
Mach number	$\mathrm{Ma} = \dfrac{U}{a}$	$\dfrac{\text{Flow speed}}{\text{Sound speed}}$	Compressible flow
Froude number	$\mathrm{Fr} = \dfrac{U^2}{gL}$	$\dfrac{\text{Inertia}}{\text{Gravity}}$	Free-surface flow
Weber number	$\mathrm{We} = \dfrac{\rho U^2 L}{\Upsilon}$	$\dfrac{\text{Inertia}}{\text{Surface tension}}$	Free-surface flow
Cavitation number (Euler number)	$\mathrm{Ca} = \dfrac{p - p_v}{\rho U^2}$	$\dfrac{\text{Pressure}}{\text{Inertia}}$	Cavitation
Prandtl number	$\mathrm{Pr} = \dfrac{\mu c_p}{k}$	$\dfrac{\text{Dissipation}}{\text{Conduction}}$	Heat convection
Eckert number	$\mathrm{Ec} = \dfrac{U^2}{c_p T_0}$	$\dfrac{\text{Kinetic energy}}{\text{Enthalpy}}$	Dissipation
Specific heat ratio	$\gamma = \dfrac{c_p}{c_v}$	$\dfrac{\text{Enthalpy}}{\text{Internal energy}}$	Compressible flow
Strouhal number	$\mathrm{St} = \dfrac{\omega L}{U}$	$\dfrac{\text{Oscillation}}{\text{Mean speed}}$	Oscillating flow
Roughness ratio	$\dfrac{\epsilon}{L}$	$\dfrac{\text{Wall roughness}}{\text{Body length}}$	Turbulent, rough walls
Grashof number	$\mathrm{Gr} = \dfrac{\beta\, \Delta T g L^3 \rho^2}{\mu^2}$	$\dfrac{\text{Buoyancy}}{\text{Viscosity}}$	Natural convection
Temperature ratio	$\dfrac{T_w}{T_0}$	$\dfrac{\text{Wall temperature}}{\text{Stream temperature}}$	Heat transfer

Suppose one knew that the force F on a particular body immersed in a stream of mud depended only on the body length L, the stream velocity U, the fluid density ρ, and the fluid viscosity μ; that is,

$$F = f(L, U, \rho, \mu) \tag{1}$$

Show the development of the following dimensionless form,

$$\frac{F}{\rho V^2 L^2} = g\left(\frac{\rho VL}{\mu}\right)$$

or

$$C_F = g(Re)$$

that is, the dimensionless force coefficient $F/\rho V^2 L^2$ is a function only of the dimensionless Reynolds number $\rho VL/\mu$.

a) Using the power-product method explain the "force coefficient."

b) Using the π-Theorem

Solution: (a) The basic function is $F = f(L, U, \rho, \mu)$. The list of dimensions can be made for these five variables:

$$\{F\} = \{MLT^{-2}\} \quad \{L\} = \{L\} \quad \{U\} = \{LT^{-1}\}$$

$$\{\rho\} = \{ML^{-3}\} \quad \{\mu\} = \{ML^{-1}T^{-1}\}$$

We expect no fewer than 5 - 3 = 2 dimensionless variables. The suggested power product is for f equal to a force

$$f_1 = (\text{const})(L)^a(U)^b(\rho)^c(\mu)^d$$

or

or $\{MLT^{-2}\} = \{L\}^a\{LT^{-1}\}^b\{ML^{-3}\}^c\{ML^{-1}T^{-1}\}^d$ (1)

Equate exponents:

Length: $1 = a + b - 3c - d$

Mass: $1 = c + d$

Time: $-2 = -b - d$

Solve for three unknowns in terms of the fourth. A variety of formulations will occur, depending upon which we choose to be the "free" exponent. If we choose d, then

$$a = 2 - d \qquad b = 2 - d \qquad c = 1 - d$$

Equation (1) becomes

$$f_1 = (\text{const})L^{2-d}U^{2-d}\rho^{1-d}\mu^{d} = (\text{const})(\rho U^2L^2)\left(\frac{\rho UL}{\mu}\right)^{-d} \quad (2)$$

As usual, the arbitrariness of d implies an arbitrary function of its argument. The original function can now be rewritten as

$$\frac{F}{\rho U^2L^2} = G\frac{\rho UL}{\mu} \qquad \text{or} \qquad C_F = G(\text{Re}) \qquad \text{Ans. } (3)$$

This is exactly Eq. (2). Since the theory of fluid forces on immersed bodies is still rather weak and qualitative, the function G(Re) is generally determined by experiment.

If we choose another free exponent, two different but related force coefficients will arise. For example, if we solve for a, b, and d in terms of c, the solution is

$$a = 1 + c \qquad b = 1 + c \qquad d = 1 - c$$

or

$$f_1 = (\text{const})(LU\mu)\left(\frac{\rho UL}{\mu}\right)^{c}$$

or

$$\frac{F}{LU\mu} = G_1\frac{\rho UL}{\mu} = G_1(\text{Re}) \quad (4)$$

The new force coefficient is not unique, but

$$\frac{F}{LU\mu} = \frac{F}{\rho U^2L^2}\,\frac{\rho UL}{\mu} \equiv C_F\,\text{Re} \quad (5)$$

The "correct" force coefficient is thus a matter of taste and custom. This particular choice $F/LU\mu$ is uncommon but very useful in highly visous, "creeping" motion, for which it equals a pure constant.

Further, if we had solved for a, c, and d in terms of b, we would have obtained

$$a = b \qquad c = b - 1 \qquad d = 2 - b$$

or

$$\frac{F\rho}{\mu^2} = G_2\,\frac{\rho UL}{\mu} = G_2(\text{Re}) \quad (6)$$

This force coefficient is not unique either

$$\frac{F\rho}{\mu^2} = \frac{F}{\rho U^2L^2}\left(\frac{\rho UL}{\mu}\right)^2 \equiv C_F\,\text{Re}^2 \quad (7)$$

(b) Step 1. Write the function and count variables:

$F = f(L, U, \rho, \mu)$ there are five variables ($n = 5$)

Step 2. List dimensions of each variable. From Table 1

Table 1
DIMENSIONS OF FLUID-MECHANICS QUANTITIES

Quantity	Symbol	Dimensions {MLTΘ}	{FLTΘ}
Length	L	L	L
Area	A	L^2	L^2
Volume	$\mathcal{V}$	L^3	L^3
Velocity	V	LT^{-1}	LT^{-1}
Speed of sound	a	LT^{-1}	LT^{-1}
Volume flux	Q	L^3T^{-1}	L^3T^{-1}
Mass flux	$\dot{m}$	MT^{-1}	FTL^{-1}
Pressure, stress	p, σ	$ML^{-1}T^{-2}$	FL^{-2}
Strain rate	$\dot{\epsilon}$	T^{-1}	T^{-1}
Angle	θ	None	None
Angular velocity	ω	T^{-1}	T^{-1}
Viscosity	μ	$ML^{-1}T^{-1}$	FTL^{-2}
Kinematic viscosity	ν	L^2T^{-1}	L^2T^{-1}
Surface tension	Υ	MT^{-2}	FL^{-1}
Force	F	MLT^{-2}	F
Moment, torque	M	ML^2T^{-2}	FL
Power	P	ML^2T^{-3}	FLT^{-1}
Density	ρ	ML^{-3}	FT^2L^{-4}
Temperature	T	Θ	Θ
Specific heat	c_p, c_v	$L^2T^{-2}\Theta^{-1}$	$L^2T^{-2}\Theta^{-1}$
Thermal conductivity	k	$MLT^{-3}\Theta^{-1}$	$FT^{-1}\Theta^{-1}$
Expansion coefficient	β	Θ^{-1}	Θ^{-1}

F	L	U	ρ	μ
$\{MLT^{-2}\}$	$\{L\}$	$\{LT^{-1}\}$	$\{ML^{-3}\}$	$\{ML^{-1}T^{-1}\}$

Step 3. Find j. No variable contains the dimension Θ, and so j is less than or equal to 3 (MLT). We inspect the list and see that L, U, and ρ cannot form a pi group because only ρ contains mass and only U contains time. Therefore j does equal 3, and $n - j = 5 - 3 = 2 = k$. The pi theorem guarantees for this problem that there will be exactly two independent dimensionless groups.

Step 4. Select j variables. The group L, U, ρ we found to prove that j = 3 will do fine.

Step 5. Combine L, U, ρ with one additional variable, in sequence, to find the two pi products.

First add force to find Π_1. You may select any exponent on this additional term as you please, to place it in the numerator or denominator to any power. Since F is the output, or dependent, variable, we select it to appear to the first power in the numerator

$$\Pi_1 = L^a U^b \rho^c F = (L)^a (LT^{-1})^b (ML^{-3}) (MLT^{-2}) = M^0 L^0 T^0$$

Equate exponents:

Length: $a + b - 3c + 1 = 0$

Mass: $c + 1 = 0$

Time: $-b \quad - 2 = 0$

We can solve explicitly for

$$a = -2 \qquad b = -2 \qquad c = -1$$

Therefore

$$\Pi_1 = L^{-2}U^{-2}\rho^{-1}F = \frac{F}{\rho U^2 L^2} = C_F \qquad \text{Ans.}$$

This is exactly the right pi group as in Eq. (2). By varying the exponent on F, we could have found other equivalent groups such as $UL\rho^{1/2}/F^{1/2}$.

Finally, add viscosity to L, U, and ρ to find Π_2. Select any power you like for viscosity. By hindsight and custom, we select the power -1 to place it in the denominator:

$$\Pi_2 = L^a U^b \rho^c \mu^{-1} = L^a (LT^{-1})^b (ML^{-3})^c (ML^{-1}T^{-1})^{-1}$$
$$= M^0 L^0 T^0$$

Equate exponents:

Length: $a + b - 3c + 1 = 0$

Mass: $c - 1 = 0$

Time: $-b \quad + 1 = 0$

from which we find

$$a = b = c = 1$$

Therefore

$$\Pi_2 = L^1 U^1 \rho^1 \mu^{-1} = \frac{\rho UL}{\mu} = \text{Re} \qquad \text{Ans.}$$

We know we are finished; this is the second and last pi group. The theorem guarantees that the functional relationship must be of the equivalent form

$$\frac{F}{\rho U^2 L^2} = g\,\frac{\rho UL}{\mu} \qquad \text{Ans.}$$

which is exactly what we found by the power-product method in part (a).

The thrust F of a screw propeller is known to depend upon the diameter d, speed of advance v, fluid density ρ, revolutions per second N, and the coefficient of viscosity μ of the fluid. Find an expression for F in terms of these quantities.

Solution: The general relationship must be $F = \phi(d, v, \rho, N, \mu)$, which can be expanded as the sum of an infinite series of terms giving

$$F = A(d^m v^p \rho^q N^r \mu^s) + B(d^{m'} v^{p'} \rho^{q'} N^{r'} \mu^{s'}) + \ldots,$$

where A, B, etc. are numerical constants and m, p, q, r, s are unknown powers. Since, for dimensional homogeneity, all terms must be dimensionally the same, this can be reduced to

$$F = K d^m v^p \rho^q N^r \mu^s, \tag{1}$$

where K is a numerical constant.

The dimensions of the dependent variable F and the independent variables d, v, ρ, N and μ are

$$[F'] = [\text{Force}] = [MLT^{-2}],$$

$$[d] = [\text{Diameter}] = [L],$$

$$[v] = [\text{Velocity}] = [LT^{-1}]$$

$$[\rho] = [\text{Mass density}] = [ML^{-3}],$$

$$[N] = [\text{Rotational speed}] = [T^{-1}],$$

$$[\mu] = [\text{Dynamic viscosity}] = [ML^{-1}T^{-1}].$$

For convenience, these can be set out in the form of a table or dimensional matrix, in which a column is provided for each variable and the power of each fundamental dimension in its dimensional formula is inserted in the corresponding row:

	F	d	v	ρ	N	μ
M	1	0	0	1	0	1
L	1	1	1	-3	0	-1
T	-2	0	-1	0	-1	-1

Substituting the dimensions for the variables in (I),

$$[MLT^{-2}] = [L]^m [LT^{-1}]^p [ML^{-3}]^q [T^{-1}]^r [ML^{-1}T^{-1}]^s$$

Equating powers of [M], [L] and [T]:

$$[M],\ 1 = q + s; \tag{2}$$

$$[L],\ 1 = m + p - 3q - s; \tag{3}$$

$$[T],\ -2 = -p - r - s. \tag{4}$$

Since there are five unknown powers and only three equations, it is impossible to obtain a complete solution, but three unknowns can be determined in terms of the remaining two. If we solve for m, p and q, we get

$$q = 1 - s \text{ from (2)}$$

$$p = 2 - r - s \text{ from (3)}$$

$$m = 1 - p + 3q + s = 2 + r - s \text{ from (4).}$$

Substituting these values in (1),

$$F = K\, d^{2+r-s} v^{2-r-s} \rho^{1-s} N^{r} \mu^{s}.$$

Regrouping the powers,

$$F = K\rho v^2 d^2 (\rho v d/\mu)^{-s} (dN/v)^{r}.$$

Since s and r are unknown this can be written

$$F = \rho v^2 d^2 \phi\{\rho v d/\mu,\ dN/v\} \tag{5}$$

where ϕ means 'a function of'. At first sight, this appears to be a rather unsatisfactory solution, (5) indicates that

$$F = C\rho v^2 d^2, \tag{6}$$

where C is a constant to be determined experimentally and the value of which is dependent on the values of $\rho v d/\mu$ and dN/v.

• PROBLEM 8-9

a. Fluid flows in a long pipe. Reduce the variables to π terms for experimental work to determine head loss.

b. Rearrange the Π terms in part (a) to the form of Darcy's equation.

$$h_L = f\,\frac{L}{D}\,\frac{V^2}{2g}$$

Solution: a. The first step is always to list all quantities that will affect the problem. This list requires a real understanding of the problem. The quantities and the reasoning behind each are listed below.

Quantity	Reasoning
Head loss, h_L	What are you solving for.
Diameter of pipe, D	Certainly a variable and it could affect head loss.
Roughness of pipe, e	Rougher pipes should cause greater losses.
Length of pipe, l	The longer the pipe, the greater the losses.
Velocity of fluid, V	Friction depends on relative velocity.
Viscosity of liquid, μ	This could seem necessary in any problem involving friction.
Gravitational acceleration, g	To take care of elevation changes in the pipe.
Density of fluid, ρ	In problems of motion add the mass for inertia forces.

Therefore,

$$F(h_L, D, e, l, g, V, \mu, \rho) = 0$$

Notice that the reasoning seems vague in some cases, the typically weakest point of the dimensional analysis approach. Selecting the proper quantities is the most difficult part of the problem.

Use the M, L, T system of fundamental dimensions in this problem, although the selection is completely arbitrary because the F, L, T system works as well. List all quantities with their proper dimensions.

$$h_L \stackrel{d}{=} L,\ D \stackrel{d}{=} L,\ e \stackrel{d}{=} L,\ V \stackrel{d}{=} L/T,\ l \stackrel{d}{=} L,\ \mu \stackrel{d}{=} \frac{M}{LT},$$

$$\rho \stackrel{d}{=} \frac{M}{L^3},\ g \stackrel{d}{=} \frac{L}{T^2}$$

For this problem write the mathematical statement of dimensional homogeneity.

$$M^0L^0T^0 \stackrel{d}{=} h_L^{a_1}D^{a_2}e^{a_3}l^{a_4}V^{a_5}\mu^{a_6}\rho^{a_7}g^{a_8} \tag{1}$$

Substitute the dimensional equivalent,

$$M^0L^0T^0 \stackrel{d}{=} L^{a_1}L^{a_2}L^{a_3}L^{a_4}\left(\frac{L}{T}\right)^{a_5}\left(\frac{M}{LT}\right)^{a_6}\left(\frac{M}{L^3}\right)^{a_7}\left(\frac{L}{T^2}\right)^{a_8}$$

Rearrange to

$$M^0L^0T^0 \stackrel{d}{=} M^{a_6+a_7}L^{a_1+a_2+a_3+a_4+a_5-a_6-3a_7+a_8}T^{-a_5-a_6-2a_8}$$

Equate the exponents of mass, length, and time to get

$$M: \quad 0 = a_6 + a_7$$

$$L: \quad 0 = a_1 + a_2 + a_3 + a_4 + a_5 - a_6 - 3a_7 + a_8$$

$$T: \quad 0 = -a_5 - a_6 - 2a_8$$

You can solve these equations for any three constants in terms of the others. In this case, solve for constant a_6, a_2, a_8.

$$a_2 = -a_1 - a_3 - a_4 - \frac{1}{2}a_5 + \frac{3}{2}a_7$$

$$a_6 = -a_7$$

$$a_8 = \frac{1}{2}a_7 - \frac{1}{2}a_5$$

The following develops by substituting these values into Equation (1).

$$M^0L^0T^0 \stackrel{d}{=} h_L^{a_1}D^{(-a_1-a_3-a_4-1/2a_5+3/2a_7)}e^{a_3}l^{a_4}V^{a_5}$$

$$\mu^{-a_7}\rho^{+a_7}g^{1/2a_7-1/2a_5}$$

Collecting terms with the same exponent gives

$$M^0L^0T^0 \stackrel{d}{=} \left(\frac{h_L}{D}\right)^{a_1}\left(\frac{e}{D}\right)^{a_3}\left(\frac{l}{D}\right)^{a_4}\left(\frac{V}{\sqrt{gD}}\right)^{a_5}\left(\frac{\rho g^{1/2}D^{3/2}}{\mu}\right)^{a_7}$$

These terms are dimensionless, π terms.

$$\pi_1 = \frac{h_L}{D}, \quad \pi_2 = \frac{e}{D}, \quad \pi_3 = \frac{l}{D}, \quad \pi_4 = \frac{V}{\sqrt{gD}}, \quad \pi_5 = \frac{\rho g^{1/2}D^{3/2}}{\mu}$$

Therefore,

$$F(h_L, D, e, l, V, \mu, \rho, g) = f(\pi_1, \pi_2, \pi_3, \pi_4, \pi_5) = 0$$

Now, you have reduced the number of unknowns from eight to five, which you could have predicted.

b. Given from part (a)

$$\pi_1 = \frac{h_L}{D}, \quad \pi_2 = \frac{e}{D}, \quad \pi_3 = \frac{l}{D}, \quad \pi_4 = \frac{V}{\sqrt{gD}}, \quad \pi_5 = \frac{\rho g^{1/2}D^{3/2}}{\mu}$$

Select new π terms.

$$\pi_6 = \frac{\pi_1}{\pi_4{}^2} = \frac{h_L}{D\,\frac{V^2}{gD}} = \frac{h_L}{\frac{V^2}{g}}$$

$$\pi_7 = \pi_4\pi_5 = \frac{\rho g^{1/2}D^{3/2}}{\mu} \quad \frac{V}{g^{1/2}D^{1/2}} = \frac{\rho VD}{\mu}$$

The new π terms are

$$\pi_2 = \frac{e}{D}, \quad \pi_3 = \frac{l}{D}, \quad \pi_4 = \frac{V}{\sqrt{gD}}, \quad \pi_6 = \frac{h_L}{V^2/g}, \quad \pi_7 = \frac{\rho VD}{\mu}$$

$$\pi_6 = f(\pi_2,\ \pi_3,\ \pi_4,\ \pi_7)$$

$$h_L = f_1(\pi_2,\ \pi_3,\ \pi_4,\ \pi_7)\frac{V^2}{2g}$$

Experiments shown that $V/\sqrt{gD}$ is not important and that the head loss varies directly with L/D. Therefore,

$$h_L = f_2(\pi_2,\ \pi_7)\ \frac{l}{D}\ \frac{V^2}{2g}$$

$$f = f_2(\pi_2,\ \pi_7) = f_2\left(\frac{e}{D},\ \frac{\rho VD}{\mu}\right)$$

Finally,

$$h_L = f\ \frac{l}{D}\ \frac{V^2}{2g}$$

• PROBLEM 8-10

A solid sphere travels under the surface of the ocean. Reduce the number of variables to π terms to be used to experimentally determine the drag force.

Solution: List all the variables that affect the drag force.

Quantity	Reasoning
Drag force, D	This is the quantity of primary interest.
Density of liquid, ρ	As the sphere moves through the water it accelerates fluid particles. Thus, the mass density of these particles contributes to the force.
Size of sphere, l	Size of the sphere certainly affects the drag force.

Viscosity of water, μ	This determines the friction forces.
Velocity of sphere, V	This affects the shear forces on the side of the sphere.

Let's use the F, L, T system in this problem.

$$D \overset{d}{=} F,\ \rho \overset{d}{=} \frac{FT^2}{L^4},\ \mu \overset{d}{=} \frac{FT}{L^2},\ l \overset{d}{=} L,\ V \overset{d}{=} \frac{L}{T}$$

For dimensional homogeneity.

$$F^0L^0T^0 \overset{d}{=} D^{a_1}\rho^{a_2}\mu^{a_3}L^{a_4}V^{a_5} \qquad (1)$$

Substitute the dimensional equivalents into Equation (1) to get

$$F^0L^0T^0 \overset{d}{=} F^{a_1}\left(\frac{FT^2}{L^4}\right)^{a_2}\left(\frac{FT}{L^2}\right)^{a_3}L^{a_4}\left(\frac{L}{T}\right)^{a_5}$$

Rearrange

$$F^0L^0T^0 = F^{a_1+a_2+a_3}L^{-4a_2-2a_3+a_4+a_5}T^{2a_2+a_3-a_5}$$

Equate the exponents of force, length, and time.

$$F: \quad 0 = a_1 + a_2 + a_3$$

$$L: \quad 0 = -4a_2 - 2a_3 + a_4 + a_5$$

$$T: \quad 0 = 2a_2 + a_3 - a_5$$

Here are three equations and five unknowns of which three can be solved in terms of the other two.

$$a_1 = -a_2 - a_3$$

$$a_4 = 2a_2 + a_3$$

$$a_5 = 2a_2 + a_3$$

Substitute these into Equation (1).

$$F^0L^0T^0 \overset{d}{=} D^{(-a_2-a_3)}\rho^{a_2}\mu^{a_3}L^{2a_2+a_3}V^{2a_2+a_3}$$

Collect terms with the same exponent.

$$F^0L^0T^0 \overset{d}{=} \left(\frac{\rho V^2 l^2}{D}\right)^{a_2}\left(\frac{\mu V l}{D}\right)^{a_3}$$

Terms to the same exponent are dimensionless and are π terms.

$$\pi_1 = \frac{\rho V^2 l^2}{D}, \qquad \pi_2 = \frac{\mu V l}{D}$$

$$F(D,\ \rho,\ \mu,\ l,\ V) = f(\pi_1,\ \pi_2) = 0$$

You have reduced the number of unknowns, therefore, from five to two.

At low velocities (laminar flow), the volume flux Q through a small-bore tube is a function only of the pipe radius r, the fluid viscosity μ, and the pressure drop per unit pipe length dp/dx. Using the power-product method, rewrite the suggested relationship Q = f(r, μ, dp/dx) in dimensionless form.

Table 1

DIMENSIONS OF FLUID-MECHANICS QUANTITIES

		Dimensions	
Quantity	*Symbol*	$\{MLT\Theta\}$	$\{FLT\Theta\}$
Length	L	L	L
Area	A	L^2	L^2
Volume	$\mathcal{V}$	L^3	L^3
Velocity	V	LT^{-1}	LT^{-1}
Speed of sound	a	LT^{-1}	LT^{-1}
Volume flux	Q	L^3T^{-1}	L^3T^{-1}
Mass flux	$\dot{m}$	MT^{-1}	FTL^{-1}
Pressure, stress	p, σ	$ML^{-1}T^{-2}$	FL^{-2}
Strain rate	$\dot{\epsilon}$	T^{-1}	T^{-1}
Angle	θ	None	None
Angular velocity	ω	T^{-1}	T^{-1}
Viscosity	μ	$ML^{-1}T^{-1}$	FTL^{-2}
Kinematic viscosity	ν	L^2T^{-1}	L^2T^{-1}
Surface tension	Υ	MT^{-2}	FL^{-1}
Force	F	MLT^{-2}	F
Moment, torque	M	ML^2T^{-2}	FL
Power	P	ML^2T^{-3}	FLT^{-1}
Density	ρ	ML^{-3}	FT^2L^{-4}
Temperature	T	Θ	Θ
Specific heat	c_p, c_v	$L^2T^{-2}\Theta^{-1}$	$L^2T^{-2}\Theta^{-1}$
Thermal conductivity	k	$MLT^{-3}\Theta^{-1}$	$FT^{-1}\Theta^{-1}$
Expansion coefficient	β	Θ^{-1}	Θ^{-1}

<u>Solution</u>: First list the dimensions of the variables (Table 1):

$$Q = \{L^3T^{-1}\} \qquad r = \{L\} \qquad \mu = \{ML^{-1}T^{-1}\} \qquad \frac{dp}{dx} = \{ML^{-2}T^{-2}\}$$

Since Q is a volume flux and we assume dimensional homogeneity, the function f must be a volume flux. Assume a power product

$$f_1 = (\text{const})\,(r)^a\,(\mu)^b\left(\frac{dp}{dx}\right)^c$$

or

$$\{L^3T^{-1}\} = \{L\}^a\{ML^{-1}T^{-1}\}^b\{ML^{-2}T^{-2}\}^c$$

Equating respective exponents, we have:

Length: $\quad 3 = a - b - 2c$

Mass: $\quad 0 = b + c$

Time: $-1 = -b - 2c$

With three equations in three unknowns, the solution is

$$a = 4 \quad b = -1 \quad c = 1$$

There is no arbitrariness; only one power product can be formed

$$Q = (\text{const}) \frac{r^4}{\mu} \frac{dp}{dx} \qquad \text{Ans.}$$

The constant is dimensionless. Laminar-flow theory shows the value of the constant to be $\pi/8$.

SIMILITUDE

• PROBLEM 8-12

> A submarine-launched missile, 1 m diameter by 5 m long, is to be studied in a water tunnel to determine the loads acting on it during its underwater launch. The maximum speed during this initial part of the missile's flight is 10 m s^{-1}. Calculate the mean water tunnel flow velocity if a 1/20 scale model is to be employed and dynamic similarity is to be achieved.

Solution: For dynamic similarity, the Reynolds number must be constant for the model and the prototype:

$$Re_m = Re_p,$$

$$\bar{v}_m l_m \rho_m / \mu_m = \bar{v}_p l_p \rho_p / \mu_p.$$

The model flow velocity is given by

$$\bar{v}_m = \bar{v}_p (l_p / l_m)(\rho_p/\rho_m)(\mu_m/\mu_p),$$

but $\rho_p = \rho_m$ and $\mu_p = \mu_m$. Therefore,

$$\bar{v}_m = 10 \times 20 \times 1 \times 1 = 200 \text{ m s}^{-1}.$$

This is a high flow velocity and illustrates the reason why a few model tests are made with completely equal Reynolds numbers. At high Re values, however, the divergences become of lesser importance.

A marine research facility uses a towing basin to test models of proposed ship hull configurations. A new hull shape utilizing a bulbous underwater bow is proposed for a nuclear-powered aircraft carrier that is to be 300 m long. A 3-m model has been tested in the towing tank and found to have a maximum practical hull speed of 1.4 m/s. What is the anticipated hull speed for the prototype?

Solution: In the study of ship hulls, surface tension and compressibility effects are not significant. Therefore, for geometrically similar bodies, dynamic similarity occurs when

$$\left(\frac{V^2}{\ell g}\right)_m = \left(\frac{V^2}{\ell g}\right)_p \quad \text{and} \quad \left(\frac{\rho V \ell}{\mu}\right)_m = \left(\frac{\rho V \ell}{\mu}\right)_p$$

Experience has shown that the Froude number is of greater significance than the Reynolds number in this particular application. Thus, the fluid used in the towing tank is generally water; the Froude number alone is maintained between model and prototype; and empirical corrections are made to compensate for the differences that exist between the Reynolds numbers.

Hence we ignore the viscous effects, as measured by the Reynolds number, and concentrate on the hull's wave-making characteristics, as measured by the Froude number.

$$\left(\frac{V^2}{\ell g}\right)_m = \left(\frac{V^2}{\ell g}\right)_p$$

Since the gravitational acceleration is the same for model and prototype, the anticipated prototype velocity becomes

$$V_p = V_m\left(\frac{\ell_p}{\ell_m}\right)^{1/2}$$

or

$$V_p = 1.4(10.0) = 14 \text{ m/s}$$

As a matter of interest, a speed of 14 m/s translates into 27.2 knots, where 1 knot = 1 nautical mile per hour and 1 nautical mile = 6080 feet.

• PROBLEM 8-14

A model of a proposed low-speed crop-dusting plane is to be tested in a wind tunnel capable of maintaining velocities of 264 ft/s, using atmospheric air. The prototype plane has a design wing span of 27 ft and a proposed speed of 88 ft/s. In order to preserve dynamic similarity, what should be the wing span of the model if the wind tunnel is to operate at full speed? Assume the air at these low velocities to be essentially incompressible. In addition, assume that the temperature of atmospheric air and the temperature of the wind-tunnel air are the same.

Solution: For low-speed flows, where viscous and inertia forces predominate, dynamic similarity requires that the Reynolds numbers be the same for both the model and the prototype, and that they be geometrically similar. Thus,

$$\left(\frac{\rho V \ell}{\mu}\right)_p = \left(\frac{\rho V \ell}{\mu}\right)_m \quad \text{and} \quad \left(\frac{b}{\ell}\right)_m = \left(\frac{b}{\ell}\right)_p$$

Here, ℓ is the average chord length and b is the span length. Now we may assume that $p_m \approx p_p$; then we may take $\rho_m \approx \rho_p$, since $T_m = T_p$ (this follows from the ideal gas law). Similarly, $\mu_m = \mu_p$, since $T_m = T_p$. Therefore,

$$b_m = \frac{b_p \ell_m}{\ell_p} = b_p\left(\frac{\rho V}{\mu}\right)_p \left(\frac{\mu}{\rho V}\right)_m = b_p\left(\frac{V_p}{V_m}\right) = \frac{27 \times 88}{264} = 9 \text{ ft}$$

• PROBLEM 8-15

A copepod is a water crustacean approximately 1 mm in diameter. We want to know the drag force on the cope pod when it moves slowly in fresh water. A scale model 100 times larger is made and tested in glycerin at V = 30 cm/s. The measured drag on the model is 1.3 N. For similar conditions, what are the velocity and drag of the actual copepod in water? Assume that the temperature is 20°C.

Solution: The fluid properties are:

Water (prototype): $\mu_p = 0.001$ kg/(m · s)

$\rho_p = 999$ kg/m³

Glycerin (model): $\mu_m = 1.5$ kg/(m · s)

$\rho_m = 1263$ kg/m³

The length scales are $L_m = 100$ mm and $L_p = 1$ mm. We are given enough model data to compute the Reynolds number and force coefficient

$$Re_m = \frac{\rho_m V_m L_m}{\mu_m} = \frac{(1263\ kg/m^3)(0.3\ m/s)(0.1\ m)}{1.5\ kg/(m \cdot s)} = 25.3$$

$$C_{F_m} = \frac{F_m}{\rho_m V_m^2 L_m^2} = \frac{1.3\ N}{(1263\ kg/m^3)(0.3\ m/s)^2(0.1\ m)^2} = 1.14$$

Both these numbers are dimensionless, as you can check. For conditions of similarity, the prototype Reynolds number must be the same, and this requires the prototype force coefficient to be the same.

$$Re_p = Re_m = 25.3 = \frac{999 V_p (0.001)}{0.001}$$

or

$$V_p = 0.0253\ m/s = 2.53\ cm/s \qquad \text{Ans.}$$

$$C_{F_p} = C_{F_m} = 1.14 = \frac{F_p}{999(0.0253)^2(0.001)^2}$$

or

$$F_p = 7.31 \times 10^{-7} N \qquad \text{Ans.}$$

It would obviously be difficult to measure such a tiny drag force.

• PROBLEM 8-16

Model studies of a sea wall are conducted at a scale of 1:25. If the wave period in the prototype is 10 sec, what should be the period of wave generation in the model to insure similarity? To what prototype force per foot of wall would a model wave force of 12 lb/ft correspond?

Solution: According to the Froude criterion for similarity, $V \propto L^{\frac{1}{2}}$. Hence, since $T \propto L/V$, $T \propto L^{\frac{1}{2}}$ and

$$\frac{T_m}{T_p} = \left(\frac{L_m}{L_p}\right)^{\frac{1}{2}} \qquad \text{or} \qquad T_m = 10\left(\frac{1}{25}\right)^{\frac{1}{2}} = 2\ sec$$

Since the Euler numbers must be the same, $F/L \propto pL^2/L \propto V^2 L \propto L^2$, and

$$\frac{(F/L)_p}{(F/L)_m} = \left(\frac{L_p}{L_m}\right)^2 \qquad \text{or} \qquad \left(\frac{F}{L}\right)_p = 12 \times \overline{25}^2 = 7500\ lb/ft$$

• PROBLEM 8-17

Tests are to be made in a wind tunnel on a 1:5 model of a submerged buoy. If the maximum water currents to be expected are 3 fps, what air speed should be used to insure similarity of the flow pattern? To what prototype drag would a model drag of 5.0 lb correspond?

Solution: From property tables of air and water at atmospheric pressure it is found that for an assumed temperature of 60° F $\nu_m/\nu_p = 1.58 \times 10^{-4}/1.21 \times 10^{-5} = 13.1$. Then, for equality of the Reynolds numbers,

$$V_m = V_p \frac{L_p}{L_m} \frac{\nu_m}{\nu_p} = 3 \times 5 \times 13.1 = 196 \text{ fps}$$

With equality of the Euler numbers (since $p \propto \rho V^2$),

$$\frac{F_p}{F_m} = \frac{p_p L_p^2}{p_m L_m^2} = \frac{\rho_p}{\rho_m} \frac{\nu_p^2}{\nu_m^2} \frac{L_p^2}{L_m^2} = \frac{1.94}{0.00237} \times \frac{3^2}{196^2} \times 5 = 4.80$$

Hence,

$$F_p = F_m \times 4.80 = 5.0 \times 4.80 = 24 \text{ lb}$$

• PROBLEM 8-18

The drag of a ship in water is assumed to depend on the Reynolds number and the Froude number so that

$$\frac{\text{drag}}{\frac{1}{2}\rho V^2 A} = f\left(\frac{\rho VD}{\mu}, \frac{V^2}{gD}\right)$$

It is proposed that a model one-tenth the size of a full-scale ship be tested in water and the results be used to predict the performance of the full-scale ship. Is this feasible?

Solution: The prediction of full-scale tests from the model tests employed to determine the form of the drag law requires that Reynolds numbers and Froude numbers of model and prototype be equal. Thus

$$R_c = \frac{\rho V_m D_M}{\mu} = \frac{\rho V_p D_p}{\mu}$$

and

$$F_r = \frac{V_m^2}{gD_m} = \frac{V_p^2}{gD_p}$$

From equality of Reynolds numbers

$$\frac{V_m}{V_p} = \frac{D_p}{D_m} = 10$$

From equality of Froude numbers

$$\frac{V_m}{V_p} = \left(\frac{D_m}{D_p}\right)^{\frac{1}{2}} = \frac{1}{\sqrt{10}}$$

These results contradict one another, and we may conclude that the attainment of dynamic similarity in model and prototype is not possible.

• PROBLEM 8-19

The nondimensional drag on a sphere (the drag coefficient) is a function of the Reynolds number and is expressed as follows,

$$C_d = \frac{\text{drag}}{\frac{1}{2}\rho V^2 A} = f\left(\frac{\rho VD}{\mu}\right)$$

where V is the upstream velocity, A is the cross-sectional area, and D is the sphere diameter. What will be the ratio of the drag force on a sphere in a flow at a given Reynolds number in 70° F air to the drag force on the same sphere in 70° F water at the same Reynolds number?

TABLE 1

Density of Air versus Temperature

Temperature, °F	ρ, lbm/ft^3	ρ, slugs/ft^3 × 10^3
0	0.0862	2.68
10	0.0846	2.63
20	0.0827	2.57
30	0.0811	2.52
40	0.0795	2.47
50	0.0780	2.42
60	0.0764	2.37
70	0.0750	2.33
80	0.0736	2.28
90	0.0722	2.24
100	0.0709	2.20
120	0.0685	2.13
140	0.0663	2.06
150	0.0651	2.02
200	0.0602	1.87
300	0.0523	1.63
400	0.0462	1.44
500	0.0414	1.29

Pressure 14.696 lbf/in.^{2}.

TABLE 2

Density and Saturation Pressure of Water

Temperature, °F	Saturation Pressure, lbf/in.²	Density ρ, lbm/ft³	Density ρ, slugs/ft³
32	0.088	62.42	1.940
40	0.122	62.42	1.940
50	0.178	62.38	1.939
60	0.256	62.34	1.938
70	0.363	62.27	1.935
80	0.507	62.19	1.933
90	0.698	62.11	1.930
100	0.949	62.00	1.927
110	1.275	61.84	1.922
120	1.692	61.73	1.919
140	2.889	61.38	1.908
160	4.741	61.01	1.896
180	7.510	60.57	1.884
200	11.526	60.13	1.869
212	14.696	59.83	1.861

Solution: Equality of Reynolds numbers implies that

$$\frac{\rho_{air} V_{air} D}{\mu_{air}} = \frac{\rho_{H_2O} V_{H_2O} D}{\mu_{H_2O}}$$

or

$$\frac{V_{air}}{V_{H_2O}} = \frac{\nu_{air}}{\nu_{H_2O}}$$

Equality of Reynolds numbers also implies equality of drag coefficients, and we have

$$\frac{(drag)_{air}}{\frac{1}{2}\rho_{air} V^2_{air} A} = \frac{(drag)_{H_2O}}{\frac{1}{2}\rho_{H_2O} V^2_{H_2O} A}$$

Therefore

$$\frac{(drag)_{air}}{(drag)_{H_2O}} = \frac{\rho_{air}}{\rho_{H_2O}} \frac{V^2_{air}}{V^2_{H_2O}} = \frac{\rho_{air}}{\rho_{H_2O}} \left(\frac{\nu_{air}}{\nu_{H_2O}}\right)^2$$

The values of density and kinematic viscosity at 70°F are

$$\rho_{air} = 2.33 \times 10^{-3} \text{ slugs/ft}^3$$

$$\rho_{H_2O} = 1.94 \text{ slugs/ft}^3$$

$$\nu_{air} = 1.64 \times 10^{-4} \text{ft}^2/\text{sec}$$

$$\nu_{H_2O} = 1.05 \times 10^{-5} \text{ft}^2/\text{sec}$$

TABLE 3

Absolute and Kinematic Viscosities of Air at Atmospheric Pressure

°F	$\mu \times 10^7$ lbf-sec/ft²	$\nu \times 10^4$ ft²/sec	°F	$\mu \times 10^7$ lbf-sec/ft²	$\nu \times 10^4$ ft²/sec
0	3.28	1.26	100	3.96	1.89
10	3.45	1.31	120	4.07	1.89
20	3.50	1.36	140	4.14	2.01
30	3.58	1.42	160	4.22	2.12
40	3.62	1.46	180	4.34	2.25
50	3.68	1.52	200	4.49	2.40
60	3.74	1.58	250	4.87	2.80
70	3.82	1.64	300	4.99	3.06
80	3.85	1.69	400	5.40	3.65
90	3.90	1.74	500	5.89	4.56

TABLE 4

Absolute and Kinematic Viscosities of Water

°F	$\mu \times 10^5$ lbf-sec/ft²	$\nu \times 10^5$ ft²/sec	°F	$\mu \times 10^5$ lbf-sec/ft²	$\nu \times 10^5$ ft²/sec
32	3.75	1.93	120	1.17	0.609
35	3.54	1.82	125	1.12	0.582
40	3.23	1.66	130	1.08	0.562
45	2.97	1.53	135	1.02	0.534
50	2.73	1.41	140	0.981	0.514
55	2.53	1.30	145	0.940	0.493
60	2.35	1.22	150	0.899	0.472
65	2.24	1.13	155	0.868	0.457
70	2.04	1.05	160	0.837	0.440
75	1.92	0.988	165	0.806	0.426
80	1.80	0.929	170	0.776	0.411
85	1.68	0.870	175	0.750	0.397
90	1.60	0.825	180	0.725	0.384
95	1.51	0.782	185	0.701	0.372
100	1.43	0.738	190	0.679	0.362
105	1.35	0.698	195	0.657	0.351
110	1.29	0.668	200	0.637	0.341
115	1.23	0.637	212	0.593	0.318

Therefore

$$\frac{(\text{drag})_{\text{air}}}{(\text{drag})_{H_2O}} = \frac{2.33 \times 10^{-3}}{1.94} \times \left(\frac{1.64 \times 10^{-4}}{1.05 \times 10^{-5}}\right)^2 = 2.92 \times 10^{-1}$$

• PROBLEM 8-20

A gravity fed lock in a proposed shipping canal is to be studied with a 1/100-scale geometrically similar model. (a) If the model lock fills in 0.8 min, estimate the time for the prototype to fill. (b) Find the ratio of corresponding mass flow rates in the model and prototype locks.

Solution: (a) We will assume that sufficient similitude is maintained by preserving constant Froude numbers for the model and the prototype. Hence we ignore the viscous effects, as measured by the Reynolds number, and concentrate on the hull's wave-making characteristics, as measured by the Froude number.

$$\left(\frac{V^2}{\ell g}\right)_m = \left(\frac{V^2}{\ell g}\right)_p$$

Since the gravitational acceleration is the same for model and prototype, the anticipated prototype velocity becomes

$$V_p = V_m\left(\frac{\ell_p}{\ell_m}\right)^{1/2}$$

Let the geometric ratio ℓ_p/ℓ_m be designated as λ (lambda). Then

$$V_p = V_m(\lambda)^{1/2}$$

Fill time in the model studies is proportional to the ratio $(\ell^3/Q)_m = (\ell^3/AV)_m$, while time in the prototype is proportional to $(\ell^3/AV)_p$. Since $A_m/A_p = \ell_m^2/\ell_p^2$, the ratio of the fill times, t_p/t_m, is given by

$$\frac{t_p}{t_m} = \left(\frac{\ell}{V}\right)_p\left(\frac{V}{\ell}\right)_m = \frac{\lambda}{\sqrt{\lambda}} = \sqrt{\lambda}$$

Therefore, $t_p = t_m\sqrt{\lambda}$, or for this specific case,

$$t_p = 0.80\sqrt{100} = 8 \text{ min}$$

(b) The mass flow rates can be represented as the product of appropriate densities, velocities, and areas. Thus,

$$\frac{\dot{m}_p}{\dot{m}_m} = \frac{(\rho VA)_p}{(\rho VA)_m}$$

or

$$\frac{\dot{m}_p}{\dot{m}_m} = \frac{\rho_p}{\rho_m} \times \sqrt{\lambda} \times \lambda^2 = \lambda^{5/2}$$

since $\rho_p/\rho_m = 1.0$. (The fluid is water in each case.)

• PROBLEM 8-21

A one-fifth scale model of an airplane is tested in (a) a wind tunnel, and (b) a water tunnel. Caculate the tunnel speeds required to correspond to a full-scale speed of 100 fps at sea level.

TABLE 1

Liquid	Sp gr	$\mu \times 10^5$ lb-sec/ft^2
Alcohol	0.8	2.4
Benzine	0.88	1.36
Gasoline	0.8	0.63
Glycerine	1.3	1,800
Linseed oil	0.95	90
Mercury	13.55	3.24
Olive oil	0.9	175
Turpentine	0.85	3.11
Water, fresh	1.0	2.09
Water, sea	1.03	3.2

TABLE 2

Gas	ρ, slug/ft^3	$\mu \times 10^5$ lb-sec/ft^2
Air	0.00238	0.0378
Carbon dioxide	0.00363	0.0305
Hydrogen	0.000166	0.0184
Nitrogen	0.00229	0.0363
Oxygen	0.00262	0.0417

Solution:

$$R_{e_{model}} = R_{e_{full\ scale}}$$

Therefore

$$\frac{v_m L_m}{\nu_m} = \frac{v_f L_f}{\nu_f}$$

or

$$v_m = v_f \frac{\nu_m}{\nu_f} \frac{L_f}{L_m}$$

$$= 5\frac{\nu_m}{\nu_f} v_f = 500 \frac{\mu_m}{\mu_f} \frac{\rho_f}{\rho_m}$$

(a) In the wind tunnel, $\mu_m = \mu_f$ and $\rho_m = \rho_f$, and so

$$v_m = 500 \text{ fps}$$

(b) In the water tunnel,

$\mu_m = 2.1 \times 10^{-5}$ lb-sec/ft^2

$\mu_f = 0.0377 \times 10^{-5}$ lb-sec/ft^2

$\rho_f = 0.00238$ slug/ft^3

$\rho_m = 1.94$ slugs/ft^3

Therefore

$$v_m = 500 \frac{2.1}{0.0377} \frac{0.00238}{1.94}$$

$$= 34.2 \text{ fps}$$

• PROBLEM 8-22

A certain submerged body is to move horizontally through oil (γ = 52 lb/ft^3, μ = 0.0006 lb·s/ft^2) at a velocity of 45 fps. To study the characteristics of this motion, an enlarged model of the body is tested in 60°F water. The model ratio λ is 8: 1. Determine the velocity at which this enlarged model should be pulled through the water to achieve dynamic similarity. If the drag force on the model is 0.80 lb, predict the drag force on the prototype. Body is submerged, hence there is no wave action. Reynolds criterion must be satisfied. ρ(water) = 1.94 slugs/ft^3.

TABLE

Temp, F	Specific weight γ, lb/ft^3	Density ρ, slugs/ft^3	Viscosity $\mu \times 10^5$, lb.s/ft^2	Kinematic viscosity $\nu \times 10^5$, ft^2/s
32	62.42	1.940	3.746	1.931
40	62.43	1.940	3.229	1.664
50	62.41	1.940	2.735	1.410
60	62.37	1.938	2.359	1.217
70	62.30	1.936	2.050	1.059
80	62.22	1.934	1.799	0.930
90	62.11	1.931	1.595	0.826
100	62.00	1.927	1.424	0.739
110	61.86	1.923	1.284	0.667
120	61.71	1.918	1.168	0.609
130	61.55	1.913	1.069	0.558
140	61.38	1.908	0.981	0.514
150	61.20	1.902	0.905	0.476
160	61.00	1.896	0.838	0.442
170	60.80	1.890	0.780	0.413
180	60.58	1.883	0.726	0,385
190	60.36	1.876	0.678	0.362
200	60.12	1.868	0.637	0.341
212	59.83	1.860	0.593	0.319

Solution: For Reynolds criterion to be satisfied,

$$Re_p = Re_m$$

$$\left(\frac{DV}{\nu}\right)_p = \left(\frac{DV}{\nu}\right)_m \quad \text{where} \quad \frac{D_m}{D_p} = \frac{8}{1}$$

$$\nu_m = 1.22 \times 10^{-5}\ \text{lb}\cdot\text{s/ft}^2 \text{ (see Table)}$$

$$\nu_p = \frac{\mu}{\rho} = \frac{0.0006}{52/32.2} = 0.000322\ \text{lb}\cdot\text{s/ft}^2$$

$$\frac{D_p(45)}{0.000322} = \frac{(8D_p)V_m}{1.22 \times 10^{-5}}$$

$$V_m = 0.213 \text{ fps}$$

$$F \propto \rho V^2 L^2 \qquad \text{Hence} \qquad \frac{F_p}{F_m} = \frac{\rho_p V_p^2 L_p^2}{\rho_m V_m^2 L_m^2}$$

$$\frac{F_p}{F_m} = \frac{(52/32.2)(45)^2 1}{1.94(0.213)^2(8)^2} = 580$$

$$F_p = 380F_m \qquad F_p = 580(0.8) = 465 \text{ lb}$$

• PROBLEM 8-23

A smooth streamline-shaped body 3 m long is to be towed at 10 m/s through atmospheric air. Design a model test in water to predict the drag of the prototype. Assume viscosity of air and water = 1.8×10^{-5} N·s/m² and .001 N·s/m², respectively, and density of air and water = .075 lb/ft³ and 62.4 lb/ft³, respectively.

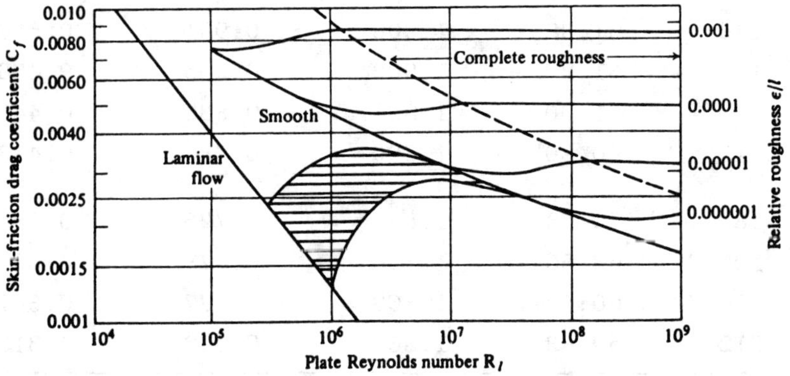

Solution: If the prototype and model flows are dynamically similar, the drag coefficients for both will be the same. For dynamic similarity, the model and prototype must be geometrically similar, and the Reynolds number of both must be the same

$$\frac{L_m V_{\infty m} \rho_m}{\mu_m} = \frac{_p V_{\infty p} \rho_p}{\mu_p}$$

where subscripts m and p refer to the model and prototype. Then

$$L_m V_{\infty m} = \frac{\mu_m}{\mu_p} \frac{\rho_p}{\rho_m} \ell_p V_{\infty p} = \frac{0.001 \text{ N·s/m}^2}{0.000018 \text{ N·s/m}^2} \frac{0.075 \text{ lb/ft}^3}{62.4 \text{ lb/ft}^3} (3\text{m})(10\text{m/s})$$

$$= 2\text{m}^2/\text{s}$$

Dynamic similarity is achieved if the product LV_∞ for the model is 2 m^2/s. A model 50 cm long tested in a 4 m/s streaming flow of water is one possible combination assuring dynamic similarity.

The Reynolds number for this flow is about 2 x 10^6. The figure shows that large surface roughness is required to promote complete turbulence at this rather low Reynolds number. Thus, because the prototype's surface is smooth, turbulence is still developing and the requirement for dynamic similarity of strict matching of Reynolds number should not be relaxed. Also, the relative roughness of the prototype and model should be matched. Since the prototype surface is smooth, a highly polished model surface is required to match the relative roughness of the larger prototype.

Finally, the drag of the prototype is obtained from the measured drag of the model by equating the drag coefficients. Since area varies as the square of the linear dimension of the body L,

$$D_p = \left[\frac{\rho_p}{\rho_m}\left(\frac{V_{\infty p}}{V_{\infty m}}\right)^2 \left(\frac{L_p}{L_m}\right)^2\right] D_m$$

Substituting numerical values, we have

$$D_p = \left[\frac{0.075 \text{ lb/ft}^3}{62.4 \text{ lb/ft}^3}\left(\frac{10 \text{ m/s}}{4 \text{ m/s}}\right)^2 \left(\frac{3 \text{ m}}{0.5 \text{ m}}\right)^2\right] D_m = 0.270 \, D_m$$

The drag of the prototype is only about one-quarter that of the model. This is caused by testing the model in water, which has such a high density compared to air.

Two drag-measurement tests are run on the body of circular cross section shown in the Figure. The results are presented in the Table. Estimate the drag of a geometrically similar body 7.5 cm in diameter moving at 0.5 m/s through SG = 0.68 gasoline at 10°C.

Given data:

$p = 101{,}000\ N/m^2$

$m_{gas} = 28.97$ kg/kg-mol

fluid	$\rho(kg/m^3)$	$\mu\ (N\cdot s/m^2)$
air	1.14	2×10^{-5}
water	1000	4.3×10^{-4}
gasoline	680	3.5×10^{-4}

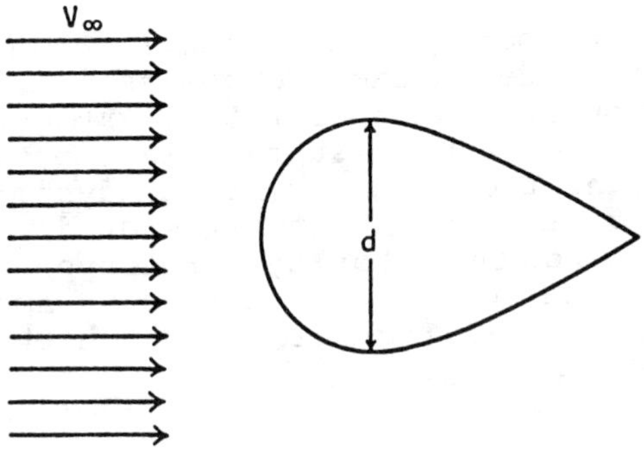

Solution: The body is a bluff shape, and so the drag is primarily pressure drag caused by boundary-layer separation. Furthermore, the body is well rounded with no sharp corners. The location of boundary-layer separation and the magnitude of the drag depend on whether the flow is laminar or turbulent.

First we calculate the Reynolds numbers of the test data. The density of the airflow is

$$\rho_{air} = \frac{p}{(R/m)T} = \frac{(101{,}000\ N/m^2)(28.97\ kg/kg\ mol)}{(8314\ N\cdot m/kg\ mol \cdot K)(273.16 + 35)K}$$

$$= 1.14\ kg/m^3$$

$$R_{d_{air}} = \frac{(0.05\ m)(30\ m/s)(1.14\ kg/m^3)}{0.00002\ N\cdot s/m^2}\ \frac{1N}{1\ kg\cdot m/s^2}$$

$$= 8.55 \times 10^4$$

TABLE

Fluid	Velocity V_∞, m/s	Body Diameter d, cm	Drag Force D, N
Air, 35 °C, atm pressure	30	5	0.575
Water, 65 °C	3	25	62.5

The Reynolds number of the water flow is

$$R_{d_{water}} = \frac{(0.25\ \text{m})(3\ \text{m/s})(1000\ \text{kg/m}^3)}{0.00043\ \text{N}\cdot\text{s/m}^2}\ \frac{1\text{N}}{1\ \text{kg}\cdot\text{m/s}^2}$$

$$= 1.74 \times 10^6$$

Transition from laminar to turbulent flow for bodies of various shapes occurs in the Reynolds-number range of about 10^5 to 10^6. The airflow is laminar, and the water flow is turbulent. We calculate the Reynolds number of the gasoline flow

$$R_{d_{gas}} = \frac{(0.075\ \text{m})(0.5\ \text{m/s})(680\ \text{kg/m}^3)}{0.00035\ \text{N}\cdot\text{s/m}^2}\ \frac{1\ \text{N}}{1\ \text{kg}\cdot\text{m/s}^2}$$

$$= 7.3 \times 10^4$$

The gasoline flow is laminar, like the airflow. The drag coefficient of a rounded body is fairly constant over a wide range of Reynolds number for laminar flow. The air and gasoline flows should have nearly the same drag coefficient. We calculate the drag coefficient for the airflow

$$C_D = \frac{(0.575\ \text{N})\ \dfrac{1\ \text{kg}\cdot\text{m/s}^2}{1\ \text{N}}}{\frac{1}{2}(1.14\ \text{kg/m}^3)(30\ \text{m/s})^2(\pi/4)(0.05\ \text{m})^2} = 0.571$$

Then, using this drag coefficient, we calculate the drag in the gasoline flow

$$D = 0.571\left(\frac{1}{2}\right)(680\ \text{kg/m}^3)(0.5\ \text{m/s})^2\ \frac{\pi}{4}\ (0.075\ \text{m})^2\ \frac{1\ \text{N}}{1\ \text{kg}\cdot\text{m/s}^2}$$

$$= 0.214\text{N}$$

A model arrangement is set up to determine the drag on an actual sphere 10" in diameter when immersed in sea water. The relative velocity between the sphere and the water is 4 knots. A wind tunnel is used for the model arrangement and the model sphere diameter is to be only 3". During the test procedure the drag on the model sphere is measured as 4.0 lbf.

(a) Find the velocity of the air blown through the wind tunnel, and

(b) the drag on the actual sphere in the sea water.

Solution: (a) To simulate dynamic conditions in the wind tunnel that are similar to the conditions of the actual sphere in sea water, the Reynolds number for flow conditions in the sea water is set equal to Reynolds number for flow conditions in the wind tunnel. Accordingly,

$$Re_{sea} = Re_{tunnel} \tag{1}$$

To compute the Reynolds number for the actual sphere in sea water,

$$\rho_{sea} = 1.98 \text{ slug/ft}^3$$

$$\nu_{sea} = 1.4 \times 10^{-5} \text{ ft}^2/\text{sec}$$

$$V_{sea} = 4 \text{ knots} \times \frac{6080 \text{ ft}}{\text{knot}} \times \frac{\text{hr}}{3600 \text{ sec}} = 6.76 \frac{\text{ft}}{\text{sec}}$$

Therefore the Reynolds number for flow conditions of the sphere immersed in sea water is

$$Re_{sea} = \left.\frac{VD}{\nu}\right]_{sea} = 6.76 \frac{\text{ft}}{\text{sec}} \times 10 \text{ in} \times \frac{\text{ft}}{12 \text{ in}}$$

$$\times \frac{1 \text{ sec}}{1.4 \times 10^{-5} \text{ ft}^2} = 4.02 \times 10^5$$

The only unknown in the Reynolds number for the flow in the tunnel is the velocity of the air. This velocity can be calculated from the relationship (1) that the Reynolds numbers are equal.

$$V_{tunnel} = \left.\frac{Re\nu}{D}\right]_{tunnel}$$

Since $\rho_{air} = .00238 \text{ slug/ft}^3$

and

$$\nu_{air} = 1.56 \times 10^{-4} \text{ ft}^2/\text{sec}$$

$$V_{tunnel} = 4.02 \times 10^5 \times 1.56 \times 10^{-4} \frac{\text{ft}^2}{\text{sec}} \times \frac{1}{3 \text{ in}} \times \frac{12 \text{ in}}{\text{ft}}$$

$$= 250.8 \text{ ft/sec}$$

(b) The drag, D, on the actual sphere and the model is given by

$$D = \left.C_D \rho \frac{V^2}{2} (\text{area})\right]_{sea}$$

$$D = \left.C_D \rho \frac{V^2}{2} (\text{area})\right]_{tunnel}$$

The drag coefficient is a function of the Reynolds number, and since these numbers are equal from (1), C_D's for the sea and tunnel flows are equal. Since we know the drag on the model, D for the sea conditions may be obtained from the ratios,

$$\frac{D_t}{D_s} = \frac{\rho_t V_t^2 d_t^2}{\rho_s V_s^2 d_s^2}$$

where the subscripts t and s designate tunnel and sea flow conditions, respectively. Therefore,

$$D_s = \frac{\rho_s V_s^2 d_s^2}{\rho_t V_t^2 d_t^2} D_t$$

$$D_s = \frac{1.9}{.00238} \times \left(\frac{6.76 \times 10}{250.8 \times 3}\right)^2 \times 4.0 \text{ lbf}$$

$$= 25.8 \text{ lbf}$$

A surface ship is 100 ft long. Design a towing-tank test in water using a 3-ft-long model to predict the power required to propel the prototype at 20 knots.

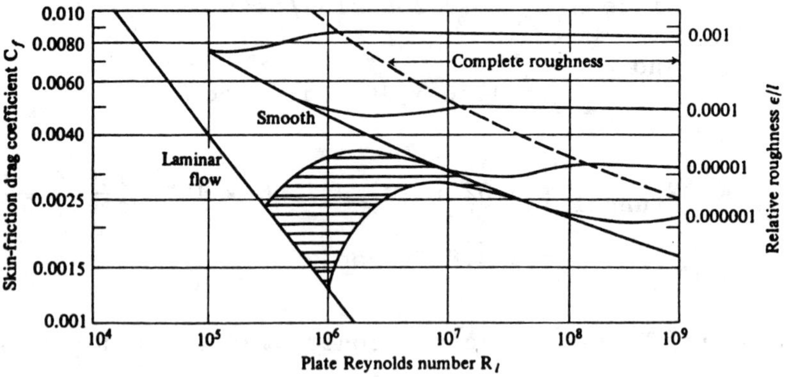

Solution: The drag of a surface ship is due predominantly to the generation of gravity waves on the water surface by the ship's motion. The viscous drag is primarily skin friction with very little boundary-layer separation. Thus, Froude-number matching is the primary criterion for dynamic similarity.

The submerged portion of the model must be geometrically similar to that of the prototype. Matching Froude numbers between model and prototype

$$\frac{V_m}{\sqrt{g l_m}} \quad \frac{V_p}{\sqrt{g l_p}}$$

results in a required model speed of

$$V_m = \sqrt{\frac{l_m}{l_p}}\, V_p = \sqrt{\frac{3\text{ ft}}{100\text{ ft}}}\ (20\text{ knots})\ \frac{1.689\text{ ft/s}}{1\text{ knot}} = 5.85\text{ ft/s}$$

Dynamic similarity requires equality of all force coefficients

$$\frac{D_m}{\frac{1}{2}\rho V_m^2 A_m} = \frac{D_p}{\frac{1}{2}\rho V_p^2 A_p}$$

The area varies as the square of the linear dimension. Then

$$D_p = \left(\frac{V_p}{V_m}\right)^2 \left(\frac{l_p}{l_m}\right)^2 D_m$$

but Froude-number matching requires

$$\frac{V_p}{V_m} = \sqrt{\frac{l_p}{l_m}}$$

so that

$$D_p = \frac{l_p}{l_m} \left(\frac{l_p}{l_m}\right)^2 D_m = \left(\frac{l_p}{l_m}\right)^3 D_m$$

In Froude-number matching, forces scale as the cube of the body size.

The power required to propel the prototype is

$$P = D_p V_p = \left(\frac{l_p}{l_m}\right)^3 D_m \sqrt{\frac{l_p}{V_m}}\, F_m = \left(\frac{l_p}{l_m}\right)^{7/2} V_m D_m$$

In Froude-number matching, power scales as the $\frac{7}{2}$ power of the body size. Substituting numerical values, for power in the horsepower we get

$$P\,(\text{hp}) = \left(\frac{100\ \text{ft}}{3\ \text{ft}}\right)^{7/2} \frac{(5.85\ \text{ft/s})(D_m\ \text{lbf})}{\dfrac{550\ \text{ft lbf/s}}{1\ \text{hp}}} = 2275\,(D_m\ \text{lbf})$$

The power required to propel the prototype at 20 knots is obtained by substituting the measured drag from the model test in the above equation.

Finally, let us see how the Reynolds numbers of model and prototype compare under Froude-number matching

$$\frac{R_m}{R_p} = \frac{l_m}{l_p}\frac{V_m}{V_p} = \frac{l_m}{l_p}\sqrt{\frac{l_m}{l_p}} = \left(\frac{l_m}{l_p}\right)^{3/2}$$

The Reynolds number in Froude-number matching scale as the $\frac{3}{2}$ power of the body size.

The Reynolds number of the prototype is

$$R_{l_p} = \frac{l_p V_p \rho}{\mu} = \frac{(100 \text{ ft})(20 \text{ knots})(62.4 \text{ lb/ft}^3)\ \dfrac{1.689 \text{ ft/s}}{1 \text{ knot}}}{(0.000023 \text{ lbf} \cdot \text{s/ft}^2)(32.17 \text{ lb} \cdot \text{ft/lbf s}^2)}$$

$$= 2.85 \times 10^8$$

The Reynolds number of the model flow is

$$R_{l_m} = \left(\frac{l_m}{l_p}\right)^{3/2} R_{l_p} = \left(\frac{3 \text{ ft}}{100 \text{ ft}}\right)^{3/2} (2.85 \times 10^8) = 1.48 \times 10^6$$

The figure shows that the prototype flow is turbulent but the model's flow is close to transition. It may be desirable to roughen the model's surface to promote turbulence to simulate the prototype flow. Once again, we are in the realm of experience--almost art--in knowing how best to proceed in considering Reynolds-number effects. If the model's flow is turbulent, its skin-friction drag coefficient will be greater than that of the prototype, making the power prediction for the prototype conservative.

• PROBLEM 8-27

A model pump has an impeller of 6 in. diameter. When running at 1200 rpm, the pump delivers 2 cfs against a head of 16 ft.

A similar pump is required to discharge 40 cfs at 600 rpm. What should be the diameter of this pump, and what head will it develop?

Solution: For a centrifugal pump, the peripheral velocity is given by

$$u = \frac{\pi D N}{60}$$

If follows that

$$u \propto DN$$

or

$$D \propto \frac{u}{N} \tag{1}$$

Since the reference quantity to which velocities are related is the square root of the head generated by the impeller, then assuming the theoretical head

$$u \propto \sqrt{H}$$

and substituting for u from this proportionality in eq. (1)

$$D \propto \frac{\sqrt{H}}{N} \tag{2}$$

The flow through the impeller is given by

$$Q = \pi D b v_f$$

and since $v_f \propto \sqrt{H}$, and $b \propto D$, where D is the reference linear dimension, then

$$Q \propto D^2\sqrt{H} \tag{3}$$

Using small letters for the model pump and capitals for the larger pump, then from eq. (2)

$$d = k\,\frac{\sqrt{h}}{n} \tag{4}$$

and

$$D = k\,\frac{\sqrt{H}}{N} \tag{5}$$

Substituting for the proportionality constant k from (4) in (5)

$$D = d\left(\frac{n}{N}\right)\frac{\sqrt{H}}{\sqrt{h}} = \left(\frac{6}{12}\right)\left(\frac{1200}{600}\right)\frac{\sqrt{H}}{\sqrt{16}}$$

$$D = 0.25\,\sqrt{H} \tag{6}$$

Similarly from eq. (3)

$$q = k'd^2\sqrt{h}$$

$$Q = k'D^2\sqrt{H}$$

where k' is another constant.

$$\frac{Q}{q} = \frac{D^2}{d^2}\sqrt{\frac{H}{h}}$$

Putting the data in this equation

$$\frac{40}{2} = \frac{D^2\sqrt{H}}{(6/12)^2\sqrt{16}}$$

$$D^2\sqrt{H} = 20$$

Substituting for $\sqrt{H} = 20/D^2$ from this equation in eq. (6)

$$D = (0.25)(20/D^2)$$

$$D^3 = 5$$

$$D = 1.71 \text{ ft}$$

Substituting this value in eq. (6)

$$1.71 = 0.25\sqrt{H}$$

from which

$$H = 46.8 \text{ ft}$$

• PROBLEM 8-28

A model of an aeroplane built $\frac{1}{10}$th scale is to be tested in a wind tunnel which operates at a pressure of 20 atmospheres. The aeroplane is expected to fly at a speed of 500 km/h. At what speed should the wind tunnel operate to give dynamic similarity between model and prototype. The drag measured on the model is 337.5 newtons. What power will be needed to propel the aircraft at 500 km/h.

Solution: A relationship between a set of physical variables can be stated as a relationship between a set of independent dimensionless groups made up from suitably chosen variables.

If the variables selected are ρ, v, l and μ the resulting group will be the ratio of inertia force to viscous force, that is Reynolds number. Thus

$$\pi_1 = \rho^a v^b l^c R$$

where R represents the resistance to motion.

The second π group can be formed similarly, the repeating variables, ρ, v, l , being combined with the other variable μ

$$\pi_2 = \rho^a v^b l^c \mu$$

The method of processing these π groups can now be demonstrated. Each group must be dimensionless, thus for π_1

$$M^a L^{-3a} L^b T^{-b} L^c MLT^{-2} = |0|$$

Equating indices of M, L and T respectively to zero gives

$$M: a + 1 = 0$$

$$T: -b-2 = 0$$

$$L: -3a + b + c + 1 = 0$$

Thus $a = -1$, $b = -2$ and $c = -2$. Hence

$$\pi_1 = \rho^{-1} v^{-2} l^{-2} R$$

Similarly

$$\pi_2 = \rho^a v^b l^c \mu$$

Thus

$$M^a L^{-3a} L^b T^{-b} L^c ML^{-1} T^{-1} = |0|$$

$$M: a + 1 = 0$$

$$T: -b-1 = 0$$

$$L: -3a + b + c - 1 = 0$$

Thus $a = -1$, $b = -1$ and $c = -1$,

so

$$\pi_2 = \rho^{-1} v^{-1} l^{-1} \mu = \mu/(\rho v l)$$

In most problems, the value of $\mu/\rho vL$ is very small. This is inconvenient and the reciprocal of this number is

used instead. This is of course quite valid as it is also dimensionless. By substituting these expressions into the equation resulting from the Buckingham Π theorem,

$$f(\Pi_1 , \Pi_2) = 0$$

$$f\left(\frac{R}{\rho v^2 l^2}, \frac{\rho v L}{\mu}\right) = 0$$

or

$$\frac{R}{\rho v^2 l^2} = f\left(\frac{\rho v l}{\mu}\right)$$

or

$$R = \rho v^2 l^2 f(\rho v\, l/\mu) = \rho v^2 l^2 f(Re)$$

This last form is called the Rayleigh equation.

For dynamic similarity

$$Re_m = Re_p$$

$$\frac{\rho_m v_m l_m}{\mu_m} = \frac{\rho_p v_p l_p}{\mu_p}$$

$$\therefore \qquad v_m = \frac{\rho_p}{\rho_m}\frac{\mu_m}{\mu_p}\frac{l_p}{l_m} v_p$$

Now the coefficient of dynamic viscosity is not significantly altered by pressure changes unless the pressure change is large, so:

$$\mu_m = \mu_p$$

Due to the compressibility of air,

$$\rho_m = 20\ \rho_p$$

$$v_m = \frac{1}{20} \times \frac{1}{1} \times 10 \times 500 = 250 \text{ km/hr}$$

Using the Rayleigh equation and

$$Re_m = Re_p ,$$

$$\frac{R_p}{R_m} = \frac{\rho_p v_p^2 l_p^2 f(Re_p)}{\rho_m v_m^2 l_m^2 f(Re_m)}$$

$$\frac{R_p}{R_m} = \frac{\rho_p}{\rho_m}\left(\frac{v_p}{v_m}\right)^2\left(\frac{l_p}{l_m}\right)^2$$

$$R_p/R_m = \frac{1}{20} \times 4 \times 10^2$$

$$R_p = 20\ R_m$$

Power needed to propel the aircraft

$$P = v_p\ R_p$$

$$= 500\ \frac{km}{hr}\left(1000\ \frac{m}{km}\right)\frac{1\ hr}{3600\ sec} \times 20\ R_m$$

$$= \frac{5000}{36}\ \frac{m}{s} \times 20\ (337.5N)\ \frac{1\ kN}{1000\ N}$$

$$P = 938\ kW$$

CHAPTER 9

POTENTIAL AND VORTEX FLOW

Basic Attacks and Strategies for Solving Problems in this Chapter. See pages 486 to 562 for step-by-step solutions to problems.

Mathematically, any vector whose curl is zero can be expressed as the gradient of a scalar function. In fluid mechanics, this identity can be applied to velocity. If a fluid flow is irrotational,

$$\text{vorticity} = \text{curl}\,(\mathbf{V}) = \nabla \times \mathbf{V} = 0,$$

and hence velocity can be written as the gradient of some scalar function, ϕ, called the velocity potential function, i.e.,

$$\mathbf{V} = \nabla\phi. \tag{1}$$

Other than the irrotationality condition (zero vorticity), no other restrictions are imposed on Equation (1). The flow can be steady or unsteady, viscous or inviscid, two- or three-dimensional, although the presence of viscosity in a fluid always leads to non-zero vorticity (and therefore rotationality) near solid walls. Far away from solid walls (i.e., outside the viscous boundary layer), the flow may generally be considered potential.

A companion function to the velocity potential can be defined for two-dimensional flows. From the differential continuity equation in Cartesian coordinates, stream function ψ is defined as

$$u = \frac{\partial\psi}{\partial y}, \quad v = -\frac{\partial\psi}{\partial x} \tag{2}$$

with similar definitions in cylindrical coordinates. ψ is more physically appealing than ϕ since ψ turns out to be constant along a streamline. In a two-dimensional potential flow, ψ and ϕ are closely related since Equations (1) and (2) must define the same velocity field. In fact, it turns out that ψ and ϕ are mutually orthogonal, and one can be found from the other by first finding velocity and then integrating, using Equations (1) and (2), or vice-versa.

The incompressible continuity equation $\nabla \cdot \mathbf{V} = 0$ and the irrotationality condition $\nabla \times \mathbf{V} = 0$ can be combined with Equations (1) and (2) to yield

$$\nabla^2 \phi = 0 \quad \text{and} \quad \nabla^2 \psi = 0$$

respectively for an irrotational two-dimensional incompressible flowfield. Since both of these Laplace equations are *linear*, the very convenient method of *superposition* may be applied to either ψ or ϕ. In particular, some fundamental plane flow solutions can be defined and then simply added together to simulate more complex flows. The three fundamental "building block" potential flows are:

1) **Uniform Stream** $\psi = U_\infty y,\ \phi = U_\infty x$

2) **Line Source** (or sink) $\psi = m\theta,\ \phi = m \ln r$

3) **Line Vortex** $\psi = -K \ln r,\ \phi = K\theta$

Note that ψ, ϕ, and $\mathbf{V}$ can be added (superposed), but pressure must be obtained from Bernoulli's equation and *cannot* be superposed because of the V^2 term in the equation. To find p, first obtain $\mathbf{V}$ for the entire (superposed) flowfield, and then apply Bernoulli's equation.

The line vortex and line source satisfy irrotationality everywhere except at the origin where there is a mathematical singularity. For the vortex, a line integration of tangential velocity surrounding the vortex center around a closed counterclockwise curve C is defined as the circulation, Γ:

$$\Gamma = \int_C \mathbf{V} \cdot d\mathbf{s}, \tag{3}$$

where ds is the differential arc length along the closed curve C. It turns out that Γ is constant for *any* closed curve C surrounding the vortex center, and $\Gamma = 2\pi K$ where K is the vortex strength.

Step-by-Step Solutions to Problems in this Chapter, "Potential and Vortex Flow"

POTENTIAL FLOW

• PROBLEM 9-1

Let $\phi = x^2 - y^2$ be the potential functions for an irrotational flow. Find the velocity components as functions of x, y, and z.

Solution: A potential flow is a flow for which the velocity $\vec{V}$ may be defined as the gradient of some scalar function ϕ (x,y,z,t) called the potential function, that is,

$$\vec{V} = \nabla\phi$$

There are no other restrictions for whether the flow is steady or unsteady, viscous or nonviscous, etc.

From $\vec{V} = \nabla\phi$

$$\hat{i}V_x + \hat{j}V_y + \hat{k}V_z = i\frac{\partial\phi}{\partial x} + j\frac{\partial\phi}{\partial y} + k\frac{\partial\phi}{\partial z}$$

Therefore

$$V_x = \frac{\partial\phi}{\partial x} = 2x$$

$$V_y = \frac{\partial\phi}{\partial y} = -2y$$

$$V_z = \frac{\partial\phi}{\partial z} = 0$$

• PROBLEM 9-2

The stream function for a particular flow is given as

$$\psi(x,y) = x^2 - y^2 .$$

Is this flow irrotational? If so, calculate the velocity potential.

Solution: The velocity components are

$$u = \frac{\partial\psi}{\partial y} = -2y$$

and

$$v = -\frac{\partial\psi}{\partial x} = -2x$$

The vorticity components are then

$$\xi = 0, \quad \eta = 0, \quad \zeta = -2 + 2 = 0$$

The flow is irrotational, since all vorticity components are zero. A particle would not rotate, it would only deform.

The velocity potential is found as follows. From the first equation,

$$u = \frac{\partial\phi}{\partial x} = -2y$$

so that

$$\phi = -2xy + f(y)$$

Differentiating the above with respect to y gives

$$\frac{\partial\phi}{\partial y} = -2x + \frac{\partial f}{\partial y}$$

Equating this to $v = -2x$ gives $\partial f/\partial y = C$ where C is a constant. The velocity potential is then

$$\phi = -2xy + C$$

The constant C is not important since it does not affect the velocity or pressure fields. Hence, it is often set equal to zero.

• PROBLEM 9-3

A flow field is represented by the potential function

$$\phi = r^{\frac{1}{2}}\cos\frac{\theta}{2}$$

(a) Determine the corresponding stream function.

(b) Check to see if the flow is incompressible.

(c) Compute the radial and tangential components of acceleration of a particle moving in the flow field.

Solution: (a) $\vec{V} = -\nabla\phi = -\hat{\imath}_r\frac{\partial\phi}{\partial r} - \hat{\imath}_\theta\frac{1}{r}\frac{\partial\phi}{\partial\theta}$ $\left[\because \text{ in cylindrical coordinate } \nabla = \hat{\imath}_r\frac{\partial}{\partial r} + \hat{\imath}_\theta\frac{\partial}{\partial\theta}\right]$

$$= \hat{\imath}_r\left(-\frac{1}{2}r^{-\frac{1}{2}}\cos\frac{\theta}{2}\right) + \hat{\imath}_\theta\left(\frac{1}{2}r^{-\frac{1}{2}}\sin\frac{\theta}{2}\right)$$

and

$$V_r = -\frac{1}{2}r^{-\frac{1}{2}}\cos\frac{\theta}{2}, \qquad V_\theta = +\frac{1}{2}r^{-\frac{1}{2}}\sin\frac{\theta}{2}$$

From the definition of the stream function, ψ

$$\begin{Bmatrix} V_r = \frac{1}{r}\frac{\partial\psi}{\partial\theta} \\ V_\theta = -\frac{\partial\psi}{\partial r} \end{Bmatrix} \Rightarrow \begin{Bmatrix} \psi = r\int V_r d\theta + C_1 \\ \psi = -\int V_\theta dr + C_2 \end{Bmatrix} \Rightarrow$$

$$\begin{Bmatrix} \psi = r\int\left(-\frac{1}{2}r^{-\frac{1}{2}}\cos\frac{\theta}{2}\right)d\theta + C_1 \\ \psi = \int\left(-\frac{1}{2}r^{-\frac{1}{2}}\sin\frac{\theta}{2}\right)dr + C_2 \end{Bmatrix} \Rightarrow \begin{Bmatrix} \psi = -r^{\frac{1}{2}}\sin\frac{\theta}{2} + C_1 \\ \psi = -r^{\frac{1}{2}}\sin\frac{\theta}{2} + C_2 \end{Bmatrix}$$

Comparing these two expressions, we see that $C_1 = C_2$.

Since these constants do not affect so much on the velocity or pressure fields, therefore they can be neglected. Therefore the stream function is

$$\psi = -r^{\frac{1}{2}}\sin\frac{\theta}{2}$$

(b) The flow is incompressible if

or
$$\frac{\partial(rV_r)}{\partial r} + \frac{\partial V_\theta}{\partial\theta} = 0$$

$$\frac{\partial\left(-\frac{1}{2}r^{\frac{1}{2}}\cos\frac{\theta}{2}\right)}{\partial r} + \frac{\partial\left(\frac{1}{2}r^{-\frac{1}{2}}\sin\frac{\theta}{2}\right)}{\partial\theta}$$

$$= -\frac{1}{4}r^{-\frac{1}{2}}\cos\frac{\theta}{2} + \frac{1}{4}r^{-\frac{1}{2}}\cos\frac{\theta}{2} = 0$$

(c) The acceleration of the particle is given by

$$a = \frac{dV}{dt} = \frac{\partial \vec{V}}{\partial r}\frac{dz}{dt}\bigg)_P + \frac{\partial \vec{V}}{\partial \theta}\frac{d\theta}{dt}\bigg)_P + \frac{\partial \vec{V}}{\partial t} = \frac{\partial \vec{V}}{\partial r}V_r + \frac{\partial \vec{V}}{\partial \theta}\frac{V_\theta}{r} + \frac{\partial \vec{V}}{\partial t}$$

Substituting $V = V_r \hat{\imath}_r + V_\theta \hat{\imath}_\theta$

$$a = V_r \frac{\partial}{\partial r}(V_r \hat{\imath}_r + V_\theta \hat{\imath}_\theta) + \frac{V_\theta}{r}\frac{\partial}{\partial \theta}(V_r \hat{\imath}_r + V_\theta \hat{\imath}_\theta) + \frac{\partial}{\partial t}(V_r \hat{\imath}_r + V_\theta \hat{\imath}_\theta)$$

$$= V_r\left(\frac{\partial V_r}{\partial r}\hat{\imath}_r + V_r \cancelto{0}{\frac{\partial \hat{\imath}_r}{\partial r}} + \frac{\partial V_\theta}{\partial r}\hat{\imath}_\theta + V_\theta \cancelto{0}{\frac{\partial \hat{\imath}_\theta}{\partial r}}\right) + \frac{V_\theta}{r}\left(\frac{\partial V_r}{\partial \theta}\hat{\imath}_r\right.$$

$$\left. + V_r \cancelto{0}{\frac{\partial \hat{\imath}_r}{\partial \theta}} + \frac{\partial V_\theta}{\partial \theta}\hat{\imath}_\theta + V_\theta \cancelto{0}{\frac{\partial \hat{\imath}_\theta}{\partial \theta}}\right) + \left(\frac{\partial V_r}{\partial t}\hat{\imath}_r + V_r \cancelto{0}{\frac{\partial \hat{\imath}_r}{\partial t}} + \frac{\partial V_\theta}{\partial t}\hat{\imath}_\theta\right.$$

$$\left. + V_\theta \cancelto{0}{\frac{\partial \hat{\imath}_\theta}{\partial t}}\right)$$

Collecting the terms,

$$a_r = V_r \frac{\partial V_r}{\partial r} + \frac{V_\theta}{r}\frac{\partial V_r}{\partial \theta} + \frac{\partial V_r}{\partial t} = \frac{1}{8r^2}$$

$$a_\theta = V_r \frac{\partial V_\theta}{\partial r} + \frac{V_\theta}{r}\frac{\partial V_\theta}{\partial \theta} + \frac{\partial V_\theta}{\partial t} = \frac{1}{8r^2}\sin\theta$$

• PROBLEM 9-4

A point source is placed at $x = -a$ and an equal strength sink at $x = +a$ (see accompanying figure). Determine the maximum radius r_0 of the axisymmetric body formed if the source and sink are placed in a uniform flow.

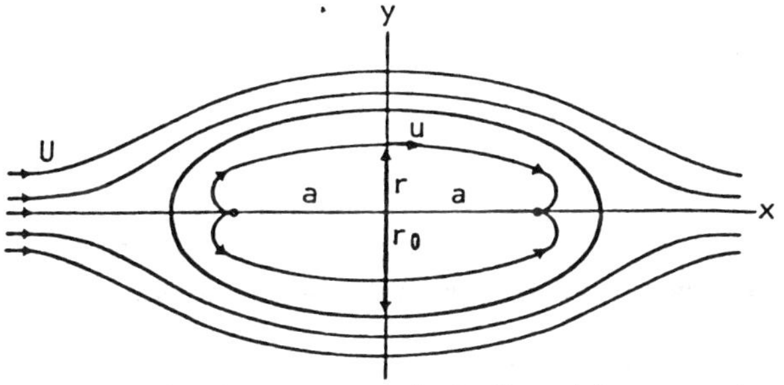

Solution: The simple potential function $\phi = \frac{A}{r}$ (1)

represents a point source with streamlines emanating along radial lines in all directions from the origin of r.

An integration of the velocity around a spherical surface allows us to introduce the source strength q and write

$$\phi = -\frac{q}{4\pi r} \tag{2}$$

for a source. Using equation 2 in cartesian coordinates, the velocity potential is

$$\phi = Ux - \frac{q}{4\pi\sqrt{(x+a)^2 + y^2 + z^2}} + \frac{q}{4\pi\sqrt{(x-a)^2 + y^2 + z^2}}$$

Because of symmetry, the body will have its maximum thickness at $x = 0$, as shown. One may find the body radius by integrating from $r = 0$ to $r = r_0$ so that the total mass flux across the area of integration is q. The x-component of velocity is

$$u = \frac{\partial \phi}{\partial x} = U + \frac{(x + a)q}{4\pi[(x + a)^2 + y^2 + z^2]^{3/2}} - \frac{(x - a)q}{4\pi[(x - a)^2 + y^2 + z^2]^{3/2}}$$

Along the x = 0 plane

$$u = U + \frac{2qa}{4\pi(a^2 + y^2 + z^2)^{3/2}}$$

$$= U + \frac{qa}{2\ (a^2 + r^2)^{3/2}}$$

The flow rate q through a circle of radius r_o is

$$q = \int_0^{r_0} \left[U + \frac{qa}{2\pi(a^2 + r^2)^{3/2}}\right] 2\pi r\ dr$$

$$= U\pi r_0^2 - qa \left[\frac{1}{\sqrt{a^2 + r_0^2}} - \frac{1}{a}\right]$$

For a particular set of flow parameters r_0 could be determined.

• PROBLEM 9-5

A flow is defined by u = 2x and v = -2y. Find the stream function and potential function for this flow.

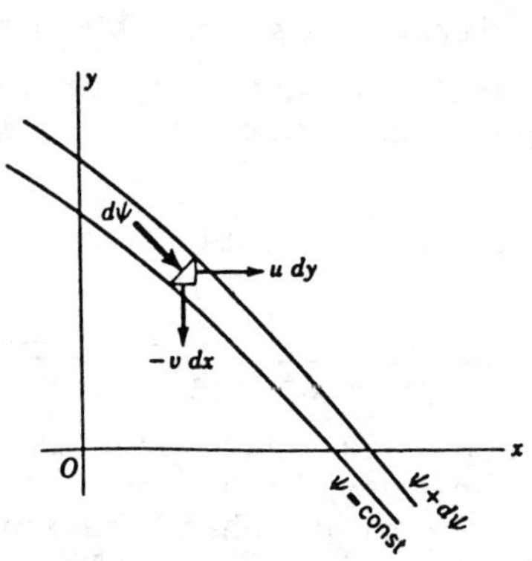

Solution: The equation of continuity for an incompressible fluid is given by

$$\frac{\partial u}{\partial x} + \frac{\partial v}{\partial y} + \frac{\partial w}{\partial z} = 0$$

The stream function ψ, based on the continuity principle, is a mathematical expression that describes a flow field. In the figure are shown two adjacent streamlines of a two-dimensional flow field. Let $\psi(x,y)$ represent the streamline nearest the origin. Then $\psi + d\psi$ is representative of the second streamline. Since there is no flow across a streamline, we can let ψ be indicative of the flow carried through the area from the origin 0 to the first streamline. And thus $d\psi$ represents the flow carried between the two streamlines. From continuity referring to the triangular fluid element in the figure, we see that for an incompressible fluid

$$d\psi = -v\,dx + u\,dy \tag{1}$$

The total derivative $d\psi$ may also be expressed as

$$d\psi = \frac{\partial \psi}{\partial x}\,dx + \frac{\partial \psi}{\partial y}\,dy \tag{2}$$

Comparing these last two equations, we note that

$$u = \frac{\partial \psi}{\partial y}$$

and (3)

$$v = -\frac{\partial \psi}{\partial x}$$

Check continuity:

$$\frac{\partial u}{\partial x} + \frac{\partial v}{\partial y} = 2 - 2 = 0$$

Hence continuity is satisfied and it is possible for a stream function to exist and using Eq. 1

$$d\psi = -v\,dx + u\,dy = 2y\,dx + 2x\,dy$$

$$\psi = 2xy + C_1$$

The vorticity ξ is defined as the circulation per unit of enclosed area, and is given by

$$\xi = \frac{\partial v}{\partial x} - \frac{\partial u}{\partial y} \tag{4}$$

It has been found that an irrotational flow is one for which the vorticity $\xi = 0$, and the flow is rotational if $\xi \neq 0$.

For two-dimensional flow the velocity potential $\phi(x,y)$ may be defined in cartesian coordinates as

$$u = -\frac{\partial \phi}{\partial x}$$

and (5)

$$v = -\frac{\partial \phi}{\partial y}$$

The corresponding expressions in polar coordinates are

$$v_r = -\frac{\partial \phi}{\partial r}$$

and (6)

$$v_t = -\frac{\partial \phi}{r\,\partial \theta}$$

If we substitute the expressions of Eq. (5) into the continuity equation, we get

$$\frac{\partial^2 \phi}{\partial x^2} + \frac{\partial^2 \phi}{\partial y^2} = 0 \qquad (7)$$

This is known as the Laplace equation; it is of importance in both solid mechanics and fluid mechanics.

If the expressions of Eq. (5) are substituted into the equation for vorticity (Eq. 4), we get

$$\xi = \frac{\partial v}{\partial x} - \frac{\partial u}{\partial y} = \frac{\partial}{\partial x}\left(-\frac{\partial \phi}{\partial y}\right) - \frac{\partial}{\partial y}\left(-\frac{\partial \phi}{\partial x}\right) =$$

$$-\frac{\partial^2 \phi}{\partial x\,\partial y} + \frac{\partial^2 \phi}{\partial y\,\partial x} = 0$$

Since $\xi = 0$, the flow is irrotational, and thus, if a velocity potential exists, the flow must be irrotational. Conversely, if the flow is rotational, the velocity potential ϕ does not exist.

It was noted in Eq. (2) that

$$d\psi = \frac{\partial \psi}{\partial x}\,dx + \frac{\partial \psi}{\partial y}\,dy$$

Similarly,

$$d\phi = \frac{\partial \psi}{\partial x} dx + \frac{\partial \phi}{\partial y} dy$$

From Eqs. (3) and (5) we can express these two equations as

$$d\psi = -v\,dx + u\,dy \tag{8}$$

and

$$d\phi = -u\,dx - v\,dy \tag{9}$$

Check to see if the flow is irrotational:

$$\frac{\partial v}{\partial x} - \frac{\partial u}{\partial y} = 0 - 0 = 0$$

Hence the flow is irrotational and a potential function exists, and using Eq. 9

$$d\phi = -u\,dx - v\,dy = -2x\,dx + 2y\,dy$$

$$\phi = -(x^2 - y^2) + C_2$$

• PROBLEM 9-6

Determine the scalar potential function of the vector function $\vec{A}$:

$$\vec{A} = 2xy\hat{i} + x^2\hat{j} + 3z^2\hat{k}$$

Solution: The Stokes' theorem is given by

$$\oint_L \vec{A} \cdot d\vec{l} = \int_S \hat{n} \cdot (\vec{\nabla} \times \vec{A})\, dS \tag{1}$$

where S may be a three-dimensional surface and is bounded by the line L, $d\vec{l}$ is a directed line element of L, and $\hat{n}$ is normal to dS.

The circulation Γ of the vector $\vec{A}$ around the line L is defined as

$$\Gamma = \oint \vec{A} \cdot d\vec{l}$$

If a vector $\vec{A}$ is given by the gradient of a scalar function ϕ, that is

$$\vec{A} = \vec{\nabla}\phi \qquad (2)$$

then Stokes' theorem (Eq. 1) shows that the circulation is zero, or

$$\oint \vec{A} \cdot d\vec{l} = 0$$

since $\vec{\nabla} \times \vec{\nabla}\phi = 0$. (The curl of a gradient is always zero.) The vector field $\vec{A}$ is called a conservative vector field, and the scalar function ϕ is a scalar potential function of the vector $\vec{A}$.

Then using Eq. (2),

$$\vec{A} = \vec{\nabla}\phi = \frac{\partial\phi}{\partial x}\,\hat{i} + \frac{\partial\phi}{\partial y}\,\hat{j} + \frac{\partial\phi}{\partial z}\,\hat{k}$$

Hence

$$\frac{\partial\phi}{\partial x} = 2xy \qquad \frac{\partial\phi}{\partial y} = x^2 \qquad \frac{\partial\phi}{\partial z} = 3z^2$$

The first of these equations gives

$$\phi = x^2y + f(y,z)$$

Substitute this in the second equation and

$$\frac{\partial\phi}{\partial y} = x^2 = x^2 + \frac{\partial f}{\partial y}$$

Thus $\partial f/\partial y = 0$ or $f = f(z)$. Substitute in the third equation and

$$\frac{\partial\phi}{\partial z} = 3z^2 = \frac{\partial f}{\partial z}$$

giving $f = z^3 + C$. We finally have

$$\phi = x^2y + z^3 + C$$

• **PROBLEM 9-7**

In a two-dimensional, incompressible flow the fluid velocity components are given by: $v_x = x - 4y$ and $v_y = -y - 4x$. Show that the flow satisfies the continuity equation and obtain the expression for the stream function. If the flow is potential obtain also the expression for the velocity potential.

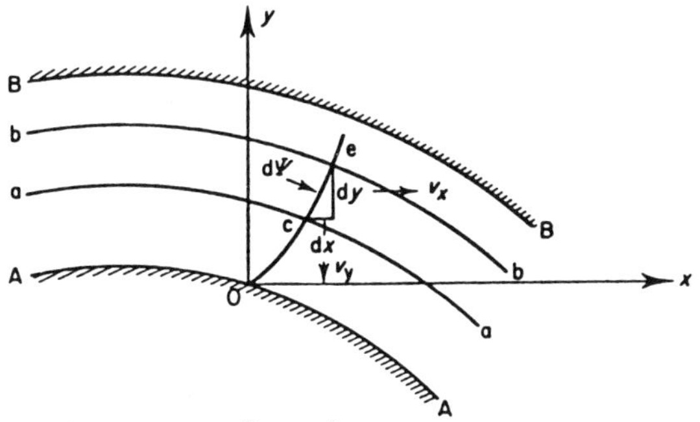

The stream function

Solution: For incompressible, two-dimensional flow, the continuity equation is

$$\frac{\partial v_x}{\partial x} + \frac{\partial v_y}{\partial y} = 0 ,$$

but

$$v_x = x - 4y$$

and

$$v_y = -y - 4x$$

and

$$\frac{\partial v_x}{\partial x} = 1 ,$$

$$\frac{\partial v_y}{\partial y} = -1;$$

therefore, 1 - 1 = 0 and the flow satisfies the continuity equation.

Referring to the accompanying figure, if the streamline aa is denoted by ψ_a , which will be labelled by a numerical value representing the flow rate per unit depth between AA and the streamline aa, then,

$$\psi_a = Q_{0c}$$

and, similarly, if

$$\psi_b = Q_{0e} ,$$

it follows that

$$d\psi = \psi_b - \psi_a = Q_{ce} ,$$

so that

$$d\psi = v_x \, dy - v_y \, dx \qquad (1)$$

and ψ, which is called the stream function, is given by

$$\psi = \int v_x \, dy - \int v_y \, dx . \qquad (2)$$

Thus, the stream function depends upon position coordinates,

$$\psi = f(x,y)$$

and, hence, the total derivative:

$$d\psi = \frac{\partial\psi}{\partial x} dx + \frac{\partial\psi}{\partial y} dy . \qquad (3)$$

Comparing equations (3) and (1), the relationships between the stream function and the velocity components are obtained:

$$v_x = \frac{\partial\psi}{\partial y}$$

and (4)

$$v_y = - \frac{\partial\psi}{\partial x}$$

Velocity potential is defined as

$$\phi = \int_A^B v_s \, ds \, , \tag{5}$$

where A and B are two points in a potential field and v_s is the velocity tangential to the elementary path s.

The condition for potential flow is

$$\frac{\partial v_y}{\partial x} - \frac{\partial v_x}{\partial y} = 0 \, . \tag{6}$$

In potential flow

$$\phi = \int_A^B v_s \, ds$$

exists, from which it follows that, if v_x and v_y are the orthogonal components of v_s , then,

$$v_x = \frac{\partial \phi}{\partial x}$$

and (7)

$$v_y = \frac{\partial \phi}{\partial y} \, ,$$

Comparing equations (4) and (7), from (4)

$$v_x = \frac{\partial \psi}{\partial y}$$

and

$$v_y = - \frac{\partial \psi}{\partial x} \, ;$$

from 7

$$v_x = \frac{\partial \phi}{\partial x}$$

and

$$v_y = \frac{\partial \phi}{\partial y} \, .$$

Thus, equating for v_x and v_y , we obtain

$$\frac{\partial \psi}{\partial y} = \frac{\partial \phi}{\partial x}$$

and

$$\frac{\partial \phi}{\partial y} = - \frac{\partial \psi}{\partial x}$$

These equations are known as Cauchy-Riemann equations and they enable the stream function to be calculated if the velocity potential is known and vice versa in a potential flow.

It is now possible to return to the condition for potential flow and to restate it in terms of the stream function. The condition is

$$\frac{\partial v_y}{\partial x} - \frac{\partial v_x}{\partial y} = 0 \, ,$$

but

$$v_y = - \frac{\partial \psi}{\partial x} \qquad \text{and} \qquad v_x = \frac{\partial \psi}{\partial y} \, ,$$

so that, by substitution,

$$\frac{\partial}{\partial x} \left(- \frac{\partial \psi}{\partial x} \right) - \frac{\partial}{\partial y} \left(\frac{\partial \psi}{\partial y} \right) = 0$$

and

$$\frac{\partial^2 \psi}{\partial x^2} + \frac{\partial^2 \psi}{\partial y^2} = 0 \, . \tag{8}$$

This is the Laplace equation for stream function, which must be satisfied for the flow to be potential.

To obtain the stream function, using equations (4)

$$v_x = \frac{\partial \psi}{\partial y} = x - 4y \, , \tag{9}$$

$$v_y = \frac{\partial \psi}{\partial x} = -(y + 4x) \ . \qquad (10)$$

Therefore, from (9),

$$\psi = (x - 4y)dy + f(x) + C$$

$$= xy - 2y^2 + f(x) + C \ .$$

But, if $\psi_0 = 0$ at $x = 0$ and $y = 0$, which means that the reference streamline passes through the origin, then $C = 0$ and

$$\psi = xy - 2y^2 + f(x) \ . \qquad (11)$$

To determine $f(x)$, differentiate partially the above expression with respect to x and equate to $-v_y$, equation (10):

$$\frac{\partial \psi}{\partial x} = y + \frac{\partial}{\partial x} f(x) = y + 4x \ ,$$

$$f(x) = \int 4x \, dx = 2x^2$$

Substitute into (11)

$$\psi = 2x^2 + xy - 2y^2 \ .$$

To check whether the flow is potential, there are two possible approaches:

(a) Since

$$\frac{\partial v_y}{\partial x} - \frac{\partial v_x}{\partial y} = 0 \ ,$$

but

$$v_y = -(4x + y) \qquad \text{and} \qquad v_x = (x - 4y) \ .$$

Therefore,

$$\frac{\partial v_y}{\partial x} = -4 \qquad \text{and} \qquad \frac{\partial v_x}{\partial y} = -4 \ ,$$

so that

$$\frac{\partial v_y}{\partial x} - \frac{\partial v_x}{\partial y} = -4 + 4 = 0$$

and flow is potential.

(b) Laplace's equation must be satisfied:

$$\frac{\partial^2 \psi}{\partial x^2} + \frac{\partial^2 \psi}{\partial y^2} = 0 \ ,$$

$$\psi = 2x^2 + xy - 2y^2 \ .$$

Therefore,

$$\frac{\partial \psi}{\partial x} = 4x + y \quad \text{and} \quad \frac{\partial \psi}{\partial y} = x - 4y,$$

$$\frac{\partial^2 \psi}{\partial x^2} = 4 \quad \text{and} \quad \frac{\partial^2 \psi}{\partial y^2} = -4 \ .$$

Therefore $4 - 4 = 0$ and so flow is potential.

Now, to obtain the velocity potential,

$$\frac{\partial \phi}{\partial x} = v_x = x - 4y \ ,$$

therefore,

$$\phi = \int (x - 4y)dx + f(y) + G \ .$$

But $\phi_0 = 0$ at $x = 0$ and $y = 0$, so that $G = 0$. Therefore

$$\phi = x^2/2 + 4yx + f(y).$$

Differentiating with respect to y and equating to v_y ,

$$\frac{\partial \phi}{\partial y} = 4x + \frac{d}{dy} f(y) = -(4x + y)$$

$$\frac{d}{dy} f(y) = -y$$

and

$$f(y) = -\frac{y^2}{2},$$

so that

$$\phi = x^2/2 + 4yx - y^2/2 .$$

CIRCULATION

• PROBLEM 9-8

> Determine the circulation of A around the triangle which has vertices at the origin (0,4,0) and (4,0,0). The vector function A is given by
>
> $$A = 4x^2\hat{i} + 2yz\hat{j} - (4y^2 + z^2)\hat{k}$$

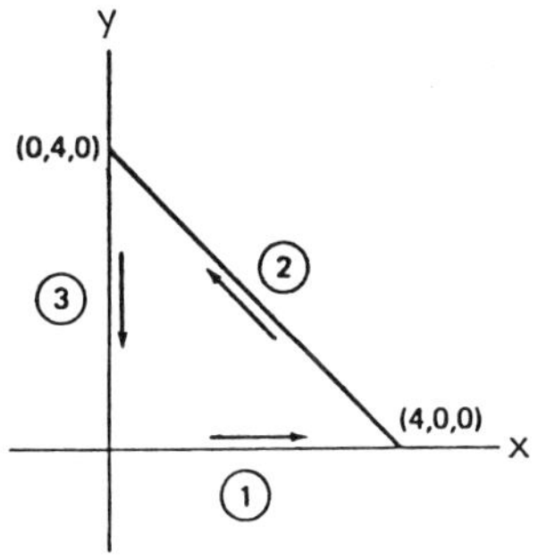

Solution: The circulation is defined as

$$\Gamma = \oint_L A \cdot dl$$

$$= \int_{①} A \cdot dl + \int_{②} A \cdot dl + \int_{③} A \cdot dl$$

Referring to the Fig. and using $A \cdot dl = A_x\, dx + A_y\, dy$

$+ A_z dz$, we have

$$\Gamma = \int_{①} A_x\,dx + \int_{②} (A_x\,dx + A_y\,dy) + \int_{③} A_y\,dy$$

$$= \int_0^4 4x^2\,dx + \int_4^0 4x^2\,dx + \int_0^4 2y\cancelto{0}{z}\,dy + \int_4^0 2y\cancelto{0}{z}\,dy$$

$$= 0$$

• PROBLEM 9-9

> A cylinder 4 ft in diameter and 25 ft long rotates at 90 rpm with its axis perpendicular to an airstream with a wind velocity of 120 fps (81.8 mph). The specific weight of the air is 0.0765 lb/ft^3. Assuming no slip between the cylinder and the circulatory flow, find (a) the value of the circulation and (b) the transverse or life force.

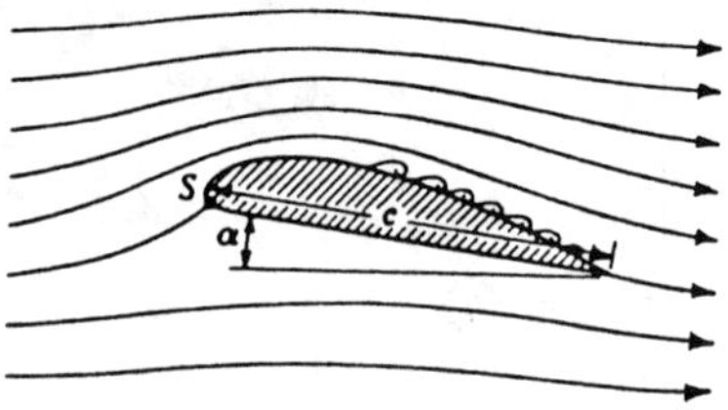

Fig. 1

Solution: (a) The increased velocity over the top of the wing of Fig. 1 and the decreased velocity around the bottom of the wing can be explained by noting that a circulation is induced as the wing moves relative to the flow field. The strength of the circulation depends, in the real case, on the shape of the wing and its velocity and orientation with respect to the flow field. A schematic diagram of the situation is presented in Fig. 2.

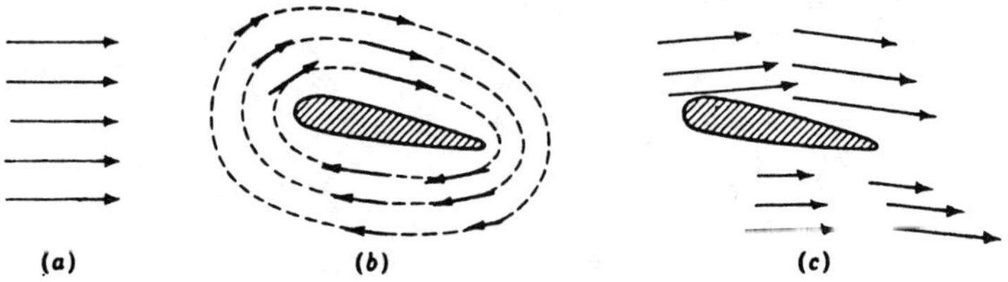

Fig. 2

Schematic superposition of circulation on uniform rectilinear flow field. (a) Uniform rectilinear flow field. (b) Circulation. (c) Net effect.

The relationship between lift and circulation is one that has been studied exhaustively for years by many investigators. To illustrate the theory of lift, consider the flow of an ideal fluid past a cylinder and assume that a circulation about the cylinder is imposed on the flow. First, though, let us consider the velocity field surrounding a free vortex (Fig. 3). The equation for this field is given as vr = C, a constant.

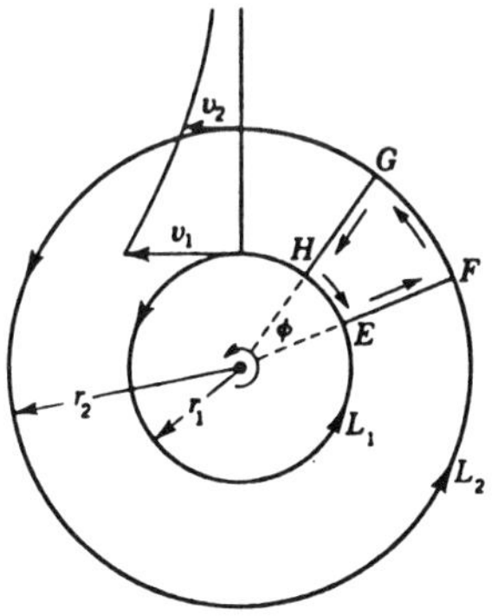

Fig. 3

Let L represent any closed path in a two-dimensional flow field. The circulation Γ is defined mathematically as a line integral of the velocity about a closed path. Thus

$$\vec{\Gamma} = \oint_L \vec{V} \cdot d\vec{L} = \oint_L V \cos \beta \, dL$$

where $\vec{V}$ is the velocity in the flow field at the element $d\vec{L}$ of the path, and β is the angle between V and the tangent to the path (in the positive direction along the path) at that point.

The circulation can be readily computed by application of the above equation if we choose the closed path as the circular streamline L_1 concentric with the center of the vortex. The velocity is evidently around this path and tangent to it ($\cos \beta = 1$), and the line integral of dL is simply the circumference of the circle. Applying the same treatment to another concentric circle L_2 , we get

$$\Gamma = v \oint_L dL = v_1 (2\pi r_1) = v_2 (2\pi r_2)$$

But from the vortex velocity field $v_1 r_1 = v_2 r_2 = C$, and hence

$$\Gamma = 2\pi C = 2\pi vr$$

which demonstrates that the circulation around two different curves, each completely enclosing the vortex center, is the same. It may be proved more rigorously that the circulation around any path enclosing the vortex center is given by $\Gamma = 2\pi C$. The circulation is seen to depend only on the vortex constant C, which is called the strength of the vortex.

First we calculate the tangential or peripheral velocity.

$$v_t = \frac{2\pi Rn}{60} = 2\pi \times 2 \times \frac{90}{60} = 18.84 \text{ fps}$$

Then

$$\Gamma = 2\pi R v_t = 2\pi \times 2 \times 18.84 = 237 \text{ ft}^2/\text{s}$$

(b) The lift force is given by the equation

$$F_L = \rho BU\Gamma$$

where F_L is the lift force, B is the length of the cylinder, and U is the uniform flow velocity.

$$F_L = \rho BU\Gamma = \frac{0.0765}{32.2} \times 25 \times 120 \times 237 = 1,685 \text{ lb}$$

• PROBLEM 9-10

A circular tank, with radius r_0, containing fluid is rotated in a large body of the same fluid at a constant angular velocity ω. Neglecting viscous effects near the solid wall, find the velocity distribution, vorticity, and circulation in the fluid inside and outside the tank.

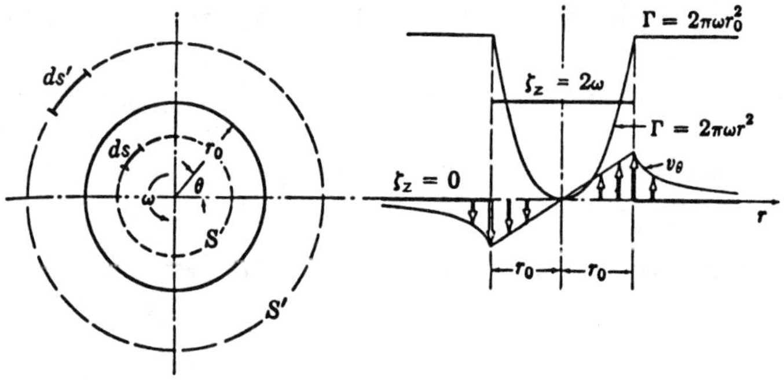

Velocity, vorticity, and circulation.

Solution: The fluid inside the tank is revolving with the tank; its angular velocity will be that of the tank, and the velocity of the fluid as a function of the radius is

$$v_\theta = \omega r$$

The differential equation for an irrotational plane motion of the free fluid outside the tank is obtained from equations 1 and 2

$$u = \frac{\partial \psi}{\partial y} \qquad v = -\frac{\partial \psi}{\partial x} \tag{1}$$

$$\xi_z = \frac{\partial v}{\partial x} - \frac{\partial u}{\partial y} \tag{2}$$

If the quantities for u and v in eq. (1) are substituted into eq. (2), the following differential equation is found when $\xi_z = 0$.

$$\frac{\partial}{\partial x}\left(-\frac{\partial \psi}{\partial x}\right) - \frac{\partial}{\partial y}\left(\frac{\partial \psi}{\partial y}\right) = 0$$

$$\frac{\partial^2 \psi}{\partial x^2} + \frac{\partial^2 \psi}{\partial y^2} = 0$$

This is called Laplace's equation. For a simple circulatory motion with axisymmetry and $v_r = 0$ and eq. (3) which expresses the vorticity component normal to the $r - \theta$ plane, the solution can be found.

$$\xi_z = \frac{\partial v_\theta}{\partial r} + \frac{v_\theta}{r} - \frac{1}{r}\frac{\partial v_r}{\partial \theta} \tag{3}$$

In this case, since v_θ is a function of r alone, for irrotational motion, eq. (3) reduces to

$$\frac{dv_\theta}{dr} + \frac{v_\theta}{r} = 0$$

Separating variables

$$\frac{dv_\theta}{v_\theta} + \frac{dr}{r} = 0$$

Integrating

$$\ln v_\theta + \ln r = \text{constant}$$

Finally,

$$v_\theta r = C$$

This type of motion is called a free vortex, where the velocity is zero and the peripheral velocity varies inversely as the radial distance.

The peripheral velocity will vary hyperbolically with the radius

$$v_\theta = \frac{C}{r}$$

Since, at both sides of the tank wall, the velocity of the fluid should have the peripheral velocity of the tank (no slip), the constant C can be evaluated.

$$\omega r_0 = \frac{C}{r_0}$$

$$C = \omega r_0^2$$

Since the fluid within the tank revolves like a rigid body, the vorticity is the same everywhere inside and equal to twice the rotation 2ω. The free-vortex flow is an irrotational flow, and therefore the vorticity everywhere outside the tank is zero.

In order to compute the circulation inside and outside the tank as a function of the radius, for convenience two concentric streamlines S and S' are chosen. The circulation inside the tank for $r < r_0$ is, according to Eq. (4),

$$\Gamma_s = \oint v_\theta \, ds = \int_0^{2\pi} v_\theta r \, d\theta$$

$$= \int_0^{2\pi} (\omega r) r \, d\theta = 2\pi\omega r^2$$

which is equal to the product of the vorticity times the area. Therefore Γ is a parabolic function of the radius for inside the tank. The circulation outside the tank along the contour S', $r > r_0$ is

$$\Gamma_{s'} = \oint v_\theta \, ds' = \int_0^{2\pi} v_\theta r \, d\theta$$

$$= \int_0^{2\pi} \frac{C}{r} r \, d\theta$$

$$= 2\pi C = 2\pi\omega r_0^2 = \text{constant}$$

This shows that, since the vorticity is zero outside the tank, there is no longer any contribution to the circulation for $r > r_0$.

VORTEX FLOW

• PROBLEM 9-11

With the aid of the Stokes theorem find the expression for the z-component of vorticity in cylindrical coordinates.

Solution:

From the figure we can see that

$$\oint \vec{q} \cdot ds = \int_s \zeta_n \cdot dS = \zeta_z r d\theta \, dr$$

$$= v_r dr + \left(v_\theta + \frac{\partial v_\theta}{\partial r} dr\right) (r + dr) d\theta -$$

$$\left(v_r + \frac{\partial v_r}{\partial \theta} d\theta\right) dr - v_\theta r d\theta$$

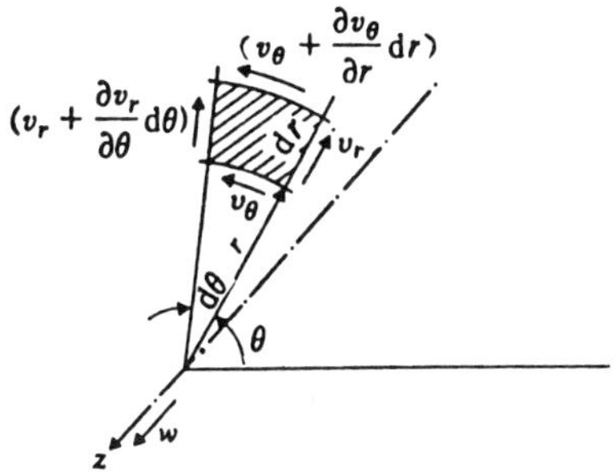

Vorticity in cylindrical coordinates.

Therefore

$$\zeta_z = \frac{1}{r}\left[\frac{\partial}{\partial r}(r v_\theta) - \frac{\partial v_r}{\partial \theta}\right]$$

• PROBLEM 9-12

A viscous, incompressible fluid flows between two plates. The flow is laminar and two-dimensional, the velocity profile is parabolic:

$$u = U_c(1 - y^2/b^2)$$

Calculate the shear stress τ and rotation ω.

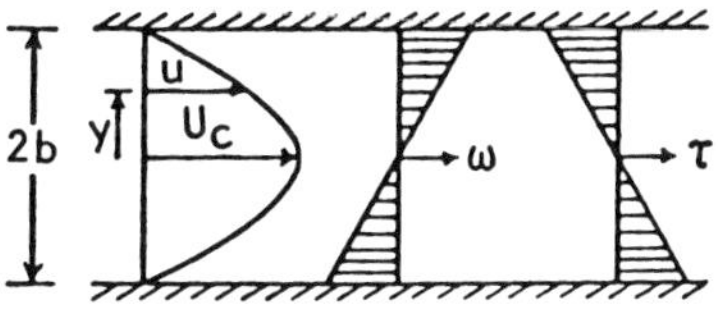

Solution: $\tau = \mu\, du/dy = -2\mu U_c y/b^2$

$$\omega = \frac{1}{2}\left(\frac{\partial v}{\partial x} - \frac{\partial u}{\partial y}\right) = -\frac{1}{2}[-2U_c y/b^2] = U_c y/b^2$$

The rotation and vorticity are large where the shear stress is large.

• PROBLEM 9-13

Check these flows for continuity and determine the vorticity of each: (a) $v_t = 6r$, $v_r = 0$; (b) $v_t = 0$, $v_r = -5/r$.

Solution: (a) For incompressible fluid flow, the differential continuity equation, in polar coordinates, is given by

$$\frac{v_r}{r} + \frac{\partial v_r}{\partial r} + \frac{\partial v_t}{r\partial\theta} = 0$$

The vorticity ξ is defined as the circulation per unit of enclosed area, and, in polar coordinates, is given by

$$\xi = \frac{\partial v_t}{\partial r} + \frac{v_t}{r} - \frac{\partial v_r}{r\partial\theta}$$

It has been found that an irrotational flow is one for which the vorticity $\xi = 0$, the flow is rotational if $\xi \neq 0$.

$$\frac{0}{r} + \frac{\partial(0)}{\partial r} + \frac{\partial(6)r}{r\,\partial\theta} = 0 \qquad \text{(continuity is satisfied)}$$

$$\xi = \frac{\partial(6r)}{\partial r} + \frac{6r}{r} - \frac{\partial(0)}{r\partial\theta} = 6+6-0 = 12 \qquad \text{(flow is rotational)}$$

(b) $$-\frac{5/r}{r} + \frac{\partial(-5r^{-1})}{\partial r} + \frac{\partial(0)}{r\partial\theta} = -\frac{5}{r^2} + \frac{5}{r^2} + 0 = 0 \qquad \text{(continuity is satisfied)}$$

$$\xi = \frac{\partial(0)}{\partial r} + \frac{0}{r} - \frac{\partial(-5/r)}{r\partial\theta} = 0 \qquad \text{(flow is irrotational)}$$

Obtain expressions for the vorticity in the following flow fields:

(a) a forced vortex
(b) a free vortex; and
(c) potential flow around a circular cylinder with its axis normal to the stream direction.

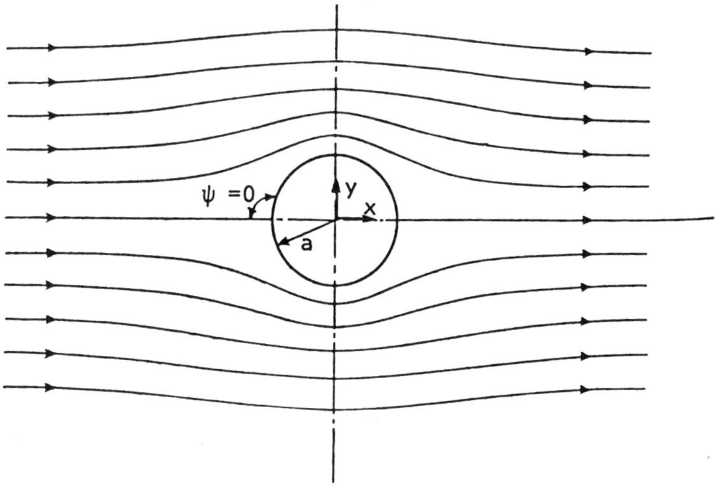

Solution: For solving (a) and (b) it is convenient to work in cylindrical coordinates. Working in a cylindrical instead of cartesian coordinates, for the case in which the local fluid velocity is everywhere normal to a radius from the axis of the flow field (i.e. the radial component of velocity is everywhere zero); the result is

$$\text{vorticity, } \xi = \frac{V}{r} + \frac{\partial V}{\partial r}.$$

This can be put into more precise form for particular velocity distributions:

(a) For a forced vortex,

$$V = kr \quad (k \text{ is a constant})$$

$$\therefore \frac{dV}{dr} = k.$$

Then $\xi = k + k = 2k.$

(b) For a free vortex,

$$V = \frac{k}{r},$$

$$\therefore \frac{dV}{dr} = -\frac{k}{r^2}.$$

Then

$$\xi = \frac{k}{r^2} - \frac{k}{r^2} = 0.$$

A quantity ψ, known as the "stream function," is defined by

$$\frac{\partial\psi}{\partial y} \equiv -V_x, \quad \frac{\partial\psi}{\partial x} \equiv V_y. \tag{1}$$

In terms of the stream function, flow around a cylinder of radius a with its axis normal to the stream is found to be represented by

$$\psi = V_\infty y \left(1 - \frac{a^2}{x^2 + y^2}\right) \tag{2}$$

and is shown diagrammatically in the Fig. The streamline defined by $\psi = 0$ forms the axis of the flow and the surface of the cylinder; for large values of x and y, equation 2 tends to

$$\psi = V_\infty y,$$

representing a uniform stream flowing in the x-direction with velocity V_∞. Differentiation of the stream-function to give the distributions of velocity-components V_x and V_y gives

$$V_x \equiv -\frac{\partial\psi}{\partial y} = -V_\infty \left[1 - \frac{a^2}{x^2 + y^2} + \frac{2a^2y^2}{(x^2 + y^2)^2}\right]$$

and

$$V_y \equiv \frac{\partial\psi}{\partial x} = \frac{2V_\infty a^2 xy}{(x^2 + y^2)^2}.$$

Differentiation of these expressions gives

$$\frac{\partial V_x}{\partial y} = \frac{2V_\infty a^2 y(y^2 - 3x^2)}{(x^2 + y^2)^3},$$

and

$$\frac{\partial V_y}{\partial x} = \frac{2V_\infty a^2 y(y^2 - 3x^2)}{(x^2 + y^2)^3}.$$

Clearly, then

$$\xi \equiv \frac{\partial V_y}{\partial x} - \frac{\partial V_x}{\partial y} = 0.$$

a. Consider the two-dimensional flowfield against a plate given by the equations: $u = 4x$, $v = -4y$. Show that this flow satisfies the equation of continuity. Calculate the vorticity of the flow. (Fig 1)

b. A two-dimensional incompressible flowfield is described by the equations $v_t = \omega r$ and $v_r = 0$, in which ω is a constant. Sketch this flow and show that it satisfies the continuity equation. Also calculate the vorticity of the flow. (Fig. 2)

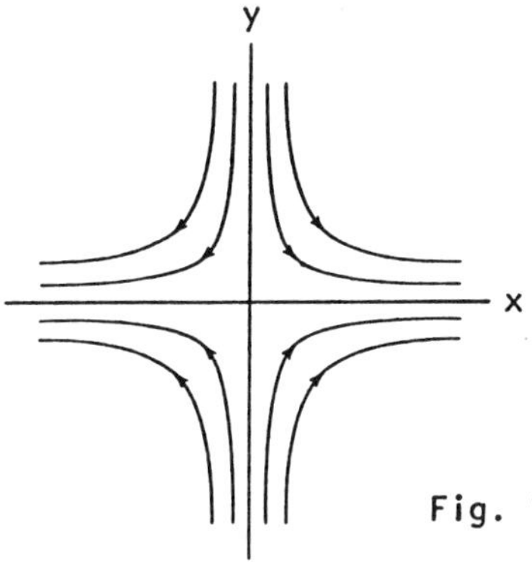

Fig. 1

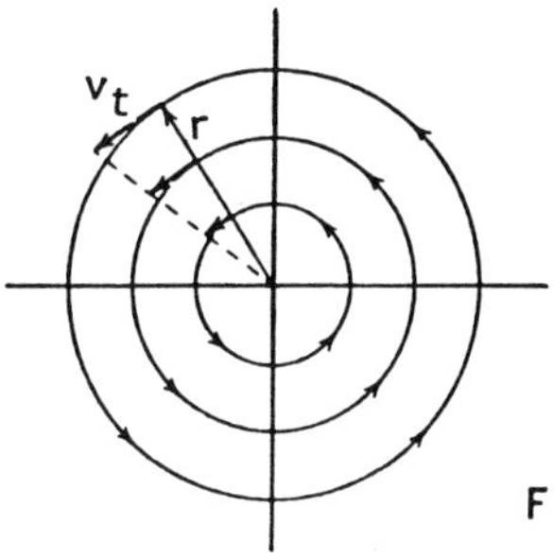

Fig. 2

Solution: a. From the equations it is seen that the x-component of velocity increases with x and the y-component increases with negative y. Accordingly the flow may actually be sketched as shown (Fig. 1) in all four quadrants. Note that the flow in any one quadrant (with the axes taken as solid boundaries) describes a possible flow in a square corner, providing that it satisfies the continuity equation. Substituting 4x and -4y for u and v in

$$\frac{\partial u}{\partial x} + \frac{\partial v}{\partial y} = 0, \qquad (1)$$

gives

$$\frac{\partial}{\partial x}(4x) + \frac{\partial}{\partial y}(-4y) = 4 - 4 = 0 \qquad (1)$$

and so shows that continuity is satisfied. The flow is thus physically possible. The equation used here assumes steady, incompressible flow. If the flow was given in polar coordinates the continuity equation would be

$$\frac{v_r}{r} + \frac{\partial v_r}{\partial r} + \frac{\partial v_t}{r\partial\theta} = 0 \qquad (2)$$

ξ: vorticity $\quad\Gamma$: circulation

$$\xi = \frac{\partial\Gamma}{\partial x\partial y} = \frac{\partial v}{\partial x} - \frac{\partial u}{\partial y} \qquad (3)$$

$$\xi = \frac{\partial}{\partial x}(-4y) - \frac{\partial}{\partial y}(4x) = 0 - 0 = 0 \tag{3}$$

Since the vorticity is zero, the flowfield of the first problem is an irrotational (or potential) flow.

b. Evidently the streamlines of this flow must be concentric circles because the radial component of velocity is everywhere zero. This flow is known as a forced vortex. Substituting $v_r = 0$ and $v_t = \omega r$ in to the continuity equation in polar coordinates gives

$$\frac{0}{r} + \frac{\partial}{\partial r}(0) + \frac{\partial}{r\partial\theta}(\omega r) = 0 + 0 + 0 = 0 \tag{2}$$

The equation of continuity is satisfied; this flow is physically possible.

Vorticity in polar coordinates is

$$\xi = \frac{\partial v_t}{\partial r} + \frac{v_t}{r} - \frac{\partial v_r}{r\partial\theta} \tag{4}$$

so

$$\xi = \frac{\partial}{\partial r}(\omega r) + \frac{\omega r}{r} - \frac{\partial}{r\partial\theta}(0) = \omega + \omega - 0 = 2\omega \tag{4}$$

Evidently this is a rotational flow possessing a constant vorticity (over the whole flowfield) of 2ω; the forced vortex is thus a rotational flowfield.

• PROBLEM 9-16

Find the circulation for an element of area enclosing the center, and an element not enclosing the center of a free vortex and a forced vortex, parallel to the z-axis.

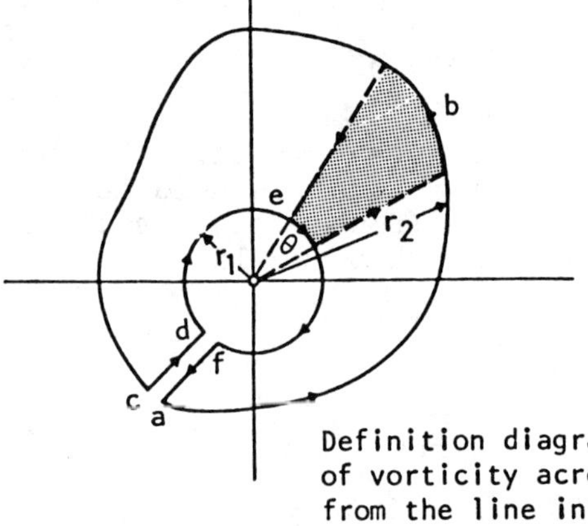

Fig. 1

Definition diagram for calculation of vorticity across an area element from the line integral of velocity around the area.

Solution: The velocity distribution for a free vortex (Fig. 1), using cylindrical polar coordinates is

$$v_r = w = 0 \qquad v_\theta = v = \frac{c}{r}$$

The circulation around the shaded area is

$$\Gamma = \oint q \cos \alpha \, ds = v_2 \theta r_2 - v_1 \theta r_1 = \frac{c}{r_2}\theta r_2 - \frac{c}{r_1}\theta r_1 = 0$$

Here, $\alpha = 0$ because v is tangential to the element θr of the curve. This is true for any area excluding the center, for example, for area bounded by abcdef. If the lines $\overline{cd}$ and $\overline{fa}$ are brought together then the integrals

$$\oint q \cdot ds = \int_{abc} q \cdot ds + \int_{cd} q \cdot ds + \int_{def} q \cdot ds +$$

$$\int_{fa} q \cdot ds$$

along those lines cancel each other and we have two integrals over closed curves aba and ded with a net result zero as for the sector. However, along the circle ded

$$\oint q \cdot ds = \int_0^{2\pi} \frac{c}{r_1} r_1 \, d\theta = 2\pi c = \Gamma$$

showing that the circulation along any closed curve including the origin is Γ. This circulation is seen to be independent of r, and we can make area as small as we like, but in the axis through the origin there is a circulation of magnitude Γ. This point is a singular point for which $c/r \to \infty$ and $c = \Gamma/2\pi$ is known as the vortex strength.

By the Stokes theorem the surface integral $\int_S \zeta \cdot dS$ for the whole surface bounded by a closed curve including the origin is also equal to Γ. The vorticity

$$\zeta_n = \zeta_z = \frac{1}{r}\left[\frac{\partial}{\partial r}(r v_\theta) - \frac{\partial v_r}{\partial \theta}\right]$$

and is zero everywhere except at $r = 0$ where it is indeterminate. That is to say, the motion in a free vortex is irrotational except at the singularity.

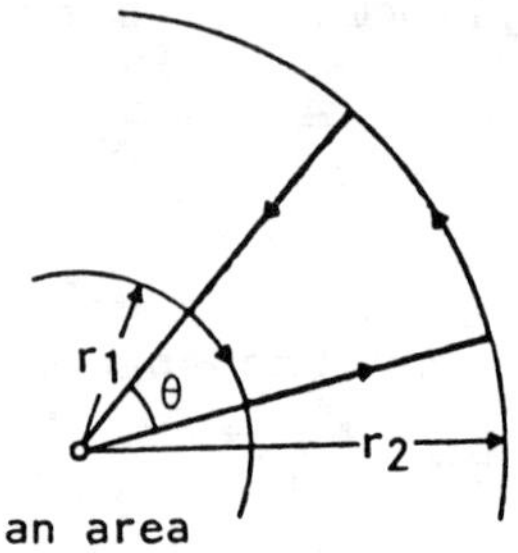

Circulation about an area element in vortex motion.

Fig. 2

A forced vortex is formed when a body of fluid rotates as a rigid body, with no relative motion between elements. The velocity distribution in cylindrical polar coordinates is

$$v_r = w = 0 \qquad v_\theta = v = \omega r$$

where ω is the angular velocity. The circulation about an element of area (Fig. 2) is

$$\Gamma = \oint q \cos\alpha \, ds = v_2 r_2 \theta - v_1 r_1 \theta = \omega\theta(r_2^2 - r_1^2)$$

The area enclosed by the bounding line is

$$A = \int_{r_1}^{r_2} (\theta r)\, dr = \frac{\theta}{2}(r_2^2 - r_1^2)$$

whence

$$\Gamma = 2\omega A$$

that is, the circulation equals twice the product of the angular velocity and the enclosed area. This is valid for any area element. With the rotation in xy plane,

$$\zeta = \zeta_z \text{ and } u = -\omega y, \; v = \omega x, \; q = \omega r. \quad \text{Then}$$

$$\zeta_z = \frac{\partial v}{\partial x} - \frac{\partial u}{\partial y} = 2\omega$$

An air duct of 2- by 2-ft-square cross section turns a bend of radius 4 ft as measured to the center line of the duct. If the measured pressure difference between the inside and outside walls of the bend is 1 in of water, estimate the rate of air flow in the duct. Assume standard sea-level conditions in the duct, assume ideal flow around the bend (free vortex flow), ρ(air) = .002377 slug/ft^3, and ρ(water) = 1.94 slugs/ft^3.

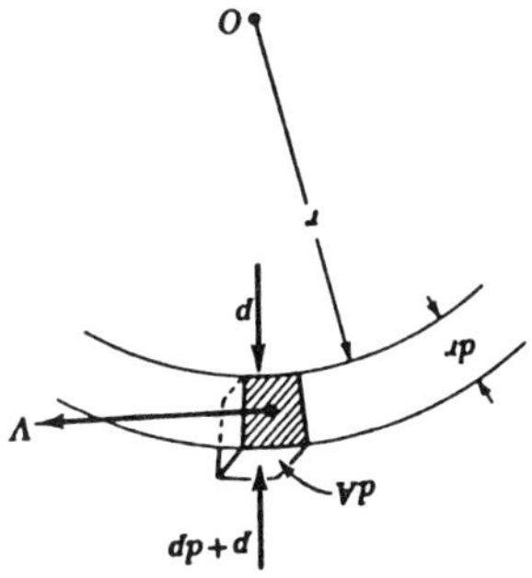

Solution: The figure shows an element of fluid moving in a horizontal plane with a velocity V along a curved path of radius r. The element has a linear dimension dr in the plane of the paper and an area dA normal to the plane of the paper. The mass of this fluid element is ρ dA dr, and the normal component of acceleration is V^2/r. Thus the centripetal force acting upon the element toward the center of curvature is ρ dA drV^2/r. As the radius increases from r to r + dr, the pressure will change from p to p + dp. Thus the resultant force in the direction of the center of curvature is dp dA. Equating these two forces,

$$dp = \rho \frac{V^2}{r}\, dr \tag{1}$$

The angular momentum with respect to the center of rotation of a particle of mass m moving along a circular path of radius r at a velocity V is mVr. Newton's second law states that, for the case of rotation, the torque is equal to the time rate of change of angular momentum. Hence, torque = d(mVr)/dt. In the case of a free vortex there is no torque applied; therefore, mVr = constant and thus Vr = C, where the value of C is determined by knowing the value of V at some radius r. Inserting V = C/r in Eq. (1), we obtain

$$dp = \rho\frac{C^2}{r^2}\frac{dr}{r} = \frac{\gamma}{g}\frac{C^2}{r^3}\, dr$$

Between any two radii r_1 and r_2 this integrates as

$$\frac{p_2}{\gamma} - \frac{p_1}{\gamma} = \frac{C^2}{2g}\left(\frac{1}{r_1^2} - \frac{1}{r_2^2}\right) \tag{2}$$

$$\frac{p_2}{\gamma} - \frac{p_1}{\gamma} = \frac{1}{12}\left(\frac{1.94}{0.002377}\right) = 68 \text{ ft of air}$$

From Eq. 2

$$\frac{p_2}{\gamma} - \frac{p_1}{\gamma} = \frac{C^2}{2g}\left(\frac{1}{r_1^2} - \frac{1}{r_2^2}\right)$$

Thus, with

$$r_1 = 3 \text{ ft} \qquad r_2 = 5 \text{ ft} \qquad 68 = \frac{C^2}{64.4}\left(\frac{1}{3^2} - \frac{1}{5^2}\right)$$

$$C = 248 \text{ ft}^2/\text{s}$$

Thus, with $Q = \int V\, dA$ and $V = C/r$, while $dA = B\, dr$, where B is the width of the duct,

$$Q = BC\int_{r_1}^{r_2} \frac{dr}{r} = BC \ln \frac{r_2}{r_1} = 2 \times 248 \ln \frac{3}{5}$$

$$= 2 \times 248 \times 0.511 = 254 \text{ cfs} = 15{,}240 \text{ ft}^3/\text{min}$$

• PROBLEM 9-18

A closed vertical cylinder, 3 ft. diameter, is full of water, which is made to rotate by paddles, 1 ft. in diameter, revolving axially at 2 revs. per sec. Find the velocity of the water at 6 in. and 12 in. from the center and the increase of pressure-head at these radii above the pressure-head at the center. If there is also an outward radial flow, whose velocity is 3 ft. per sec. at the circumference of the wheel, find the resultant velocity and its inclination to the radius at 12 in. radius.

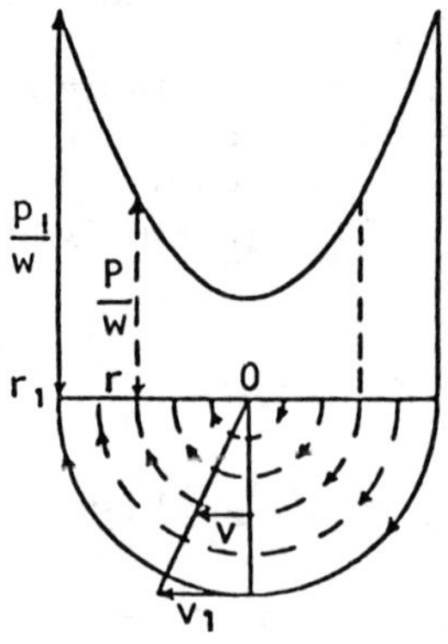

Solution: If fluid is forced to rotate about a vertical axis by revolving paddles, or by the friction of a revolving casing in which it is contained, its angular velocity ω will be constant for all points, like a revolving solid, the stream-lines will be circular, and the velocity at any point will be proportional to the radius, or $v = \omega r$. The radius of curvature, R, of a stream-line = r, the radius of the circle (Figure)

$$\therefore \quad \delta H = \frac{v}{g}\left(\frac{dv}{dr} + \frac{v}{r}\right)\delta r = \frac{v}{g}(\omega + \omega)\delta r = \frac{2\omega^2 r\delta r}{g} ;$$

$$\therefore \quad H_1 - H = \frac{\omega^2(r_1^2 - r^2)}{g} = \frac{v_1^2 - v^2}{g}$$

= increase of total head from r to r_1 .

But

$$H_1 - H = \frac{v_1^2 - v^2}{2g} + \frac{p_1 - p}{w}$$

as z is constant;

$$\therefore \quad \frac{p_1 - p}{w} = \frac{v_1^2 - v^2}{g} - \frac{v_1^2 - v^2}{2g} = \frac{v_1^2 - v^2}{2g} = \frac{\omega^2(r_1^2 - r^2)}{2g} .$$

The pressure, therefore, decreases from the outside towards the center, as shown in the figure, and if the vortex is uncovered on the top, this curve will form the free surface. This is the principle of the centrifugal pump; the pressure of the fluid, entering at the center and leaving at the circumference, is increased at the circumference by the forced rotation of the fluid.

(a) $v_6 = 2\pi \times \frac{1}{2} \times 2 = 2\pi = 6.2832$ ft. per sec.

= velocity on entering the free vortex at 6 in. radius;

$\therefore$ in the free vortex

$$vr = C = 2\pi \times \frac{1}{2} = \pi ;$$

$\therefore \quad v_{12} = \frac{C}{1} = \pi = 3.1416$ ft. per sec. at 12 in. radius

Forced vortex,

$$\frac{p_6 - p_0}{\gamma} = \frac{v_6{}^2 - v_0{}^2}{\gamma} = \frac{4\pi^2 - 0}{\gamma} = .613 \text{ ft.} =$$

increase of pressure-head from center to 6 in. radius.

Free vortex,

$$\frac{p_{12} - p_6}{\gamma} = \frac{v_6{}^2 - v_{12}{}^2}{2g} = \frac{4\pi^2 - \pi^2}{2g} = \frac{3\pi^2}{64.4} = .460 \text{ ft.} =$$

increase from 6 in. to 12 in. radius;

$\therefore$ the pressure-head at 6 in. radius is .613 ft., and at 12 in. radius is 1.073 ft. greater than that at the center.

(b) The radial velocity u at 12 in. radius $= 3 \times \frac{6}{12} =$ 1.5 ft. per sec.

The inclination of the resultant velocity to the radius

$$= \tan^{-1} \frac{\pi}{1.5} = 64° \ 29'.$$

The resultant velocity = 1.5 sec. 64° 29' = 3.482 ft. per sec.

• PROBLEM 9-19

A point A on the free surface of a free vortex is at a radius r_A = 200 mm and a height z_A = 125 mm above datum. If the free surface at a distance from the axis of the vortex, which is sufficient for its effect to be negligible, is 180 mm above datum, what will be the height above datum of a point B on the free surface at a radius of 100 mm?

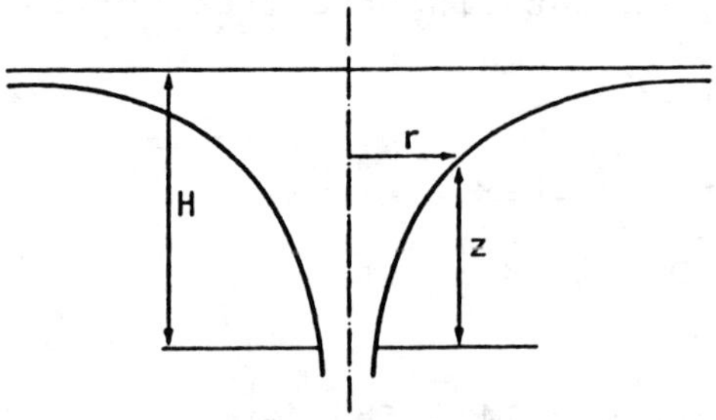

Free Vortex

Solution: For point A, the profile of the free surface of the free vortex is given by

$$H - z_A = C^2/2gr^2_A,$$

therefore,

$$C^2/2g = r^2(H - z_A).$$

Now H is the head above datum at an infinite distance from the axis of rotation, where the effect of the vortex is negligible, so that H = 180 mm = 0.18 m. Also, z_A = 0.125 m and r_A = 0.2 m. Substituting,

$$\frac{C^2}{2g} = 0.2^2(0.18 - 0.125) = 2.2 \times 10^{-3} m^3.$$

For point B,

$$H - z_B = C^2/2gr^2_B$$

$$z_B = H - C^2/2gr^2_B$$

$$= 0.18 - (2.2 \times 10^{-3})/0.1^2 = -0.04 \text{ m}$$

$$= 40 \text{ mm below datum.}$$

• PROBLEM 9-20

A closed vertical cylinder 400 mm diameter and 500 mm high is filled with oil of relative density 0.9 to a depth of 340 mm, the remaining volume containing air at atmospheric pressure. The cylinder revolves about its vertical axis at such a speed that the oil just begins to uncover the base. Calculate (a) the speed of rotation for this condition and (b) the upward force on the cover.

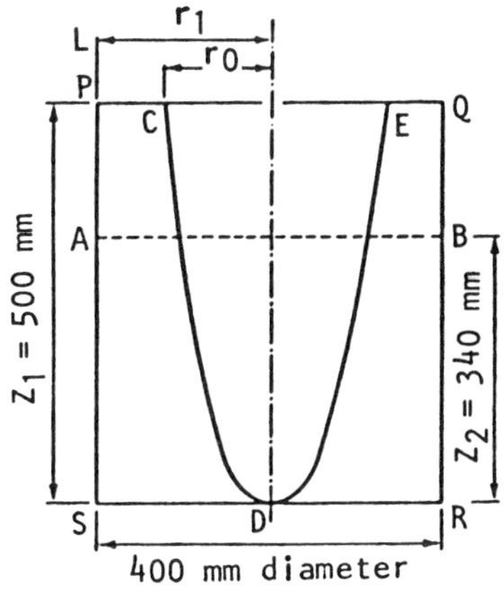

Solution: (a) When stationary, the free surface will be at AB (see figure), a height Z_2 above the base.

$$\text{Volume of oil} = \pi r_1^2 Z_2.$$

When rotating at the required speed ω, a forced vortex is formed and the free surface will be the paraboloid CDE

Volume of oil = Volume of cylinder PQRS -

Volume of paraboloid CDE

$$= \pi r_1^2 Z_1 - \frac{1}{2}\pi r_o^2 Z_1$$

since the volume of a paraboloid is equal to half the volume of the circumscribing cylinder.

No oil is lost from the container, therefore,

$$\pi r_1^2 Z_2 = \pi r_1^2 Z_1 - \frac{1}{2}\pi r_0^2 Z_1,$$

$$r_0^2 = 2r_1^2(1 - Z_2/Z_1),$$

$$r_0 = r_1\sqrt{\{2(1 - Z_2/Z_1)\}} = r_1\sqrt{\{2(1 - 340/500)\}}$$

$$= 0.8r_1 = 0.8 \times 200 = 160 \text{ mm}.$$

The profile of the parabolic free surface of the forced vortex is given by

$$z = \omega^2 r^2/2g + C. \tag{1}$$

or, between points C and D, taking D as datum level,

$$Z_D = 0$$

when $r = 0$

and

$$Z_C = Z_1$$

when $r = r_0$

giving

$$Z_1 - 0 = \omega^2 r_0^2/2g,$$

$$\omega = \sqrt{(2gZ_1/r_0^2)}$$

$$= \sqrt{(2 \times 9.81 \times 0.5/0.16^2)} = 19.6 \text{ rad s}^{-1}.$$

(b) The oil will be in contact with the top cover from radius $r = r_0$ to $r = r_1$. If p is the pressure at any radius r, the force on an annulus of radius r and width δr is given by

$$\delta F = p \times 2\pi r \delta r.$$

Integrating from $r = r_0$ to $r = r_1$,

$$\text{Force on top cover, } F = 2\pi \int_{r_0}^{r_1} pr\delta r \tag{2}$$

For any horizontal plane, for which Z will be constant, the pressure distribution will be given by

$$p/\rho g = \omega^2 r^2/2g + C.$$

Since the pressure at r_0 is atmospheric, $p = 0$ when $r = r_0$, so that

$$C = -\omega^2 r_0^2/2g$$

and

$$p = \rho g \left\{\frac{\omega^2 r^2}{2g} - \frac{\omega^2 r_0^2}{2g}\right\} = \frac{\rho\omega^2}{2g}(r^2 - r_0^2).$$

Substituting in (2),

$$F = 2\pi \frac{\rho\omega^2}{2} \int_{r_0}^{r_1} (r^2 - r_0^2) r \, dr$$

$$= \rho\omega^2\pi \int_{r_0}^{r_1} (r^3 - r_0^2 r) dr$$

$$= \rho\omega^2\pi\left[\frac{1}{4}r^4 - \frac{1}{2}r_0^2 r^2\right]_{r_0}^{r_1}$$

$$= \rho\omega^2\pi\left\{\frac{1}{4}r_1^4 - \frac{1}{4}r_0^4 - \frac{1}{2}r_0^2 r_1^2 + \frac{1}{2}r_0^4\right\}$$

$$= \frac{\rho\omega^2\pi}{4}(r_1^4 + r_0^4 - 2r_0^2 r_1^2) = \frac{\pi}{4}\rho\omega^2(r_1^2 - r_0^2)^2$$

$$= \frac{\pi}{4} \times (0.9 \times 1000) \times 19.6^2(0.2^2 - 0.16^2)^2 N$$

$$= 56.3 \text{ N}.$$

• PROBLEM 9-21

"A forced vortex" is the name given to a flow pattern in which the streamlines are concentric circles and the velocity is proportional to radius; in such a situation the fluid is moving exactly like a rotating rigid body. Forced vortex motion rarely occurs in practice, but it can readily be demonstrated by spinning a cylindrical container of liquid (with its axis vertical) on a turntable (see Fig.).

(a) Obtain a general expression relating pressure to radius r and elevation z in a forced vortex; take the fluid density to be constant and the axis of the vortex vertical.

(b) From the result of (a) derive an equation describing the shape of a constant-pressure surface in the vortex.

(c) A flow pattern in which the streamlines are concentric circles, and in which the fluid velocity V varies inversely as the radius r, is called a free vortex.

Obtain a general expression relating pressure with radius in such a vortex in a uniform-density fluid.

(d) Using the result of (c), calculate the difference in pressure between two points which are 10 ft and 0.2 miles from the center of a tornado, given that the velocity at the latter point is 1 ft/sec. Take air density ρ to be 0.0765 lbm/ft^3.

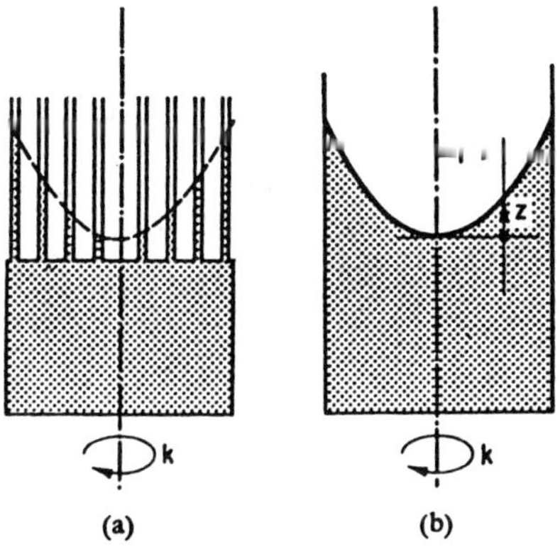

Solution: (a) The force-momentum relationship when applied in the direction perpendicular to streamlines is given by

$$\frac{\partial p}{\partial r} = \frac{\rho V^2}{r g_c} \tag{1}$$

In this problem, let V = kr, where V represents fluid velocity and k is a constant. Then from equation 1,

$$\frac{\partial p}{\partial r} = \frac{\rho V}{r g_c} = \frac{\rho k^2 r}{g_c}$$

$$\therefore \qquad p - p_0 = \frac{\rho k^2}{2g_c}(r^2 - r_0^2) ,$$

where p_0 is the pressure at a reference radius r_0.

This equation accounts for pressure variation in the r-direction only. Since the r-direction is always in a horizontal plane in the given flow, the variation of pressure with elevation can be added directly, thus giving as the required equation

$$p - p_0 = \frac{\rho k^2}{2g_c}(r^2 - r_0^2) - \frac{\rho g}{g_c}(z - z_0) .$$

(b) From the last equation the shape of a constant-pressure surface can easily be derived by putting $p/p_0 = j$ (a constant).

Then

$$p_0(j - 1) = \frac{\rho k^2}{2g_c}(r^2 - r_0^2) - \frac{\rho g}{g_c}(z - z_0). \tag{2}$$

Since the left-hand side is constant, this equation represents a parabola. The demonstration mentioned above does, in fact, result in the liquid surface (which is essentially a constant-pressure surface), assuming a parabolic shape. If the reference position is taken at the intersection of the axis of rotation and the liquid surface, then equation 2 simplifies to

$$0 = \frac{\rho k^2 r^2}{2g_c} - \frac{\rho g z}{g_c},$$

$$\therefore \quad z = \frac{k^2 r^2}{2g}.$$

(c) Let the velocity distribution be represented by the equation

$$V = k/r.$$

Substitution of this into equation 1, and subsequent integration, gives the required result:

$$\frac{dp}{dr} = \frac{\rho V^2}{r g_c} = \frac{\rho k^2}{r^3 g_c}$$

$$\therefore \quad p_1 - p_2 = \frac{\rho k^2}{2g_c}\left(\frac{1}{r_2^2} - \frac{1}{r_1^2}\right)$$

(d) If the motion can be treated as a free vortex, the constant k can be found from the conditions at the distant point,

$$k = V_1 r_1 = 1 \times (5280 \times 0.2)\,\text{ft}^2/\text{sec} = 1056\ \text{ft}^2/\text{sec}.$$

Then, inserting numerical values into the equation just derived,

$$p_1 - p_2 = \frac{0.0765 \times (1056)^2}{2 \times 32.2}\left[\frac{1}{(10)^2} - \frac{1}{(1056)^2}\right]\text{lbf/ft}^2$$

$$\simeq \frac{0.0765}{64.4} \times (105.6)^2\ \text{lbf/ft}^2 = 132.5\ \text{lbf/ft}^2.$$

The eye of a tornado has a radius of 75 ft. Find the variation of pressure in the flow field of the tornado if the maximum wind velocity is 150 fps. Assume that the core or so-called eye of the tornado behaves closely to that of solid body rotation, while the flow outside the eye is well represented by a free vortex field. (See Fig. 1).

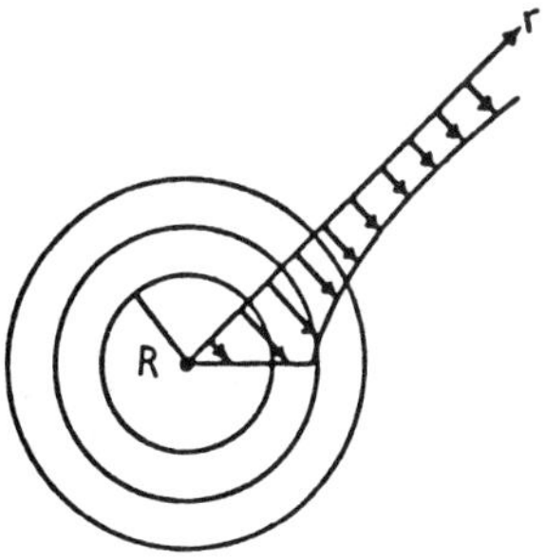

Fig. 1

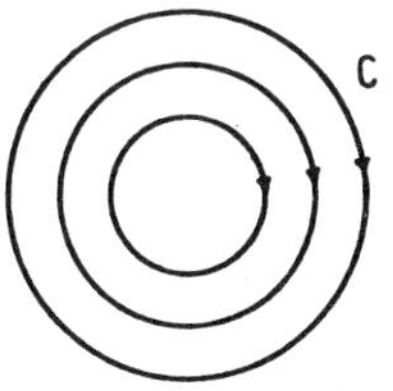

Fig. 2

Solution: As shown in Figure 2, a flow continuously circles the origin. This type of flow is called free vortex flow. In free vortex flow, we have the velocity components

$$v_r = \frac{\partial \phi}{\partial r} = 0$$

and

$$v_\theta = \frac{1}{r}\frac{\partial \phi}{\partial \theta} = -\frac{C}{r}$$

The flow velocity is thus in the tangential direction only and varies inversely as the distance from the origin. For positive values of the constant C, the flow is in the clockwise direction, since v_θ is taken as positive in the counter-clockwise direction. The tangential velocity distribution (there are no radial velocities in this case) will be a function of the radial distance from the center of the tornado, as shown in Figure 1. In the analysis, we neglect any translational motion of the tornado and consider motion only with respect to the center of the tornado. The streamlines of flow are circular, as shown in the figure. In the eye ($r \leq R$), we have rigid body rotation with velocity distribution

$$v_\theta = \omega r$$

Outside the eye ($r \geq R$), the flow is represented by a free vortex, and hence the velocity distribution is

$$v_\theta = \frac{C}{r}$$

Since the tangential velocity as given by the two equations must be equal at the edge of the eye (r = R), we can evaluate the constant C:

$$v_\theta = \omega R = \frac{C}{R}$$

hence

$$C = \omega R^2$$

Therefore the velocity distribution is

$$v_\theta = \omega r \qquad r \leq R$$

and

$$v_\theta = \frac{\omega R^2}{r} \qquad r \geq R$$

The maximum wind velocity occurs at the edge of the eye. If the maximum velocity is 150 fps at a radius of 75 ft, we obtain

$$\omega = \frac{v_\theta}{R} = \frac{150 \text{ fps}}{75 \text{ ft}} = 2 \text{ rad/sec}$$

and the velocity variation is given by

$$v_\theta = 2r\,(\text{fps}) \qquad r \leq 75 \text{ ft}$$

and

$$v_\theta = \frac{10{,}250}{r}(\text{fps}) \qquad r \geq 75 \text{ ft}$$

Since the flow outside the eye is potential, we can obtain its pressure distribution from Bernoulli's equation given by the equation

$$\frac{1}{2}V^2 + \frac{g_c}{\rho}p + gy = C$$

where $V^2 = u^2 + v^2$ is the total velocity with u and v as the x- and y- components, and C is constant throughout the region of irrotational flow, giving us

$$\frac{1}{2}v_\theta^2 + \frac{g_c}{\rho}p = \frac{g_c}{\rho}p_a$$

where p_a is the atmospheric pressure far from the eye and where the flow velocity vanishes. Rewrite to obtain the pressure change with respect to the atmospheric pressure

$$(p_a - p)\frac{g_c}{\rho} = \frac{1}{2}v_\theta^2 = \frac{1}{2}\,\frac{\omega^2 R^4}{r^2} = \frac{1}{2}\,\frac{V_\theta^2 R^2}{r^2}$$

where V_θ is the magnitude of the velocity at r = R, the edge of the eye. Since v_θ^2 is always positive, the pressure external to the eye will always be less than atmospheric pressure. This underpressure is in part responsible for the damage caused by tornadoes.

To evaluate the pressure distribution within the eye, we must refer to Euler's equations, for the flow is rotational there. Euler's equations, in rectangular coordinates, are given by

$$-\frac{g_c}{\rho}\frac{\partial p}{\partial x} = u\frac{\partial u}{\partial x} + v\frac{\partial u}{\partial y}$$

$$-\frac{g_c}{\rho}\frac{\partial p}{\partial y} - g = u\frac{\partial v}{\partial x} + v\frac{\partial v}{\partial y}$$

In steady, two-dimensional cylindrical coordinates and in the absence of gravity effects, Euler's equations in cylindrical coordinates can be shown to be

$$-\frac{g_c}{\rho}\frac{\partial p}{\partial r} = v_r\frac{\partial v_r}{\partial r} + \frac{v_\theta}{r}\frac{\partial v_r}{\partial \theta} - \frac{v_\theta^2}{r}$$

and

$$-\frac{\rho}{g_c}\frac{\partial p}{r\partial \theta} = v_r\frac{\partial v_\theta}{\partial r} + \frac{v_\theta}{r}\frac{\partial v_\theta}{\partial \theta} + \frac{v_\theta v_r}{r}$$

In the case under consideration here, these equations simplify, since $v_r = 0$ (no radial velocities) and v_θ is independent of the variable θ

$$\frac{-g_c}{\rho}\frac{\partial p}{\partial r} = \frac{-v_\theta^2}{r}$$

and

$$\frac{-g_c}{\rho}\frac{\partial p}{r\partial \theta} = 0.$$

The second equation indicates that the pressure p is a function of the radial distance r only, and the first equation can thus be written as an ordinary differential equation

$$\frac{g_c}{\rho}\frac{dp}{dr} = \frac{\omega^2 r^2}{r} = \omega^2 r$$

Integrating, we obtain

$$p = \frac{1}{2g_c}\omega^2 r^2 \quad + \text{constant}$$

The constant is evaluated from the previously obtained pressure relation in the potential region of the flow by equating the two pressures at the edge of the eye (r = R). Thus

$$p = \frac{1}{2g_c}\rho\omega^2 R^2 + \text{constant} = p_a - \frac{1}{2g_c}\rho V_\theta^2$$

Hence the constant is

$$\text{constant} = p_a - \frac{1}{2g_c}\rho V_\theta^2 - \frac{1}{2g_c}\frac{\rho V_\theta^2 R^2}{R^2}$$

$$= p_a - \frac{\rho}{g_c}V_\theta^2$$

and

$$p = \frac{\rho}{2g_c}V_\theta^2\frac{r^2}{R^2} - \frac{\rho}{g_c}V_\theta^2 + p_a$$

Simplifying, we obtain

$$(p_a - p)\frac{g_c}{\rho} = V_\theta^2\left(1 - \frac{1}{2}\frac{r^2}{R^2}\right)$$

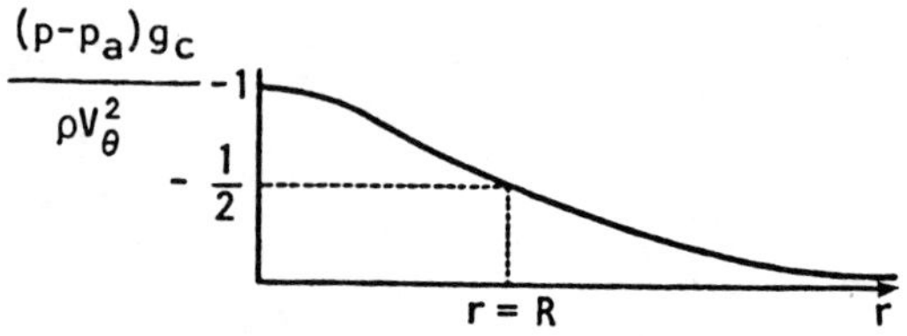

Fig. 3

We can now sketch the variation of pressure in the entire flow field as in Figure 3. From the figure it is seen that the minimum pressure occurs at the center of the eye and that the entire pressure field of the tornado is below atmospheric.

VELOCITY POTENTIAL AND STREAM FUNCTION

• PROBLEM 9-23

Find the velocity potential of a uniform stream with velocity V inclined at an angle α to the x axis (see Figure).

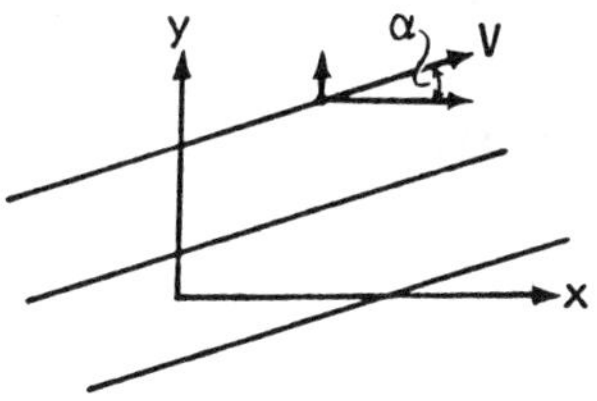

Solution: The velocity potential $\phi(x, y)$ is a function of the space coordinates x, y and is related to the velocity components as follows:

$$u = \frac{\partial \phi}{\partial x}$$
$$v = \frac{\partial \phi}{\partial y} \tag{1}$$

That is, the velocity component u is given by the partial derivative of ϕ in the x direction and the velocity component v by the partial derivative of ϕ in the y direction. With $u = V \cos \alpha$ and $v = V \sin \alpha$ in (1), we obtain

$$\frac{\partial \phi}{\partial x} = V \cos \alpha$$

and

$$\frac{\partial \phi}{\partial y} = V \sin \alpha$$

Integrating, we obtain

$$\phi = V \cos \alpha \cdot x + V \sin \alpha \cdot y + C$$

where C is the constant of integration, which can be set equal to zero, since the velocity components are related to the spatial derivatives of the potential and not to the value of the potential.

• PROBLEM 9-24

Assume that $\phi = \ln (x^2 + y^2)^{\frac{1}{2}}$ is the velocity potential of a two-dimensional irrotational incompressible flow defined everywhere but at the origin. Determine the stream function for this flow.

Solution:

$$\psi = \int \frac{\partial \phi}{\partial x} dy + f(x) = \int \frac{x}{x^2 + y^2} dy + f(x) \qquad \text{(a)}$$

$$\psi = -\int \frac{\partial \phi}{\partial y} dx + g(y) = -\int \frac{y}{x^2 + y^2} dx + g(y) \qquad \text{(b)}$$

Carrying out the partial integrations,

$$\psi = \tan^{-1} \frac{y}{x} + f(x) \qquad \text{(c)}$$

$$\psi = -\tan^{-1} \frac{x}{y} + g(y) \qquad \text{(d)}$$

From simple trigonometric relations we may say

$$\tan^{-1} \frac{x}{y} = \frac{\pi}{2} - \tan^{-1} \frac{y}{x} \qquad \text{(e)}$$

Substituting the above expression into (d) and equating the right-hand sides of (c) and (d) gives us

$$\tan^{-1} \frac{y}{x} + f(x) = -\frac{\pi}{2} + \tan^{-1} \frac{y}{x} + g(y) \qquad \text{(f)}$$

From this it is clear that $f(x) - g(y) = -\pi/2$. Since $f(x)$ is a function only of x and $g(y)$ is a function only of y, it is necessary that each be constant in order that the expression $f(x) - g(y)$ be constant over the entire range of independent variables x and y. It is usually permissible to drop the additive constants in the stream function. Hence, the stream function is seen to be

$$\psi = \tan^{-1} \frac{y}{x} \qquad \text{(g)}$$

If the function ψ is known, the velocity potential may be ascertained by a similar procedure.

• PROBLEM 9-25

Given a flow field in the x-y plane, which is defined by the relationship

$$h = cx^2 - cy^2 \qquad (1)$$

(1) Determine whether the flow is irrotational, and

(2) find the velocity potential, g. Thereafter

(3) show that g and h are orthogonal.

Solution: (1) The flow is irrotational if the angular velocity component, $\omega_z = 0$. Now

$$\omega_z = \frac{\partial v}{\partial x} - \frac{\partial m}{\partial y} \qquad (2)$$

Since

$$m = \frac{\partial h}{\partial y} \quad \text{and} \quad v = -\frac{\partial h}{\partial x}$$

then m and v can be found by operating on equation (1). Consequently,

$$m = \frac{\partial}{\partial y}(cx^2 - cy^2) = -2cy \qquad (3)$$

$$v = -\frac{\partial}{\partial x}(cx^2 - cy^2) = -2cx \qquad (4)$$

Finding now the partial derivatives in equation (2) for ω_z,

$$\frac{\partial v}{\partial x} = -2c \; ; \qquad \frac{\partial m}{\partial y} = -2c$$

Substituting into (2),

$$\omega_z = -2c - (-2c) = 0$$

The flow is consequently irrotational.

(2) The velocity potential, g, can be found from

$$m = -\frac{\partial g}{\partial x} \qquad (5)$$

and

$$v = -\frac{\partial g}{\partial y} \qquad (6)$$

Now, from equation (3) we found that $m = -2cy$. Equating this expression to the right side of equation (5),

$$-\frac{\partial g}{\partial x} = -2cy$$

Integrating this relationship with respect to x, we obtain

$$g = 2cyx + q(y) \qquad (7)$$

where q(y) is an arbitrary function of y.

Applying now equation (6) to (7),

$$v = -\frac{\partial g}{\partial y} = -2cx - \frac{\partial q(y)}{\partial y} \qquad (8)$$

But v has been found to be $-2cx$ from equation (4). Equating this expression to (8),

$$-2cx - \frac{\partial q(y)}{\partial y} = -2cx$$

from which we obtain that

$$\frac{\partial q(y)}{\partial y} = 0$$

and carrying out the integration tells us that q must be a constant, k. Consequently,

$$g = 2cyx + k$$

(3) For g and h to be orthogonal, the slope of one function must equal the negative reciprocal of the slope of the other function, or

$$\left.\frac{dy}{dx}\right)_{h=\text{const}} = -\left.\frac{dx}{dy}\right)_{g=\text{const}}$$

If h is constant, then dh = 0, and from (1),

$$dh = 0 = 2cxdx - 2cydy$$

Solving for the slope dy/dx,

$$\left.\frac{dy}{dx}\right)_{h=\text{const}} = \frac{x}{y}$$

Similarly,

$$dg = 0 = 2cydy + 2cxdx$$

From which

$$\left.\frac{dx}{dy}\right)_{g=\text{const}} = -\frac{y}{x}$$

The orthogonality property has therefore been shown.

• PROBLEM 9-26

If ϕ_1 and ϕ_2 represent two different solutions of Laplace's equation

$$\nabla^2 \phi = 0 \tag{1}$$

show that

$$\phi_3 = \phi_1 \pm \phi_2 \tag{2}$$

also satisfies Laplace's equation, and

$$\phi_4 = \phi_1 \phi_2 \tag{3}$$

does not satisfy Laplace's equation.

Solution: Substituting the velocity potential ϕ_3 given by Equation (2) into Laplace's equation (1) gives

$$\frac{\partial^2\phi_3}{\partial x^2} + \frac{\partial^2\phi_3}{\partial y^2} + \frac{\partial^2\phi_3}{\partial z^2} = \frac{\partial^2\phi_1}{\partial x^2} + \frac{\partial^2\phi_1}{\partial y^2} + \frac{\partial^2\phi_1}{\partial z^2} +$$

$$\frac{\partial^2 \phi_2}{\partial x^2} + \frac{\partial^2 \phi_2}{\partial y^2} + \frac{\partial^2 \phi_2}{\partial z^2} = 0$$

since $\nabla^2 \phi_1 = 0$ and $\nabla^2 \phi_2 = 0$.

Consider next the velocity potential ϕ_4. From Equation (3), since

$$\frac{\partial \phi_4}{\partial x} = \phi_1 \frac{\partial \phi_2}{\partial x} + \phi_2 \frac{\partial \phi_1}{\partial x}$$

and

$$\frac{\partial^2 \phi_4}{\partial x^2} = \phi_1 \frac{\partial^2 \phi_2}{\partial x^2} + \phi_2 \frac{\partial^2 \phi_1}{\partial x^2} + 2 \frac{\partial \phi_1}{\partial x} \frac{\partial \phi_2}{\partial x}$$

with similar expression for $\frac{\partial^2 \phi_4}{\partial y^2}$ and $\frac{\partial^2 \phi_4}{\partial z^2}$, then

$$\frac{\partial^2 \phi_4}{\partial x^2} + \frac{\partial^2 \phi_4}{\partial y^2} + \frac{\partial^2 \phi_4}{\partial z^2} = \phi_1\left(\frac{\partial^2 \phi_2}{\partial x^2} + \frac{\partial^2 \phi_2}{\partial y^2} + \frac{\partial^2 \phi_2}{\partial z^2}\right) + \phi_2\left(\frac{\partial^2 \phi_1}{\partial x^2} + \frac{\partial^2 \phi_1}{\partial y^2} + \frac{\partial^2 \phi_1}{\partial z^2}\right) + 2\left(\frac{\partial \phi_1}{\partial x}\frac{\partial \phi_2}{\partial x} + \frac{\partial \phi_1}{\partial y}\frac{\partial \phi_2}{\partial y} + \frac{\partial \phi_1}{\partial z}\frac{\partial \phi_2}{\partial z}\right) \qquad (4)$$

But $\nabla^2 \phi_1 = 0$ and $\nabla^2 \phi_2 = 0$ by Equation (1), so that Equation (4) becomes

$$\nabla^2 \phi_4 = 2\left(\frac{\partial \phi_1}{\partial x}\frac{\partial \phi_2}{\partial x} + \frac{\partial \phi_1}{\partial y}\frac{\partial \phi_2}{\partial y} + \frac{\partial \phi_1}{\partial z}\frac{\partial \phi_2}{\partial z}\right)$$

which is not necessarily zero.

• PROBLEM 9-27

A source and sink of equal strength of $Q = 20\ ft^3/sec$ is shown in Fig. 1. What is the potential at the location $(x,y) = (15,15)$?

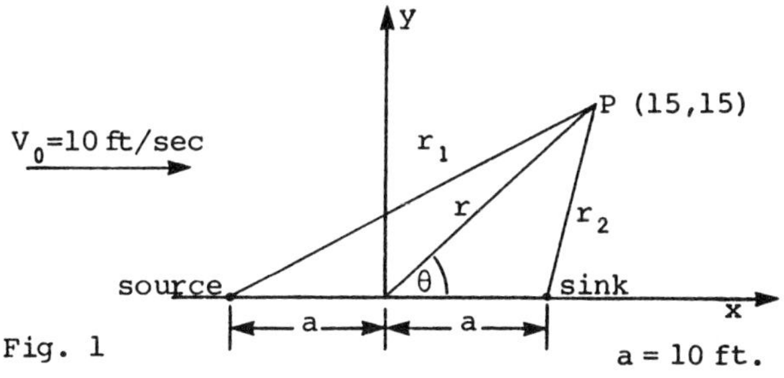

Fig. 1

Solution: At point P(15,15) the complete potential ϕ is a combination of three different flows from the source, sink and the uniform velocity parallel to the line joining the source and the sink. Therefore,

$$\phi = V_o x + \frac{Q}{2\pi}(\ln r_1 - \ln r_2) \tag{1}$$

where, from the figure, $x = r\cos\theta$.

From the law of cosines,

$$r_1 = (r^2 + a^2 + 2ra\cos\theta)^{\frac{1}{2}} \tag{2}$$

$$r_2 = (r^2 + a^2 - 2ra\cos\theta)^{\frac{1}{2}} \tag{3}$$

Substituting equations (2) and (3) into (1), we have

$$\phi = V_o r\cos\theta + \frac{Q}{4\pi}\left[\ln(r^2 + a^2 + 2ra\cos\theta) - \ln(r^2 + a^2 - 2ra\cos\theta)\right]$$

The radial velocity at point P is

$$V_r = -\frac{\partial\phi}{\partial r} = -\left\{V_o\cos\theta + \frac{Q}{2\pi}\left[\frac{(r + a\cos\theta)}{(r^2 + a^2 + 2ra\cos\theta)} - \frac{(r - a\cos\theta)}{(r^2 + a^2 - 2ra\cos\theta)}\right]\right\} \tag{4}$$

The angular velocity at point P is

$$V_\theta = -\frac{1}{r}\frac{\partial\phi}{\partial\theta} = -\frac{1}{r}\left[-V_o r\sin\theta + \frac{Q}{2\pi}\left[\frac{-(ra\sin\theta)}{(r^2 + a^2 + 2ra\cos\theta)} - \frac{(ra\sin\theta)}{(r^2 + a^2 - 2ra\cos\theta)}\right]\right] \tag{5}$$

For,

$$r = \frac{15}{\cos 45^0} = 21.21, \quad \theta = 45^0, \quad a = 10 \text{ ft.}$$

$$V_r = -7.019 \text{ ft/sec}$$

$$V_\theta = -7.0126 \text{ ft/sec}$$

• PROBLEM 9-28

A source with the strength 6 cfs/ft and a vortex with strength 12 ft^2/sec are located at the origin. Determine the equation for velocity potential and stream function. What are the velocity components at x = 2, y = 3?

$$u = -\frac{\partial\phi}{\partial x} \tag{1}$$

$$v = -\frac{\partial\phi}{\partial y} \tag{2}$$

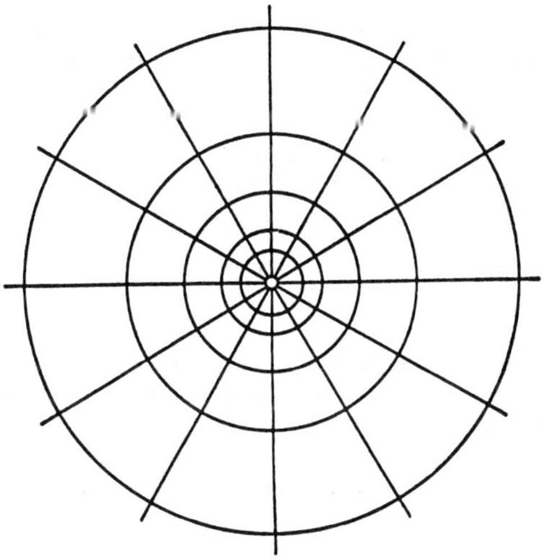

FLOW NET FOR SOURCE OR VORTEX.

Solution: A line normal to the xy-plane, from which fluid is imagined to flow uniformly in all directions at right angles to it, is a source. It appears as a point in the customary two-dimensional flow diagram. The total flow per unit time per unit length of line is called the strength of the source. As the flow is in radial lines from the source, the velocity a distance r from the source is determined by the strength divided by the flow area of the cylinder, or $2\pi\mu/2\pi r$, in which the strength is $2\pi\mu$. Then, since by Eq.1 the velocity in any direction is given by the negative derivative of the velocity potential with respect to the direction,

$$-\frac{\partial \phi}{\partial r} = \frac{\mu}{r}, \qquad \frac{\partial \phi}{\partial \theta} = 0$$

and

$$\phi = -\mu \ln r$$

is the velocity potential, in which ln indicates the natural logarithm and r is the distance from the source. This value of ϕ satisfies the Laplace equation in two dimensions.

The streamlines are radial lines from the source, i.e.,

$$\frac{\partial \psi}{\partial r} = 0, \qquad -\frac{1}{r}\frac{\partial \psi}{\partial \theta} = \frac{\mu}{r}$$

From the second equation

$$\psi = -\mu\theta$$

Lines of constant ϕ (equipotential lines) and constant ψ are shown in the Fig. A sink is a negative source, a line into which fluid is flowing.

Vortex: In examining the flow case given by selecting the stream function for the source as a velocity potential,

$$\phi = -\mu\theta, \qquad \psi = \mu \ln r$$

which also satisfies the Laplace equation, it is seen that the equipotential lines are radial lines and the streamlines are circles.

The velocity potential for the source is

$$\phi = -\frac{6}{2\pi} \ln r$$

and the corresponding stream function is

$$\psi = -\frac{6}{2\pi}\theta$$

The velocity potential for the vortex is

$$\phi = -\frac{12}{2\pi}\theta$$

and the corresponding stream function is

$$\psi = \frac{12}{2\pi} \ln r$$

By adding the respective functions

$$\phi = -\frac{3}{\pi}(\ln r + 2\theta)$$

and

$$\psi = -\frac{3}{\pi}(\theta - 2 \ln r)$$

The radial and tangential velocity components are

$$v_r = -\frac{\partial\phi}{\partial r} = \frac{3}{\pi r}, \qquad v_\theta = -\frac{1}{r}\frac{\partial\phi}{\partial\theta} = \frac{6}{\pi r}$$

At (2,3), $r = \sqrt{2^2 + 3^2} = 3.61$, $v_r = 0.265$, $v_\theta = 0.53$

Determine which of the following functions represent a possible velocity potential, and sketch the flow pattern for these possible cases.

(a) $f = Ux$

(b) $f = Vy$

(c) $f = Kx^3$

(d) $f = \sin(x + y)$

(e) Parts a and b were solutions to the Laplace equation and, hence, were velocity potentials. Prove that the sum of these two velocity potentials is a third velocity potential and explain the flow pattern that this new velocity potential represents.

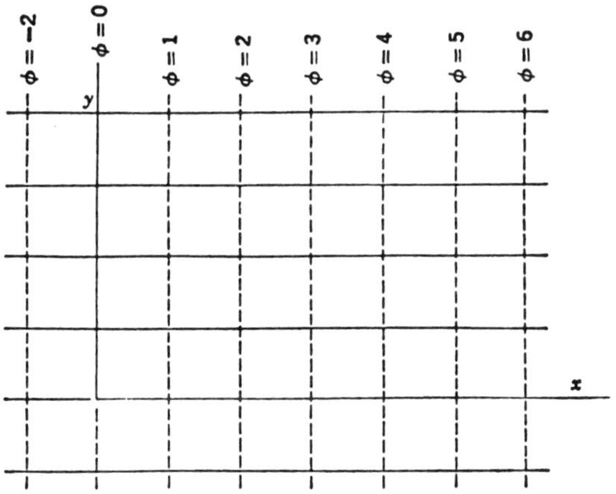

Fig. 1

Solution: (a)

$f = Ux$

This function must represent a flow field that satisfies the equation of continuity in order to be a possible velocity potential, which means it must satisfy Laplace's equation.

$$\frac{\partial^2 f}{\partial x^2} + \frac{\partial^2 f}{\partial y^2} = 0$$

$$\frac{\partial^2 f}{\partial x^2} = \frac{\partial^2 (Ux)}{\partial x^2} = 0$$

$$\frac{\partial^2 f}{\partial y^2} = \frac{\partial^2 (Ux)}{\partial y^2} = 0$$

or

$$0 + 0 = 0$$

This function is a velocity potential.

(b) $f = Vy$

This function also must satisfy the Laplace equation.

$$\frac{\partial^2 f}{\partial x^2} = \frac{\partial^2 (Vy)}{\partial x^2} = 0;$$

$$\frac{\partial^2 f}{\partial y^2} = \frac{\partial^2 (Vy)}{\partial y^2} = 0;$$

The Laplace equation is

$$\frac{\partial^2 f}{\partial x^2} + \frac{\partial^2 f}{\partial y^2} = 0$$

Substitute for f to get

$$\frac{\partial^2 (Vy)}{\partial x^2} + \frac{\partial^2 (Vy)}{\partial y^2} = 0 + 0 = 0$$

This function is a velocity potential.

(c) $f = Kx^3$

$$\frac{\partial^2 f}{\partial x^2} + \frac{\partial^2 f}{\partial y^2} = \frac{\partial^2 (Kx^3)}{\partial x^2} + \frac{\partial^2 (Kx^3)}{\partial y^2} = 6\ Kx$$

This does not satisfy the Laplace equation for it does not equal zero, and, accordingly, it is not a possible velocity potential.

(d) $f = \sin(x + y)$

$$\sin(x + y) = \sin x \cos y + \cos x \sin y$$

$$\frac{\partial^2 f}{\partial x^2} = \frac{\partial^2(\sin x \cos y + \cos x \sin y)}{\partial x^2}$$

$$= - \sin x \cos y - \cos x \sin y$$

$$\frac{\partial^2 f}{\partial y^2} = \frac{\partial^2(\sin x \cos y + \cos x \sin y)}{\partial y^2}$$

$$= - \sin x \cos y - \cos x \sin y$$

By substituting these values of $\partial^2 f/\partial x^2$ and $\partial^2 f/\partial y^2$ into the Laplace equation, you get

$$(- \sin x \cos y - \cos x \sin y)$$

$$+ (- \sin x \cos y - \cos x \sin y) = 0$$

$$- 2 \sin x \cos y - 2 \cos x \sin y = 0$$

$$2 \sin(x + y) = 0$$

which is not necessarily true, and so $f = \sin(x + y)$ is not a velocity potential.

The functions

$$f = Ux$$

$$f = Vy$$

are velocity potentials. Figure 1 is a sketch of the flow pattern represented by $\phi = Ux$. The vertical dotted lines are lines of constant ϕ called equipotential lines. The horizontal lines are everywhere normal to the equipotential lines and, hence, are streamlines. You compute the velocity at any point from the definition of the velocity potential.

$$u = - \frac{\partial \phi}{\partial x}, \qquad v = - \frac{\partial \phi}{\partial y}$$

Consequently,

$$u = - \frac{\partial (Ux)}{\partial x} = - U$$

$$v = -\frac{\partial(Ux)}{\partial y} = 0$$

The velocity potential $\phi = Ux$ represents horizontal flow of an ideal fluid from higher to lower values of x at a constant velocity U.

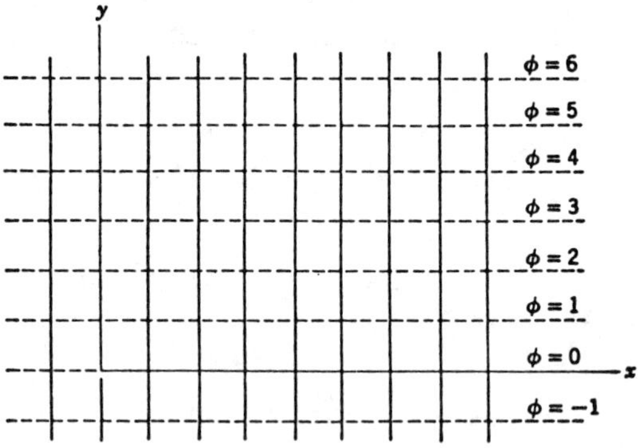

Fig. 2

Figure 2 is a sketch of the flow pattern represented by $\phi = Vy$. In this case, the horizontal lines are equipotential lines and the vertical lines are streamlines. Compute velocity as before.

$$u = -\frac{\partial\phi}{\partial x} = -\frac{\partial(Vy)}{\partial x} = 0$$

$$v = -\frac{\partial\phi}{\partial y} = -\frac{\partial(Vy)}{\partial y} = -V$$

Therefore, the flow, with a constant velocity V, is in the negative y direction.

e) From part (a)

$$\phi_1 = Ux$$

and from part (b)

$$\phi_2 = Vy$$

New velocity potential ϕ is

$$\phi = Ux + Vy$$

You know that the sum of two velocity potentials is a third one, and, therefore, ϕ is a solution to the Laplace equation. Check this statement once to demonstrate that it is correct.

$$\frac{\partial^2 (Ux + Vy)}{\partial x^2} + \frac{\partial^2 (Ux + Vy)}{\partial y^2} = 0$$

$$\frac{\partial U}{\partial x} + \frac{\partial V}{\partial y} = 0 + 0 = 0$$

It checks as you know it must.

To explain the flow pattern corresponding to $\phi = Ux + Vy$, first sketch the equipotential lines.

$$\phi = Ux + Vy = c_1$$

$$y = -\frac{U}{V}x + \frac{c_1}{V}$$

This is the equation of a family of straight lines with a slope $-U/V$. The streamlines are normal to the equipotential lines and must have a slope equal to the negative reciprocal of the slope of the equipotential lines. The streamlines also are a family of straight lines.

$$y = \frac{V}{U}x + c_2$$

where c_2 is a different constant for each straight line.

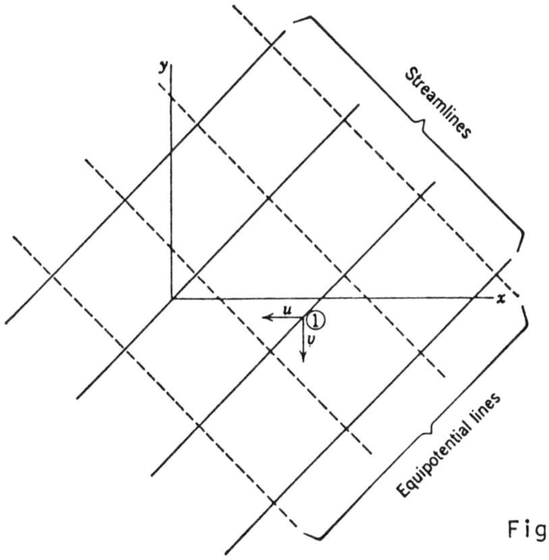

Fig. 3

Figure (3) is a sketch of the streamlines and equipotential lines.

From the velocity potential definition

$$u = -\frac{\partial \phi}{\partial x}$$

$$v = -\frac{\partial\phi}{\partial y}$$

Thus,

$$u = -\frac{\partial(Ux + Vy)}{\partial x} = -U$$

$$v = -\frac{\partial(Ux + Vy)}{\partial y} = -V$$

Point 1 (Fig. 3) is arbitrary and shows the components of the velocity which are constant through out the field. The velocity is the vector sum of the components which means that the magnitude $|q|$ is

$$|q| = \sqrt{U^2 + V^2}$$

Figure 3 also shows that the flow is downward to the left at a slope V/U.

• PROBLEM 9-30

Determine the velocity components, u, v, and w for

a) $\phi = ax + by + cz$

b) $\phi = U_0 x$

c) $\phi = -1/2(ax^2 + by^2 + cz^2)$

Solution: Any linear function of (x,y,z), $\phi = ax + by + cz$ for example, is a solution of the Laplace equation. It will represent a translatory motion of constant velocity

$$u = -\frac{\partial\phi}{\partial x} = -a \qquad v = -\frac{\partial\phi}{\partial y} = -b$$

$$w = -\frac{\partial\phi}{\partial z} = -c$$

Here the simplest case is that of uniform flow in a given coordinate direction, that is

$$\phi = U_0 x \quad \text{and} \quad u = -\frac{\partial\phi}{\partial x} = -U_0$$

giving flow in the negative x-direction. The quadratic form $\phi = -1/2(ax^2 + by^2 + cz^2)$ is a solution of the Laplace equation only if $\nabla^2\phi = a + b + c = 0$. Take $a = -b$ and $c = 0$, then

$$\phi = -\frac{a}{2}(x^2 - y^2)$$

and

$$u = ax \qquad v = -ay \qquad w = 0$$

• PROBLEM 9-31

A flow field is defined by the relationship

$$h(x,y) = 3x^2y - y^3$$

(a) Determine whether the flow is irrotational.

(b) Determine whether the magnitude of the velocity at an arbitrary point in the flow field is a function of only the distance of the point from the origin of the x-y coordinates.

Solution: (a) In examining the function h(x,y), it is seen that this function and hence the flow field is confined to the x-y plane. Consequently, the flow is irrotational if $\omega_z = 0$.

Now

$$\omega_z = \frac{1}{2}\left(\frac{\partial v}{\partial x} - \frac{\partial m}{\partial y}\right) \qquad (1)$$

where

$$v = -\frac{\partial h}{\partial x} \qquad (2)$$

and

$$m = +\frac{\partial h}{\partial y} \qquad (3)$$

From (2)

$$\frac{\partial v}{\partial x} = -\frac{\partial^2 h}{\partial x^2}$$

and $-\frac{\partial^2 h}{\partial x^2} = -6y$ obtained from the flow field relationship.

Similarly,

$$\frac{\partial m}{\partial y} = +\frac{\partial^2 h}{\partial y^2} = +6y$$

Substituting into equation (1),

$$\omega_z = \frac{1}{2}\left(-\frac{\partial^2 h}{\partial x^2} + \frac{\partial^2 h}{\partial y^2}\right) \tag{4}$$

Inserting values into (4),

$$\omega_z = \frac{1}{2}(-6y + 6y) = 0$$

Consequently, the flow is irrotational.

To investigate the magnitude of the velocity at any point in the flow field, we take note that velocity is a vector quantity having both magnitude and direction. The velocity vector $\bar{R}$ is determined by its components m and v taken along a set of coordinate axes. The resultant of those components m and v gives the magnitude and direction. We are interested in the magnitude only.

Therefore

$$R = \sqrt{v^2 + m^2}$$

Now

$$v = -\frac{\partial h}{\partial x}$$

and

$$m = \frac{\partial h}{\partial y}$$

Finding these derivatives from the given equation for the flow field,

$$-\frac{\partial h}{\partial x} = -6yx = v$$

and

$$\frac{\partial h}{\partial y} = 3x^2 - 3y^2 = m$$

$$v^2 = 36y^2x^2$$

$$m^2 = 9x^4 - 18x^2y^2 + 9y^4$$

$$m^2 + v^2 = 9x^4 + 18x^2y^2 + 9y^4$$

$$\sqrt{m^2 + v^2} = \sqrt{9(x^4 + 2x^2y^2 + y^4)}$$

factoring,

$$= \sqrt{9(x^2 + y^2)(x^2 + y^2)}$$

$$= \sqrt{3^2(x^2 + y^2)^2}$$

from which

$$\sqrt{m^2 + v^2} = 3(x^2 + y^2) = R$$

But x and y are the position coordinates of the point, and the distance S of the point from the origin of the x-y axes is given by

$$S = \sqrt{x^2 + y^2}$$

so that $x^2 + y^2 = S^2$.

Substituting in the last expression for R, there is obtained

$$R = 3S^2$$

and consequently the magnitude of the velocity R in the flow field at an arbitrary point is a function of only the distance of the point from the origin - in particular the square of that distance.

• PROBLEM 9-32

The complex potential

$$w(z) = V_0\left[z + y_0 e^{2\pi z\iota/\lambda}\right]$$

represents the flow over a sinusoidally corrugated wall where y_0 is the amplitude of the corrugation and λ is the wavelength. Find the amplitude y corresponding to the streamline, $\psi = 0$.

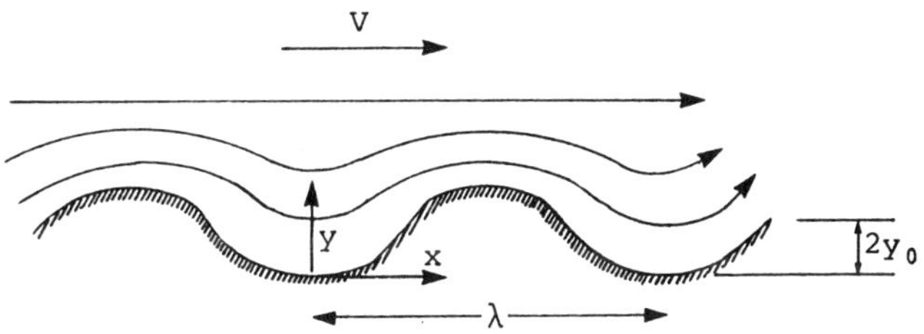

Solution: $$w(z) = V_0\left[z + y_0 e^{2\pi z\iota/\lambda}\right]$$
$$= V_0\left[(x + \iota y) + y_0 e^{2\pi(x + \iota y)\iota/\lambda}\right]$$

or

$$w(z) = \phi + \iota\psi = V_0\left[x + \iota y + y_0 e^{-2\pi y/\lambda + 2\pi x\iota/\lambda}\right]$$

$$w(z) = \phi + \iota\psi = V_0\left[x + \iota y + y_0 e^{-(2\pi y/\lambda)}\left(\cos\frac{2\pi x}{\lambda} + \iota\sin\frac{2\pi x}{\lambda}\right)\right]$$

$$w(z) = \phi + \iota\psi = V_0\left[x + y_0 e^{-(2\pi y/\lambda)}\cos\frac{2\pi x}{\lambda}\right.$$
$$\left. + \iota\left(y + y_0 e^{-(2\pi y/\lambda)}\sin\frac{2\pi x}{\lambda}\right)\right]$$

Therefore the stream function is

$$\psi = V_o \left\{ y + y_o e^{-(2\pi y/\lambda)} \sin \frac{2\pi x}{\lambda} \right\}$$

The streamline $\psi = 0$ is then given by

$$y + y_o e^{-(2\pi y/\lambda)} \sin \frac{2\pi x}{\lambda} = 0$$

or

$$y = -y_o \sin \frac{2\pi x}{\lambda} \left\{ 1 - \frac{2\pi y}{\lambda} + \frac{(2\pi y/\lambda)^2}{2!} - \frac{(2\pi y/\lambda)^3}{3!} + \dots \right\}$$

When $2\pi y/\lambda$ is small compared to unity, the above equation becomes

$$y = -y_o \sin \frac{2\pi x}{\lambda}$$

• **PROBLEM 9-33**

Investigate the stream function in polar coordinates

$$\psi = U \sin \theta \left(r - \frac{R^2}{r} \right) \qquad (1)$$

Where U and R are constants: a velocity and a length respectively. Plot the streamline. What does the flow represent? Is it a realistic solution to the basic equations?

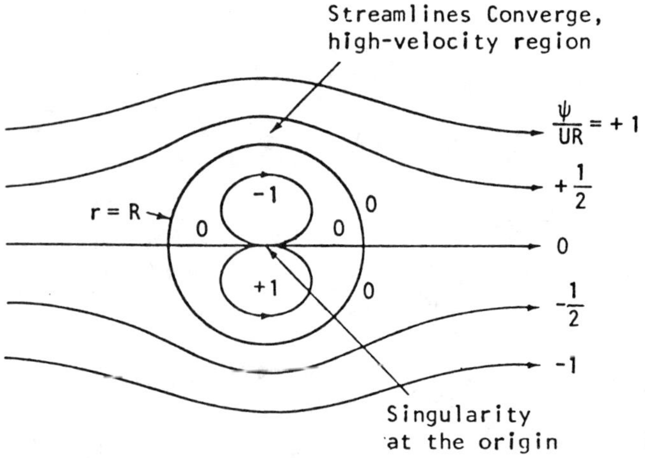

Fig. 1

Solution: The streamlines are lines of constant ψ, which has units of square meters per second. Note that ψ/UR is dimensionless. Rewrite Eq. (1) in dimensionless form

$$\frac{\psi}{UR} = \sin\theta\left(\eta - \frac{1}{\eta}\right) \qquad \eta = \frac{r}{R} \tag{2}$$

Of particular interest is the special line $\psi = 0$. From Eq. (1) or (2) this occurs when (a) $\theta = 0°$ or $180°$ and (b) $r = R$. Case (a) is the x axis and case (b) is a circle of radius R, both of which are plotted in Fig. 1.

For any other nonzero value of ψ, it is easiest to pick a value of r and solve for θ:

$$\sin\theta = \frac{\psi/UR}{r/R - R/r}$$

In general, there will be two solutions for θ because of the symmetry about the y axis. For example $\psi/UR = +1.0$;

Guess r/R	3.0	2.5	2.0	1.8	1.7	1.618
Compute	22°	28°	42°	54°	64°	90°
	158°	152°	138°	156°	116°	

This line is plotted in Fig. 1 and passes over the circle r = R. You have to watch it, though, because there is a second curve for $\psi/UR = +1.0$ for small r < R below the x axis.

Guess r/R	0.618	0.6	0.5	0.4	0.3	0.2	0.1
Compute	-90°	-70°	-42°	-28°	-19°	-12°	-6°
		-110°	-138°	-152°	-161°	-168°	-174°

This second curve plots as a closed curve inside the circle r = R. there is a singularity of infinite velocity and indeterminate flow direction at the origin. Figure 1 shows the full pattern.

The given stream function, Eq. (1) is an exact and classic solution to the momentum equation for frictionless flow. Outside the circle r = R represents two dimensional inviscid flow of a uniform stream past a circular cylinder. Inside the circle it represents a rather ugly and unrealistic trapped circulating motion of what is called a line doublet.

A revolving paddle-wheel, 12 in. diameter, rotating at 2 revolutions per sec. generates a free vortex in a vertical closed cylinder, 36 in. diameter and 6 in. deep, in which it is mounted axially. There is also an outward radial flow, whose velocity at the circumference of the wheel is 3 ft. per sec. Draw the stream-lines for the free spiral vortex formed.

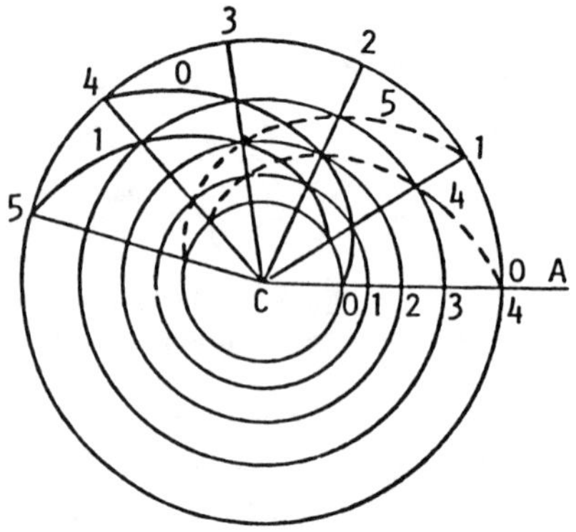

Fig. 1

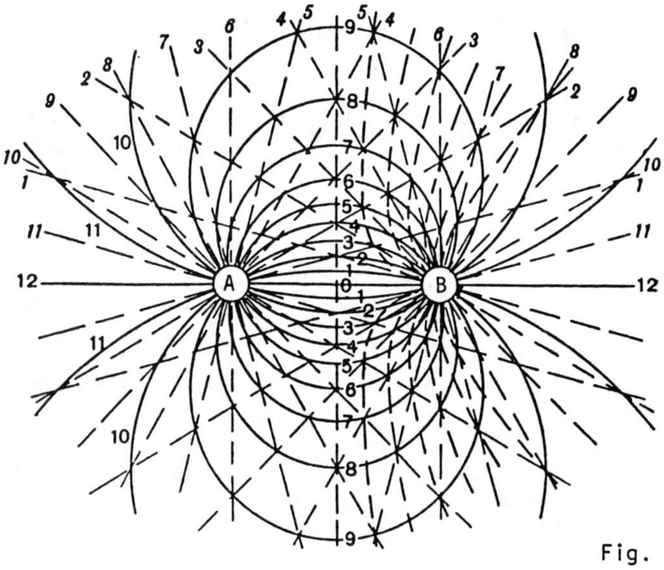

Fig. 2

Solution: First calculate the radii of three ψ_1 lines at equal increments of flux between the 0.5 ft. radius of the wheel and the 1.50 ft. radius of the cylinder. The value of v at

$$.5 \text{ radius} = 2\pi \times .5 \times 2 = 2\pi.$$

The value of v at any radius $r = \frac{\pi}{r}$. The flux ψ_1 across any line between the wheel and a point at radius r,

$$= \int_{\frac{1}{2}}^{r} v \cdot \frac{1}{2} dr = \frac{1}{2}\int_{\frac{1}{2}}^{r} \frac{\pi}{r} \cdot dr = \frac{\pi}{2} \left(\log_e r - \log_e \frac{1}{2}\right)$$

$$= \frac{\pi}{2} \log_e 2r .$$

The total flux

$$= \frac{\pi}{2} \log_e 3 = \frac{\pi}{2} \times 1.0986 = 1.7255 \text{ ft.}^3 \text{ per sec.}$$

$$\frac{1}{4} \text{ of this} = .4314 \text{ ft.}^3 \text{ per. sec.} = \delta\psi.$$

Then

$$\frac{\pi}{2} \log_e 2r_1 = .4314, \text{ or } \log_e 2r_1 = .2746$$

$$\therefore\ 2r_1 = 1.316;\ \therefore\ r_1 = .658 \text{ft.}$$

$$\frac{\pi}{2}\log_e 2r_2 = .8627, \text{ or } \log_e 2r_2 = .5492$$

$$\therefore\ 2r_2 = 1.732;\ \therefore\ r_2 = .866 \text{ ft.}$$

$$\frac{\pi}{2} \log_e 2r_3 = 1.2941, \text{ or } \log_e 2r_3 = .8238$$

$$\therefore\ 2r_3 = 2.279;\ \therefore\ r_3 = 1.139 \text{ ft.}$$

We draw circles of these radii, also of .500 and 1.500 ft. rad. (Fig. 1). The total radial flow

$$= 3 \times \pi \times 1 \times \frac{1}{2} = 4.7124 \text{ ft.}^3 \text{ per sec.}$$

The ψ_2 lines must be drawn so as to have the same increment of flux (.4314) as the ψ_1 lines, i.e., the angle between them must be $\frac{.4314}{4.7124} \times 360° = 32.96°$. From OA(say) we draw five straight radial lines at this angle apart to act as ψ_2 lines, and draw fair curves through the intersections of the ψ_1 and ψ_2 curves as shown. The full lines $(\psi_1 - \psi_2)$ are for anti-clockwise rotation, the dotted ones $(\psi_1 + \psi_2)$ for clockwise rotation. It is obvious that the angle made by the radius with the initial radius increases proportionally to the logarithm of the radius.

• PROBLEM 9-35

Sketch the flow pattern represented by the stream function

$$\psi = Uy\left(1 - \frac{a^2}{x^2 + y^2}\right)$$

The pressure at $x = \pm\infty$ is $p = p_\infty$. Compute the pressure along the line $y = 0$. What is the maximum pressure and where does it occur.

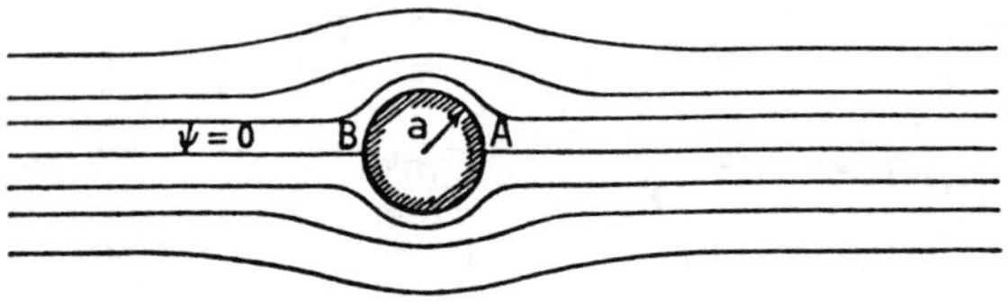

Solution: The figure is a sketch of the family of curves

$$\psi = \text{constant}$$

or

$$Uy\left(1 - \frac{a^2}{x^2 + y^2}\right) = C$$

The curve $\psi = 0$ is $y = 0$ and $x^2 + y^2 = a^2$.

Take the streamline $\psi = 0$ as a solid boundary and choose the inside of the streamline $\psi = 0$ as solid rather than above or below the streamline $\psi = 0$. A line has zero thickness, and so the only inside section of the streamline that can be considered solid is the inside of the circle $x^2 + y^2 = a^2$. The streamlines outside the circle represent the flow around the cylinder. Do not sketch any streamlines within the circle because this region is considered a solid cylinder.

From the work so far, you do not know what the flow was before the cylinder was placed in its position. The streamlines may have been parallel or in some other pattern. The velocity profile may have been uniform or some function of x and y.

To determine the type of flow into which the cylinder has been placed, examine the flow well upstream of the cylinder where it has not been disturbed, by examining the stream function ψ for large values of x where the $a^2/(x^2 + y^2)$ term becomes very small compared to one. As an approximation for large values of x, write the stream function as

$$\psi = Uy$$

which is a straight line parallel to the x axis. Consequently, in the limit as x approaches ∞, the streamlines become parallel to the x axis.

Make use of the definition of the stream function to determine the velocity profile.

$$u = -\frac{\partial\psi}{\partial y}, \qquad v = \frac{\partial\psi}{\partial x} \tag{1}$$

In this case,

$$u = -\frac{\partial Uy\left(1 - \frac{a^2}{x^2 + y^2}\right)}{\partial y} = -U\left(1 - \frac{a^2}{x^2 + y^2}\right) - \frac{2Uy^2a^2}{(x^2 + y^2)^2}$$

$$= -U\left[1 - \frac{a^2}{x^2 + y^2} + \frac{2a^2y^2}{(x^2 + y^2)^2}\right]$$

$$v = \frac{\partial Uy\left(1 - \frac{a^2}{x^2 + y^2}\right)}{\partial x} = (-1)\,\frac{[-Uya^2(2x)]}{(x^2 + y^2)^2}$$

$$= \frac{2Ua^2xy}{(x^2 + y^2)^2}$$

In the limit as $x \to \pm \infty$,

$$\lim_{x \to \pm\infty} u = -U$$

$$\lim_{x \to \pm\infty} v = 0$$

At plus infinity, therefore, the flow is parallel at uniform velocity in the negative direction.

The stream function $\psi = Uy[1 - (a^2/x^2 + y^2)]$ represents the flow around a circular cylinder placed in a stream of uniform parallel flow of an ideal fluid.

For irrotational flow the stream function ψ must satisfy the Laplace equation. Therefore, differentiate ψ twice with respect to each variable.

$$\frac{\partial \psi}{\partial x} = \frac{2Ua^2xy}{(x^2 + y^2)^2}$$

$$\frac{\partial^2 \psi}{\partial x^2} = \frac{2Ua^2y}{(x^2 + y^2)^2} - \frac{8\,Ua^2x^2y^2}{(x^2 + y^2)^3}$$

$$\frac{\partial \psi}{\partial y} = +U\left(1 - \frac{a^2}{x^2 + y^2} + \frac{2a^2y^2}{(x^2 + y^2)^2}\right)$$

$$\frac{\partial^2 \psi}{\partial y^2} = +\left(\frac{6ya^2U}{(x^2 + y^2)^2} - \frac{8a^2Uy^3}{(x^2 + y^2)^3}\right)$$

Substitute these into the Laplace equation.

$$\frac{\partial^2 \psi}{\partial x^2} + \frac{\partial^2 \psi}{\partial y^2} = 0$$

$$\frac{2Ua^2y}{(x^2 + y^2)^2} - \frac{8Ua^2x^2y}{(x^2 + y^2)^3} + \frac{6Ua^2y}{(x^2 + y^2)^2} - \frac{8Ua^2y^3}{(x^2 + y^2)^3} = 0$$

$$\frac{8Ua^2y}{(x^2 + y^2)^2}\left[1 - \frac{x^2 + y^2}{x^2 + y^2}\right] = \frac{8Ua^2y}{(x^2 + y^2)^2}[1 - 1] = 0$$

Thus, the flow is irrotational.

The Bernoulli equation is valid for irrotational flow, and you can use it, therefore, to compute the pressure along $\psi = 0$, which is $y = 0$.

$$\frac{p}{\rho} + \frac{q^2}{2} + gz = \text{constant}$$

Assume that the xy plane is horizontal with no change in z throughout. At x = +∞ the velocity is q = U, the pressure $p = p_\infty$, and the elevation $z = z_c$. Substitute these values into the Bernoulli equation and calculate the value of the constant.

$$\frac{p_\infty}{\rho} + \frac{U^2}{2} + gz_c = \text{constant}$$

You can write the Bernoulli equation, then, for any point in the flow as

$$\frac{p}{\rho} + \frac{q^2}{2} + gz_c = \frac{p_\infty}{\rho} + \frac{U^2}{2} + gz_c$$

Solve for p to get

$$p = \rho\left(\frac{U^2 - q^2}{2}\right) + p_\infty \qquad (2)$$

You previously computed the equations for u and v

$$u = -U\left(1 - \frac{a^2}{x^2 + y^2} + \frac{2a^2y^2}{(x^2 + y^2)^2}\right)$$

$$v = +\frac{2Ua^2xy}{(x^2 + y^2)^2}$$

Substitute y = 0 into these equations to get the values of u and v on the x axis.

$$u = -U\left[1 - \frac{a^2}{x^2}\right]$$

$$v = 0$$

Therefore,

$$q = \sqrt{u^2 + v^2} = -U\left[1 - \frac{a^2}{x^2}\right] \quad \text{on the x axis}$$

Substitute this value of q into Equation (2) for p to get the equation for p on the x axis.

$$p = \left\{\rho\,\frac{U^2 - U^2\left[1 - \frac{a^2}{x^2}\right]^2}{2}\right\} + p_\infty$$

$$= p_\infty + \frac{\rho U^2}{2}\left\{1 - 1 + \frac{2a^2}{x^2} - \frac{a^4}{x^4}\right\}$$

$$p = p_\infty + \frac{\rho U^2}{2}\left\{\frac{a^2}{x^2}\left(2 - \frac{a^2}{x^2}\right)\right\} \qquad (3)$$

From Equation (2) you can see that the pressure p will be a maximum when the velocity q is zero. The equation for the velocity on the x axis is

$$q = -U\left(1 - \frac{a^2}{x^2}\right)$$

Set q equal to zero and solve for x.

$$0 = U\left(1 - \frac{a^2}{x^2}\right)$$

$$x = \pm a$$

From this, note that the velocity is zero at $x = \pm a$, the two points A and B (Figure) are stagnation points, and the pressure at these points can be computed by substituting $q = 0$ into Equation (2).

$$p = p_\infty + \frac{\rho U^2}{2}$$

In this problem you computed the pressure variation along the streamline $\psi = 0$. It can be computed along any streamline in a similar manner. Therefore, you can compute the pressure at any point in the fluid.

• PROBLEM 9-36

a. An incompressible velocity field is given by

$$u = a(x^2 - y^2) \qquad v \text{ unknown} \qquad w = b$$

where a and b are constants. What must the form of the velocity component v be?

b. Take the velocity field of part (a) with $b = 0$ for algebraic convenience

$$u = a(x^2 - y^2) \qquad v = -2axy \qquad w = 0$$

and determine under what conditions it is a solution to the Navier-Stokes momentum equation. Assuming that these conditions are met, determine the resulting pressure distribution when z is "up" ($g_x = 0$, $g_y = 0$, $g_z = -g$).

c. Does a stream function exist for the velocity field of part (a)

$$u = a(x^2 - y^2) \qquad v = -2axy \qquad w = 0$$

If so, find it and plot it and interpret it.

d. Does a velocity potential exist for the velocity field of part (a)

$$u = a(x^2 - y^2) \qquad v = -2axy \qquad w = 0$$

If so, find it and plot it and compare with part (c).

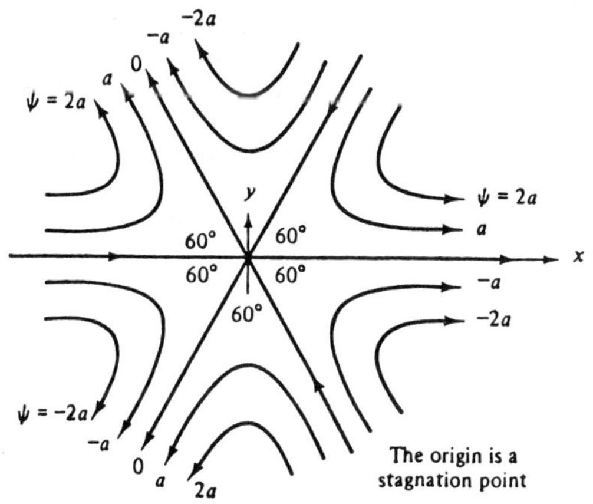

Fig. 1

Solution: a) Since the flow is incompressible, the following continuity equation applies:

Cartesian: $$\frac{\partial u}{\partial x} + \frac{\partial v}{\partial y} + \frac{\partial w}{\partial z} = 0$$

Cylindrical: $$\frac{1}{r}\frac{\partial}{\partial r}(rv_r) + \frac{1}{r}\frac{\partial}{\partial \theta}(v_\theta) + \frac{\partial}{\partial z}(v_z) = 0$$

$$\frac{\partial}{\partial x}(ax^2 - ay^2) + \frac{\partial v}{\partial y} + \frac{\partial b}{\partial z} = 0$$

or

$$\frac{\partial v}{\partial y} = -2ax$$

This is easily integrated partially with respect to y

$$v(x, y, z, t) = -2axy + f(x, z, t) \qquad \text{Ans.}$$

This is the only possible form for v which satisfies the incompressible continuity equation. The function of integration f is entirely arbitrary since it vanishes when v is differentiated with respect to y.

b) Make a direct substitution of u, v, w into the Navier-Stokes Eq.

$$\rho(0) - \frac{\partial p}{\partial x} + \mu(2a - 2a) = 2a^2\rho(x^3 + xy^2) \qquad (1)$$

$$\rho(0) - \frac{\partial p}{\partial y} + \mu(0) = 2a^2\rho(x^2y + y^3) \qquad (2)$$

$$\rho(-g) - \frac{\partial p}{\partial z} + \mu(0) = 0 \qquad (3)$$

The viscous terms vanish identically (although μ is not zero). Equation (3) can be integrated partially to obtain

$$p = -\rho gz + f_1(x, y) \qquad (4)$$

i.e., the pressure is hydrostatic in the z direction, which follows anyway from the fact that the flow is two-dimensional ($w = 0$). Now the question is: Do equations (1) and (2) show that the given velocity field is a solution? One way to find out is to form the mixed derivative $\partial^2 p/(\partial x\, \partial y)$ from (1) and (2) separately and then compare them.

Differentiate Eq. (1) with respect to y

$$\frac{\partial^2 p}{\partial x \partial y} = -4a^2 \rho xy \tag{5}$$

Now differentiate Eq. (2) with respect to x

$$\frac{\partial^2 p}{\partial x \partial y} = \frac{\partial}{\partial x}[2a^2\rho(x^2y + y^3)] = -4a^2\rho xy \tag{6}$$

These are identical. Therefore the given velocity field is an exact solution to the Navier-Stokes equation.

To find the pressure distribution, substitute Eq. (4) into Eqs. (1) and (2), which will enable us to find f (x, y)

$$\frac{\partial f_1}{\partial x} = -2a^2\rho(x^3 + xy^2) \tag{7}$$

$$\frac{\partial f_1}{\partial y} = -2a^2\rho(x^2y + y^3) \tag{8}$$

Integrate Eq. (7) partially with respect to x

$$f_1 = -\tfrac{1}{2}a^2\rho(x^4 + 2x^2y^2) + f_2(y) \tag{9}$$

Differentiate this with respect to y and compare with Eq. (8)

$$\frac{\partial f_1}{\partial y} = -2a^2\rho x^2 y + f'_2(y) \tag{10}$$

Comparing (8) and (10), we see they are equivalent if

$$f'_2(y) = -2a^2\rho y^3$$

or

$$f_2(y) = \tfrac{1}{2}a^2\rho y^4 + C \tag{11}$$

where C is a constant. Combine Eqs. (4), (9), and (11) to give the complete expression for pressure distribution

$$p(x, y, z) = -\rho gz - \tfrac{1}{2}a^2\rho(x^4 + y^4 + 2x^2y^2) + C \quad \text{Ans.} \tag{12}$$

This is the desired solution. Do you recognize it? Not unless you go back to the beginning and square the velocity components:

$$u^2 + v^2 + w^2 = V^2 = a^2(x^4 + y^4 + 2x^2y^2) \tag{13}$$

Comparing with Eq. (12), we can rewrite the pressure distribution as

$$p + \frac{1}{2}\rho V^2 + \rho g z = C \qquad (14)$$

This is Bernoulli's equation. That is no accident, because the velocity distribution given in this problem is one of a family of flows which are solutions to the Navier-Stokes equation and which satisfy Bernoulli's incompressible equation everywhere in the flow field. They are called irrotational flows, for which curl $V = \nabla \times V \equiv 0$.

Since this flow field was shown in part (a) to satisfy the equation of continuity, we are pretty sure that stream function does exist. We can check again to see if

$$\frac{\partial u}{\partial x} + \frac{\partial v}{\partial y} = 0$$

Substitute:

$$2ax + (-2ax) = 0 \qquad \text{checks}$$

Therefore we are certain that a stream function exists. To find ψ we simply set

$$u = \frac{\partial \psi}{\partial y} = ax^2 - ay^2 \qquad (15)$$

$$v = -\frac{\partial \psi}{\partial x} = -2axy \qquad (16)$$

and work from either one toward the other. Integrate (15) partially

$$\psi = ax^2y - \frac{ay^3}{3} + f(x) \qquad (17)$$

Differentiate (17) with respect to x and compare with (16)

$$\frac{\partial \psi}{\partial x} = 2axy + f'(x) = 2axy \qquad (18)$$

Therefore $f'(x) = 0$, or f = constant. The complete stream function is thus found

$$\psi = a\left(x^2y - \frac{y^3}{3}\right) + C \qquad \text{Ans.} \quad (19)$$

To plot this, set $C = 0$ for convenience and plot the function

$$3x^2y - y^3 = \frac{3\psi}{a} \qquad (20)$$

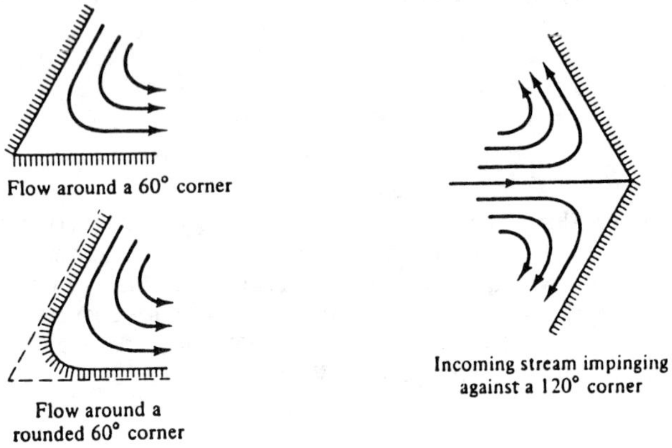

Fig. 2

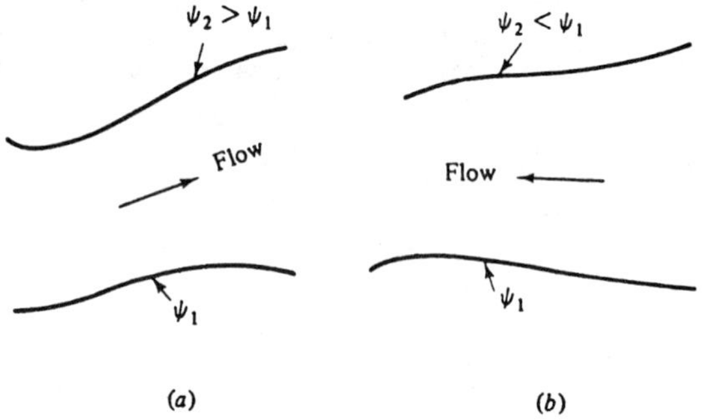

Fig. 3

Sign convention for flow direction in terms of the change in stream function: (*a*) flow to the right if ψ_U is greater; (*b*) flow to the left if ψ_L is greater.

for constant values of ψ. The result is shown in Fig. 2 to be six 60° wedges of circulating motion, each with identical flow patterns except for the arrows. Once the streamlines are labeled, the flow directions follow from the sign convention of Fig. 3. How can the flow be interpreted? Since there is slip along all streamlines, no streamline can truly represent a solid surface in a viscous flow. However, the flow could represent the impingement of three incoming streams at 60, 180, and 300°. This would be a rather unrealistic yet exact solution to the Navier-Stokes equation, as we showed in part (a).

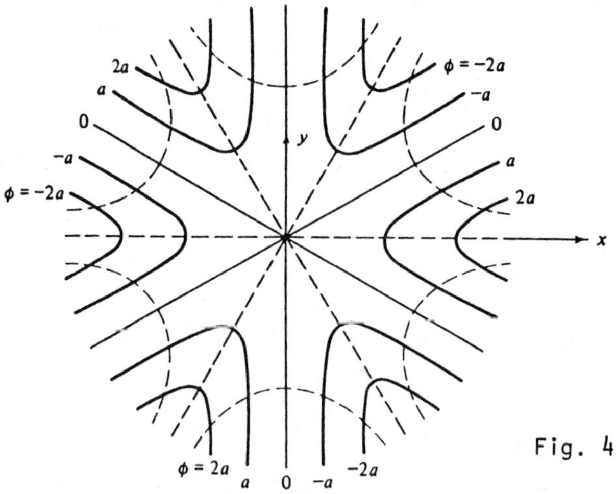

Fig. 4

By allowing the flow to slip as a frictionless approximation, we could let any given streamline be a body shape. Some examples are shown in Fig. 2.

d) Since w = 0, the curl of V has only one (z) component, and we must show that it is zero

$$(\nabla \times V)_z = \omega_z = \frac{\partial v}{\partial x} - \frac{\partial u}{\partial y} = \frac{\partial}{\partial x}(-2axy) - \frac{\partial}{\partial y}(ax^2 - ay^2)$$

$$= -2ay + 2ay = 0 \quad \text{checks}$$

The flow is indeed irrotational. A potential exists. Ans.

To find $\phi(x, y)$, set

$$u = \frac{\partial \phi}{\partial x} = ax^2 - ay^2 \tag{21}$$

$$v = \frac{\partial \phi}{\partial y} = -2axy \tag{22}$$

Integrate (21)

$$\phi = \frac{ax^3}{3} - axy^2 + f(y) \tag{23}$$

Differentiate (23) and compare with (22)

$$\frac{\partial \phi}{\partial y} = -2axy + f'(y) = -2axy$$

Therefore f' = 0, or f = constant. The velocity potential is

$$\phi = \frac{ax^3}{3} - axy^2 + C \qquad \text{Ans.}$$

Letting C = 0, we can plot the ϕ lines in the same fashion as in part (c). The result is shown in Fig. 4 (no arrows on ϕ). For this particular problem, the ϕ lines form the same pattern as the ψ lines of Part (c) (which are shown here as dotted lines) but are displaced 30°. The ϕ and ψ lines are everywhere perpendicular except at the origin, a stagnation point, where they are 30° apart. We expected trouble at the stagnation point, and there is no general rule for determining the behavior of the lines at that point.

IRROTATIONAL FLOW

• PROBLEM 9-37

The flow in a laminar boundary layer is described by the expression

$$u = 1000y - \frac{10^9 y^3}{9} \text{ m/s}$$

where y is in meters and v = 0. At y = 1 mm, is the flow rotational or is it irrotational? Refer to Figure.

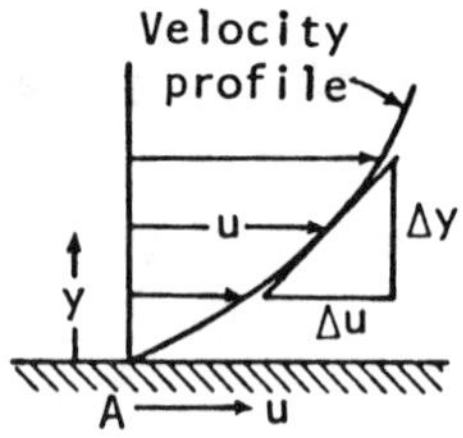

Velocity profile and velocity gradient.

Solution: It is necessary to determine whether $\partial u/\partial y = \partial v/\partial x$. If so, the flow is irrotational; if not, the flow is rotational:

$$\frac{\partial u}{\partial y} = 1000 - 3 \times \frac{10^9 y^2}{9} = 1000 - 3 \times \frac{10^9 (0.001)^2}{9}$$

$$= 667 \text{ (m/s)/m}$$

Since $\partial v/\partial x = 0$, $\partial u/\partial y \neq \partial v/\partial x$ and the flow at y = 1 mm is rotational.

• PROBLEM 9-38

The flow of a nonviscous fluid past a two-dimensional half body (Figure) is described by the x and y components of velocity as

$$u = u_s + \frac{mx}{x^2 + y^2} \quad \text{and} \quad v = \frac{my}{x^2 + y^2}$$

where u_s and m are the strengths of the free-stream velocity and source which form the flow pattern. Show that the flow is irrotational.

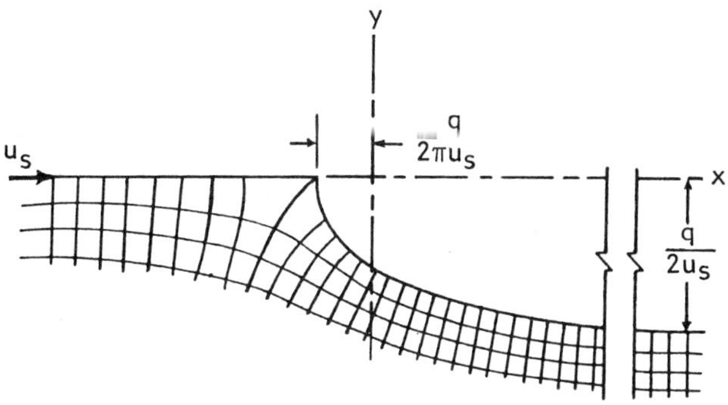

Flow net for one side of a half-body.

Solution: To prove the flow is irrotational, it must be shown that $\partial u/\partial y = \partial v/\partial x$:

$$\frac{\partial u}{\partial y} = \frac{-2mxy}{(x^2 + y^2)^2} \qquad \frac{\partial v}{\partial x} = \frac{-2mxy}{(x^2 + y^2)^2}$$

and thus the flow is irrotational.

• PROBLEM 9-39

The whirlpool above an open drain is essentially an irrotational vortex (Fig.). If the velocity at a radius of 20 cm is 15 cm/s, what is the velocity at 5 cm? At 80 cm? What is the drawdown Δy at r = 5 cm? At r = 80 cm?

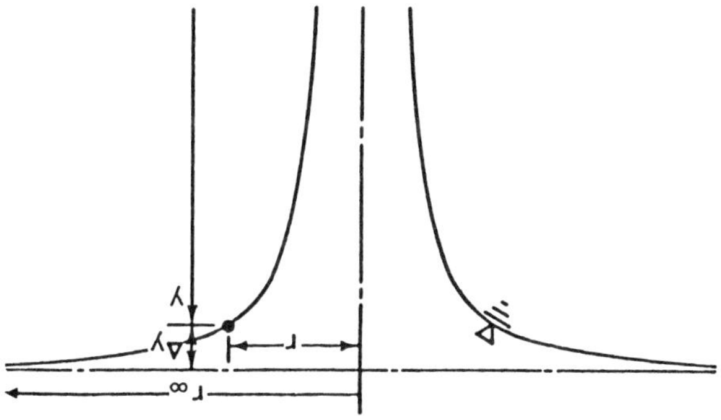

Solution: Since the velocity in an irrotational vortex is given by $v \sim 1/r$, we may write

$$v = k/r$$

or

$$k = vr\ (15\ \text{cm/s})(20\ \text{cm}) = 300\ \text{cm}^2/\text{s}$$

Hence at

$$r = 5\ \text{cm} \qquad v = \frac{300}{5} = 60\ \text{cm/s}$$

and at

$$r = 80 \text{ cm} \qquad v = \frac{300}{80} = 3.75 \text{ cm/s}$$

Since the flow is irrotational we may write the Bernoulli equation between a point at infinity and any other point even though the stream lines are approximately concentric circles. Along the free surface the pressure is zero so that between r = ∞ and r = 5 cm we have

$$y_\infty + \frac{v_\infty^2}{2g} = y_5 + \frac{v_5^2}{2g}$$

Thus

$$\Delta y_5 = y_\infty - y_5 = \frac{v_5^2}{2g} - \frac{v_\infty^2}{2g} = \frac{(60)^2}{(2)(981)} = 1.83 \text{ cm}$$

Likewise

$$\Delta y_{80} = \frac{(3.75)^2}{(2)(981)} = 0.007 \text{ cm}$$

CHAPTER 10

DRAG AND LIFT

Basic Attacks and Strategies for Solving Problems in this Chapter. See pages 563-628 for step-by-step solutions to problems.

The *no-slip* boundary condition requires that the velocity of a fluid immediately adjacent to a solid wall be the same as the velocity of the wall itself. In the usual frame of reference where the wall is stationary with fluid flowing over it, the fluid velocity right next to the wall must be zero. This no-slip condition leads to what is referred to as a *boundary layer*. A boundary layer is the thin fluid layer in which viscous effects are important near a wall. Inside a boundary layer, the fluid velocity goes from some finite value at the boundary layer edge to zero at the wall in a very short distance. Since viscous shear stress is proportional to viscosity μ and velocity gradient, the shear stress is quite large in a boundary layer, especially very close to the wall where the velocity gradient is steepest. This shear stress, by Newton's third law, imposes a frictional drag force on the wall in the same direction as the flow above the boundary layer.

Boundary layers can be either laminar (smooth and steady) or turbulent (quite unsteady and irregular). For uniform flow along a semi-infinite flat plate, the laminar boundary layer solution can be obtained exactly (albeit with the help of a digital computer), but turbulent boundary layers are too complex to solve exactly, even with the fastest computers; turbulent boundary layer results are found empirically or semi-empirically.

For engineering analyses, the three quantities of most significance are the boundary layer thickness δ, the skin friction coefficient C_f, and the displacement thickness δ^*. δ is usually taken as the distance from the wall where the velocity u has increased to 99% of the freestream velocity U. Letting τ_w denote the shear stress acting on the wall by the fluid, C_f is non-dimensionalized wall shear stress,

$$C_f = \frac{2\tau_w}{\rho U^2}. \tag{1}$$

Displacement thickness is defined as the distance to which streamlines outside the boundary layer are displaced away from the wall, and results from the fact that the fluid inside the boundary layer carries less mass flow than it would have in the absence of the wall. It turns out that

$$\delta^* = \int_0^\infty \left(1 - \frac{u}{U}\right) dy. \tag{2}$$

For laminar flow on a flat plate, expressions for δ, C_f, and δ^* are provided in Table 1 of the first problem.

At a Reynolds number greater than about 300,000 based on U and x (the distance along the plate in the flow direction), the laminar boundary layer begins to oscillate and transitions to turbulence. Empirical expressions for δ, C_f, and δ^* can be found in the literature for turbulent flow.

Skin friction is not the only source of drag on bodies—such as automobiles, baseballs, submarines, and airplanes—moving through a fluid. The uneven distribution of pressure forces along the body surface can produce significant (often dominating) drag forces as well. For non-streamlined or blunt bodies in particular, the boundary layer along the body surface cannot remain attached and separates off the surface. This leads to a gross imbalance of pressure (pressure being very high on the front end and very low on the back end of the body) and a large pressure drag.

The total aerodynamic drag on a body usually must be found by experimentation. Drag is expressed non-dimensionally by a *drag coefficient*, C_D, defined as

$$C_D = \frac{\text{Drag Force}}{\frac{1}{2}\rho U^2 A}, \tag{3}$$

where U is the freestream velocity and

A is an area,

typically the projected frontal area, but sometimes (as in the case of airplane wings or flat plates) the planform area. Drag coefficients for several body shapes are provided in the tables supplied with Problems 10.10, 10.13, and 10.18. To find the drag force, simply multiply C_D by the dynamic pressure ($\frac{1}{2}\rho U^2$) and the appropriate area A. The power required to overcome aerodynamic drag is simply the drag force multiplied by velocity.

In general, a flow with a laminar boundary layer produces much less skin friction drag than a flow with a turbulent boundary layer. However, turbulent boundary layers are much more resilient to flow separation, and hence can lead to less pressure drag. In engineering analysis, one can sometimes force the flow to be turbulent in order to decrease the overall drag. The dimples on a golf ball are one such example. The dimples force the boundary layer to be turbulent, which delays separation and decreases the pressure drag; since pressure drag dominates on bluff bodies such as spheres, the net effect is a decrease in total drag.

Lift can be analyzed in much the same way as drag. Namely, a lift coefficient, C_L, is defined as

$$C_L = \frac{\text{Lift Force}}{\frac{1}{2}\rho U^2 A}. \tag{4}$$

Lift coefficient is obtained empirically or semi-empirically for airplane wings and other lifting bodies as a function of the angle of attack, α.

Step-by-Step Solutions to Problems in this Chapter, "Drag and Lift"

BOUNDARY LAYER

• PROBLEM 10-1

Standard air flows past a flat plate at a free-stream velocity of 10 m/s. What is the thickness of the boundary layer and the displacement thickness 30 cm from the leading edge?

TABLE 1
RESULTS OF CALCULATIONS FOR LAMINAR BOUNDARY LAYER

VELOCITY PROFILE	$\frac{\delta}{x}$	C_f	$\frac{\delta^*}{x}$
Momentum analysis			
$\frac{u}{u_s} = \frac{y}{\delta}$	$\frac{3.46}{Re_x^{1/2}}$	$\frac{1.156}{Re_x^{1/2}}$	$\frac{1.73}{Re_x^{1/2}}$
$\frac{u}{u_s} = 2\left(\frac{y}{\delta}\right) - \left(\frac{y}{\delta}\right)^2$	$\frac{5.48}{Re_x^{1/2}}$	$\frac{1.462}{Re_x^{1/2}}$	$\frac{1.83}{Re_x^{1/2}}$
$\frac{u}{u_s} = \frac{3}{2}\left(\frac{y}{\delta}\right) - \frac{1}{2}\left(\frac{y}{\delta}\right)^3$	$\frac{4.64}{Re_x^{1/2}}$	$\frac{1.292}{Re_x^{1/2}}$	$\frac{1.74}{Re_x^{1/2}}$
$\frac{u}{u_s} = \sin\frac{\pi y}{2\delta}$	$\frac{4.80}{Re_x^{1/2}}$	$\frac{1.310}{Re_x^{1/2}}$	$\frac{1.74}{Re_x^{1/2}}$

(TABLE 1, continued on the following page.)

VELOCITY PROFILE	$\frac{\delta}{x}$	C_f	$\frac{\delta^*}{x}$
Exact solution[a]			
(Blasius)	$\frac{4.91}{Re_x^{1/2}}$	$\frac{1.328}{Re_x^{1/2}}$	$\frac{1.73}{Re_x^{1/2}}$

[a] To be used for engineering calculations.

Solution: The Reynolds number is $Re_x = (10)(0.3)/1.46 \times 10^{-5} = 2.05 \times 10^5$, and thus the boundary layer may be assumed to be laminar:

$$\frac{\delta}{x} = \frac{4.91}{Re_x^{1/2}} = \frac{4.91}{453} = 0.0108$$

The boundary layer thickness is then

$$\delta = (0.0108)(300) = 3.2 \text{ mm}$$

and the displacement thickness is $\delta^* = \frac{3}{8}\delta = 1.2$ mm.

• PROBLEM 10-2

Air at 100°F is flowing over a flat plate 1 ft wide. Estimate the boundary layer thickness 1 ft from the leading edge and the drag. The air speed is 7.2 ft/sec.

Table 1

Absolute and Kinematic Viscosities of Air at Atmospheric Pressure

°F	$\mu \times 10^7$ lbf-sec/ft²	$\nu \times 10^4$ ft²/sec	°F	$\mu \times 10^7$ lbf-sec/ft²	$\nu \times 10^4$ ft²/sec
0	3.28	1.26	100	3.96	1.89
10	3.45	1.31	120	4.07	1.89
20	3.50	1.36	140	4.14	2.01
30	3.58	1.42	160	4.22	2.12
40	3.62	1.46	180	4.34	2.25
50	3.68	1.52	200	4.49	2.40
60	3.74	1.58	250	4.87	2.80
70	3.82	1.64	300	4.99	3.06
80	3.85	1.69	400	5.40	3.65
90	3.90	1.74	500	5.89	4.56

Solution: The boundary layer thickness δ can be estimated from the equation

$$\delta = \sqrt{5\ \nu x/V_0} \tag{1}$$

We may also write equation 1 in the form

$$\frac{\delta}{x} = \frac{5}{\sqrt{R_{e,x}}} \tag{2}$$

where $R_{e,x} = V_0 x/\nu$ is the Reynolds number based on length from the plate leading edge.

Table 2
Density of Air versus Temperature

Temperature, °F	ρ, lbm/ft³	ρ, slugs/ft³ × 10³
0	0.0862	2.68
10	0.0846	2.63
20	0.0827	2.57
30	0.0811	2.52
40	0.0795	2.47
50	0.0780	2.42
60	0.0764	2.37
70	0.0750	2.33
80	0.0736	2.28
90	0.0722	2.24
100	0.0709	2.20
120	0.0685	2.13
140	0.0663	2.06
150	0.0651	2.02
200	0.0602	1.87
300	0.0523	1.63
400	0.0462	1.44
500	0.0414	1.29

Pressure 14.696 lbf/in.².

The kinematic viscosity of air at 100°F is $\nu = 1.8 \times 10^{-4}$ ft²/sec. The density is $\rho = 2.20 \times 10^{-3}$ slug/ft³. Therefore

$$R_{e,x} = \frac{V_0 x}{\nu} = \frac{7.2 \times 1}{1.8 \times 10^{-4}} = 4 \times 10^4$$

and

$$\delta = \frac{5 \times 1}{2 \times 10^2} \text{ ft} = 0.0025 \text{ ft.}$$

In general, the drag coefficient is given by

$$C_d = \frac{\text{drag}}{\frac{1}{2}\rho V^2 A} \tag{1}$$

where V is the upstream velocity and A is the cross-sectional area.

For laminar boundary layer flow over a flat plate, the drag coefficient has been found to be given by

$$C_D = \frac{1.328}{\sqrt{R_{e,x}}} \qquad (2)$$

Combining Eqs. 1 and 2 gives the equation for drag to be

$$\text{drag} = \frac{\text{area} \times \frac{1}{2}\rho U_0{}^2 \times 1.328}{\sqrt{R_{e,x}}}$$

$$\text{Drag} = \frac{1\ \text{ft}^2/2\times 2.20\times 10^{-3}\text{slug/ft}^3\times(7.2)^2\text{ft}^2/\text{sec}^2\times 1.328}{2 \times 10^2}$$

$$= 3.8 \times 10^{-4}\text{lb/ft}$$

• PROBLEM 10-3

a) Crude oil at 70°F ($\nu = 10^{-4}\text{ft}^2/\text{sec}$, sp. gr. = 0.86) with a free-stream velocity of 10 ft/sec flows past a thin flat plate which is 4 ft wide and 6 ft long in a direction parallel to the flow. Determine and plot the boundary-layer thickness and the shear-stress distribution along the plate. b) For the conditions of part (a), determine the resistance of one side of the plate.

Table

RESULTS-δ AND τ_0 FOR DIFFERENT VALUES OF x

	x = 0.1 ft	x = 1.0 ft	x = 2 ft	x = 4 ft	x = 6 ft
$x^{1/2}$	0.316	1.00	1.414	2.00	2.45
τ_δ, psf	0.552	0.174	0.123	0.087	0.071
δ, ft.	0.005	0.016	0.022	0.031	0.039
δ, in.	0.060	0.189	0.270	0.380	0.460

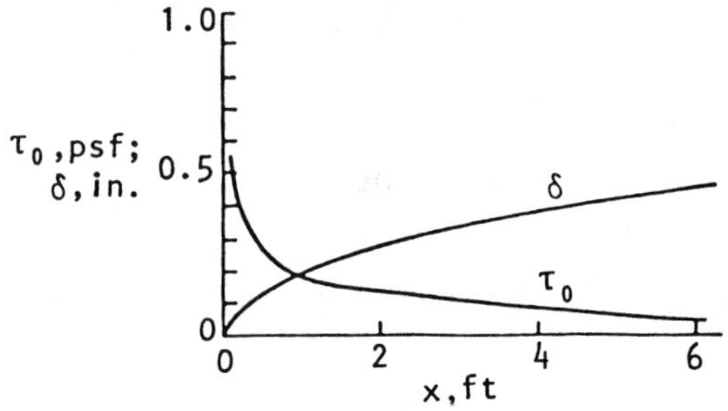

Figure

Solution: a)

$$Re_x = \frac{U_0 x}{\nu} = \frac{10x}{10^{-4}} = 10^5 x$$

$$Re_x^{1/2} = 3.16(10^2 x^{1/2})$$

The shear stress is given by

$$\tau_0 = 0.332\mu \frac{U_0}{x} Re_x^{1/2}$$

where $\mu = \rho\nu = 1.94 \times 0.86 \times 10^{-4}$

$$= 1.67 \times 10^{-4} \text{ lbf-sec/ft}^2$$

Then $\tau_0 = 0.332(1.67 \times 10^{-4}) \frac{10}{x} (3.16)(10^2 x^{1/2})$

$$= \frac{0.175}{x^{1/2}} \text{ psf}$$

The thickness of the boundary layer is

$$\delta = \frac{5x}{Re_x^{1/2}} = \frac{5x}{3.16(10^2 x^{1/2})} = 1.58(10^{-2} x^{1/2}) \text{ ft}$$

$$= 1.58(12)(10^{-2} x^{1/2}) = 0.190 x^{1/2} \text{ in.}$$

The results are plotted in the figure and listed in the table.

b)

$$F_s = \frac{C_f B L \rho U_0^2}{2}$$

Here $C_f = \frac{1.33}{Re_L^{1/2}} = \frac{1.33}{(3.16)(10^2)(6^{1/2})} = 0.0017$

Then $F_s = 0.0017(4)(6)(0.86)(1.94) \quad \frac{10^2}{2}$

$$= 3.40 \text{ lbf}$$

SI units: For part (a) the significant variables in SI units are ρ = 860 kg/m^3, U_0 = 3.05 m/s, B = 1.22 m, and L = 1.88 m. Then

$$F = 0.0017(1.22 \text{ m})(1.88 \text{ m})(860 \text{ kg/m}^3)\left(\frac{3.05^2 \text{m}^2/\text{s}^2}{2}\right)$$

$$= 15.6 \text{ kg} \cdot \text{m/s}^2 = 15.6 \text{ N}$$

• PROBLEM 10-4

Assume that a boundary layer over a smooth, flat plate is first laminar and then goes turbulent at a critical Reynolds number of 5×10^5. If we have a plate that is 3 m long by 1 m wide and if air, 20°C, at normal atmospheric pressure flows past this plate with a velocity of 30 m/s, what will be the total shearing resistance of one side, the resistance due to the turbulent and laminar part of the boundary layer, and the average resistance coefficient C_f for the plate?

Solution: The total resistance $F_s = C_f BL\rho V_0{}^2/2$, where from Eq. (1) we get C_f:

For $Re_{cr} = 500{,}000$

$$C_f = \frac{0.074}{Re_L{}^{1/5}} - \frac{1{,}740}{Re_L} \qquad (1)$$

Also, $Re_L = VL/\nu$, or

$$Re_L = \frac{30 \text{ m/s} \times 3 \text{ m}}{(1.49)(10^{-5})\text{m}^2/\text{s}} = 6.04 \times 10^6$$

Then solving Eq. (1), we have $C_f = 0.00326 - 0.000290$

$$= 0.00297$$

The total resistance is now calculated:

$$F_s = C_f BL\rho \frac{V_0{}^2}{2}$$

$$F_s = 0.00297 \times 1 \times 3 \times 1.2 \times \frac{30^2}{2} = 4.81 \text{ N}$$

Then x_{cr} is determined:

$$Re_{cr} = \frac{Vx_{cr}}{\nu} = 500{,}000$$

or $$x_{cr} = \frac{500{,}000 \times 1.49 \times 10^{-5}}{30} = 0.248 \text{ m}$$

Thus, the laminar resistance will be

$$F_{s,lam} = \frac{1.33}{(5 \times 10^5)^{1/2}} 1 \times 0.248 \times 1.2 \times \frac{30^2}{2} = 0.252 \text{ N}$$

Then $F_{s,turb} = 4.81 \text{ N} - 0.25 \text{ N} = 4.56 \text{ N}$

Determine the distance downstream from the bow of a ship moving at 7.59 kts relative to still water at which the boundary layer becomes turbulent. Also, determine the boundary layer thickness at this point and the total friction drag coefficient for this portion of the surface of the ship. Let the maximum value of Reynolds number $R_x = 5 \times 10^5$ be where the flow ceases to be laminar, so that the flow is turbulent for $R_x > 5 \times 10^5$. Assume $\rho = 1.94$ slugs/ft^3 and $\nu = 1.21 \times 10^{-5}$ ft^2/sec.

Solution:

$$U = 7.59 \text{ knots } \frac{5280 \text{ ft}}{3600 \text{ sec}} \times \frac{1.15 \text{ mile/hr}}{\text{knot}} = 12.8 \text{ fps}$$

$$\nu = 1.21 \times 10^{-5} \frac{\text{ft}^2}{\text{sec}}$$

$$\rho = 1.94 \text{ slugs/ft}^3$$

Given that the lower critical Reynolds number, below which the flow is laminar, is

$$R_x = \frac{Ux}{\nu} = 5 \times 10^5$$

so that the length along the flat plate measured from the bow is

$$x = \frac{5 \times 10^5 \nu}{U}$$

$$= \frac{(5 \times 10^5)(1.21 \times 10^{-5})}{12.8} = 0.47 \text{ ft.}$$

This means that the boundary layer becomes turbulent at a distance of only 0.47 ft downstream from the bow and for x > .47 ft our solution will be invalid.
Define the boundary layer thickness as that place where the velocity component u is 99% of the value of the free-stream velocity U, i.e. u = .99U. Hence the boundary layer thickness is, according to Blasius' calculations and the above definition,

$$\delta \approx 5.0\sqrt{\frac{\nu x}{U}}$$

$$\delta\Big|_{x=0.47 \text{ ft}} = 5\sqrt{\frac{\nu x}{U}}\Bigg|_{x=0.47 \text{ ft}}$$

or

$$= 0.0102 \text{ ft}$$

The total friction drag coefficient is given by the equation

$$C_{D_f} = \frac{1.328}{\sqrt{R_\ell}} = \frac{1.328}{\sqrt{5 \times 10^5}} = 1.92 \times 10^{-3}$$

• PROBLEM 10-6

A flat surface 40 ft wide by 200 ft long is exposed to a 80-ft/s wind (80°F) blowing in the direction of the 200-ft length. Assuming an initial laminar boundary layer followed by a turbulent boundary layer, determine:
(a) The length of the laminar boundary layer.
(b) The shear stress and boundary layer thickness at the end of the laminar region.
(c) The shear stress at the end of the 200 ft.
(d) The total force on the first 100 ft.
(e) The total force on the final 100 ft.

Solution: At 80°F, $\rho = 0.00228$ slug/ft^3 and $\nu = 1.69 \times 10^{-4}$ ft^2/s.

(a) Assuming that the laminar portion ends at $Re_x = 400{,}000$,

$$Re_x = \frac{Vx}{\nu} \tag{1}$$

$$Re_x = \frac{(80)x}{1.69 \times 10^{-4}} = 400{,}000$$

and x = 0.845 ft. Thus, less than 1 ft of the total length will be laminar.

(b) At the point x = 0.845 ft, $Re_x = 400{,}000$, the shear stress is given by

$$\tau_0 = \frac{.664}{Re_x^{1/2}} \frac{\rho U^2}{2} \tag{2}$$

Hence, $$\tau_0 = \frac{0.664}{(400{,}000)^{1/2}} \frac{(0.00228)(80)^2}{2} = 0.00766 \text{ lb/ft}^2$$

while boundary layer thickness is

$$\delta = \frac{5.2x}{Re_x^{1/2}} \tag{3}$$

$$\delta = \frac{(0.845)(5.2)}{(400{,}000)^{1/2}} = 0.00695 \text{ ft}$$

(c) At the end of the surface, x = 200 ft and

$$Re_{200} = \frac{(80)(200)}{1.69 \times 10^{-4}} = 9.47 \times 10^7$$

Although this is beyond the range of validity of Eq. 4 below, using it as a first approximation

$$\tau_0 = \frac{.059}{Re_x^{1/5}} \frac{\rho U^2}{2} \tag{4}$$

$$\tau_0 = \frac{0.059}{(9.47 \times 10^7)^{1/5}} \frac{(0.00228)(80)^2}{2} = 0.01093 \text{ lb/ft}^2$$

(d) For the first 100 ft,

$$Re_{100} = \frac{(80)(100)}{1.69 \times 10^{-4}} = 4.73 \times 10^7$$

and

$$C_F = \frac{0.059}{Re^{1/5}}$$

$$C_f = 0.0023$$

Thus, the total force equation

$$F = C_f A \frac{\rho U^2}{2} \qquad (6)$$

yields:

$$F_{100} = (0.0023)(40)(100)\frac{(0.00228)(80)^2}{2} = 67.1 \text{ lb}$$

(e) Repeating part (d) for the entire length where $Re_{200} = 9.47 \times 10^7$ and $C_f = 0.0021$ gives

$$F_{200} = (0.0021)(40)(200)\frac{(0.00228)(80)^2}{2} = 122.7 \text{ lb}$$

The force on the total length is a little less than twice the force on the first 100 ft. Finally, the force on the last 100 ft is the difference

$$F_{200} - F_{100} = 122.7 - 67.1 = 55.6 \text{ lb}$$

• PROBLEM 10-7

If the wetted area of a 200-ft ship is 7500 ft^2, approximately how great is the surface drag when the ship is traveling at a speed of 15 knots (1 knot = 1.69 fps)? What is the thickness of the boundary layer at the stern?

Solution: If the flow is turbulent at the leading edge itself, the equations for the relative boundary-layer

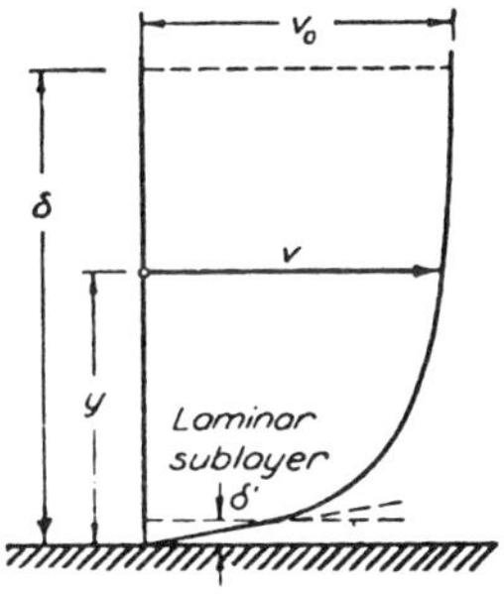

Fig. 1

Velocity distribution in the turbulent boundary layer.

thickness and the coefficient of resistance for smooth boundaries may be generalized in the forms

$$\frac{\delta}{x} = \frac{0.38}{R^{1/5}} \qquad (1)$$

$$C_f = \frac{0.074}{R^{1/5}} \qquad (2)$$

the latter corresponding to the Blasius equation for pipes. At high Reynolds numbers the Kármán-Schoenherr counterpart of the Kármán-Prandtl equation for pipes is applicable:

$$\frac{1}{\sqrt{C_f}} = 4.15 \log RC_f \qquad (3)$$

Equations (2) and (3) are plotted in Fig. 2.

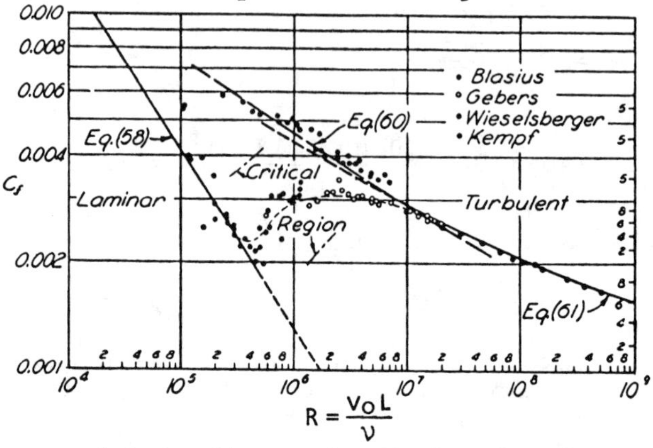

Fig. 2

Resistance diagram for the boundary layer on a smooth surface.

The Reynolds number is, for a water temperature of 60°F,

$$R = \frac{VL}{\nu} = \frac{15 \times 1.69 \times 200}{1.2 \times 10^{-5}} = 4.23 \times 10^8$$

From Fig. 2 for this value of R, $C_f = 0.00175$. Hence,

$$F = C_f A \frac{\rho v_0^2}{2} = 0.00175 \times 7500 \times \frac{1.94(15\times1.69)^2}{2} = 8180 \text{ lb}$$

From Eq. (1),
$$\delta = \frac{0.38L}{R^{1/5}} = 1.43 \text{ ft}$$

• PROBLEM 10-8

a) Air at a temperature of 20°C (68°F) and with a free-stream velocity of 30 m/s (98 ft/sec) flows past a smooth, thin plate, which is 3 m wide (9.8 ft) by 6 m long (19.2 ft), in the direction of flow. Assuming that the boundary layer is forced to be turbulent from the leading edge, determine the shear stress, the thickness of the laminar sublayer, and the thickness of the boundary layer 5 m (16.4 ft) downstream of the leading edge.
b) Determine the total drag of the plate given in part a. Solve in English units, as well as metric.

Solution: a) First compute Re_x at a distance 5 m from the leading edge:

$$Re_x = U_0 \frac{x}{\nu} = \frac{(30 \text{ m/s})(5 \text{ m})}{1.49(10^{-5})\text{m}^2/\text{s}} = 10^7$$

$$Re_x^{1/5} = (10^7)^{1/5} = 25$$

Compute τ_0, where $\tau_0 = c_f \rho U_0^2/2$. Here

$$c_f = 0.058\, Re_x^{-(1/5)} = \frac{0.058}{25}$$

Also $\rho = 1.20 \text{ kg/m}^3 (0.00232 \text{ slugs/ft}^3)$

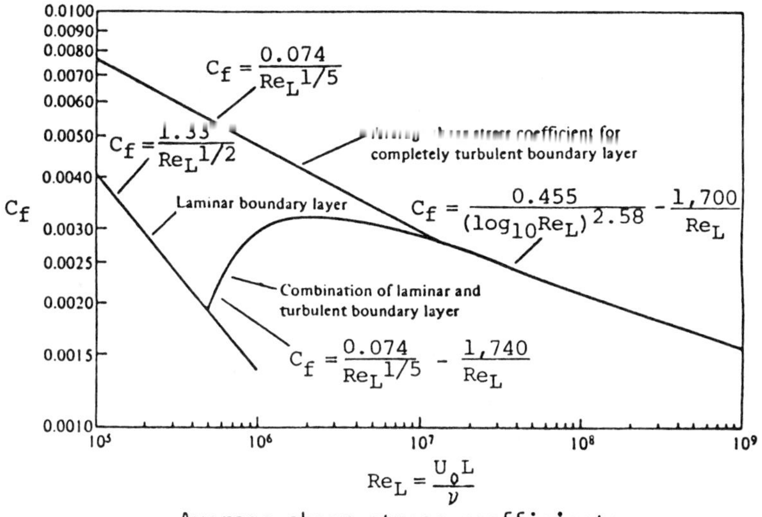

Average-shear-stress coefficients.

Then $\tau_0 = \dfrac{0.058}{25}(1.20\ \text{kg/m}^3)\ \dfrac{30^2}{2}\ \text{m}^2/\text{s}^2$

$= 1.25\ \text{kg/m}\cdot\text{s}^2 = 1.25\ \text{N/m}^2$

English units $\tau_0 = \dfrac{0.058}{25}(0.00232)\dfrac{98^2}{2} = 0.0259\ \text{lbf/ft}^2$

Now compute $u_* = \sqrt{\tau_0/\rho}$ and the thickness of the boundary layer and laminar sublayer.

$$u_* = \left(\frac{\tau_0}{\rho}\right)^{1/2} = \left(\frac{1.25\ \text{N/m}^2}{1.20\text{kg/m}^3}\right)^{1/2} = 1.02\ \text{m/s}$$

English units
$$u_* = \left(\frac{0.0259\ \text{lbf/ft}^2}{0.00232\ \text{slugs/ft}^3}\right)^{1/2} = 3.34\ \text{ft/sec}$$

The thickness of the laminar sublayer is given by

$$\delta' = \frac{5\nu}{u_*}$$

but $\nu = 1.49 \times 10^{-5}\text{m}^2/\text{s}$, so

$$\delta' = \frac{5(1.49)(10^{-5})\text{m}^2/\text{s}}{1.02\ \text{m/s}} = 7.30 \times 10^{-5}\text{m} = 0.07\ \text{mm}$$

English units $\delta' = \dfrac{5(1.61)(10^{-4})\text{ft}^2/\text{sec}}{3.33\ \text{ft/sec}}$

$= 2.42 \times 10^{-4}\ \text{ft}$

Compute the thickness of the boundary layer:

$$\delta = 0.37\ \frac{x}{\text{Re}_x^{1/5}} = \frac{(0.37)(5)\text{m}}{25} = 0.074\ \text{m} = 74\ \text{mm}$$

English units $\delta = \dfrac{(0.37)(16.4)\text{ft}}{25} = 0.24\ \text{ft} = 2.9\ \text{in.}$

b) The shearing resistance of one side will be given as $F_s = C_f BL\rho U_0^2/2$; therefore, the total drag will be twice this for two sides of the plate:

$$F_s = \frac{2C_f BL\rho U_0^2}{2}$$

Since C_f is a function of Re_L, we compute that as

$$Re_L = U_0\frac{L}{\nu} = \frac{(30\ m/s)(6m)}{(1.49)(10^{-5}m^2/s)} = 1.21(10^7)$$

From the figure, $C_f = 0.0028$. Then

$$F_s = 0.0028(3m)(6m)(1.20N\cdot s^2/m^4)(30^2m^2/s^2) = 54.4\ N$$

English units

$$F_s = (0.0028)(9.8)(19.2)(0.00232)(98^2) = 11.7\ lbf$$

• PROBLEM 10-9

(a) With a smooth horizontal flow of air over a flat plate of length L and width w, a laminar boundary layer is developed on the plate surface. The expression for velocity profile in the boundary layer is

$$\frac{u}{U} = \cos\pi\left(\frac{y}{\delta} - \frac{1}{2}\right)$$

Assuming an incompressible flow of air, develop relationships for

(i) the boundary layer thickness δ and the displacement thickness δ^* in terms of x and the Reynolds No. at point x.

(ii) the resistance to the flow of air offered by the plate surface.

(b) Air at 15 psi and at a free stream velocity $U_\infty = 90$ ft/sec enters a passage formed by two parallel, flat rectangular plates separated by a distance, d = 1.2 ft. However, inspection of the air flow inside the passage reveals that a turbulent boundary layer occurs on each plate surface, growing from the leading edge. A reasonable formulation for the velocity profile in a turbulent boundary layer and the boundary layer thickness δ can be assumed as, respectively,

$$\frac{u}{U(x)} = \left(\frac{y}{\delta}\right)^{\frac{1}{7}} \quad \text{and} \quad \frac{\delta}{x} = \frac{0.366}{Re_x^{\frac{1}{5}}}$$

where U(x) is the flow velocity in the center of passage and is a function of x.

Assuming an incompressible, steady, streamline and a non-viscous flow outside the boundary layer, calculate the pressure at a point 20 ft. downstream of the entrance. Ignore all the end effects, since the plate width is b >> d.

Solution: (a) (i) Boundary layer thickness:

The integral momentum equation for a general compressible flow is written as,

$$-\delta\frac{dp}{dx} - \tau_s = \frac{d}{dx}\int_0^\delta \rho u^2 dy - U\frac{\partial}{\partial x}\int_0^\delta \rho u\ dy$$

where $\frac{dp}{dx}$ is the pressure gradient in the direction of flow and τ_s is the shear stress at the surface along which a fluid flows.

But, for an incompressible flow over a flat plate $\frac{dp}{dx} = 0$, ρ = constant and letting $\eta = \frac{y}{\delta}$, the above equation reduces to

$$\tau_s = \rho U^2 \frac{d\delta}{dx} \int_0^1 \frac{u}{U}\left(1 - \frac{u}{U}\right) d\eta \tag{1}$$

The velocity profile in the laminar boundary layer for this problem is

$$\frac{u}{U} = \cos\pi\left(\frac{y}{\delta} - \frac{1}{2}\right)$$

Simplifying the R.H.S.,

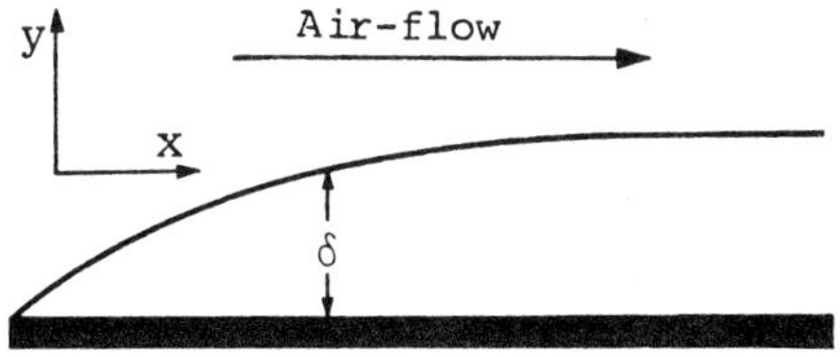

Boundary layer flow over a flat plate.

$$\frac{u}{U} = \cos\left(\frac{\pi y}{\delta} - \frac{\pi}{2}\right)$$

$$= \cos\frac{\pi y}{\delta}\cos\frac{\pi}{2} + \sin\frac{\pi y}{\delta}\sin\frac{\pi}{2} \quad \left(\cos\frac{\pi}{2} = 0,\ \sin\frac{\pi}{2} = 1\right)$$

or $$\frac{u}{U} = \sin\frac{\pi y}{\delta} = \sin\pi\eta$$

Thus, the velocity profile of the flow inside the boundary layer is sinusoidal.

Replacing $\frac{u}{U}$ by $\sin\pi\eta$ in equation (1)

$$\tau_s = \rho U^2 \frac{d\delta}{dx} \int_0^1 \sin\pi\eta(1 - \sin\pi\eta)d\eta$$

or

$$\tau_s = \rho U^2 \frac{d\delta}{dx}\left[\int_0^1 \sin\pi\eta d\eta - \int_0^1 \sin^2\pi\eta d\eta\right] \tag{2}$$

The integration gives,

$$\int_0^1 \sin\pi\eta d\eta = -\frac{1}{\pi}\Big[\cos\pi\eta\Big]_0^1 = -\frac{1}{\pi}(\cos\pi - \cos 0) = \frac{2}{\pi} \quad (\cos\pi = -1,\ \cos 0 = 1)$$

$$\int_0^1 \sin^2\pi\eta d\eta = \frac{1}{2}\int_0^1 (1 - \cos 2\pi\eta)d\eta$$

$$= \frac{1}{2}\left[\int_0^1 d\eta - \int_0^1 \cos 2\pi\eta d\eta\right]$$

$$= \frac{1}{2}\left\{\left[\eta\right]_0^1 - \left[\frac{\sin 2\pi\eta}{2}\right]_0^1\right\}$$

$$= \frac{1}{2}\left\{(1 - 0) - \frac{1}{2\pi}(\sin 2\pi - \sin 0)\right\} = \frac{1}{2}$$

Substituting the results of integration in equation (2), we get

$$\tau_s = \rho U^2 \frac{d\delta}{dx}\left[\frac{2}{\pi} - \frac{1}{2}\right] = 0.1366\rho U^2 \frac{d\delta}{dx}$$

Let $\alpha = 0.1366$ so that $\tau_s = \alpha\rho U^2 \frac{d\delta}{dx}$

Now, the shear stress in the flow at the surface of the flat plate can also be given by,

$$\tau_s = \mu \left.\frac{\partial u}{\partial y}\right|_{y=0}$$

Using the expression for velocity profile in the boundary layer,

$$u = U\sin\frac{\pi y}{\delta}$$

The velocity gradient is,

$$\frac{\partial u}{\partial y} = \frac{\pi U}{\delta}\cos\frac{\pi y}{\delta}$$

At $y = 0$, $\frac{\partial u}{\partial y} = \frac{\pi U}{\delta}\cos 0 = \frac{\pi U}{\delta}$

Therefore, $\tau_s = \frac{\mu\pi U}{\delta}$

Also, $\tau_s = \alpha\rho U^2 \frac{d\delta}{dx}$, so that $\alpha\rho U^2 \frac{d\delta}{dx} = \frac{\mu\pi U}{\delta}$

or, $\left(\frac{\alpha}{\pi}\right)\left(\frac{\rho U}{\mu}\right)\delta d\delta = \frac{\mu\pi U}{\delta} dx$

Integrating both sides, $\left(\frac{\alpha}{\pi}\right)\rho U \frac{\delta^2}{2} = x + C$

Applying the boundary condition:

At $x = 0$, $\delta = 0 \Rightarrow C = 0$

$$\therefore \left(\frac{\alpha}{\pi}\right)\rho U \frac{\delta^2}{2} = x$$

or,

$$\delta = \left[\frac{2\pi}{\alpha}\left(\frac{\mu x}{\rho U}\right)\right]^{\frac{1}{2}} = \left(\frac{2\pi}{\alpha}\right)^{\frac{1}{2}}\left[\frac{x^2}{\rho\frac{Ux}{\mu}}\right]^{\frac{1}{2}}$$

Thus, $\frac{\delta}{x} = \frac{6.78}{\sqrt{Re_x}}$

<u>Displacement Thickness δ^*:</u>

δ^* is expressed as,

$$\delta^* = \int_0^1 \rho\left(1 - \frac{u}{U}\right) d\eta$$

$$= \delta \int_0^1 (1 - \sin\pi\eta)d\eta$$

$$= \delta \left[\eta + \frac{\cos\pi\eta}{\pi}\right]_0^1 = \delta\left[(1 - 0) + \frac{1}{\pi}(\cos\pi - \cos 0)\right]$$

$$= \delta\left[1 + \frac{1}{\pi}(-1 - 1)\right] = \delta\,\frac{(\pi - 2)}{\pi}$$

$$\text{or,} \quad \delta^* = x\left(\frac{\delta}{x}\right)\frac{(\pi - 2)}{\pi}$$

In this equation, substituting the expression for boundary layer thickness, i.e. $\frac{\delta}{x} = \frac{6.78}{\sqrt{Re_x}}$

We get, $$\frac{\delta^*}{x} = \frac{6.78}{\sqrt{Re_x}} \times \frac{(\pi - 2)}{\pi} = \frac{2.46}{\sqrt{Re_x}}$$

(ii) Consider an infinitesimal area dA = bdx of the plate surface over which the air flows. Then the total resistance to the flow of air offered by the plate surface is expressed as,

$$P = \int_A \tau_s dA = \int_0^L \tau_s b dx$$

where $$\tau_s = \alpha\rho U^2 \frac{d\delta}{dx}$$

Therefore $$P = \int_0^L \alpha\rho U^2 \frac{d\delta}{dx} b dx$$

$$P = \alpha\rho b U^2 \int_0^L d\delta = \alpha\rho b U^2 \delta_{(x=L)}$$

From the boundary layer relation, $$\frac{\delta_{(x=L)}}{L} = \frac{6.78}{\sqrt{Re_L}}$$

$$\Rightarrow \quad \delta_{(x=L)} = \frac{6.78L}{\sqrt{Re_L}}$$

Also $\alpha = 0.1366$.

Therefore, $$P = (0.1366)\rho b U^2 \left(\frac{6.78L}{\sqrt{Re_L}}\right)$$

$$= \frac{0.926\rho b L U^2}{\sqrt{Re_L}}$$

(b) The pressure at a point downstream of the leading edge of the plate can be obtained by using Bernoulli's equation which is applicable when the flow is not fully developed. Recall, that a fully developed flow is the one in which the boundary layer is equal to half of the depth of the passage, i.e. $\delta = \frac{d}{2}$. Therefore, a check for the fully developed flow is made as follows:

On each flat plate exposed to the flow, the boundary layer thickness is

$$\frac{\delta}{x} = \frac{0.37}{Re_x^{1/5}}$$

where

$$Re_x = \frac{U_\infty x}{\nu}$$

we have, $U_\infty = 90$ ft/sec, $x = 20$ ft

kinematic viscosity $= 1.64 \times 10^{-4} \frac{ft^2}{sec}$

$$\therefore \quad Re_{(x=20ft)} = \frac{90 \times 20}{1.64 \times 10^{-4}} = 10,975,610$$

Thus, at point (2),

$$\delta_2 = \frac{0.37 \times 20}{(10,975,610)^{\frac{1}{5}}} = 0.289 \text{ ft}$$

The separation between the plates, d = 1.2 ft.

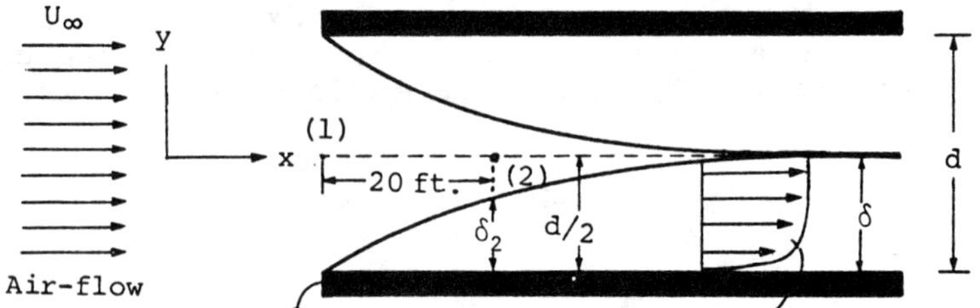

Thus, $\delta_2 < \frac{d}{2}$ showing that the flow is not fully developed and the Bernoulli's equation can be applied between points (1) and (2) located on the streamline in the center of the flow between the plates, as follows:

$$p_1 + \frac{\gamma V_1^2}{2g} + \gamma h_1 = p_2 + \frac{\gamma V_2^2}{2g} + \gamma h_2$$

As points (1) and (2) are on the same streamline, then $h_1 = h_2$ and we get,

$$p_1 + \frac{\gamma V_1^2}{2g} = p_2 + \frac{\gamma V_2^2}{2g}$$

or

$$p_2 - p_1 = \frac{\gamma}{2g}(V_1^2 - V_2^2)$$

$$= \frac{\rho V_1^2}{2}\left[1 - \left(\frac{V_2}{V_1}\right)^2\right] \qquad (\because \gamma = \rho g)$$

where, $V_1 = U_\infty$ and $V_2 = U_2$, the core velocity at point (2).

$$\therefore \quad p_2 - p_1 = \frac{\rho U_\infty^2}{2}\left[1 - \left(\frac{U_2}{U_\infty}\right)^2\right]$$

The velocity U_2 can be evaluated by applying the relation of displacement thickness δ^* at point (2) which is given by,

$$\delta_2^* = \int_0^\delta \left(1 - \frac{u}{U(x)}\right) dy$$

Let $\eta = \frac{y}{\delta}$ or at point (2) $\eta = \frac{y}{\delta_2}$. Therefore, $dy = \delta_2 d\eta$.

Also, as $y = 0$, $\eta = 0$

$$y = \delta_2, \quad \eta = \frac{\delta_2}{\delta_2} = 1.$$

Hence,

$$\delta_2^* = \delta_2 \int_0^1 \left(1 - \frac{u}{U(x)}\right) d\eta$$

Now, using the expression for velocity profile in the boundary layer, i.e.

$$\frac{u}{U(x)} = \left(\frac{y}{\delta}\right)^{\frac{1}{7}} = \eta^{\frac{1}{7}}$$

$$\delta^* = \delta_2 \int_0^1 (1 - \eta^{\frac{1}{7}}) d\eta$$

$$= \delta_2 \left[\int_0^1 d\eta - \int_0^1 \eta^{\frac{1}{7}} d\eta\right]$$

$$= \delta_2 \left\{[\eta]_0^1 - \frac{7}{8}\left[\eta^{\frac{8}{7}}\right]_0^1\right\}$$

or $$\delta_2^* = \frac{\delta_2}{8} = \frac{0.289}{8} = 0.036 \text{ ft}$$

Also, using the equation of continuity for the points (1) and (2),

$$\rho A_1 V_1 = \rho A_2 V_2 = \text{constant}$$

Since, $V_1 = U_\infty$ and $V_2 = U_2$

$$\therefore \quad \rho A_1 U_\infty = \rho A_2 U_2$$

or $$\frac{U_2}{U_\infty} = \frac{A_1}{A_2}$$

Now, the area at the entrance, $A_1 = b \times d$.

Area at point (2), $A_2 = b(d - 2\delta^*)$.

Thus,

$$\frac{U_2}{U_\infty} = \frac{b \times d}{b(d - 2\delta^*)} = \frac{d}{(d - 2\delta^*)}$$

Substituting $d = 20$ ft and $\delta^* = 0.036$

$$\frac{U_2}{U_\infty} = \frac{20}{[20 - 2(0.036)]} = 1.0036$$

Using the equation for change in pressure,

$$p_2 - p_1 = \frac{\rho U_\infty^2}{2}\left[1 - \left(\frac{U_2}{U_\infty}\right)^2\right]$$

$$= \frac{1}{2} \times 2.42 \times 10^{-3} \frac{\text{slug}}{\text{ft}^3} \times (90)^2 \frac{\text{ft}^2}{\text{sec}^2}\left[1 - (1.0036)^2\right]$$

$$= -0.07 \frac{\text{slug}}{\text{ft-sec}^2}$$

As $lb_f = 1 \text{ slug} \times 1 \frac{\text{ft}}{\text{sec}^2} \Rightarrow 1 \text{ slug} = \frac{1\, lb_f\text{-sec}^2}{\text{ft}}$

Therefore,

$$p_2 - p_1 = -0.07 \frac{lb_f\text{-sec}^2}{\text{ft}} \times \frac{1}{\text{ft-sec}^2}$$

or

$$p_2 - p_1 = -0.07 \frac{lb_f}{ft^2} \times \frac{1}{144 \frac{in^2}{ft^2}} = -4.86 \times 10^{-4} \text{ psi}$$

from which we get,

$$p_2 = p_1 - 4.86 \times 10^{-4} = 15 - 4.86 \times 10^{-4} = 14.999 \text{ psi}$$

DRAG

• PROBLEM 10-10

Determine the drag force on a cylindrical chimney that is 80 m high by 10 m in diameter when the wind (T = 15°C) velocity is 120 km/hr.

Solution: Assuming standard atmospheric pressure, $\rho = 1.226 \text{ kg/m}^3$, and for T = 15°C = 59°F,

$$\nu = (1.57 \times 10^{-4} ft^2/s)(0.0929 \text{ m}^2/ft^2)$$

$$= 1.46 \times 10^{-5} \text{ m}^2/s$$

The drag on many objects is a combination of both boundary layer and separation effects at moderate to high velocities and dependent on a different mechanism in the lower range. In the Fig., the various conditions for flow past a cylinder are shown.

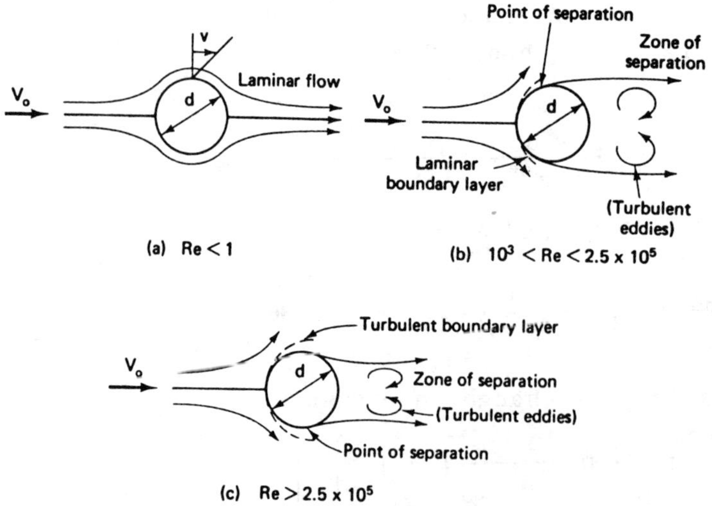

(a) Re < 1

(b) $10^3 < Re < 2.5 \times 10^5$

(c) $Re > 2.5 \times 10^5$

The flow is best characterized by a Reynolds number. In this case the reference length will be the sphere diameter and therefore, in terms of the approach velocity and kinematic viscosity,

$$Re = \frac{V_0 d}{\nu}$$

Therefore, the Reynolds number is

$$Re = \frac{V_0 d}{\nu}$$

$$= \frac{(120\ \text{km/hr})(1000\ \text{m/km})(10\ \text{m})}{(3600\ \text{s/hr})(1.46\times10^{-5}\text{m}^2/\text{s})} = 2.28 \times 10^7$$

Consider the total drag force on a body for which a length dimension L can be defined. Assuming that the approach velocity V_0 is not so high as to cause compressibility effects and that the surface is smooth, we may write

$$F_D = f(\text{geometry}, L, V_0, \rho, \mu)$$

The geometry term is included to avoid the number of length ratios required to define a complicated shape. It is sufficient since we are interested in a specific shape (which means the geometry variable is constant) in this problem.

Selecting L, V_0, and ρ as repeating variables, dimensional analysis gives

$$\frac{F_D}{\rho L^2 V_0^2} = \phi\left(\text{geometry}, \frac{V_0 L \rho}{\mu}\right)$$

Since L^2 is proportional to an area A, the equation may be rewritten as

$$\frac{F_D}{\rho A V_0^2/2} = \phi_1(\text{geometry}, Re) \tag{1}$$

where the factor 2 has been added arbitrarily to the denominator.

For a specific shape, both the area A and the length dimension in the Reynolds number must be properly defined. Nevertheless, Eq. 1 is a general drag equation and for a specific shape, ϕ_1, can be replaced by a drag coefficient C_D so that

$$F_D = C_D A \frac{\rho V_0^2}{2}$$

where C_D is a function only of the Reynolds number. From the Table, for

$$\frac{L}{d} = \frac{80}{10} = 8$$

$$C_D = 0.34 \text{ (approximately)}$$

Hence, the drag force is

$$F_D = \frac{C_D A \rho V_0^2}{2}$$

$$= \frac{(0.34)(80\text{m})(10\text{m})(1.226\text{kg/m}^3)(33.3\text{m/s})^2}{2}$$

$$= 1.85 \times 10^5 \text{ N}$$

TABLE

BODY (flow from left to right)	L/d	Re = Vd/ν	C_D
Bodies of revolutions			
1) Sphere:		10^5 $>3 \times 10^5$	0.50 0.20
2) Hemispherical Shell:		$>10^3$	1.33 0.34
3) Ellipsoid: 2, 1 = d		$>2 \times 10^5$	0.07
4) Circular cylinder axis ∥ to flow:	0 1 2 4 7	$>10^3$	1.12 0.91 0.85 0.87 0.99
5) Circular disk ⊥ to flow:		$>10^3$	1.12
6) Tandem disks ⊥ to flow (L = Spacing)	0 1 2 3	$>10^3$	1.12 0.93 1.04 1.54
"Two-dimensional" flow (finite and infinite length)			
7) Circular cylinder axis ⊥ to flow:	1 5 20 ∞ 5 ∞	10^5 $>5 \times 10^5$	0.63 0.74 0.90 1.20 0.35 0.33
8) Elliptical cylinder: 2:1, 4:1, 8:1		4×10^4 10^5 $2.5 \times 10^4 - 10^5$ 2.5×10^4 2×10^5	0.6 0.46 0.32 0.29 0.20
9) Rectangular Plate: L = length, d = width	1 5 20 ∞	$>10^3$	1.16 1.20 1.50 1.90
10) Square cylinder:		3.5×10^4 $10^4 - 10^5$	2.0 1.6

Evaluate the surface drag force on a 1-m by 1-m plate as water ($\nu = 10^{-6}m^2/s$) flows over it with a velocity of 1 m/s. Assume that:
(a) The boundary layer remains laminar.
(b) The boundary layer is entirely turbulent.
(c) The boundary layer is transitory.

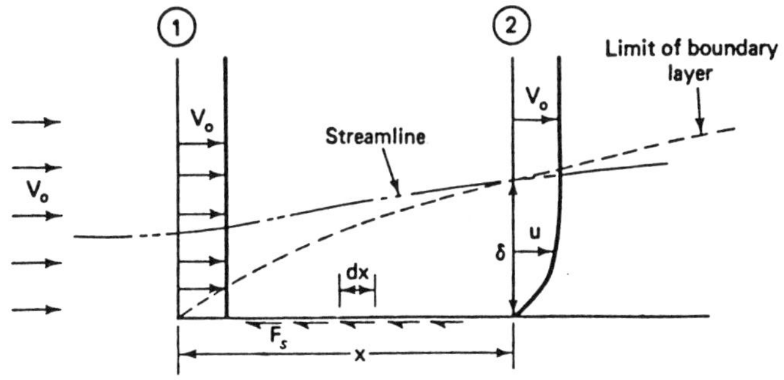

Fig. 1

Solution: In Fig. 1, for purposes of momentum analysis, a control volume can be identified of length x. This volume is bounded below by the plate and above by the streamline passing through the outer limit of the boundary layer at the downstream end of the region. The expansion of the streamline in the downstream direction is due to the ever-increasing amount of retarded fluid between it and the plate. Because of the thinness of the boundary layer, it is assumed that there is no pressure gradient across it, and the constant free stream velocity precludes a pressure gradient in the flow direction. Therefore, there is no pressure variation in any direction, and the only force is the resisting surface drag F_s of the plate on the fluid.

Applying the momentum principle with β values included, on a per unit width basis between the sections identified as 1 and 2, gives

$$-F_s = \rho q(V_2\beta_2 - V_1\beta_1)$$

Here q is the discharge per unit width and F_s is the drag force exerted by the plate on an area x units long by a unit width. At section 1 the average velocity V_1 equals V_0 and $\beta_1 = 1$. At the downstream section the velocity profile can be assumed parabolic and hence the average velocity $V_2 = 2V_0/3$, $\beta_2 = \frac{6}{5}$, and $q = V_2\delta = \frac{2}{3}V_0\delta$. Thus, the momentum equation becomes

$$-F_s = \rho \frac{2V_0\delta}{3}\left[\left(\frac{2V_0}{3}\right)\left(\frac{6}{5}\right) - V_0\right]$$

or

$$F_s = 0.133\rho V_0^2\delta \tag{1}$$

That portion of F_s acting on a differential length dx (shown in Fig. 1) may be written

$$dF_s = \tau_0 dx$$

where τ_0 is the shear stress at the plate and therefore is a function of x.

The shear stress at the plate is given by

$$\tau_0 = \mu\left(\frac{du}{dy}\right)_{y=0}$$

and the parabolic velocity distribution can be expressed as

$$u = V_0 - \frac{V_0}{\delta^2}(\delta - y)^2$$

Thus, upon differentiating, the shear stress is given by

$$\tau = \mu\frac{2V_0}{\delta^2}(\delta - y)$$

and therefore at the plate

$$\tau_0 = \frac{2\mu V_0}{\delta} \tag{2}$$

Continuing by differentiating Eq. 1,

$$\tau_0 = \frac{dF_s}{dx} = 0.133\rho V_0 \frac{d\delta}{dx}$$

which when combined with Eq. 2 yields

$$\delta d\delta = \frac{15\mu}{\rho V_0} dx$$

Integrating over the length x, during which the boundary layer thickness will range from zero at the leading edge to δ at the other limit, gives

$$\frac{\delta^2}{2} = \frac{15\mu x}{\rho V_0}$$

Solving for the ratio of δ to x,

$$\frac{\delta}{x} = \frac{5.48}{\sqrt{V_0 x\rho/\mu}}$$

This introduces a significant Reynolds number defined in terms of the distance from the leading edge,

$$Re_x = \frac{V_0 x\rho}{\mu}$$

whereupon

$$\frac{\delta}{x} = \frac{5.48}{Re_x^{1/2}} \tag{3}$$

This equation, which gives the growth of the boundary layer, may now be substituted into Eq. 1 to give, after some rearrangement,

$$F_s = \frac{1.463}{Re_x^{1/2}} \frac{x\rho V_0^2}{2} \tag{4}$$

where a drag coefficient

$$C_f = \frac{1.463}{Re_x^{1/2}} \tag{5}$$

may conveniently be introduced. Substituting Eq. 3 into Eq. 2 for δ gives another useful equation,

$$\tau_0 = \frac{0.732}{Re_x^{1/2}} \frac{\rho V_0^2}{2} \tag{6}$$

and a local drag coefficient

$$c_f = \frac{0.732}{Re_x^{1/2}} \tag{7}$$

may be defined.

On the basis of more rigorous analysis, coupled with experimental verification, the constants in Eqs. 3,4,5, 6 and 7 should be changed from 5.48 to 5.2, 1.463 to 1.328, and 0.732 to 0.664, respectively. With the inclusion of the improved values for the numerical constants, the laminar boundary layer thickness is given by

$$\frac{\delta}{x} = \frac{5.2}{Re_x^{1/2}}$$

Further, the drag force on a flat plate of breadth B and length L is

$$F_s = C_f BL \frac{\rho V_0^2}{2} \tag{8}$$

where the mean drag coefficient is

$$C_f = \frac{1.328}{Re_L^{1/2}}$$

The Reynolds number Re_L is given by

$$Re_L = \frac{V_0 L}{\nu} = \frac{(1 \text{ m/s})(1 \text{ m})}{10^{-6}\text{m}^2/\text{s}} = 10^6$$

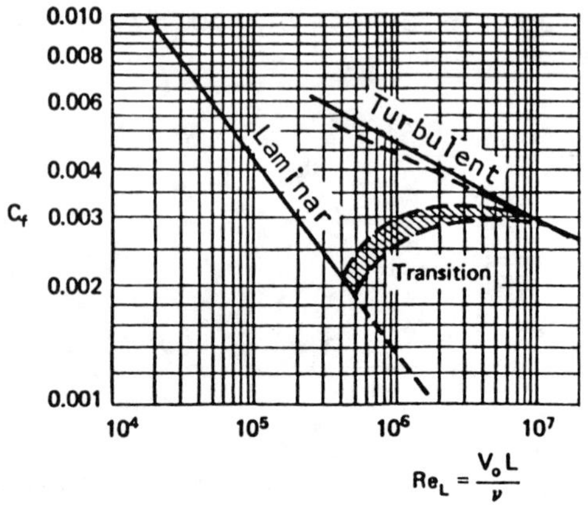

Fig. 2

From Fig. 2 the mean drag coefficient in each of the specified cases is given as follows:

(a) Laminar curve: $C_f = 0.00137$

(b) Turbulent curve: $C_f = 0.0047$

(c) Transition curve: $C_f = 0.0028$ (approximate)

Since the force is given by Eq. 8

$$F_s = C_f BL \frac{\rho V_0^2}{2} = \frac{(C_f)(1 \text{ m})(1 \text{ m})(998 \text{ kg/m}^3)(1 \text{ m/s})^2}{2}$$

$$= 449 C_f$$

and force in each case is as follows:

(a) Laminar boundary layer: $F_s = (499)(0.00137) = 0.68\text{N}$

(b) Turbulent boundary layer: $F_s = (499)(0.0047) = 2.35\text{N}$

(c) Transition: $F_s = (499)(0.0028) = 1.40\text{N}$

• PROBLEM 10-12

Calculate the drag force on a 3-m billboard 30 m wide at ground level in a 25-m/s wind normal to the bill-board. Assume standard air.

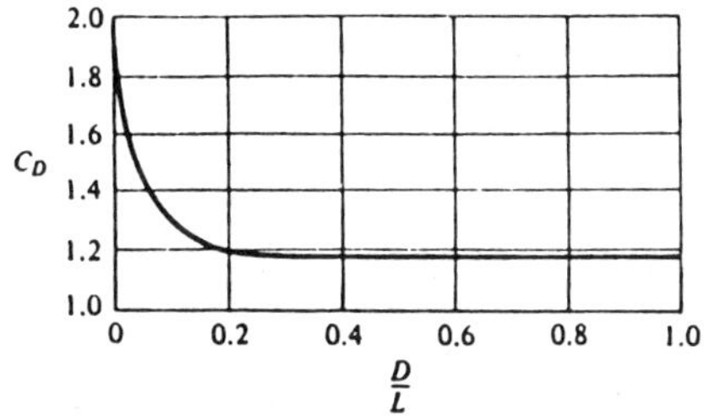

Effect of aspect ratio on drag coefficient for rectangular plate normal to the flow.

Solution: The drag will be one-half that for a 6- by 30-m rectangle, since the flow is essentially the same as that past the top half of a 6- by 30-m rectangle. Thus;

$$Re_D = \frac{u_s D}{\nu} = \frac{(25)(6)}{1.46 \times 10^{-5}} = 1.0 \times 10^7 \text{ which is greater than } 10^3$$

From the figure, $C_D = 1.2$ for $D/L = 0.2$. Thus the drag is

$$\text{drag} = C_D \frac{\rho u_s^2}{2} A = \frac{(1.2)(1.225)(25)^2(3)(30)}{2}$$

$$= 41 \text{ kN}$$

• PROBLEM 10-13

Two tandem disks, each with a diameter of 0.7 m spaced 1.6 m apart, are towed behind a ship through 18°C fresh water at 6 m/s. What power is consumed in pulling the disks?

Solution: At T = 18°C (from Table 1),

$$\nu = 1.061 \times 10^{-6} \text{ m}^2/\text{s}$$

In this case, the Reynolds number is given by the equation

$$Re = \frac{V_0 d}{\nu}$$

where V_0 is the approach velocity, d is the diameter of the disks, and ν is the kinematic viscosity.

TABLE 1

Temperature (°C)	Specific Weight, γ (N/m^3)	Density, ρ (kg/m^3)	Dynamic Viscosity, $\mu \times 10^3$ ($N \cdot s/m^2$)	Kinematic Viscosity, $\nu \times 10^6$ (m^2/s)	Surface Tension, $\sigma \times 10^2$ (N/m)	Vapor Pressure, p_v (kN/m^2)	Modulus of Compressibility, $E \times 10^{-9}$ (N/m^2)
0	9805	999.8	1.794	1.794	7.62	0.61	2.02
5	9806	1000.0	1.519	1.519	7.54	0.87	2.06
10	9802	999.7	1.308	1.308	7.48	1.23	2.11
15	9797	999.1	1.140	1.141	7.41	1.70	2.14
20	9786	998.2	1.005	1.007	7.36	2.34	2.20
25	9777	997.1	0.894	0.897	7.26	3.17	2.22
30	9762	995.7	0.801	0.804	7.18	4.24	2.23
35	9747	994.1	0.723	0.727	7.10	5.61	2.24
40	9730	992.2	0.656	0.661	7.01	7.38	2.27
45	9711	990.2	0.599	0.605	6.92	9.55	2.29
50	9689	988.1	0.549	0.556	6.82	12.33	2.30
55	9665	985.7	0.506	0.513	6.74	15.78	2.31
60	9642	983.2	0.469	0.477	6.68	19.92	2.28
65	9616	980.6	0.436	0.444	6.58	25.02	2.26
70	9588	977.8	0.406	0.415	6.50	31.16	2.25
75	9560	974.9	0.380	0.390	6.40	38.57	2.23
80	9528	971.8	0.357	0.367	6.30	47.34	2.21
85	9497	968.6	0.336	0.347	6.20	57.83	2.17
90	9473	965.3	0.317	0.328	6.12	70.10	2.16
95	9431	961.9	0.299	0.311	6.02	84.36	2.11
100	9398	958.4	0.284	0.296	5.94	101.33	2.07

Consider the total drag force on a body for which a length dimension L can be defined. Assuming that the approach velocity V_0 is not so high as to cause compressibility effects and that the surface is smooth, we may write

$$F_D = f(\text{geometry}, L, V_0, \rho, \mu)$$

The geometry term is included to avoid the number of length ratios required to define a complicated shape. It is sufficient since we are interested in a specific shape (which means the geometry variable is constant) in this problem.

Selecting L, V_0, and ρ as repeating variables, dimensional analysis gives

$$\frac{F_D}{\rho L^2 V_0^2} = \phi\left(\text{geometry}, \frac{V_0 L \rho}{\mu}\right)$$

Since L^2 is proportional to an area A, the equation may be rewritten as

$$\frac{F_D}{\rho A V_0^2/2} = \phi_1(\text{geometry}, Re) \qquad (1)$$

where the factor 2 has been added arbitrarily to the denominator.

For a specific shape, both the area A and the length dimension in the Reynolds number must be properly defined. Nevertheless, Eq. 1 is a general drag equation and for a specific shape, ϕ_1, can be replaced by a drag coefficient C_D so that

$$F_D = C_D A \frac{\rho V_0^2}{2}$$

where C_D is a function only of the Reynolds number.

$$Re = \frac{(6 \text{ m/s})(0.7 \text{ m})}{1.061 \times 10^{-6} \text{ m}^2/\text{s}} = 3.96 \times 10^6$$

TABLE 2

BODY (flow from left to right)	L/d	Re = Vd/ν	C_D
Bodies of revolutions			
1) Sphere:		10^5 $> 3 \times 10^5$	0.50 0.20
2) Hemispherical Shell:		$> 10^3$	1.33 0.34
3) Ellipsoid: 2, 1 = d		$> 2 \times 10^5$	0.07
4) Circular cylinder axis ‖ to flow:	0 1 2 4 7	$> 10^3$	1.12 0.91 0.85 0.87 0.99
5) Circular disk ⊥ to flow:		$> 10^3$	1.12
6) Tandem disks ⊥ to flow (L = Spacing)	0 1 2 3	$> 10^3$	1.12 0.93 1.04 1.54
"Two-dimensional" flow (finite and infinite length)			
7) Circular cylinder axis ⊥ to flow:	1 5 20 ∞ 5 ∞	10^5 $> 5 \times 10^5$	0.63 0.74 0.90 1.20 0.35 0.33
8) Elliptical cylinder: 2:1, 4:1, 8:1		4×10^4 10^5 $2.5 \times 10^4 - 10^5$ 2.5×10^4 2×10^5	0.6 0.46 0.32 0.29 0.20
9) Rectangular Plate: L = length d = width	1 5 20 ∞	$> 10^3$	1.16 1.20 1.50 1.90
10) Square cylinder:		3.5×10^4 $10^4 - 10^5$	2.0 1.6

Since this is well in excess of $Re = 10^3$, Table 2, may be used. For $L/d = 1.6/0.7 = 2.29$, $C_D = 1.19$ and

$$F_D = \frac{C_D A \rho V_0^2}{2}$$

$$= \frac{(1.19)(\pi/4)(0.7\text{m})^2(998\text{kg/m}^3)(6\text{m/s})^2}{2}$$

$= 8230 \text{ N}$

Thus, the power required becomes

$$P = F_D V = (8230)(6) = 49{,}400 \text{ N·m/s}$$

• PROBLEM 10-14

Find the "free-fall" velocity of a 8.5-in-diameter sphere weighing 16 lb when falling through the following fluids under the action of gravity: (a) through the standard atmosphere at sea level; (b) through the standard atmosphere at 10,000-ft elevation; (c) through water at 60°F; (d) through crude oil (s = 0.925) at 60°F.

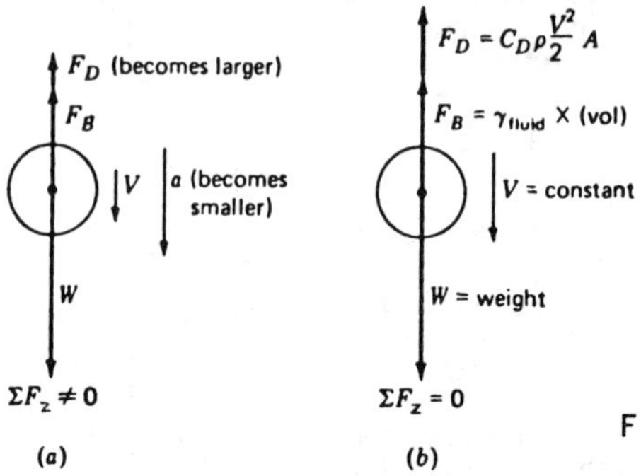

Fig. 1

TABLE

Fluid	lb/ft^3	slug/ft^3	ft^2/s	F(lb)
Air (sea level)	0.0765	0.00238	1.57×10^{-4}	0.0142
Air (10,000 ft)	0.0564	0.00175	1.57×10^{-4}	0.0105
Water, 60°F	62.4	1.94	1.22×10^{-5}	11.6
Oil, 60°F	57.6	1.79	0.001	10.7

Solution: (a) When first released the sphere will accelerate because the forces acting on it are out of balance. This acceleration results in a buildup of velocity which causes an increase in the drag force. After a while the drag force will increase to the point where the forces acting on the sphere are in balance, as indicated in Figure (1). When that point is reached the sphere will attain a constant or terminal (free-fall) velocity. Thus for free-fall conditions,

$$\sum F_z = W - F_B - F_D = \text{mass} \times \text{acceleration} = 0$$

where W is the weight, F_B the buoyant force, and F_D the drag force. The buoyant force is equal to the unit weight of the fluid multiplied by the volume ($\pi D^3/6$ =

0.186 ft^3) of the sphere. The detailed analysis for the sphere falling through the standard sea-level atmosphere is as follows:

$$16 - 0.01 - C_D \rho \frac{V^2}{2} A = 0$$

where

$$\rho = 1.94 \text{ slugs/ft}^3 \quad \text{and} \quad A = \frac{\pi(8.5/12)^2}{4}$$

$$= 0.394 \text{ ft}^2$$

or $$15.99 = C_D(0.00238)\frac{V^2}{2}(0.394) = 0.00047\, C_D V^2$$

A trial-and-error solution is required. Let $C_D \approx 0.2$, then V = 412 fps.

$$N_R = \frac{DV}{\nu} = \frac{(8.5/12)412}{1.57 \times 10^{-4}} = 1.86 \times 10^6$$

The values of C_D and N_R check Fig. 2; hence $C_D = 0.2$ and V = 412 fps.

Following a similar procedure for the other three fluids gives the following free-fall velocities:

(b) Standard atmosphere at 10,000 ft = 470 fps
(c) Water at 60°F = 7.4 fps
(c) Crude oil (s = 0.925) at 60°F = 6.15 fps

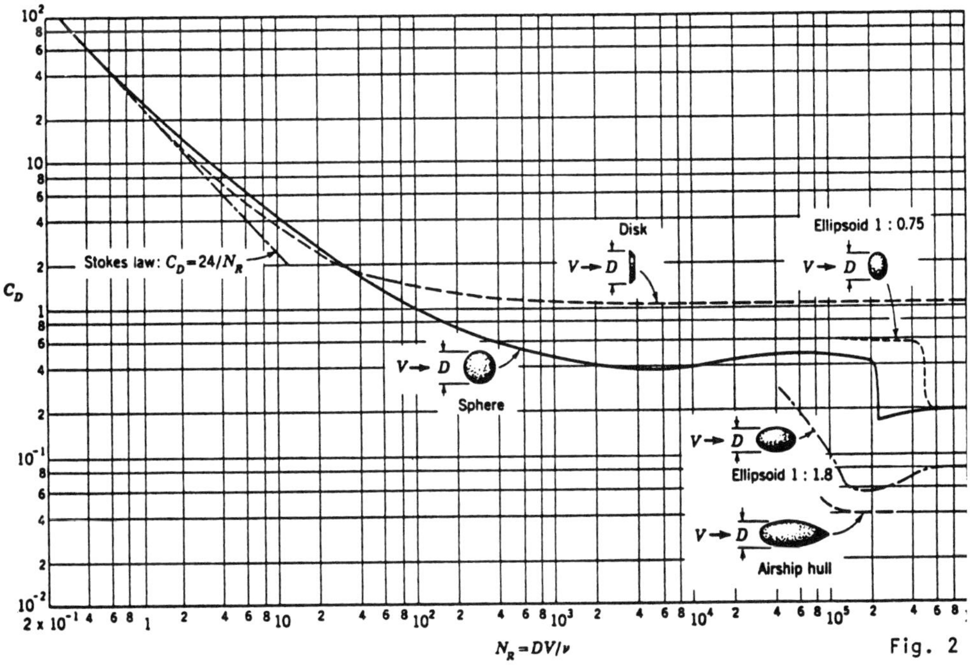

Fig. 2

A 3-mph wind blows over a flat roof (see Figure). If the air temperature is 32°F and the roof is 10 ft wide, calculate the wind drag force on the roof.

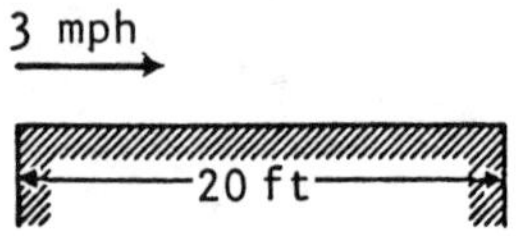

Solution: At this temperature, $\nu = 1.43 \times 10^{-4}\,ft^2/sec$ and $\rho = 0.0808\ lb_m/ft^3$. (See Tables 1 and 2.) Therefore, the Reynold's number at the trailing edge is given by

$$Re_L = \frac{VL}{\nu} = \frac{[3 \times (5280/3600)]20}{1.43 \times 10^{-4}} = 6.15 \times 10^5$$

TABLE 1

Absolute and Kinematic Viscosities of Air at Atmospheric Pressure

°F	$\mu \times 10^7$ lbf-sec/ft²	$\nu \times 10^4$ ft²/sec	°F	$\mu \times 10^7$ lbf-sec/ft²	$\nu \times 10^4$ ft²/sec
0	3.28	1.26	100	3.96	1.89
10	3.45	1.31	120	4.07	1.89
20	3.50	1.36	140	4.14	2.01
30	3.58	1.42	160	4.22	2.12
40	3.62	1.46	180	4.34	2.25
50	3.68	1.52	200	4.49	2.40
60	3.74	1.58	250	4.87	2.80
70	3.82	1.64	300	4.99	3.06
80	3.85	1.69	400	5.40	3.65
90	3.90	1.74	500	5.89	4.56

TABLE 2

Density of Air versus Temperature

Temperature, °F	ρ, lbm/ft³	ρ, slugs/ft³ × 10³
0	0.0862	2.68
10	0.0846	2.63
20	0.0827	2.57
30	0.0811	2.52
40	0.0795	2.47
50	0.0780	2.42
60	0.0764	2.37
70	0.0750	2.33
80	0.0736	2.28
90	0.0722	2.24
100	0.0709	2.20
120	0.0685	2.13
140	0.0663	2.06
150	0.0651	2.02
200	0.0602	1.87
300	0.0523	1.63
400	0.0462	1.44
500	0.0414	1.29

Pressure 14.696 lbf/in.²

This signifies a turbulent boundary layer flowing over the roof, and so, the skin friction drag coefficient is given by the equation

$$C_D = \frac{0.074}{\sqrt[5]{Re_L}} - \frac{1700}{Re_L}$$

$$= 0.00514 - 0.00276$$

$$= 0.00238$$

The total drag is given by the equation

$$\text{Total drag } D = \frac{1}{2}\frac{\rho}{g_c} V^2 C_D A$$

$$= \frac{1}{2}\frac{0.0808}{32.2}(4.4)^2(0.00238)200$$

$$= 0.0116 \text{ lb}_f$$

• PROBLEM 10-16

A parachutist weighs 175 lb and has a projected frontal area of 2 ft^2 in free fall. His drag coefficient based on frontal area is found to be 0.80. If the air temperature is 70°F, determine his terminal velocity.

Density of Air Versus Temperature

Temperature, °F	ρ, lbm/ft^3	ρ, slugs/ft^3 × 10^3
0	0.0862	2.68
10	0.0846	2.63
20	0.0827	2.57
30	0.0811	2.52
40	0.0795	2.47
50	0.0780	2.42
60	0.0764	2.37
70	0.0750	2.33
80	0.0736	2.28
90	0.0722	2.24
100	0.0709	2.20
120	0.0685	2.13
140	0.0663	2.06
150	0.0651	2.02
200	0.0602	1.87
300	0.0523	1.63
400	0.0462	1.44
500	0.0414	1.29

Pressure 14.696 lbf/in.^{2}.

Solution: It is convenient to express the drag of a bluff body in terms of a non-dimensional parameter, C_D, called drag coefficient:

$$C_D = \frac{D}{\frac{1}{2}(\rho/g_c)V^2A}$$

or

$$D = C_D \frac{1}{2} \frac{\rho}{g_c} V^2A$$

with A the projected frontal area of the bluff body normal to the flow direction.

At terminal velocity, the parachutist's weight is balanced by his drag:

$$W = C_D \frac{1}{2} \frac{\rho}{g_c} V^2A$$

The density of air at normal atmospheric pressure and 70°F is 0.075 lb_m/ft^3. Therefore,

$$175 = (0.8) \frac{1}{2} \frac{0.075}{32.2} V^2 2$$

and

$$V_{terminal} = 307 \text{ fps}$$

• PROBLEM 10-17

A large percentage of the aerodynamic drag on an automobile is due to pressure drag. Whereas a convertible open car, vintage 1920, with boxlike structure may have $C_D = 0.9$, a modern streamlined car with rounded front and tapered rear will have $C_D = 0.30$ (see Figure). For a car traveling at 60 mph, calculate the horsepower required to overcome drag for the two cases. Assume a frontal area of 18 ft^2 for both cars, and ρ for air to be .075 lb_m/ft^3.

Solution: It is convenient to express the drag of a bluff body in terms of a non-dimensional parameter, C_D, called drag coefficient:

$$C_D = \frac{D}{\frac{1}{2}(\rho/g_c)V^2A}$$

or

$$D = C_D A\frac{1}{2} \frac{\rho}{g_c} V^2$$

with A the projected frontal area of the bluff body normal to the flow direction.

For $C_D = 0.9$,

$$D = (0.9)(18)\frac{1}{2}\,\frac{0.075}{32.2}\left(\frac{60\times 5280}{3600}\right)^2 = 146.3\ lb_f$$

For $C_D = 0.3$,

$$D = 48.8\ lb_f$$

For the 1920 car, the horsepower required is

$$\text{Horsepower} = \frac{D\times V}{550} \quad (\text{with D in } lb_f \text{ and V in fps})$$

$$= \frac{146.3 \times (60 \times 5280/3600)}{550}$$

$$= 23.4\ hp$$

For the streamlined car, the horsepower required is 7.8.

• PROBLEM 10-18

Compute the overturning moment which is exerted by a 60-mph wind upon a chimney having a diameter of 8 ft and a height of 100 ft.

Solution: The Reynolds number of the flow, for air at 60°F, is

$$R = \frac{VD}{\nu} = \frac{\frac{60 \times 5280}{3600} \times 8}{1.58 \times 10^{-4}} = 4.45 \times 10^6$$

From Figure or Table, $C_D = 0.35$. Hence

$$F = C_D A\,\frac{\rho v_0{}^2}{2} = 0.35 \times 8 \times 100 \times \frac{0.0024 \times \overline{88}^2}{2} = 2600\ lb$$

and

$$M = F\,\frac{L}{2} = 2600 \times \frac{100}{2} = 130{,}000\ lb\text{-}ft.$$

Table

Approximate Values of the Drag Coefficient for Various Body Forms

Form of Body	L/D	R	C_D
Circular disk		$>10^3$	1.12
Tandem disks (L = spacing)	0	$>10^3$	1.12
	1		0.93
	2		1.04
	3		1.54
Rectangular plate (L = length, D = width)	1	$>10^3$	1.16
	5		1.20
	20		1.50
	∞		1.90
Circular cylinder (axis ‖ to flow)	0	$>10^3$	1.12
	1		0.91
	2		0.85
	4		0.87
	7		0.99
Circular cylinder (axis ⊥ to flow)	1	10^5	0.63
	5		0.74
	20		0.90
	∞		1.20
	5	$>5 \times 10^5$	0.35
	∞		0.33
Streamlined foil (1:3 airplane strut)	∞	$>4 \times 10^4$	0.07
Hemisphere: Hollow upstream		$>10^3$	1.33
Hollow downstream			0.34
Sphere		10^5	0.50
		$>3 \times 10^5$	0.20
Ellipsoid (1:2, major axis ‖ to flow)		$>2 \times 10^5$	0.07
Airship hull (model)		$>2 \times 10^5$	0.05

Figure

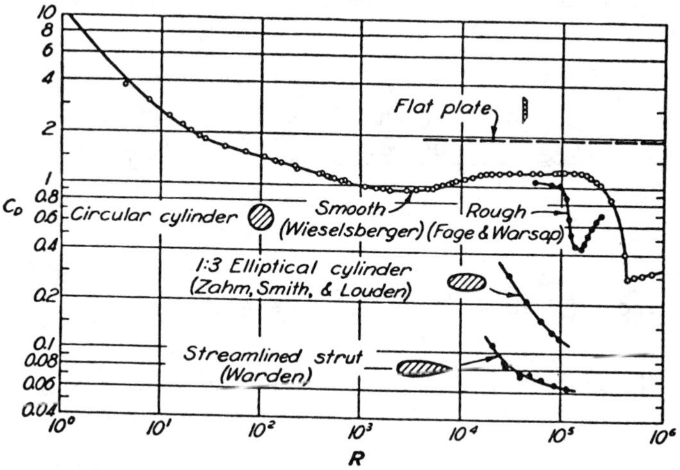

Resistance diagram for cylindrical bodies

• PROBLEM 10-19

A bathysphere of 10 ft diameter moves through sea water of density ρ = 2 slugs/ft^3 and viscosity $\nu = 1.5 \times 10^{-5}$ ft^2/sec at a uniform velocity U = 1 fps and is to be modeled by a 6 in diameter sphere moving through fresh water of density ρ = 1.935 slugs/ft^3 and viscosity ν = 1.0×10^{-5} ft^2/sec. Calculate a) the velocity of the model sphere for dynamically similar flow, b) the Reynolds number for the flow, c) the total drag coefficient, and d) the drag force on the prototype sphere.

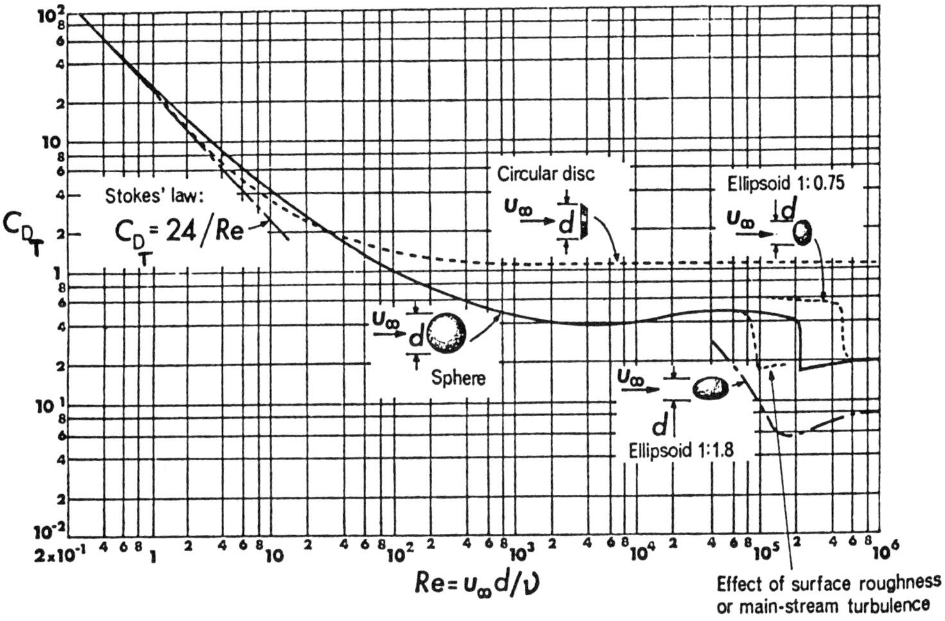

Solution: (a) To obtain the velocity of the model sphere, the Reynolds number of the prototype is equated to the Reynolds number of the model using the diameter of the sphere as the characteristic length

$$\left(\frac{UD}{\nu}\right)_1 = \left(\frac{UD}{\nu}\right)_2$$

so that the velocity of the model sphere is

$$U_1 = U_2\left(\frac{D_2}{D_1}\right)\left(\frac{\nu_1}{\nu_2}\right)$$

such that

$$U_1 = 1.0\left(\frac{10}{0.5}\right)\left(\frac{1.0 \times 10^{-5}}{1.5 \times 10^{-5}}\right)$$

$$= 13.33 \text{ fps}$$

(b) The Reynolds number for either the prototype flow or model flow is

$$R = \frac{UD}{\nu}$$

$$= \frac{1 \times 10}{1.5 \times 10^{-5}}$$

$$= 6.67 \times 10^5$$

(c) The total drag coefficient C_{D_T} is obtained from the Figure as

$$C_{D_T} = 0.2$$

(d) The drag force on the prototype sphere is obtained from the equation for drag

$$D = \frac{1}{2}\rho U^2 A\ C_D$$

Putting in the given and calculated date one obtains

$$D = \frac{1}{2} \times 2 \times 1 \times \pi\ (25) \times 0.2 \quad = 15.7 \text{ lbf}$$

as the drag force on the sphere moving through sea water at 1 fps.

• PROBLEM 10-20

A 4.60-m smooth model of an ocean vessel is towed in fresh water at 2.83 m/s with a total measured drag of 75 N. The wetted surface of the hull is 3.50 m^2.

a. Estimate the skin-friction drag.

b. Estimate the wave drag.

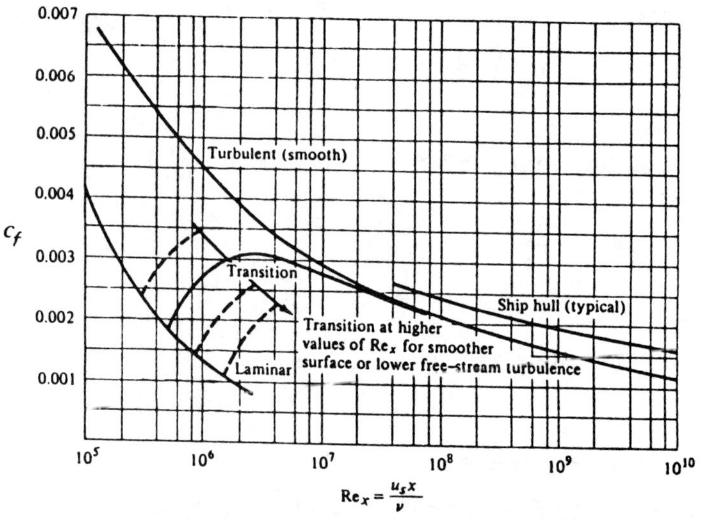

Drag coefficients for plane surfaces parallel to flow.

Solution: a. The skin-friction drag may be estimated by considering the wetted surface of the hull as equivalent to a flat plate 4.60 m long with a total area of 3.50 m^2:

$$Re_x = \frac{u_s x}{\nu} = \frac{(2.83)(4.60)}{1.1 \times 10^{-6}} = 1.2 \times 10^7$$

From the figure, $C_f = 0.0029$. Then the skin-friction drag is

$$\text{drag}_{\text{friction}} = C_f\left(\frac{\rho u_s^2}{2}\right)A = \frac{(0.0029)(1000)(2.83)^2(3.50)}{2}$$

$$= 40.6 \text{ N}$$

b. In part (a), the total measured drag was given as 75 N. The skin-friction drag was calculated to be 40.6 N. The wave drag, then, is the difference between the two, namely 75 - 40.6 = 34.4 N.

• PROBLEM 10-21

A barge is traveling down a river with a velocity of 3 mph (see Figure). Calculate the total drag on the flat bottom surface of the barge. The barge is 20 ft long and 10 ft wide. Take the kinematic viscosity of water to be 10^{-5} ft^2/sec, density 62.2 lb_m/ft

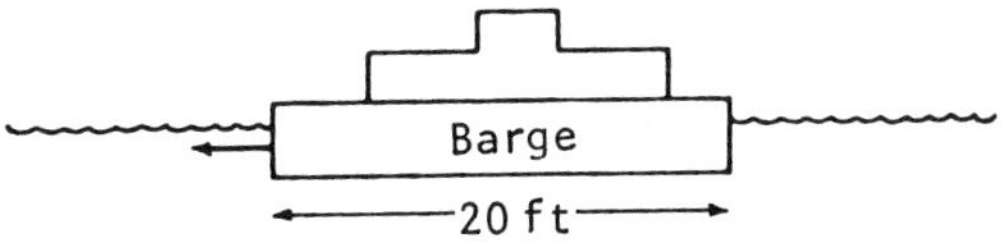

Solution: We shall first calculate Re_L in order to determine whether or not transition has occurred on the surface.

$$Re_L = \frac{VL}{\nu} = (3 \times \frac{5280}{3600})\ \frac{20}{10^{-5}} = 8.8 \times 10^6$$

Since the value is greater than 5×10^5, this indicates that a turbulent boundary layer has formed on the bottom surface of the barge. The equation for the skin friction drag coefficient is given by

$$C_{D_f} = \frac{0.074}{\sqrt[5]{Re_L}} - \frac{1700}{Re_L}$$

$$= 0.00302 - 0.00019$$

$$= 0.00283$$

Therefore the total drag on the bottom surface of the barge is given by

$$\text{total drag } D = \frac{1}{2}\frac{\rho}{g_c} V^2 C_{D_f} A$$

with $A = (20 \times 10) = 200 \text{ ft}^2$ and $V = 3\left(\frac{5280}{3600}\right) = 4.4$ fps

so

$$D = 10.6 \text{ lb}_f$$

• PROBLEM 10-22

Assuming the form resistance of an automobile to govern its speed for a given power expenditure, determine the ratio of speeds for conditons of very good and very poor streamlining.

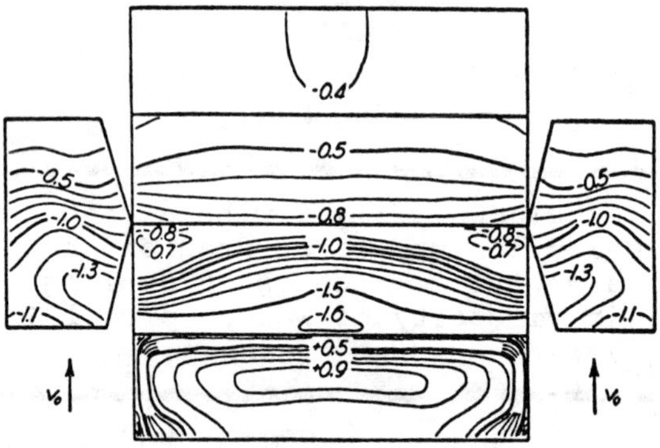

Distribution of wind pressure on a building. (Contours pass through points having equal values of $\Delta p/\rho v_0^2/2$.)

Table

APPROXIMATE VALUES OF THE DRAG COEFFICIENT FOR VARIOUS BODY FORMS

Form of Body	L/D	R	C_D
Circular disk		$>10^3$	1.12
Tandem disks (L = spacing)	0	$>10^3$	1.12
	1		0.93
	2		1.04
	3		1.54
Rectangular plate (L = length, D = width)	1	$>10^3$	1.16
	5		1.20
	20		1.50
	∞		1.90
Circular cylinder (axis ∥ to flow)	0	$>10^3$	1.12
	1		0.91
	2		0.85
	4		0.87
	7		0.99
Circular cylinder (axis ⊥ to flow)	1	10^5	0.63
	5		0.74
	20		0.90
	∞		1.20
	5	$>5 \times 10^5$	0.35
	∞		0.33

APPROXIMATE VALUES OF THE DRAG COEFFICIENT FOR VARIOUS BODY FORMS

Form of Body	L/D	R	C_D
Streamlined foil (1:3 airplane strut)	∞	$>4 \times 10^4$	0.07
Hemisphere: Hollow upstream Hollow downstream		$>10^3$	1.33 0.34
Sphere		10^5 $>3 \times 10^5$	0.50 0.20
Ellipsoid (1:2, major axis ‖ to flow)		$>2 \times 10^5$	0.07
Airship hull (model)		$>2 \times 10^5$	0.05

Solution: From either the figure or Table, the highest and the lowest values of C_D are found to be about 1.1 and 0.05. Since the power is the product of F and v_0, and since the densities and projected areas are the same,

$$P = Fv_0 = C_D A \frac{\rho v_0{}^3}{2}$$

and

$$\frac{(v_0)_{max}}{(v_0)_{min}} = \sqrt[3]{\frac{(C_D)_{max}}{(C_D)_{min}}} = \sqrt[3]{\frac{1.1}{0.05}} = 2.8$$

• PROBLEM 10-23

A smooth thin square plate 1 m on a side is attached to one end of a rod(see Figure 1). A hemispherical cup with its concave surface to the flow is attached to the other end. The rod is free to rotate about a frictionless shaft at its midpoint. The entire apparatus is immersed in a uniform streaming flow of water at 5°C moving at 5 m/s. Estimate the diameter of the cup if the rod is not to rotate. Assume viscosity of water = $1.5 \times 10^{-3} N \cdot s/m^2$.

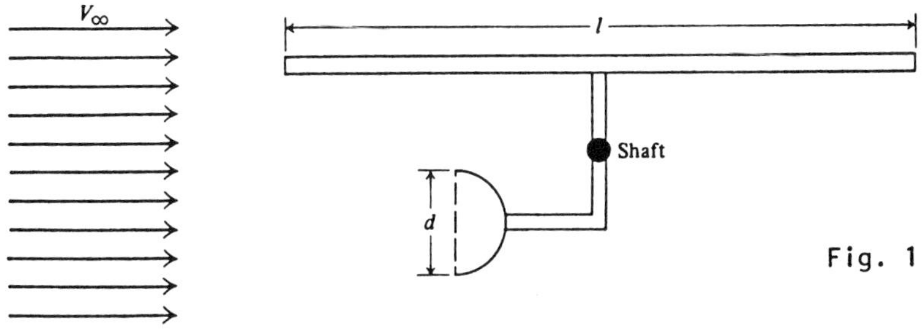

Fig. 1

Solution: For no rotation of the rod, the sum of the moments acting about the shaft must be zero. Fluid drag forces act on the rod, but the rod has the same length on either side of the shaft and the moments mutually cancel. Thus, the moments due to the plate and cup must be equal and opposite; or since their moment arms are the same, the drag forces must be equal.

The drag on the plate is all skin friction and is caused equally by the top and bottom surfaces and is given by

$$D_{plate} = 2C_f \tfrac{1}{2}\rho V_\infty^2 \, lw$$

The drag on the hemispherical cup is nearly all pressure drag given by

$$D_{cup} = C_D \tfrac{1}{2}\rho V_\infty^2 \frac{\pi d^2}{4}$$

Equating the drag forces on the plate and cup and solving for d, we get

$$d = \sqrt{\frac{8}{\pi}\frac{C_f}{C_D} l\,w}$$

C_f depends on the plate Reynolds number Re_l

$$Re_l = \frac{l\ V_\infty\ \rho}{\mu}$$

$$Re = \frac{(1\ \text{m})(5\ \text{m/s})(1000\ \text{kg/m}^3)}{0.0015\ \text{N}\cdot\text{s/m}^2}\ \frac{1\ \text{N}}{1\cdot\text{kg}\cdot\text{m/s}^2}$$

$$= 3.33 \times 10^6$$

Now consider what happens above a Reynolds number of about 300,000. In the Reynolds-number range of 300,000 to 1 million, depending upon the surface roughness and any external disturbances, a region of turbulent-boundary-layer flow develops at the trailing edge of the plate. At higher Reynolds numbers the boundary layer becomes turbulent farther upstream. Figure 2 shows these two modes of boundary-layer flow on a flat plate; x_{crit} is the distance from the leading edge, where the laminar boundary layer ends and the turbulent boundary layer starts.

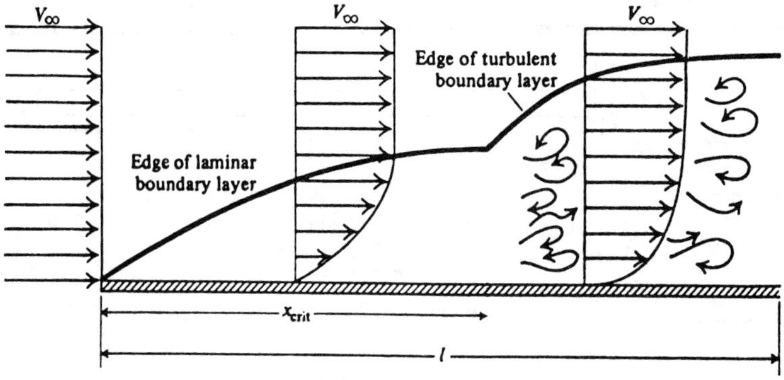

Fig. 2

The plate Reynolds number at which boundary-layer turbulence first appears at the trailing edge of the plate is called Re_{crit}. As the plate Reynolds number Re_l is increased to a higher value than Re_{crit}, x_{crit} moves upstream. However, x_{crit} is always located at a local Reynolds number equal to Re_{crit}. When Re_{crit} is unknown for a specific flat-plate flow situation, a representative value of 500,000 is usually used.

Table

DRAG COEFFICIENTS FOR THREE-DIMENSIONAL SHAPES

Hollow hemisphere or cup		
Convex face to flow	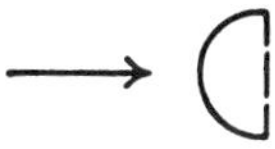	**0.38**
Concave face to flow	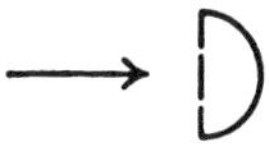	**1.42**

Assuming an average Re_{crit} of 500,000 there will be a transition from laminar to turbulent flow.

C_f for a combined laminar and turbulent-boundary-layer flow on a smooth plate is given by

$$C_f = \frac{0.455}{(\log Re_1)^{2.58}} - \frac{A}{Re_1}$$

where A depends on Re_{crit} as follows:

Re_{crit}	300,000	500,000	1,000,000	3,000,000
A	1050	1700	3300	8700

The C_f for combined laminar and turbulent flow is

$$C_f = \frac{0.455}{(\log 3.33 \times 10^6)^{2.58}} - \frac{1700}{3.33 \times 10^6}$$

$$= 0.0031$$

C_D for the cup is about the same for laminar or turbulent flow, since the boundary layer separates at the sharp edge. From the Table, $C_D = 1.42$. Now substituting back into the equation for the cup diameter d, we find

$$d = \sqrt{\frac{8}{\pi}\,\frac{0.0031}{1.42}}\,(1\ \text{m})^2 = 0.075\ \text{m} = 7.5\ \text{cm}$$

• PROBLEM 10-24

A 30-in-diameter cylindrical body floats in still water with a depth of submergence of 4 ft. The body has a stabilizing weight on its bottom surface to keep it upright (Fig. 1). If the water level suddenly increases by a small amount, what is the subsequent motion of the body?

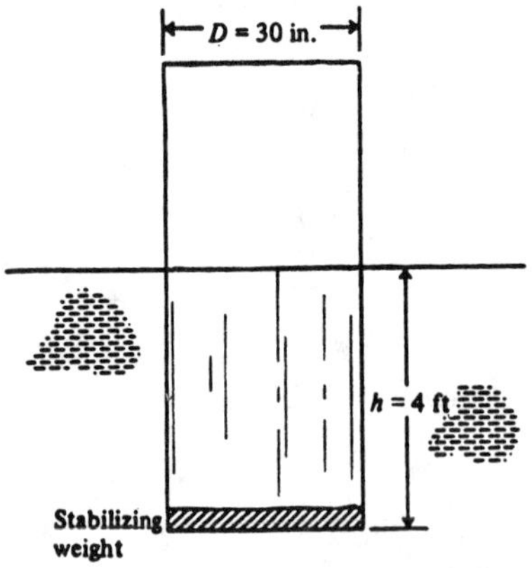

Fig. 1

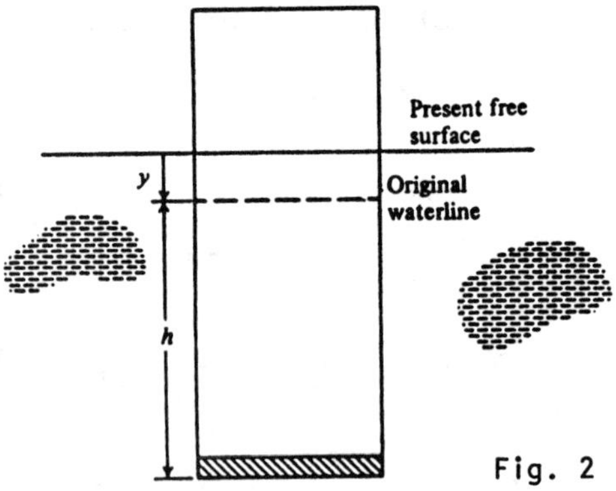

Fig. 2

Solution: Initially the body floating in still water is in static equilibrium and displaces a volume of water whose weight is equal to its weight W_{body}

$$W_{body} = \rho g \frac{\pi D^2}{4} h$$

Now consider what happens when the water level suddenly rises. The body feels an increased buoyant force, causing it to accelerate upward. We apply Newton's second law of motion to the body and measure y down from the present free surface to the original waterline on the body (Fig. 2). The net upward propelling force is the buoyant force less the weight and drag of the body. The propelling force must overcome the inertia of the body as well as that of the added mass of water associated with the shape of the body

$$W_{body} + D - F_b = (m_{body} + m_h) \frac{d^2y}{dt^2}$$

The buoyant force F_b is

$$F_b = \rho g \frac{\pi D^2}{4} (h + y)$$

Then

$$-\rho g \frac{\pi D^2 y}{4} + D - (m_{body} + m_h) \frac{d^2y}{dt^2}$$

The drag force D is

$$D = C_D \tfrac{1}{2} \rho V^2 A$$

If the change in water level is small, the body does not travel very far before reaching its new position of static equilibrium. The acceleration is not necessarily small, but the body must travel some distance before the acceleration can develop a substantial velocity. We shall assume that the body's velocity never becomes very large and that the drag force always remains small. Neglecting the drag force, we have

$$\frac{d^2y}{dt^2} = \frac{\rho g \pi D^2}{4(m_{body} + m_h)} y$$

The net force increases with displacement y from the new static-equilibrium position or free surface and always opposes the motion: it is a restoring force. The differential equation is immediately recognized as the one which describes an undamped single-degree-of-freedom oscillatory motion.

The solution is obtained by writing the differential equation as a polynomial, where the derivatives are represented by a variable, say r, raised to a power, equal to the order of the derivative. The polynomial corresponding to our differential equation is

$$r^2 + \frac{\rho g \pi D^2}{4(m_{body} + m_h)} = 0$$

The two roots of the polynomial are imaginary

$$r = \pm\, i \sqrt{\frac{\rho g \pi D^2}{4(m_{body} + m_h)}}$$

The solution to the differential equation is

$$y = c_1 \exp\left[i \sqrt{\frac{\rho g \pi D^2}{4(m_{body} + m_h)}}\, t\right] + c_2 \exp\left[-i \sqrt{\frac{\rho g \pi D^2}{4(m_{body} + m_h)}}\, t\right]$$

The oscillatory nature of the solution is better shown by transforming from the exponential form to

$$y = A \sin\left(\sqrt{\frac{\rho g \pi D^2}{4(m_{body} + m_h)}}\, t + \phi\right)$$

The motion is oscillatory with amplitude A, phase angle ϕ, and frequency $[\rho g \pi D^2/4(m_{body} + m_h)]^{1/2}$. According to the solution, the oscillatory motion continues forever with no attenuation in its amplitude. This is because we neglected the drag force, which would ultimately have damped out the motion. The frequency we found is called the undamped natural frequency

$$\omega_n = \sqrt{\frac{\rho g \pi D^2}{4(m_{body} + m_h)}}$$

The amplitude and phase angle are obtained by applying the boundary conditions $t = 0$, $y = y_0$ and $V = dy/dt = 0$; y_0 is simply equal to the change in elevation of the water surface. Applying the two boundary conditions gives $y_0 = A \sin \phi$ and $A\omega_n \cos \phi = 0$. Then $A = y_0$ and $\phi = \pi/2$, giving

$$y = y_0 \sin\left(\omega_n t + \frac{\pi}{2}\right) = y_0 \cos \omega_n t$$

The body oscillates about the new water level or new static-equilibrium position, continually overshooting it by its original displacement from it.

• PROBLEM 10-25

A 1.5-mm steel sphere ($\rho_s = 7800$ kg/m³) falls steadily at a velocity of 3.20 mm/s in an oil of density $\rho_f = 870$ kg/m³ contained in a vertical cylinder 100 mm in diameter. What is the oil viscosity?

Missouri River sand has a mean diameter of about 0.17 mm. What is the settling velocity of this sand in water at 15°C? Assume the particles are spheres with a specific gravity 2.65. Then $\nu = 1.14 \times 10^{-6}$ m²/s from the Table.

VISCOSITY OF WATER

TEMPERATURE (°C)	DYNAMIC μ (kg/m s)	KINEMATIC ν (m²/s)
0	1.787×10^{-3}	1.787×10^{-6}
5	1.519	1.519
10	1.307	1.307
15	1.139	1.140
20	1.002	1.004
25	0.890	0.893
30	0.798	0.801
35	0.719	0.723
40	0.653	0.658
45	0.596	0.602
50	0.547	0.553
55	0,504	0.511
60	0.466	0.474
65	0.433	0.441
70	0.404	0.409

Solution: If a sphere falls through a fluid of infinite extent (fluid dimensions much greater than the sphere diameter), the buoyant and drag forces at terminal or steady velocity are equal to the gravity force on the sphere. Thus for $Re_D < 0.1$ Stokes' law will apply and

$$\gamma_f \frac{\pi D^3}{6} + 3\mu u_s \pi D = \gamma_s \frac{\pi D^3}{6}$$

Thus, if the fall velocity u_s, tne fluid and sphere specific weights γ_f and γ_s, respectively, and the sphere diameter D are known, the fluid viscosity is

$$\mu = \frac{D^2(\gamma_s - \gamma_f)}{18\ u_s} \tag{1}$$

and this equation provides an extremely simple method for measuring dynamic viscosity. If the fluid is of finite extent, the influence of the boundaries of the container is such as to indicate an apparent drag coefficient higher than that in an infinite fluid. If, for example, the sphere falls in the center of a vertical cylinder of diameter D_c, the relative velocity of the fluid adjacent to the sphere increases, the drag increases, and the sphere will fall slower than in an infinite fluid. The measured velocity u_m should be corrected to its equivalent velocity in an infinite fluid u_s by the equation

$$u_s = \left(1 + 2.4\frac{D}{D_c}\right) u_m \tag{2}$$

For a number of particles uniformly distributed in a fluid, mutual interference will cause them to fall more slowly than if each particle fell alone. The settling velocity of natural particles such as sand and gravel is less than that for equivalent spheres, since the drag coefficient increases with increasing departure from a spherical shape.

From Eq. (2),

$$u_s = \left(1 + 2.4\frac{1.5}{100}\right)(3.20) = 3.32 \text{ mm/s}$$

From Eq. (1)

$$\mu = \frac{(0.0015)^2(7800 - 870)(9.81)}{(18)(0.00332)} = 2.56 \text{ kg/m s}$$

Equation (1) is valid, since $Re_D = u_s D \rho_f/\mu = (0.00332) \times (0.0015)(870)/2.56 = 0.0016$, which is less than 0.1 and is well within the range of validity for the Stokes equation.

Equating drag = weight - buoyancy gives

$$C_D u_s^2 = \frac{4dg}{d}\left(\frac{\rho_s}{\rho_f} - 1\right)$$

$$= \frac{(4)(1.7 \times 10^{-4})(9.807)}{3}(2.65 - 1)$$

$$= 3.66 \times 10^{-3}$$

Oseen in 1910 made an improvement in the Stokes solution by including, in part, the inertia terms Stokes omitted. Oseen's solution gave

$$C_D = \frac{24}{Re_D}\left(1 + \frac{3}{16}Re_D\right) \tag{3}$$

which is valid for $Re_D < 1$.

It is interesting to note that experimental results lie midway between the Stokes and Oseen curves in the figure.

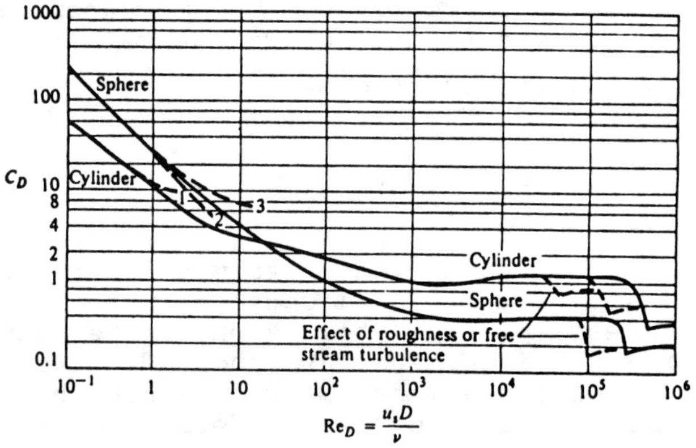

Drag coefficients for spheres and infinite circular cylinders. Curve 1: Lamb's solution for cylinder. Curve 2: Stokes' solution for a sphere Curve 3: Oseen's solution for a sphere

Thus, since the plot is logarithmic, the experimental data follow the equation

$$C_D = \frac{24}{Re_D}\left(1 + \frac{3}{16}Re_D\right)^{1/2} \tag{4}$$

quite accurately for Reynolds numbers up to 100.

To solve for u_s, assume a C_D, then calculate u_s, Re, and C_D from Eq. (4). Repeat until C_D values agree:

C_D	10	10.8	10.7	
u_s	0.0191	0.0184	0.0185	Thus u_s = 1.85cm/s
Re	2.85	2.75	2.76	
C_D	10.4	10.7	10.7	

• PROBLEM 10-26

(a) Find the frictional drag on the top and sides of a box-shaped moving van 8 ft wide, 10 ft high, and 35 ft long, traveling at 60 mph through air ($\gamma = 0.0725$ lb/ft^3) at 50°F. Assume that the vehicle has a rounded nose so that the flow does not separate from the top and sides (see Figure 1). Assume also that even though the top and sides of the van are relatively smooth there is enough roughness so that for all practical purposes a turbulent boundary layer starts immediately at the leading edge. Assume $\gamma = .00015$ ft^2/s.

(b) Using the data of (a) determine the total drag exerted by the air on the van. Assume that $C_D \approx 0.45$ (see Figs. 1 and 2).

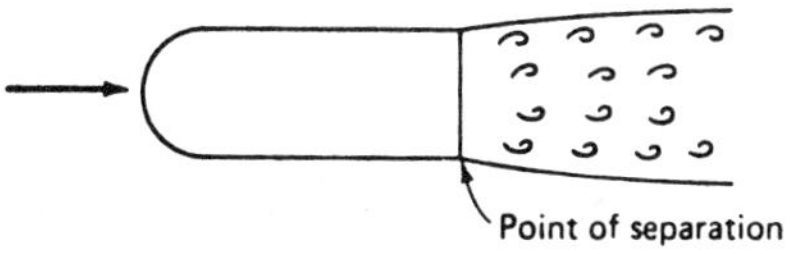

Fig. 1

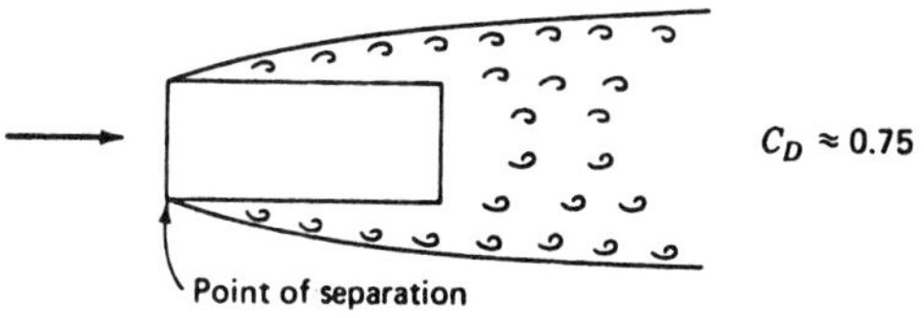

Fig. 2

Solution: (a) Calculate the Reynolds number based on the length of the van, with U = 60 mph = 88 ft/s,

$$\text{Re} = \frac{LU}{\nu} = \frac{35 \times 88}{0.00015} = 20{,}550{,}000$$

For Re > 10^7 (turbulent boundary layer), the friction drag coefficient C_f is given by the empirical equation

$$C_f = \frac{0.455}{(\log \text{Re})^{2.58}}$$

$$C_f = \frac{0.455}{(7.31)^{2.58}} = 0.00268$$

The friction drag F_f is given by the equation

$$F_f = C_f \rho \frac{V^2}{2} BL$$

where C_f = friction-drag coefficient, dependent on viscosity, among other factors

L = length of surface parallel to flow

B = transverse width, conveniently approximated for irregular shapes by dividing total surface area by L

It is important to note that the above equation gives the drag for one side only. In this problem, since we need to find the friction drag on the top and sides of the van, B = width of top + 2·(width of side) = 8 + 10 + 10 ft.

$$F_f = 0.00268 \times \frac{0.0725}{32.2} \times \frac{(88)^2}{2} \times (10 + 8 + 10)35 = 22.9 \text{ lb}$$

(b) To find the total drag force F_D, we can use the equation, where A is the cross sectional area normal to the flow and C_D is an overall drag coefficient,

$$F_D = C_D \rho \frac{V^2}{2} A = 0.45 \left(\frac{0.0725}{32.2}\right) \frac{(88)^2}{2} (8 \times 10)$$

$$F_D = 314 \text{ lb}$$

Thus the pressure drag = 314 - 23 = 291 lb; in this case the pressure drag is responsible for about 93 percent of the total drag while the friction drag comprises only 7 percent of the total.

• PROBLEM 10-27

A thin rectangular plate 10 m long and 1 m wide is to be towed through 15°C salt water at a steady speed of 5 m/s. Both surfaces of the plate are covered with a regular pattern of convex glass reflectors, as shown in Fig. 1. Estimate the power required to tow the plate. Assume density = 1025 kg/m^3 and viscosity = 1.7×10^{-3} $N \cdot s/m^2$.

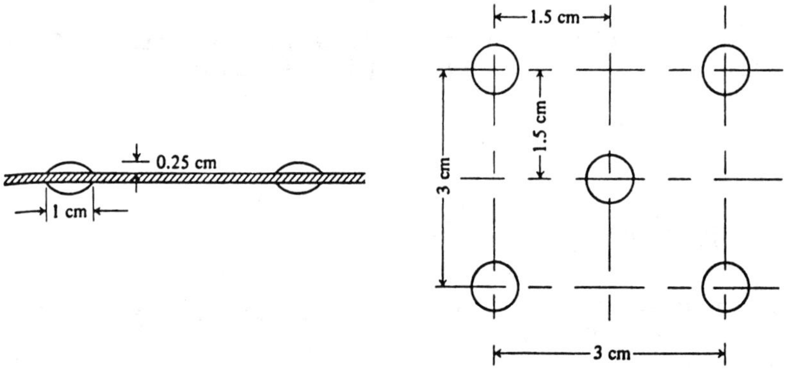

Fig. 1

Solution: The drag on the plate is all skin friction. The plate Reynolds number is

$$Re_l = \frac{l V_\infty \rho}{\mu}$$

$$Re_l = \frac{(10 \text{ m})(5 \text{ m/s})(1025 \text{ kg/m}^3)}{0.0017 \text{ N} \cdot \text{s/m}^2} \; \frac{1 \text{ N}}{1 \text{ kg} \cdot \text{m/s}^2} = 3 \times 10^7$$

In the Reynolds-number range of 300,000 to 1 million, depend-ing upon the surface roughness and any external disturbance, a region of turbulent-boundary-layer flow develops at the trailing edge of the plate. At higher Reynolds numbers the boundary layer becomes turbulent farther upstream. Figure 2 shows these two modes of boundary-layer flow on a flat plate; x_{crit} is the distance from the leading edge, where the laminar boundary layer ends and the turbulent boundary layer starts.

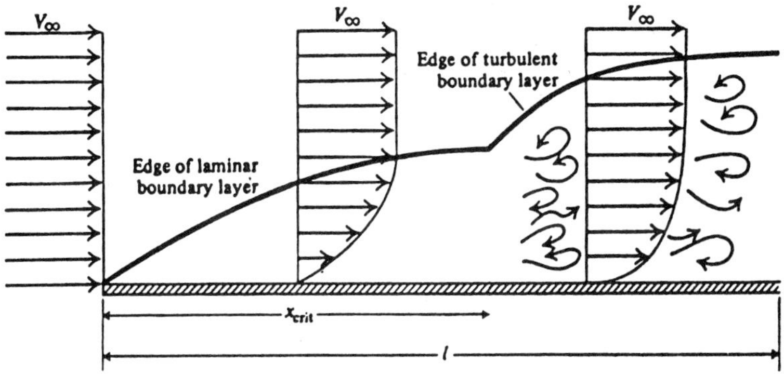

Fig. 2

item	dimensions	D [cm]	d [cm]	k [cm]	ϵ [cm]
spheres	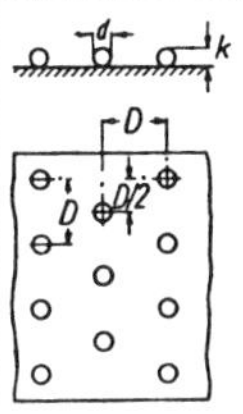	4	0.41	0.41	0.093
		2	0.41	0.41	0.344
		1	0.41	0.41	1.26
		0.6	0.41	0.41	1.56
		densest arrgt.	0.41	0.41	0.257
		1	0.21	0.21	0.172
		0.5	0.21	0.21	0.759
spherical segments	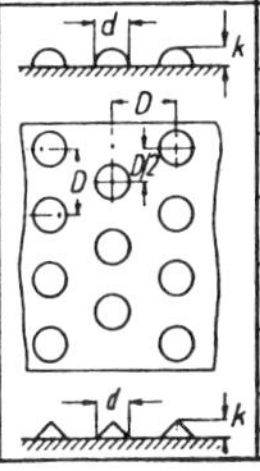	4	0.8	0.26	0.031
		3	0.8	0.26	0.049
		2	0.8	0.26	0.149
		densest arrgt.	0.8	0.26	0.365
cones		4	0.8	0.375	0.059
		3	0.8	0.375	0.164
		2	0.8	0.375	0.374

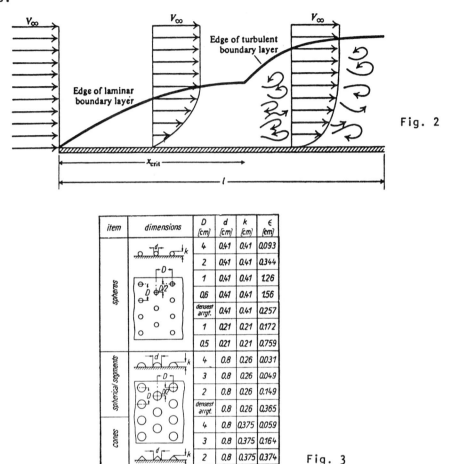

Fig. 3

The plate Reynolds number at which boundary-layer turbulence first appears at the trailing edge of the plate is called Re_{crit}. As the plate Reynolds number Re_l is increased to a higher value than Re_{crit}, x_{crit} moves upstream. However, x_{crit} is always located at a local Reynolds number equal to Re_{crit}. When Re_{crit} is unknown for a specific flat-plate flow situation, a representative value of 500,000 is usually used. Re_{crit} may be calculated using the equation

$$Re_{crit} = \frac{x_{crit} V_\infty \rho}{\mu}$$

There will be a transition from laminar to turbulent flow on the plate since $Re_l > Re_{crit}$. Assuming that transition occurs at $Re_{crit} \approx 500{,}000$, the location of the transition is

$$x_{crit} = \frac{500{,}000\mu}{\rho V_\infty} = 0.166 \text{ m}$$

Less than 2 percent of the plate surface experiences laminar flow. We shall ignore the small laminar portion and assume that the flow is turbulent over the whole plate. In turbulent flow the skin-friction drag coefficient depends on Reynolds number and the relative roughness ε/l Matching the roughness pattern of the plate to Fig. 3 gives d = 1 cm, k = 0.25 cm, and D = 3 cm, which gives an equivalent sand roughness of about $\varepsilon \approx 0.05$ cm. The relative roughness ε/l is then

$$\frac{\varepsilon}{l} = \frac{0.05 \text{ cm}}{10 \text{ m}} = \frac{1 \text{ m}}{100 \text{ cm}} = 0.00005$$

From Fig. 4

$$C_f\left(Re_l = 3 \times 10^7, \frac{\varepsilon}{l} = 0.00005\right) = 0.0045$$

When both sides of the plate are considered, the drag force is

$$D = 2C_f(\tfrac{1}{2}\rho V_\infty^2)(lw) = 2(0.0045)[\tfrac{1}{2}(1025 \text{ kg/m}^3)(5 \text{ m/s})^2]$$

$$\times [(10 \text{ m})(1 \text{ m})] \frac{1 \text{ N}}{1 \text{ kg} \cdot \text{m/s}^2} = 1153 \text{ N}$$

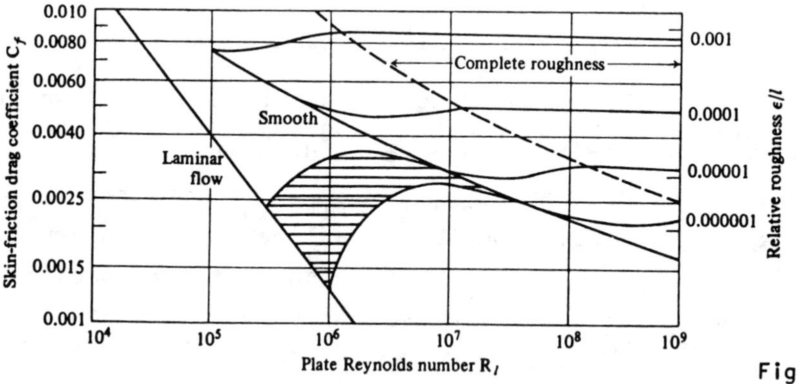

Fig. 4

The power τ is simply the product of the drag force and velocity

$$\tau = DV_\infty = (1153 \text{ N})(5 \text{ m/s}) = 5765 \text{ N} \cdot \text{m/s} = 5765 \text{ J/s}$$

$$= 5765 \text{ W} = 7.7 \text{ hp}$$

• PROBLEM 10-28

Determine the drag and lift forces acting on a cylinder as the result of a flow with free-stream velocity V_0 and free-stream pressure P_0. The cylinder is assumed to be infinitely long and exposed at right angle to the flow. Neglect friction losses.

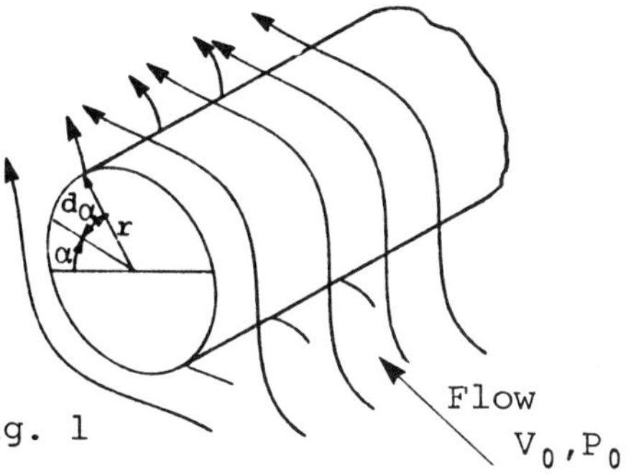

Fig. 1

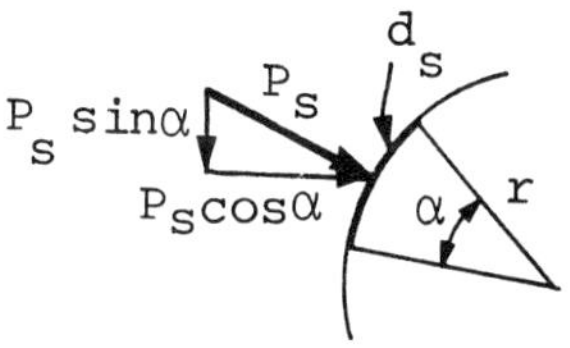

Fig. 2

Solution: Assumptions: (1) No frictional losses
(2) Incompressible flow
(3) Two-dimensional uniform flow

The velocity at the surface is given by

$$V_s = -2\, V_0 \sin\alpha \tag{1}$$

The force exerted at the differential surface ds, is

$$F_s = P_s ds \quad \text{(no shear stress)}$$

$$ds = r d\alpha$$

$$\therefore F_s = P_s r d\alpha \tag{2}$$

The drag force is the component parallel to the flow, thus,

$$D = -rP_s \cos\alpha d\alpha \tag{3}$$

The drag force over the entire surface will be

$$D = \int_0^{2\pi} rP_s \cos\alpha d\alpha \tag{4}$$

To ascertain the pressure at each point on the surface, Bernoulli's equation is considered between the flow far from the surface and the points of the surface (assume $z_0 = z_s$), thus

$$\frac{V_0^2}{2g} + \frac{P_0}{\gamma} = \frac{V_s^2}{2g} + \frac{P_s}{\gamma} \tag{5}$$

$$P_s = \frac{\rho}{2} (V_0^2 - V_s^2) + P_0 \tag{6}$$

Finally, introducing eq.(1) in eq.(6), we get

$$P_s = \frac{\rho V_0^2}{2}(1-4\sin^2\alpha) + P_0$$

The above equation in Eq.(4)

$$D = \int_0^{2\pi} \left[\frac{\rho V_0^2}{2}(1-4\sin^2\alpha)+P_0\right] r\cos\alpha d\alpha$$

from which D = 0.

Following the same procedure, the lift force (L) is found to be zero.

Note: The calculations are carried out in terms of unit length of the cylinder.

• PROBLEM 10-29

(a) A 5-mm-diameter spherical iron pellet is dropped into a tank of SG = 0.680 gasoline at 5°C. Estimate how fast the pellet falls.

(b) A 5-mm-diameter spherical iron pellet is dropped in a tank of 30°C glycerin. Estimate how fast the pellet falls.

(c) Estimate how long it takes for the 5-mm-diameter spherical iron pellet of (a) to attain terminal velocity.

Assume density of iron = 7900 kg/m^3, density of glycerine = 1260 kg/m^3, and viscosity of glycerin = .35 $N \cdot s/m^2$.

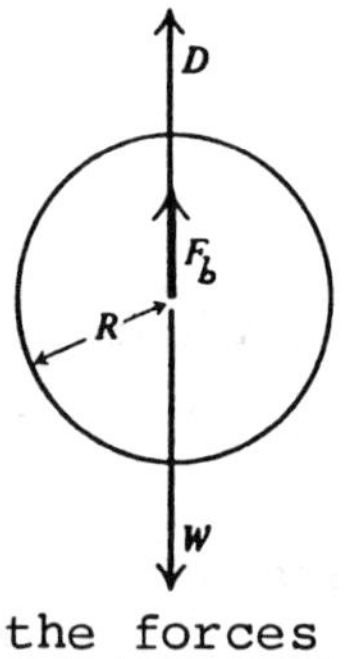

Fig. 1

Solution: (a) Consider the forces acting on the sphere as it falls through the gasoline (Fig. 1), F_b is the buoyant force, which is equal to the weight of gasoline displaced by the sphere $\rho_{gas} g \nu_{sphere}$, D is the fluid drag force $D = C_D \frac{1}{2}\rho V^2 A_{proj}$, and W is the weight of the sphere $\rho_{iron} g \nu_{sphere}$. Newton's second law of motion applied to the sphere is

$$W - (F_b + D) = m\frac{dV}{dt}$$

As the sphere accelerates from rest, the drag force increases as the square of the velocity. After a very short time of acceleration, the drag force increases to $D = W - F_b$, resulting in dV/dt = 0. The sphere now falls at a constant velocity, called its terminal or settling velocity V_T. Then

$$C_D \tfrac{1}{2}\rho_{gas} V_T^2 A_{proj} = \rho_{iron} g \nu_{sphere} - \rho_{gas} g \nu_{sphere}$$

Solving for the terminal velocity and noting that $\nu_{sphere}/A_{proj} = \frac{4}{3}R$, we get

$$V_T = \sqrt{\frac{8}{3}\frac{gR}{C_D}\left(\frac{\rho_{iron}}{\rho_{gas}} - 1\right)}$$

C_D depends on whether the flow over the sphere is laminar or turbulent, but we need to know V_T to calculate the Reynolds number to see whether the flow is laminar or turbulent. We shall guess that the flow is turbulent, calculate V_T using the turbulent value of $C_D = 0.10$ (See Table 1), and then calculate the Reynolds number with the V_T obtained to see whether the flow is turbulent or not:

TABLE 1

Drag coefficients for three-dimensional shapes

Shape		Laminar flow	Turbulent flow
Sphere	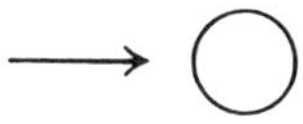	0.47	0.10

$$V_T = \left[\frac{\frac{8}{3}(9.8\text{m/s}^2)(2.5\text{mm})}{0.10}\left(\frac{7900\ \text{kg/m}^3}{680\ \text{kg/m}^3} - 1\right)\frac{1\ \text{m}}{1000\ \text{mm}}\right]^{1/2}$$

$$= 2.63\ \text{m/s}$$

The Reynolds number is then $Re_D = \dfrac{D V_T \rho_{gas}}{\mu_{gas}}$

$$Re_D = \frac{(5\text{mm})(2.63\text{m/s})(680\text{kg/m}^3)}{0.00035\text{N}\cdot\text{s/m}^2}\ \frac{1\text{m}}{1000\text{mm}}\ \frac{1\text{N}}{1\text{kg}\cdot\text{m/s}^2}$$

$$= 2.55 \times 10^4$$

A Reynolds number of 2.55×10^4 is too low for turbulent flow. The flow must be laminar. Recalculating V_T using the laminar value of $C_D = 0.47$ (see Table 1) results in a terminal velocity of 1.21 m/s. Laminar flow causes a larger drag coefficient and thus a lower terminal velocity.

(b) Since glycerin has such a high viscosity, it may be suspected that the spherical pellet experiences a low-Reynolds-number flow. Just as in a high-Reynolds-number flow,in (a)the pellet quickly accelerates to a constant settling speed. However, in a low-Reynolds-number flow the drag force on the sphere is given by Stokes' drag law

$$D = 6\pi\mu RV$$

Then the balance of forces becomes

$$6\pi\mu RV_T = \rho_{iron} g \nu_{sphere} - \rho_{glyc} g \nu_{sphere}$$

Solving for the settling velocity, we have

$$V_T = \frac{2}{9}\frac{R^2 g}{\mu}\rho_{glyc}\left(\frac{\rho_{iron}}{\rho_{glyc}} - 1\right)$$

Substituting numerical values gives

$$V_T = \frac{2}{9}\frac{(2.5\text{mm})^2(9.8\text{m/s}^2)(1260\text{kg/m}^3)}{(0.35\ \text{N}\cdot\text{s/m}^2)\left(\frac{1000\text{mm}}{1\text{m}}\right)^2}\left(\frac{7900}{1260}\right) - 1\ \frac{1\text{N}}{1\text{kg}\cdot\text{m/s}^2}$$

$$= 0.26\text{m/s}$$

Now calculate the Reynolds number to check that the flow is a low-Reynolds-number

$$Re_D = \frac{(5\text{mm})(0.26\text{m/s})(1260\text{kg/m}^3)}{(0.35\text{N}\cdot\text{s/m}^2)\frac{1000\text{mm}}{1\text{m}}}\ \frac{1\text{N}}{1\text{kg}\cdot\text{m/s}^2} = 4.7$$

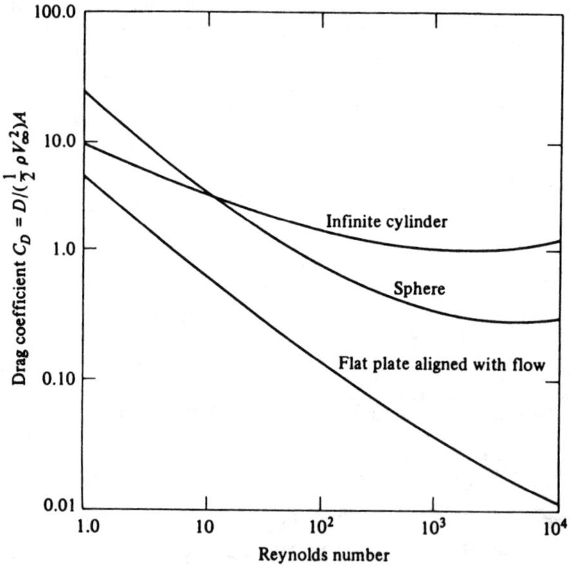

Fig. 2

Since a Reynolds number of 4.7 is a little high for a low-Reynolds-number flow, we try for a better estimate by using Fig. 2 for intermediate-Reynolds-number flows. Figure 2 gives $C_D \approx 8$ for $Re_D = 4.7$. Using the equation for V_T in terms of the drag coefficient developed in (a), we have

$$V_T = \sqrt{\frac{8}{3}\frac{gR}{C_D}\left(\frac{\rho_{iron}}{\rho_{glyc}} - 1\right)} = 0.21\text{m/s}$$

resulting in a Reynolds number of 3.8.

(c) Referring back to (a), we see that the equation of motion of the pellet is

$$W = (F_b + D) = m\frac{dV}{dt}$$

The mass m must include both the mass of the pellet and the hydrodynamic mass associated with the pellet, $m = m_p + m_h$. At terminal velocity, $V = V_T$ and $dV/dt = 0$,

$$W - F_b = D = C_D \tfrac{1}{2}\rho V_T^2 A$$

The equation of motion then becomes

$$C_D \tfrac{1}{2}\rho A(V_T^2 - V^2) = (m_p + m_h)\frac{dV}{dt}$$

We have made the assumption that the drag coefficient is constant over the acceleration of the pellet to terminal velocity. This assumption is reasonable except right at the very beginning of the pellet's motion, when the flow is low-Reynolds-number. It would greatly complicate the problem mathematically to introduce the variation of the drag coefficient with Reynolds number.

A further complication also arises. (a) states that the pellet is dropped in a tank: during its initial motion the pellet is at and near the free surface, and its hydrodynamic mass is changing. We shall ignore this hydrodynamic-mass variation in the interest of obtaining an analytical solution. Now we can integrate the equation. Separating variables, we get

$$\frac{C_D \tfrac{1}{2}\rho A}{m_p + m_h}dt = \frac{dV}{V_T^2 - V^2}$$

The integral on the right is a common one tabulated in most tables of integrals

$$\int \frac{dx}{p^2 - x^2} = \frac{1}{2p}\ln\frac{p + x}{p - x}$$

Then

$$\frac{C_D \tfrac{1}{2}\rho A}{m_p + m_h}t = \frac{1}{2V_T}\ln\frac{V_T + V}{V_T - V} + \text{const}$$

The boundary conditions state that the pellet is dropped from rest at the surface of the gasoline, $t = 0$, $V = 0$, giving const = 0. Then

$$\frac{V_T - V}{V_T + V} = \exp -\left(\frac{C_D \rho V_T A}{m_p + m_h}t\right)$$

The quantity $(m_p + m_h)/C_D\rho V_T A$ has units of time and can be considered as a time constant τ

$$\tau \equiv \frac{m_p + m_h}{C_D \rho V_T A}, \quad \frac{V_T - V}{V_T + V} = e^{-t/\tau}$$

The dimensionless velocity quantity $(V_T - V)/(V_T + V)$ has the familiar exponential time decay shown in Fig. 3.

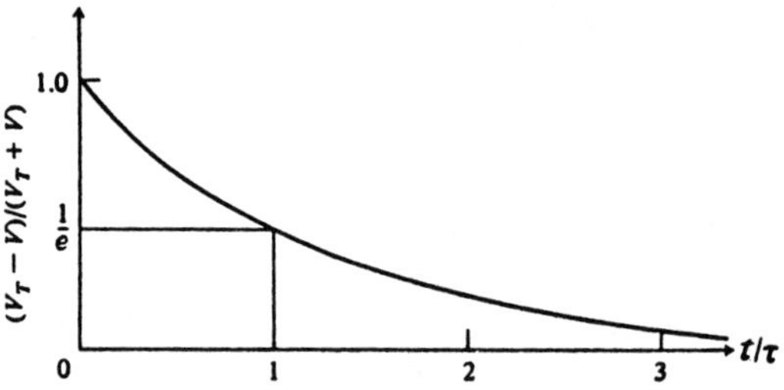

Fig. 3

In one time constant the quantity $(V_T - V)/(V_T + V)$ has been reduced to $1/e \approx 0.368$ of its original value. In two time constants it has been reduced to $1/e^2 \approx 0.135$ of its original value. The terminal velocity is a limiting value and is precisely attained only after infinite time.

TABLE 2

Hydrodynamic mass of three-dimensional bodies

Shape		Hydrodynamic mass
Sphere of diameter D	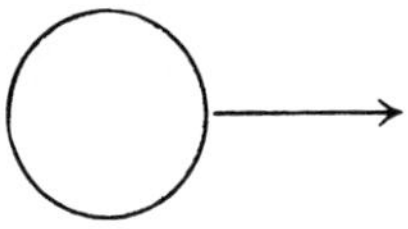	$\frac{\pi}{12}\rho D^3$

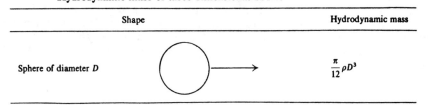

To evaluate the time constant for this motion numerically we use the hydrodynamic mass of a sphere accelerating in an infinite fluid, Table 2,

$$m_h = \frac{\pi}{12}\rho D^3$$

$$\tau = \frac{m_p + m_h}{C_D \rho V_T A} = \frac{\frac{1}{6}\pi D^3 \rho_{iron} + \frac{1}{12}\pi D^3 \rho_{gas}}{C_D \rho_{gas} V_T (\pi D^2/4)} = \frac{2}{3}\frac{D}{C_D V_T}\left[(\rho_{iron}/\rho_{gas}) + \frac{1}{2}\right]$$

Referring back to (a) we have

$$\frac{\rho_{iron}}{\rho_{gas}} = \frac{7900 \text{ kg/m}^3}{680 \text{ kg/m}^3} = 11.62$$

The effect of hydrodynamic mass in this problem is seen to be small, ½ compared with ρ_{iron}/ρ_{gas} = 11.62, or less than 5 percent of the time constant. Continuing, we have

$$\tau = \frac{2(0.005m)(11.62 + 0.5)}{3(0.47)(1.21m/s)} = 0.071s$$

After three time constants, 0.213 s,

$$\frac{V_T - V}{V_T + V} = \frac{1}{e^3} \quad \text{or} \quad V = \frac{e^3 - 1}{e^3 + 1}V_T = 0.905V_T$$

After three time constants the actual velocity is more than 90 percent of the terminal velocity.

LIFT

• PROBLEM 10-30

The wings of an aeroplane have a total plan area of 320 ft^2 The plane weighs 18,000 lbf, and the normal speed in level flight is 230 ft/sec when the atmospheric density is 0.076 lbm/ft^3.

(a) If the tailplane contributes 5 percent of the total lift force, what must be the lift coefficient of the wings at this operating condition?

(b) If, at the same flight condition, the drag coefficient of the wings is 0.055, and the total drag is 1.75 times that caused by the wings alone, what forward thrust is required? Ignore interference between the wings and the fuselage.

Solution: (a) In level flight the weight of the plane is exactly supported by the wings and tailplane. The lift force can be expressed in non-dimensional form in the same way as that used for drag, i.e.

$$C_D \equiv \frac{2Dg_0}{\rho V_\infty^2 A},$$

$$C_L \equiv \frac{2Lg_0}{\rho V_\infty^2 A},$$

where L is the lift force, and A, as before, is a reference area; in aeronautical work it is usual to take for A the plan area of the wing, since this is easily obtained from the dimensions of the aircraft; it is usual, too, to use this in the definitions for both lift and drag, so that the ratio of the coefficients is also the ratio of the forces.

For the conditions specified the lift force L provided by the wings is 0.95 of the total, i.e. 0.95 x 18,000 lbf = 17,100 lbf. Then from the definition of lift coefficient given in the previous section,

$$C_L = \frac{2 \times 17{,}100 \times 32.2}{0.0765 \times (230)^2 \times 320} = 0.85.$$

(b) The drag force on the wings, D, can be calculated using the definition of C_D; alternatively, by proportion,

$$D = L \times \frac{C_D}{C_L} = 17{,}100 \times \frac{0.055}{0.85} \text{lbf} = 1106 \text{ lbf},$$

so that the total drag force is 1.75 x 1106 lbf = 1937 lbf.

The thrust required is equal (and opposite in direction) to the total drag, i.e. forward thrust = 1937 lbf.

• PROBLEM 10-31

A kite, which may be assumed to be a flat plate of face area 1.2m^2 and mass 1.0kg, soars at an angle to the horizontal. The tension in the string holding the kite is 50N when the wind velocity is 40km h^{-1} horizontally and the angle of the string to the horizontal direction is 35°. The density of air is 1.2 kg m^{-3}. Calculate the lift and the drag coefficients for the kite in the given position indicating the definitions adopted for these coefficients.

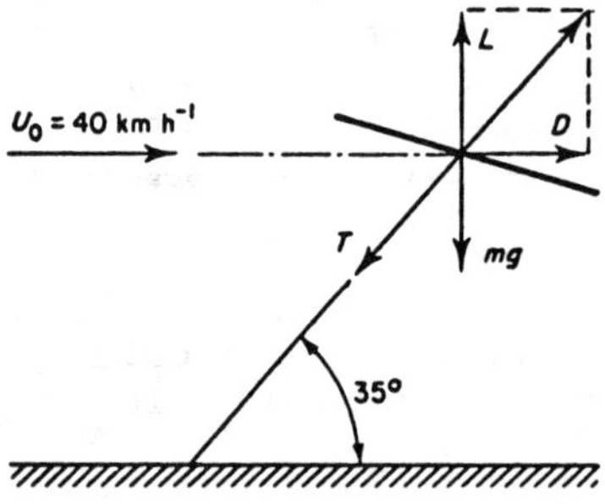

Solution: Since the wind is horizontal, the drag, by definition, will also be horizontal and the lift vertical. The kite is in equilibrium and, therefore, lift and drag must be balanced by the string tension and the weight of the kite. Resolving forces into horizontal and vertical components,

$$L = T \sin 35° + mg = 50 \sin 35 + 1.0 \times 9.81$$

$$= 38.49 \text{ N},$$

$$D = T \cos 35° = 50 \cos 35 = 40.95 \text{ N}.$$

But, lift,

$$L = \frac{1}{2} C_L \rho U_0^2 A$$

and, therefore,

$$C_L = 2L/\rho U_0^2 A = 2 \times 38.49/1.2\left(\frac{40 \times 1000}{3600}\right)^2 1.2$$

$$= 0.432.$$

Similarly, the drag coefficient,

$$C_D = \frac{2D}{\rho U_0^2 A} = \frac{2 \times 40.95}{1.2(40 \times 1000/3600)^2 1.2} = 0.460.$$

Both coefficients have been based on the full area of the kite, because the projected area varies with incidence. This is also the accepted practice in the case of aerofoils.

• PROBLEM 10-32

In the design of an aircraft weighing about 17,000 lbs. (including passengers and freight) and having a wing area of 400 ft², it is desired to estimate the required engine output for a takeoff speed of 180 ft/sec. Wind tunnel experiments on the wings have provided relationships between the angle of attack and the coefficients of lift and drag. These are as follows:

$$C_L = 0.4 + .072\alpha \quad (1)$$

and

$$C_D = .007 + .0085\alpha \quad (2)$$

where α is expressed in degrees. Assume the atmospheric density is .0025 slug/ft^3.

Solution: The required power P for takeoff is given by

$$P = DV \quad (3)$$

where D is the drag and V is the takeoff velocity.

Now the drag D is given by

$$D = C_D \frac{\rho V^2 A}{2} \quad (4)$$

To find C_D we need to know the angle of attack α in this problem. We can find this angle α if we involve C_L and its corresponding function of α. This can be obtained from the condition that at takeoff, the lift L must support the weight of the aircraft. The lift is given by

$$L = C_L \frac{\rho V^2 A}{2} \quad (5)$$

If we solve C_L from (5) and substitute in (1), we can find α. Carrying out these computations,

$$C_L = \frac{2L}{\rho V^2 A}$$

$$C_L = 17{,}000 \text{ lbs}(2) \times \frac{ft^3}{.0025 \text{ slug}} \times \frac{sec^2}{(180)^2 ft^2} \times \frac{1}{400 \text{ ft}^2}$$

$$\times \frac{\text{slug ft}}{\text{lb sec}^2} = 1.05$$

Substituting the value for C_L into equation (1) and solving for α,

$$\alpha = \frac{C_L - 0.4}{.072} = \frac{1.05 - 0.4}{.072} = 9.0$$

Having found α, we can now solve for C_D from (2)

$$C_D = .007 + .0085(9.0) = .0835$$

Substituting into (4) and solving for D,

$$D = \frac{.0835}{2} \times .0025 \frac{\text{slug}}{ft^3} \times (180)^2 \frac{ft^2}{sec^2} \times 400 \text{ ft}^2 \times \frac{\text{lb sec}^2}{\text{slug ft}}$$

$$= 1353 \text{ lbs}$$

To find now the power from (3)

$$P = 1353 \text{ lbs} \times 180 \frac{ft}{sec} \times \frac{\text{HP sec}}{550 \text{ ft lbs}} = 443 \text{ HP}$$

• PROBLEM 10-33

Determine the smallest angle at which a sailplane with a wing having the characteristics of Fig. 2 with an aspect ratio of 8 could glide continously without power. If the plane weighs 1500 lb and has a wing area of 150 ft², what velocity would it attain?

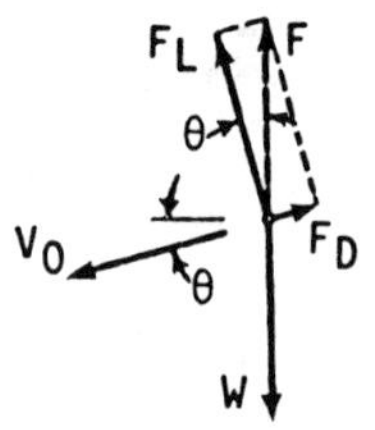

Fig. 1

Solution: In a glide at constant speed, the resultant of the lift and drag must be equal and opposite to the weight of the plane. If the drag of the fuselage can be ignored, the maximum lift-drag ratio of the wing thus determines the co-tangent of the minimum glide angle. From Fig. 2 it is seen that for $\frac{l}{c} = 8$ the maximum value of C_L/C_D is 0.40/0.017 = 23.5 at α = 0°. The minimum glide angle

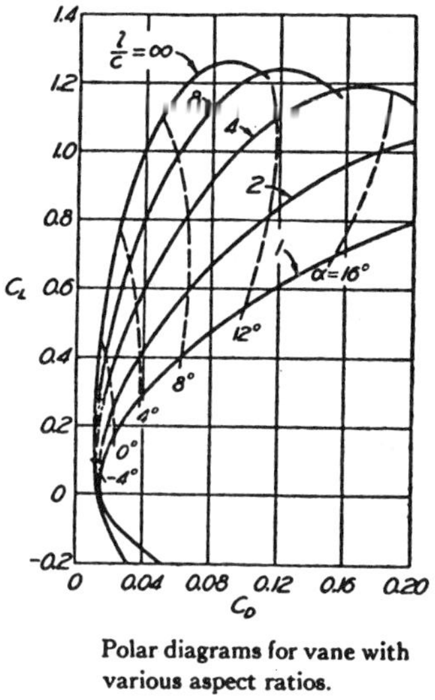

Polar diagrams for vane with various aspect ratios.

Fig. 2

is thus $\theta = \cot^{-1} 23.5 = 2.45°$. Then

$$W = F = \sqrt{C_L{}^2 + C_D{}^2}\, lc\, \frac{\rho v_0{}^2}{2}$$

$$1500 = \sqrt{\overline{0.40}^2 + \overline{0.017}^2} \times 150 \times \frac{0.0024 v_0{}^2}{2}$$

and

$$v_0 = \sqrt{\frac{1500 \times 2}{0.40 \times 150 \times 0.0024}} = 145 \text{ fps}$$

• PROBLEM 10-34

The cruising speed of the DC-6 is 310 mph at an altitude of 20,000 ft (ambient pressure 973 psf, ambient temperature 448°R). Using the data of Figure 1, determine the wing lift and drag under cruise conditions at an angle of attack of 8 deg. Assume the wing has an aspect ratio (ratio of span to average chord) of 10, with a total span b of 115 ft. Approximate the wing as a rectangle of area bc, with C_L and C_D invariant across the entire span.

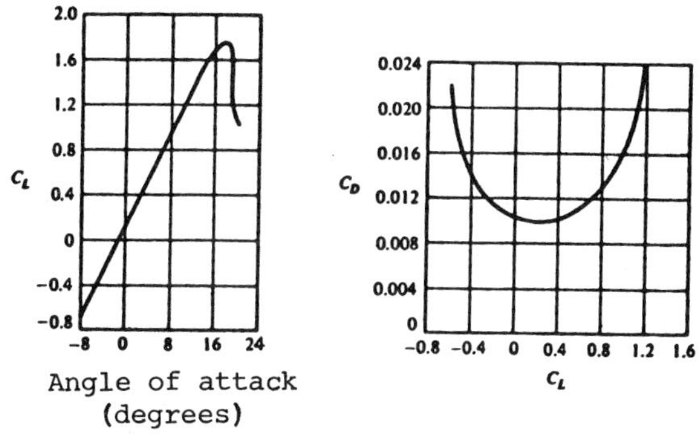

Fig. 1

Solution: Consider the symmetrical airfoil shown in Figure 2. The potential flow pressure distribution is given in Figure 3. Stagnation points are located at points A and C of Fig. 2, with minimum pressure at B, the point of maximum thickness. The body is reduced in thickness very gradually, from B to the trailing edge, so that no large adverse pressure gradients are incurred. In this case, the point of separation can be delayed almost until the trailing edge itself is reached and the resultant pressure drag reduced to a minimum.

Streamlined body shape Fig. 2

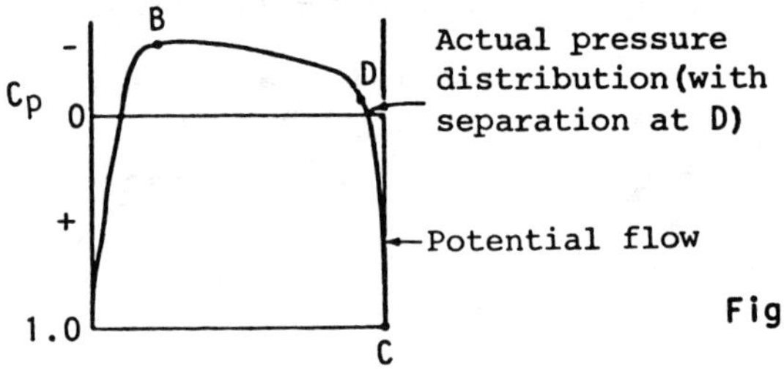

Fig. 3

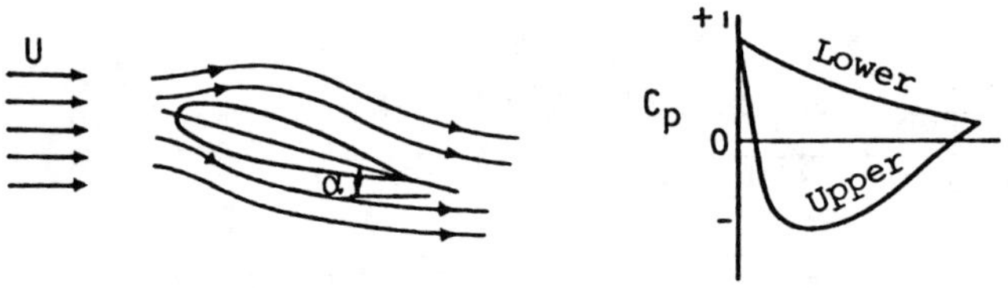

Fig. 4

As the angle of attack of the foil is increased from zero, the fluid moving over the top surface must accelerate more rapidly, yielding a more negative pressure coefficient (see Figure 4). The fluid traveling over the lower surface undergoes a much more gradual acceleration. The resultant difference in pressure between upper and lower surfaces yields a positive lift force on the foil, generally expressed in terms of a lift coefficient C_L with

$$C_L = \frac{L}{\frac{1}{2}(\rho/g_c)V^2 A}$$

Using the perfect gas law, at 20,000 feet

$$\rho = \frac{p}{RT} = \frac{973}{53.3 \times 448} = 0.0408 \ \frac{lb_m}{ft^3}$$

From Figure 1, $C_L = 1.0$ and $C_D = 0.017$. The wing area $A = b^2/\mathit{R\!\!\!/}$, with $\mathit{R\!\!\!/}$ the aspect ratio; $A = (115)^2/10 = 1320 \ ft^2$.

$$L = C_L \frac{1}{2} \frac{\rho}{g_c} V^2 \ A = (1.0) \ \frac{1}{2} \ \frac{0.0408}{32.2} \left(\frac{310 \times 5280}{3600}\right)^2 1320$$

$$= 17.3 \times 10^4 \text{ lb}_f$$

$$D = C_D\left(\frac{1}{2}\frac{\rho}{g_c}V^2A\right) = 2940 \text{ lb}_f$$

• PROBLEM 10-35

> An aircraft is flying in level flight at a speed of 250 km/hr through air at standard conditions. The lift coefficient at this speed is 0.4 and the drag coefficient is 0.0065. The mass of the aircraft is 850 kg. Calculate the effective lift area for the craft.

Solution: Apply definition of lift coefficient.

$$C_L \equiv \frac{F_D}{\frac{1}{2}\rho V^2 A_p}$$

Assume lift equals weight in level flight. Then

$$F_L = mg = C_L \tfrac{1}{2}\rho V^2 A_p$$

Solving for A_p

$$A_p = \frac{2mg}{C_L \rho V^2}$$

$$A_p = \frac{2}{0.4} \times 850\text{Kg} \times 9.81 \frac{\text{m}}{\text{sec}^2} \times \frac{\text{m}^3}{1.23\text{Kg}}\left(\frac{\text{hr}}{250 \times 10^3\text{m}} \times \frac{3600 \text{ sec}}{\text{hr}}\right)^2 = 7.03\text{m}^2$$

• PROBLEM 10-36

> A tennis ball weighs 2 oz and is 2.5 in. in diameter. A typical forehand shot is hit at a speed of 60 ft/sec, with top spin of 5500 rpm. Determine the aerodynamic lift acting on the ball and compute the radius of curvature of the path of the spinning ball.

Solution: A) Using Newton's law of motion

$$\Sigma F_r = ma_r; \quad C_L = \frac{F_L}{\frac{1}{2}\rho V^2 A} = f\left(R_e, \frac{\omega D}{2V}\right)$$

Computing R_e,

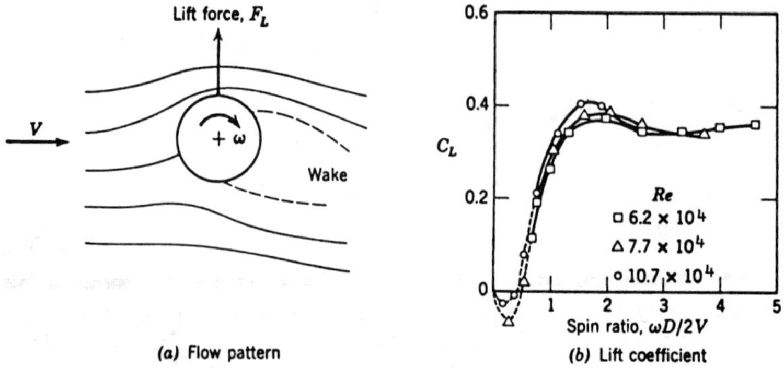

Flow pattern and lift coefficient for a spinning sphere in uniform flow. Fig. 1

$$\nu = 1.6 \times 10^{-4} \text{ ft}^2/\text{sec}$$

$$R_e = \frac{VD}{\nu} = \frac{60 \text{ ft}}{\text{sec}} \times 2.5 \text{ in} \times \frac{\text{sec}}{1.6 \times 10^{-4}\text{ft}^2} \times \frac{\text{ft}}{12 \text{ in.}}$$

$$= 7.81 \times 10^4$$

Then

$$\frac{\omega D}{2V} = \frac{1}{2} \times 5500 \frac{\text{rev}}{\text{min}} \times 2\pi \frac{\text{rad}}{\text{rev}} \times \frac{\text{min}}{60 \text{ sec}} \times 2.5 \text{ in} \times \frac{\text{ft}}{12 \text{ in}}$$

$$\times \frac{\text{sec}}{60 \text{ ft}} \simeq 1$$

Then from Figure (1b),

$$C_L = 0.3,$$

so with

$$A = \frac{\pi}{4} D^2 = 0.0341 \text{ ft}^2,$$

$$F_L = C_L \tfrac{1}{2}\rho V^2 A = \left(\frac{0.30}{2}\right) 0.00238 \frac{\text{slug}}{\text{ft}^3} \times (60)^2 \frac{\text{ft}^2}{\text{sec}^2}$$

$$\times 0.0341 \text{ ft}^2 \times \frac{\text{lbf} \cdot \text{sec}^2}{\text{slug} \cdot \text{ft}}$$

$$F_L = 0.0438 \text{ lbf} = 0.701 \text{ oz}$$

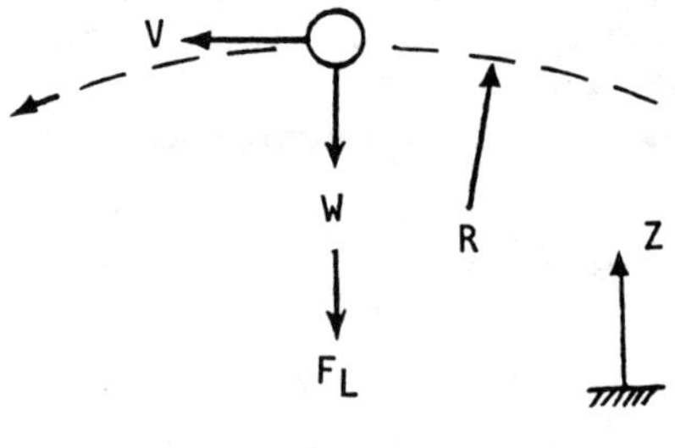

Fig. 2

B) Applying the equation of motion (see Fig. (2))

$$\Sigma F_r = -W - F_L = ma_r = \frac{-W}{g}\frac{V^2}{R}$$

Thus

$$R = \frac{W}{W + F_L}\frac{V^2}{g}$$

$$= \frac{2 \text{ oz}}{(2.00 + 0.701) \text{ oz}} \times (60)^2 \frac{\text{ft}^2}{\text{sec}^2} \times \frac{\text{sec}^2}{32.2 \text{ ft}}$$

$$R = 82.8 \text{ ft}$$

• PROBLEM 10-37

A spherical balloon contains helium and ascends through standard air. The mass of the balloon and its payload is 150 kg. Determine the required diameter if it is to ascend at 3 m/sec. Assumption: Balloon at terminal speed; $a_y = 0$

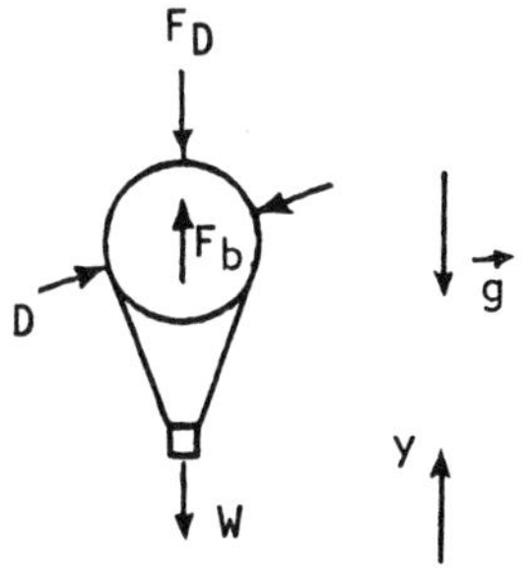

Fig. 1

Solution: Apply Newton's second law of motion, definition of C_D

$$\Sigma F_y = ma_y; \quad C_D = \frac{F_D}{\frac{1}{2}\rho V^2 A}$$

Summing forces, $F_b - F_D - mg = 0$ (1)

where

$$F_b = \rho g \forall = \rho g \frac{\pi}{6} D^3$$

$$F_D = C_D \tfrac{1}{2}\rho V^2 A = \rho V^2 C_D \frac{\pi}{8} D^2$$

Substituting into equation (1)

$$\frac{\pi}{6}\rho g D^3 - \frac{\pi}{8} C_D \rho V^2 D^2 - mg = 0$$

or

$$D^3 - \frac{3}{4}\frac{C_D V^2}{g} D^2 - \frac{6m}{\pi\rho} = 0 \qquad (2)$$

But C_D is a function of Reynolds number. For standard air, $\nu = 1.5 \times 10^{-5} m^2/sec$ with $D \simeq 5m$,

$$R_e = \frac{VD}{\nu} = \frac{3m}{sec} \times 5m \times \frac{sec}{1.5 \times 10^{-5} m^2} = 10^6$$

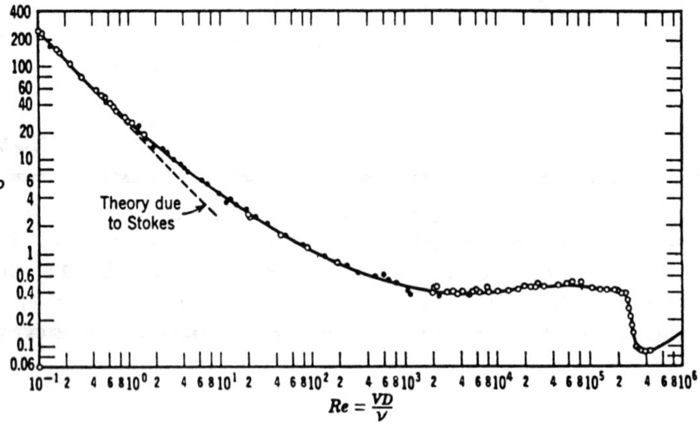

Drag coefficient of a sphere as a function of Reynolds number. Fig. 2

Then from Figure (2),

$$C_D \simeq 0.2$$

Solving for the coefficients,

$$\frac{3}{4}\frac{C_D V^2}{g} = \frac{3}{4} \times 0.2 \times (3)^2 \frac{m^2}{sec^2} \times \frac{sec^2}{9.81m} = 0.138m$$

$$\frac{6m}{\pi\rho} = \frac{6}{\pi} \times 150Kg \times \frac{m^3}{1.23\,Kg} = 233m^3$$

Substituting back into eq. (2)

$$D^3 - 0.138D^2 - 233 = 0$$

Iterating gives:

D(m)	$D^3 - 0.138D^2 - 233(m^3)$
5.0	-111
6.0	- 22
6.1	- 11.2
6.2	- 0.0233

Thus $D \simeq 6.2m$

CHAPTER 11

CHANNEL FLOW

Basic Attacks and Strategies for Solving Problems in this Chapter. See pages 629 to 677 for step-by-step solutions to problems.

Open channel flow is somewhat similar to pipe flow except that, with a free surface exposed to atmospheric pressure, there can be no streamwise pressure gradient. The fluid flows due to gravity alone, with the flow rate determined by a balance between gravitational and frictional forces. A reasonable analysis can be obtained by using the Moody Chart for pipe flow (see Chapter 7) with the hydraulic diameter of the channel. Instead, engineers prefer the *hydraulic radius*, defined as one-fourth of the hydraulic diameter,

$$\text{Hydraulic radius} = R = \frac{A}{P} = \frac{D_h}{4}, \qquad (1)$$

where A is the cross-sectional area of the fluid in the channel and

P is the wetted perimeter (which does *not* include the free surface).

For uniform flow in long, straight, inclined channels of a constant shape, the average velocity V at any streamwise location remains constant and the energy equation between two streamwise locations 1 and 2 reduces to

$$h_f = y_1 - y_2 = L \sin \alpha, \qquad (2)$$

where h_f is the frictional head loss which is exactly balanced by the change in surface height $y_1 - y_2$. In the above, L is the streamwise distance from 1 to 2 and α is the inclination angle with respect to the horizontal. Frictional head loss can also be expressed in terms of the Darcy friction factor f as

$$h_f = f\frac{LV^2}{8gR}. \qquad (3)$$

Combining Equations (2) and (3), and introducing the Chezy constant $C = (8g/f)^{1/2}$, the velocity in the channel is found:

$$V = C(R \sin \alpha)^{1/2}. \qquad (4)$$

Typically, volume flow rate Q is the unknown in a channel flow problem. The procedure is to find the Chezy constant C for the given channel, and then to use Equation (4) to find V. Finally, Q = volume flow rate = VA.

In some cases, direct empirical relationships for C are available, but in most cases, C is found by the *Manning correlation,*

$$C = \frac{1.49}{n} R^{1/6} \text{ (}R\text{ in units of feet),} \tag{5}$$

when n is the Manning coefficient, a non-dimensional roughness coefficient which can be obtained from empirical tables such as that supplied with Problem 11.5 or 11.6. In practice, the calculation of Chezy's coefficient can be bypassed by combining Equations (4) and (5):

$$V = \frac{1.49}{n} R^{2/3} (\sin\alpha)^{1/2} \text{ (}R\text{ in ft, }V\text{ in ft/s)} \tag{6a}$$

or

$$V = \frac{1.0}{n} R^{2/3} (\sin\alpha)^{1/2} \text{ (}R\text{ in m, }V\text{ in m/s).} \tag{6b}$$

The procedure for finding the volume flow rate is to look up n for the channel (n depends greatly on the amount of roughness on the walls of the channel), calculate V using Equation (6) for a known inclination angle α and hydraulic radius R, and finally calculate $Q = VA$.

Another feature unique to open channel flows is the hydraulic jump, which is quite analogous to a normal shock wave in a compressible duct flow (see Chapter 12). Here, a high-velocity supercritical flow (the Froude number is greater than one) is suddenly slowed to subcritical ($F_r < 1$) as the fluid depth increases and the velocity decreases across the hydraulic jump. (See the discussion of hydraulic jumps in Problem 11.4.) The ratio of fluid heights downstream (y_2) and upstream (y_1) of a stationary hydraulic jump is

$$\frac{y_2}{y_1} = \frac{1}{2}\left[\left(1 + 8F_{r_1}^2\right)^{1/2} - 1\right], \tag{7}$$

where F_{r_1} is the Froude number upstream,

$$F_{r_1} = \left[\frac{V_1^2}{(y_1 g)}\right]^{1/2}.$$

In problems where the hydraulic jump is not stationary, as, for example, with a surge caused by a sudden gate closure, it is most convenient to transform the frame of reference to that of a stationary hydraulic jump, apply Equation (7), and then transform back to the original frame of reference.

Step-by-Step Solutions to Problems in this Chapter, "Channel Flow"

OPEN CHANNEL FLOW

• PROBLEM 11-1

An open channel conveying water is of trapezoidal cross section. The base width is 1.5 m and the side slopes are at 60° to the horizontal. The channel bed slope is 1 in 400 and the depth is constant at 1 metre. Calculate the discharge in cubic metres per second if the Chezy C is calculated from the Bazin relationship $C = 87/(1 + 0.2/\sqrt{m})$ where m is the hydraulic mean depth.

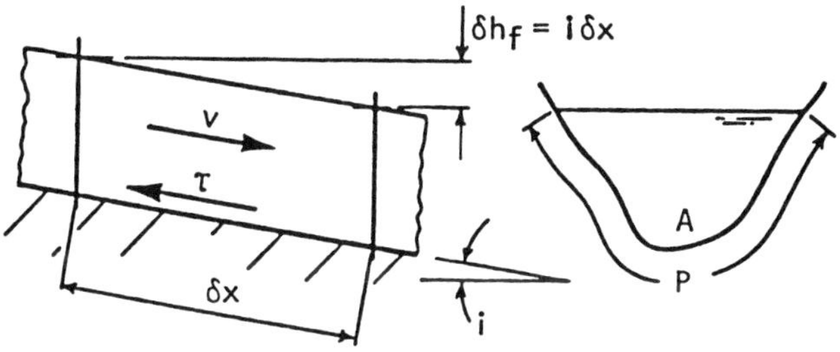

Solution: The energy loss due to friction per unit weight δh_f over a length δx (see Fig.) must be exactly balanced by conversion of potential energy. Thus

$$\delta h_f = i\delta x$$

$\therefore$ $$i = \delta h_f/\delta x$$

but $$\frac{dh_f}{dx} = \frac{fv^2}{2gm} = i$$

$\therefore$ $$v = \sqrt{(2g/f)}\sqrt{(mi)}$$

$\therefore$ $$v = C\sqrt{(mi)}$$

and $$C = \sqrt{(2g/f)}$$

This approach links the Chezy C and the Darcy f. The flow in a channel is given by $Q = CA\sqrt{(mi)}$.

$$m = A/P$$

$$A = 1 \times (3.0 + 2 \times \tan 30°)/2 = 2.077$$

$$P = 1.5 + 2 \times \sec 30° \times 1 = 3.809$$

$$m = 0.545 \text{ m}$$

$$C = \frac{87}{\left(1 + \frac{0.2}{\sqrt{0.545}}\right)}$$

$$C = 68.46$$

$$Q = CA\sqrt{(mi)}$$

$$= 68.46 \times 2.077\sqrt{0.545} \times \frac{1}{20}$$

$$Q = 5.25 \text{ m}^3/\text{s}$$

• PROBLEM 11-2

A heat-exchange conduit of extruded aluminum has a cross section of the form shown. What are the magnitudes of f and C for a Reynolds number $4VR/\nu$ of 50,000? (Fig. 1)

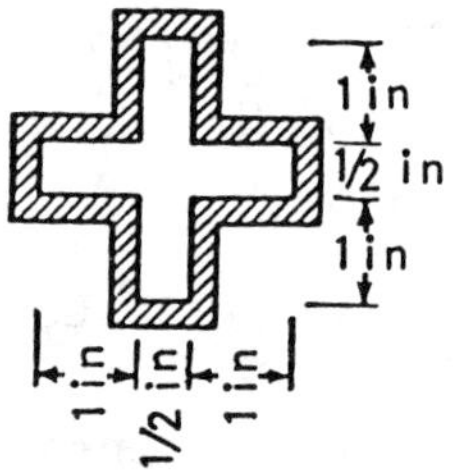

Fig. 1

Solution: The hydraulic radius is computed as follows:

$$R = \frac{A}{P} = \frac{4 \times 1 \times \frac{1}{2} + \frac{1}{2} \times \frac{1}{2}}{4(1 + \frac{1}{2} + 1)} = 0.225 \text{ in.} = 0.0188 \text{ ft}$$

From the right-hand scale of Figure 2, for drawn material, $f = 0.021$. From the left-hand scale, $1/\sqrt{f} = 6.9$, and $C = 110 \text{ ft}^{1/2}/\text{sec}$.

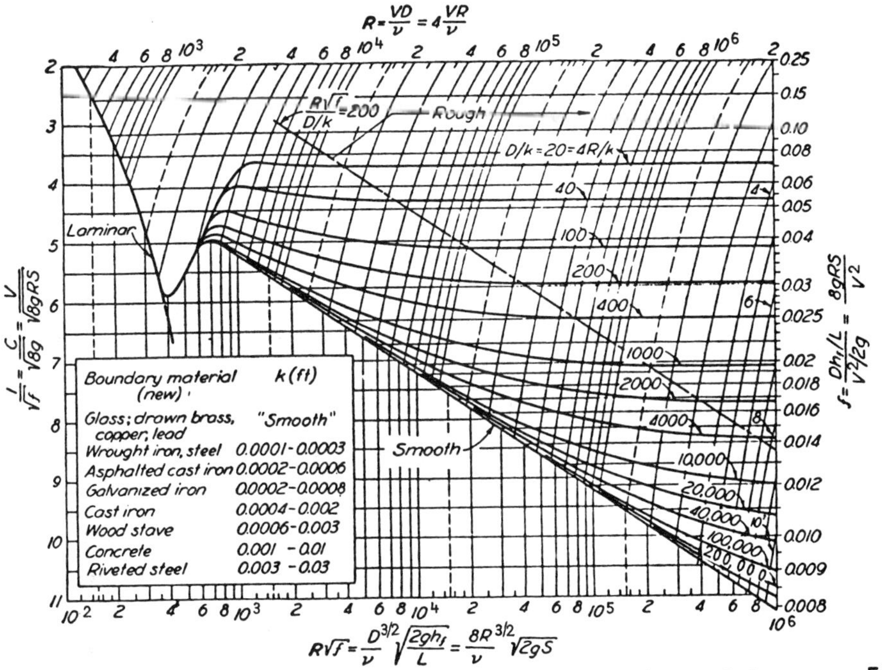

General resistance diagram for uniform flow in conduits.

Fig. 2

• PROBLEM 11-3

The flow rate and water depth just upstream of the dam shown in Figure 1 are 400 cfs and 10 ft, respectively. The channel cross section is rectangular with a width of 20 ft. The channel has a slope of 0.001 and a roughness coefficient n of 0.020. Determine the variation of water depth along the channel.

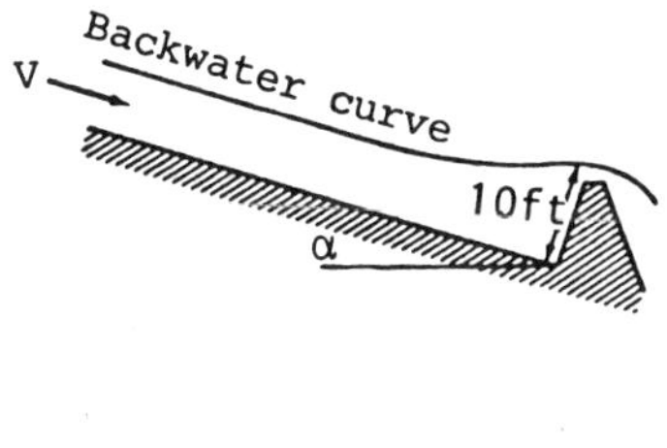

Fig. 1

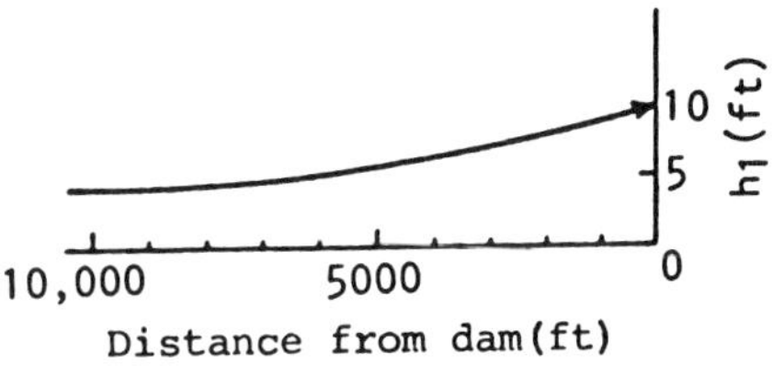

Fig. 2

Solution: At a large distance upstream from the dam, the velocity of the flow will be that corresponding to uniform flow. Using the Manning equation

$$V = \frac{1.486}{n}\left(\frac{A}{P}\right)^{2/3}\sqrt{\sin\alpha}$$

we obtain

$$V = \frac{1.486}{0.02}\left(\frac{20h}{20 + 2h}\right)^{2/3} \sqrt{0.001} = \frac{Q}{hb} = \frac{400}{20h}$$

Rewriting, we obtain an equation for h:

$$0.117h = \left(\frac{20 + 2h}{20h}\right)^{2/3}$$

This equation is solved by trial and error as indicated in Table 1.

Table 1

h	$\frac{P}{A} = \frac{20 + 2h}{20h}$	$\left(\frac{20 + 2h}{20h}\right)^{2/3}$	$0.117h$
7	0.243	0.390	0.819
6	0.267	0.415	0.702
5	0.300	0.450	0.585
4	0.350	0.496	0.468
4.1	0.344	0.492	0.480
4.2	0.338	0.485	0.491
4.16	0.340	0.488	0.487

Thus water depth for uniform flow is 4.16 ft and the water depth in the nonuniform flow will vary from 4.16 to 10 ft. The depth variation is written as a difference equation:

$$\left(1 - \frac{V^2_{av}}{gh_{av}}\right)\Delta h = \left(\sin \alpha - \frac{n^2}{2.21} \frac{V^2_{av}}{(A/P)^{4/3}_{av}}\right)\Delta x$$

For evaluation of the preceding equation, divide the depth range 4.16 to 10 ft into six increments as indicated in Table 2. The value of the quantities with subscript av are taken as the average over the interval. The velocity of the flow at each assumed depth is obtained from the continuity equation

$$V = \frac{Q}{bh} = \frac{20}{h}$$

The other quantities are evaluated as indicated in the table, resulting in an incremental distance Δx along the channel over which the selected change in depth occurs. Summing all Δx's, we obtain the total length of the backwater equal to 9555 ft. A plot of the backwater curve behind the dam is given in Figure 2.

Table 2

h(ft)	A(ft^2)	P(ft)	$\frac{A}{P}$(ft)	V(fps)	$\frac{V^2}{g}$	$\left(\frac{A}{P}\right)_{av}$	$\left(\frac{V^2}{g}\right)_{av}$
10	200	40	5.00	2.00	0.124		
						4.88	0.139
9	180	38	4.75	2.22	0.153		
						4.60	0.174
8	160	36	4.45	2.50	0.194		
						4.29	0.228
7	140	34	4.12	2.86	0.254		
						3.99	0.300
6	120	32	3.75	3.34	0.347		
						3.54	0.423
5	100	30	3.33	4.00	0.498		
						3.14	0.610
4.16	83.2	28.32	2.94	4.82	0.721		

h(ft)	h_{av}	$\left(\frac{V^2}{hg}\right)_{av}$	$1-\frac{V^2}{hg}$	$\left(1-\frac{V^2}{hg}\right)\Delta h$	$\left(\frac{A}{P}\right)^{4/3}_{av}$	$\frac{n^2}{2.21}\frac{V^2_{av}}{\left(\frac{A}{P}\right)^{4/3}_{av}}$	$\sin\alpha - \frac{n^2}{2.21}\frac{V^2_{av}}{\left(\frac{A}{P}\right)^{4/3}_{av}}$	Δx(ft)
10								
	9.5	0.0146	0.985	0.985	8.30	0.0971×10^{-3}	0.903×10^{-3}	1090
9								
	8.5	0.0205	0.980	0.980	7.68	0.131×10^{-3}	0.869×10^{-3}	1130
8								
	7.5	0.0305	0.970	0.970	7.00	0.188×10^{-3}	0.812×10^{-3}	1195
7								
	6.5	0.461	0.954	0.954	6.37	0.273×10^{-3}	0.727×10^{-3}	1310
6								
	5.5	0.0770	0.923	0.923	5.40	0.455×10^{-3}	0.545×10^{-3}	1690
5								
	4.58	0.133	0.867	0.728	4.60	0.768×10^{-3}	0.232×10^{-3}	3140
4.16								$L = 9555$ ft

A 9-ft-wide rectangular channel carries 180 cfs at a depth of 2 ft. If a downstream gate is abruptly closed, at what velocity will the resulting surge move upstream and what will be the consequent depth?

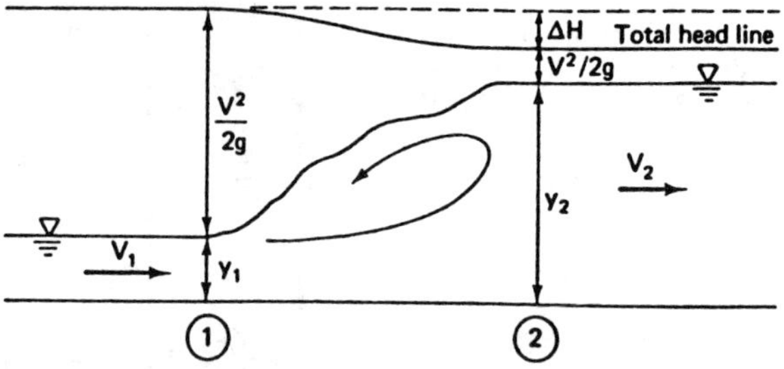

(a) Two-dimensional hydraulic jump.

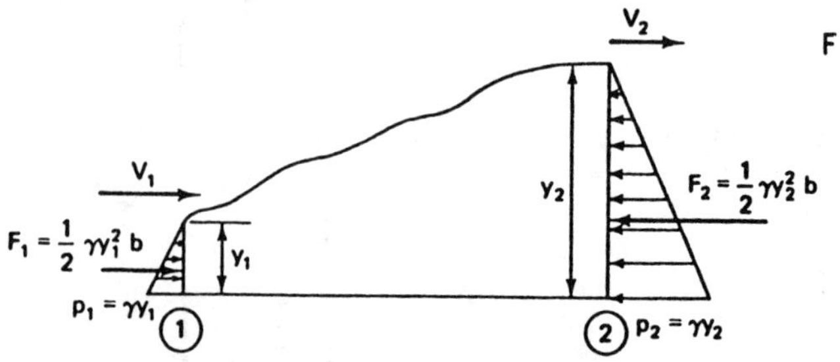

Fig. 1

(b) Control volume.

Solution: The hydraulic jump is a phenomenon which occurs frequently in open channel flows when a high-velocity flow is suddenly retarded to become a much slower, deeper flow. The foot of a spillway below a dam is one location where the hydraulic jump is frequently encountered. The hydraulic jump, in a rectangular channel of width b, is shown on Fig. 1a, and the corresponding control volume and force diagram are given in Fig. 1b.

The region between sections 1 and 2 is highly nonuniform with a considerable roller formed, as shown. Since sections 1 and 2 are in regions of uniform flow, the pressure distribution is hydrostatic at both sections, as shown on the force diagram. The momentum equation is, accordingly,

$$\frac{1}{2}\gamma y_1^2 b - \frac{1}{2}\gamma y_2^2 b = \rho Q(V_2 - V_1)$$

From continuity,

$$Q = V_1 y_1 b = V_2 y_2 b$$

Combining the two equations,

$$y_1^2 - y_2^2 = \frac{2y_1V_1^2}{g}\left(\frac{y_1}{y_2} - 1\right)$$

$$y_1 + y_2 = 2\,\frac{y_1}{y_2}\,\frac{V_1^2}{g}$$

$$\frac{y_2}{y_1}\left(\frac{y_2}{y_1} + 1\right) = \frac{2V_1^2}{gy_1}$$

and finally, by solution of the preceding quadratic equation in y_2/y_1, we get

$$\frac{y_2}{y_1} = \frac{1}{2}\left(\sqrt{1 + \frac{8V_1^2}{gy_1}} - 1\right) \tag{1}$$

Thus, the downstream depth, more particularly the downstream/upstream depth ratio, is given by upstream (and assumed known) conditions alone.

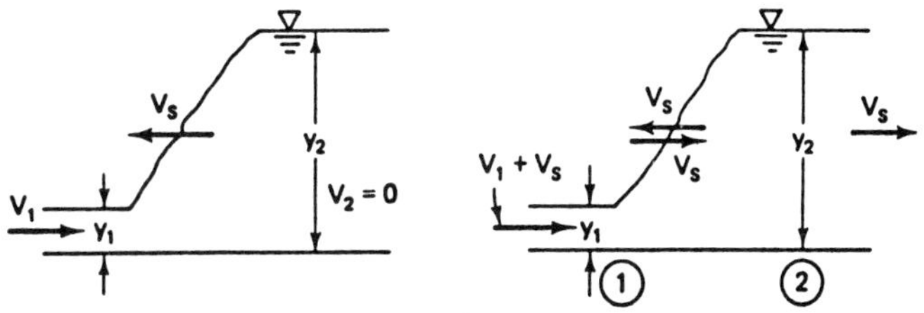

(a) Physical problem. (b) Steady state conditions.

Fig. 2

The physical problem is shown in Fig. 2a, and the steady state that results upon adding the surge velocity V_s in the opposite direction is given in Fig. 2b. The hydraulic jump equations may now be applied to Fig. 2b. From Eq. 1,

$$\frac{y_2}{y_1} = \frac{1}{2}\left(\sqrt{1 + \frac{8(V_1 + V_s)^2}{gy_1}} - 1\right)$$

or

$$\frac{y_2}{2} = \frac{1}{2}\left(\sqrt{1 + \frac{(8)(10 + V_s)^2}{(32.2)(2)}} - 1\right)$$

and from continuity

$$(V_1 + V_s)y_1 = V_s y_2$$

or

$$(10 + V_s)(2) = V_s y_2$$

Thus, we have two equations which must be solved simultaneously for the two unknowns. Upon solving, we obtain

$$y_2 = 4.98 \text{ ft}$$

and

$$V_s = 6.72 \text{ ft/s}$$

Upon abruptly closing the gate, a surge will form and move upstream at 6.72 ft/s, leaving stationary water with a depth of 4.98 ft.

• PROBLEM 11-5

A long channel with a rectangular cross section and an unfinished concrete surface is 35 ft wide and has a constant slope of 0.5°. What is the water depth when the channel carries 3500 ft^3/s?

TABLE

VALUE OF MANNING COEFFICIENT *n* FOR VARIOUS SURFACES

Type of channel surface	Manning coefficient *n*
Welded steel	0.012
Riveted or spiral steel, galvanized iron	0.016
Uncoated cast iron, black wrought iron	0.014
Corrugated storm drain	0.024
Concrete culvert with bends, connections, debris	0.013
Straight sewer with manholes, inlets, etc.	0.015
Common clay drainage tile	0.013
Sanitary sewers with slime, bends, connections	0.013
Concrete surface, trowel-finished	0.013
Float-finished	0.015
Unfinished	0.017
Earth channel, clean, dredged, straight, uniform	0.018
Stony bottom, weedy banks, winding	0.035
Unmaintained, dense high weeds	0.080
Small natural streams (width less than 100 ft)	
clean, straight, no deep pools or rifts	0.030
Some weeds, stones; winding	0.045
Very weedy, deep pools, underbrush	0.100
Flood, over high grass pastures	0.035
Over light brush and trees in summer	0.060
Rivers, with no boulders or brush	0.025–0.060
Irregular, with rough sections	0.035–0.100

Solution: Since the channel is long and has a constant slope and surface roughness, we assume that the flow is uniform, $y_0 = y_\infty$. We also assume that the flow is completely turbulent and use the Manning equation to calculate the flow rate

$$\dot{Q} = A\bar{V} = A\frac{1}{n}\left[\frac{A}{P}\right]^{2/3} S_0^{1/2} \qquad \frac{A}{P} \text{ in meters}$$

where A is the cross-sectional area of the flow, P is the wetted perimeter of the flow, $S_0 = \tan\theta$, where θ is the angle made by the channel slope and the horizontal, and n is the Manning coefficient.

The area and wetted perimeter both depend on the water depth y_∞. For a channel of rectangular cross section $A = wy_\infty$ and $P = w + 2y_\infty$. Then

$$\dot{Q} = wy_\infty \frac{1}{n}\left(\frac{wy_\infty}{w + 2y_\infty}\right)^{2/3} S_0^{1/2}$$

The equation cannot be solved explicitly for y_∞. We must use either trial and error or iteration to find y_∞. Let us find y_∞ using simple iteration. Arrange the equation in the form $y_\infty = G(y_\infty)$. One possible form is

$$y_\infty = \frac{1}{w}\left(\frac{\dot{Q}n}{S_0{}^{1/2}}\right)^{3/5}(w + 2y_\infty)^{2/5}$$

Now substitute numerical values. The Manning equation requires A/P in meters, and the average velocity $\bar{V}$ in meters per second. Remember that the Manning equation is not dimensionally consistent, so that we cannot check the units in the calculation. To be safe, we shall work in SI units and then convert back into British engineering units.

From the Table, $n = 0.017$ for an unfinished concrete surface.

$$\dot{Q} = (3500\ ft^3/s)[(1\ m)/(3.28\ ft)]^3 = 99.185\ m^3/s:$$

$$w = (35\ ft)[(1\ m)/(3.28\ ft)] = 10.671\ m:$$

and $S_0 = \tan 0.5 = 0.5°$ (in radians) $= 0.5(\pi/180)$

$= 0.008727$ rad. Then

$$y_\infty = 0.5317(10.671 + 2y_\infty)^{2/5}$$

We start the iteration with an initial estimate for y_∞ of 5 m. The simple iteration sequence is then

$$y_{\infty,0} = 5\ m$$

$$y_{\infty,1} = 1.786\ m$$

$$y_{\infty,2} = 1.538\ m$$

$$y_{\infty,3} = 1.517\ m$$

$$y_{\infty,4} = 1.515\ m$$

$$y_{\infty,5} = 1.515\ m$$

The water depth is then $y_\infty = 1.52\ m \approx 5\ ft$.

What slope is required to produce a flow of 400 cfs at a uniform depth of 4 ft in a trapezoidal earth channel with a base width of 6 ft and side slopes of 1 vertical on 2 horizontal?

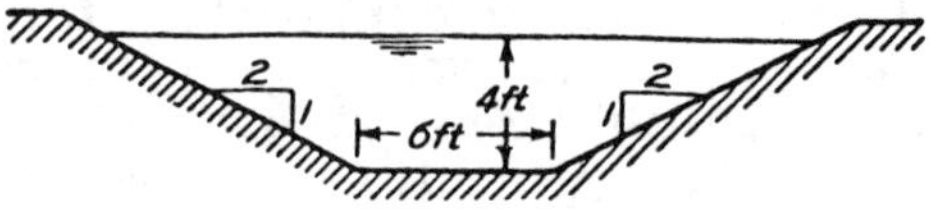

Fig. 1

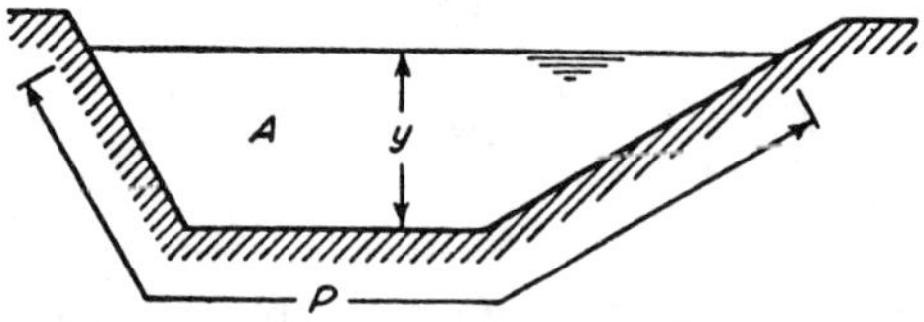

Fig. 2

Definition sketch for hydraulic radius.

Solution: The Chézy formula is

$$Q = AC\sqrt{RS} \tag{1}$$

Herein R is a quantity known as the hydraulic radius, which is (see Fig. 2) simply the ratio of the cross-sectional area A to the wetted perimeter P (the length of the cross-sectional line of contact between fluid and boundary); S is the channel slope; and C is a variable discharge coefficient.

The Manning formula is

$$Q = 1.5\,\frac{AR^{2/3}S^{1/2}}{n}$$

in which n is a roughness factor having the values for particular boundaries listed in Table I.

TABLE 1

VALUE OF THE MANNING ROUGHNESS FACTOR FOR VARIOUS BOUNDARY MATERIALS

Boundary Surface	Manning *n* $(ft)^{1/6}$
Planed wood	0.010–0.014
Unplaned wood	0.011–0.015
Finished concrete	0.011–0.013
Unfinished concrete	0.013–0.016
Cast iron	0.013–0.017
Riveted steel	0.017–0.020
Brick	0.012–0.020
Rubble	0.020–0.030
Earth	0.020–0.030
Gravel	0.022–0.035
Earth with weeds	0.025–0.040

If the Chézy and Manning expressions are solved simultaneously for C, it will be found that

$$C = 1.5 \frac{R^{1/6}}{n} \qquad (2)$$

The hydraulic radius is

$$R = \frac{A}{P} = \frac{4(22 + 6)/2}{6 + 2\sqrt{4^2 + 8^2}} = \frac{56}{23.9} = 2.34 \text{ ft}$$

From Eq. (2), for an average value of n from Table I,

$$C = 1.5 \frac{R^{1/6}}{n} = 1.5 \frac{\overline{2.34}^{1/6}}{0.025} = 69 \text{ ft}^{1/2}/\text{sec}$$

Hence, upon solving Eq. (1) for S,

$$S = \frac{Q^2}{C^2 A^2 R} = \frac{\overline{400}^2}{\overline{69}^2 \times \overline{56}^2 \times 2.34} = 0.0046$$

• PROBLEM 11-7

A large reservoir 5 m deep has a rectangular sluice gate 1 m wide. How does the volume flow rate change as the sluice gate is raised (see Figure)?

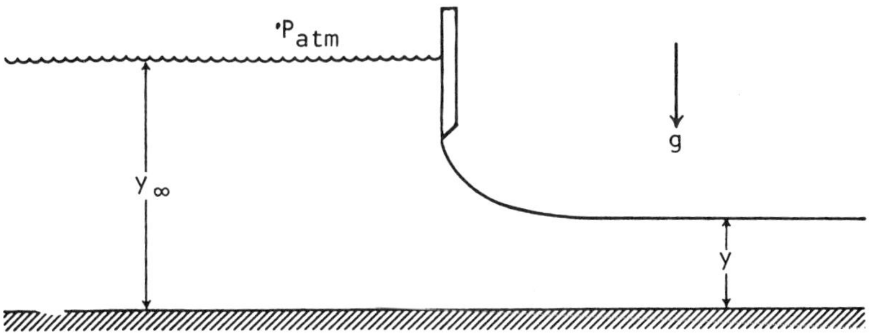

Solution: Apply the ideal Bernoulli equation in head form at the water surface and far enough up and downstream of the gate for the streamlines to be parallel

$$\frac{P_{atm}}{\rho g} + \frac{V_\infty^2}{2g} + y_\infty = \frac{P_{atm}}{\rho g} + \frac{V^2}{2g} + y$$

In terms of specific head $H_\infty = H$. This is a constant-specific-head flow. The sluice gate is not an obstacle which changes the specific head of the flow; it simply changes the distribution of energy between kinetic and potential.

Now we examine H_∞, $H_\infty = V_\infty^2/2g + y_\infty$. Assume that the reservoir is so large that y_∞ upstream of the gate does not change with changes in sluice-gate opening. Furthermore, V_∞ is very small. Then, $H_\infty \approx y_\infty$ and is the same for all sluice-gate openings. Thus, the outflow has the same specific head regardless of the height of the sluice-gate opening.

The flow rate is then given by

$$\frac{\dot{Q}}{w} = y\sqrt{2g(H - y)}$$

with $H = y_\infty$

$$\dot{Q} = wy\sqrt{2g(y_\infty - y)}$$

The minimum value of H occurs at y_c, called the critical depth. The critical depth y_c changes with the flow rate and consequently with y

$$y_c = \left[\frac{(\dot{Q}/w)^2}{g}\right]^{1/3} = [2y^2(y_\infty - y)]^{1/3}$$

The corresponding Froude number is given by

$$F = \left(\frac{y_c}{y}\right)^{3/2} = \sqrt{2\left(\frac{y_\infty}{y} - 1\right)}$$

What happens as the sluice gate is raised? First, consider a small opening which results in a y of, say, 0.5 m. The flow rate is

$$\dot{Q} = (1\text{ m})(0.5\text{ m})\sqrt{2(9.8\text{ m/s}^2)(5\text{ m} - 0.5\text{ m})}$$

$$= 4.696\text{ m}^3/\text{s}$$

and the corresponding Froude number is

$$F = \sqrt{2\left(\frac{5\text{ m}}{0.5\text{ m}} - 1\right)} = 4.243$$

As the gate is further raised, y increases, causing the Froude number to decrease and the flow rate to increase. Finally, at a gate height which gives $y = \frac{2}{3}y_\infty = \frac{2}{3}(5\text{ m}) =$ 3.333 m, the Froude number is 1 and the maximum flow rate is passed

$$\dot{Q}_{max} = w\sqrt{\frac{8}{27}gy_\infty^3} = (1\text{ m})\sqrt{\frac{8}{27}(9.8\text{ m/s}^2)(5\text{ m})^3}$$

$$= 19.05\text{ m}^3/\text{s}$$

Find the ratio of the depth of flow D to the breadth B in a rectangular channel which will make the discharge a maximum for a given cross-sectional area, slope and surface roughness. Assume steady uniform flow and that the coefficient C in the Chezy formula is constant.

A channel of rectangular cross-section conveys water at a rate of 9 $m^3 s^{-1}$ with a velocity of 1.2 ms^{-1}. Find the gradient required for uniform steady flow if the proportions are those for maximum discharge. Take C = 50 SI units.

Solution: The Darcy formula for the loss of head h_f in a length l can be written

$$h_f = \frac{fl}{m} \cdot \frac{\bar{v}^2}{2g}, \qquad (1)$$

since, for a pipe running full, $m = A/P = (\pi/4)d^2/\pi d = d/4$ and, in the form of equation (1), the Darcy formula can be applied directly to open channels of any form.

Many problems involve steady flow at uniform depth and constant cross-sectional area. Under these conditions, the bed slope s is equal to the slope of the total energy line i, which is the loss of energy per unit weight per unit length h_f/l. It is, therefore, more convenient for resistance formulae for channels to be written in terms of i. From equation (1),

$$\bar{v}^2 = \frac{2g}{f} m \frac{h_f}{l}$$

so that

$$\bar{v} = \sqrt{(2g/f)}\sqrt{(mi)}$$

or, putting $\sqrt{(2g/f)} = C$,

$$\bar{v} = C\sqrt{(mi)}. \qquad (2)$$

This is the Chezy formula.

$$Q = AC\sqrt{(mi)}.$$

Putting m = A/P and, since i = bed slope s for uniform flow,

$$Q = AC\sqrt{\{(A/P)s\}}.$$

Since A, C and s are fixed, Q will be a maximum for the proportions which make P a minimum. For a rectangular section,

$$\text{Wetted perimeter, } P = B + 2D, \tag{3}$$

$$\text{Area, } A = BD$$

and so $$B = A/D.$$

Substituting in equation (3),

$$P = A/D + 2D. \tag{4}$$

For a given value of A, the minimum value of P occurs when $dP/dD = 0$. Differentiating equation (4),

$$\frac{dP}{dD} = -A/D^2 + 2 = 0,$$

$$A = 2D^2$$

or, since $A = BD$,

$$BD = 2D^2.$$

Thus, for maximum discharge in a rectangular section, $B = 2D$.

$$\text{Discharge, } Q = A\bar{v},$$

$$\text{Cross-sectional area, } A = Q/\bar{v} = 9/12 = 7.5 \text{ m}^2.$$

For maximum discharge, $B = 2D$,

$$A = BD = 2D^2,$$

$$D = \sqrt{\tfrac{1}{2}A} = \sqrt{(3.75)} = 1.936 \text{ m},$$

$$B = 2D = 3.872 \text{ m},$$

$$m = A/P = A/(B + 2D) = 7.5/(3.872 + 3.872)$$

$$= 0.968 \text{ m}.$$

Using the Chezy formula, $\bar{v} = C\sqrt{(mi)}$,

$$\text{Gradient, } i = \bar{v}^2/C^2m = 1.2^2/50^2 \times 0.968 = 1/1681.$$

• PROBLEM 11-9

A rectangular channel, 10 ft wide, and laid on a slope of 3 in 800, carries water at a rate of 125 cfs. A weir raises the depth of the water to 5 ft, measured shortly upstream from the weir. If the Chèzy coefficient is 100 $ft^{1/2}$/sec, what is the depth 100 ft upstream from the weir?

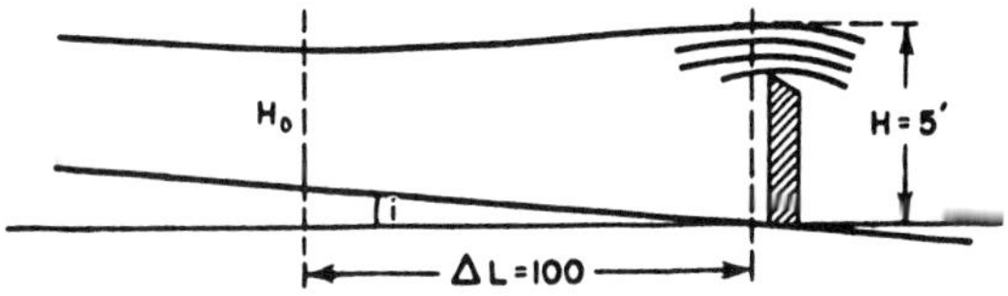

Solution: Referring to the section shortly upstream from the weir

$$A = (10)(5) = 50 \text{ ft}^2$$

$$P = 10 + 2(5) = 20 \text{ ft}$$

$$m = AP = 2.5 \text{ ft}$$

$$v = \frac{Q}{A} = \frac{125}{50} = 2.5 \text{ fps}$$

By analogy to flow in pipes

$$dh_f = \alpha(dL) = 4f\,\frac{(dL)}{D}\left(\frac{v^2}{2g}\right)$$

where f is the Darcy friction factor.

For a pipe of diameter D

$$m = \frac{A}{P} = \frac{\pi/4D^2}{\pi D} = \frac{D}{4}$$

$$D = 4m$$

$$\alpha(dL) = 4f\,\frac{(dL)}{4m}\left(\frac{v^2}{2g}\right)$$

from which

$$\alpha = \frac{fv^2}{2mg}$$

From the Chézy equation

$$v = C\sqrt{(mi)}$$

in which $C = \sqrt{(g/C_D)} = \sqrt{(2g/f)}$

$$f = \frac{2g}{C^2}$$

$$\alpha = \frac{fv^2}{2mg} = \left(\frac{2g}{C^2}\right)\left(\frac{v^2}{2mg}\right)$$

$$\alpha = \frac{v^2}{mC^2}$$

Using the above equation

$$\alpha = \frac{v^2}{mC^2} = \frac{(2.5)^2}{(2.5)(10,000)}$$

$$\alpha = 0.00025$$

$$i = 3/800 = 0.00375$$

$$\therefore i - \alpha = 0.0035$$

$$\frac{v^2}{gH} = \frac{(2.5)^2}{(32.2)(5)} = \frac{2.5}{64.4}$$

The rate of change of the slope of the free surface is then given by

$$\frac{dH}{dL} = \frac{i - \alpha}{1 - v^2/gH}$$

$$\frac{dH}{dL} = \frac{i - \alpha}{1 - v^2/gH} = \frac{0.0035}{1 - 2.5/64.4} = 0.00365$$

For small changes in the slope, the free surface may be assumed to curve uniformly, and

$$\frac{dH}{dL} = \frac{\Delta H}{\Delta L} = 0.00365$$

For $\Delta L = 100$ ft

$$\Delta H = 0.365 \text{ ft}$$

The depth, 100 ft upstream from the weir, is then

$$H_0 = 4.635 \text{ ft}$$

• PROBLEM 11-10

A rectangular channel carries 7 cfs of water per unit width. If a submerged obstruction with maximum height of 0.25 ft is placed across the channel floor, what will be the increase (or decrease) in water level due to the obstruction? Assume frictionless flow to determine the change in water level for two cases: (a) $h_1 = 0.75$ ft and (b) $h_1 = 2.0$ ft.

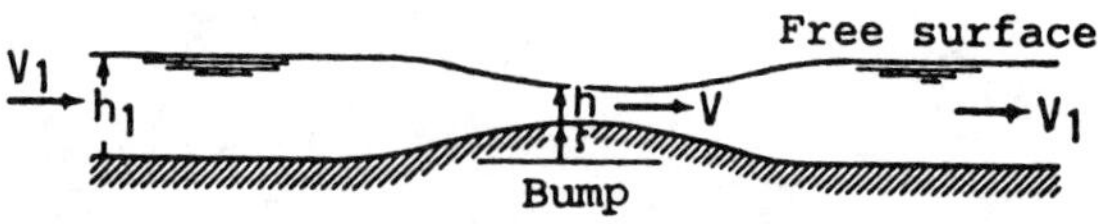

Fig. 1

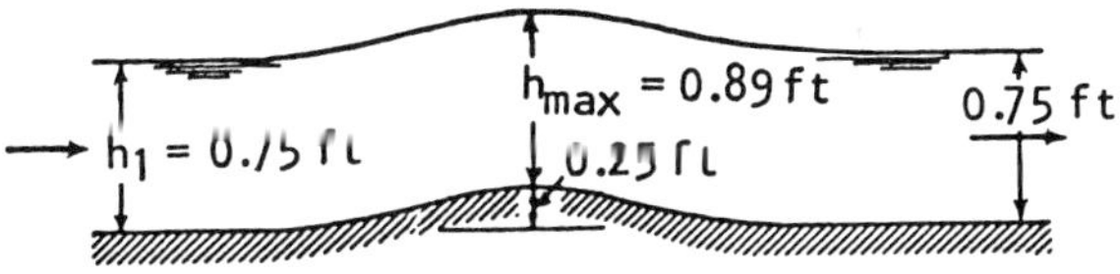

Fig. 2

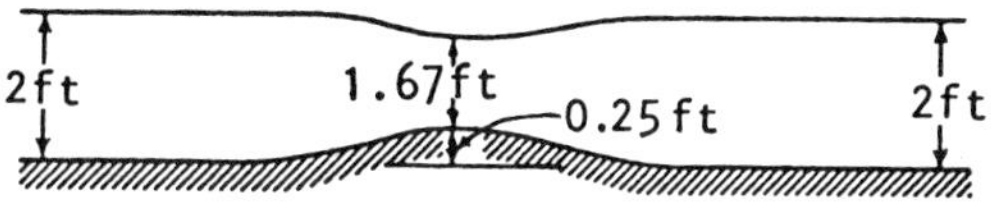

Fig. 3

Solution: (a) Consider the case of frictionless flow along a horizontal rectangular channel floor with a small bump in the floor, as indicated in Figure 1. Taking Bernoulli's equation along the free surface, we obtain

$$\frac{V_1^2}{2g_c} + \frac{g}{g_c} h_1 = \frac{V^2}{2g_c} + \frac{g}{g_c}(h_1 + \zeta)$$

where ζ is the elevation of the bump above the channel floor. The velocities can be eliminated from the preceding equation by using the continuity equation

$$Q = bhV$$

where b is the width of the rectangular channel. Thus we obtain

$$\frac{Q^2}{2g_c b^2 h_1^2} + \frac{g}{g_c} h_1 = \frac{Q^2}{2g_c b^2 h^2} + \frac{g}{g_c}(h + \zeta)$$

or multiplying g_c/g, we have

$$\frac{Q^2}{2gb^2h_1^2} + h_1 = \frac{Q^2}{2gb^2h^2} + (h + \zeta) \tag{1}$$

In this case h_1 = 0.75 ft and Q/b = 7 cfs/ft. Hence

$$\frac{Q^2}{b^2 g} = 1.52 \quad \text{and} \quad \frac{Q^2}{2b^2 g h_1^2} = 1.35 \text{ ft}$$

Therefore,

$$\frac{Q^2}{2gb^2h_1^2} + h_1 = 2.10 \text{ ft}$$

Using equation (1), we write

$$\frac{Q^2}{2gb^2h_1^2} + h_1 = 2.10 = \frac{Q^2}{2gb^2h^2} + h + \zeta_{max}$$

where ζ_{max} is the maximum height of the bump.
Therefore,

$$\frac{Q^2}{2gb^2h^2} + h = 1.85 \text{ ft}$$

and after trial and error, we obtain

$$h = 0.89 \text{ ft}$$

Thus the water level (0.89 ft) at the maximum height of the obstruction is above the upstream water level (0.75 ft) as shown in Figure 2.

(b) In this case, $h_1 = 2.0$ ft; $Q/b = 7$ cfs/ft and

$$\frac{Q^2}{2b^2gh_1^2} = 0.190 \text{ ft}$$

Therefore,

$$\frac{Q^2}{2gb^2h_1^2} + h_1 = 2.190 \text{ ft}$$

We thus obtain

$$2.19 = \frac{Q^2}{2gb^2h^2} + h + 0.25$$

and

$$\frac{Q^2}{2gb^2h^2} + h = 1.94 \text{ ft}$$

Solving for h, we obtain

$$h = 1.67 \text{ ft}$$

Here the water level (1.67 ft) above the maximum height of the obstruction is less than the upstream water level (2.0 ft), as shown in Figure 3.

• PROBLEM 11-11

A trapezoidal channel that was cut through bare earth has side slopes of 2 on 1 and a bottom width of 8 ft (see Fig.). If the channel slope is 0.001 and the flow rate is 340 cfs, determine the normal depth, y_n.

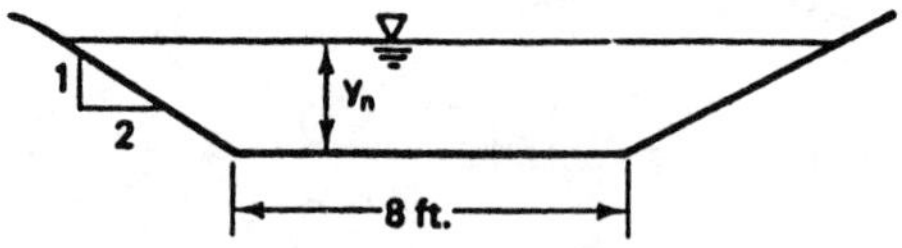

Solution: Substituting S for the head loss per unit length (h/L) and solving for V, the Darcy equation becomes

$$V = \sqrt{\frac{8g}{f}}\sqrt{RS}$$

Replacing the first radical by a coefficient C, we get the Chezy equation,

$$V = C\sqrt{RS}$$

The value of the Chezy C can be obtained directly from the two foregoing equations which give C as a function of f, namely,

$$C = \sqrt{\frac{8g}{f}}$$

Noting that the Chezy C has dimensions of $g^{1/2}$, a dimensionless constant $C' = C/\sqrt{g}$ is sometimes introduced such that

$$V = C'\sqrt{gRS}$$

TABLE

Boundary Surface	*Manning n*
Very smooth surface (glass, plastic, machined metal)	0.010
Planed timber	0.011
Unplaned wood	0.012–0.015
Smooth concrete	0.012–0.013
Unfinished concrete	0.013–0.016
Brickwork	0.014
Vitrified clay	0.015
Rubble masonry	0.017
Earth channels smooth, no weeds	0.020
Firm gravel	0.020
Earth channels with some stones and weeds	0.025
Earth channels in bad condition, winding natural streams	0.035
Mountain streams	0.040–0.050
Sand channels (with flat bed) (d = mean grain diameter, ft)	$0.034d^{1/6}$

An empirical equation known as the Manning equation is currently the most common resistance equation used in the analysis of open-channel flows. It essentially assumes that the Chezy C is proportional to $R^{1/6}$ rather than a pure constant, and introduces a resistance coefficient rather than a discharge coefficient. The Manning equation originally presented in the metric system was as follows:

$$V = \frac{1}{n} R^{2/3}S^{1/2}$$

When the equation was converted to English units it was felt desirable to keep the magnitude of the roughness coefficient n values unchanged. Thus, it now appears as

$$V = \frac{1.49}{n} R^{2/3} S^{1/2} \tag{1}$$

Here V, R, and S are again the average velocity, hydraulic radius, and slope and the additional variable known as the Manning n.

Assuming that n = 0.025, the Manning equation (Eq. 1) may be written in terms of the discharge as

$$Q = AV = \frac{1.49}{n} AR^{2/3} S^{1/2} = \frac{1.49}{n} \frac{A^{5/3}}{P^{2/3}} S^{1/2}$$

where for the trapezoid

$$A = 8y_n + 2y_n^2$$

and

$$P = 8 + 2\sqrt{y_n^2 + 2^2 y_n^2}$$

$$= 8 + 2y_n\sqrt{5}$$

Separating the known from the unknown quantities and substituting,

$$\frac{Qn}{1.49S^{1/2}} = \frac{(340)(0.025)}{(1.49)(0.001)^{1/2}} = \frac{(8y_n + 2y_n^2)^{5/3}}{(8 + 2y_n\sqrt{5})^{2/3}}$$

Solving by trial and error, the normal depth is

$$y_n = 4.93 \text{ ft}$$

• PROBLEM 11-12

At a certain section in a very smooth 6-ft-wide rectangular channel the depth is 3.00 ft when the flow rate is 160 cfs. Compute the distance to the section where the depth is 3.20 ft if S_0 = 0.002 and n = 0.012.

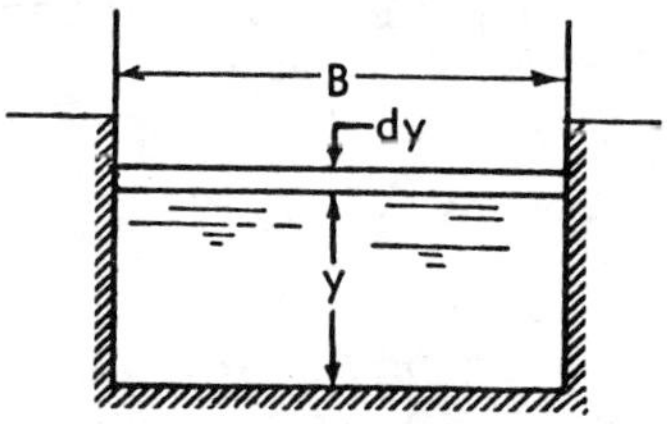

Solution: For conduits having noncircular cross sections, some value other than the diameter must be used for the linear dimension in the Reynolds number. Such a characteristic is the hydraulic radius, defined as

$$R_h = \frac{A}{P}$$

where A is the cross-sectional area of the flowing fluid, and P is the wetted perimeter, that portion of the perimeter of the cross section where there is contact between fluid and solid boundary.

One of the best as well as one of the most widely used formulas for open-channel flow is that of Robert Manning. In English units, the Manning formula is

$$V(\text{fps}) = \frac{1.49}{n} R_h^{2/3} S^{1/2} \tag{1}$$

where 1.49 is the cube root of 3.28, the number of feet in a meter, S is the slope of the channel, and n is known as the Manning roughness coefficient.

In terms of flow rate Eq. (1) may be expressed as

$$Q(\text{cfs}) = \frac{1.49}{n} A R_h^{2/3} S^{1/2} \tag{2}$$

Letting y_0 represent the height of the channel,

$$Q = \frac{1.49}{n} A R_h^{2/3} S^{1/2} = \left(\frac{1.49}{0.012}\right) 6y_0 \left(\frac{6y_0}{6 + 2y_0}\right)^{2/3} (0.002)^{1/2} = 160 \text{ cfs}$$

By trial,

$$y_0 = 3.5 \text{ ft}$$

Totally separate from the concept of energy gradient, or energy difference, between two sections of a stream is the matter of the energy at a single section with reference to the channel bed (see Fig.). This is called the specific energy at that section, and its value is given by the following equation:

$$E = y + \frac{V^2}{2g}$$

or

$$E = y + \frac{Q^2}{2gA^2}$$

Differentiating with respect to y,

$$\frac{dE}{dy} = 1 - \frac{Q^2}{2g}\left(\frac{2}{A^3}\frac{dA}{dy}\right)$$

This may now be set equal to zero and solved for the value of the critical depth for the given flow. As A may or may not be a reasonable function of y, it is helpful to observe that dA = B dy, and thus dA/dy = B, the width of the water surface. Substituting this in the above expression results in

$$\frac{Q^2}{g} = \left(\frac{A^3}{B}\right)_{y=y_c}$$

as the equation which must be satisfied for critical flow. For a given cross section the right-hand side is a function of y only.

Critical depth occurs when

$$\frac{Q^2}{g} = \frac{A^3}{B} \quad \text{or} \quad \frac{(160)^2}{32.2} = \frac{(6y_c)^3}{6}$$

from which $y_c = 2.8$ ft. Since $y_0 > y_c$, the flow is subcritical and the depth increases toward normal depth as one proceeds upstream. With S_0 and n known and the depth and velocity at one end of the reach given, the length L to the end corresponding to the other depth can be computed from

$$L = \frac{(y_1 + V_1^2/2g) - (y_2 + V_2^2/2g)}{S - S_0}$$

The calculations are shown in the following table. The total distance is calculated to be 73 ft. The accuracy could be improved by taking more steps. In computing $S - S_0$, a slight error in the calculated value of S will introduce a sizable error in the calculated value of L. Thus it is important that S be calculated as accurately as possible, preferably by use of an electric calculator.

TABLE

y, ft	A, ft^2	P $(6 + 2y)$, ft	R_h, ft	V, fps	$V^2/2g$, ft	$y + \frac{V^2}{2g}$	Numerator $\Delta\left(y + \frac{V^2}{2g}\right)$	V_{avg}, fps	R_{avg}, ft	S	Denominator $S - S_0$	L, ft
3.00	18.00	12.00	1.500	8.89	1.227	4.227						
							0.022	8.74	1.512	0.00284	0.00084	26
3.10	18.60	12.20	1.525	8.60	1.149	4.249						
							0.029	8.47	1.536	0.00262	0.00062	47
3.20	19.20	12.40	1.548	8.33	1.078	4.278						

$\Sigma = 73$

A trapezoidal channel with a bottom width of 4 m and side slopes of 4:1 (four horizontal to one vertical) carries water at a depth of 2 m (Figure 1).
(A) Compute the average depth.
(B) If the discharge in the channel is 50 m^3/s, calculate the average velocity.
(C) If the discharge in the trapezoidal channel described in (A) is 30 m^2/s what is the critical depth, the critical velocity, and the critical discharge of the flow?

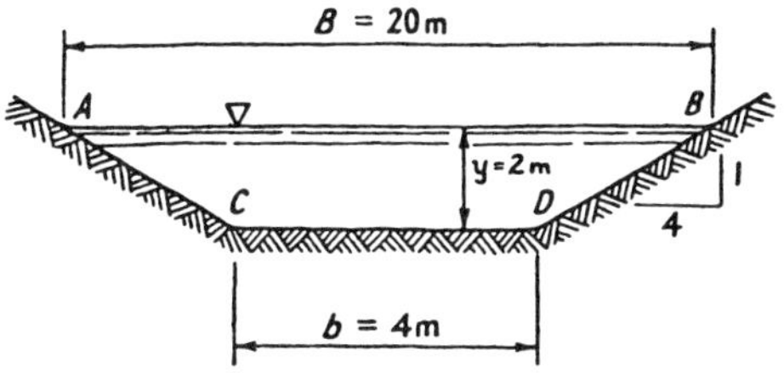

Fig. 1

Solution: (A) The cross-sectional area of the channel is

$$A = by + zy^2 = 4 \times 2 + 4 \times 2^2 = 24 \text{ m}^2$$

The width of water surface, B = 20 m

$$\text{The wetted perimeter, } P = 2\overline{AC} + \overline{CD}$$

$$= 2\sqrt{2^2 + 8^2} + 4$$

$$= 20.49 \text{ m}$$

$$z = 4$$

$$b = 4 \text{ m}$$

$$B = 20 \text{ m}$$

In channels other than those of rectangular shape the depth of the water is variable in each cross section along the flow line. For this reason it is convenient for computational reasons to introduce the concept of average depth. Average depth is computed by dividing the cross-sectional area of the flow by the width of the water surface, B. Both of these values are to be measured in a plane that is perpendicular to the flow line, that is, to the direction of the velocity. Hence, the average depth is defined as

$$y_{ave} = \frac{A}{B} \tag{1}$$

$$= \frac{24}{20} = 1.2 \text{ m}$$

(B) The velocity in any channel is variable across the flow area because of the boundary layer effects along the channel boundaries. In practice the magnitude and direction of the velocity measured at various points in the cross-sectional area are rarely found to be uniform and parallel with the direction of the channel. For the purposes of hydraulic computations an average flow velocity is used; this is defined as the discharge of the flow divided by its cross-sectional area, that is,

$$v_{ave} = \frac{Q}{A} \tag{2}$$

where

$$v = \frac{Q}{A}$$

$$Q = 50 \ \frac{m^3}{s}$$

$$A = 24 \ m^2$$

therefore

$$v = \frac{50}{24}$$

$$= 2.08 \ \frac{m}{s}$$

(C) The area of the channel was 24 m^2. Hence the average velocity is $v = Q/A = 30/24 = 1.25$ m/s. The critical depth y_c may be given by the equation

$$y_c = cE$$

where c may be obtained from Fig. 2.

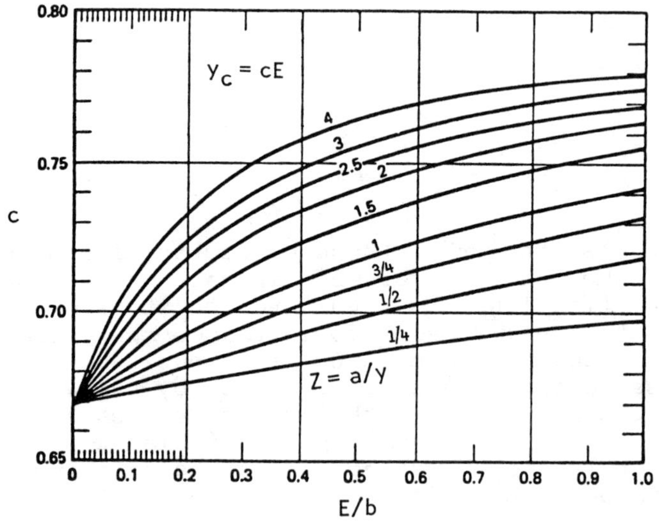

Fig. 2

The kinetic energy is

$$\frac{v^2}{2g} = \frac{1.25^2}{2(9.81)} = 0.0796 \ m$$

The total available energy is then given by

$$E = \frac{v^2}{2g} + y = 0.0796 + 2.0 = 2.0796 \text{ m}$$

We need to express E/b, which is

$$\frac{E}{b} = \frac{2.0796}{4} = 0.52$$

which is in Figure 2 and for z = 4 gives c = 0.768. Hence

$$y_c = cE = 0.768(2.0796)$$

$$y_c = 1.59 \text{ m}$$

The critical velocity for trapezoidal channels is given by the equation

$$v_c = \sqrt{\frac{b + zy_c}{b + 2zy_c} gy_c} = \sqrt{\frac{4 + 4(1.59)}{4 + 2(4)(1.59)} 9.81(1.59)}$$

$$= 9.85$$

$$v_c = 9.85 \frac{m}{s}$$

To obtain the critical discharge we first compute the area corresponding to the critical depth

$$A_c = b \cdot y_c + 2\left(\frac{zy_c^2}{2}\right) = 4(1.59) + 2\left(\frac{4(1.59)^2}{2}\right)$$

$$= 5.96 + 2.53 = 8.49 \text{ m}^2$$

The critical discharge is then equal to

$$Q_c = A_c v_c = 8.49(9.85) = 83.6 \frac{m^3}{s}$$

• PROBLEM 11-14

The cross-section of an open channel is shown in Fig. 1. If the slope of the channel bed is 1 ÷ 3600, what is the flow for a centre-line depth of 4 ft? What would be the depth if the flow was increased by 20 per cent? In each case, the Chèzy coefficient may be assumed to have the same value of C = 100 $ft^{1/2}$/sec.

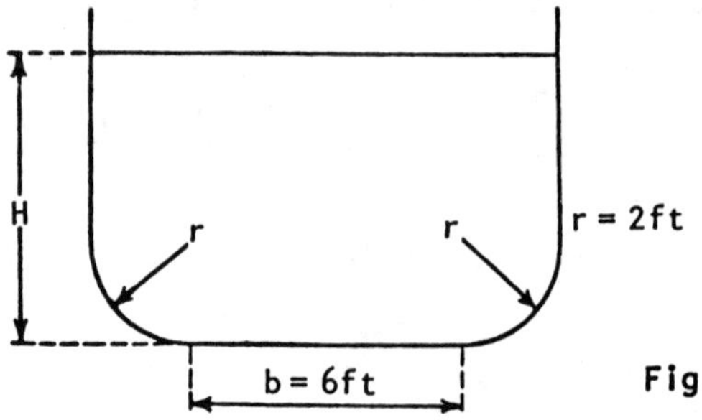

Fig. 1

Solution: Consider, in Fig. 2, a section of an open channel of uniform flow area laid on a slope i. Let L be a length of the channel numerically equal to the average velocity v of the liquid flowing in it, then the quantity of the liquid crossing any section of the channel in unit time is AL_ρ, when A is the flow area and ρ is the density of the liquid.

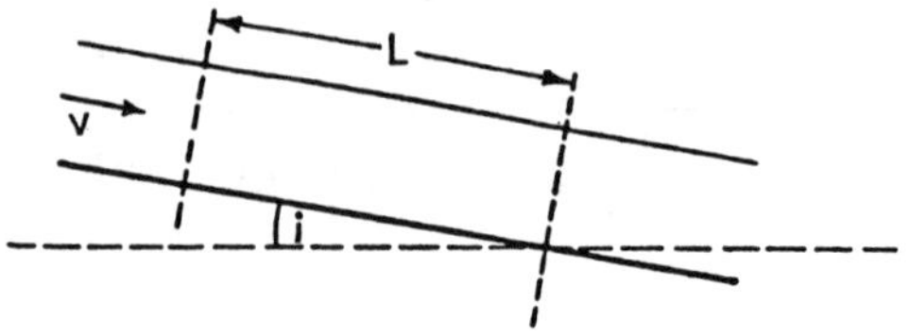

Fig. 2

The only force causing flow in open channels is the force of gravity, then, if we ignore losses other than friction, this force will be equal to the resistance experienced in the flow. This resistance may be related to the velocity v by an equation of the form

$$R = SC_D\rho v^2$$

Let P be the wetted perimeter of the channel, then the wetted surface area in the length considered is

$$S = PL$$

and

$$R = (PL)C_D\rho v^2$$

The work done per unit time in overcoming friction is therefore

$$RL = PL^2C_D\rho v^2$$

This work equals the loss in potential energy of AL_ρ mass units of the liquid when it changes its altitude by h units of length, then by an energy balance

$$AL\rho gh = PL^2C_D\rho v^2$$

where g is the gravitational acceleration.

But h = L sin i, and since for very small slopes normally encountered in open channel practice, we can safely take sin i = i, the above equation becomes

$$AL\rho g(Li) = PL^2 C_D \rho v^2$$

Simplifying

$$Agi = PC_D v^2 \tag{1}$$

Let

$$m = \frac{A}{P} \tag{2}$$

where m is called the hydraulic radius, or hydraulic mean depth. Equation (1) may then be written

$$v^2 = \frac{gmi}{C_D}$$

This equation is normally presented in the form known as the Chèzy equation.

$$v = C\sqrt{(mi)} \tag{3}$$

The coefficient C which appears in this equation is known as the Chèzy coefficient.

With reference to Fig. 1, for a depth of H = 4 ft

$$A = 10(4 - 2) + (6)(2) + 2\pi$$

$$A = 38.28 \text{ ft}^2$$

$$P = 6 + (2)(2) + 2\pi$$

$$P = 16.28 \text{ ft}$$

Using Eq. (3), and taking m = A/P (from eq. (2))

$$v = 100\sqrt{\frac{Ai}{P}}$$

$$Q = vA = 100\sqrt{\frac{A^3 i}{P}} = 100\sqrt{\frac{(38.28)^3}{(16.28)(3600)}}$$

$$Q = 97.83 \text{ cfs}$$

For Q = (1.2)(97.83), let the unknown depth be H, then

$$A = 10(H - 2) + (6)(2) + 2\pi$$

$$A = 10\,H - 1.72$$

$$P = 6 + 2(H - 2) + 2\pi$$

$$P = 2H + 8.28$$

Again

$$Q = 100\sqrt{\frac{A^3 i}{P}}$$

$$(1.2)(97.83) = 100\sqrt{\frac{(10H - 1.72)^3}{(2H + 8.28)(3600)}}$$

$$\frac{(10H - 1.72)^3}{2H + 8.28} = 4960$$

By trial

$$H = 4.6 \text{ ft}$$

• PROBLEM 11-15

Find the proportions of a trapezoidal channel (see Figure) which will make the discharge a maximum for a given cross-sectional area of flow and given side slopes. Show also that if the side slopes can be varied the most efficient of all trapezoidal sections is a half hexagon.

A trapezoidal channel has side slopes of 3 horizontal to 4 vertical and the slope of its bed is 1 in 2000. Determine the optimum dimensions of the channel if it is to carry water at 0.5 m^3s^{-1}. Use the Chezy formula assuming that C = 80 $m^{1/2}s^{-1}$.

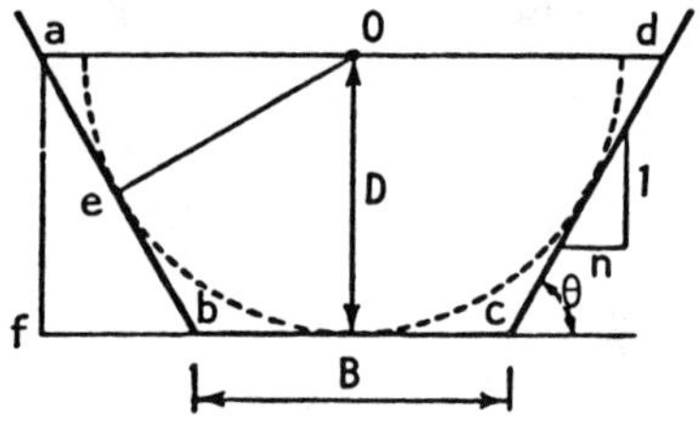

Solution: Using the Chezy formula,

$$Q = ACm^{1/2}i^{1/2} = AC(A/P)^{1/2}i^{1/2}$$

Maximum discharge for given values of A, C and i will, therefore, occur when P is a minimum.

In the Figure, base width = B, depth = D and the side slope is 1 vertical to n horizontal.

$$\text{Area of flow, } A = (B + nD)D, \qquad (1)$$

from which

$$B = A/D - nD. \qquad (2)$$

Wetted perimeter, P = bc + 2 cd

$$= B + 2D\sqrt{(n^2 + 1)}.$$

Substituting from equation (2),

$$P = A/D + (2\sqrt{(n^2 + 1)} - n)D. \quad (3)$$

If A and n are fixed, P will be a minimum when dP/dD = 0. Differentiating equation (3),

$$\frac{dP}{dD} = -A/D^2 + 2\sqrt{(n^2 + 1)} - n = 0,$$

$$A = D^2(2\sqrt{(n^2 + 1)} - n).$$

Substituting for A from equation (1),

$$BD + nD^2 = D^2(2\sqrt{(n^2 + 1)} - n).$$

For maximum discharge,

$$B = 2D(\sqrt{(n^2 + 1)} - n). \quad (4)$$

We now find the side slopes for the section which will have the greatest possible efficiency. The value of B will be that for optimum efficiency, given by equation (4),

$$\text{Area, } A = BD + nD^2 = 2D^2(\sqrt{(n^2 + 1)} - n) + nD^2$$

$$= D^2(2\sqrt{(n^2 + 1)} - n).$$

So, for maximum efficiency,

$$D = A^{1/2}/(2\sqrt{(n^2 + 1)} - n)^{1/2}.$$

Substituting in equation (3),

$$P = 2A^{1/2}(2\sqrt{(n^2 + 1)} - n)^{1/2}. \quad (5)$$

Since P^2 will also be a minimum when P is a minimum, it is convenient to square equation (5):

$$P^2 = 4A(2\sqrt{(n^2 + 1)} - n).$$

Differentiating and equating to zero,

$$\frac{d(P^2)}{dn} = 4A\left(\frac{2n}{\sqrt{(n^2 + 1)}} - 1\right) = 0,$$

$$2n = \sqrt{(n^2 + 1)}.$$

Squaring,

$$4n^2 = n^2 + 1 \quad \text{and} \quad n = 1/\sqrt{3}.$$

If θ is the angle of the side of the horizontal, tan θ = $1/n = \sqrt{3}$ and $\theta = 60°$. Thus, the cross-section of flow of greatest possible efficiency will be a half-hexagon.

For the given cross-section in this problem, $n = \frac{3}{4}$; therefore, substituting in equation (4) for maximum discharge,

$$B = 2D\left(\sqrt{\left(\frac{3}{4}\right)^2 + 1} - \frac{3}{4}\right) = 2D\left(\frac{5}{4} - \frac{3}{4}\right) = D,$$

$$\text{Area of cross-section, } A = BD + \frac{3}{4}D^2$$

$$= D^2 + \frac{3}{4}D^2 = \frac{7}{4}D^2,$$

$$\text{Wetted perimeter, } P = B2 + \frac{5}{4}D = \frac{7}{2}D,$$

$$\text{Hydraulic mean depth, } m = A/P = \frac{1}{2}D.$$

Substituting in the Chezy formula,

$$Q = ACm^{1/2}i^{1/2} = \frac{7}{4}D^2 \times C\left(\frac{1}{2}D\right)^2 i^{1/2}.$$

Putting $Q = 0.5\ m^3s^{-1}$, $C = 80\ m^{1/2}s^{-1}$ and $i = 1/2000$,

$$0.5 = \frac{7}{4} \times \frac{80\ D^{5/2}}{(2 \times 2000)^{1/2}}$$

$$\text{Depth, } D = \left(\frac{4 \times 0.5 \times 63.25}{7 \times 80}\right)^{2/5} = 0.552\ m,$$

$$\text{Base width, } B = D = 0.552\ m.$$

It can also be shown that for a channel of optimum proportions the sides and base are tangential to a semicircle with its centre O in the free surface.

Drawing Oe perpendicular to ab from the midpoint of the free surface:

$$\sin \widehat{Oae} = \frac{Oe}{Oa} = \frac{Oe}{\frac{1}{2}(B + 2nD)},$$

$$\sin \widehat{abf} = \frac{af}{ab} = \frac{D}{D\sqrt{(n^2 + 1)}}.$$

But $\widehat{Oae} = \widehat{abf}$ and so

$$\frac{Oe}{\frac{1}{2}(B + 2nD)} = \frac{D}{D\sqrt{(n^2 + 1)}},$$

$$Oe = \frac{B + 2nD}{2\sqrt{(n^2 + 1)}} .$$

From equation (4),

$$B = 2D(\sqrt{(n^2 + 1)} - n) = 2D\sqrt{(n^2 + 1)} - 2nD \qquad (6)$$

and so

$$Oe = \frac{2D\sqrt{(n^2 + 1)} - 2nD + 2nD}{2\sqrt{(n^2 + 1)}} = D.$$

As ab is perpendicular to Oe it will be a tangent to a semicircle of centre O to which bc and cd will also be tangential.

The hydraulic mean depth for an optimum trapezoidal section will be

$$m = \frac{BD + nD^2}{B + 2D\sqrt{(n^2 + 1)}} = \frac{B + nD}{B/D + 2\sqrt{(n^2 + 1)}} .$$

Substituting for B from equation (6),

$$m = \frac{2D\sqrt{(n^2 + 1)} - nD}{2\sqrt{(n^2 + 1)} - 2n + 2\sqrt{(n^2 + 1)}} = \frac{D}{2} .$$

• PROBLEM 11-16

An open channel having a trapezoidal cross section with side slopes of 1 vertical to 2 horizontal is required to discharge 15 m^3/s of water when running full. The longitudinal slope is 1 in 2000. Determine the depth d and the base width b to give minimum cross sectional area. $C = 70\ m^{1/6}$.

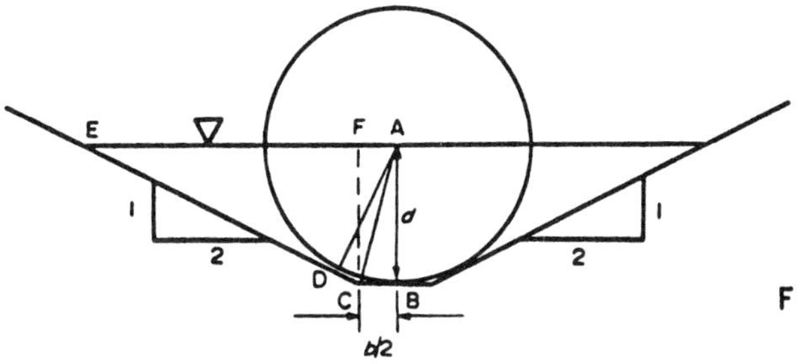

Fig. 1

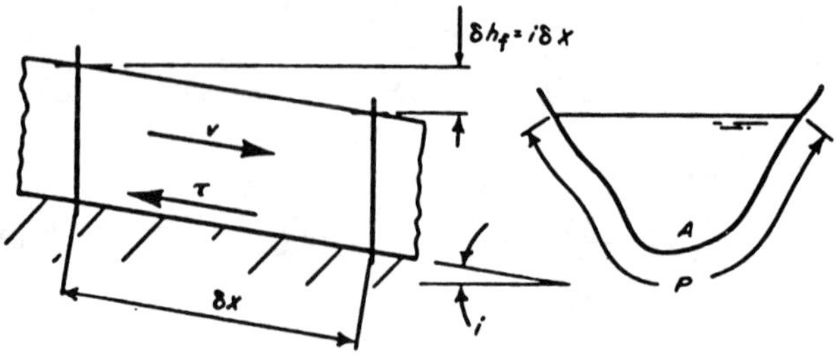

Fig. 2

Solution: The energy loss due to friction per unit weight δh_f over a length δx (see Fig. 2) must be exactly balanced by conversion of potential energy. Thus

$$\delta h_f = i\delta x$$

$$\therefore \quad i = \delta h_f/\delta x$$

but

$$\frac{dh_f}{dx} = \frac{fv^2}{2gm} = i$$

$$\therefore \quad v = \sqrt{(2g/f)}\sqrt{(mi)}$$

$$\therefore \quad v = C\sqrt{(mi)}$$

and

$$C = \sqrt{(2g/f)}$$

This approach links the Chezy C and the Darcy f. The flow in a channel is given by $Q = CA\sqrt{(mi)}$. Consider a trapezoidal channel for which the side slopes are 1 in s and the bottom width is 2b (Fig. 3). Then

$$A = 2bd + sd^2$$

$$P = 2b + 2d\sqrt{(1 + s^2)}$$

$$\therefore \quad P = A/d - sd + 2d\sqrt{(1 + s^2)}$$

$$dP/dd = 0 = -(A/d^2) - s + 2\sqrt{(1 + s^2)}$$

$$\therefore \quad (A/d) + sd = 2d\sqrt{(1 + s^2)}$$

and from before

$$A/d = 2b + sd$$

$$\therefore \quad b + sd = d\sqrt{(1 + s^2)}$$

This condition must be satisfied if discharge is to be maximum. It is now necessary to examine what this condition implies as regards the shape of the cross section. Referring to Fig. 3, $b + sd = AB$ and $d\sqrt{(1 + s^2)} = BC$ so the condition is that AB = BC.

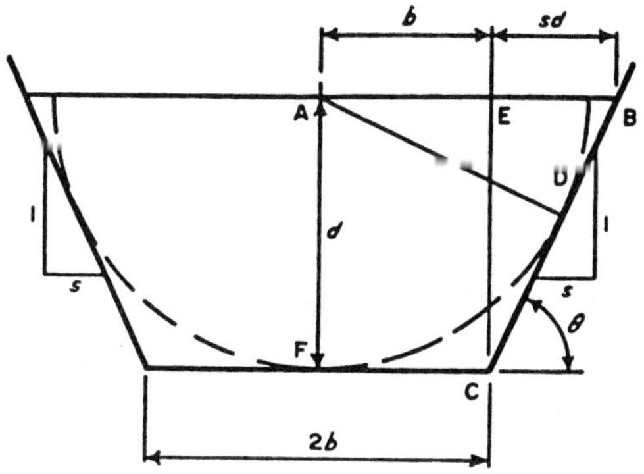

Fig. 3

If this is so then $\Delta ABD \equiv \Delta EBC$ as AB = BC, $\angle ADB = \angle BEC$ = 1 right angle and $\angle EBC$ is common to both.

$\therefore$ AD = EC and EC = AF

If AD = AF it must be possible to inscribe a semicircle within the channel cross section which will be tangential to all three sides and have its centre in the surface of the flow. An economic channel is defined by this condition.

The value of the hydraulic mean depth for an economic channel is always half the depth.

$$m = \frac{A}{P} = \frac{2bd + sd^2}{2b + 2d(1 + s^2)^{1/2}}$$

$$m = \frac{2bd + sd^2}{2b + 2(b + sd)} = \frac{2bd + sd^2}{2(2b + sd)} = \frac{d}{2}$$

In the case of a rectangular channel the economic channel breadth is equal to twice the depth.

For an economic section, a semicircle must be inscribable in the channel cross section. Under these circumstances, the hydraulic mean radius equals half the depth, d. The area A = (2b + 4d)d/2; m = d/2.

$$Q = AC\sqrt{(mi)} = A \times 70\ m^{2/3}i^{1/2}$$

$$15/(70i^{1/2}) = (b + 2d)d \times (d/2)^{2/3}$$

$$\Delta AED \equiv \Delta CFE$$

so

$$AE = CE$$

$$b/2 + 2d = \sqrt{[d^2 + (2d)^2]}$$

$$b = 2(\sqrt{5}d - 2d) = 0.472d$$

$$9.583 = 2.472d^2(d/2)^{2/3}$$

$$d = 1.977 \text{ m}$$

$$b = 0.472 \times 1.977 = 0.933 \text{ m}$$

• PROBLEM 11-17

Consider a rectangular flume 4.5 m wide, built of unplaned planks (n = 0.014), leading from a reservoir in which the water surface is maintained constant at a height of 1.8 m above the bed of the flume at entrance (see Fig. 1). The flume is on a slope of 0.001. The depth 300 m downstream from the head end of the flume is 1.20 m. Assuming an entrance loss of $0.2\ V_1^2/2g$, find the flow rate for the given conditions.

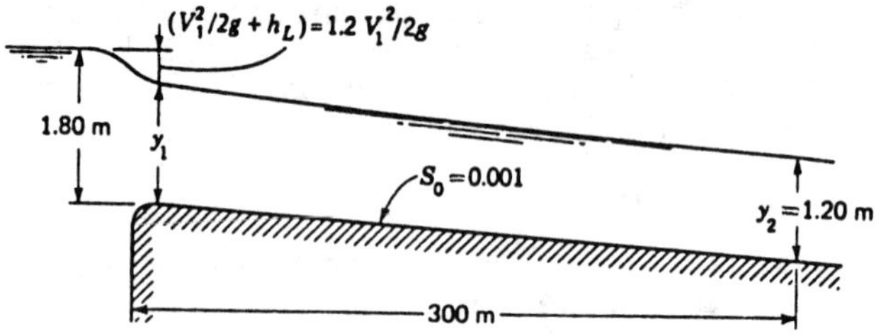

Fig. 1

Solution: The principal forces involved in flow in an open channel are inertia, gravity, hydrostatic force due to change in depth, and friction. The first three represent the useful kinetic and potential energies of the liquid, while the fourth dissipates useful energy into the useless kinetic energy of turbulence and eventually into heat because of the action of viscosity. Referring to Fig. 2, the total energy of the elementary volume of liquid shown is proportional to

$$H = z + y + \alpha \frac{V^2}{2g} \tag{1}$$

where z + y is the potential energy head above the arbitrary datum, and $\alpha V^2/2g$ is the kinetic energy head, V being the mean velocity in the section. Each term of the equation represents energy in foot-pounds per pound of fluid.

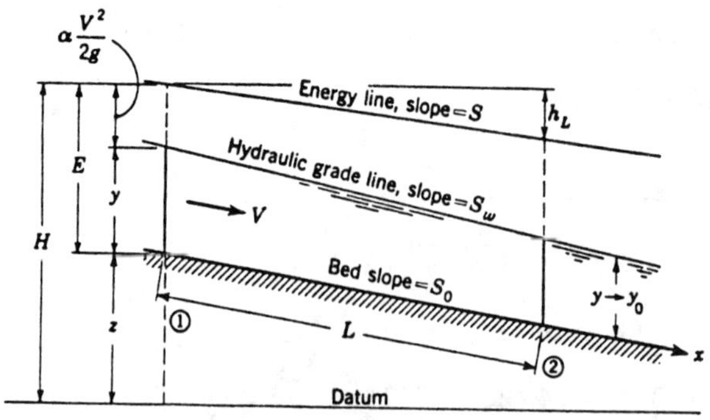

Fig. 2

The value of α is generally found to be higher in open channels than in pipes. It may range from 1.05 to 1.40, and in the case of a channel with an obstruction the value of α just upstream may be as high as 2.00 or even more. As the value of α is not known unless the velocity distribution is determined, it is often omitted, i.e. it is assumed to be unity.

Differentiating Eq. (1) with respect to x, the distance along the channel, the rate of energy dissipation is found to be (with $\alpha = 1$)

$$\frac{dH}{dx} = \frac{dz}{dx} + \frac{dy}{dx} + \frac{1}{2g}\frac{d(V^2)}{dx} \tag{2}$$

The slope of the energy line is defined as $S = -dH/dx$, while the slope of the channel bed is $S_0 = -dz/dx$, and the slope of the hydraulic grade line or water surface is given by $S_w = -dz/dx - dy/dx$.

The energy equation for steady flow between two sections (1) and (2) of Fig. 2 a distance L apart is

$$z_1 + y_1 + \alpha_1 \frac{V_1^2}{2g} = z_2 + y_2 + \alpha_2 \frac{V_2^2}{2g} + h_L \tag{3}$$

As $z_1 - z_2 = S_0 L$ and $h_L = SL$, the energy equation may also be written in the form (with $\alpha_1 = \alpha_2 = 1$)

$$y_1 + \frac{V_1^2}{2g} = y_2 + \frac{V_2^2}{2g} + (S - S_0)L \tag{4}$$

The Manning equation for uniform flow can be applied to nonuniform flow with an accuracy dependent on the length of the reach taken. Thus a long stream will be divided into several reaches of varying length such that the change in depth is roughly the same within each reach. Then, within a reach, the Manning formula gives

In SI units: $$S = \left(\frac{nV_m}{R_m^{2/3}}\right)^2 \tag{5}$$

where V_m and R_m are the means of the respective values at the two ends of the reach.

For a first approximate answer we shall consider the entire flume as one reach. The equations to be satisfied are

Energy at entrance:

$$y_1 + \frac{1.2V_1^2}{2g} = 1.80 \tag{6}$$

Energy equation (4) for the entire reach:

$$y_1 + \frac{V_1^2}{2g} = 1.20 + \frac{V_2^2}{2g} + (S - 0.001)L \qquad (7)$$

where S is given by Eq.:

$$S = \left(\frac{nV_m}{R_m^{2/3}}\right)^2 \qquad (8)$$

The procedure is to make successive trials of the upstream depth y_1. This determines corresponding values of V_1, q, V_2, V_m, R_m, and S. The trials are repeated until the value of L from Eq. (7) is close to 300 m. The solution is conveniently set in tabular form as follows:

Trial y_1, m	V_1, Eq. (6), m/s	$q = y_1 V_1$, m^2/s	$V_2 = q/1.20$, m/s	V_m, m/s	R_{h1}, m	R_{h2}, m	R_m, m	S, Eq. (c)	L, Eq. (7), m
1.50	2.22	3.33	2.78	2.50	0.90	0.78	0.89	0.00143	358
1.48	2.29	3.39	2.82	2.56	0.89	0.78	0.835	0.00163	226

Thus $y_1 \approx 1.49$ m and the flow rate $Q = qB \approx 3.36 \times 4.5 \approx 15.1\ m^3/s$.

• PROBLEM 11-18

Water flows under a sluice gate into a channel with a horizontal base 12 ft wide, at the rate of 300 ft³/sec. Calculate the stream depth h_1, assuming the flow to be one-dimensional and reversible, and given that the depth upstream of the gate, h_0, is 10 ft. Take gravitational acceleration to be 32.2 ft/sec².

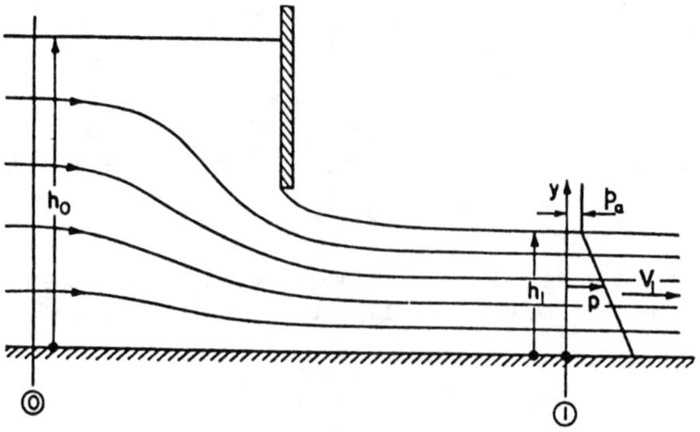

Fig. 1

Solution: Consider the situation illustrated in Fig. 1, in which a liquid flows steadily under a sluice-gate at one side of a reservoir into a horizontal channel having a rectangular cross-section of uniform width. We assume that the downstream conditions are such that the flow in the channel becomes uniform, and can be represented by fluid properties at a single station 1; the streamlines at this station are parallel straight lines and the pressure variation is the simple hydrostatic one, which we can write in the form

$$p - p_a = w(h_1 - y), \tag{1}$$

where p is the fluid pressure at a point y above the bottom of the channel, h_1 is the height of the surface, and w is the specific weight of the fluid. Since equation (1) can be rearranged to show that

$$\frac{p}{w} + y = \frac{p_a}{w} + h_1 = \text{constant for given } p_a,\ h_1 \text{ and } w,$$

it is clear that if the velocity is uniform at the value V_1 across the stream at station 1, then the flow there is one-dimensional in the sense that the Bernoulli constant has the same value for all streamlines. This, together with the assumptions that the flow is reversible and the velocity at station 0 is negligible, permits us to write Bernoulli's equation in the form

$$\frac{p_a}{w} + h_0 = \frac{p_a}{w} + h_1 + \frac{V_1^2}{2g},$$

i.e.

$$h_0 = h_1 + \frac{V_1^2}{2g}. \tag{2}$$

The volume flow rate per unit width of the channel is $V_1 h_1$, and if this quantity is denoted by q, equation (2) becomes

$$h_0 = h_1 + \frac{q^2}{2h_1^2 g} \tag{3}$$

$$\therefore \quad q^2 = h_1^2[2g(h_0 - h_1)]$$

This expression shows that q is zero when $h_1 = 0$ and when $h_1 = h_0$; by differentiating it with respect to h_1, keeping h_0 constant, it is possible to determine the condition for maximum discharge from a reservoir of given depth:

$$2q\frac{dq}{dh_1} = 2g(2h_1h_0 - 3h_1^2)$$

$$\therefore \quad \frac{dq}{dh_1} = \frac{g(2h_1h_0 - 3h_1^2)}{h_1[2g(h_0 - h_1)]^{1/2}}$$

$$= 0 \quad \text{when} \quad h_1 = \frac{2h_0}{3}.$$

By substituting back into equation (2) we find that at the maximum discharge condition

$$V_1 = (gh_1)^{1/2}$$

or

$$\frac{V_1^2}{gh_1} = 1. \tag{4}$$

Now $(gh)^{1/2}$ [$\equiv$ c, say] is the velocity of small-amplitude surface waves. It is frequently called the "critical velocity." Flows at velocities greater and less than the critical velocity have different characteristics, and c is somewhat analogous to the acoustic velocity a in compressible fluid flow. The group $V^2/gh(\equiv V^2/c^2)$, which is dimensionless, is called the Froude number. Froude numbers greater or less than unity therefore correspond to flows at velocities greater or less than c; these are called respectively "supercritical," or "shooting," and "subcritical," or "tranquil."

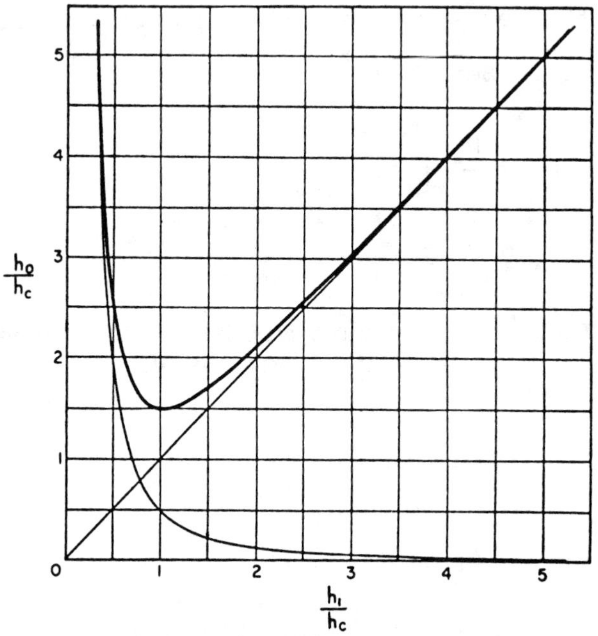

Fig. 2

A second way of determining which flow régime exists in a given channel is to compare the actual depth with the "critical depth" (h_c, say)--the depth that the stream would have for the same flow rate if moving at the critical velocity. At the critical conditions

$$q = ch_c$$

$$= g^{1/2}h_c^{3/2}$$

$$\therefore \quad h_c = \left(\frac{q^2}{g}\right)^{1/3}. \tag{5}$$

From equation (5) the critical depth is found,

$$h_c = \left(\frac{q^2}{g}\right)^{1/3} = \left[\left(\frac{300}{12}\right)^2 \times \frac{1}{32.2}\right]^{1/3} = 2.69 \text{ ft.}$$

Given $h_0 = 10$ ft, then $h_0/h_c = 3.72$.

By inspection of Fig. 2, values of 3.69 and 0.387 are obtained for h_1/h_c, giving $h_1 = 9.92$ ft and $h_1 = 1.04$ ft respectively. Of these two values the second, which corresponds, of course, to supercritical flow, is the more reading admissible; the stream would have to decelerate in order to reach the state corresponding to a stream depth of 9.92 ft, and the accompanying irreversibility would largely invalidate the basis of the solution. Consequently the only acceptable solution is

$$h_1 = 1.04 \text{ ft}$$

• PROBLEM 11-19

Determine the downstream depth in a 3-m-wide rectangular channel in which the channel bottom or invert drops 0.1 m. The flow rate is 5.5 m^3/s and the upstream depth 0.4 m.

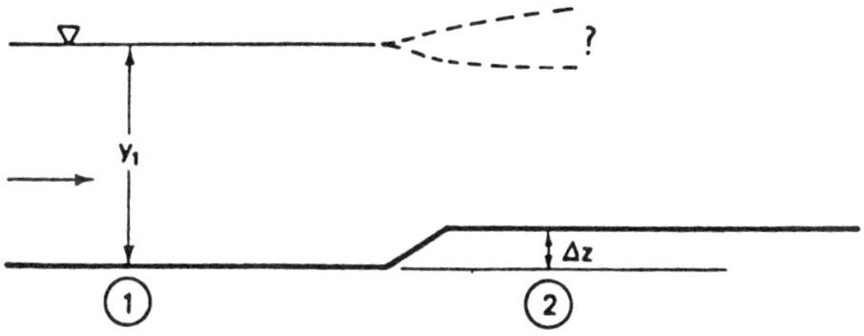

Fig. 1

Solution: Consider flow at depth y in a rectangular channel of constant width b. It is desired to determine what will happen to the depth or water surface elevation if it encounters a change in bottom elevation Δz as shown on Fig. 1.

Flow is from left to right, and assuming that section 2 is reasonably close to section 1, channel friction may be ignored in this type of analysis. Note that if flow at sections 1 and 2 is uniform (or nearly uniform) the pressure distribution vertically will be hydrostatic and the pressure and elevation heads may be replaced by the water surface elevation relative to a horizontal datum. Immediately,

$$y_1 + \frac{V_1^2}{2g} = (y_2 + \Delta z) + \frac{V_2^2}{2g}$$

Assuming that the flow rate is known, this expression can be put in terms of only one unknown, y_2, through use of continuity in its two-dimensional form,

$$q = y_1V_1 = y_2V_2$$

The result is

$$y_1 + \frac{q^2}{2gy_1^2} = \Delta z + y_2 + \frac{q^2}{2gy_2^2}$$

Although this equation can be solved by trial and error, it is a cubic equation and we cannot even say which is the correct root, as there will usually be two positive roots. To circumvent this impasse, we define a quality known as the specific energy, H_0. By definition the specific energy is the total head relative to the channel bottom. Thus, in general,

$$H_0 = y + \frac{V^2}{2g}$$

and in a rectangular channel,

$$H_0 = y + \frac{q^2}{2gy^2} \quad (1)$$

Note at this point that the values of specific energy at the two sections of Fig. 1 are

$$H_{0_1} = y_1 + \frac{q^2}{2gy_1^2}$$

and

$$H_{0_2} = y_2 + \frac{q^2}{2gy_2^2}$$

and that they are related, in this case, through the conventional energy equation as follows:

$$H_{0_1} = H_{0_2} + \Delta z$$

The two possible flow states for a given value of H_0 consist of a relatively deep flow (y_B) and low velocity (as indicated by $V_B^2/2g$) and a relatively shallow flow (y_A) and high velocity (V_A). The former is called subcritical flow and the latter supercritical flow.

Below a certain minimum value of specific energy, the flow is not possible. It is this minimum that will be considered next. If Eq. (1) is differentiated with respect to y and the result set equal to zero, the minimum point of the curve can be located. Namely,

$$\frac{dH_0}{dy} = 1 - \frac{q^2}{gy^3} = 0$$

We will refer to this depth as the critical depth y_c. Hence, upon solving,

$$y_c = \sqrt[3]{\frac{q^2}{g}}$$

With critical depth defined in this manner it will be noted that for

Subcritical flow: $y > y_c$

Supercritical flow: $y < y_c$

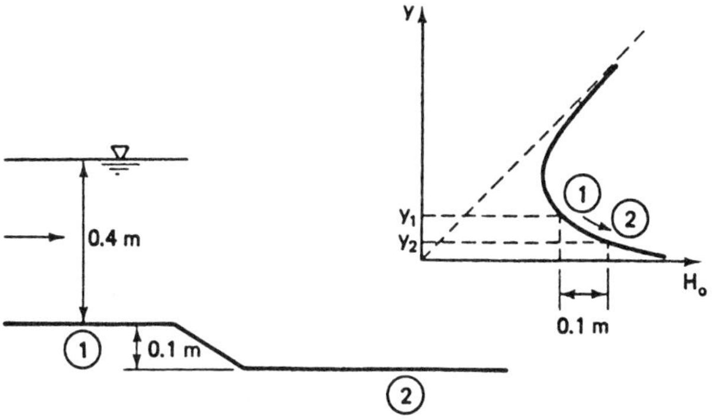

Fig. 2

The channel and specific energy diagram (not to same scale) are shown in Fig. 2. The discharge per unit width is

$$q = \frac{Q}{b} = \frac{5.5\ m^3/s}{3\ m} = 1.83\ (m^3/s)/m$$

and the critical depth is

$$y_c = \sqrt[3]{\frac{q^2}{g}}$$

$$= \sqrt[3]{\frac{(1.83\ m^2/s)^2}{9.81\ m/s^2}} = 0.70\ m$$

Since $y = 0.40\ m < 0.70\ m$, the flow is supercritical and y_1 is on the lower limb, as shown on the specific energy diagram. In passing from section 1 to section 2 the specific energy must increase by 0.1 m and hence y_2 must be less than y_1. Solving,

$$H_{0_1} = y_1 + \frac{q^2}{2gy_1^2} = (0.4) + \frac{(1.83)^2}{(2)(9.81)(0.4)^2}$$

$$= 1.467\ m$$

$$H_{0_2} = H_{0_1} + 0.1 \text{ m} = 1.567 \text{ m}$$

and finally

$$H_{0_2} = y_2 + \frac{q^2}{2gy_2^2}$$

or

$$1.567 = y_2 + \frac{(1.83)^2}{(2)(9.81)y_2^2} = y_2 + \frac{0.171}{y_2^2}$$

The downstream depth must lie in the range $0 < y_2 < y_1$. Upon solving by trial and error, the solution is found to be $y_2 = 0.38$ m.

• PROBLEM 11-20

A hydraulic jump occurs in a rectangular channel having a discharge of 30 cfs of water per foot width. The approach depth is 2.5 ft. Calculate the depth in the channel after the jump, as well as the energy dissipated per foot width. Assume density of water is 62.4 lbm/ft^3.

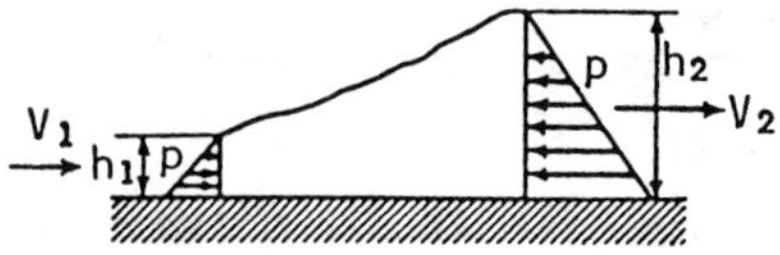

Solution: Consider the control volume as shown in the figure extending over the channel with b. Here it can be assumed that the friction forces are negligible due to the short length of channel over which the jump takes place. Furthermore, since we have taken the channel bed to be horizontal, the component of the gravity forces in the flow direction is zero. The momentum equation for one-dimensional flow

$$\sum F_x = \frac{\rho Q}{g_c}(V_2 - V_1)$$

becomes, using the average hydrostatic pressures acting on the cross sections as the only forces,

$$\frac{1}{2}\rho\frac{g}{g_c}h_1h_1b - \frac{1}{2}\rho\frac{g}{g_c}h_2h_2b = \frac{\rho Q}{g_c}(V_2 - V_1)$$

From the continuity equation, we have

$$Q = V_1h_1b = V_2h_2b$$

Substituting the expressions for $V_1 = Q/bh_1$ and $V_2 = Q/bh_2$ into the momentum equation results in

$$\frac{h_2^2 - h_1^2}{2} = \frac{Q^2}{b^2 g}\left(\frac{1}{h_1} - \frac{1}{h_2}\right) = \frac{Q^2}{b^2 g}\,\frac{h_2 - h_1}{h_1 h_2}$$

Divide by $h_2 - h_1$, multiply by 2, and rearrange terms to obtain the quadratic equation for h_2:

$$h_2^2 + h_1 h_2 - \frac{2Q^2}{b^2 g h_1} = 0$$

Finding the root of this equation gives the relation between the two depths of the jump and the channel discharge:

$$h_2 = -\frac{h_1}{2} + \sqrt{\frac{2Q^2}{b^2 g h_1} + \frac{h_1^2}{4}} \qquad (1)$$

where the negative sign in front of the radical has been deleted, for a negative h_2 has no physical meaning. The value of the downstream velocity V_2 can now be evaluated from

$$V_2 = \frac{Q}{bh_2}$$

The energy loss (ΔE) across the hydraulic jump can be calculated using a modified Bernoulli equation taken along the free surface:

$$\frac{g}{g_c} h_1 + \frac{V_1^2}{2g_c} = \frac{g}{g_c} h_2 + \frac{V_2^2}{2g_c} + \Delta E$$

whence

$$\Delta E = -\frac{g}{g_c}(h_2 - h_1) + \frac{V_1^2 - V_2^2}{2g_c}$$

where ΔE is given in ft-lb_f/lb_m.

By using the continuity equation and the relationship between h_1 and h_2 given in (1), we obtain, after appropriate simplification, an expression for the energy loss at the jump in terms of the liquid depths h_1 and h_2:

$$\Delta E = \frac{g}{g_c}\,\frac{(h_2 - h_1)^3}{4h_1 h_2} \qquad (2)$$

For a given discharge Q(cfs), the rate of energy loss can be computed:

$$\rho Q \Delta E \quad (\text{ft-lb}_f/\text{sec})$$

The channel depth after the jump is obtained from (1):

$$h_2 = -\frac{2.5}{2} + \sqrt{\frac{2 \times (30)^2}{32.2 \times 2.5} + \frac{(2.5)^2}{4}} = 3.65 \text{ ft}$$

The energy dissipated per pound mass of water is obtained from (2):

$$\Delta E = \frac{g}{g_c} \frac{(3.65 - 2.5)^3}{4 \times 2.5 \times 3.65} = 0.0417 \text{ ft-lb}_f/\text{lb}_m$$

For a discharge of 30 cfs, the rate of energy loss therefore becomes

$$\rho Q \Delta E = 62.4 \times 30 \times 0.0417 = 78 \text{ ft-lb}_f/\text{sec}$$

The velocity

$$V_1 = \frac{Q}{bh_1} = \frac{30}{2.5} = 12 \text{ fps}$$

while velocity

$$V_2 = \frac{Q}{bh_2} = \frac{30}{3.65} = 7.52 \text{ fps}$$

• PROBLEM 11-21

Analyze the water-surface profile in a long rectangular channel with concrete lining (n = 0.013). The channel is 10 ft wide, the flow rate is 400 cfs, and there is an abrupt change in channel slope from 0.0150 to 0.0016. Find also the horsepower loss in the jump.

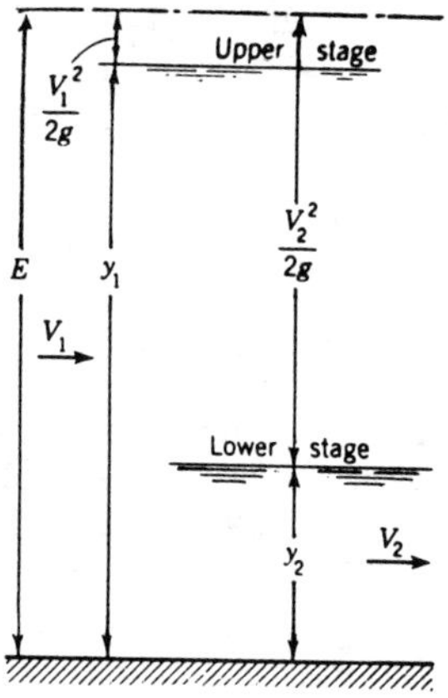

Fig. 1

Solution: Using the Manning formula, given by

$$Q(\text{cfs}) = \frac{1.49}{n} AR_h^{2/3} S^{1/2}$$

$$400 = \frac{1.49}{0.013} (10y_{0_1}) \left(\frac{10y_{0_1}}{10 + 2y_{0_1}}\right)^{2/3} (0.015)^{1/2}$$

By trial, $y_{0_1} = 2.17$ ft (normal depth on upper slope)

Using a similar procedure, the normal depth y_{0_2} on the lower slope is found to be 4.80 ft. Totally separate from the concept of energy gradient, or energy difference, between two sections of a stream is the matter of the energy at a single section with reference to the channel bed (Fig. 1). This is called the specific energy at that section, and its value is given by the following equation:

$$E = y + \frac{V^2}{2g} \tag{1}$$

In this section, for simplicity, we shall confine the discussion to wide rectangular channels.

If q denotes the flow per unit width of a wide rectangular channel, then $V = Q/A = qb/yb = q/y$ and

$$E = y + \frac{q^2}{2gy^2} \tag{2}$$

or

$$q = y\sqrt{2g(E - y)} \tag{3}$$

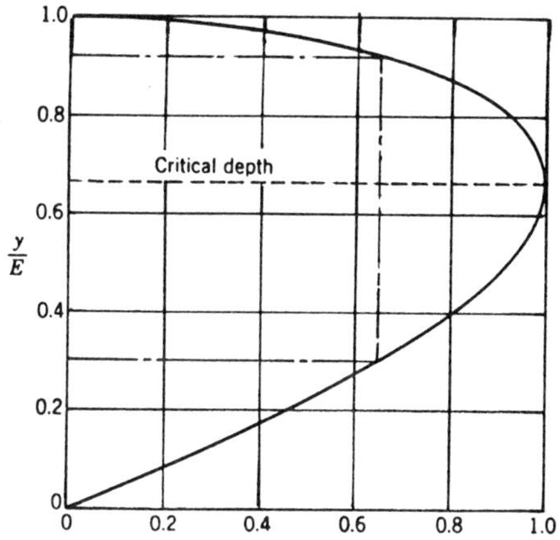

Fig. 2

This is the equation of the curve which is shown in dimensionless form in Fig. 2. It is seen that the maximum discharge for a given specific energy occurs when the depth is between 0.6E and 0.7E. This may be established more exactly by differentiating Eq. (3) with respect to y and equating to zero. Thus

$$\frac{dq}{dy} = \sqrt{2g}\left(\sqrt{E - y} - \frac{1}{2}\,\frac{y}{\sqrt{E - y}}\right) = 0$$

from which

$$y_c = \frac{2}{3}E \tag{4}$$

where y_c is called the critical depth for the given specific energy. Substituting from Eq. (4) into Eq. (3):

$$q_{max} = \sqrt{g}\left(\frac{2}{3}E\right)^{3/2} = \sqrt{gy_c^3} \qquad (5)$$

or

$$y_c = \left(\frac{q^2}{g}\right)^{1/3} \qquad (6)$$

$$y_c = \left(\frac{q^2}{g}\right)^{1/3} = \left[\frac{\left(\frac{400}{10}\right)^2}{32.2}\right]^{1/3} = 3.68 \text{ ft}$$

Thus flow is supercritical ($y_{0_1} < y_c$) before break in slope and subcritical ($y_{0_2} > y_c$) after break, so a hydraulic jump must occur.

For channels on a gradual slope (i.e., less than about 3°) the gravity component of the weight is relatively small and may be neglected without introducing significant error. The friction forces acting are negligible because of the short length of channel involved and because the shock losses are large in comparison. Applying Newton's second law to the element of fluid contained between sections 1 and 2 of Fig. 3 we have

$$\sum F_x = \gamma h_{c_1} A_1 - \gamma h_{c_2} A_2 = \frac{\gamma}{g} Q(V_2 - V_1)$$

which can be reordered to give

$$\frac{\gamma}{g} QV_1 + \gamma h_{c_1} A_1 = \frac{\gamma}{g} QV_2 + \gamma h_{c_2} A_2$$

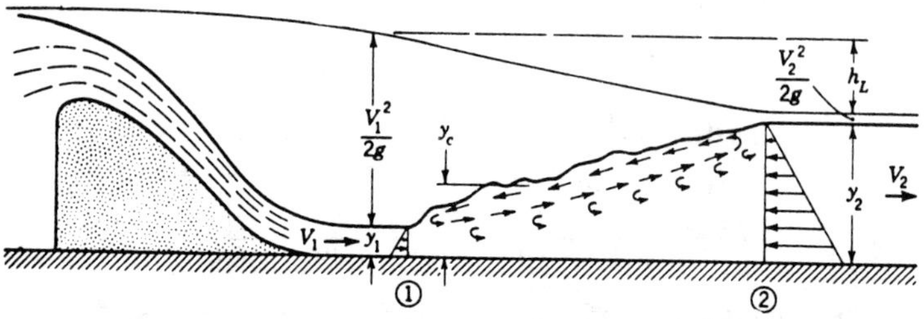

Fig. 3

This states that the momentum plus the pressure force on the cross-sectional area is constant, or dividing by γ and observing that $V = Q/A$,

$$F_m = \frac{Q^2}{Ag} + Ah_c = \text{constant}$$

In the case of a rectangular channel, this reduces for a unit width to:

$$f_m = \frac{q^2}{y_1 g} + \frac{y_1^2}{2} = \frac{q^2}{y_2 g} + \frac{y_2^2}{2} \qquad (7)$$

A curve of values of f_m for different values of y is

plotted to the right of the specific-energy diagram shown in Fig. 4. Both curves are plotted for the condition of 2 cfs/ft of width. As the loss of energy in the jump does not affect the "force" quantity f_m, the latter is the same after the jump as before, and therefore any vertical line intersecting the f_m curve serves to locate two conjugate depths y_1 and y_2. These depths represent possible combinations of depth that could occur before and after the jump.

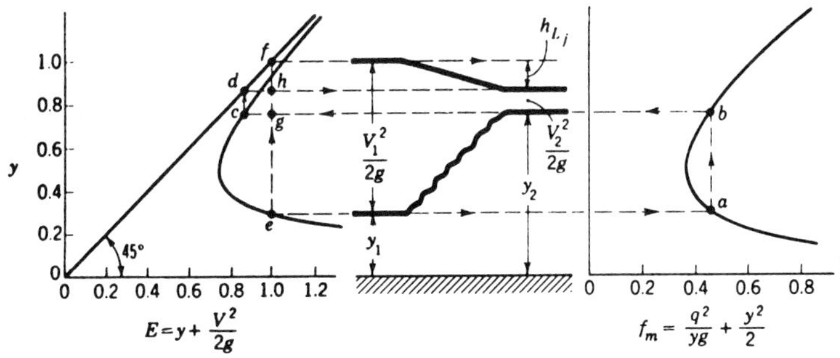

Fig. 4

Thus, in Fig. 4, the line for the initial water level y_1 intersects the f_m curve at a as shown, giving the value of f_m, which must be the same after the jump. The vertical line ab then fixes the value of y_2. This depth is then transposed to the specific-energy diagram to determine the value cd of $V_2^2/2g$. The value of $V_1^2/2g$ is the vertical distance ef, and the head loss h_{L_j} caused by the jump is the drop in energy from 1 to 2. Or

$$h_{L_j} = \left(y_1 + \frac{V_1^2}{2g}\right) - \left(y_2 + \frac{V_2^2}{2g}\right)$$

On Fig. 4 the head loss in the hydraulic jump is given by either the horizontal distance hd or the vertical distance fh.

When the rate of flow and the depth before or after the jump are given, it is seen that Eq. (7) becomes a cubic equation when solving for the other depth. This may readily be reduced to a quadratic, however, by observing that $y_2^2 - y_1^2 = (y_2 + y_1)(y_2 - y_1)$ and substituting the known depth in the resulting expression:

$$\frac{q^2}{g} = y_1 y_2 \frac{y_1 + y_2}{2}$$

This equation can be rearranged to give an explicit expression for the depth after jump in a rectangular channel,

$$y_2 = \frac{y_1}{2}\left(-1 + \sqrt{1 + \frac{8q^2}{gy_1^3}}\right) \qquad (8)$$

Applying Eq. (7) to determine the depth conjugate to the 2.17-ft (upper-slope) normal depth, we get

$$y_2' = \frac{2.17}{2}\left\{-1 + 1 + \left[\frac{8(40)^2}{32.2(2.17)^3}\right]^{1/2}\right\}$$

$$= 5.75 \text{ ft}$$

Therefore the depth conjugate to the upper-slope normal depth of 2.17 ft is 5.75 ft. This jump cannot occur because the normal depth y_{0_2} on the lower slope is less than 5.75 ft.

Applying Eq. (7) to determine the depth conjugate to the 4.80-ft (lower-slope) normal depth, we get

$$4.8 = \frac{y_1'}{2}\left[-1 + 1 + \left(\frac{8(40)^2}{32.2y_1^3}\right)^{1/2}\right]$$

$$y_1' = 2.76 \text{ ft}$$

The lower conjugate depth of 2.76 ft will occur downstream of the break in slope.

The Manning equation for uniform flow can be applied to nonuniform flow with an accuracy dependent on the length of the reach taken. Thus a long stream will be divided into several reaches of varying length such that the change in depth is roughly the same within each reach.

Then, within a reach, the Manning formula gives

$$S = \left(\frac{nV_m}{1.49R_m^{2/3}}\right)^2 \qquad (9)$$

where V_m and R_m are the means of the respective values at the two ends of the reach. With S_0 and n known and the depth and velocity at one end of the reach given, the length L to the end corresponding to the other depth can be computed from

$$L = \frac{(y_1 + V_1^2/2g) - (y_2 + V_2^2/2g)}{S - S_0} \qquad (10)$$

or

$$L = \frac{E_1 - E_2}{S - S_0}$$

$$E_1 = 2.17 + \frac{(400/2.17)^2}{64.4} = 7.53 \text{ ft}$$

$$E_2 = 2.76 + \frac{(400/27.6)^2}{64.4} = 6.02 \text{ ft}$$

$$V_m = \frac{1}{2}\left(\frac{400}{21.7} + \frac{400}{27.6}\right) = 16.45 \text{ fps}$$

$$R_m = \frac{1}{2}\left(\frac{21.7}{14.34} + \frac{27.6}{15.52}\right) = 1.645 \text{ ft}$$

From Eq. (9),

$$S = \left[\frac{(0.013)(16.45)}{1.49(1.645)^{2/3}}\right]^2 = 0.0107$$

Finally,

$$L = \frac{7.53 - 6.02}{0.0107 - 0.0016} = 165 \text{ ft}$$

Thus depth on the upper slope is 2.17 ft; downstream of break the depth increases gradually to 2.76 ft over a distance of approximately 165 ft; then a hydraulic jump occurs to depth 4.80 ft; downstream of jump the depth remains constant at 4.8 ft.

$$\text{HP loss} = \frac{\gamma Q h_{L_j}}{550}$$

$$h_{L_j} = E_1 - E_2 = 1.51 \text{ ft}$$

$$\text{HP loss} = \frac{62.4(400)(1.51)}{550} = 68$$

CHAPTER 12

COMPRESSIBLE FLOW

Basic Attacks and Strategies for Solving Problems in this Chapter. See pages 678 to 806 for step-by-step solutions to problems.

There are several flow phenomena, such as choking, shock waves, etc., which occur only when a fluid flow is highly compressible. Compressibility becomes important when the Mach number, $M = V/a$, becomes greater than about 0.3. For a perfect gas, the speed of sound

$$a = (\alpha RT)^{1/2},$$

where α (or k in alternate notation) is the ratio of specific heats C_p/C_v,

R is the gas constant, and

T is the *absolute* temperature.

When M is less than one, the flow is subsonic, while supersonic flows are those with Mach numbers greater than one.

Most compressible flow problems encountered by engineers involve the flow of a gas in a duct. Of these, three different simplifications enable the analysis of three primary categories of compressible duct flow: isentropic flow in a duct of a changing area, adiabatic flow in a duct of a constant area (with friction), and frictionless flow in a constant area duct with heat transfer. When none of these simplifications can be made, the problem is much more complicated. Isothermal duct flow with friction is one case where there is heat transfer as well as frictional effects, but this can be analyzed. (See Problem 12.22 for a good discussion of this type of flow.)

For the case of adiabatic flow in a constant area duct with friction, the stagnation enthalpy of the fluid must remain constant since no energy is added and no work is done. This leads to the rather unique result that the Mach number always approaches unity towards the end of the duct. This applies to both subsonic flow (where the Mach number will increase towards one) and supersonic flow (where M will decrease towards one). Problems of this type are attacked by utilizing L^*, the *sonic length*, defined as the duct length required to develop from some initial Mach number to $M = 1$. The dimensionless parameter fL^*/D is tabulated as a function of the Mach number (see, for example, Table 1 of Problem 12.9), where f is the Darcy friction factor obtainable from the Moody Chart (see the discussion in Chapter 7) and D is the diameter of a round duct or the hydraulic diameter of a non-round duct. For cases where the duct is not long enough to reach sonic conditions at its exit, the relationship between the duct length and the Mach numbers M_1 and M_2 at the inlet and outlet of the duct respectively is

$$\frac{fL}{D} = \left(\frac{fL^*}{D}\right)_{M_1} - \left(\frac{fL^*}{D}\right)_{M_2}. \tag{1}$$

Problems 12.9 and 12.22 demonstrate the use of this equation.

Constant area duct flow problems, where friction can be ignored but heat is added or subtracted, must be attacked from the integral (control volume) conservation laws of mass, momentum, and energy, where heat transfer rate $\dot{Q}$ appears in the energy equation. Ratios of temperatures, stagnation temperatures, pressures, stagnation pressures, and velocities can be found as functions of the Mach number and are tabulated. These tables aid in the solution as illustrated in Problem 12.27.

The most common category of flows encountered is adiabatic, isentropic flow in a duct of a varying area. Thermodynamic relationships, combined with the integral conservation laws of mass and energy, lead to expressions of temperature, pressure, and density ratios as functions of the Mach number. These are tabulated in the form T/T_o, p/p_o, ρ/ρ_o, and A/A^* as functions of M, where the subscript o denotes stagnation conditions and the asterisk denotes the value at $M = 1$. The only way to attain supersonic flow in a duct of this kind is by first passing through a converging section, then a throat, followed by a diverging section. The flow upstream of the throat will be subsonic and the flow downstream may reach supersonic conditions if the pressure drop is sufficient. $M = 1$ can only occur at the throat, and the throat area is then defined as A^*. If the flow remains isentropic, one can find all desired quantities for a given A/A^* by using the isentropic tables or equations to find M, and then using M to find all other quantities.

If the downstream pressure is not low enough to attain supersonic flow throughout the entire length of the diverging nozzle section, a normal shock wave will form in the diverging section. A normal shock leads to a sudden rise in temperature and pressure, and the flow abruptly changes from supersonic to subsonic. Tables are available for the ratios of temperature, pressure, etc., across a normal shock wave. These, combined with the isentropic tables, are used to find the Mach numbers upstream and downstream of the shock. The flow is assumed to be isentropic everywhere except across the shock where entropy increases significantly, resulting in a great loss of stagnation pressure.

Oblique shocks are also possible. Problem 12.39 provides a good discussion of these.

Step-by-Step Solutions to Problems in this Chapter, "Compressible Flow"

ACOUSTIC VELOCITY AND MACH NUMBER

• PROBLEM 12-1

Calculate the value of the acoustic velocity in

(a) air at 32°F (γ = 1.40, R = 53.3 ft lbf/lbm °R),

(b) air at 2000°R (γ = 1.32, R = 53.3 ft lbf/lbm °R),

(c) steam at 480°F, 400 lbf/in^2, using the information from the Table and

(d) water at 60°F (adiabatic-reversible bulk modulus $\rho(\partial p/\partial \rho)_{ad,rev} = 3.13 \times 10^5$ lbf/in^2, density ρ = 62.4 lbm/ft^3).

Extract from steam tables — TABLE

Pressure (lbf/in²)		Temperature (°F) 470	480	490
395	Specific volume, v (ft³/lbm)	1·2385	1·2607	1·2824
	Specific enthalpy, h (Btu/lbm)	1224·8	1232·0	1239·0
	Specific entropy, s (Btu/lbm °R)	1·5079	1·5156	1·5230
400	v	1·2205	1·2426	1·2641
	h	1224·0	1231·3	1238·3
	s	1·5058	1·5135	1·5210
405	v	1·2030	1·2249	1·2463
	h	1223·2	1230·5	1237·6
	s	1·5037	1·5115	1·5190

Solution: Consider a planar pressure wave; it moves with velocity a in the direction perpendicular to its own plane. The Figure shows the conditions on either side of the wave, relative to a control surface moving with it. The small change in pressure is accompanied by changes in the other properties ρ, a, etc.

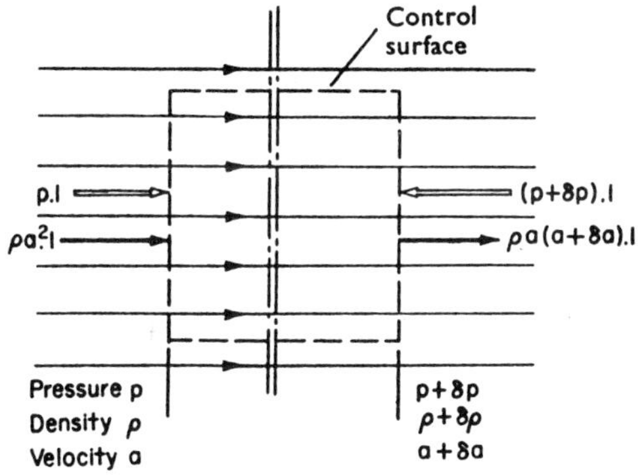

Applying the mass-continuity equation to the control volume shown, we have, taking the inlet and outlet areas to be unity,

$$\rho . a . 1 = (\rho + \delta\rho)(a + \delta a).1,$$

$$\therefore \quad a\,\delta\rho + \rho\,\delta a = 0. \tag{1}$$

The force-momentum relation, applied in the flow direction, is

$$-g_c\,\delta p.1 = \rho . a.1(a + \delta a - a),$$

$$g_c\,\delta p + \rho a\,\delta a = 0. \tag{2}$$

Combining equations 1 and 2 we have, in the limit,

$$a^2 = g_c\left(\frac{\partial p}{\partial \rho}\right)_{ad,rev.} \tag{3}$$

In the derivation of Eq. 3 no assumption has been made about the character of the fluid except that it is compressible. Since we have already assumed that the pressure wave has a negligible magnitude, it follows that its passage through the fluid is accomplished without irreversibility or heat transfer; this accounts for the suffix to the differential, indicating that it is the rate of change of pressure with density in an adiabatic and reversible process.

When the relationships between fluid properties can be expressed in algebraic form, equation 3 can be rewritten in more explicit form. For example, in the case of a perfect gas, for an adiabatic, reversible process,

$$\frac{p}{\rho^{\gamma}} = \text{constant},$$

from which we see that

$$\left(\frac{\partial p}{\partial \rho}\right)_{ad,rev} = \frac{\gamma p}{\rho} = \gamma RT.$$

By substituting these expressions into equation 3, we obtain the relations

$$a^2 = \frac{\gamma p g_c}{\rho} = \gamma R g_c T. \qquad (4)$$

Since the process undergone by the gas produces a vanishingly small change of state, equation 4 applies to a "semi-perfect" gas also, provided that the value of γ appropriate to the local gas conditions is used.

Over a wide range of temperatures, air behaves as a semi-perfect, rather than a perfect, gas; it is therefore necessary to use appropriate values of γ when calculating a.

(a) Using equation 4, for air at 32°F, we obtain

$$a = \gamma (R g_c T)^{1/2}$$

$$= [1.40 \times 53.3 \times 32.2 \times (460 + 32)]^{1/2} \text{ ft/sec} =$$

$$1087 \text{ ft/sec}.$$

(b) Similarly, for air at 2000°R,

$$a = (1.32 \times 53.3 \times 32.2 \times 2000)^{1/2} \text{ ft/sec} = 2128 \text{ ft/sec}.$$

(c) At the conditions specified, steam does not obey the ideal gas rule, and it is necessary to refer back to equation 3. For convenience, this can be written in terms of specific volume v instead of density $\rho (v \equiv 1/\rho)$:

$$a^2 = g_c\left(\frac{\partial p}{\partial \rho}\right)_{ad,rev} = - g_c v^2 \left(\frac{\partial p}{\partial v}\right)_{ad,rev}$$

The specific volumes at the beginning and end of an adiabatic, reversible process between 395 lbf/in^2 and 405 lbf/in^2, which passes through the specified state (480°F, 400 lbf/in^2) can be found by interpolation among the tabulated values; they are 1.2546 ft^3/lbm and 1.2306 ft^3/lbm respectively. The mean value of $(\partial p/\partial v)_{ad,rev}$ in this range is, therefore,

$$\left(\frac{\partial p}{\partial v}\right)_{ad,rev} = \left(\frac{\Delta p}{\Delta v}\right)_{ad,rev}$$

$$= \frac{1440}{1.2306 - 1.2546} \text{ lbf lbm/ft}^3$$

$$= - 60{,}000 \text{ lbf lbm/ft}^3.$$

Then by substituting into equation 3, we find

$$a = \left[- g_c v^2 \left(\frac{\partial p}{\partial v}\right)_{ad,rev}\right]^{1/2}$$

$$= [32.2 \times (1.2426)^2 \times 60{,}000]^{1/2} \text{ ft/sec} = 1727 \text{ ft/sec}.$$

(d) In a fluid of precisely constant density, the acoustic velocity would be infinite; all real liquids are, however, compressible, and in consequence the adiabatic-reversible bulk modulus, for which the definition is given above, also has a finite value. From the data we find that

$$\left(\frac{\partial p}{\partial \rho}\right)_{ad,rev} = \frac{3.13 \times 10^5 \times 144}{62.4} \text{ lbf ft/lbm} = 7.228 \times 10^5 \text{ lbf ft/lbm}.$$

The corresponding value of a can be calculated from equation 3:

$$a = (32.2 \times 7.228 \times 10^5)^{1/2} \text{ ft/sec} = 4823 \text{ ft/sec}.$$

a. What is the speed of sound in water at 20°C and at atmospheric pressure?

b. What is the speed of sound in air at 20°C at sea level and at an altitude where the pressure is less than that at sea level? Refer to Table 1 for gas properties.

c. Calculate the speed of sound in a mixture of liquid and tiny gas bubbles. (Results valid for nonresonant sound frequencies.) Let subscripts m, 1, and g refer to property values for the mixture, liquid, and gas, respectively.

TABLE 1
TYPICAL PROPERTIES OF GASES AT ROOM TEMPERATURE

GAS	k	SI UNITS (m N/kg °K) R	c_p
Air	1.40	287.1	1,005
Helium	1.66	2077	5,224
Hydrogen	1.40	4124	14,434
Methane	1.31	518	2,190
Xenon	1.66	63.3	159

Solution: a. Fluids may be deformed by viscous shear or compressed by an external pressure applied to a volume of fluid. All fluids are compressible by this method, liquids to a much smaller degree, however, than gases.

The compressibility is defined in terms of an average bulk modulus of elasticity

$$\bar{K} = -\frac{p_2 - p_1}{(\forall_2 - \forall_1)/\forall_1} = \frac{\Delta p}{\Delta\forall/\forall} \quad (1)$$

and the elastic modulus equals the absolute pressure times the ratio of specific heat capacities ($k = c_p/c_v = 1.4$ for air) during an isentropic compression.

The bulk modulus of elasticity K is of interest in acoustics as well as in fluid mechanics. The velocity of sound in any medium is

$$c = \sqrt{\frac{K}{\rho}} \quad (2)$$

and for a gas, sound waves are transmitted essentially isentropically, so that the velocity of sound in a perfect gas is

$$c = \sqrt{\frac{kp}{\rho}} = \sqrt{kRT} \quad (3)$$

TABLE 2

DENSITY OF WATER

TEMPERATURE (°C)	DENSITY ρ (kg/m^3)
0	999.8
5	999.9
10	999.7
15	999.1
20	998.2
25	997.1
30	995.6
35	994.1
40	992.2
45	990.2
50	988.1
55	985.7
60	983.2
65	980.6
70	977.8
75	974.9
80	971.8
85	968.6
90	965.3
95	961.9
100	958.4

Determined by applying the continuity and momentum principles in Eq. (2), $K = 2.18 \times 10^9$ Pa and $\rho = 998$ kg/m^3 from Table 2 so that

$$c = \sqrt{(2.18 \times 10^9)/998} = 1478 \text{ m/s}$$

b. Equation (3) indicates that the speed of sound is independent of pressure, and thus

$$c = \sqrt{kRT} = \sqrt{(1.4)(287.1)(293)} = 343 \text{ m/s}$$

c. For a nonresonant mixture, sound is considered to be propagated at constant temperature at low gas concentrations, and the sound speed is given by

$c_m = \sqrt{K_m/\rho_m}$. If x is the proportion of gas by

volume, the density of the mixture is

$$\rho_m = x\rho_g + (1 - x)\rho_1$$

and the elastic modulus K_m for the mixture is given by

$$\frac{1}{K_m} = \frac{x}{K_g} + \frac{1 - x}{K_1}$$

and thus

$$c_m = \sqrt{\frac{p_g K_1}{[xK_1 + (1 - x)p_g][x\rho_g + (1 - x)\rho_1]}}$$

A 20-cm-diameter round-nosed projectile whose drag coefficient is shown in the Figure travels at 600 m/s through the standard atmosphere at an altitude of 6,000 m. Find the drag. Assume $\gamma = 1.4$.

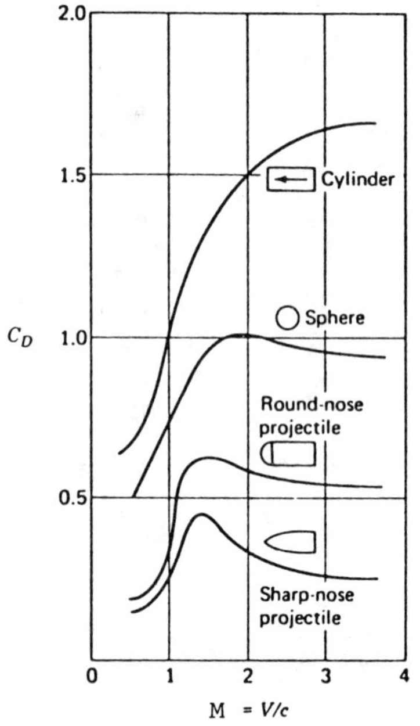

Solution: The acoustic velocity is given by $c = \sqrt{kRT}$.

From the Table, the temperature is -24°C, or 249 K.

TABLE 1

Altitude, km	Temp, °C	Pressure, kN/m², abs	Specific weight γ, N/m³	Density ρ, kg/m³	Viscosity μ × 10⁵, N·s/m²
0	15.0	101.33	12.01	1.225	1.79
2	2.0	79.50	9.86	1.007	1.73
4	−4.5	60.12	8.02	0.909	1.66
6	−24.0	47.22	6.46	0.660	1.60
8	−36.9	35.65	5.14	0.526	1.53
10	−49.9	26.50	4.04	0.414	1.46
12	−56.5	19.40	3.05	0.312	1.42
14	−56.5	14.20	2.22	0.228	1.42
16	−56.5	10.35	1.62	0.166	1.42
18	−56.5	7.57	1.19	0.122	1.42
20	−56.5	5.53	0.87	0.089	1.42
25	−51.6	2.64	0.41	0.042	1.45
30	−40.2	1.20	0.18	0.018	1.51

Then with $\gamma = 1.4$ and $R \cong 287\ m^2/(s^2)(K)$.

$$c = \sqrt{1.4 \times 287 \times 249} = 317\ m/s$$

and so

$$M = \frac{600}{317} = 1.89$$

From the Fig., $C_D = 0.62$

$$\frac{p}{\rho} = RT \qquad \rho = \frac{p}{RT}$$

$$\rho = \frac{47.22\ kN/m^2,abs}{[287\ m^2/(s^2)(K)]249K} = 0.00066\ \frac{kN.s^2}{m^4} = 0.66\ kg/m^3$$

Now using the equation for the total drag force

$$F_D = C_D \rho \frac{V^2}{2} A$$

$$F_D = 0.62 \times 0.66 \times \frac{(600)^2}{2} \times \frac{\pi(0.20)^2}{4} = 2,310\ N$$

• PROBLEM 12-4

Air ($k = 1.4$, $M_W = 29$) flows at a velocity of 224 fps under a pressure of 14.7 psia. If the termperature of the air is 17°C, at what Mach number does the flow take place?

Solution: Sonic velocity is defined as the speed with which a small pressure disturbance is transmitted through a fluid medium. The algebraic form of the definition is

$$\frac{E}{\rho} = \frac{dp}{d\rho}$$

where E is the bulk modulus for the fluid (analogous to the Young's modulus of elasticity for solid rods), and ρ as its density.

The square root of the modulus-density ratio, i.e. the term $\sqrt{(E/\rho)}$, has the dimension of velocity, and it can be proved that this is the sonic velocity, so that

$$v_s = \sqrt{\frac{E}{\rho}}$$

or

$$v_s = \sqrt{\frac{dp}{d\rho}} \qquad (1)$$

Equation (1) shows that the sonic velocity is inversely proportional to the ratio $d\rho/dp$ which represents the rate of change of density with pressure. This ratio also defines an important property of the fluid, namely its compressibility. For an isentropic expansion

$$pV^k = C \qquad (2)$$

where C and k are constants, the latter being the ratio of the specific heats at constant pressure and constant volume, respectively.

Since the specific volume $V = 1/\rho$, then from eq. (2)

$$p = C\rho^k \qquad (3)$$

Presenting this equation in a differential form

$$\frac{dp}{d\rho} = kC\rho^{k-1}$$

Substituting in this equation, from eq. (3), for

$$C = \frac{p}{\rho^k}$$

$$\frac{dp}{d\rho} = k\left(\frac{p}{k}\right)(\rho^{k-1})$$

Simplifying

$$\frac{dp}{d\rho} = k\frac{p}{\rho}$$

or

$$\frac{dp}{d\rho} = kpV$$

Substituting for $dp/d\rho$ from this equation in eq. (1)

$$v_s = \sqrt{(kpV)} \qquad (4)$$

At 17°C, and normal pressure, the specific volume of air is

(Note: At STP, the specific volume of an ideal gas is 359 ft^3/pound mole.

let

$$V = \frac{V_m}{M_W} \qquad V_m: \text{ specific volume on a mole basis}$$

M_W: molecular weight

V: specific volume)

$$V = \frac{(359)(273 + 17)}{(29)(273)}$$

$$V = 13.15 \text{ ft}^3/\text{lb}$$

$$p = (14.7)(144)(32.2) = 68{,}170 \text{ poundal/ft}^2$$

Using eq. (4)

$$v_s = \sqrt{\{(1.4)(68{,}170)(13.15)\}}$$

$$v_s = 1120 \text{ fps}$$

(It will be noted that the NACA standard sonic velocity of air at 59°F is 1117 fps.)

The Mach number, M, is given by the equation

$$M = \frac{v}{v_s} = \frac{224}{1120}$$

$$M = 0.2$$

• PROBLEM 12-5

An airflow at M = 3 enters a diverging passage (Fig. 1). The passage has an exit-to-inlet area ratio A_e/A_i of 3. Determine the exit-to-inlet pressure ratio p_e/p_i which causes a shock at an area equal to twice the inlet area, $A_s = 2A_i$. Assume steady inviscid quasi-one-dimensional, one-directional flow of an ideal gas with constant specific heats.

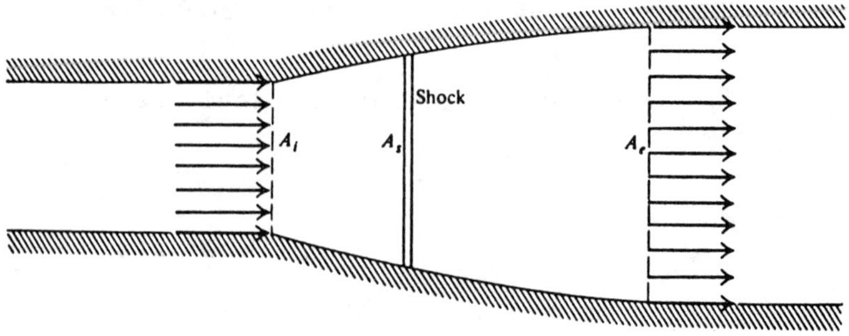

Fig. 1

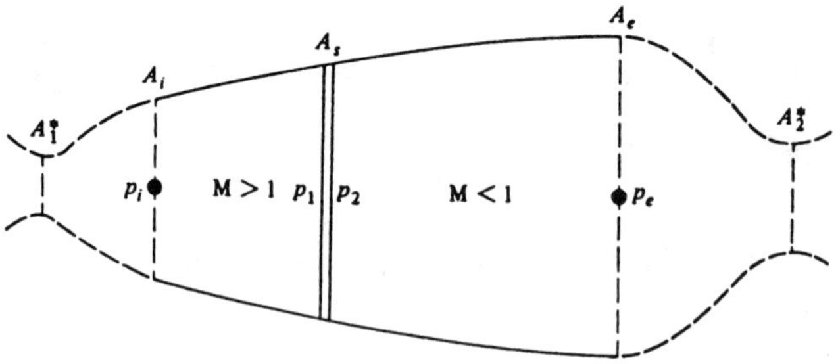

Fig. 2

Solution: Assume that the flow is adiabatic. The flow is then isentropic from A_i to the shock and from the shock to A_e. The flow is irreversible across the shock.

It is helpful to introduce two imaginary throats (Fig. 2), A_1^* and A_2^*, associated with the isentropic flow upstream and downstream of the shock, respectively. Note that $A_1^* \neq A_2^*$ because the isentropic flows on either side of the shock have different stagnation pressures. We write the pressure ratio p_e/p_i as the product of the pressure ratios across each section of the flow

$$\frac{p_e}{p_i} = \frac{p_e}{p_2}\frac{p_2}{p_1}\frac{p_1}{p_i}$$

We shall evaluate each pressure ratio separately. The sequence of calculations to obtain the solution is outlined below. The required numerical values can be obtained from the isentropic flow and normal shock tables.

Table 1

M	A/A^*	p/p_0	ρ/ρ_0	T/T_0
2.12	1.87	0.1060	0.2012	0.5266
2.14	1.90	0.1027	0.1968	0.5219
2.16	1.94	0.0996	0.1925	0.5173
2.18	1.97	0.0965	0.1882	0.5127
2.20	2.00	0.0935	0.1841	0.5081
2.22	2.04	0.0906	0.1800	0.5036
2.24	2.08	0.0878	0.1760	0.4991
2.26	2.12	0.0851	0.1721	0.4947
2.28	2.15	0.0825	0.1683	0.4903
2.30	2.19	0.0800	0.1646	0.4859
2.32	2.23	0.0775	0.1609	0.4816
2.34	2.27	0.0751	0.1574	0.4773
2.36	2.32	0.0728	0.1539	0.4731
2.38	2.36	0.0706	0.1505	0.4688
2.40	2.40	0.0684	0.1472	0.4647
2.42	2.45	0.0663	0.1439	0.4606
2.44	2.49	0.0643	0.1408	0.4565
2.46	2.54	0.0623	0.1377	0.4524
2.48	2.59	0.0604	0.1346	0.4484
2.50	2.64	0.0585	0.1317	0.4444
2.52	2.69	0.0567	0.1288	0.4405
2.54	2.74	0.0550	0.1260	0.4366
2.56	2.79	0.0533	0.1232	0.4328
2.58	2.84	0.0517	0.1205	0.4289
2.60	2.90	0.0501	0.1179	0.4252
2.62	2.95	0.0486	0.1153	0.4214
2.64	3.01	0.0471	0.1128	0.4177
2.66	3.06	0.0457	0.1103	0.4141
2.68	3.12	0.0443	0.1079	0.4104
2.70	3.18	0.0430	0.1056	0.4068
2.72	3.24	0.0417	0.1033	0.4033
2.74	3.31	0.0404	0.1010	0.3998
2.76	3.37	0.0392	0.0989	0.3963
2.78	3.43	0.0380	0.0967	0.3928
2.80	3.50	0.0368	0.0946	0.3894
2.82	3.57	0.0357	0.0926	0.3860
2.84	3.64	0.0347	0.0906	0.3827
2.86	3.71	0.0336	0.0886	0.3794
2.88	3.78	0.0326	0.0867	0.3761
2.90	3.85	0.0317	0.0849	0.3729
2.92	3.92	0.0307	0.0831	0.3696
2.94	4.00	0.0298	0.0813	0.3665
2.96	4.08	0.0289	0.0796	0.3633
2.98	4.15	0.0281	0.0779	0.3602
3.00	4.23	0.0272	0.0762	0.3571
3.02	4.32	0.0264	0.0746	0.3541

Calculate p_1/p_i: From Table 1

$$\frac{A_i}{A^*_1} = G(M_i) = 4.235$$

$$\frac{A_s}{A_i} = 2 \qquad \text{given in statement of problem}$$

Then

$$\frac{A_s}{A^*_1} = \frac{A_s}{A_i}\frac{A_i}{A^*_1} = 2(4.235) = 8.47$$

Table 2

M	A/A^*	p/p_0	ρ/ρ_0	T/T_0
3.04	4.40	0.0256	0.0730	0.3511
3.06	4.48	0.0249	0.0715	0.3481
3.08	4.57	0.0242	0.0700	0.3452
3.10	4.66	0.0234	0.0685	0.3422
3.12	4.75	0.0228	0.0671	0.3393
3.14	4.84	0.0221	0.0657	0.3365
3.16	4.93	0.0215	0.0643	0.3337
3.18	5.02	0.0208	0.0630	0.3309
3.20	5.12	0.0202	0.0617	0.3281
3.22	5.22	0.0196	0.0604	0.3253
3.24	5.32	0.0191	0.0591	0.3226
3.26	5.42	0.0185	0.0579	0.3199
3.28	5.52	0.0180	0.0567	0.3173
3.30	5.63	0.0175	0.0555	0.3147
3.32	5.74	0.0170	0.0544	0.3121
3.34	5.84	0.0165	0.0533	0.3095
3.36	5.96	0.0160	0.0522	0.3069
3.38	6.07	0.0156	0.0511	0.3044
3.40	6.18	0.0151	0.0501	0.3019
3.42	6.30	0.0147	0.0491	0.2995
3.44	6.42	0.0143	0.0481	0.2970
3.46	6.54	0.0139	0.0471	0.2946
3.48	6.66	0.0135	0.0462	0.2922
3.50	6.79	0.0131	0.0452	0.2899
3.52	6.92	0.0127	0.0443	0.2875
3.54	7.05	0.0124	0.0434	0.2852
3.56	7.18	0.0120	0.0426	0.2829
3.58	7.31	0.0117	0.0417	0.2806
3.60	7.45	0.0114	0.0409	0.2784
3.62	7.59	0.0111	0.0401	0.2762
3.64	7.73	0.0108	0.0393	0.2740
3.66	7.87	0.0105	0.0385	0.2718
3.68	8.02	0.0102	0.0378	0.2697
3.70	8.17	0.0099	0.0370	0.2675
3.72	8.32	0.0096	0.0363	0.2654
3.74	8.47	0.0094	0.0356	0.2633
3.76	8.63	0.0091	0.0349	0.2613
3.78	8.79	0.0089	0.0342	0.2592
3.80	8.95	0.0086	0.0335	0.2572
3.82	9.11	0.0084	0.0329	0.2552
3.84	9.28	0.0082	0.0323	0.2532
3.86	9.45	0.0080	0.0316	0.2513
3.88	9.62	0.0077	0.0310	0.2493
3.90	9.80	0.0075	0.0304	0.2474
3.92	9.98	0.0073	0.0299	0.2455
3.94	10.16	0.0071	0.0293	0.2436

From Table 2

$$M_1 = G\left(\frac{A_s}{A_i^*}\right) = 3.74$$

From Table 2

$$\frac{p_1}{p_{01}} = G(M_1) = 0.0094$$

From Table 1

$$\frac{p_i}{p_{01}} = G(M_i) = 0.0272$$

Then

$$\frac{p_1}{p_i} = \frac{p_1/p_{01}}{p_i/p_{01}} = \frac{0.0094}{0.0272} = 0.3445$$

Table 3

M_1	M_2	p_2/p_1	p_{02}/p_{01}	ρ_2/ρ_1	T_2/T_1
3.32	0.4587	12.6928	0.2489	4.1276	3.0751
3.34	0.4578	12.8482	0.2446	4.1431	3.1011
3.36	0.4569	13.0045	0.2404	4.1583	3.1273
3.38	0.4560	13.1618	0.2363	4.1734	3.1537
3.40	0.4552	13.3200	0.2322	4.1884	3.1802
3.42	0.4544	13.4791	0.2282	4.2032	3.2069
3.44	0.4535	13.6392	0.2243	4.2178	3.2337
3.46	0.4527	13.8002	0.2205	4.2323	3.2607
3.48	0.4519	13.9621	0.2167	4.2467	3.2878
3.50	0.4512	14.1250	0.2129	4.2609	3.3151
3.52	0.4504	14.2888	0.2093	4.2749	3.3425
3.54	0.4496	14.4535	0.2057	4.2888	3.3701
3.56	0.4489	14.6192	0.2022	4.3026	3.3978
3.58	0.4481	14.7858	0.1987	4.3162	3.4257
3.60	0.4474	14.9533	0.1953	4.3296	3.4537
3.62	0.4467	15.1218	0.1920	4.3429	3.4819
3.64	0.4460	15.2912	0.1887	4.3561	3.5103
3.66	0.4453	15.4615	0.1855	4.3692	3.5388
3.68	0.4446	15.6328	0.1823	4.3821	3.5674
3.70	0.4439	15.8050	0.1792	4.3949	3.5962
3.72	0.4433	15.9781	0.1761	4.4075	3.6252
3.74	0.4426	16.1522	0.1731	4.4200	3.6543
3.76	0.4420	16.3272	0.1702	4.4324	3.6836
3.78	0.4414	16.5031	0.1673	4.4447	3.7130
3.80	0.4407	16.6800	0.1645	4.4568	3.7426
3.82	0.4401	16.8578	0.1617	4.4688	3.7723
3.84	0.4395	17.0365	0.1589	4.4807	3.8022
3.86	0.4389	17.2162	0.1563	4.4924	3.8323
3.88	0.4383	17.3968	0.1536	4.5041	3.8625
3.90	0.4377	17.5783	0.1510	4.5156	3.8928
3.92	0.4372	17.7608	0.1485	4.5270	3.9233
3.94	0.4366	17.9442	0.1460	4.5383	3.9540
3.96	0.4360	18.1285	0.1435	4.5494	3.9848
3.98	0.4355	18.3138	0.1411	4.5605	4.0158
4.00	0.4350	18.5000	0.1388	4.5714	4.0469
4.10	0.4324	19.4450	0.1276	4.6245	4.2048
4.20	0.4299	20.4133	0.1173	4.6749	4.3666
4.30	0.4277	21.4050	0.1080	4.7229	4.5322
4.40	0.4255	22.4200	0.0995	4.7685	4.7017
4.50	0.4236	23.4583	0.0917	4.8119	4.8751
4.60	0.4217	24.5200	0.0846	4.8532	5.0523
4.70	0.4199	25.6050	0.0781	4.8926	5.2334
4.80	0.4183	26.7133	0.0721	4.9301	5.4184
4.90	0.4167	27.8450	0.0667	4.9659	5.6073
5.00	0.4152	29.0000	0.0617	5.0000	5.8000
5.10	0.4138	30.1783	0.0572	5.0326	5.9966

Calculate p_2/p_1: From Table 3

$$\frac{p_2}{p_1} = G(M_1) = 16.15$$

Calculate p_e/p_2: Both throats pass the same mass flow rate, $\dot{m}_{max}$, which is given by

$$\dot{m}_{max} = A_1^* p_{01} \sqrt{\frac{\gamma}{(R/m)T_{01}}} \left(\frac{\gamma + 1}{2}\right)^{-(\gamma+1)/2(\gamma-1)}$$

$$= A_2^* p_{02} \sqrt{\frac{\gamma}{(R/m)T_{02}}} \left(\frac{\gamma + 1}{2}\right)^{-(\gamma+1)/2(\gamma-1)}$$

Since the flow throughout the passage including that across the shock is adiabatic, $T_{01} = T_{02}$,

$$A_1^* p_{01} = A_2^* p_{02}$$

From Table 3

$$\frac{p_{02}}{p_{01}} = G(M_1) = 0.173$$

Then

$$\frac{A_1^*}{A_2^*} = \frac{p_{02}}{p_{01}} = 0.173$$

$$\frac{A_e}{A_i} = 3, \text{ given in statement of problem}$$

$$\frac{A_e}{A_2^*} = \frac{A_e}{A_i}\frac{A_i}{A_1^*}\frac{A_1^*}{A_2^*} = 3(4.235)(0.173) = 2.198,$$

Table 4

M	A/A^*	p/p_0	ρ/ρ_0	T/T_0
0.00	∞	1.0000	1.0000	1.0000
0.02	28.94	0.9997	0.9998	0.9999
0.04	14.48	0.9989	0.9992	0.9997
0.06	9.67	0.9975	0.9982	0.9993
0.08	7.26	0.9955	0.9968	0.9987
0.10	5.82	0.9930	0.9950	0.9980
0.12	4.86	0.9900	0.9928	0.9971
0.14	4.18	0.9864	0.9903	0.9961
0.16	3.67	0.9823	0.9873	0.9949
0.18	3.28	0.9776	0.9840	0.9936
0.20	2.96	0.9725	0.9803	0.9921
0.22	2.71	0.9668	0.9762	0.9904
0.24	2.50	0.9607	0.9718	0.9886
0.26	2.32	0.9541	0.9670	0.9867

From Tables 4 and 5

$$M_e = G\left(\frac{A_e}{A_2^*}\right) = 0.275$$

From Tables 4 and 5

$$\frac{p_e}{p_{02}} = G(M_e) = 0.949$$

From Table 3

$$M_2 = G(M_1) = 0.443$$

Table 5

M	A/A^*	p/p_0	ρ/ρ_0	T/T_0
0.28	2.17	0.9470	0.9619	0.9846
0.30	2.04	0.9395	0.9564	0.9823
0.32	1.92	0.9315	0.9506	0.9799
0.34	1.82	0.9231	0.9445	0.9774
0.36	1.74	0.9143	0.9380	0.9747
0.38	1.66	0.9052	0.9313	0.9719
0.40	1.59	0.8956	0.9243	0.9690
0.42	1.53	0.8857	0.9170	0.9659
0.44	1.47	0.8755	0.9094	0.9627
0.46	1.42	0.8650	0.9016	0.9594
0.48	1.38	0.8541	0.8935	0.9559
0.50	1.34	0.8430	0.8852	0.9524
0.52	1.30	0.8317	0.8766	0.9487
0.54	1.27	0.8201	0.8679	0.9449
0.56	1.24	0.8082	0.8589	0.9410
0.58	1.21	0.7962	0.8498	0.9370
0.60	1.19	0.7840	0.8405	0.9328
0.62	1.17	0.7716	0.8310	0.9286
0.64	1.15	0.7591	0.8213	0.9243
0.66	1.13	0.7465	0.8115	0.9199
0.68	1.11	0.7338	0.8016	0.9153
0.70	1.09	0.7209	0.7916	0.9107
0.72	1.08	0.7080	0.7814	0.9061
0.74	1.07	0.6951	0.7712	0.9013
0.76	1.06	0.6821	0.7609	0.8964
0.78	1.05	0.6691	0.7505	0.8915
0.80	1.04	0.6560	0.7400	0.8865
0.82	1.03	0.6430	0.7295	0.8815
0.84	1.02	0.6300	0.7189	0.8763
0.86	1.02	0.6170	0.7083	0.8711
0.88	1.01	0.6041	0.6977	0.8659
0.90	1.01	0.5913	0.6870	0.8606
0.92	1.01	0.5785	0.6764	0.8552
0.94	1.00	0.5658	0.6658	0.8498
0.96	1.00	0.5532	0.6551	0.8444
0.98	1.00	0.5407	0.6445	0.8389
1.00	1.00	0.5283	0.6339	0.8333
1.02	1.00	0.5160	0.6234	0.8278
1.04	1.00	0.5039	0.6129	0.8222
1.06	1.00	0.4919	0.6024	0.8165
1.08	1.01	0.4800	0.5920	0.8108
1.10	1.01	0.4684	0.5817	0.8052
1.12	1.01	0.4568	0.5714	0.7994
1.14	1.02	0.4455	0.5612	0.7937
1.16	1.02	0.4343	0.5511	0.7879
1.18	1.02	0.4232	0.5411	0.7822

From Table 5

$$\frac{p_2}{p_{02}} = G(M_2) = 0.874$$

Then

$$\frac{p_e}{p_2} = \frac{p_e/p_{02}}{p_2/p_{02}} = \frac{0.949}{0.874} = 1.086$$

Finally

$$\frac{p_e}{p_i} = (1.086)(16.15)(0.3445) = 6.04$$

• PROBLEM 12-6

What diameter would a hailstone (s = 0.9) have to attain to fall through air with the speed of sound?

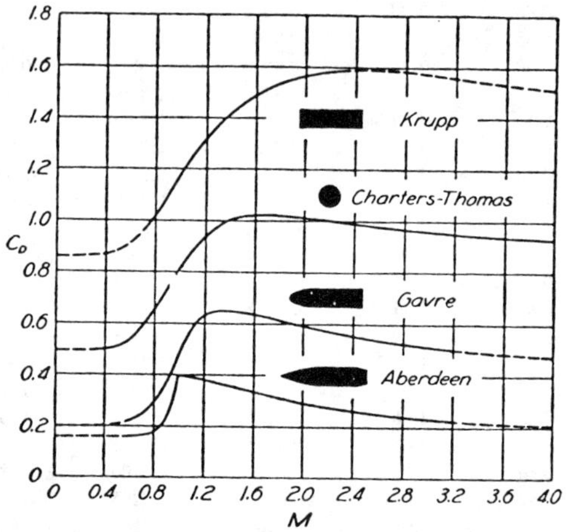

Resistance diagram for bodies having supersonic velocities.

Solution: From the figure C_D for a sphere is 0.8 when M = 1. The drag equation thus indicates that

$$F_D = 62.4 \times 0.9 \frac{\pi d^3}{6} = C_D A \frac{\rho v_0^2}{2} = 0.8 \frac{\pi d^2}{4} \frac{0.0024 \times \overline{1100}^2}{2}$$

Solution for d then yields

$$d = \frac{6}{62.4 \times 0.9} \times \frac{0.8}{4} \times \frac{0.0024 \times \overline{1100}^2}{2} = 31 \text{ ft}$$

Evidently, a normal hailstone inherently lacks the size, the shape, and the density to fall with sonic speed.

• PROBLEM 12-7

A perfect gas with $\gamma = 1.4$ flows steadily through the nozzle shown in Figure. The entropy of the gas remains constant, and $R = 53.3$ ft-lb_f/lb_m-°R. The inlet temperature and pressure are 200°F and 100 lb_f/in^2, respectively, and the exit pressure is 20 lb_f/in^2. The inlet is relatively large in comparison with the exit area which is 12 in^2. Obtain the exit Mach number, temperature, density, the mass flow rate through the nozzle, and the velocity at the smallest cross-sectional area.

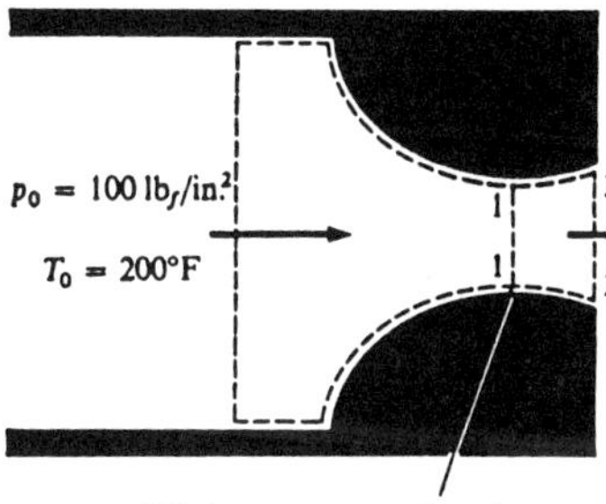

Solution: In a large inlet, the velocity is low and the inlet pressure and temperature may be taken as stagnation values. Then, $p_0 = 100$ lb_f/in^2 and $T_0 = 200°F = 660°R$.

These are constants for the flow through the nozzle since the flow is isentropic and without shaft work.

$$\frac{p_2}{p_{02}} = \frac{20 \ lb_f/in^2}{100 \ lb_f/in^2} = 0.20$$

TABLE 1

ISENTROPIC FLOW ($\gamma = 1.4$)

N_M	A/A^*	p/p_0	ρ/ρ_0	T/T_0
1.20	1.03	0.4124	0.5311	0.7764
1.22	1.04	0.4017	0.5213	0.7706
1.24	1.04	0.3912	0.5115	0.7648
1.26	1.05	0.3809	0.5019	0.7590
1.28	1.06	0.3708	0.4923	0.7532
1.30	1.07	0.3609	0.4829	0.7474
1.32	1.08	0.3512	0.4736	0.7416
1.34	1.08	0.3417	0.4644	0.7358
1.36	1.09	0.3323	0.4553	0.7300
1.38	1.10	0.3232	0.4463	0.7242
1.40	1.11	0.3142	0.4374	0.7184
1.42	1.13	0.3055	0.4287	0.7126
1.44	1.14	0.2969	0.4201	0.7069
1.46	1.15	0.2886	0.4116	0.7011
1.48	1.16	0.2804	0.4032	0.6954
1.50	1.18	0.2724	0.3950	0.6897
1.52	1.19	0.2646	0.3869	0.6840
1.54	1.20	0.2570	0.3789	0.6783
1.56	1.22	0.2496	0.3710	0.6726
1.58	1.23	0.2423	0.3633	0.6670
1.60	1.25	0.2353	0.3557	0.6614
1.62	1.27	0.2284	0.3483	0.6558
1.64	1.28	0.2217	0.3409	0.6502
1.66	1.30	0.2151	0.3337	0.6447
1.68	1.32	0.2088	0.3266	0.6392
1.70	1.34	0.2026	0.3197	0.6337
1.72	1.36	0.1966	0.3129	0.6283
1.74	1.38	0.1907	0.3062	0.6229
1.76	1.40	0.1850	0.2996	0.6175
1.78	1.42	0.1794	0.2931	0.6121
1.80	1.44	0.1740	0.2868	0.6068
1.82	1.46	0.1688	0.2806	0.6015
1.84	1.48	0.1637	0.2745	0.5963
1.86	1.51	0.1587	0.2686	0.5910
1.88	1.53	0.1539	0.2627	0.5859
1.90	1.56	0.1492	0.2570	0.5807
1.92	1.58	0.1447	0.2514	0.5756
1.94	1.61	0.1403	0.2459	0.5705
1.96	1.63	0.1360	0.2405	0.5655
1.98	1.66	0.1318	0.2352	0.5605
2.00	1.69	0.1278	0.2300	0.5556
2.02	1.72	0.1239	0.2250	0.5506
2.04	1.75	0.1201	0.2200	0.5458
2.06	1.78	0.1164	0.2152	0.5409
2.08	1.81	0.1128	0.2104	0.5361
2.10	1.84	0.1094	0.2058	0.5313

From Table .1,

$$N_{M_2} = 1.71$$

and

$$\frac{T_2}{T_{02}} = 0.631 \quad \text{and} \quad \frac{\rho_2}{\rho_{02}} = 0.316$$

$$T_2 = T_0(0.631) = (660)(0.631) = 416°R$$

$$V_2 = N_{M_2} c_2 = N_{M_2}(\gamma R T_2)^{1/2}$$

$$= (1.71)[(1.4)(53.3)(416)(32.2)]^{1/2} = 1718 \text{ ft/sec}$$

$$\rho_0 = \frac{p_0}{RT_0} = \frac{(100)(144)}{(53.3)(660)} = 0.410 \text{ lb}_m/\text{ft}^3$$

$$\rho_2 = \rho_{02}(0.316) = (0.410)(0.316) = 0.130 \text{ lb}_m/\text{ft}^3$$

$$m = \rho_2 A_2 V_2$$

$$= (0.130)\left(\frac{12}{144}\right)(1718) = 18.6 \text{ lb}_m/\text{sec}$$

Since the flow passes from the subsonic to the supersonic condition through the nozzle, the Mach number at the minimum cross-sectional area must be 1. With $N_{M_1} = 1.00$,

$$\frac{T_1}{T_{01}} = 0.833 \quad \text{from Table .1}$$

$$T_1 = T_0(0.833) = (660)(0.833) = 550°\text{R}$$

$$V_1 = N_{M_1}(\gamma RT_1)^{1/2} = [(1.4)(53.3)(550)(32.2)]^{1/2}$$

$$= 1151 \text{ ft/sec}$$

• PROBLEM 12-8

An airflow at M = 3 enters a diverging passage (Fig. 1). The passage has an exit-to-inlet area ratio A_e/A_i of 3. Determine the exit-to-inlet pressure ratio p_e/p_i which results in isentropic flow through the passage. Assume steady inviscid quasi-one-dimensional, one-directional flow of an ideal gas with constant specific heats.

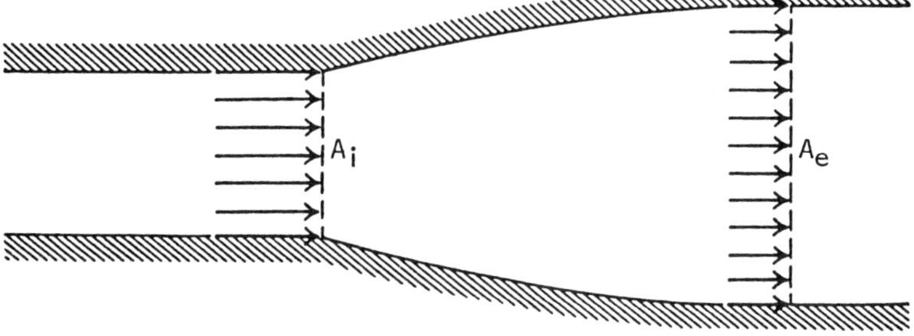

Fig. 1

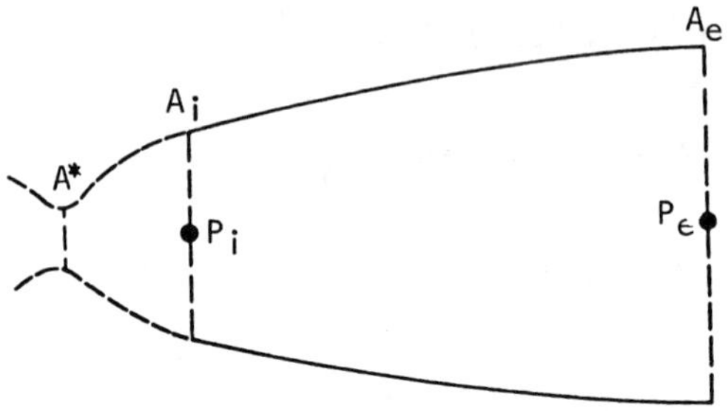

Fig. 2

Solution: It is useful to introduce an imaginary throat A* (Fig. 2), which acts as a reference area and allows us to work in terms of the area ratio A/A*. We shall work the problem in terms of ratios of properties wherever possible. We write the pressure ratio as the product of two pressure ratios

$$\frac{p_e}{p_i} = \left(\frac{p_e}{p_0}\right)\left(\frac{p_0}{p_i}\right)$$

and evaluate each pressure ratio separately. The sequence of calculations necessary to obtain the solution is outlined below. The required numerical values can be obtained from the isentropic flow tables.

Calculate p_e/p_0:

$$\frac{A_e}{A_i} = 3$$

given in statement of problem

From Table 1

$$\frac{A_i}{A^*} = G(M_i) = 4.235$$

Then

$$\frac{A_e}{A^*} = \frac{A_e}{A_i}\frac{A_i}{A8} = 3(4.235) = 12.70$$

From Table 2

$$M_e = G\left(\frac{A_e}{A^*}\right) = 4.20$$

From

$$\frac{p_e}{p_0} = G(M_e) = 0.0051$$

TABLE 1

M	A/A^*	p/p_0	ρ/ρ_0	T/T_0
2.12	1.87	0.1060	0.2013	0.5266
2.14	1.90	0.1027	0.1968	0.5219
2.16	1.94	0.0996	0.1925	0.5173
2.18	1.97	0.0965	0.1882	0.5127
2.20	2.00	0.0935	0.1841	0.5081
2.22	2.04	0.0906	0.1800	0.5036
2.24	2.08	0.0878	0.1760	0.4991
2.26	2.12	0.0851	0.1721	0.4947
2.28	2.15	0.0825	0.1683	0.4903
2.30	2.19	0.0800	0.1646	0.4859
2.32	2.23	0.0775	0.1609	0.4816
2.34	2.27	0.0751	0.1574	0.4773
2.36	2.32	0.0728	0.1539	0.4731
2.38	2.36	0.0706	0.1505	0.4688
2.40	2.40	0.0684	0.1472	0.4647
2.42	2.45	0.0663	0.1439	0.4606
2.44	2.49	0.0643	0.1408	0.4565
2.46	2.54	0.0623	0.1377	0.4524
2.48	2.59	0.0604	0.1346	0.4484
2.50	2.64	0.0585	0.1317	0.4444
2.52	2.69	0.0567	0.1288	0.4405
2.54	2.74	0.0550	0.1260	0.4366
2.56	2.79	0.0533	0.1232	0.4328
2.58	2.84	0.0517	0.1205	0.4289
2.60	2.90	0.0501	0.1179	0.4252
2.62	2.95	0.0486	0.1153	0.4214
2.64	3.01	0.0471	0.1128	0.4177
2.66	3.06	0.0457	0.1103	0.4141
2.68	3.12	0.0443	0.1079	0.4104
2.70	3,18	0.0430	0.1056	0.4068
2.72	3.24	0.0417	0.1033	0.4033
2.74	3.31	0.0404	0.1010	0.3998
2.76	3.37	0.0392	0.0989	0.3963
2.78	3.43	0.0380	0.0967	0.3928
2.80	3.50	0.0368	0.0946	0.3894
2.82	3.57	0.0357	0.0926	0.3860
2.84	3.64	0.0347	0.0906	0.3827
2.86	3.71	0.0336	0.0886	0.3794
2.88	3.78	0.0326	0.0867	0.3761
2.90	3.85	0.0317	0.0849	0.3729
2.92	3.92	0.0307	0.0831	0.3696
2.94	4.00	0.0298	0.0813	0.3665
2.96	4.08	0.0289	0.0796	0.3633
2.98	4.15	0.0281	0.0779	0.3602
3.00	4.23	0.0272	0.0762	0.3571
3.02	4.32	0.0264	0.0746	0.3541

TABLE 2

M	A/A^*	p/p_0	ρ/ρ_0	T/T_0
3.96	10.34	0.0069	0.0287	0.2418
3.98	10.53	0.0068	0.0282	0.2399
4.00	10.72	0.0066	0.0277	0.2381
4.10	11.71	0.0058	0.0252	0.2293
4.20	12.79	0.0051	0.0229	0.2208
4.30	13.95	0.0044	0.0209	0.2129
4.40	15.21	0.0039	0.0191	0.2053
4.50	16.56	0.0035	0.0174	0.1980
4.60	18.02	0.0031	0.0160	0.1911
4.70	19.58	0.0027	0.0146	0.1846
4.80	21.26	0.0024	0.0134	0.1783
4.90	23.07	0.0021	0.0123	0.1724
5.00	25.00	0.0019	0.0113	0.1667

Calculate p_0/p_i:

From Table 1

$$\frac{p_i}{p_0} = G(M_i) = 0.0272$$

Then

$$\frac{p_e}{p_i} = \frac{0.0051}{0.0272} = 0.186$$

• PROBLEM 12-9

Air flows steadily through a circular duct of constant cross-section. The air enters the duct with a velocity of 500 ft/sec at a pressure of p_1 = 30 psia and a temperature T_1 = 80°F. As a result of heat transfer and friction, the air at the exit is at p_2 = 20 psia and T_2 = 90°F.

(a) Determine the stagnation pressures and temperatures (p_{01}, p_{02}, T_{01}, T_{02}) at the inlet and outlet of the duct.

(b) If the duct is made of cast iron with 6 in. inside diameter, what must be the length of the duct, L_{12}, in order to satisfy the inlet and outlet conditions? Assume that the flow is adiabatic.

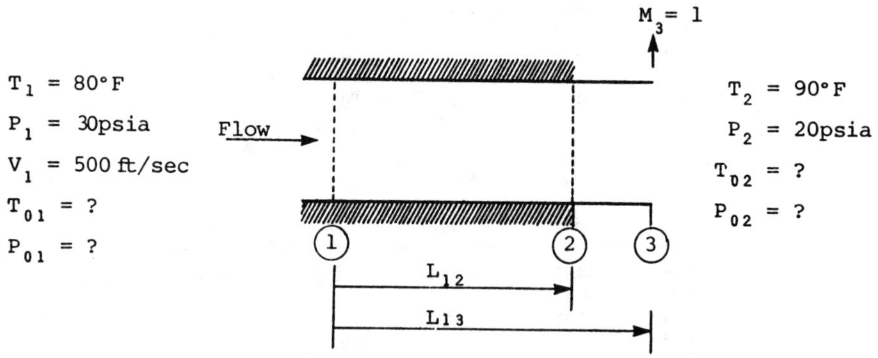

Fig. 1

Solution: (a) Assuming that the air behaves as an ideal gas, the speed of sound at section 1 is

$$C_1 = (\gamma R T_1)^{\frac{1}{2}} = \left[(1.4)(32.2)(53.3)(540)\right]^{\frac{1}{2}} = 1,140 \text{ ft/sec}$$

The inlet Mach number, M_1, is

$$M_1 = \frac{V_1}{C_1} = \frac{500}{1,140} = 0.4386$$

The stagnation temperature at point 1 is given by the formula

$$T_{01} = T_1\left[1 + \left(\frac{\gamma - 1}{2}\right)M_1^2\right] = (540)\left[1 + \left(\frac{1.4 - 1}{2}\right)(0.4386)^2\right]$$

$$= 560.77^0\text{R}$$

The stagnation pressure at point 1 is

$$P_{01} = P_1\left[1 + \left(\frac{\gamma - 1}{2}\right)M_1^2\right]^{\frac{\gamma}{\gamma-1}} = (60)\left[1 + \left(\frac{1.4 - 1}{2}\right)(0.4386)^2\right]^{\frac{1.4}{1.4-1}}$$

$$= 68.47 \text{ psi}$$

From the continuity equation

$$\frac{\partial}{\partial t}\int_{CV} \rho d\overline{V} \,(\to 0) + \int_{CS} \rho\overline{V}dA = 0$$

$$\Rightarrow \quad -\rho_1 V_1 A + \rho_2 V_2 A = 0 \Rightarrow \quad \rho_1 V_1 = \rho_2 V_2$$

$$\Rightarrow \quad V_2 = \frac{\rho_1}{\rho_2} V_1$$

Also,

$$\left.\begin{aligned} p_1 &= \rho_1 R T_1 \\ p_2 &= \rho_2 R T_2 \end{aligned}\right\} \Rightarrow \frac{\rho_1}{\rho_2} = \frac{p_1}{p_2}\frac{T_2}{T_1}$$

Therefore,

$$V_2 = \frac{p_1}{p_2}\frac{T_2}{T_1} V_1 = \frac{(30)(550)}{(20)(540)} (500) = 763.43 \text{ ft/sec}$$

The speed of sound at section 2 is

$$C_2 = (\gamma R T_2)^{\frac{1}{2}} = \left[(1.4)(32.2)(53.3)(55.0)\right]^{\frac{1}{2}}$$
$$= 1,149.57 \text{ ft/sec}$$

The outlet Mach number M_2 is

$$M_2 = \frac{V_2}{C_2} = \frac{(763.43)}{(1,149.51)} = 0.6641$$

The stagnation temperature at point 2 is

$$T_{02} = T_2\left[1 + \left(\frac{\gamma - 1}{2}\right)M_2^2\right] = (550)\left[1 + \frac{1.4 - 1}{2}(0.6641)^2\right]$$
$$= 598.51^0\text{R}$$

and the stagnation pressure is

$$P_{02} = P_2\left[1 + \left(\frac{\gamma - 1}{2}\right)M_2^2\right]^{\frac{\gamma}{\gamma - 1}} = 20\left[1 + \left(\frac{1.4 - 1}{2}\right)(0.6641)^2\right]^{\frac{1.4}{1.4-1}}$$

$$= 26.88 \text{ psia}$$

(b) The duct length L required to change the Mach number from an initial value M_1, to a final value M_2, may be found from the equation

$$\frac{fL}{D} = \left(\frac{fL_{max}}{D}\right)_{M_1} - \left(\frac{fL_{max}}{D}\right)_{M_2} \qquad (1)$$

From fig. 2 (Problem 12-14) and e/D = 0.0012 the friction factor is f = 0.023, from table 1

$$(L_{13})_{max} = \left(\frac{fL_{max}}{D}\right)_{M_1 = 0.4386} = 2.401 \qquad (2)$$

and

$$(L_{23})_{max} = \left(\frac{fL_{max}}{D}\right)_{M_2 = 0.6641} = 0.271 \qquad (3)$$

Combining equation (1), (2) and (3), we have

$$L = \left[(L_{13})_{max} - (L_{23})_{max}\right]\frac{D}{f} = (2.401 - 0.271)\,\frac{(0.5)}{0.023}$$

$$\Rightarrow \quad L = 46.30 \text{ ft.}$$

Table 1 Change in properties with Mach No. in an adiabatic flow of air in a duct with friction. ($\gamma = 1.4$)

M_1	T/T^*	p/p^*	V/V^*	p_o/p_o^0	$fl\,max/D$
0.00	1.200	∞	0.000	∞	∞
0.05	1.199	21.903	0.055	11.591	280.020
0.10	1.198	10.944	0.109	5.822	66.922
0.15	1.195	7.287	0.164	3.910	27.932
0.20	1.190	5.455	0.218	2.964	14.533
0.25	1.185	4.355	0.272	2.403	8.483
0.30	1.179	3.619	0.326	2.035	5.299
0.35	1.171	3.092	0.379	1.778	3.452
0.40	1.163	2.696	0.431	0.590	2.308
0.45	1.153	2.386	0.483	1.449	1.566
0.50	1.143	2.138	0.535	1.340	1.069
0.55	1.132	1.934	0.585	1.255	0.728
0.60	1.119	1.763	0.635	1.188	0.491
0.65	1.107	1.618	0.684	1.136	0.325
0.70	1.093	1.493	0.732	1.094	0.208
0.75	1.079	1.385	0.779	1.062	0.127
0.80	1.064	1.289	0.825	1.038	0.072
0.85	1.048	1.205	0.870	1.021	0.036
0.90	1.033	1.129	0.915	1.009	0.015
1.00	1.000	1.000	1.000	1.000	0.000

• **PROBLEM 12-10**

Air flows isentropically through a duct. At a given point the area is 0.5 ft^2 and the Mach number is $M_1 = 0.4$. At another point in the duct the area is 0.4 ft^2. What is the Mach number at the second point? What would the area be at a point where M = 1. Choose $\gamma = 1.4$.

Solution: The Mach number at the second area A_2 can be found if the ratio A_2/A^* is known. This can be found, however, from the relation

$$\frac{A_2}{A^*} = \frac{A_2}{A_1} \times \frac{A}{A^*}$$

where A_1/A^* is known as M_1 is specified and A_1/A_2 can be found.

From the table and M = 0.4,

$$\frac{A_1}{A^*} = 1.59$$

$$\frac{A_2}{A^*} = 1.59 \times \frac{A_2}{A_1} = 1.59 \times \frac{0.4}{0.5} = 1.27$$

Turning to the table we see that a value $A_2/A^* = 1.27$ leads to two possible values of M_2, one subsonic and the other supersonic:

$$M_2 \cong 0.54$$

$M_2 \cong 1.63$

Since no further information is given, both values are possible answers.

The area where M = 1 can be found from

$$A^* = \frac{A^*}{A_1} \times A_1$$

$$A^* = \left(\frac{A^*}{A_1}\right)_{M_1} A_1 = \frac{0.5}{1.59} = 0.31 \text{ ft}^2$$

If $M_2 = 1.63$, it would be necessary that a throat exist with an area equal to A^*. If $M_2 = 0.54$, no throat need exist. The area may decrease steadily from $A_1 = 0.5$ to $A_2 = 0.4$.

One-Dimensional Isentropic Compressible-Flow Functions

For an ideal gas with constant specific heat and molecular weight, $\gamma = 1.4$

M	$\frac{A}{A^*}$	$\frac{P}{P_0}$	$\frac{\rho}{\rho_0}$	$\frac{T}{T_0}$	$\frac{A}{A^*}\frac{P}{P_0}$
0	∞	1.00000	1.00000	1.00000	∞
0.05	11.592	0.99825	0.99875	0.99950	11.571
0.10	5.8218	0.99303	0.99502	0.99800	5.7812
0.15	3.9103	0.98441	0.98884	0.99552	3.8493
0.20	2.9635	0.97250	0.98027	0.99206	2.8820
0.25	2.4027	0.95745	0.96942	0.98765	2.3005
0.30	2.0351	0.93947	0.95638	0.98232	1.9119
0.35	1.7780	0.91877	0.94128	0.97608	1.6336
0.40	1.5901	0.89562	0.92428	0.96899	1.4241
0.45	1.4487	0.87027	0.90552	0.96108	1.2607
0.50	1.3398	0.84302	0.88517	0.95238	1.12951
0.55	1.2550	0.81416	0.86342	0.94295	1.02174
0.60	1.1882	0.78400	0.84045	0.93284	0.93155
0.65	1.1356	0.75283	0.81644	0.92208	0.85493
0.70	1.09437	0.72092	0.97158	0.91075	0.78896
0.75	1.06242	0.68857	0.76603	0.89888	0.73155
0.80	1.03823	0.65602	0.74000	0.88652	0.68110
0.85	1.02067	0.62351	0.71361	0.87374	0.63640
0.90	1.00886	0.59126	0.68704	0.86058	0.59650
0.95	1.00214	0.55946	0.66044	0.84710	0.56066
1.00	1.00000	0.52828	0.63394	0.83333	0.52828
1.05	1.00202	0.49787	0.60765	0.81933	0.49888
1.10	1.00793	0.46835	0.58169	0.80515	0.47206
1.15	1.01746	0.43983	0.55616	0.79083	0.44751
1.20	1.03044	0.41238	0.53114	0.77640	0.42493
1.25	1.04676	0.38606	0.50670	0.76190	0.40411
1.30	1.06631	0.36092	0.48291	0.74738	0.38484
1.35	1.08904	0.33697	0.45980	0.73287	0.36697
1.40	1.1149	0.31424	0.43742	0.71839	0.35036
1.45	1.1440	0.29272	0.41581	0.70397	0.33486
1.50	1.1762	0.27240	0.39498	0.68965	0.32039
1.55	1.2115	0.25326	0.37496	0.67545	0.30685
1.60	1.2502	0.23527	0.35573	0.66138	0.29414
1.65	1.2922	0.21839	0.33731	0.64746	0.28221
1.70	1.3376	0.20259	0.31969	0.63372	0.27099

• PROBLEM 12-11

The pressure and Mach number of helium flowing through a channel at a certain point are found to be P = 30 psia and M = 1.5. At a point further downstream the pressure is 15 psia. Assuming isentropic flow through the channel, determine the Mach number at the second point. Assume γ = 1.67.

One-Dimensional Isentropic Compressible-Flow Functions

For an ideal gas with constant specific heat and molecular weight, $\gamma = 1.67$

M	$\frac{A}{A^*}$	$\frac{P}{P_0}$	M	$\frac{A}{A^*}$	$\frac{P}{P_0}$
0	∞	1.0000	1.75	1.312	0.1721
0.05	11.265	0.9979	1.80	1.351	0.1601
0.10	5.661	0.9917	1.85	1.392	0.1490
0.15	3.805	0.9815	1.90	1.436	0.1386
0.20	2.887	0.9674	1.95	1.482	0.1290
0.25	2.344	0.9497	2.00	1.530	0.1201
0.30	1.989	0.9286	2.05	1.580	0.1119
0.35	1.741	0.9046	2.10	1.632	0.1042
0.40	1.560	0.8780	2.15	1.687	0.09712
0.45	1.424	0.8491	2.20	1.744	0.09053
0.50	1.320	0.8184	2.25	1.803	0.08442
0.55	1.239	0.7862	2.30	1.865	0.07875
0.60	1.176	0.7529	2.35	1.929	0.07349
0.65	1.126	0.7190	2.40	1.995	0.06862
0.70	1.0874	0.6847	2.45	2.064	0.06410
0.75	1.0576	0.6503	2.50	2.135	0.05990
0.80	1.0351	0.6162	2.55	2.209	0.05601
0.85	1.0189	0.5826	2.60	2.285	0.05239
0.90	1.0080	0.5497	2.65	2.364	0.04903
0.95	1.0019	0.5177	2.70	2.445	0.04591
1.00	1.0000	0.4867	2.75	2.529	0.04301
1.05	1.0018	0.4568	2.80	2.616	0.04032
1.10	1.0071	0.4282	2.85	2.705	0.03781
1.15	1.0154	0.4009	2.90	2.797	0.03547
1.20	1.0266	0.3749	2.95	2.892	0.03330
1.25	1.0406	0.3502	3.0	2.990	0.03128
1.30	1.0573	0.3269	3.5	4.134	0.01720
1.35	1.0765	0.3049	4.0	5.608	0.009939
1.40	1.0981	0.2842	4.5	7.456	0.006007
1.45	1.122	0.2647	5.0	9.721	0.003778
1.50	1.148	0.2465	6.0	15.68	$165(10)^{-5}$
1.55	1.176	0.2295	7.0	23.85	$807(10)^{-6}$
1.60	1.207	0.2136	8.0	34.58	$429(10)^{-6}$
1.65	1.240	0.1988	9.0	48.24	$244(10)^{-6}$
1.70	1.275	0.1850	10.0	65.18	$147(10)^{-6}$
			∞	∞	0

Solution: Because the flow is isentropic, the Mach number will be determined at a point as soon as a static-to-stagnation-pressure ratio is determined. Conversely, if M is known, P/P_0 is known. This is true at the first point.

Letting P_1 be the pressure at the first point and P_2 be the pressure at the second point, we may write

$$\frac{P_2}{P_0} = \frac{P_1}{P_0} \times \frac{P_2}{P_1}$$

Because the ratios P_1/P_0 and P_2/P_1 can be found, P_2/P_0 and M_2 can be found.

From the table and a value $M_1 = 1.5$ we have

$$\frac{P_1}{P_0} = 0.247$$

Therefore

$$\frac{p_2}{p_0} = 0.247 \times \frac{15}{30} = 0.124$$

The Mach number corresponding to this pressure ratio is $M_2 = 1.98$.

COMPRESSIBLE FLOW

• PROBLEM 12-12

Atmospheric air is allowed to fill an initially evacuated insulated bottle. What is the temperature of the air in the bottle when it is full? Assume air is at 70°F and $\gamma = 1.4$.

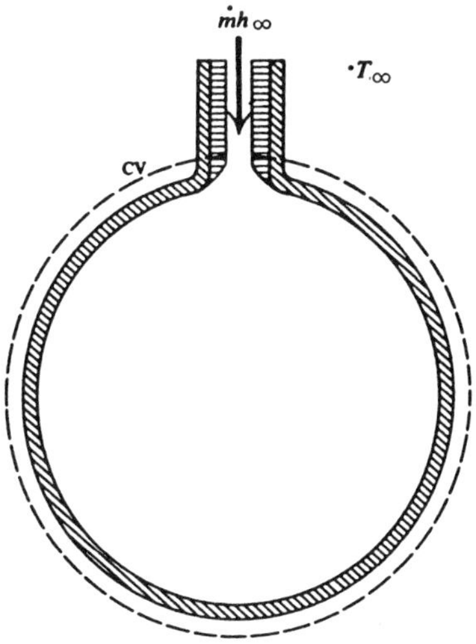

Solution: Select a control volume enclosing the bottle and apply the first law of thermodynamics (see Figure).

The first law for a control volume has the form

$$\dot{Q}_{CV} + \dot{W}_{CV} = \frac{\partial}{\partial t}\iiint_{CV} \rho e \, d\upsilon + \oiint_{CS} \rho\left(h + \frac{V^2}{2} + gy\right)V \cdot dA$$

where $\dot{Q}_{CV}$ = rate at which heat is added to control volume

$\dot{W}_{CV}$ = rate at which work is done on control volume

e = energy per unit mass of fluid in control volume

$= u + V^2/2 + gy$

$V^2/2$ = kinetic energy per unit mass

gy = potential energy per unit mass

The entering and leaving flow must do work against the resisting pressure of the adjoining fluid, which results in the extra energy term pv, called the flow work or transmitted energy. The flow-work term is conveniently added to the internal energy to form the property called enthalpy, h = u + pv.

There is no heat transfer because the bottle is insulated, $\dot{Q}_{CV} = 0$. No work is done on the fluid inside the control volume because the walls of the bottle are rigid, $W_{CV} = 0$. Since potential- and kinetic-energy changes across the control surface are negligible,

$$0 = \frac{\partial}{\partial t}\iiint_{CV} \rho e \, d\upsilon + \oiint_{CS} \rho h V \cdot dA$$

$\iiint \rho e \, d\upsilon$ is the energy of the fluid inside the control volume. The fluid inside the control volume is motionless and has only internal energy

$$\iiint_{CV} \rho e \, dV = mu$$

where m is the mass of fluid inside the control volume at any time and u is its internal energy per unit mass. The enthalpy of the entering air h_∞ is constant because it is at atmospheric conditions. Then

$$\oiint_{cs} \rho h V \cdot dA = -h_\infty \dot{m} = -h_\infty \frac{dm}{dt}$$

where dm/dt is the rate at which the mass inside the control volume is changing. The first law becomes

$$0 = \frac{d(mu)}{dt} - h_\infty \frac{dm}{dt}$$

Integrating gives

$$mu - h_\infty m = \text{const}$$

The tank is initially evacuated, m = 0 at t = 0, giving const = 0. Then

$$u = h_\infty$$

The internal energy of the air in the tank after filling is equal to the enthalpy of the atmospheric air; the flow work of the entering air has been converted into internal energy.

Since air at atmospheric conditions acts as an ideal gas with constant specific heats.

$$\frac{1}{\gamma - 1} \frac{R}{m} T = \frac{\gamma}{\gamma - 1} \frac{R}{m} T_\infty \quad \text{or } T = \gamma T_\infty$$

The temperature of the air in the tank is γ times atmospheric temperature. For atmospheric air at 70°F = 530°R

$$T = 1.4(530°R) = 742°R = 282°F$$

This is a substantial increase in the temperature of the air.

• PROBLEM 12-13

Determine the pressure, density, temperature, and velocity in the reduced section of the circular duct of the Fig. The mass flow rate of air is 6.0 slugs/s. A pressure gage at section 1 reads 100 psi, and the temperature at that section is 80°F.

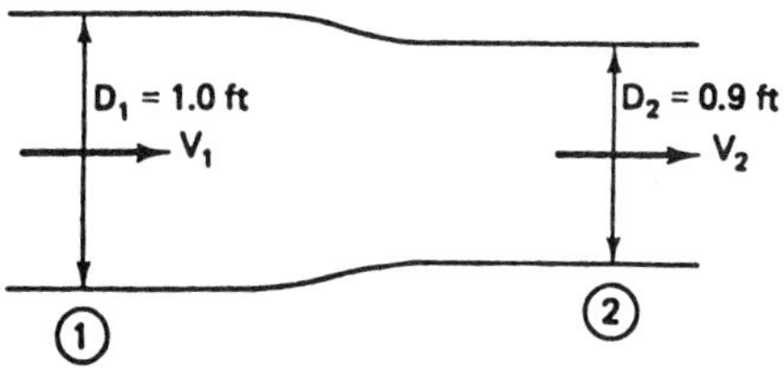

Solution: Since $p_1 = 100 + 14.7 = 114.7$ psia and $T_1 = 80 + 460 = 540°R$, the density is

$$\rho_1 = \frac{p_1}{gRT_1} = \frac{(114.7)(144)}{(32.2(53.3)(540)} = 0.0178 \text{ slugs/ft}^3$$

and the velocity accordingly is

$$V_1 = \frac{6.0}{\rho_1 A_1} = \frac{6.0}{(0.0178)(0.785)} = 429 \text{ ft/s}$$

Since

$$c_1 = \sqrt{kgRT_1} = \sqrt{(1.4)(32.2)(53.3)(540)} = 1139 \text{ ft/s}$$

the Mach number becomes

$$Ma_1 = \frac{V_1}{c_1} = \frac{429}{1139} = 0.377$$

TABLE

Gas	*Chemical Formula*	*Specific Weight, γ* lb/ft³	N/m³	*Density, ρ* slugs/ft³	kg/m³	*Gas Constant, R* ft/°R	m/°K	*Adiabatic Constant, k*
Air	—	0.0753	11.8	0.00234	1.206	53.3	29.2	1.40
Ammonia	NH_3	0.0448	7.04	0.00139	0.716	89.5	49.1	1.32
Carbon dioxide	CO_2	0.115	18.1	0.00357	1.840	34.9	19.1	1.29
Helium	He	0.0104	1.63	0.000323	0.166	386.	212.	1.66
Hydrogen	H_2	0.00522	0.820	0.000162	0.0835	767.	421.	1.40
Methane	CH_4	0.0416	6.53	0.00129	0.665	96.4	52.9	1.32
Nitrogen	N_2	0.0726	11.4	0.00225	1.160	55.2	30.3	1.40
Oxygen	O_2	0.0830	13.0	0.00258	1.330	48.3	26.5	1.40
Sulfur dioxide	SO_2	0.170	26.7	0.00528	2.721	23.6	12.9	1.26

Within the constraints of isentropic (frictionless-adiabatic) flow, we are frequently concerned with flows exposed to relatively abrupt changes in cross-sectional geometry; that is, changes in cross section that cause pressure, velocity, and in the case of compressible flow, density and temperature changes, over a short length so that friction can be reasonably ignored. As with incompressible flow, the fundamental equations are continuity and energy. Between two sections they may be written as

$$\rho_1 V_1 A_1 = \rho_2 V_2 A_2 \tag{1}$$

$$\frac{V_1^2}{2} + \frac{k}{k-1}\frac{p_1}{\rho_1} = \frac{V_2^2}{2} + \frac{k}{k-1}\frac{p_2}{\rho_2} \tag{2}$$

In a duct where V, p, and ρ are all subject to change with changing cross-sectional area A, we may best anticipate the results by differentiating the equations as follows. From continuity,

$$\rho VA = \text{constant}$$

Thus, changes in the flow direction s are related according to

$$\frac{\partial(\rho VA)}{\partial s} = VA\frac{\partial \rho}{\partial s} + \rho V\frac{\partial A}{\partial s} + \rho A\frac{\partial V}{\partial s} = 0$$

Treating the energy equation (Eq. 2) in a similar manner,

$$\frac{\partial}{\partial s}\left(\frac{V^2}{2} + \frac{k}{k-1}\frac{p}{\rho}\right) = V\frac{\partial V}{\partial s} + \frac{k}{k-1}\left(\frac{1}{\rho}\frac{\partial p}{\partial s} - \frac{p}{\rho^2}\frac{\partial \rho}{\partial s}\right) \quad (3)$$

The former may be arranged to yield

$$\frac{1}{A}\frac{\partial A}{\partial s} = -\frac{1}{V}\frac{\partial V}{\partial s} - \frac{1}{\rho}\frac{\partial \rho}{\partial s} \quad (4)$$

If Eq. 3 is now solved for the pressure gradient $\partial p/\partial s$ and the result substituted into Eq. 4 along with

$$p = \rho gRT$$

and

$$c^2 = kgRT$$

we can obtain

$$\frac{1}{A}\frac{\partial A}{\partial s} = -\frac{1}{V}\frac{\partial V}{\partial s}\left(1 - \frac{V^2}{c^2}\right)$$

or

$$\frac{1}{V}\frac{\partial V}{\partial s} = \frac{(1/A)(\partial A/\partial s)}{Ma^2 - 1} \quad (5)$$

Equation 5 now permits the prediction of velocity changes due to area changes. If a stream-tube expands in the flow direction ($\partial A/\partial s > 0$), and if $Ma > 1$ (supersonic), then from Eq. 5, $\partial V/\partial s > 0$ and the velocity will increase as it goes through the expansion. Conversely, if $\partial A/\partial s > 0$ while $Ma < 1$ (subsonic), then $\partial V/\partial s < 0$ and the velocity decreases as it passes through an expansion. Similarly, if $\partial A/\partial s < 0$ (a converging section), the velocity will increase with s if the flow is subsonic and decrease if supersonic. Finally, if $Ma = 1$, Eq. 5

requires that $\partial A/\partial s$ also equal zero. This may be satisfied only in a region of minimum cross section known as a throat, such as would occur between converging and diverging sections. This does not imply, however, that Ma = 1 in a throat. Rather, if $Ma \neq 1$ when $\partial A/\partial s = 0$ (the geometric condition at the throat), then $\partial V/\partial s = 0$ also. Thus, if the flow is subsonic ($Ma < 1$) in the throat, the velocity that has increased toward the throat will be a maximum in the throat, and if the throat velocity is supersonic ($Ma > 1$), the velocity in the throat will be a minimum.

With reference to Eq. 5, the contracting section ($\partial A/\partial s < 0$) and the subsonic flow ($Ma_1 < 0$) require that the velocity must increase, and therefore V_2 will be greater than V_1. From Eq. 1,

$$6.0 = \rho_2 V_2 \left(\frac{\pi}{4}\right)(0.9)^2 = 0.636\rho_2 V_2$$

while from Eq. 2,

$$\frac{(427)^2}{2} + \frac{1.4}{0.4}\,\frac{(114.7)(144)}{0.0178} = \frac{V_2^2}{2} + \frac{1.4}{0.4}\,\frac{p_2}{\rho_2}$$

Using the frictionless adiabatic relationship

$$\frac{p_1}{\gamma_1^k} = \frac{p_2}{\gamma_2^k}$$

$$\frac{(114.7)(144)}{[(0.0178)(32.2)]^{1.4}} = \frac{p_2}{[(32.2)\rho_2]^{1.4}}$$

Solving the three equations simultaneously,

$$V_2 = 554 \text{ ft/s}$$

Density and pressure may now be calculated from the preceding equations. They are

$$\rho_2 = \frac{6.0}{(0.636)(554)} = 0.0170 \text{ slugs/ft}^3$$

$$p_2 = \frac{(114.7)(0.0170)^{1.4}}{(0.0178)^{1.4}} = 107.5 \text{ psia} = 92.8 \text{ psi (gage)}$$

Using the equation of state, the final temperature may also be calculated as

$$T_2 = \frac{p_2}{\rho_2 gR} = \frac{(107.5)(144)}{(0.0170)(32.2)(53.3)} = 531\,^\circ\text{R} = 71\,^\circ\text{F}$$

Natural gas is to be piped along an 18 in. diameter main at a uniform temperature of 80°F. In what length of pipe would the pressure drop from 50 lbf/in^2 to 20 lbf/in^2 if the flow rate were 50,000 lbm/hr?

Take R = 96.3 ft lbf/lbm°R, $\gamma = 1.31$, $\mu = 2.34 \times 10^{-7}$ lbf sec/ft^2? and e = 0.00005 ft.

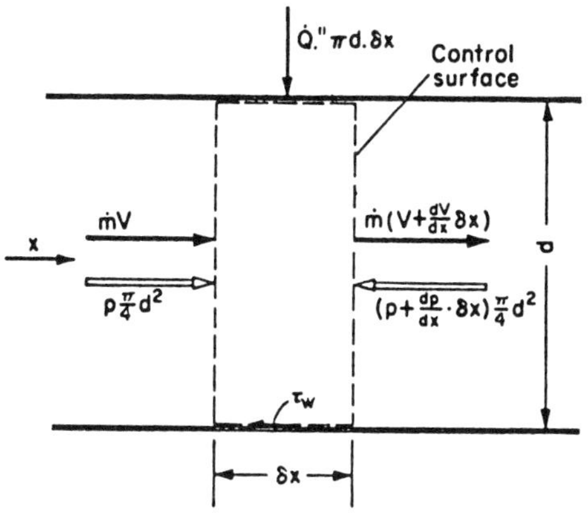

Fig. 1

Solution: An equation can be derived by applying the force-momentum relation to the control-volume defined in Fig. 1.

Ignoring the weight of the fluid (as is usual in the analysis of gas flows), and equating the net force in the x-direction to the increase of the momentum flow rate, we obtain

$$- \frac{\pi d^2}{4} \frac{dp}{dx} \delta x - \pi d \ \delta x . \tau_w = \frac{\dot{m}}{g} \frac{dV}{dx} \delta x, \tag{1}$$

where $\dot{m}$ is the mass flow rate, and the remaining symbols are defined in the diagram.

Now $\dot{m} = \rho V \pi d^2/4$, so that equation 1 may be rewritten as

$$\frac{1}{\rho} \frac{dp}{dx} + \frac{4\tau_w}{\rho d} + \frac{V}{g} \frac{dV}{dx} = 0 \tag{2}$$

or, alternatively, using the definition of friction factor as

$$\frac{1}{\rho} \frac{dp}{dx} + \frac{fV^2}{2dg} + \frac{V}{g} \frac{dV}{dx} = 0. \tag{3}$$

For isothermal processes of a perfect gas

$$\frac{p}{\rho} = RT = \text{constant},$$

so that

$$\frac{p}{p_1} = \frac{\rho}{\rho_1} \tag{4}$$

and

$$\frac{dp}{dx} = \frac{p_1}{\rho_1}\frac{d\rho}{dx} \quad , \tag{5}$$

where suffix 1 refers to a reference state; let this be at a station in the pipe defined by $x = x_1$, and let symbols without suffixes represent quantities at another station in the pipe. The simple statement of mass-continuity,

$$\dot{m} = \frac{\pi}{4} d^2 \rho V,$$

gives, for constant $\dot{m}$ and d,

$$\rho V = \text{constant} = \rho_1 V_1, \tag{6}$$

$$\therefore \quad \rho V^2 = \frac{(\rho_1 V_1)^2}{\rho};$$

and by introducing equation 4

$$\rho V^2 = \frac{p_1 \rho_1 V_1^2}{p} \tag{7}$$

By using equation 4 again, and differentiating, we see that

$$V^2 = \frac{p_1^2 V_1^2}{p^2} \tag{8}$$

$$\therefore \quad 2V \frac{dV}{dx} = - \frac{2p_1^2 V_1^2}{p^3}\frac{dp}{dx} \tag{9}$$

$$\therefore \quad \rho V \frac{dV}{dx} = - \frac{p_1 \rho_1 V_1^2}{p^2}\frac{dp}{dx} \; . \tag{10}$$

By multiplying equation 3 through by ρ, and substituting for ρV^2 and $\rho V dV/dx$ from equations 7 and 10 respectively, we obtain

$$\frac{dp}{dx} + \frac{f}{2dg}\frac{p_1 \rho_1 V_1^2}{p} - \frac{1}{g}\frac{p_1 \rho_1 V_1^2}{p^2}\frac{dp}{dx} = 0,$$

which can easily be transformed to

$$2\,\frac{dp}{dx}\left(\frac{p}{\gamma p_1^2 Ma_1^2} - \frac{1}{p}\right) = -\,\frac{f}{d}\,, \qquad (11)$$

where Ma_1 is the Mach number of velocity V_1.

With the assumption that f is constant, equation 11 can be integrated to yield the following relation between p and x:

$$\frac{1}{\gamma Ma_1^2}\left[1 - \left(\frac{p}{p_1}\right)^2\right] + \ln\left(\frac{p}{p_1}\right)^2 = \frac{f}{d}\,(x-x_1). \qquad (12)$$

In this problem, let station 1 be the place at which the pressure is 50 lbf/in^2. The Mach number at this station, Ma_1, and the Reynolds number of the flow, Re, can be calculated from the data. The value of friction factor f corresponding to Re can be found from the Moody chart, Fig. 2, and known values then substituted in equation 12.

At station 1, the density,

$$\rho_1 = \frac{p_1}{RT} = \frac{50 \times 244}{96.3 \times 540}\ \text{lbm/ft}^3$$

$$= 0.138\ \text{lbm/ft}^3;$$

the mean flow velocity, $\quad V_1 = \dfrac{4\dot{m}}{\pi \rho d^2}$

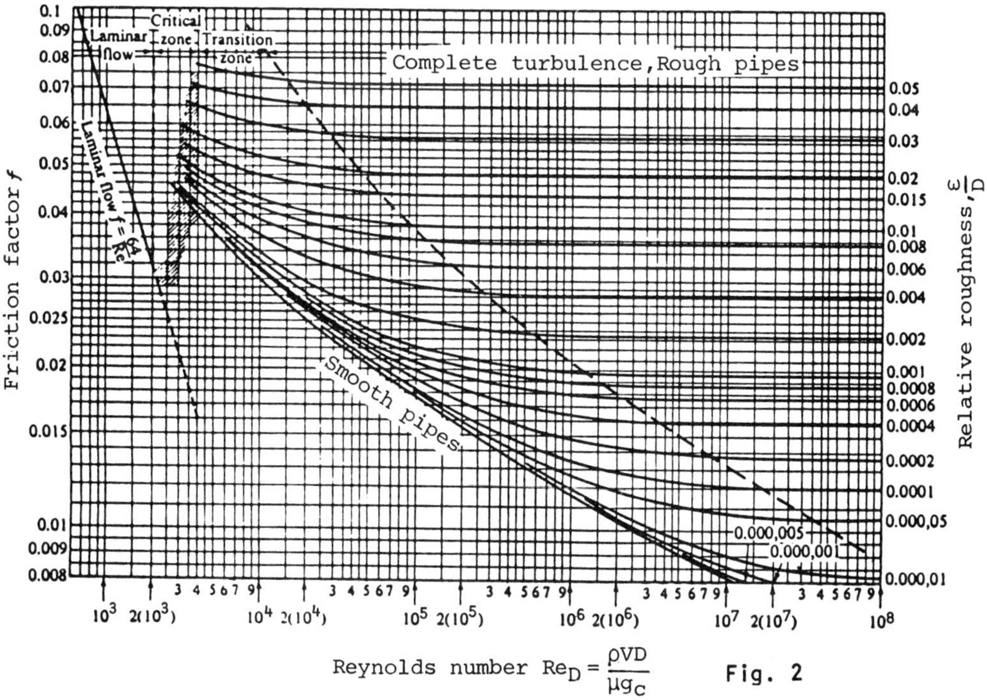

Fig. 2

$$= \frac{4}{\pi \times 0.138 \times (1.5)^2} \left(\frac{5 \times 10^4}{3600}\right) \text{ ft/sec}$$

$$= 56.8 \text{ ft/sec;}$$

the Mach number, $Ma_1 = \dfrac{V_1}{(\gamma R g T)^{1/2}} = \dfrac{56.8}{(1.31 \times 96.3 \times 32.2 \times 540)^{1/2}}$

$$= 0.0383;$$

the Reynolds number,

$$Re = \frac{\rho V_1 d}{\mu g} = \frac{0.138 \times 56.8 \times 1.5}{2.34 \times 10^{-7} \times 32.2}$$

$$= 1.56 \times 10^6.$$

The value of f corresponding to this Reynolds number is 0.0113 (for e/d = 0.00003); putting the appropriate values into equation 12, we have

$$\frac{1}{1.31(0.0383)^2}\left[1 - \left(\frac{20}{50}\right)^2\right] + \ln\left(\frac{20}{50}\right)^2 = \frac{0.0113}{1.5}(x - x_1)$$

$$x - x_1 = \frac{1.5}{0.0113}\left[\frac{0.84}{1.31(0.0383)^2} - 1.83\right]$$

$$= 5.77 \times 10^4 \text{ ft.}$$

• PROBLEM 12-15

(a) Design a theoretical nozzle in order to expand isentropically superheated steam (P_0 = 280 psia, T_0 = 900°F) to a pressure of 120 psia at a rate of 1 lbm/sec.

(b) If the nozzle efficiency is to be 80 percent, compute the exit area at section 8 $(A_e)_{actual}$, of the nozzle.

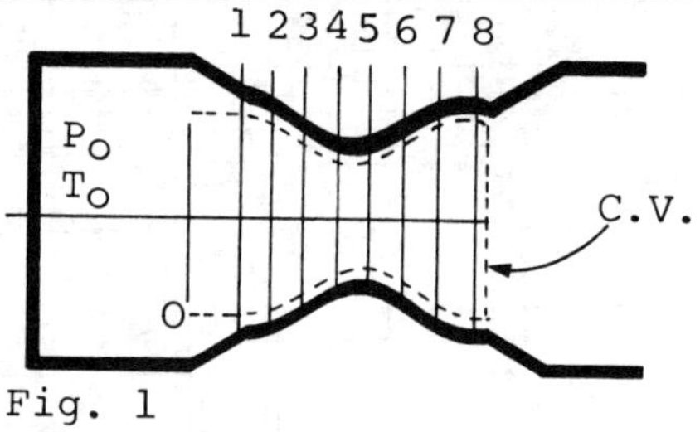

Fig. 1

Solution: The nozzle (Fig. 1) is divided into eight different sections. These sections represent a set of pressure decrements of 20 psia and a set of temperature decrements of 40°F. (A better accuracy can be achieved by separating the nozzle into a larger number of sections). Using the Mollier Diagram (Appendix A) the enthalpy of the superheated steam can be determined (Table 1 below).

By using the steam tables, at given pressure and temperature, the specific volume, v, of the superheated steam is determined.

TABLE 1

Section	Pressure lb/m^2	Temperature °F	Entholpy btu/lbm	Specific Volume ft^3/lbm
0	280	860	1451	2.754
1	260	820	1432	2.985
2	240	780	1411	3.013
3	220	740	1395	3.178
4	200	700	1375	3.378
5	180	660	1354	3.621
6	160	620	1334	3.926
7	140	580	1320	4.320
8	120	540	1295	4.845

The average velocities for each section may be computed from the first law of thermodynamics in the following manner:

$$\cancelto{0}{\dot{Q}} + \cancelto{0}{\dot{W}} = \cancelto{0}{\frac{\partial}{\partial t}\int_{cv} e\rho\, d\forall} + \int_{cs} (e+pv)\rho\vec{V}d\vec{A}$$

or

$$(u_0+p_0v_0+\frac{V_0^2}{2})\{-\rho_0V_0A_0\} + (u_n+p_nv_n+\frac{V_n^2}{2})\{\rho_nV_nA_n\} = 0$$

or

$$h_0 + \cancelto{0}{\frac{V_0^2}{2}} = h_n + \frac{V_n^2}{2}$$

or

$$V_n = \sqrt{2(h_0-h_n)} \quad \text{where } n=1,2,3,\ldots,8 \qquad (1)$$

The areas for each section may be evaluated by using the continuity equation

$$\frac{\partial}{\partial t}\cancelto{0}{\int_{cv}\rho\, d\forall} + \int_{cs} \rho\vec{V}d\vec{A} = 0$$

or

$$\rho_nV_nA_n = \rho_{n-1}V_{n-1}A_{n-1} = \dot{m}$$

or

$$A_n = \dot{m}\frac{1}{\rho_nV_n}$$

or

$$A_n = \frac{v_n}{V_n} \qquad (2)$$

where v_n = the specific volume of the steam at each section, and

V_n = the velocity of the steam at each section.

By substituting numerical values into equations 1 and 2, the following table is obtained.

For example,

$$V_1 = \sqrt{2(h_0 - h_1)}$$

$$= \sqrt{(2)(31.93)(778)(1451-1432)}$$

$$= 971.58 \text{ ft/sec}$$

and

$$A_1 = \frac{v_1}{V_1} = \frac{2.985}{911.58} = 0.0031 \text{ ft}^2$$

$$= 0.4424 \text{ in}^2.$$

Knowing the areas at each section, the nozzle is designed.

(B) Assuming that the losses are taking place between the throat (section 4) and the exit (section 8) the actual exit velocity, Ve, can be calculated from the definition of the nozzle efficiency, or

$$\eta = \frac{(Ve^2/2)}{(h_0 - h_e)} \quad \text{or}$$

$$Ve = \sqrt{2\eta(h_0 - h_e)}$$

$$Ve = \sqrt{(2)(31.93)(778)(0.80)(1472-1277)}$$

$$= 2,784 \text{ ft/sec}$$

But

$$\eta = \frac{h_0 - (h_e)_{actual}}{h_0 - (h_e)_{ideal}} \quad \text{or}$$

$$(h_e)_{out} = h_0 - \eta(h_0 - h_e)_{ideal}$$

$$= (1472)-(0.80)(1472-1277)$$

$$= 1,316 \text{ Btu/lbm}$$

The actual specific volume at the exit (section 8) can be obtained by using the steam tables. For an exit pressure of p_e = 120 psia and an actual enthalpy of $(h_e)_{out}$ = 1,316 Btu/lbm, the actual specific volume, $(v_e)_{out}$ at section 8 is

$$(v_e)_{out} = 6.21 \text{ ft}^3/\text{lbm}.$$

Using the equation 2, the actual exit area, Ae, is computed as

$$(A_e)_{actual} = \frac{(v_e)_{actual}}{(V_e)_{actual}} = \frac{6.21}{2,490} = 0.00249 \text{ ft}^2$$

A blunt-nosed object is travelling through standard air at a velocity of 400 mph. Compute the pressure on the nose of the object for each of the following assumptions: (a) the fluid is incompressible; (b) the fluid is a perfect gas and the process is isothermal; (c) the process is isentropic. The density of standard air is 0.00238 slug/ft^3, and the pressure is 14.7 psia. There is no friction in the flow in front of the body.

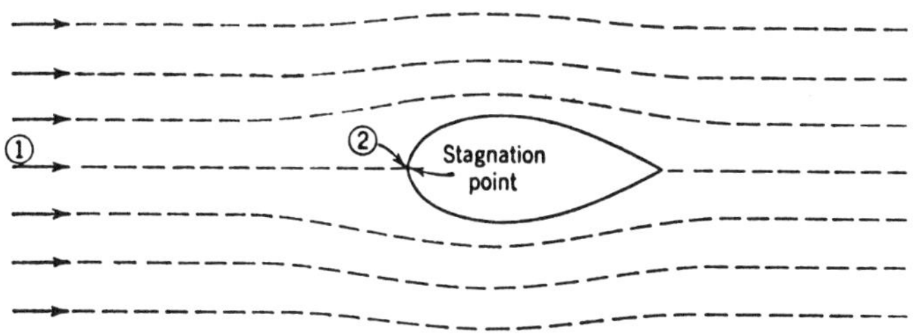

Solution: The easiest way to solve this problem is to consider the object as stationary and the air as flowing at a uniform velocity of 400 mph well upstream of the object. At first glance this would appear to be an entirely different problem, but it is really the same problem as an observer on the object would see it. The pressures will not change just because the flow is seen from a different point. Another way to look at this is to add a velocity of 400 mph in the opposite direction to the velocity of the body. The 400 mph velocity is added to all the particles of the flow. This will give the same flow pattern as shown in the figure, but because the velocity is constant, the pressures will not change. Consider the undisturbed stream of air (Figure) at standard pressure and temperature. The dotted lines are streamlines. The center streamline which hits the nose of the object has a 400 mph velocity at the undistrubed point 1 and a zero velocity at the nose which is point 2.

(a) For an incompressible fluid

ρ = constant

Write the modified Euler equation between points 1 and 2.

$$\frac{V_2^2 - V_1^2}{2} + g(z_2 - z_1) + \int_1^2 \frac{dp}{\rho} = 0 \ . \ . \ . \qquad (1)$$

Since ρ is a constant, take it outside the integral sign. Then integrate to get

$$\frac{V_2^2 - V_1^2}{2} + g(z_2 - z_1) + \frac{p_2 - p_1}{\rho} = 0 \qquad (2)$$

Points 1 and 2 are at the same elevation; therefore, $z_2 = z_1$. Substitute the given values of $V_1 = 400$ mph, $V_2 = 0$, $p_1 = 14.7$ psia into Equation (2)

$$0 - \frac{1}{2}\left(\frac{(400)(5280)}{3600}\right)^2 + 0 + \frac{p_2(144) - (14.7)(144)}{0.00238} = 0$$

$$-172{,}600 + \frac{p_2(144) - (14.7)(144)}{0.00238} = 0$$

$$p_2 = 14.7 + \frac{(172{,}600)(0.00238)}{144} = 17.55 \text{ psia}$$

(b) For a perfect gas and an isothermal process

$$\frac{p}{\rho} = g_c RT$$

where T = constant

$$\frac{p}{\rho} = \text{constant}$$

Again write Equation (1) between points 1 and 2.

$$\frac{V_2^2 - V_1^2}{2} + g(z_2 - z_1) + \int_{p1}^{p2} \frac{dp}{\rho} = 0 \qquad (1)$$

The Bernoulli equation was derived from this equation for incompressible fluids or flow when we could neglect the compressibility effects.

All the terms in the equation are simple and straightforward except the term that includes the integral sign. To carry out this integration, we must know the relation between p and ρ. We assume that this relation is known and can be written as

$$\rho = f(p)$$

where f is a known function. This function is then substituted for ρ in Equation (1) to get

$$\frac{V_2^2 - V_1^2}{2} + g(z_2 - z_1) + \int_{p1}^{p2} \frac{dp}{f(p)} = 0 \qquad (3)$$

Now we can carry out the integration and use the equation for compressible flow in the same way the Bernoulli equation is used for incompressible flow. Equation (3) has limitations as follows: the flow must be along a streamline, it must be steady and frictionless, and the relationship between ρ and p must be written $\rho = f(p)$.

$$\frac{V_2^2 - V_1^2}{2} + g(z_2 - z_1) + \int_1^2 \frac{dp}{f(p)} = 0$$

Given

$$V_2 = 0, \quad V_1 = \frac{(400)(5280)}{3600} = 588 \text{ fps}$$

$$z_1 = z_2$$

which when substituted in the equation between points 1 and 2 gives

$$\frac{0 - (588)^2}{2} + 0 + \int_{p1}^{p2} \frac{dp}{\rho} = 0 \qquad (4)$$

$$\frac{p}{\rho} = \frac{p}{\rho_1} \quad \text{as } \frac{p}{\rho} = \text{constant}$$

therefore $\quad \dfrac{1}{\rho} = \dfrac{p_1}{\rho_1 p}$

which you substitute in equation (4).

$$-172{,}600 + \frac{p_1}{\rho_1}\int_{p_1}^{p_2} \frac{dp}{p} = 0$$

$$-172{,}600 + \frac{(14.7)(144)}{0.00238} \log_e p \Big|_{p_1}^{p_2} = 0$$

$$-172{,}600 + 890{,}000 \log_e \frac{p_2}{p_1} = 0$$

$$\log_e \frac{p_2}{p_1} = \frac{172{,}600}{890{,}000}$$

but $p_1 = (14.7)(144)$

therefore,

$$\log_e \left[\frac{p_2(144)}{(14.7)(144)}\right] = 0.194$$

and

$$p_2 = (14.7)e^{0.194} = (14.7)(1.214) = 17.85 \text{ psia}$$

During the pressure increase, the fluid temperature probably will increase and so this result is not the true case.

(c) For an isentropic process

$$\frac{p}{\rho^k} = \text{constant}$$

where k = 1.4 for air.

Substitute the appropriate values into Equation (3) to obtain

$$-\ 172{,}600 + \int_{p1}^{p2} \frac{dp}{\rho} = 0 \qquad (5)$$

Now,

$$\frac{p}{\rho^{1.4}} = \frac{p_1}{\rho_1^{1.4}} = \frac{p_2}{\rho_2^{1.4}}$$

Rearrange this to

$$\frac{1}{\rho} = \left(\frac{p_1}{p}\right)^{1/1.4} \frac{1}{\rho_1}$$

and substitute it into Equation (5)

$$-\ 172{,}600 + \frac{p_1^{1/1.4}}{\rho_1} \int_{p1}^{p2} \frac{dp}{p^{1/1.4}} = 0$$

$$-\ 172{,}600 + \frac{p_1^{1/1.4}}{\rho_1} \left. \frac{p^{1-1/1.4}}{1 - \frac{1}{1.4}} \right|_{p1}^{p2} = 0$$

$$-\ 172{,}600 + \frac{p_1^{5/7}}{\rho_1} \frac{7}{2} (p_2^{2/7} - p_1^{2/7}) = 0$$

Substitute

$$p_1 = (14.7)(144) \text{ psia}$$

and

$$\rho_1 = 0.00238 \text{ slug/ft}^3$$

Thus,

$$-\ 172{,}600 + \frac{[(14.7)(144)]^{5/7}}{0.00238} \left(\frac{7}{2}\right) \left\{ p_2^{2/7} - [(14.7)(144)]^{2/7} \right\} = 0$$

$$-\ 172{,}600 + \frac{236}{0.00238} \frac{7}{2} p_2^{2/7} - \frac{7}{2} \frac{(14.7)(144)}{0.00238} = 0$$

$$347{,}000 p_2^{2/7} = 172{,}600 + \frac{(14.7)(144)\left(\frac{7}{2}\right)}{0.00238}$$

$$p_2^{2/7} 347{,}000 = 3{,}287{,}100$$

$$p_2 = \left(\frac{3{,}287{,}100}{347{,}000}\right)^{7/2} = 2600 \text{ psfa}$$

$$= \frac{2600}{144} = 18.05 \text{ psia}$$

• PROBLEM 12-17

An airflow ($\gamma = 1.4$) is expanded isentropically in a nozzle from $M_1 = 0.3$, $A_1 = 1.0$ ft^2, to a Mach number M_2 of 3.0. Determine (a) the minimum nozzle area, (b) A_2, (c) p_2/p_1 and (d) T_2/T_1 (See Figure).

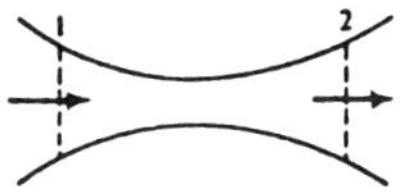

Solution: (a) Since flow in the nozzle goes from subsonic to supersonic speeds, the flow must pass through a minimum area A* at which M = 1. From the Table , at $M_1 = 0.3$, $A_1/A^* = 2.0351$ so that the minimum area = 1/2.0351 = 0.491 ft^2.

TABLE FOR $\frac{p}{p_t}$

$\gamma = 1.4$

M	$\frac{p}{p_t}$	$\frac{T}{T_t}$	$\frac{A}{A_*}$	M	$\frac{p}{p_t}$	$\frac{T}{T_t}$	$\frac{A}{A_*}$
0	1.0000	1.0000	∞	0.30	0.9395	0.9823	2.0351
.01	.9999	1.0000	57.8738	.31	.9355	.9811	1.9765
.02	.9997	.9999	28.9421	.32	.9315	.9799	1.9219
.03	.9994	.9998	19.3005	.33	.9274	.9787	1.8707
.04	.9989	.9997	14.4815	.34	.9231	.9774	1.8229
.05	.9983	.9995	11.5914	.35	.9188	.9761	1.7780
.06	.9975	.9993	9.6659	.36	.9143	.9747	1.7358
.07	.9966	.9990	8.2915	.37	.9098	.9733	1.6961
.08	.9955	.9987	7.2616	.38	.9052	.9719	1.6587
.09	.9944	.9984	6.4613	.39	.9004	.9705	1.6234
.10	.9930	.9980	5.8218	.40	.8956	.9690	1.5901
.11	.9916	.9976	5.2992	.41	.8907	.9675	1.5587
.12	.9900	.9971	4.8643	.42	.8857	.9659	1.5289
.13	.9883	.9966	4.4969	.43	.8807	.9643	1.5007
.14	.9864	.9961	4.1824	.44	.8755	.9627	1.4740

TABLE FOR $\frac{T}{T_t}$

M or M_1	$\frac{p}{p_t}$	$\frac{T}{T_t}$	$\frac{A}{A_*}$
2.90	0.3165 -1	0.3729	3.850
2.91	.3118 -1	.3712	3.887
2.92	.3071 -1	.3696	3.924
2.93	.3025 -1	.3681	3.961
2.94	.2980 -1	.3665	3.999
2.95	.2935 -1	.3649	4.038
2.96	.2891 -1	.3633	4.076
2.97	.2848 -1	.3618	4.115
2.98	.2805 -1	.3602	4.155
2.99	.2764 -1	.3587	4.194
3.00	.2722 -1	.3571	4.235
3.01	.2682 -1	.3556	4.275
3.02	.2642 -1	.3541	4.316
3.03	.2603 -1	.3526	4.357
3.04	.2564 -1	.3511	4.399

(b) At $M_2 = 3.0$, $A_2/A^* = 4.235$ so $A_2 = 2.08$ ft^2.

(c) For this isentropic flow, T_t and p_t are constants.

$$\frac{p_2}{p_1} = \frac{p_2/p_{t_2}}{p_1/p_{t_1}} = \frac{0.0272}{0.9395} = 0.0290 \qquad \text{(see the Table for } \frac{p}{p_t}\text{)}$$

(d) $$\frac{T_2}{T_1} = \frac{T_2/T_{t_2}}{T_1/T_{t_1}} = \frac{0.3571}{0.9823} = 0.363 \qquad \text{(see the Table for } \frac{T}{T_t}\text{)}$$

• PROBLEM 12-18

At one point in an air duct the temperature of the flow is 200° F and the local pressure is 30 psia. At this point the cross-sectional area of the duct is 1 ft^2. Downstream of this point the flow temperature is 30° F at a point where the pressure is 15 psia and the area of flow is 0.3 ft^2. Calculate the velocity of flow at the second point and the mass flow rate.

Solution: Bernouilli's equation for incompressible ideal flow was established from

$$\frac{dp}{\gamma} + \frac{vdv}{g} + dz = 0 \qquad (1)$$

assuming γ to be constant. The compressible form may be established from the same root, remembering that γ is now a variable.

Integrating Eq. (1) between limits gives

$$\int_1^2 \frac{dp}{\gamma} = \int_2^1 \frac{v\,dv}{g} + \int_2^1 dz$$

$$= \frac{v_2^2 - v_1^2}{2g} + (z_2 - z_1) \qquad (2)$$

Now it is usual when dealing with compressible gas flow problems to assume that all processes are rapid and therefore adiabatic. Then the following equation is applicable

$$\frac{p}{\gamma^k} = c \qquad \text{a constant}$$

or

$$p = c\gamma^k$$

and so

$$dp = ck\gamma^{k-1}\,d\gamma$$

Substituting these in Eq. (2)

$$\frac{v_2^2 - v_1^2}{2g} + (z_2 - z_1) = \int_1^2 \frac{ck\gamma^{k-1}\,d\gamma}{\gamma}$$

$$= \int_1^2 ck\gamma^{k-2}\,d\gamma$$

$$= \frac{ck}{k-1}(\gamma_1^{k-1} - \gamma_2^{k-1})$$

$$= \frac{k}{k-1} c\gamma_2^{k-1}\left[\left(\frac{\gamma_1}{\gamma_2}\right)^{k-1} - 1\right]$$

But

$$c\gamma_2^{k-1} + c\frac{\gamma_2^k}{\gamma_2} = \frac{p}{\gamma_2}$$

Therefore

$$\frac{v_2^2 - v_1^2}{2g} + (z_2 - z_1) = \frac{p_2 k}{\gamma_2 (k-1)} \left[\left(\frac{\gamma_1}{\gamma_2} \right)^{k-1} - 1 \right]$$

Very often in problems involving the use of this equation, the term $(z_2 - z_1)$ is negligible, in which case

$$\frac{v_2^2 - v_1^2}{2g} = \frac{p_2 k}{\gamma_2 (k-1)} \left[\left(\frac{\gamma_1}{\gamma_2} \right)^{k-1} - 1 \right] \qquad (3)$$

This is one form of Bernoulli's equation for ideal compressible flow. By making the substitution

$$\frac{\gamma_1}{\gamma_2} = \left(\frac{p_1}{p_2} \right)^{1/k}$$

Eq. (3) can be written as

$$\frac{v_2^2 - v_1^2}{2g} = \frac{p_2}{\gamma_2} \frac{k}{k - 1} \left[\left(\frac{p_1}{p_2} \right)^{(k-1)/k} - 1 \right] \qquad (4)$$

By considering the relationship

$$\frac{\gamma_2}{\gamma_1} = \frac{p_2 T_1}{p_1 T_2} = \left(\frac{p_2}{p_1} \right)^{1/k}$$

and dividing by p_2/p_1 ,

$$\frac{T_1}{T_2} = \left(\frac{p_2}{p_1} \right)^{(1/k)-1} = \left(\frac{p_2}{p_1} \right)^{(1-k)/k} = \left(\frac{p_1}{p_2} \right)^{(k-1)/k}$$

Eq. (4) is reduced to

$$\frac{v_2^2 - v_1^2}{2g} = \frac{p_2}{\gamma_2} \frac{k}{k - 1} \left(\frac{T_1}{T_2} - 1\right)$$

$$= RT_2 \frac{k}{k - 1} \left(\frac{T_1}{T_2} - 1\right)$$

$$= \frac{Rk}{k - 1} (T_1 - T_2) \tag{5}$$

In this problem

$$\gamma_1 = \frac{p_1}{RT_1} = \frac{30 \times 144}{53.3 \times 660} = 0.1225 \text{ lb/ft}^3$$

and

$$\gamma_2 = \frac{p_2}{RT_2} = \frac{15 \times 144}{53.3 \times 490} = 0.0825 \text{ lb/ft}^3$$

Now

$$G = \frac{\gamma_1 A_1 v_1}{g} = \frac{\gamma_2 A_2 v_2}{g}$$

Therefore

$$v_1 = \frac{\gamma_2 A_2 v_2}{\gamma_1 A_1} = \frac{0.0825}{0.1225} \times 0.3v_2 = 0.202v_2$$

Applying Eq. (5),

$$\frac{v_2^2 - v_1^2}{2g} = \frac{Rk}{k - 1} (T_1 - T_2)$$

$$\frac{v_2^2}{2g} (1 - 0.202^2) = \frac{53.3 \times 1.4 \times 170}{0.4}$$

$$v = \sqrt{64.4 \times 33,100} = 1,460 \text{ fps}$$

Hence the mass flow rate is

$$\frac{\gamma_2 A_2 v_2}{g} = \frac{0.0825 \times 0.3 \times 1,460}{32.2} = 1.13 \text{ slugs/sec}$$

A high-speed aeroplane flies steadily at 2500 ft/sec through an atmosphere where the temperature and pressure are 410° F and 628 lbf/ft^2 respectively. Calculate the stagnation temperature and pressure on the nose of the aircraft. Treat air as a perfect gas (γ = 1.40, R = 53.3 ft lbf/lbm °R.)

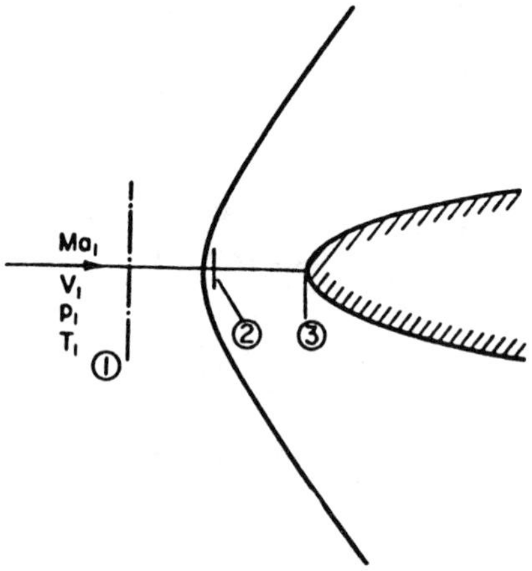

Fig. 1

Solution: The flow pattern in the region of the aircraft nose is rather complicated, but along the central streamline, as indicated in Fig. 1, we may treat the flow as adiabatic, and reversible except in the shock wave, which may be taken as perpendicular to the streamline. For convenience we consider motion relative to the aircraft.

At 410 °R the acoustic velocity is given by

$$a_1 = (\gamma R g_0 T_1)^{\frac{1}{2}}$$

$$= (1.4 \times 53.3 \times 32.2 \times 410)^{\frac{1}{2}} \text{ ft/sec} = 992 \text{ ft/sec},$$

The ratio of a fluid velocity V to the local acoustic velocity is called the Mach number of that velocity, ie.,

Mach number, $$Ma \equiv \frac{V}{a}$$

so that Mach number of the oncoming air is

$$Ma_1 = \frac{2500}{992} = 2.52;$$

In terms of the Mach number the expressions for the ratios of stagnation and static properties derived in the previous section take rather simpler forms, thus

$$\frac{T_s}{T} = \left(1 + \frac{\gamma - 1}{2} Ma^2\right) \quad (1)$$

$$\frac{p_s}{p} = \left(1 + \frac{\gamma - 1}{2} Ma^2\right)^{\gamma/(\gamma-1)} \quad (2)$$

Therefore, its stagnation temperature is (using equation 1)

$$T_{s1} = T_1 \left(1 + \frac{\gamma - 1}{2} Ma_1^2\right)$$

$$= 410[1 + 0.2 \times (2.52)^2]°R = 410 \times 2.270°R = 930°R$$

and its stagnation pressure (using equation 2) is

$$P_{s1} = p_1 \left(1 + \frac{\gamma - 1}{2} Ma_1^2\right)^{\gamma/(\gamma-1)}$$

$$= 628 \times (2.270)^{3.5} \text{ lbf/ft}^2 = 11,060 \text{ lbf/ft}^2.$$

On the assumption that the flow is adiabatic, the stagnation temperature has the same value for all points on the streamline so that

$$T_{s3} = T_{s2} = T_{s1} = 930° R.$$

For the control volume defined by Fig. 2, the continuity and force-momentum equations have the following forms:

$$\rho_1 V_1 = \rho_2 V_2 \;,$$

and

$$g_0(p_1 - p_2) = \rho_2 V_2^2 - \rho_1 V_1^2 \;.$$

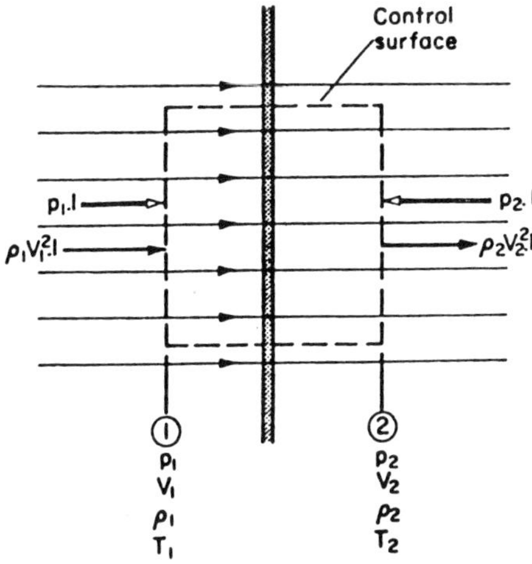

Fig. 2

By substitution for ρ from the perfect-gas relation $p = \rho RT$, and for

$$V[=Ma(\gamma R g_0 T)^{\frac{1}{2}}],$$

these may be rewritten respectively as

$$\frac{p_1 Ma_1}{T_1^{\frac{1}{2}}} = \frac{p_2 Ma_2}{T_2^{\frac{1}{2}}} \tag{3}$$

and

$$p_1(1 + \gamma Ma_1^2) = p_2(1 + \gamma Ma_2^2) \ . \tag{4}$$

The elimination of p_1/p_2 from these two equations gives

$$\frac{T_1}{T_2} = \left[\frac{Ma_1(1 + \gamma Ma_2^2)}{Ma_2(1 - \gamma Ma_1^2)}\right]^2 \ . \tag{5}$$

But from the steady-flow energy equation,

$$\frac{T_1}{T_2} = \frac{1 + \frac{1}{2}(\gamma - 1)Ma_2^2}{1 + \frac{1}{2}(\gamma - 1)Ma_1^2} \tag{6}$$

so that

$$\left[\frac{Ma_1(1 + \gamma Ma_2^2)}{Ma_2(1 + \gamma Ma_1^2)}\right]^2 = \frac{1 + \frac{1}{2}(\gamma - 1)Ma_2^2}{1 + \frac{1}{2}(\gamma - 1)Ma_1^2} \tag{7}$$

Equation 7 is effectively a quadratic relation between Ma_1^2 and Ma_2^2, and can be solved by the standard procedure to give

$$Ma_2^2 = \frac{1 + \frac{1}{2}(\gamma - 1)Ma_1^2}{\gamma Ma_1^2 - \frac{1}{2}(\gamma - 1)} \tag{8}$$

or

$$Ma_2^2 = Ma_1^2 \tag{9}$$

From the expression for Ma_2 the ratios of fluid properties upstream and downstream of the shock can be found by

substitution in the appropriate preceding equations:

$$\frac{T_2}{T_1} = \frac{2[1 + \frac{1}{2}(\gamma - 1)Ma_1^2](2\gamma Ma_1^2 - \gamma + 1)}{(\gamma + 1)Ma_1^2} \tag{10}$$

$$\frac{p_2}{p_1} = \frac{2\gamma Ma_1^2 - \gamma + 1}{\gamma + 1}, \tag{11}$$

$$\frac{\rho_2}{\rho_1} = \frac{V_1}{V_2} = \frac{(\gamma + 1)Ma_1^2}{2[1 + \frac{1}{2}(\gamma - 1)Ma_1^2} \tag{12}$$

Further, by making use of the relations between stagnation and static pressure, we obtain

$$\frac{p_{s_2}}{p_{s_1}} = \left[\frac{\gamma + 1}{2\gamma Ma_1^2 - \gamma + 1}\right]^{1/(\gamma-1)} \left[\frac{\frac{1}{2}(\gamma + 1)Ma_1^2}{1 + \frac{1}{2}(\gamma - 1)Ma_1^2}\right]^{\gamma/(\gamma-1)} \tag{13}$$

Calculations of the stagnation pressure on the downstream side of the shock requires the use of equation 13; if the flow between the shock wave and the stagnation point is reversible no change in stagnation pressure occurs except in the shock wave itself, and

$$p_{s_3} = p_{s_2} = p_{s_1}\left[\frac{\gamma + 1}{2\gamma Ma_1^2 - \gamma + 1}\right]^{1/(\gamma-1)} \times$$

$$\left[\frac{\frac{1}{2}(\gamma + 1)Ma_1^2}{1 + \frac{1}{2}(\gamma - 1)Ma_1^2}\right]^{\gamma/(\gamma-1)}$$

$$= 11060 \left(\frac{2.40}{17.38}\right)^{2.5} \left(\frac{7.62}{2.270}\right)^{3.5} \text{lbf/ft}^2$$

$$= 11060 \times 0.491 \text{ lbf/ft}^2 = 5420 \text{ lbf/ft}^2.$$

• PROBLEM 12-20

What is the maximum air velocity that will not exceed a 2°F temperature rise at a stagnation point? Assume air at 70°F and 14.7 psia.

TABLE

Gas	Chemical Formula	Specific Weight, γ		Density, ρ		Gas Constant, R		Adiabatic Constant, k
		lb/ft³	N/m³	slugs/ft³	kg/m³	ft/°R	m/°K	
Air	—	0.0753	11.8	0.00234	1.206	53.3	29.2	1.40
Ammonia	NH_3	0.0448	7.04	0.00139	0.716	89.5	49.1	1.32
Carbon dioxide	CO_2	0.115	18.1	0.00357	1.840	34.9	19.1	1.29
Helium	He	0.0104	1.63	0.000323	0.166	386.	212.	1.66
Hydrogen	H_2	0.00522	0.820	0.000162	0.0835	767.	421.	1.40
Methane	CH_4	0.0416	6.53	0.00129	0.665	96.4	52.9	1.32
Nitrogen	N_2	0.0726	11.4	0.00225	1.160	55.2	30.3	1.40
Oxygen	O_2	0.0830	13.0	0.00258	1.330	48.3	26.5	1.40
Sulfur dioxide	SO_2	0.170	26.7	0.00528	2.721	23.6	12.9	1.26

Solution: At 70°F, $\rho = 0.00233$ slug/ft^3,

so

$$c = \sqrt{\frac{kp}{\rho}} = \sqrt{\frac{(1.4)(14.7)(144)}{0.00344}} = 1128 \text{ ft/s}$$

Also,

$$T_1 = 70 + 460 = 530°R$$

and the stagnation temperature T_2 is

$$T_2 = 72 + 460 = 532°R$$

The energy equation, with $h = u + p/\gamma$, may be expressed in the form

$$\frac{V_1^2}{2g} + y_1 + h_1 = \frac{V_2^2}{2g} + y_2 + h_2$$

For purposes of application, certain thermodynamic relationships for perfect gases are required. Introducing c_p and c_v as the coefficients of specific heat at constant pressure and constant volume, respectively, we have, from thermodynamics,

$$k = c_p/c_v$$

and

$$c_p - c_v = R \qquad \left(\frac{\text{ft-lb}}{\text{lb-°R}} \quad \text{or} \quad \frac{\text{m}\cdot\text{N}}{\text{N.°K}}\right) \tag{1}$$

Combining, we obtain

$$c_p = \frac{Rk}{k-1} \tag{2}$$

Additionally, the thermodynamic relationship for internal energy,

$$u = c_v T$$

may be combined with the perfect gas law yielding for the enthalpy

$$h = u + p/\gamma$$

$$= c_v T + RT$$

$$= (c_v + R)T$$

and therefore from Eq. 1

$$h = c_p T$$

or, from Eq. 2,

$$h = \frac{RkT}{k - 1}$$

Thus, the energy equation becomes alternatively,

$$\frac{V_1^2}{2g} + y_1 + c_p T_1 = \frac{V_2^2}{2g} + y_2 + c_p T_2$$

or

$$\frac{V_1^2}{2g} + y_1 + \frac{RkT_1}{k - 1} = \frac{V_2^2}{2g} + y_2 + \frac{RkT_2}{k - 1}$$

The flow will be frictionless and adiabatic, a condition referred to as isentropic flow. We will also ignore the effects of elevation differences. The energy equation may immediately be espressed as either

$$\frac{V_1^2}{2g} + c_p T_1 = \frac{V_2^2}{2g} + c_p T_2 \qquad (3)$$

or

$$\frac{V_1^2}{2g} + \frac{RkT_1}{k - 1} = \frac{V_2^2}{2g} + \frac{RkT_2}{k - 1} \qquad (4)$$

For isentropic flow, Eq. 4 may be arranged to

$$\frac{V_2^2}{2g} - \frac{V_1^2}{2g} = \frac{Rk}{k - 1} (T_1 - T_2) \qquad (5)$$

Since the temperature change at a stagnation point is of interest, Eq. 5 gives

$$T_2 - T_1 = \frac{V_1^2}{2g} \frac{k - 1}{Rk}$$

Thus, the stagnation temperature (T_2) is

$$T_2 = T_1 + \frac{V_1^2}{2g} \frac{k - 1}{Rk}$$

and therefore

$$\frac{T_2}{T_1} = 1 + \frac{k - 1}{2} \frac{V_1^2}{gKRT_1}$$

or

$$\frac{T_2}{T_1} = 1 + \frac{k - 1}{2} Ma_1^2 \tag{6}$$

The maximum Mach number is given by Eq. 6:

$$\frac{523}{530} = 1 + \frac{1.4 - 1}{2} (Ma_1^2)$$

Solving, we obtain

$$Ma_1^2 = 0.0189$$

and

$$Ma_1 = 0.137$$

Finally,

$$Ma_1 = 0.137 = \frac{V_1}{1128}$$

and

$$V_1 = 153 \text{ ft/s}$$

• PROBLEM 12-21

A supersonic wind-tunnel consists of a large reservoir containing gas under high pressure which is discharged through a convergent-divergent nozzle to a test section of constant cross-sectional area. The cross-sectional area of the throat of the nozzle is 500 mm^2 and the Mach number in the test section is 4. Calculate the cross-sectional area of the test section assuming $\gamma = 1.4$.

Solution: If ρ, $\bar{v}$ and A are the density, velocity and cross-sectional area at any section of the nozzle and ρ_t, $\bar{v}_t$, A_t are the critical values at the throat, then, since the mass flow rate is the same at each cross-section,

$$\rho\bar{v}A = \rho_t\bar{v}_tA_t ,$$

$$A/A_t = \rho_t\bar{v}_t/\rho\bar{v} . \qquad (1)$$

The velocity at any point can be expressed in terms of the Mach number at that point and the local speed of sound,

$$\bar{v} = \text{Ma } c = \text{Ma} \sqrt{(\gamma RT)}$$

for adiabatic conditions.

At the throat, Ma = 1 and $T = T_t$, so that $\bar{v}_t = \sqrt{(\gamma RT_t)}$. Substituting in equation (1),

$$\frac{A}{A_t} = \frac{\rho_t}{\rho}\left(\frac{T_t}{T}\right)\frac{1}{\text{Ma}} . \qquad (2)$$

Now, for the isentropic flow from a large reservoir in which the conditions are given by p_0, ρ_0 and T_0 and $\bar{v}_0$ is zero, from Bernoulli's equation at any section of the nozzle,

$$\frac{\bar{v}^2}{2} = \left(\frac{\gamma}{\gamma - 1}\right) R(T_0 - T) .$$

Dividing by c^2, where $c = \sqrt{(\gamma RT)}$, the local velocity of sound, and rearranging,

$$\frac{\bar{v}^2}{c^2} = \text{Ma}^2 = \frac{2}{\gamma - 1}\left(\frac{T_0}{T} - 1\right) .$$

$$\frac{T_0}{T} = 1 + \left(\frac{\gamma - 1}{2}\right)\text{Ma}^2 \qquad (3)$$

and since, for isentropic flow,

$$\frac{T_0}{T} = \left(\frac{p_0}{p}\right)^{(\gamma-1)/\gamma} = \left(\frac{\rho_0}{\rho}\right)^{(\gamma-1)}$$

$$\frac{p_0}{p} = \left(1 + \frac{\gamma - 1}{2}\text{Ma}^2\right)^{\gamma/(\gamma-1)} \qquad (4)$$

and

$$\frac{\rho_0}{\rho} = \left(1 + \frac{\gamma - 1}{2} Ma^2\right)^{1/(\gamma-1)} \qquad (5)$$

we have,

$$\frac{\rho_t}{\rho} = \frac{\rho_t}{\rho_0} \times \frac{\rho_0}{\rho} = \left\{\frac{1 + [(\gamma - 1)/2]Ma^2}{(\gamma + 1)/2}\right\}^{1/(\gamma-1)}$$

and

$$\frac{T_t}{T} = \frac{T_t}{T_0} \times \frac{T_0}{T} = \left\{\frac{1 + [\gamma - 1)/2]Ma^2}{(\gamma + 1)/2}\right\} .$$

Substituting these values in equation (2),

$$\frac{A}{A_t} = \frac{1}{Ma}\left\{\frac{1 + [(\gamma - 1)/2]Ma^2}{(\gamma + 1)/2}\right\}^{(\gamma+1)/2(\gamma-1)} . \qquad (6)$$

From equation (6), putting $\gamma = 1.4$ and Ma = 4,

$$\frac{A}{A_t} = \frac{1}{4}\left\{\frac{1 + 0.2 \times 4^2}{1.2}\right\}^{2.4/0.8} = 10.72,$$

Area of test section = 10.72 × 500 = 5360 mm^2.

• PROBLEM 12-22

(a) It is desired to pump methane (k = 1.31 and R = 518 J/kg °K) through a 30-cm commercial steel pipe. The discharge from a compressor is at a pressure of 2000 kPa abs, a temperature of 60°C (dynamic viscosity is 1.3×10^{-5} kg/m s), and a velocity of 15 m/s. For both isothermal and adiabatic flow with friction, find (a) the minimum pressure possible; (b) the maximum length of pipe possible; (c) the maximum velocity possible; (d) the pressure, temperature, velocity, and Mach number at a distance equal to one-half the maximum length from the compressor; and (e) the location of a second compressor if the compressor inlet pressure is 500 kPa abs.

(b) Carbon dioxide (R = 188.5 J/kg °K, k = 1.30 and $\mu = 2.2 \times 10^{-5}$ kg/m s) enters a 60-mm smooth tube at a pressure of 1200 kPa abs, a velocity of 30 m/s, and a density of 20 kg/m^3. The tube discharges the gas into the atmosphere. Compare the maximum allowable tube length for isothermal flow with that for adiabatic flow.

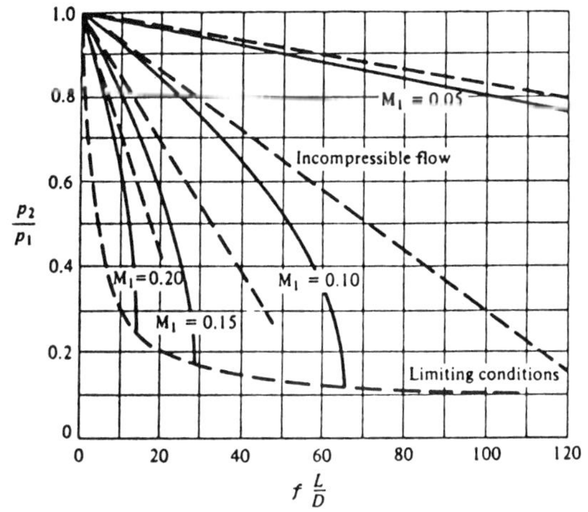

Fig. 1

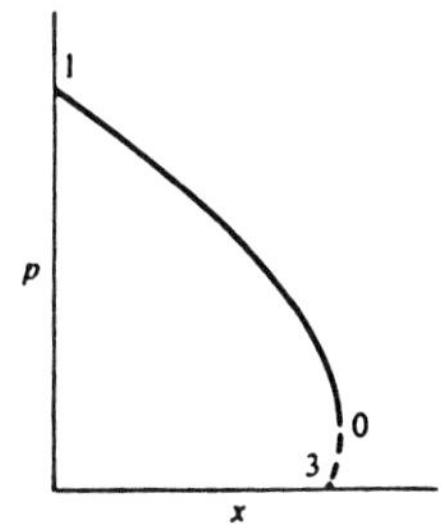

Fig. 2

Solution: Subsonic gas flow in pipes differs from liquid flow in pipes primarily in that the gas density decreases and hence the velocity increases in the direction of flow. The wall shear stess may be expressed in terms of the friction factor, so that the pressure gradient is

$$-\frac{dp}{dx} = \frac{f}{D}\frac{\rho V^2}{2} + \beta\rho V\frac{dV}{dx} \tag{1}$$

This equation is similar to the Darcy-Weisbach equation except for the last term on the right, which represents the pressure drop required to increase the flow momentum. The value of β is about 1.03 for fully developed incompressible turbulent flow. Thus a one-dimensional analysis, with $\beta = 1.0$, is justified.

A dimensional analysis would indicate that the friction factor for compressible flow could depend upon the Mach number as well as the relative roughness of the pipe and the Reynolds number. Experiments have shown, however, that the dependence on Mach number for subsonic flow is negligible and that the friction factor may be obtained in the same manner as for incompressible flow.

Equation (1) with $\beta = 1$ may be written in dimensionless form as

$$\frac{dp}{\rho V^2/2} + f\frac{dx}{D} + 2\frac{dV}{V} = 0 \tag{2}$$

This general equation will be integrated directly for isothermal flow and used in the analysis for adiabatic flow in pipes with friction.

ISOTHERMAL FLOW

In order to integrate Eq. (2), the variable density and velosity will have to be expressed in terms of the variable pressure, and the variation in the friction factor f will have to be investigated.

If all conditions are known at some upstream section (section 1), those at any arbitrary section downstream can be expressed in terms of known values at section 1.

From the perfect gas equation of state,

$$\frac{p}{\rho} = \frac{p_1}{\rho_1} = RT$$

a constant, from which $\rho = p\rho_1/p_1$.

From continuity,

$$V\rho = V_1\rho_1$$

then $V = V_1p_1/p$. The differential form of this last equation (for later use) is

$$\frac{dV}{V} + \frac{dp}{p} = 0$$

or

$$\frac{dV}{V} = -\frac{dp}{p}$$

The first term in Eq. (2) can then be written as

$$\frac{2}{\rho_1V_1^2p_1}\ p\ dp$$

The friction factor depends on the relative roughness of the pipe, which is assumbed to be constant, and on the Reynolds number $Re = VD\rho/\mu$, which is also constant, since $V\rho$ is constant, from continuity, and the dynamic viscosity depends only on the temperature, which is constant. Hence the friction factor is constant and can be found in the usual manner for known conditions at any section.

If $L = x_2 - x_1$, Eq. (2) can now be integrated to obtain

$$p_1^2 - p_2^2 = \rho_1V_1^2p_1\left(f\frac{L}{D} - 2\ \ln\frac{p_2}{p_1}\right)$$

or

$$p_1^2 - p_2^2 = kM_1^2 p_1^2 \left(f \frac{L}{D} - 2 \ln \frac{p_2}{p_1} \right) \tag{3}$$

in terms of the initial Mach number. These equations give the pressure at some distance L downstream of any initial section 1 where conditions are known. The logarithmic term is often small compared to fL/D and may be neglected for a first approximation. Then this first approximation for p_2 should be used to calculate the magnitude of the logarithmic term, and an iterative process used to calculate p_2 precisely.

Equation (3) may be solved for fL/D to give

$$f \frac{L}{D} = \frac{1}{kM_1^2} \left[1 - \left(\frac{p_2}{p_1} \right)^2 \right] - 2 \ln \frac{p_1}{p_2} \tag{4}$$

This equation gives the distance L from section 1, where conditions are known, to some downstream section where p_2 is specified. It is a dimensionless equation and indicates that the dimensionless length fL/D is related to the pressure ratio p_2/p_1 by a family of curves, one for each initial Mach number, for a given gas (given k). These relationships are shown in Fig. 1

A dimensional plot of pressure vs. length would have the same general appearance as the curves in Fig. 1. Equation (2), by letting $dV/V = -dp/p$, contains only two differentials, dp and dx. Thus the pressure gradient, which is the slope of the curves of Figs. 1 and 2, can be shown to be

$$\frac{dp}{dx} = \frac{\frac{pf}{2D}}{1 - \frac{p}{\rho V^2}} = \frac{\frac{f}{D} \frac{\rho V^2}{2}}{kM^2 - 1} \tag{5}$$

As $M \rightarrow 0$, this is essentially the Darcy equation for liquid pipe flow; thus at low Mach numbers, isothermal flow may be considered as incompressible. The pressure gradients will be within 5 percent at Mach numbers up to 0.184 for air. A thermodynamic analysis will indicate the entropy increases as the fluid goes from 1 to 0 in Fig. 2, and decreases from 0 to 3. This latter decrease is impossible since it contradicts the second law of thermodynamics. Thus flow can exist only up to 0, which is called the limiting point. Associated with this limiting point are a limiting Mach number, a limiting or minimum pressure, and a limiting or maximum length of pipe. These will be indicated by asterisks, (M^*). Flow from 3 to 0 represents supersonic flow.

* Limiting or Maximum Mach Number. At the limiting point the pressure gradient is infinite. Thus for dp/dx to be infinite, the denominator of Eq. (3) must be zero. Then

$$M^* = \frac{1}{\sqrt{k}}$$

and the velocity can increase along the pipe only until the Mach number equals $1/\sqrt{k} = 0.845$ for air, for example.

* Limiting or Minimum Pressure. Recall that from continuity

$$V_1 p_1 = Vp = V^*p^*$$

so that

$$\frac{p^*}{p_1} = \frac{V_1}{V^*} = \frac{M_1 c_1}{M^* c^*} = \frac{M_1}{M^*} = M_1\sqrt{k}$$

since the acoustic velocity remains constant for isothermal conditions. Therefore,

$$\frac{p^*}{p_1} = M_1\sqrt{k} \qquad (6)$$

and the pressure drops along the pipe to a minimum which depends on the initial Mach number.

* Limiting or Maximum Length. The limiting or maximum length of pipe can be found by inserting the expression for the minimum pressure, Eq. 6, into the general expression for fL/D, Eq. (4), to get

$$\frac{fL^*}{D} = \frac{1}{kM_1^2} - 1 - \ln\frac{1}{kM_1^2} \qquad (7)$$

A number of conclusions may be drawn concerning subsonic isothermal flow:

1. The pressure drops at an increasing rate along the pipe, in contrast with a constant pressure gradient for fully developed liquid flow.

2. The velocity and Mach number increase up to a maximum Mach number of $1/\sqrt{k}$. This increase ceases at a limiting or maximum length, which must be at the end of the pipe. If assumed initial conditions at some upstream section result in a maximum length of pipe which is less than a given pipe length, then the initial conditions will have to be adjusted (lower M_1) so that the given pipe length becomes at least equal to or less than the maximum length for the adjusted initial conditions.

3. The limiting pressure ratio and the limiting or maximum length depend only on the gas (k value) and the initial Mach number. Thus tables may be prepared and used for making pipe calculations. Then

$$\frac{p_2}{p_1} = \frac{M_1}{M_2}$$

and from Eq. (7)

$$\frac{fL}{D} = \left(\frac{fL^*}{D}\right)_{M_1} - \left(\frac{fL^*}{D}\right)_{M_2}$$

These equations relate pressures and lengths between two sections for which the Mach numbers are given.

It is interesting to speculate whether heat is added or removed from the gas as it flows through a pipe. The general energy equation becomes

$$\frac{V_1^2}{2} + q = \frac{V_2^2}{2}$$

and since $V_2 > V_1$ for subsonic flow, q must be positive and heat added to the gas. The tendency to cool as a result of expansion is greater than the tendency to heat as a result of wall friction, and therefore heat must be added to maintain a constant temperature. For supersonic flow, $V_2 < V_1$, q is negative, and heat is removed from the gas.

ADIBATIC FLOW

Integration of Eq. (2) again requires that the variable density and velocity be expressed in terms of the variable pressure, and the variation of the friction factor must again be investigated.

The momentum equation [Eq. (2)] when multiplied by $kM^2/2$ may be written as

$$\frac{dp}{\rho} + \frac{kM^2}{2} f \frac{dx}{D} + kM^2 \frac{dV}{V} = 0$$

The continuity equation in differential form is

$$\frac{dV}{V} + \frac{d\rho}{\rho} = 0$$

The perfect gas equation in differential form is

$$\frac{dp}{p} = \frac{d\rho}{\rho} + \frac{dT}{T}$$

From the definition of the Mach number $M = V/c = V/\sqrt{kRT}$

$$\frac{dM}{M} = \frac{dV}{V} - \frac{dT}{2T}$$

The energy equation $h + V^2/2$ = constant may be written in differential form as

$$\frac{dT}{T} + (k - 1)M^2 \frac{dV}{V} = 0$$

We have five simultaneous equations in six differential variables: dp/p, dV/V, $d\rho/\rho$, dT/T, dM/M, and $f\ dx/D$.

The first five (dependent) variables may be expressed in terms of $f\ dx/D$, the independent variable. The results are

$$\frac{dp}{p} = -\frac{1 + (k - 1)M^2}{1 - M^2} \frac{kM^2}{2} f \frac{dx}{D}$$

$$\frac{dV}{V} = \frac{1}{1 - M^2} \frac{kM^2}{2} f \frac{dx}{D}$$

$$\frac{d\rho}{\rho} = -\frac{1}{1 - M^2} \frac{kM^2}{2} f \frac{dx}{D}$$

$$\frac{dT}{T} = \frac{M^2(k - 1)}{1 - M^2} \frac{kM^2}{2} f \frac{dx}{D}$$

and

$$\frac{dM}{M} = \frac{1 + \frac{1}{2}(k - 1)M^2}{1 - M^2} \frac{kM^2}{2} f \frac{dx}{D}$$

A thermodynamic analysis shows that since dx is positive in the direction of flow, f must be positive. Therefore the preceding equations indicate that for subsonic flow ($M < 1$) the velocity and Mach number increase, while the pressure, density, and temperature decrease in the direction of flow. The opposite is true in each instance for supersonice flow ($M > 1$).

* Limiting or Maximum Mach Number. The equation for dp/p can be solved for the pressure gradient:

$$\frac{dp}{dx} = -\frac{fkp}{2D} M^2 \left[\frac{1 + (k - 1)M^2}{1 - M^2}\right] = -\frac{f}{D}\frac{\rho V^2}{2} \times$$

$$\left[\frac{1 + (k - 1)M^2}{1 - M^2}\right] \qquad (8)$$

As $M \to 0$, this is essentially the Darcy equation for liquid pipe flow; thus for low Mach numbers, adiabatic gas flow may also be treated as incompressible. The pressure gradients will be within 5 percent at Mach numbers up to 0.185 for air. Figure 2, which is applicable for adiabatic flow as well as isothermal, indicates that the pressure drops to the limiting point 0, and at that point dp/dx is infinite. From Eq. 8, the limiting Mach number is 1 for adiabatic flow.

* Limiting or Minimum Pressure. To obtain the limiting pressure, we combine the equations for dp/p and dM/M to obtain

$$\frac{dp}{p} = -\frac{1 + (k - 1)M^2}{M[1 + \frac{1}{2}(k - 1)M^2]} dM$$

which, when integrated between some given section and the point where $M = 1$ (the limiting point), gives

$$\frac{p^*}{p_1} = M \sqrt{\frac{2[1 + \frac{1}{2}(k - 1)M_1^2]}{k + 1}} \qquad (9)$$

The integration is evident if $dM = dM^2/2M$.

* Limiting or Maximum Length. The expression for dM/M can be rearranged to give

$$f\frac{dx}{D} = \frac{1 - M^2}{kM^4[1 + \frac{1}{2}(k - 1)M^2]} dM^2$$

by letting $dM = dM^2/2M$. Intergrating between a given section and the point where M = 1,

$$\frac{\bar{f}L^*}{D} = \frac{1 - M_1^2}{kM_1^2} + \frac{k + 1}{2k} \ln \frac{(k + 1)M_1^2}{2[1 + \frac{1}{2}(k - 1)M_1^2]} \quad (10)$$

where $\bar{f}$ is the average friction factor. At High Reynolds numbers, especially for rough pipes, the friction factor dependes only on the pipe roughness, and thus the friction factor could conceivably be constant for adiabatic flow.

If not,

$$\bar{f} = \frac{1}{L^*} \int_0^{L^*} f\,dx$$

The value of f at section 1 may be assumed to equal $\bar{f}$, and after solving for conditions at some downstream point, the value of f throughout the entire pipe can be examined and an appropriate average value used to recalculate downstream conditions.

* Limiting Temperature and Velocity. The preceding expressions for dT/T, dV/V, and dM/M can be solved simultaneously by eliminating f dx/D to obtain, after integration,

$$\frac{T^*}{T_1} = \frac{2[1 + \frac{1}{2}(k - 1)M_1^2]}{k + 1}$$

and

$$\frac{V^*}{V_1} = \frac{1}{M_1}\sqrt{\frac{2[1 + \frac{1}{2}(k - 1)M_1^2]}{k + 1}}$$

These equations may be used in the following manner: for the flow conditions known at some arbitrary section 1, the limitng conditions can be calculated. These limited conditions are the same for corresponding flow conditions at any other section. Therefore

$$\frac{p_2}{p_1} = \frac{p^*/p_1}{p^*/p_2}$$

$$\left(\frac{\bar{f}L}{D}\right)_{1-2} = \left(\frac{\bar{f}L^*}{D}\right)_1 - \left(\frac{\bar{f}L^*}{D}\right)_2$$

$$\frac{T_2}{T_1} = \frac{T^*/T_1}{T^*/T_2}$$

$$\frac{V_2}{V_1} = \frac{V^*/V_1}{V^*/V_2}$$

Flow near the limiting condition would be very nearly adiabatic, because the high heat-transfer rates required to maintain isothermal flow would be very difficult to achieve.

If adiabatic flow is assumed for a given initial subsonic condition, the pressure at any downstream point where the Mach number is M_2 is always slightly less than it would be if the flow were assumed to be isothermal. The ratio is

$$\frac{p_{2\,ad}}{p_{2iso}} = \left[\frac{1 + \frac{1}{2}(k - 1)M_1^2}{1 + \frac{1}{2}(k - 1)M_2^2}\right]^{1/2}$$

which is less than 1 for subsonic flow and depends on the ratio of specific heat capacities k for the gas and on the initial and downstream Mach numbers. For air with $M_2 = 0.854$ (the limiting Mach number for isothermal flow), $p_{2\,ad}/p_{2\,iso} \geq 0.934$. Similarly, if adiabatic flow is assumed for a given initial subsonic condition, the temperature is always slightly less at any prescribed point where the Mach number is M_2 than if isothermal flow is assumed. The ratio is

$$\frac{T_{2\,ad}}{T_{2iso}} = \frac{1 + \frac{1}{2}(k - 1)M_1^2}{1 + \frac{1}{2}(k - 1)M_2^2}$$

which is less than 1 for subsonic flow. For air with $M_2 = 0.854$, $T_{2\,ad}/T_{2\,iso} \geq 0.875$; and for $M_2 = 0.5$, this ratio ≥ 0.952.

Therefore, except at high Mach numbers, subsonic isothermal and subsonic adiabatic flow do not differ appreciably.

(a) $p_1 = 2000$ kPa abs $\qquad \rho_1 = \frac{p_1}{RT_1} = 11.59 \text{kg/m}^3$

$T_1 = 333°K$

$c_1 = \sqrt{kRT_1} = \sqrt{(1.31)(518)(333)} = 475$ m/s

$$M_1 = \frac{V_1}{c_1} = \frac{15}{475} = 0.0316$$

$$Re_1 = \frac{V_1 D \rho_1}{\mu_1} = \frac{(15)(0.30)(11.59)}{1.3 \times 10^{-5}} = 4.0 \times 10^6$$

$$\frac{k}{D} = \frac{0.000045}{0.30} = 0.00015$$

so that

$$f_1 = 0.0134$$

See the accompanying table.

ISOTHERMAL FLOW	ADIABATIC FLOW
(a) $\frac{p^*}{p_1} = M_1\sqrt{k} = 0.0316\sqrt{1.31}$	(a) $\frac{p^*}{p_1} = M_1\sqrt{\frac{2[1 + \frac{1}{2}(k-1)M_1^2]}{k+1}}$
$= 0.0362$	$= 0.0316\sqrt{\frac{2[1 + (0.155)(0.0316)^2]}{2.31}}$
$p^* = (0.0362)(2000) = 72.3$ kPa abs	$= 0.0294$
	$p^* = (0.0294)(2000) = 58.8$ kPa abs
(b) $\frac{fL^*}{D} = \frac{1}{kM_1^2} - 1 - \ln\frac{1}{kM_1^2}$	(b) $\frac{\bar{f}L^*}{D} = \frac{1 - M_1^2}{kM_1^2}$
$= \frac{1}{(1.31)(0.0316)^2} - 1$	$+ \frac{k+1}{2k}\ln\frac{(k+1)M_1^2}{2[1 + \frac{1}{2}(k-1)M_1^2]}$
$- \ln\frac{1}{(1.31)(0.0316)^2}$	$= \frac{1 - 0.000999}{1/764}$
$= 764 - 1 - 6.6$	$+ \frac{2.31}{2.62}\ln\frac{(2.31)(0.000999)}{2(1.00015)}$
$= 756$	$= 763 - 6$
$L^* = \frac{756D}{f} = \frac{(756)(0.30)}{0.0134}$	$= 757$
$L^* = 16{,}925$ m	$L^* = \frac{757D}{f} = 16{,}950$ m
	When T^* is found, Re^* will be calculated and f^* determined. A new $\bar{f}$ may then give a slightly different L^*.
(c) $V^* = c^*M^* = \frac{475}{\sqrt{1.31}} = 415$ m/s	(c) $V^* = c^* = \sqrt{kRT^*}$, where
	$T^* = T_1\frac{2[1 + \frac{1}{2}(k-1)M_1^2]}{k+1}$
	$= 333\frac{2(1.00015)}{2.31} = 288°K$
	$V^* = \sqrt{(1.31)(518)(288)} = 442$ m/s
	Thus
	$Re^* = \frac{V^*\rho^*D}{\mu^*} = \frac{V_1\rho_1 D}{\mu^*}$
	$Re^* = \frac{(15)(0.30)(11.59)}{1.2 \times 10^{-5}} = 4.35 \times 10^6$
	$f^* = 0.0133$ and $\bar{f} = 0.01335$
	$L^* = \frac{757D}{\bar{f}} = 17{,}010$ m, a truer value

ISOTHERMAL FLOW

(d) At $L_2 = 8460$ m,

$$\frac{fL_2}{D} = \frac{756}{2} = 378$$

$$1 - \left(\frac{p_2}{p_1}\right)^2 = kM_1^2\left(f\frac{L_2}{D} + 2\ln\frac{p_1}{p_2}\right)$$

For a first approximation,

$$1 - \left(\frac{p_2}{p_1}\right)^2 = \frac{1}{764}(378) = 0.495$$

$$\frac{p_2}{p_1} = 0.711$$

and $+2\ln(p_1/p_2) = +0.68$, and a recalculation gives

$$\frac{p_2}{p_1} = 0.710$$

$$p_2 = (0.710)(2000) = 1420 \text{ kPa abs}$$

$$T_2 = 60°\text{C} = 333°\text{K}$$

$$V_2 = \frac{V_1\rho_1}{\rho_2} = V_1\frac{p_1}{p_2} = 15\frac{(11.59)}{8.23}$$

$$= 21.1 \text{ m/s}$$

(e) If $p_3 = 500$ kPa abs,

$$\left(\frac{fL}{D}\right)_{1-3} = \frac{1}{kM_1^2}\left[1 - \left(\frac{p_3}{p_1}\right)^2\right] - 2\ln\frac{p_1}{p_3}$$

$$= 764\left[1 - \frac{1}{16}\right] - 2\ln 4$$

$$= 713$$

$$L_3 = \frac{713D}{f} = 15{,}960 \text{ m}$$

ADIABATIC FLOW

(d) At $L_2 = 8505$ m,

$$\frac{p_2}{p_1} = \frac{p^*/p_1}{p^*/p_2} = \frac{0.0294}{\text{function of } M_2}$$

Eq. 10 with $fL_2/D = 378.5$ is solved for $M_2 = 0.0446$. Then

$$\frac{p^*}{p_2} = M_2\sqrt{\frac{2[1 + \frac{1}{2}(k-1)M_2^2]}{k+1}}$$

$$p_2 = 1417 \text{ kPa abs}$$

$$T_2 = \frac{T^*}{2[1 + \frac{1}{2}(k-1)M_2^2]/(k+1)}$$

$$= \frac{288}{0.866} = 333°\text{K}$$

and the temperature has not dropped noticeably. Since $c_1 \approx c_2$,

$$V_2 = V_1\frac{M_2}{M_1}$$

$$= 60\frac{0.0446}{0.0316}$$

$$= 21.2 \text{ m/s}$$

(e) If $p_3 = 500$ kPa abs, $(fL/D)_3$ is a function of M_3, which can be found by trial from Eq. 9 to be $M_3 = 0.126$:

$$\left(\frac{fL^*}{D}\right)_3 = \frac{1 - M_3^2}{kM_3^2} - \frac{k+1}{2k}\ln\frac{2[1 + \frac{1}{2}(k-1)M_3^2]}{(k+1)M_3^2}$$

$$= 47.3 - 3.5 = 43.8$$

$$L_3^* = \frac{(43.8)(0.30)}{(0.01335)} = 984 \text{ m}$$

$$L_3 = (L^*)_1 - (L^*)_3$$
$$= 17{,}010 - 984$$
$$= 16{,}020 \text{ m}$$

The results of (a) indicate that for given initial conditions: (a) the minimum pressure for adiabatic flow is slightly less than for isothermal flow; (b) the maximum length for both types of flow is essentially the same, that for adiabatic flow being less than 1 percent greater than for isothermal flow; (c) the pressure and temperature at a point halfway to the limiting point are essentially the same for either type of flow; (d) the rate of pressure drop is very large near the limiting point, and practical considerations rule out the advisability of having pipes longer than 80-90 percent of the maximum length; and (3) for practical purposes, since there is always some uncertainty in the values of the friction factor, and purely isothermal or purely adiabatic flow in pipes is rarely if ever achieved, either isothermal or adiabatic flow may be assumed in making engineering calculations.

(b) The initial Mach number and the friction factor are both needed in Eqs. (7) and (10).

$$M_1 = V_1 \sqrt{kp_1/\rho_1} = 30/\sqrt{1.30)(1.20 \times 10^6)/(20)} = 0.1074$$

$$Re = Vd\rho/\mu = (30)(0.06)(20)/2.2 \times 10^{-5} = 1.64 \times 10^6$$

so

$$f = 0.0107$$

For isothermal flow, Eq. (7) gives

$$fL^*/D = \frac{1}{(1.30)(0.1074)^2} - 1 - \ln \frac{1}{(1.30)(0.1074)^2}$$

$$= 66.8 - 1 - 4.20 = 61.6$$

and

$$L^* = (61.6)(0.06)/0.0107$$

$$= 345\text{m.}$$

For adiabatic flow, Eq (10) gives

$$fL^*/D = 66.03 + \frac{2.3}{2.6} \ln 0.01322 = 66.03 - \frac{2.3}{2.6} \ln 75.6$$

$$= 62.2.$$

and

$$L^* = (62.2)(0.06)/0.0107 = 349 \text{ m}$$

As in (a), the maximum or limiting length for isothermal and adiabatic flow is essentially the same, that for adiabatic flow being slightly greater by about 1 percent. The higher the initial Mach number, the greater is the difference.

• PROBLEM 12-23

Air is moving as a steady flow through a duct having a constant rectangular cross section measuring 2 by 1 ft. At a position 20 ft from the end, the pressure is 18 psia, and the temperature is 500°F. The fluid leaves the duct subsonically at a pressure of 14.7 psia. If there is 40 lbm of fluid flow/sec, what is the heat transfer per pound mass of fluid between the afore-mentioned section and the exit? Assume a constant specific head c_p of 0.26 Btu/lbm/°F and neglect friction.

TABLE 1

RAYLEIGH LINE
(For a perfect gas with $k = 1.4$)

M	$\frac{T_0}{T_0^*}$	$\frac{T}{T^*}$	$\frac{p}{p^*}$	$\frac{p_0}{p_0^*}$	$\frac{V}{V^*}$
0.22	0.206	0.244	2.25	1.23	0.109
0.24	0.239	0.284	2.22	1.22	0.128
0.26	0.274	0.325	2.19	1.21	0.148
0.28	0.310	0.367	2.16	1.21	0.170
0.46	0.630	0.725	1.85	1.13	0.392
0.48	0.661	0.759	1.81	1.12	0.418
0.50	0.691	0.790	1.78	1.11	0.444
0.52	0.720	0.820	1.74	1.10	0.471

TABLE 2

ONE DIMENSIONAL ISENTROPIC RELATIONS
(For a perfect gas with $k = 1.4$)

M	A/A^*	p/p_0	ρ/ρ_0	T/T_0
0.22	2.71	0.967	0.976	0.990
0.24	2.50	0.961	0.972	0.989
0.26	2.32	0.954	0.967	0.987
0.28	2.17	0.947	0.962	0.985
0.46	1.42	0.865	0.902	0.959
0.48	1.38	0.854	0.893	0.956
0.50	1.34	0.843	0.885	0.952
0.52	1.30	0.832	0.877	0.949

Solution: The first step will be to ascertain the Mach number at section 1, where we have known data. We need c_1 and V_1. Thus

$$c_1 = \sqrt{kRT} = \sqrt{(1.4)(53.3)(32.2)(500 + 460)} = 1{,}515 \text{ fps}$$

Using the continuity equation, we have

$$\rho_1 V_1 A_1 = 40 \text{ lbm/sec}$$

Therefore

$$V_1 = \frac{40}{(\rho_1)(2)(1)}$$

Since $\rho_1 = p_1/RT_1$, we have for V_1

$$V_1 = \frac{(40)(53.3)(500 + 460)}{(18)(144)(2)} = 394 \text{ fps}$$

Thus:

$$M_1 = \frac{V_1}{c_1} = 0.26$$

Now use the table of Rayleigh lines (Table 1). Thus for the above Mach number

$$\frac{p_1}{p^*} = 2.19$$

Therefore

$$p^* = 8.22 \text{ psia}$$

Since we know p* and also p_c at the exit, it is then a simple matter to ascertain M_e from the table. Thus for $p_e/p^* = 1.79$ we get $M_e = 0.48$.

Next, we must compute T_e so that we may establish the stagnation temperature at the exit. From conditions at section 1 we can get T_1/T^* from the tables as 0.325 so that $T^* = 2950°R$, and using our solved value of M_e, we can likewise determine T_e as 2280°R.

Now with M_1, T_1 and M_e, T_e known, all we need is to consult the one dimensional isentropic table (Table 2) to find $(T_0)_1$ and $(T_0)_2$. The desired value for the computation is then

$$\frac{dQ}{dm} = c_p(T_{02} - T_{01}) = 0.26(2,380 - 972) = 366 \text{ Btu/lbm}$$

• PROBLEM 12-24

Air in a wind tunnel used to study flow past a body has a temperature of 25°C and a pressure of 100 kN/m² (abs). The tunnel velocity is 150 m/s and the air in the vicinity of the object reaches a maximum of 220 m/s. What is the temperature of the air at this point?

TABLE

Gas	Chemical Formula	Specific Weight, γ		Density, ρ		Gas Constant, R		Adiabatic Constant, k
		lb/ft³	N/m³	slugs/ft³	kg/m³	ft/°R	m/°K	
Air	—	0.0753	11.8	0.00234	1.206	53.3	29.2	1.40
Ammonia	NH_3	0.0448	7.04	0.00139	0.716	89.5	49.1	1.32
Carbon dioxide	CO_2	0.115	18.1	0.00357	1.840	34.9	19.1	1.29
Helium	He	0.0104	1.63	0.000323	0.166	386.	212.	1.66
Hydrogen	H_2	0.00522	0.820	0.000162	0.0835	767.	421.	1.40
Methane	CH_4	0.0416	6.53	0.00129	0.665	96.4	52.9	1.32
Nitrogen	N_2	0.0726	11.4	0.00225	1.160	55.2	30.3	1.40
Oxygen	O_2	0.0830	13.0	0.00258	1.330	48.3	26.5	1.40
Sulfur dioxide	SO_2	0.170	26.7	0.00528	2.721	23.6	12.9	1.26

Solution: The energy equation, with $h = u + p/\gamma$, may be expressed in the form

$$\frac{V_1^2}{2g} + y_1 + h_1 = \frac{V_2^2}{2g} + y_2 + h_2$$

For purposes of application, certain thermodynamic relationships for perfect gases are required. Introducing c_p and c_v as the coefficients of specific heat at constant pressure and constant volume, respectively, we have, from thermodynamics,

$$k = c_p/c_v$$

and

$$c_p - c_v = R\left(\frac{\text{ft-lb}}{\text{lb-°R}} \text{ or } \frac{\text{m.N}}{\text{N.°K}}\right) \tag{1}$$

Combining, we obtain

$$c_p = \frac{Rk}{k - 1} \tag{2}$$

Additionally, the thermodynamic relationship for internal energy,

$$u = c_v T$$

may be combined with the perfect gas law yielding for the enthalpy

$$h = u + p/\gamma$$

$$= c_v T + RT$$

$$= (c_v + R)T$$

and therefore from Eq. 1,

$$h = c_p T$$

or, from Eq. 2,

$$h = \frac{RkT}{k - 1}$$

Thus, the energy equation becomes alternatively,

$$\frac{V_1^2}{2g} + y_1 + c_p T_1 = \frac{V_2^2}{2g} + y_2 + c_p T_2$$

or

$$\frac{V_1^2}{2g} + y_1 + \frac{RkT_1}{k - 1} = \frac{V_2^2}{2g} + y_2 + \frac{RkT_2}{k - 1}$$

The flow will be frictionless and adiabatic, a condition referred to as isentropic flow. We will also ignore the effects of elevation differences. The energy equation may immediately be expressed as either

$$\frac{V_1^2}{2g} + c_p T_1 = \frac{V_2^2}{2g} + c_p T_2 \tag{3}$$

or

$$\frac{V_1^2}{2g} + \frac{RkT_1}{k - 1} = \frac{V_2^2}{2g} + \frac{RkT_2}{k - 1} \tag{4}$$

For isentropic flow, Eq. 4 may be arranged to

$$\frac{V_2^2}{2g} - \frac{V_1^2}{2g} = \frac{Rk}{k - 1} (T_1 - T_2) \tag{5}$$

Because the velocity change is the product of an isentropic flow, Eq. 5 may be used to calculate the temperature change.

$$T_1 - T_2 = \frac{k - 1}{Rk}\left(\frac{V_2^2}{2g} - \frac{V_1^2}{2g}\right)$$

or

$$25 - T_2 = \frac{1.4 - 1}{(29.2)(1.4)}\left[\frac{(220)^2}{(2)(9.81)} - \frac{(150)^2}{(2)(9.81)}\right]$$

Solving, we obtain $T_2 = 12.1°C$.

• PROBLEM 12-25

Air flows from a large pressure vessel, in which the velocity can be assumed negligible, through a well-rounded nozzle into the working section of a wind-tunnel, as indicated in the Fig. The pressure and temperature in the supply vessel are 60 lbf/in^2 and 760°R respectively. Calculate the velocity in the working section when the pressure there is 40 lbf/in^2, assuming air to be a perfect gas (R = 53.3 lbf ft/lbm °R, γ = 1.40), and the flow to be adiabatic and reversible.

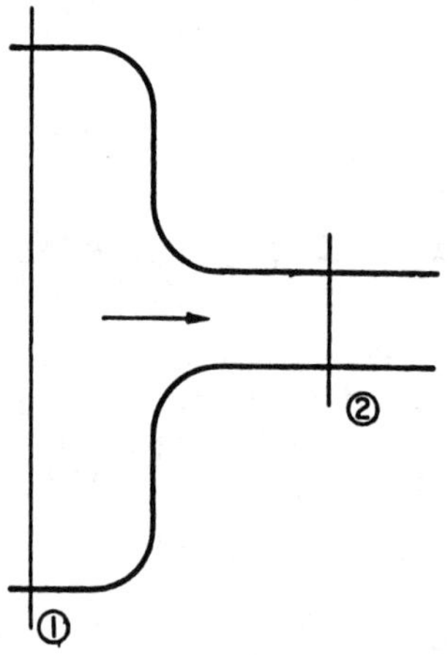

Solution: The ideal gas law pv = RT, on differentiation, yields the equation

$$\frac{dp}{p} + \frac{dv}{v} = \frac{dT}{T}, \qquad (1)$$

where v, T are specific volume and absolute temperature respectively. For a perfect gas $u = c_v T$, and the general thermodynamic relation

$$Tds = du + \frac{p}{J}\, dv$$

(where s and u are specific entropy and specific internal energy respectively) can be rearranged as

$$ds = c_v \frac{dT}{T} + \frac{p}{JT}\, dv$$

$$= c_v \frac{dT}{T} + \frac{R}{J}\frac{dv}{v}. \qquad (2)$$

Substitution for dT/T from equation 1 gives

$$ds = c_v \left(\frac{dp}{p} + \frac{dv}{v}\right) + \frac{R}{J}\frac{dv}{v}$$

$$= c_v \left(\frac{dp}{p} + \frac{dv}{v}\right) + (c_p - c_v)\frac{dv}{v}$$

$$= c_v \frac{dp}{p} + c_p \frac{dv}{v}. \qquad (3)$$

Now an adiabatic reversible process is isentropic, so that the left-hand side of equation 3 may be put equal to zero, and then

$$0 = d(\ln p) + \gamma\, d(\ln v) \quad (\text{where } \gamma \equiv c_p/c_v),$$

$$\therefore \quad pv^{\gamma} = \text{constant, C say}$$

or

$$\frac{p}{\rho^{\gamma}} = C. \qquad (4)$$

It follows that

$$\int \frac{dp}{\rho} = \int \left(\frac{C}{p}\right)^{1/\gamma} dp = C^{1/\gamma} \frac{\gamma}{\gamma - 1}\, p^{(\gamma-1)/\gamma} + \text{constant.}$$

Integration of the Euler equation, in the form

$$\frac{1}{\rho}\frac{dp}{ds} + \frac{V}{g_c}\frac{dV}{ds} = 0$$

leads to

$$c^{1/\gamma}\,\frac{\gamma}{\gamma - 1}\,p^{(\gamma-1)/\gamma} + \frac{V^2}{2g_c J} = \text{constant for all s} \qquad (5)$$

Equation 5 may be written in the alternative form

$$\frac{\gamma}{\gamma - 1}\frac{p_1}{\rho_1} + \frac{V_1^2}{2g_c} = \frac{\gamma}{\gamma - 1}\frac{p_1}{\rho_1}\left(\frac{p_2}{p_1}\right)^{(\gamma-1)/\gamma} + \frac{V_2^2}{2g_c} \qquad (6)$$

where suffixes 1 and 2 refer to two stations in the streamtube.

In this problem, with $V_1 = 0$, Eq. 6 becomes

$$\frac{\gamma}{\gamma - 1}\frac{p_1}{\rho_1} = \frac{\gamma}{\gamma - 1}\frac{p_1}{\rho_1}\left(\frac{p_2}{p_1}\right)^{(\gamma-1)/\gamma} + \frac{V_2^2}{2g_c},$$

$$\therefore \quad \frac{V_2^2}{2g_c} = \frac{\gamma}{\gamma - 1}\frac{p_1}{\rho_1}\left[1 - \left(\frac{p_2}{p_1}\right)^{(\gamma-1)/\gamma}\right]$$

$$= \frac{\gamma}{\gamma - 1}RT_1\left[1 - \left(\frac{p_2}{p_1}\right)^{(\gamma-1)/\gamma}\right],$$

$$\therefore \quad V_2 = \left\{\frac{2\gamma g_c RT_1}{\gamma - 1}\left[1 - \left(\frac{p_2}{p_1}\right)^{(\gamma-1)/\gamma}\right]\right\}^{\frac{1}{2}}$$

$$= \left\{\frac{2 \times 1.4 \times 32.2 \times 53.3 \times 760}{0.4}\left[1 - \left(\frac{2}{3}\right)^{0.286}\right]\right\}^{\frac{1}{2}} \text{ ft/sec}$$

$$= 999 \text{ ft/sec.}$$

Air flows through a variable cross-sectional passage with a Mach No. $M_1 = 0.25$, temperature $T_1 = 160°F$, and pressure $p_1 = 100$ psia at a given point 1 in the passage. At a point 2, further downstream in the passage, the Mach No. is $M_2 = 0.7$.

Assuming an isentropic flow of air,

(a) Draw the passage and show the isentropic flow process on T-S diagram.

(b) Calculate the area of cross-section at point 2, if the area at point 1 is $A_1 = 0.01\ ft^2$.

(c) Check whether the stagnation pressures at points 1 and 2 satisfy the isentropic flow condition.

Solution: (a) Since the Mach No. is increasing from 0.25 to 0.7, the air flow is subsonically accelerating. In this case, the passage must be a converging nozzle.

During an isentropic process, the entropy remains constant and the process is represented by a vertical straight line on a T-S diagram. For the given problem, a solid line joining the points 1 and 2 shows the isentropic expansion through the nozzle.

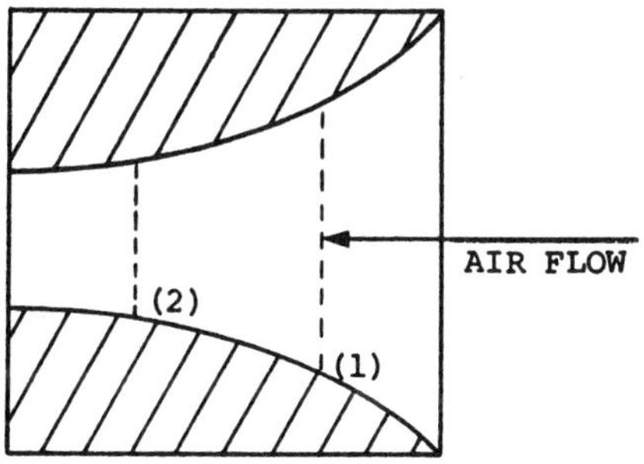

Fig. 1

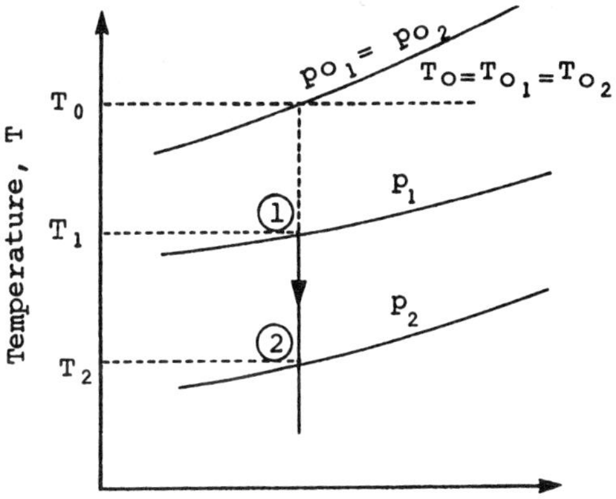

Fig. 2 T-S diagram

(b) The area of cross-section A_2 at point 2 can be calculated by applying the continuity equation between points 1 and 2, so that

$$\text{mass flow rate} = \rho_1 A_1 V_1 = \rho_2 A_2 V_2 = \text{constant}$$

or

$$A_2 = \frac{\rho_1 A_1 V_1}{\rho_2 V_2} \tag{1}$$

where V_1, the velocity at point 1, can be defined in terms of Mach No. M_1 and the local sonic velocity C_1 as,

$$V_1 = M_1 C_1 = 0.25 C_1$$

and for a perfect gas,

$$C_1 = \sqrt{\gamma R T_1}$$

where $T_1 = 160^0 F + 460 = 620^0 R$,

Specific heat ratio γ for air = 1.4,

Gas constant R = 53.3 lbf-ft/lbm-^{0}R

or

$$C_1 = \sqrt{1.4 \times 53.3 \text{ lbf-ft/lbm-}^0\text{R} \times 32.2 \text{ lbm-ft/lbf-sec}^2 \times 620^0\text{R}}$$

$$= 1220.5 \text{ ft/sec}$$

$$\therefore \quad V_1 = 0.25 \times 1220.5 = 305.1 \text{ ft/sec}$$

The density of air at point 1 i.e., ρ_1 can be computed by writing equation of state for a perfect gas, i.e.,

$$p_1 = \rho_1 R T_1$$

or

$$\rho_1 = \frac{p_1}{RT_1} = \frac{100 \text{ lbf/in}^2 \times 144 \text{ in}^2/\text{ft}^2}{53.3 \text{ lbf-ft/lbm-}^0\text{R} \times 620^0\text{R}} = 0.436 \text{ lbm/ft}^3$$

Now, the velocity V_2 and the density ρ_2 can be computed by using the isentropic flow relations.

As each state on the isentropic line in the T-S diagram has the same isentropic stagnation temperature T_0 and pressure p_0, therefore

$$T_{0_2} = T_{0_1}$$

For point 1, the isentropic stagnation temperature relation is,

$$\frac{T_{0_1}}{T_1} = 1 + \frac{\gamma - 1}{2} M_1^2$$

or

$$T_{0_1} = T_1 \left[1 + \frac{\gamma - 1}{2} M_1^2\right]$$

Substituting the values of γ, T_1 and M_1,

$$T_{0_1} = 620 \times \left[1 + \frac{1.4 - 1}{2}(0.25)^2\right] = 628^0\text{R}$$

$$\therefore \quad T_{0_2} = T_{0_1} = 628^0\text{R}$$

Again, for point 2,

$$\frac{T_{0_2}}{T_2} = 1 + \frac{\gamma - 1}{2} M_2^2$$

from which

$$T_2 = \frac{T_{0_2}}{\left[1 + \frac{\gamma - 1}{2} M_2^2\right]}$$

or,

$$T_2 = \frac{628}{\left[1 + \frac{(1.4 - 1)}{2}(0.7)^2\right]} = 572\,^0R$$

Again using the relations for Mach No. and local sonic velocity at point 2,

$$M_2 = \frac{V_2}{C_2}$$

from which

$$V_2 = M_2 C_2 = 0.7C_2$$

where

$$C_2 = \sqrt{\gamma R T_2}$$

or

$$C_2 = \sqrt{1.4 \times 53.3 \text{ lbf-ft/lbm-}^0\text{R} \times 32.2 \text{ lbm-ft/lbf-sec}^2 \times 572\,^0\text{R}}$$

$$= 1172.4 \text{ ft/sec}$$

$$\therefore \quad V_2 = 0.7 \times 1172.4 = 821 \text{ ft/sec}$$

From the well known p-v relation for an isentropic process,

$$p_1 v_1^\gamma = p_2 v_2^\gamma = \text{constant}$$

which yields,

$$\frac{p_2}{p_1} = \left(\frac{v_1}{v_2}\right)^\gamma \qquad (2)$$

Also, from the equation of state for a perfect gas,

$$\frac{p_1 v_1}{T_1} = \frac{p_2 v_2}{T_2}$$

or

$$\frac{v_1}{v_2} = \frac{p_2}{p_1} \times \frac{T_1}{T_2} \qquad (3)$$

Substituting the R.H.S. of (3) for the R.H.S. of (2),

$$\frac{p_2}{p_1} = \left(\frac{T_2}{T_1}\right)^{\frac{\gamma}{\gamma-1}}$$

or

$$p_2 = p_1 \left(\frac{T_2}{T_1}\right)^{\frac{\gamma}{\gamma-1}}$$

$$= 100 \times \left(\frac{572}{620}\right)^{\frac{1.4}{(1.4 - 1)}}$$

$$= 75.4 \text{ psi}$$

Finally, applying the perfect gas equation at point 2,

$$\rho_2 = \frac{p_2}{RT_2} = \frac{75.4 \text{ lbf/in}^2 \times 144 \text{ in}^2/\text{ft}^2}{53.3 \text{ lbf-ft/lbm-}^\circ\text{R} \times 572}$$

$$\therefore \quad \rho_2 = 0.356 \text{ lbm/ft}^3$$

Substituting the values of A_1, ρ_1, ρ_2, V_1 and V_2 back into equation (1),

$$A_2 = \frac{0.436 \times 0.01 \times 305.1}{0.356 \times 821} = 4.55 \times 10^{-3} \text{ ft}^2$$

Thus $A_1 > A_2$ which is obvious for a converging nozzle.

(c) A necessary condition for isentropic flow implies that the stagnation pressure should be the same for each state on the isentropic line. Thus for point 1, the isentropic relation holds,

$$\frac{p_{0_1}}{p_1} = \left(\frac{T_{0_1}}{T_1}\right)^{\frac{\gamma}{\gamma-1}} = \left(\frac{628}{620}\right)^{\frac{1.4}{(1.4-1)}} = 1.0459$$

$$\therefore \quad p_{0_1} = p_1 \times 1.0459 = 100 \times 1.0459 = 104.59 \text{ psi}$$

Similarly, for point 2,

$$\frac{p_{0_2}}{p_2} = \left(\frac{T_{0_2}}{T_2}\right)^{\frac{\gamma}{\gamma-1}} = \left(\frac{628}{572}\right)^{\frac{1.4}{(1.4-1)}} = 1.3867$$

$$\therefore \quad p_{0_2} = p_2 \times 1.3867 = 75.4 \times 1.3867 = 104.56 \text{ psi}$$

Thus, the stagnation pressures for points 1 and 2 are of the same values, satisfying a necessary condition for isentropic flow. A slight difference in the values of p_{0_1} and p_{0_2} arises from the approximations made in the calculations of T_{0_1} and T_2.

• PROBLEM 12-27

Air is heated during a frictionless flow between two cross-sections (1) and (2) of a pipe of uniform internal diameter, figure 1.

At section (1), $T_1 = 120\,^\circ$F, $\rho_1 = 0.0699 \frac{\text{ft}^3}{\text{lb}}$ and $V_1 = 295 \frac{\text{ft}}{\text{sec}}$ whereas at (2), $p_2 = 7$ psia. Between the two cross-sections, calculate

(i) the change in stagnation temperature, $(T_{O_2} - T_{O_1})$ and the stagnation pressure $(p_{O_2} - p_{O_1})$,

(ii) rate of heat addition per unit mass flow rate of air,

(iii) change in entropy, ΔS_{12}.

If the conditions at section (1) are changed to $T_1 = 80\,^\circ$F, $p_1 = 10$ psia and $V_1 = 2220 \frac{\text{ft}}{\text{sec}}$ whereas $V_2 = 1700 \frac{\text{ft}}{\text{sec}}$ is the only known quantity at section (2), calculate (i), (ii) and (iii). In both the above cases, show the processes on a T-S diagram.

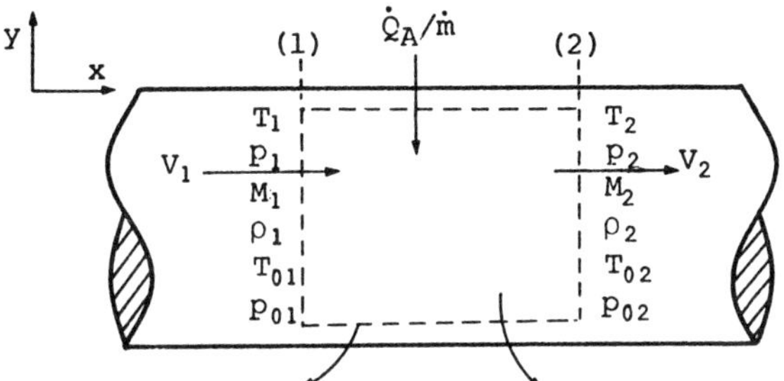

Fig. 1

Solution:

The local isentropic stagnation temperature T_{o_1} at section (1) is,

$$T_{o_1} = T_1\left[1 + \frac{(k-1)}{2}M_1^2\right] \tag{1}$$

The isentropic temperature at section (2) is,

$$T_{o_2} = T_2\left[1 + \frac{(k-1)}{2}M_2^2\right] \tag{2}$$

Now,

$$M_1 = \frac{V_1}{c_1}$$

where M_1 = Mach No. at section (1)

and c_1 = local sonic velocity at section (1)

$$= \sqrt{kRT_1 g_c}$$

$$\therefore \quad M_1 = \frac{V_1}{\sqrt{kRT_1 g_c}} = \frac{295\ \frac{\text{ft}}{\text{sec}}}{\sqrt{1.4 \times 53.3\ \frac{\text{lb}_f\text{-ft}}{\text{lb}_m\text{-}^0\text{R}} \times 580^0\text{R} \times 32.2}}$$

$$= 0.25$$

Thus, from equation (1)

$$T_{o_1} = 580\left[1 + \frac{(1.4-1)}{2}(0.25)^2\right] = 587^0\text{R}$$

Now, to find out T_{o_2}, V_2 and T_2 should be known. In order to find V_2, let us make use of the x-component of the equation of momentum for a control volume $d\forall$, as follows:

$$\Sigma F_x = \frac{\partial}{\partial t}\int_{\text{C.V.}} \rho V_x d\forall + \int_{\text{C.S.}} \rho(\vec{V} \cdot d\vec{A})V_x$$

where

$$\Sigma F_x = F_{bx} + F_{sx}$$

and

$$F_{bx} = \text{Body Force}, \quad F_{sx} = \text{Surface Force}$$

Assuming that

(1) the flow is steady and uniform at each section so that

$$\frac{\partial}{\partial t}\int_{C.V.} \rho V_x dV = 0$$

(2) F_{bx} is so small compared to F_{sx} so that $F_{bx} \approx 0$,

we have $$F_{sx} = \int_{C.S.} \rho(\vec{V} \cdot d\vec{A})V_x$$

But,

$$F_{sx} = F_{sx_1} - F_{sx_2}$$

$$= p_1 A - p_2 A \qquad (\because \ A_1 = A_2 = A)$$

$$= A(p_1 - p_2)$$

Also,

$$\int_{C.S.} \rho(\vec{V} \cdot d\vec{A})V_x = V_2(\rho_2 V_2)A - V_1(\rho_1 V_1)A$$

$$= V_2\dot{m} - V_1\dot{m} \qquad (\because \ \rho_1 AV_1 = \rho_2 AV_2 = \dot{m})$$

$$= \dot{m}(V_2 - V_1)$$

Therefore $$A(p_1 - p_2) = \dot{m}(V_2 - V_1)$$

or $$\dot{m}(V_2 - V_1) = A(p_1 - p_2)$$

$$\frac{\dot{m}}{A}(V_2 - V_1) = (p_1 - p_2)$$

$$\rho_1 V_1(V_2 - V_1) = (p_1 - p_2)$$

$$V_2 - V_1 = \frac{(p_1 - p_2)}{\rho_1 V_1}$$

$$V_2 = V_1 + \frac{(p_1 - p_2)}{\rho_1 V_1} \qquad (3)$$

Applying the equation of state for an ideal gas to section (1)

$$p_1 = \rho_1 RT_1$$

$$\therefore \quad p_1 = 0.0699 \times 53.3 \times 580 = 2160 \text{ lb/ft}^2 = 15 \text{ psia}$$

Substituting the values of ρ_1, p_1, p_2, and V_1 in equation (3)

$$V_2 = 295 \frac{\text{ft}}{\text{sec}} + \frac{(15-7)\frac{\text{lb}_f}{\text{in}^2} \times 144 \frac{\text{in}^2}{\text{ft}^2} \times 32.2 \frac{\text{lb}_m}{\text{slug}} \times \frac{\text{slug} \times \text{ft}}{\text{lbf} \times \text{sec}^2}}{0.0699 \frac{\text{lb}_m}{\text{ft}^3} \times 295 \frac{\text{ft}}{\text{sec}}}$$

$= 2094$ ft/sec

Using the equation of state for section (2),

$$T_2 = \frac{p_2}{\rho_2 R}$$

The continuity equation gives,

$$\rho_2 = \left(\frac{V_1}{V_2}\right)\rho_1 = \left(\frac{295}{2094}\right) \times 0.0699$$

$$\therefore \quad \rho_2 = 9.85 \times 10^{-3} \text{ lbm/ft}^3$$

giving

$$T_2 = \frac{7 \times 144}{9.85 \times 10^{-3} \times 53.3} = 1920^0\text{R}$$

Using the definition of Mach No. at section (2),

$$M_2 = \frac{V_2}{\sqrt{kRT_2 g_c}} = \frac{2094}{\sqrt{1.4 \times 53.3 \times 1920 \times 32.2}} = 0.975$$

Substitution for M_2 and T_2 in equation (2) yields,

$$T_{O2} = 1920\left[1 + \frac{(1.4 - 1)}{2}(0.975)^2\right] = 2285^0\text{R}$$

Thus, the increase in stagnation temperature is

$$T_{O2} - T_{O1} = 2285 - 587 = 1698^0\text{R}$$

Now, to calculate $(p_{O2} - p_{O1})$, let us find out p_{O1} first. The isentropic law holds,

$$\frac{p_{O1}}{p_1} = \left(\frac{T_{O1}}{T_1}\right)^{\frac{k}{k-1}}$$

$$\Rightarrow \quad p_{O1} = p_1\left[1 + \frac{k-1}{2}M_1^2\right]^{\frac{k}{k-1}} = 15\left[1 + \frac{(1.4-1)}{2}(0.25)^2\right]^{\frac{1.4}{(1.4-1)}}$$

$$= 15.67 \text{ psia}$$

Similarly,

$$p_{O_2} = p_2\left[1 + \frac{k-1}{2}M_2^2\right]^{\frac{k}{k-1}} = 7\left[1 + \frac{(1.4-1)}{2}(0.975)^2\right]^{\frac{1.4}{1.4-1}}$$

$$= 12.9 \text{ psia}$$

$$p_{O_2} - p_{O_1} = 12.9 - 15.67 = -2.77 \text{ psia}$$

Therefore stagnation pressure decreases when heat is added to a subsonic flow.

(ii) Rate of heat addition per unit mass flow:

The heat transfer rate per unit mass flow can be evaluated by writing the First law of Thermodynamics in the form of basic energy equation which can be stated as,

$$\int_{C.S.} (E + pv)\rho(\vec{V} \cdot d\vec{A}) + \frac{\partial}{\partial t}\int_{C.V.} E\rho d\forall$$

$$= \dot{Q}_A + \dot{W}_{shaft} + \dot{W}_{shear}$$

where

$$E = u + \frac{V^2}{2} + gz$$

Assuming that,

(1) no mechanical work is being drawn so that $\dot{W}_{shaft} = 0$

(2) the flow is non-viscous so that $\dot{W}_{shear} = 0$

(3) the flow is steady, therefore

$$\frac{\partial}{\partial t}\int_{C.V.} E\rho d\forall = 0,$$

the basic energy equation is reduced to,

$$\dot{Q}_A = (u_2 + p_2v_2 + \frac{V_2^2}{2})(\rho_2V_2A) - (u_1 + p_1v_1 + \frac{V_1^2}{2})(\rho_1V_1A) + g(z_2 - z_1)$$

Assuming $z_1 \simeq z_2$ and writing $h_1 = u_1 + p_1v_1$ and $h_2 = u_2 + p_2v_2$, where h_1 and h_2 are the specific enthalpies of air at cross-sections (1) and (2) respectively,

$$\dot{Q}_A = \left(h_2 + \frac{V_2^2}{2}\right)\dot{m} - \left(h_1 + \frac{V_1^2}{2}\right)\dot{m}$$

Defining the stagnation enthalpy, $h_o = h + \frac{V^2}{2}$ we have,

$$\dot{Q}_A = \dot{m}(h_{o_2} - h_{o_1})$$

or

$$\dot{Q}_A = \dot{m}C_p(T_{o_2} - T_{o_1})$$

where

$$\frac{\dot{Q}}{\dot{m}} = \frac{dQ/dt}{dm/dt} = \frac{dQ}{dm}$$

$$\therefore \quad \frac{\dot{Q}_A}{\dot{m}} = \frac{dQ_A}{dm} = C_p(T_{o_2} - T_{o_1}) \tag{4}$$

$$= 0.24 \frac{\text{Btu}}{\text{lbm} \times {}^0\text{R}}(1698^0\text{R}) = 407.5 \text{ Btu/lbm}$$

Entropy Change, ΔS_{12}:

For air as a perfect gas, using a relation governed by second law of thermodynamics,

$$dS = \frac{du}{T} + \frac{pdv}{T}$$

But $\quad h = u + pv$

Differentiating,

$$dh = du + d(pv) = du + pdv + vdp$$

$$\Rightarrow \quad du = dh - pdv - vdp$$

Thus we have,

$$dS = \frac{dh}{T} - \frac{pdv}{T} - \frac{vdp}{T} + \frac{pdv}{T}$$

or

$$dS = \frac{dh}{T} - v\,\frac{dp}{T}$$

where $\quad dh = C_p dT$ and from ideal gas law, $v = \frac{RT}{p}$, hence,

$$dS = \frac{C_p dT}{T} - \frac{RT}{p}\frac{dp}{T}$$

On integration,

$$\Delta S_{12} = S_2 - S_1 = \int_{T_1}^{T_2} \frac{C_p dT}{T} - R\int_{p_1}^{p_2} \frac{dp}{p}$$

$$= C_p \ln T\Big]_{T_1}^{T_2} - R\ln p\Big]_{p_1}^{p_2}$$

$$= C_p\left[\ln T_2 - \ln T_1\right] - R\left[\ln p_2 - \ln p_1\right]$$

$$\therefore \quad \Delta S_{12} = C_p \ln\frac{T_2}{T_1} - R\ln\frac{p_2}{p_1} \qquad (5)$$

Substituting the values of

$$\Delta S_{12} = \frac{0.24\ \text{Btu}}{\text{lbm} \times {}^0\text{R}} \times \ln\left(\frac{2285}{587}\right) - \frac{53.3\ \dfrac{\text{lbf} \times \text{ft}}{\text{lbm} \times {}^0\text{R}}}{778\ \dfrac{\text{lbf} \times \text{ft}}{\text{Btu}}} \times \ln\left(\frac{7}{15}\right)$$

or

$$\Delta S_{12} = 0.378\ \frac{\text{Btu}}{\text{lbm} \times {}^0\text{R}}$$

The above changes in stagnation temperatures, stagnation pressure and entropy is shown in Fig. 2.

Now, if the conditions are changed at both the sections of the control volume, we will proceed in the following manner:

Mach No. at section (1) $= M_1 = \dfrac{V_1}{\sqrt{kRT_1 g_c}}$

or

$$M_1 = \frac{2220}{\sqrt{1.4 \times 53.3 \times 540 \times 32.2}} = 1.95$$

From equation (1),

$$T_{O_1} = (540)\left[1 + \frac{(1.4 - 1)}{2}(1.95)^2\right] = 951^\circ R$$

In order to find T_{O_2}, we need to know T_2 and M_2. Use of working tables can be made to find T_2 and M_2. The working table consists of ratio of properties at a section to the critical properties.

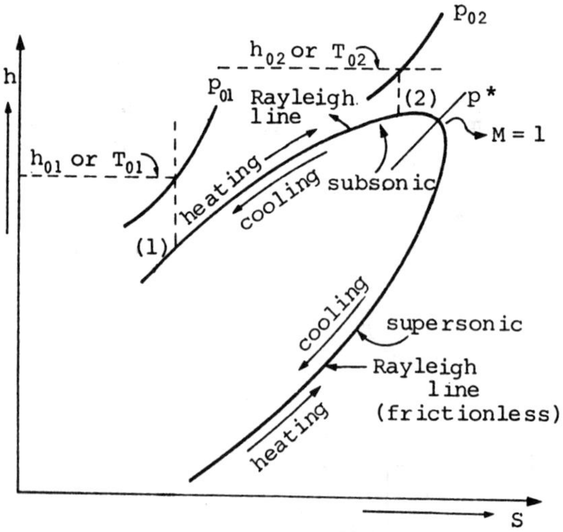

Fig.(2) Heating of a subsonic frictionless
flow from point (1) to (2).

WORKING TABLE
Frictionless Flow Through a Uniform Cross-Sectional
Duct With Heating and Cooling

M	T_o/T_o*	T/T*	p/p*	p_o/p_o*	V/V*
1.15	.98721	.93685	.84166	1.01092	1.1131
1.95	.80359	.54774	.37954	1.4516	1.4432

We have $V_1 = 2220$ ft/sec and $V_2 = 1700$ ft/sec. Since the critical properties are the same for every point in the flow, therefore,

$$\frac{V_2}{V_1} = \frac{(V_2/V^*)}{(V_1/V^*)} \implies \frac{1700}{2220} = \frac{V_2/V^*}{1.4432}$$

where V_1/V* has been found against the value of $M_1 = 1.95$ from the table.

$$\therefore \quad \frac{V_2}{V^*} = \frac{1700}{2220} \times 1.4432 \approx 1.11$$

From the working table, for $\frac{V_2}{V^*} = 1.11$ the nearest value of Mach No. $M_2 = 1.15$.

Again, from the working table, for $M_1 = 1.95$, $\frac{T_1}{T^*} = 0.54774$ and for $M_2 = 1.15$, $\frac{T_2}{T^*} = 0.93685$. With these values, we have

$$\frac{T_2}{T_1} = \frac{(T_2/T^*)}{(T_1/T^*)} = \frac{0.93685}{0.54774} = 1.71$$

or

$$T_2 = 1.71T_1 = 1.71 \times 540 = 923.4^0R$$

Thus, from equation (2)

$$T_{O_2} = T_2\left[1 + \frac{k-1}{2}M_2^2\right] = (923.4)\left[1 + \frac{(1.4-1)}{2}(1.15)^2\right]$$

$$= 1167.6^0R$$

This allows us to calculate,

$$T_{O_2} - T_{O_1} = 1167.6 - 951 = 216.6^0R$$

Thus

$$T_{O_2} > T_{O_1} \quad \text{as shown in Fig. 3.}$$

Also, from working tables,

$$\frac{p_2}{p_1} = \frac{p_2/p^*}{p_1/p^*} = \frac{0.84166}{0.37954} = 2.22$$

$$\Rightarrow \quad p_2 = 2.22p_1 = 2.22 \times 10 = 22.2 \text{ psia}$$

From the isentropic flow tables,

at $M_1 = 1.95$, $p_1/p_{O_1} = 0.13813 \Rightarrow p_{O_1} = 7.24p_1$

at $M_2 = 1.15$, $p_2/p_{O_2} = 0.43983 \Rightarrow p_{O_2} = 2.274p_2$

Therefore $p_{O_2} - p_{O_1} = (2.274 \times 22.2) - (7.24)(10)$

$$= -21.5 \text{ psia}$$

This shows that the stagnation pressure decreases during heating of a supersonic flow.

Heat transfer per unit mass flow:

Using equation (4), the heat transfer is

$$\frac{\dot{Q}_A}{\dot{m}} = C_p(T_{O_2} - T_{O_1}) = 0.24 \times 216.6 = 52 \frac{\text{Btu}}{\text{lbm}}$$

Finally, from equation (5), the change in entropy ΔS_{12} is,

$$\Delta S_{12} = C_p \ln \frac{T_2}{T_1} - R \ln \frac{p_2}{p_1}$$

from which,

$$\Delta S_{12} = 0.24 \ln\left(\frac{923}{540}\right) - \left(\frac{53.3}{778}\right) \ln\left(\frac{22.2}{10}\right)$$

or

$$\Delta S_{12} = 0.074 \frac{\text{Btu}}{\text{lbm}^0\text{R}}$$

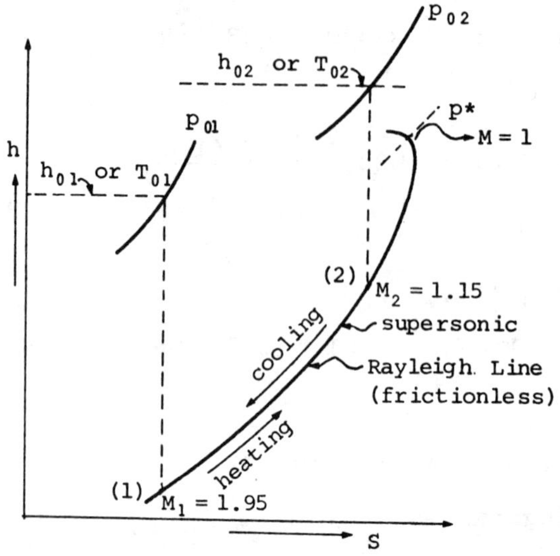

Fig. (3) Mach No. of the supersonic Rayleigh flow decreases with the addition of heat.

NOZZLE

• PROBLEM 12-28

Air at 70° F is stored in a large reservoir. If the stored air is allowed to escape through a converging nozzle, with the exit area equal to 1 in^2, into a region having a pressure equal to 14.5 psia, find the mass rate through the nozzle for the following conditions: (a) reservoir pressure equals 20 psia and (b) reservoir pressure equals 50 psia.

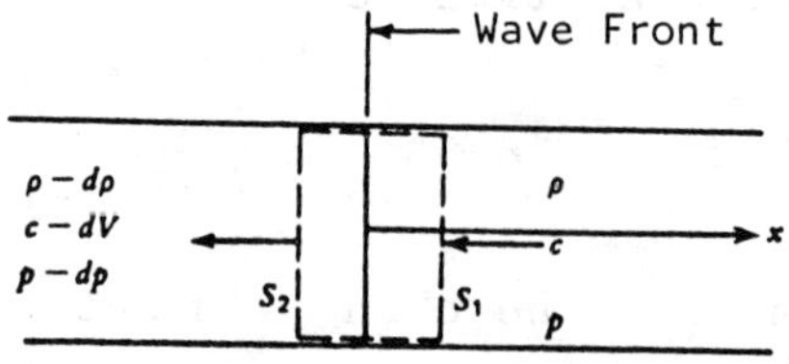

Solution: (a) First we determine the critical pressure ratio from the isentropic tables. From Table 1, $p/p_0 = 0.528$ for $M = 1$. Thus, $p_c/p_0 = 0.528$. We know that if $p_b/p_0 > 0.528$, then the nozzle is not choked and the exit pressure, p_e, equals p_b.

$$\frac{p_b}{p_0} = \frac{14.5}{20.0} = 0.725 > 0.528$$

Thus, $p_e/p_0 = 0.725$. The Mach number at the exit, M_e, can also be determined from the isentropic tables. For $p_e/p_0 = 0.725$, we can determine, by interpolation, that $M_e = 0.694$ and that $T_e/T_0 = 0.9122$. Thus, the temperature at the exit, T_e, equals 483.2R.

We must now determine C_e.

Select a control volue around the wave front, as shown in the figure. Now apply equations of conservation of mass and Newton's second law to the control volume. Conservation of mass gives

$$-\rho cA + (\rho - d\rho)(c - dV)A = 0$$

which, after neglecting the higher order terms and simplifying, reduces to

$$c\, d\rho + \rho\, dV = 0 \tag{1}$$

Newton's second law gives

Substituting these expressions into Eq. 3, we obtain

$$c = \sqrt{g_c kRT}$$

$$c_e = \sqrt{g_c kRT_e} = \sqrt{32.17(1.4)(53.35)(483.2)} = 1077.5 \text{ ft/s}$$

The mass flow rate can be determined by

$$m = \rho_e A_e V_e = \frac{p_e}{RT_e} A_e M_e c_e = \frac{14.5(1)(0.694)(1077.5)}{53.35(483.2)} = 0.421 \text{ lb}_m/\text{S}$$

(b)

$$\frac{p_b}{p_0} = \frac{14.5}{50.0} = 0.29 < 0.528$$

Therefore, the nozzle is chocked, giving $M_e = 1$ and $p_e = 0.528(50) = 26.4$ psia. From the isentropic table, $T_e T_0 = 0.8333$, giving $T_e = 441.4$R. Thus

TABLE 1

M	A/A^*	p/p_0	ρ/ρ_0	T/T_0
0.00	—	1.000	1.000	1.000
0.01	57.87	0.9999	0.9999	0.9999
0.02	28.94	0.9997	0.9999	0.9999
0.04	14.48	0.999	0.999	0.9996
0.06	9.67	0.997	0.998	0.999
0.08	7.26	0.996	0.997	0.999
0.10	5.82	0.993	0.995	0.998
0.12	4.86	0.990	9.993	0.997
0.14	4.18	0.986	0.990	0.996
0.16	3.67	0.982	0.987	0.995
0.18	3.28	0.978	0.984	0.994
0.20	2.96	0.973	0.980	0.992
0.22	2.71	0.967	0.976	0.990
0.24	2.50	0.961	0.972	0.989
0.26	2.32	0.954	0.967	0.987
0.28	2.17	0.947	0.962	0.985
0.30	2.04	0.939	0.956	0.982
0.32	1.92	0.932	0.951	0.980
0.34	1.82	0.923	0.944	0.977
0.36	1.74	0.914	0.938	0.975
0.38	1.66	0.905	0.931	0.972
0.40	1.59	0.896	0.924	0.969
0.42	1.53	0.886	0.917	0.966
0.44	1.47	0.876	0.909	0.963
0.46	1.42	0.865	0.902	0.959
0.48	1.38	0.854	0.893	0.956
0.50	1.34	0.843	0.885	0.952
0.52	1.30	0.832	0.877	0.949
0.54	1.27	0.820	0.868	0.945
0.56	1.24	0.808	0.859	0.941
0.58	1.21	0.796	0.850	0.937
0.60	1.19	0.784	0.840	0.933
0.62	1.17	0.772	0.831	0.929
0.64	1.16	0.759	0.821	0.924
0.66	1.13	0.747	0.812	0.920
0.68	1.12	0.734	0.802	0.915
0.70	1.09	0.721	0.792	0.911
0.72	1.08	0.708	0.781	0.906
0.74	1.07	0.695	0.771	0.901
0.76	1.06	0.682	0.761	0.896
0.78	1.05	0.669	0.750	0.891
0.80	1.04	0.656	0.740	0.886
0.82	1.03	0.643	0.729	0.881
0.84	1.02	0.630	0.719	0.876
0.86	1.02	0.617	0.708	0.871
0.88	1.01	0.604	0.698	0.865
0.90	1.01	0.591	0.687	0.860
0.92	1.01	0.578	0.676	0.855
0.94	1.00	0.566	0.666	0.850
0.96	1.00	0.553	0.655	0.844
0.98	1.00	0.541	0.645	0.839
1.00	1.00	0.528	0.632	0.833

$$c_e = \sqrt{g_c kRT_e} = \sqrt{32.17(1.4)(53.35)(441.4)} = 1029.8 \text{ ft/S}$$

Then the mass flow rate is determined by

$$\dot{m} = \frac{p_e}{RT_e} A_e M_e c_e = \frac{26.4(1)(1)(1029.8)}{53.35(441.4)} = 1.15 \text{ lb}_m\text{/S}$$

• PROBLEM 12-29

In a converging-diverging nozzle shown, the air flow at the entrance has a stagnation temperature $T_o = 300\,^0\text{K}$ and absolute stagnation pressure $P_o = 1.5\text{MPa}$. At the throat the velocity corresponds to 0.55 Mach, and at the exit the pressure $P_2 = 800\text{kPa}$ (abs). The exit area is $.002\text{m}^2$. Find (a) the temperature, pressure, density, and the velocity of the flow at the throat; and (b) the velocity of the flow leaving at the exit. (c) The nozzle is operated at a back pressure of 45.5kPa (abs) although the design back pressure is 77.0kPa (abs). Assuming the flow is isentropic, find the flow velocity and the mass flow rate under these conditions at the exit.

Solution: The basic flow relationships for converging-diverging nozzles from which it is possible to determine characteristics at the throat are

$$\frac{T_o}{T_1} = 1 + \frac{k-1}{2} M_1^2 \qquad (1)$$

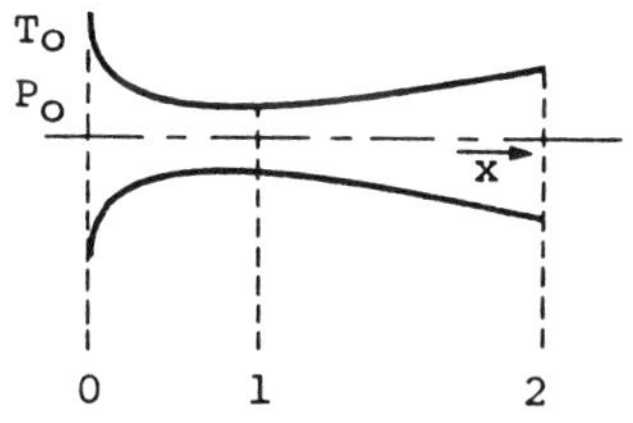

$$P_1 = P_o \left(\frac{T_1}{T_o}\right)^{k/k-1} \qquad (2)$$

$$\rho_1 = \frac{P_1}{RT_1} \qquad (3)$$

$$V_1 = M_1 c_1 = M_1 \sqrt{kRT_1} \qquad (4)$$

Substituting values into (1) to (4),

$$\frac{300^{\circ}K}{T} = 1 + \frac{1.4 - 1}{2}(.55)^2 = 1.0605$$

from which $T_2 = 283^{\circ}K$.

Substituting in equation (2),

$$P_1 = 1.5\left(\frac{283}{300}\right)^{\frac{1.4}{.4}} = 0.978\text{MPa}$$

and substituting into equations (3) and (4), respectively,

$$\rho_1 = \frac{0.978(10^6)N}{m^2} \times \frac{kg \times K}{287Nm} \times \frac{1}{300^{\circ}K} = 11.4\ \frac{kg}{m^3}$$

$$V_1 = .55\sqrt{1.4(287)(283)} = 185\text{m/sec}$$

(b) Since $M_1 < 1$, the flow at the exit must be subsonic, and the exit flow Mach No. M_2 may be obtained from

$$\frac{P_o}{P_2} = \left[1 + \frac{k-1}{2}M_2^2\right]^{k/k-1}$$

Solving for M_2,

$$\left(\frac{P_o}{P_2}\right)^{k-1/k} = 1 + \frac{k-1}{2}M_2^2$$

and

$$M_2 = \left\{\left[\left(\frac{P_o}{P_2}\right)^{k-1/k} - 1\right]\left[\frac{2}{k-1}\right]\right\}^{\frac{1}{2}} \qquad (5)$$

Substituting values,

$$M_2 = \left\{\left[\left(\frac{1.5 \times 10^6}{0.8 \times 10^6}\right)^{.286} - 1\right]\left[\frac{2}{1.4-1}\right]\right\}^{\frac{1}{2}} = 0.44$$

(c) Since the nozzle is operated at a back pressure substantially below the design back pressure, the flow through the nozzle is underexpanded. Whereas the flow within the nozzle is isentropic, the expansion from the design exit pressure to the operated back pressure is irreversible, resulting in an increase in entropy. The expansion curve of flow through the nozzle appears as follows:

For isentropic flow the stagnation properties are constant, and therefore the Mach number at the exit can be computed from

$$M_2 = \left\{\left[\left(\frac{P_o}{P_2}\right)^{k-1/k} - 1\right]\frac{2}{k-1}\right\}^{\frac{1}{2}} \qquad (6)$$

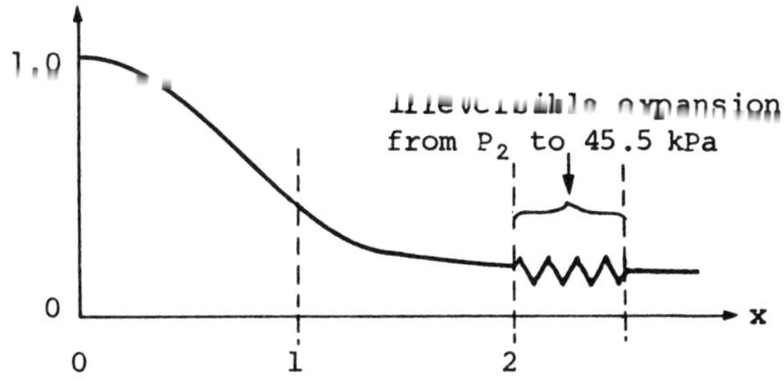

which is derived from

$$\frac{P_o}{P_2} = \left[1 + \frac{(k - 1)}{2}M_2^2\right]^{k/k-1} \tag{7}$$

Now the mass flow rate $\frac{dm}{dt}$ at the exit may be obtained from

$$\frac{dm}{dt} = \rho VA\Big]_{exit}$$

$$= \rho AcM\Big]_{exit}$$

$$= \rho AM\sqrt{kRT}\Big]_{exit}$$

Since $\rho = \frac{P}{RT}$

$$\frac{dm}{dt} = \frac{PAM}{RT}\sqrt{kRT}\Big]_{exit}$$

$$= PAM\sqrt{\frac{k}{RT}}\Big]_{exit} \tag{8}$$

where

$$\frac{T_o}{T_2} = 1 + \frac{(k - 1)}{2}M_2^2$$

from which

$$T_2 = \frac{T_o}{1 + \frac{(k-1)}{2}M_2^2} \tag{9}$$

Substituting values into equation (6),

$$M_2 = \left\{\left[\left(\frac{1.5 \times 10^6}{.077 \times 10^6}\right)^{.286} - 1\right]\frac{2}{1.4 - 1}\right\}^{\frac{1}{2}} = 2.59$$

To compute $\frac{dm}{dt}$, we first find T_2 from equation (9) and substitute into (8).

$$T_2 = \frac{300}{1 + \frac{(1.4-1)}{2}(2.59)^2} = 128^0\text{K}$$

and

$$\frac{dm}{dt} = 77 \times 10^3 \frac{N}{m^2} \times .002m^2 \times 2.59\sqrt{\frac{1.4}{287 \times 128}}$$

$$\frac{dm}{dt} = 2.46 \text{ kg/sec}$$

• PROBLEM 12-30

(a) Air is supplied to a converging nozzle installed at the end of a duct of large cross-sectional area. The temperature and pressure at the inlet of the nozzle are 70 ^{0}F and 30 psia respectively. The nozzle has a throat area of 0.017 ft^2 and it discharges to a back pressure of 25 psia. Calculate the Mach No. at the exit and the mass flow rate in the nozzle using

(i) isentropic flow equations

(ii) isentropic flow tables.

(b) Suppose that the stagnation temperature and pressure are not known for a converging nozzle having an isentropic flow of air. However, at a certain cross-section in the nozzle, the local Mach No., temperature and pressure are 0.44, 80 ^{0}F and 70 psia respectively, and the nozzle diameter is 1 3/4 in. at this section. The nozzle exhausts to a back pressure of 20 psia. Calculate, using (i) isentropic flow equations and (ii) isentropic flow tables, the throat Mach No., area at the throat and the mass flow rate.

Fig. (1)

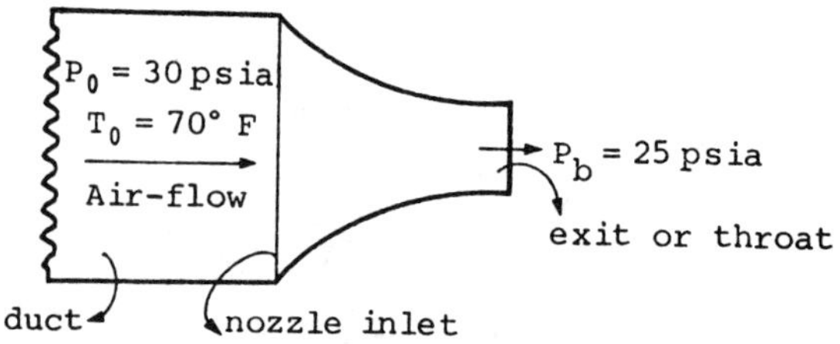

Solution: (a) (i)

Given:

Back Pressure, P_b = 25 psia

Stagnation or Total Pressure, P_o = 30 psia

Check for the choking condition of nozzle:

$$\frac{P_b}{P_o} = \frac{25}{30} = 0.833$$

Since, a nozzle is choked for $\frac{P_b}{P_o} \leqq 0.5283$, therefore, the nozzle in our case is not choked, the flow is subsonic and hence $P_b = P_e$, where P_e is the exit plane pressure. We will use e as the subscript for all the properties in the exit plane of nozzle.

Now, from the isentropic relation between P_o and P_e

$$\frac{P_o}{P_e} = \left[1 + \frac{(k - 1)}{2}M_e^2\right]^{\frac{k}{k-1}}$$

or

$$\left(\frac{P_o}{P_e}\right)^{(k-1)/k} = 1 + \frac{(k - 1)}{2}M_e^2$$

or

$$\left(\frac{P_o}{P_e}\right)^{(k-1)/k} - 1 = \frac{(k - 1)}{2}M_e^2$$

$$M_e^2 = \frac{2}{k - 1}\left[\left(\frac{P_o}{P_e}\right)^{(k-1)/k} - 1\right]$$

Substituting the values of P_o, P_e and k,

$$M_e^2 = \frac{2}{(1.4 - 1)}\left[\left(\frac{30}{25}\right)^{(1.4-1)/1.4} - 1\right] = 0.2673$$

$$\therefore \quad M_e = 0.52$$

The mass flow rate '$\dot{m}$' through the nozzle is given by the following equation,

$$\dot{m} = \rho_e A_e V_e$$

$$\frac{T_e}{T_o} = \left(\frac{P_e}{P_o}\right)^{(k-1)/k} = \left(\frac{25}{30}\right)^{(1.4-1)/1.4} = 0.95$$

where $T_o = 70^\circ F + 460 = 530^\circ R$.

$$\therefore \quad T_e = 0.95T_o = 0.95 \times 530 = 503.5^\circ R$$

From perfect gas equation,

$$\rho_e = \frac{P_e}{RT_e} = \frac{25 \times 144}{53.3 \times 503.5} = 0.134 \text{ lbm/ft}^3$$

Using the definition of Mach No.,

$$M_e = \frac{V_e}{C_e}$$

or

$$V_e = M_e C_e$$

where the velocity of sound at the nozzle exit is given by

$$C_e = \sqrt{kRT_e g_c}$$

$$= \sqrt{1.4 \times 53.3 \frac{\text{lbf-ft}}{\text{lbm-}^0\text{R}} \times 503.5^0\text{R} \times 32.2 \frac{\text{ft-lbm}}{\text{lbf-sec}^2}}$$

$$= 1100 \text{ ft/sec}$$

Therefore, $V_e = M_e C_e = 0.52 \times 1100 = 572$ ft/sec.

Also, in the converging nozzle, the smallest cross-sectional area (which is at the exit) is called throat, Fig. 1.

$$\therefore \quad A_e = 0.017 \text{ ft}^2$$

Substituting the values of P_e, A_e and V_e in the equation for $\dot{m}$,

$$\dot{m} = 0.134 \frac{\text{lbm}}{\text{ft}^3} \times 0.017 \text{ ft}^2 \times 572 \text{ ft/sec}$$

$$= 1.3 \text{ lbm/sec}$$

(ii) Since $\frac{P_b}{P_o} = 0.833 \geq 0.5283$, therefore, the nozzle is not choked and the flow is subsonic. From the isentropic tables, for $\frac{P_b}{P_o} = 0.833$ the nearest number is 0.832 corresponding to which, $M_e = 0.52$.

Now, from the continuity equation, the mass flow rate across the exit plane of the nozzle is given by,

$$\dot{m} = \rho_e A_e V_e$$

where, $V_e = M_e C_e$

and $C_e = \sqrt{kRT_e g_c}$

Therefore, from the isentropic table, for

$$M = 0.52, \quad \frac{T_e}{T_o} = 0.949$$

from which, $T_e = 0.949T_o = 0.949 \times 530 = 503^0\text{R}$ giving,

$$C_e = \sqrt{1.4 \times 53.3 \times 503 \times 32.2} = 1099.4 \frac{\text{ft}}{\text{sec}}$$

from which $V_e = M_eC_e = 0.52 \times 1099.4 = 571.7$ ft/sec.

From the ideal gas equation,

$$\rho_e = \frac{p_e}{RT_e} = \frac{25 \times 144}{53.3 \times 503} = 0.134 \text{ lbm/ft}^3$$

$$\therefore \quad \dot{m} = \rho_e A_e V_e = 0.134 \times 0.017 \times 571.7 = 1.3$$

We can also find out $\dot{m}$ by proceeding in the following manner:

$$\dot{m} = \rho_e A_e V_e$$

$$= \left(\frac{\rho_e}{\rho_o}\right) \rho_o A_e M_e \sqrt{kRT_e g_c}$$

$$= \left(\frac{\rho_e}{\rho_o}\right)\left(\frac{P_o}{RT_o}\right) A_e M_e \sqrt{kRT_e g_c}$$

$$= \left(\frac{\rho_e}{\rho_o}\right)\left(\frac{P_o}{R\sqrt{T_o}}\right) A_e M_e \sqrt{kR} \sqrt{\frac{T_e}{T_o}}$$

$$\dot{m} = p_o A_e M_e \left(\frac{\rho_e}{\rho_o}\right) \sqrt{\frac{T_e}{T_o}} \sqrt{\frac{k}{RT_o}}$$

From the isentropic flow tables, for $M_e = 0.52$, $\frac{\rho_e}{\rho_o} = 0.877$, $\frac{T_e}{T_o} = 0.949$. Substituting these values in the above equation for $\dot{m}$,

$$\dot{m} = \left[30 \frac{\text{lbf}}{\text{in}^2} \times 144 \frac{\text{in}^2}{\text{ft}^2} \times 0.017 \text{ ft}^2 \times 0.52(0.877)\sqrt{0.949}\right] \times \sqrt{\frac{1.4 \times 32.2 \frac{\text{lbm-ft}}{\text{lbf-sec}^2}}{53.3 \frac{\text{lbf-ft}}{\text{lbm-}^\circ\text{R}} \times 530^\circ \text{R}}}$$

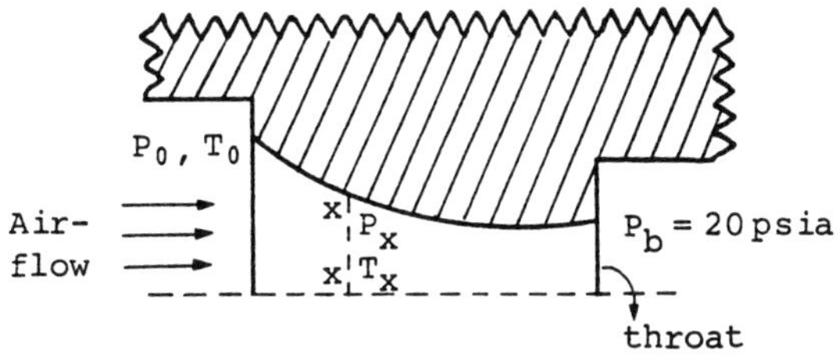

Fig. (2)

$$= (32.63\ lb_f)\sqrt{1.6 \times 10^{-3}\ \frac{lbm^2}{lbm^2\text{-}sec^2}}$$

$$\dot{m} = 1.3\ \frac{lbm}{sec}$$

(b)

(i) To see whether the nozzle is choked or not, we again use the ratio $\frac{P_b}{P_o}$, but P_o is not known this time.

Now, for the cross-section x-x, Fig. 2, the isentropic relation between T_x and T_o in terms of M_x can be described as

$$\frac{T_o}{T_x} = 1 + \frac{(k - 1)}{2}M_x^2$$

For $M_x = 0.44$

$$\frac{T_o}{T_x} = 1 + \frac{(1.4 - 1)}{2}(0.44)^2 = 1.038$$

Therefore,

$$\frac{P_o}{P_x} = \left(\frac{T_o}{T_x}\right)^{\frac{k}{k-1}} = (1.038)^{\frac{1.4}{(1.4-1)}} = 1.14$$

or

$$P_o = 1.14P_x = 1.14 \times 70 = 80 \text{ psia}$$

Thus,

$$\frac{P_b}{P_o} = \frac{20}{80} = 0.25 < 0.5283$$

which means that the nozzle is choked. Therefore, critical conditions exist at the throat. Note that the minimum area in the nozzle is called throat. Now, for the critical conditions, the Mach No. at the throat is unity, i.e. M* = 1. We will use an asterisk '*' to denote the critical conditions at the throat.

Now in the isentropic relation,

$$\frac{P_o}{P} = \left(1 + \frac{(k - 1)}{2}M^2\right)^{\frac{k}{k-1}}$$

M = 1 for P = P*, the critical pressure

$$\therefore \quad \frac{P_o}{P^*} = \left(1 + \frac{(k - 1)}{2}\right)^{\frac{k}{k-1}} = \left(\frac{2 + k - 1}{2}\right)^{\frac{k}{k-1}} = \left(\frac{k + 1}{2}\right)^{\frac{k}{k-1}}$$

For air, k = 1.4

$$\therefore \quad \frac{P_o}{P^*} = \left(\frac{2 + 1.4 - 1}{2}\right)^{\frac{1.4}{1.4 - 1}} = 1.893$$

or
$$\frac{P^*}{P_o} = 0.5283$$

giving $P^* = 0.5283P_o = 0.5283 \times 80 = 42.3$ psia

Similarly for T*,

$$\frac{T_o}{T^*} = 1 + \frac{k - 1}{2} = 1 + \frac{(1.4 - 1)}{2} = 1.2$$

or
$$\frac{T^*}{T_o} = 0.833$$

$$\therefore \quad T^* = 0.833T_o = 0.833(1.038T_x)$$

$$T^* = 0.833 \times 1.038 \times 540 = 467^0R$$

Mass flow rate can be calculated by applying equation of continuity to the properties of air at section x-x as follows:

$$\dot{m} = \rho_x A_x V_x$$

$$A = \frac{\pi d_x^2}{4} = \frac{\pi}{4} \times (0.146)^2 = 0.0167 \text{ ft}^2$$

Using the definition of Mach No.

$$M_x = \frac{V_x}{C_x}$$

or
$$V_x = M_x C_x = M_x \sqrt{kRT_x g_c}$$

where, $T_x = 540^0R$, $M_x = 0.44$

$$\therefore \quad V_x = 0.44\sqrt{1.4\left(53.3 \frac{\text{lbf-ft}}{\text{lbm-}^0\text{R}}\right)(540^0R)\left(32.2 \frac{\text{lbm-ft}}{\text{lbf-sec}^2}\right)}$$

$$= 501.2 \text{ ft/sec}$$

Also, from the ideal gas equation,

$$\rho_x = \frac{P_x}{RT_x} = \frac{70(\text{lbf/in}^2)(144 \text{ in}^2/\text{ft}^2)}{53.3 \frac{\text{lbf-ft}}{\text{lbm-}^0\text{R}} \times 540^0R} = 0.35 \frac{\text{lbm}}{\text{ft}^3}$$

$$\therefore \quad \dot{m} = \rho_x A_x V_x$$

$$= 0.35 \frac{\text{lbm}}{\text{ft}^3} \times 0.0167 \text{ ft}^2 \times 501.2 \frac{\text{ft}}{\text{sec}}$$

$$= 2.93 \frac{\text{lbm}}{\text{sec}}$$

Area at the throat, A*:

Using the equation of continuity between the section x-x and the throat,

$$\dot{m} = \rho_x A_x V_x = \rho^* A^* V^*$$

from which

$$A^* = \frac{\dot{m}}{\rho^* V^*}$$

where, for critical conditions,

$$V^* = C^* = \sqrt{kRT^* g_c}$$

Calculated previously, $T^* = 467^0R$, $R = 53.3 \frac{\text{lbf-ft}}{\text{lbm-}^0\text{R}}$ Substitution gives,

$$V^* = \sqrt{1.4 \times 53.3 \frac{\text{lbf-ft}}{\text{lbm-}^0\text{R}} \times 467^0\text{R} \times 32.2 \frac{\text{lbm-ft}}{\text{lbf-sec}^2}}$$

or

$$V^* = 1059.3 \text{ ft/sec}$$

Also, applying perfect gas equation at the throat,

$$\rho^* = \frac{P^*}{RT^*} = \frac{42.3 \times 144}{53.3 \times 467} = 0.245 \frac{\text{lbm}}{\text{ft}^3}$$

Substituting the values for ρ^* and V^* in the above equation for A^*,

$$A^* = \frac{\dot{m}}{\rho^* V^*} = \frac{2.93 \frac{\text{lbm}}{\text{sec}}}{\left(0.245 \frac{\text{lbm}}{\text{ft}^3}\right)\left(1059.3 \frac{\text{ft}}{\text{sec}}\right)} = 0.0113 \text{ ft}^2$$

(ii) <u>Calculations of M^*, A^* and $\dot{m}$ using isentropic flow tables</u>:

In order to conform that the nozzle is operating under choking condition, we use the following procedure.

For M = 0.44, from the isentropic table in the book, $\frac{P}{P_o}$ = 0.909 from which,

$$\frac{P_b}{P_o} = \left(\frac{P_b}{P_x}\right)\left(\frac{P_x}{P_o}\right) = \frac{20}{70} \times 0.909 = 0.26$$

For a nozzle, operating under choked (or critical condition), $\frac{P_b}{P_o} \leq \frac{P^*}{P_o}$ where $\frac{P^*}{P_o} = 0.5283$.

Since, in our case, $\frac{P_b}{P_o} < \frac{P^*}{P_o}$, therefore the nozzle is choked.

As stated previously, for a choked nozzle, critical conditions exist at the throat which means that M = M* = 1 at the throat.

Now to find the area at the throat, we can obtain from isentropic tables, the value of $\frac{A_x}{A^*} = 1.47$ for $M_x = 0.44$.
Therefore,

$$A^* = 0.68A_x = 0.68 \times 0.0167 = 0.011356 \text{ ft}^2$$

Similar procedure is followed for finding out mass flow rate yielding the same value for $\dot{m}$ as before. This time we will apply continuity equation on the throat.

From the isentropic tables:

for M = 1, $\frac{T^*}{T_o} = 0.833$

$\frac{\rho^*}{\rho_o} = 0.632$

For M = 0.44, $\frac{P_x}{P_o} = 0.867$, $\frac{T_x}{T_o} = 0.963$

$\therefore$ $P_o = 1.15P_x = 1.15 \times 70 = 80.5$ psia

$T_o = 1.038T_x = 1.038 \times 540 = 560.5^\circ R$

$$\rho_o = \frac{P_o}{RT_o} = \frac{80.5 \times 144}{53.3 \times 560.5} = 0.388 \frac{\text{lbm}}{\text{ft}^3}$$

With this, we can calculate,

$T^* = 0.833T_o = 0.833 \times 560.5 = 467^\circ R$

$$\rho^* = 0.632\rho_o = 0.632 \times 0.388 = 0.245 \frac{\text{lbm}}{\text{ft}^3}$$

The sonic velocity c* at the throat is given by,

$$c^* = (kRT^*g_c)^{\frac{1}{2}}$$

$$\therefore \quad c^* = (1.4 \times 53.3 \times 467 \times 32.2)^{\frac{1}{2}} = 1059.3 \frac{\text{ft}}{\text{sec}}$$

Substituting the values of ρ*, A* and c* in the following equation for $\dot{m}$,

$$\dot{m} = \rho^* A^* c^* = 0.245 \frac{\text{lbm}}{\text{ft}^3} \times 0.011356 \text{ ft}^2 \times 1059.3 \frac{\text{ft}}{\text{sec}}$$

or

$$\dot{m} = 2.95 \frac{\text{lbm}}{\text{sec}}$$

SHOCK

• PROBLEM 12-31

A normal shock wave occurs in the flow of air where p_1 = 10 psia (70 N/m^2), T_1 = 40°F(5°C), and V_1 = 1400 fps (425 m/s). Find p_2, V_2, and T_2. Use Tables 1 and 2 for gas properties.

TABLE 1

Physical properties of common gases at standard sea-level atmosphere and 68°F in English units

Gas	Chemical formula	Molecular weight	Specific weight, γ, lb/ft³	Viscosity μ × 10, lb·s/ft²	Gas constant R, ft·lb/(slug)(°R) [= ft²/(s²)(°R)]	Specific heat, ft·lb/(slug)(°R) [= ft²/(s²)(°R)] c_p	c_v	Specific heat ratio k
Air		29.0	0.0753	3.76	1,715	6,000	4,285	1.40
Carbon dioxide	CO_2	44.0	0.114	3.10	1,123	5,132	4,009	1.28
Carbon monoxide	CO	28.0	0.0726	3.80	1,778	6,218	4,440	1.40
Helium	He	4.00	0.0104	4.11	12,420	31,230	18,810	1.66
Hydrogen	H_2	2.02	0.00522	1.89	24,680	86,390	61,710	1.40
Methane	CH_4	16.0	0.0416	2.80	3,100	13,400	10,300	1.30
Nitrogen	N_2	28.0	0.0728	3.68	1,773	6,210	4,437	1.40
Oxygen	O_2	32.0	0.0830	4.18	1,554	5,437	3,883	1.40
Water vapor	H_2O	18.0	0.0467	2.12	2,760	11,110	8,350	1.33

TABLE 2

Physical properties of common gases at standard sea-level and 68°F in SI units

Gas	Chemical formula	Molecular weight	Density ρ, kg/m³	Viscosity, μ × 10⁵ N·s/m²	Gas constant R, N·m/(kg)(K) [= m²/(s²)(K)]	Specific heat, N·m/(kg)(K) [(= m²/(s²)(K)] c_p	c_v	Specific heat ratio $k = c_p/c_v$
Air		29.0	1.205	1.80	287	1,003	716	1.40
Carbon dioxide	CO_2	44.0	1.84	1.48	188	858	670	1.28
Carbon monoxide	CO	28.0	1.16	1.82	297	1,040	743	1.40
Helium	He	4.00	0.166	1.97	2,077	5,220	3143	1.66
Hydrogen	H_2	2.02	0.0839	0.90	4,120	14,450	10,330	1.40
Methane	CH_4	16.0	0.668	1.34	520	2,250	1,730	1.30
Nitrogen	N_2	28.0	1.16	1.76	297	1,040	743	1.40
Oxygen	O_2	32.0	1.33	2.00	260	909	649	1.40
Water vapor	H_2O	18.0	0.747	1.01	462	1,862	1,400	1.33

Solution:

$$\rho_1 = \frac{p_1}{RT_1} = \frac{10(144)}{1{,}715(460 + 40)} = 0.00168 \text{ slug/ft}^3$$

The acoustic velocity c is given by

$$c_1 = \sqrt{kRT_1} = \sqrt{1.4 \times 1{,}715 \times (460 + 40)} = 1096 \text{ fps}$$

The Mach number Ma is given by

$$Ma_1 = \frac{V_1}{c_1} = \frac{1{,}400}{1{,}096} = 1.28$$

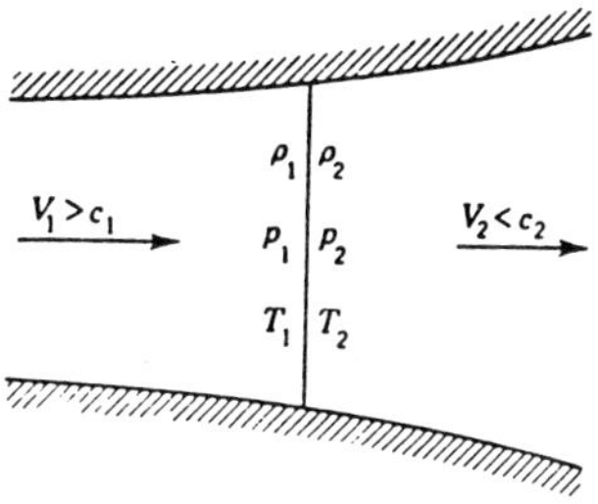

In the figure shown a one-dimensional shock wave where the approaching supersonic flow changes to subsonic flow. This phenomenon is accompanied by a sudden rise in pressure, density, and temperature. Applying the impulse-momentum principle to the fluid in the shock wave, we get

$$\sum F_x = p_1A_1 - p_2A_2 = \frac{G}{g}(V_2 - V_1) \qquad (1)$$

Substituting the continuity conditions ($G = \gamma_1A_1V_1 = \gamma_2A_2V_2$) and noting that $A_1 = A_2$, we get

$$p_2 - p_1 = \frac{1}{g}(\gamma_1V_1^2 - \gamma_2V_2^2) \qquad (2)$$

which is the pressure jump across the wave.

The flow across the shock wave may be considered adiabatic and can be expressed as

$$V_2^2 - V_1^2 = \frac{2k}{k-1}(p_1v_1 - p_2v_2) \qquad (3)$$

Equations (2) and (3) may be solved simultaneously and rearranged algebraically to give some significant relationships. Several such relations are as follows:

$$\frac{p_2}{p_1} = \frac{2kMa_1^2 - (k - 1)}{k + 1} \tag{4}$$

$$\frac{V_2}{V_1} = \frac{(k - 1)Ma_1^2 + 2}{(k + 1)Ma_1^2} \tag{5}$$

From Eq. (4)

$$\frac{p_2}{p_1} = 1.75 \qquad p_2 = 17.5 \text{ psia}$$

From Eq. (5)

$$\frac{V_2}{V_1} = 0.675 \qquad V_2 = 945 \text{ fps}$$

$$\rho_1 V_1 = \rho_2 V_2 \qquad \rho_2 = \frac{0.00168}{0.675} = 0.00249 \text{ slug/ft}^3$$

$$pv = \frac{p}{\rho} = RT \qquad T_2 = \frac{p_2}{\rho_2 R} = 590°R = 130°F$$

In SI units:

$$\rho_1 = \frac{p_1}{RT_1} = \frac{70}{287(273 + 5)} = 8.8 \times 10^{-4} \text{ kg/m}^3$$

$$c = \sqrt{kRT_1} = \sqrt{1.4 \times 287 \times (273 + 5)} = 334 \text{ m/s}$$

$$N_{M_1} = \frac{V_1}{c_1} = \frac{425}{334} = 1.27$$

From Eq. (4)

$$\frac{p_2}{p_1} = 1.75 \qquad p = 122.5 \text{ N/m}^3$$

From Eq. (5)

$$\frac{V_2}{V_1} = 0.675 \qquad V_2 = 225 \text{ m/s}$$

$$\rho_1 V_1 = \rho_1 V_1 \qquad \rho_2 = 8.8 \times \frac{10^{-4}}{0.675} = 1.3 \times 10^{-3} \text{ kg/m}^3$$

$$T_2 = \frac{p_2}{\rho_2 R} = \frac{122.5}{1.3 \times 10^{-3}(287)} = 328 \text{ K} = 55°\text{C}$$

• PROBLEM 12-32

Air at 40° flows through a normal shock with an approach velocity of 1600 fps. Determine the pressure downstream of the shock wave if the pressure upstream of the shock is 10 psia. Use (a) the basic equations of continuity, momentum, and energy and (b) the Mach number relationships. R_{air} = 1716 ft-lb/slug-°R; γ_{air} = 1.4.

Solution: (a) First, determine the density ρ_1 to be

$$\rho_1 = \frac{p_1}{RT_1} = \frac{10 \times 144}{1716 \times 500} = 0.00168 \text{ slug/ft}^3$$

Continuity requires that

$$\rho_1 V_1 = \rho_2 V_2$$

Using V_1 = 1600 fps,

$$0.00168 \times 1600 = \rho_2 V_2$$

Momentum requires that

$$p_1 - p_2 = \rho_1 V_1 (V_2 - V_1)$$

Using $p_1 = 10 \times 144$ psf,

$$1440 - p_2 = 1600 \times 0.00168(V_2 - 1600)$$

The energy equation demands that

$$\frac{V_2^2 - V_1^2}{2} + \frac{\gamma}{\gamma - 1}\left(\frac{p_2}{\rho_2} - \frac{p_1}{\rho_1}\right) = 0$$

or

$$\frac{V_2^2 - 1600^2}{2} + \frac{1.4}{0.4}\left(\frac{p_2}{\rho_2} - 1716 \times 500\right) = 0$$

There are three unknowns, ρ_2, p_2, V_2, in these three independent equations. They are found as follows: substitute continuity and momentum in the energy equation and find that

$$\frac{V_2^2 - 1600^2}{2} + 3.5\left(\frac{1440 - 2.69(V_2 - 1600)}{2.69/V_2} - 858{,}000\right)$$

$$= 0$$

This is a quadratic equation and can be solved to give

$$V_2 = 892 \text{ fps}$$

The momentum equation then yields

$$p_2 = 3340 \text{ psf}$$

or

$$23.2 \text{ psi}$$

(b) To use the Mach-number relationships, first determine M_1. It is

$$M_1 = \frac{V_1}{a_1} = \frac{V_1}{\sqrt{\gamma R T_1}} = \frac{1600}{\sqrt{1.4 \times 1716 \times 500}} = 1.46$$

The momentum equation may be given in the form

$$\frac{p_2}{p_1} = \frac{2\gamma}{\gamma + 1} M_1^2 - \frac{\gamma - 1}{\gamma + 1}$$

$$\text{or } p_2 = p_1\left[\frac{2\gamma}{\gamma + 1} M_1^2 - \frac{\gamma - 1}{\gamma + 1}\right] = 10\left(\frac{2.8}{2.4} 1.46^2 - \frac{0.4}{2.4}\right)$$

$= 23.2$ psi

This is a much easier solution than that of part (a); however, the fundamental equations are hidden in such a solution. It is important to note that continuity, momentum, and energy, with all the accompanying simplifications have been used.

• PROBLEM 12-33

Air flowing through a nozzle encounters a shock. The Mach number upstream of the shock is $M_x = 1.8$, and the static temperature downstream of the shock is $T_y = 800°R$. How much has the velocity changed across the shock? Assume $\gamma = 1.4$.

Solution: We seek $V_x - V_y$, which may be expressed as

$$V_x - V_y = V_x\left(1 - \frac{V_y}{V_x}\right)$$

V_y/V_x can be found from the Table because M_x is given.

From the Table and $M_x = 1.8$,

$$\frac{V_y}{V_x} = \frac{\rho_x}{\rho_y} = \frac{1}{\rho_y/\rho_x} = \frac{1}{2.36} = 0.425$$

$$\frac{T_x}{T_y} = \frac{1}{T_y/T_x} = \frac{1}{1.53} = 0.653$$

V_x can be determined from

$$V_x = M_x c_x = M_x\sqrt{kRT_x} = M_x\sqrt{kR(T_x/T_y)}\sqrt{T_y}$$

where T_y/T_x is known as a function of M_x.

$$V_x = M_x \cdot 49.02\sqrt{T_x/T_y}\sqrt{T_y} = 1.8 \times 49.02\sqrt{0.653}$$

$\times \sqrt{800}$

$= 2020$ ft/sec

Therefore

$$V_x - V_y = 2020 \text{ ft/sec } (1 - 0.425) = 1160 \text{ ft/sec}$$

TABLE

One-Dimensional Normal-Shock Functions

For an ideal gas with constant specific heat and molecular weight, $k = 1.4$

M_x	M_y	$\frac{P_y}{P_x}$	$\frac{\rho_y}{\rho_x}$	$\frac{T_y}{T_x}$	$\frac{P_{oy}}{P_{ox}}$
1.00	1.00000	1.00000	1.00000	1.00000	1.00000
1.05	0.95312	1.1196	1.08398	1.03284	0.99987
1.10	0.91177	1.2450	1.1691	1.06494	0.99892
1.15	0.87502	1.3762	1.2550	1.09657	0.99669
1.20	0.84217	1.5133	1.3416	1.1280	0.99280
1.25	0.81264	1.6562	1.4286	1.1594	0.98706
1.30	0.78596	1.8050	1.5157	1.1909	0.97935
1.35	0.76175	1.9596	1.6028	1.2226	0.96972
1.40	0.73971	2.1200	1.6896	1.2547	0.95819
1.45	0.71956	2.2862	1.7761	1.2872	0.94483
1.50	0.70109	2.4583	1.8621	1.3202	0.92978
1.55	0.68410	2.6363	1.9473	1.3538	0.91319
1.60	0.66844	2.8201	2.0317	1.3880	0.89520
1.65	0.65396	3.0096	2.1152	1.4228	0.87598
1.70	0.64055	3.2050	2.1977	1.4583	0.85573
1.75	0.62809	3.4062	2.2791	1.4946	0.83456
1.80	0.61650	3.6133	2.3592	1.5316	0.81268
1.85	0.60570	3.8262	2.4381	1.5694	0.79021
1.90	0.59562	4.0450	2.5157	1.6079	0.76735
1.95	0.58618	4.2696	2.5919	1.6473	0.74418
2.00	0.57735	4.5000	2.6666	1.6875	0.72088
2.05	0.56907	4.7363	2.7400	1.7286	0.69752
2.10	0.56128	4.9784	2.8119	1.7704	0.67422
2.15	0.55395	5.2262	2.8823	1.8132	0.65105
2.20	0.54706	5.4800	2.9512	1.8569	0.62812
2.25	0.54055	5.7396	3.0186	1.9014	0.60554
2.30	0.53441	6.0050	3.0846	1.9468	0.58331
2.35	0.52861	6.2762	3.1490	1.9931	0.56148
2.40	0.52312	6.5533	3.2119	2.0403	0.54015
2.45	0.51792	6.8362	3.2733	2.0885	0.51932
2.50	0.51299	7.1250	3.3333	2.1375	0.49902
2.55	0.50831	7.4196	3.3918	2.1875	0.47927
2.60	0.50387	7.7200	3.4489	2.2383	0.46012
2.65	0.49965	8.0262	3.5047	2.2901	0.44155
2.70	0.49563	8.3383	3.5590	2.3429	0.42359

• PROBLEM 12-34

A normal shock wave moves down a tube at 1695 ft/s. The undisturbed air has a temperature of 59F and a pressure of 14.7 psia. Determine (a) the Mach number on both sides of the shock as seen from a coordinate system attached to the shock and (b) the temperature and pressure of the air after the shock passes by.

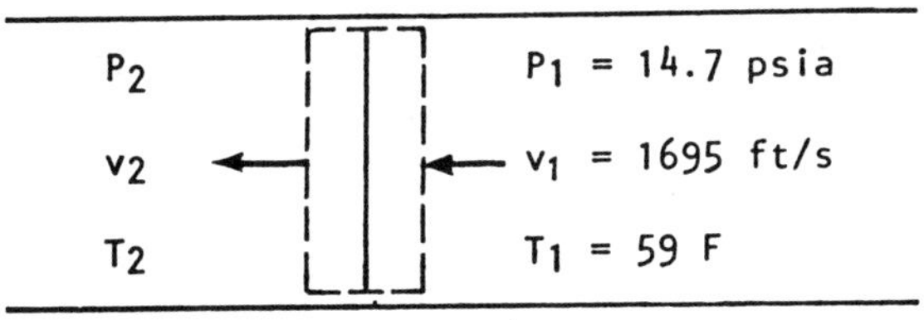

Solution: As seen from a coordinate system attached to the shock, upstream conditions are

$$v_1 = 1695 \text{ ft/s}, \quad T_1 = 59\text{F}, \quad \text{and} \quad p_1 = 1.47 \text{ psia},$$

as shown in the figure. Then

$$c_1 = \sqrt{kg_cRT} = \sqrt{1.4(32.17)(53.35)(519)}$$

$$= 1116.7 \text{ ft/s}$$

and

$$M_1 = \frac{v_1}{c_1} = \frac{1695}{1116.7} = 1.518$$

From the normal shock table, for $M_1 = 1.518$, we determine by interpolation that $M_2 = 0.6948$, $p_2/p_1 = 2.522$, and $T_2/T_1 = 1.333$. The answers to part (a) are:

$$M_1 = 1.518 \text{ and } M_2 = 0.6948.$$

The answers to part (b) are: $T = 1.333(518.6) = 691.4\text{R}$ and $p_2 = 2.522(14.7) = 37.1$ psia.

TABLE

M_1	M_2	$\frac{p_2}{p_1}$	$\frac{T_2}{T_1}$	$\frac{p_{0,2}}{p_{0,1}}$	M_1	M_2	$\frac{p_2}{p_1}$	$\frac{T_2}{T_1}$	$\frac{p_{0,2}}{p_{0,1}}$
1.00	1.000	1.000	1.000	1.000	1.40	0.740	2.120	1.255	0.958
1.02	0.980	1.047	1.013	1.000	1.42	0.731	2.186	1.268	0.953
1.04	0.962	1.095	1.026	1.000	1.44	0.723	2.253	1.281	0.948
1.06	0.944	1.144	1.039	1.000	1.46	0.716	2.320	1.294	0.942
1.08	0.928	1.194	1.052	0.999	1.48	0.708	2.389	1.307	0.936
1.10	0.912	1.245	1.065	0.999	1.50	0.701	2.458	1.320	0.930
1.12	0.896	1.297	1.078	0.998	1.52	0.694	2.529	1.334	0.923
1.14	0.882	1.350	1.090	0.997	1.54	0.687	2.600	1.347	0.917
1.16	0.868	1.403	1.103	0.996	1.56	0.681	2.673	1.361	0.910
1.18	0.855	1.458	1.115	0.995	1.58	0.675	2.746	1.374	0.903
1.20	0.842	1.513	1.128	0.993	1.60	0.668	2.820	1.388	0.895
1.22	0.830	1.570	1.140	0.991	1.62	0.663	2.895	1.402	0.888
1.24	0.818	1.627	1.153	0.988	1.64	0.657	2.971	1.416	0.880
1.26	0.807	1.686	1.166	0.986	1.66	0.651	3.048	1.430	0.872
1.28	0.796	1.745	1.178	0.983	1.68	0.646	3.126	1.444	0.864
1.30	0.786	1.805	1.191	0.979	1.70	0.641	3.205	1.458	0.856
1.32	0.776	1.866	1.204	0.976	1.72	0.635	3.285	1.473	0.847
1.34	0.766	1.928	1.216	0.972	1.74	0.631	3.366	1.487	0.839
1.36	0.757	1.991	1.229	0.968	1.76	0.626	3.447	1.502	0.830
1.38	0.748	2.055	1.242	0.963	1.78	0.621	3.530	1.517	0.821

• PROBLEM 12-35

A converging diverging nozzle with a throat area of $2m^2$ is attached to a high pressure tank of air in which the pressure is 20 psia and the temperature is 120^0F.

A normal shock occurs in the divergent part of the nozzle where the cross-sectional area is $2.5\ in^2$.

If the nozzle is to operate at off-design conditions, calculate the exit pressure as well as the Mach number at the exit.

Solution: $T_0 = 120^0F$

$P_0 = 20$ psia

$A_t = 2\ in^2$

$A_e = 5 in^2$

$A_s = 2.5 in$

$p_e = ?$

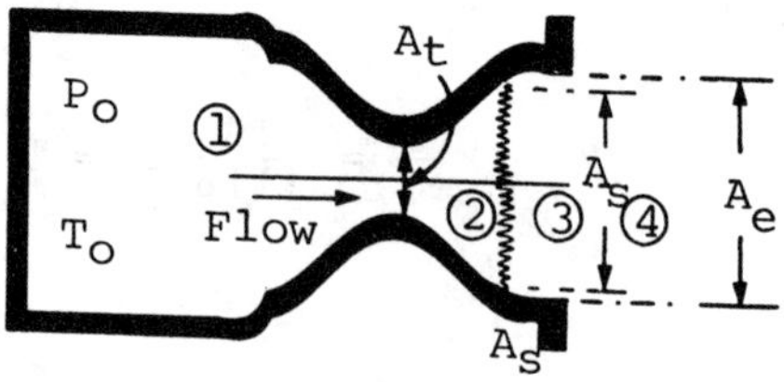

Assumptions:

(a) Isentropic flow everywhere except through the normal shock.

(b) The fluid is a perfect gas having constant specific heat.

To use the tables, the ratio $\frac{A_s}{A_t}$ is computed first, or

$$\frac{A_s}{A_t} = \frac{A_s}{A^*} = \frac{2.5}{2} = 1.25$$

TABLE 1. ISENTROPIC FLOW FUNCTIONS

M	A/A*	ρ/ρ_o	P/P_o	T/T_o
⋮	⋮	⋮	⋮	⋮
0.66	1.127	0.7465	0.8115	0.9199
	⋮	⋮	⋮	
1.58	1.234	0.7423	0.3633	0.6670
1.60	1.250	0.2353	0.3557	0.6614
1.62	1.267	0.2284	0.3483	0.6558
⋮	⋮	⋮	⋮	⋮
2.30	2.193	0.07997	0.1646	0.4859
2.32	2.274	0.07751	0.1610	0.4816
2.34	2.274	0.07513	0.1574	0.4773
⋮	⋮	⋮	⋮	⋮

From Table 1 on isentropic flow, we obtain the Mach number at section 2, or at $\frac{A_s}{A^*} = 1.25$

$$M_2 = 1.60$$

Also from this table we obtain the ratio p/p_0,

$$= p_2/p_0 = 0.2353,$$

so that the pressure at section 2 is

$$p_2 = (0.2353)(p_0) = (0.2353)(20) = 4.706 \text{ psia}$$

The properties at section 3 can be computed by using the normal shock table 2. Thus for $M_2 = 1.60$

$$M_3 = 0.6684$$

$$\frac{p_2}{p_1} = p_3/p_2 = 2.820 = p_3 = (2.820)p_2 = (2.820)(4.706)$$
$$= 13.270 \text{ psia}$$

$$\frac{p_{02}}{p_{01}} = p_{03}/p_{02} = 0.8952 \Rightarrow p_{03} = (0.8952)p_{02} = (0.8952)(20)$$
$$= 17.904 \text{ psia}$$

From table 1 at $M_3 = 0.6684$ we get,

$$\frac{A}{A^*} = \frac{A_s}{A^*} = 1.127 \Rightarrow A^* = \frac{A_s}{1.127} = \frac{2.5}{1.127} = 2.218 \text{ in}^2$$

Then at the exit $\frac{A_e}{A^*} = \frac{5}{2.218} = 2.254$

For $\frac{A_e}{A^*} = \frac{A}{A^*} = 2.254$, from the table 1 we get that

$$Me = 2.33$$

and $\frac{p}{p_0} = \frac{p_4}{p_{03}} = 0.0740 \Rightarrow p_4 = (0.0740)(p_{03})$

$$p_4 = (0.0740)(17.904) = 1.324 \text{ psia.}$$

Therefore the pressure exit is pe = 1.324 psia, and the Mach number at the exit is Me = 2.33

TABLE 2. NORMAL SHOCK FUNCTIONS

M_1	M_2	b_{o2}/b_{o1}	T_2/T_1	$b_2 b_1$	P_2/P_1
.	.	.	.	.	.
.	.	.	.	.	.
.	.	.	.	.	.
1.58	0.6746	0.9026	1.374	2.746	1.998
1.60	0.6684	0.8952	1.388	2.820	2.032
1.62	0.6625	0.8876	1.402	2.895	2.065
.	.	.	.	.	.
.	.	.	.	.	.
.	.	.	.	.	.

• **PROBLEM 12-36**

A normal shock moves through still air (14.7 psi, 60°F) with a steady velocity of 4000 fps (Figure 1). Calculate the velocity of the air behind the wave and the static pressure and temperature in this air stream. Assume $\gamma = 1.4$.

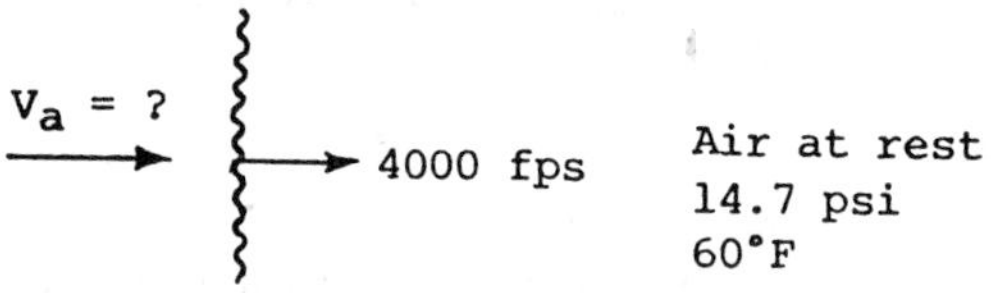

Fig. 1

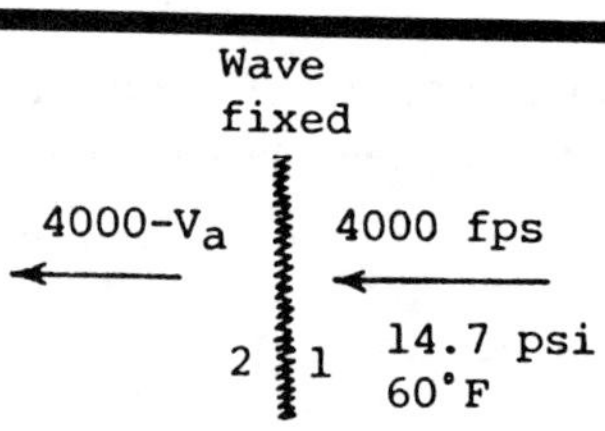

Fig. 2

Solution: In order to treat the shock with steady state equations, we shall take all velocities with respect to an observer "sitting" on the wave. In other words, impose a velocity of 4000 fps from right to left on the wave and air flow shown (See Figure 2).

The Mach number, M, is given by

$$M = \frac{V}{\sqrt{\gamma g c R T}}$$

with $T_1 = 60 + 460 = 520°R$

and

$$R = 53.3 \text{ ft-lb}_f/\text{lb}_m\text{-°R} \; ,$$

$$M_1 = \frac{4000}{\sqrt{1.4 \times 32.2 \times 53.3 \times 520}} = 3.58$$

From the Table,

$$\frac{p_2}{p_1} = 14.79, \qquad \frac{T_2}{T_1} = 3.426 , \qquad \frac{p_2}{p_1} = \frac{V_1}{V_2} = 4.316$$

Therefore the pressure behind the wave is 14.79(14.7) = 218 psi, temperature behind the wave is 3,426(520) = 1780°R, and $4000 - V_g = 4000/4.316 = 925$ fps, so that

$V_g = 4000 - 925 = 3075$ fps.

• PROBLEM 12-37

A normal shock occurs in an airflow with the following upstream conditions: $V_1 = 600$ m/s, $p_1 = 100$ kN/m^2 (abs), and $T_1 = 5°C$. Determine the downstream velocity, pressure, and temperature, and both the upstream and downstream densities and Mach numbers.

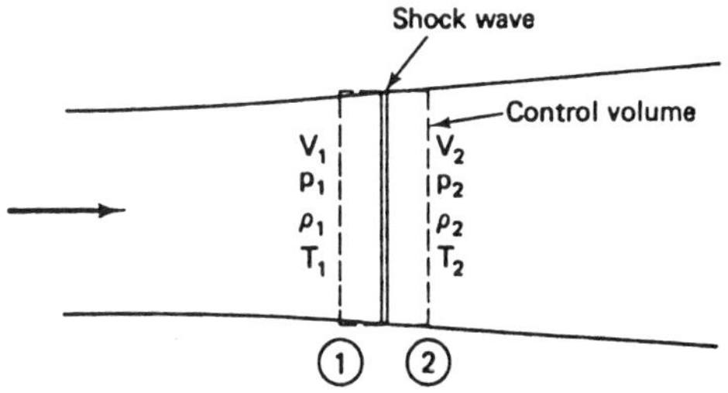

Solution: The analysis of normal shock waves follows from direct application of the momentum equation. The control volume is shown in the Fig. The shock may be treated as abrupt, since the shock takes place over a very short length which is on the order of a few times the mean free path of the molecules (approximately 10^{-3} mm). This means that the cross-sectional area will not change through the length of the shock and for sections 1 and 2 just upstream and downstream of the shock, respectively, $A_1 = A_2 = A$. Further, we will have V_1, p_1, ρ_1, and T_1 as the average velocity, pressure, density, and temperature at section 1, with corresponding quantities V_2, p_2, ρ_2, and T_2 at section 2. As in previous situations where a one-dimensional approach was applied, variation of each quantity across the section is ignored and the average value is taken to represent the section as a reasonable approximation. Applying the momentum equation to the control volume of Fig. 1 gives

$$(p_1 - p_2)A = G(v_2 - V_1)$$

where G, the mass rate of flow, is given by

$$G = \rho_1 V_1 A_1 = \rho_2 V_2 A_2 \tag{1}$$

Combining the two equations gives

$$p_2 - p_1 = \rho_1 V_1^2 - \rho_2 V_2^2 \tag{2}$$

which relates the pressure increase $p_2 - p_1$ to the upstream and downstream velocities. Making use of

$$Ma^2 = \left(\frac{V}{c}\right)^2 = \frac{V^2}{kp/\rho}$$

we may rewrite Eq. 2 as

$$p_2 - p_1 = kp_1 Ma_1^2 - kp_2 Ma_2^2$$

which, when rearranged, gives the pressure ratio p_2/p_1 in terms of the Mach number:

$$\frac{p_2}{p_1} = \frac{1 + kMa_1^2}{1 + kMa_2^2} \tag{3}$$

An additional relationship is needed at this point. We may note that although considerable internal friction occurs in the shock wave, no heat is added to or removed from the system. Hence, the adiabatic energy equation still applies. It may be arranged as follows:

$$\frac{V_1^2}{kgRT_1} + \frac{2}{k - 1} = \frac{V_2^2}{kgRT_1}\ \frac{T_2}{T_2} + \frac{2}{k - 1}\ \frac{T_2}{T_1}$$

or

$$Ma_1^2 + \frac{2}{k - 1} = \left(Ma_2^2 + \frac{2}{k - 1}\right)\frac{T_2}{T_1}$$

When solved for T_2/T_1 we have an additional equation relating the temperature ratio to the upstream and downstream Mach numbers,

$$\frac{T_2}{T_1} = \frac{Ma_1^2 + \dfrac{2}{k-1}}{Ma_2^2 + \dfrac{2}{k-1}} \qquad (4)$$

In a similar manner, the density ratio may be determined from the pressure and temperature ratios through the equation of state. Noting also that densities and velocities are related through Eq. 1, we have

$$\frac{\rho_2}{\rho_1} = \frac{V_1}{V_2} = \frac{p_2/gRT_2}{p_1/gRT_1} = \frac{p_2}{p_1}\frac{T_1}{T_2}$$

and hence

$$\frac{\rho_2}{\rho_1} = \frac{V_1}{V_2} = \frac{1 + kMa_1^2}{1 + kMa_2^2}\;\frac{Ma_2^2 + \dfrac{2}{k-1}}{Ma_1^2 + \dfrac{2}{k-1}} \qquad (5)$$

In order to obtain a relationship between the Mach numbers, the equation of state and Eq. 1 may be combined as follows:

$$\frac{p_2}{p_1} = \frac{\rho_2}{\rho_1}\frac{T_2}{T_1} = \frac{V_1}{V_2}\frac{T_2}{T_1}$$

But

$$\frac{V_1}{V_2} = \frac{V_1}{\sqrt{kgRT_1}}\;\frac{\sqrt{kgRT_1}}{V_2}\;\frac{\sqrt{T_2}}{\sqrt{T_2}} = \frac{Ma_1}{Ma_2}\left(\frac{T_1}{T_2}\right)^{1/2}$$

Substituting this equation for V_1/V_2 into the previous equation for p_2/p_1 and solving for T_2/T_1 yields

$$\frac{T_2}{T_1} = \left(\frac{p_2}{p_1}\right)^2\left(\frac{Ma_2}{Ma_1}\right)^2$$

Finally, if Eqs. 3 and 4 are substituted for T_2/T_1 and p_2/p_1, respectively, there results an equation in Ma_1 and Ma_2 only:

$$\frac{Ma_1^2 + \frac{2}{k-1}}{Ma_2^2 + \frac{2}{k-1}} = \left(\frac{1 + kMa_1^2}{1 + kMa_2^2}\right)^2 \frac{Ma_1^2}{Ma_1^2}$$

After considerable algebraic manipulation, this equation may be solved to give

$$Ma_2^2 = \frac{Ma_1^2 + \frac{2}{k-1}}{\left(\frac{2k}{k-1}\right)Ma_1^2 - 1} \qquad (6)$$

TABLE

Gas	Chemical Formula	Specific Weight, γ		Density, ρ		Gas Constant, R		Adiabatic Constant, k
		lb/ft³	N/m³	slugs/ft³	kg/m³	ft/°R	m/°K	
Air	—	0.0753	11.8	0.00234	1.206	53.3	29.2	1.40
Ammonia	NH_3	0.0448	7.04	0.00139	0.716	89.5	49.1	1.32
Carbon dioxide	CO_2	0.115	18.1	0.00357	1.840	34.9	19.1	1.29
Helium	He	0.0104	1.63	0.000323	0.166	386.	212.	1.66
Hydrogen	H_2	0.00522	0.820	0.000162	0.0835	767.	421.	1.40
Methane	CH_4	0.0416	6.53	0.00129	0.665	96.4	52.9	1.32
Nitrogen	N_2	0.0726	11.4	0.00225	1.160	55.2	30.3	1.40
Oxygen	O_2	0.0830	13.0	0.00258	1.330	48.3	26.5	1.40
Sulfur dioxide	SO_2	0.170	26.7	0.00528	2.721	23.6	12.9	1.26

To use Eq. 6, we first need Ma_1, which is given by

$$Ma_1 = \frac{V_1}{\sqrt{kgRT_1}} = \frac{600 \text{ m/s}}{\sqrt{(1.4)(9.81 \text{ m/s})(29.2 \text{ m/°K})(273 + 5\text{°K}}}$$

$$= 1.797$$

Then, from Eq. 6,

$$Ma_2^2 = \frac{(1.797)^2 + \frac{2}{1.4 - 1}}{\frac{(2)(1.4)}{1.4 - 1}(1.797)^2 - 1} = 0.3809$$

from which $Ma_2 = 0.6172$.

The pressure is obtained from Eq. 3:

$$\frac{p_2}{100 \times 10^3 \text{ N/m}} = \frac{1 + (1.4)(1.797)^2}{1 + (1.4)(0.6172)^2}$$

which gives

$$p_2 = 360 \times 10^3 \text{ N/m}^2\text{(abs)} = 360 \text{ kN/m}^2\text{(abs)}$$

Next, using Eq. 4,

$$\frac{T_2}{273 + 5} = \frac{(1.797)^2 + \frac{2}{1.4 - 1}}{(0.6172)^2 + \frac{2}{1.4 - 1}}$$

and

$$T_2 = 425°K = 152°C$$

By Eq. 5,

$$\frac{600 \text{ m/s}}{V_2} = \frac{1 + (1.4)(1.797)^2}{1 + (1.4)(0.6172)^2} \; \frac{(0.6172)^2 + \frac{2}{1.4 - 1}}{(1.797)^2 + \frac{2}{1.4 - 1}}$$

and $V_2 = 254.8$ m/s. Finally, from the equation of state,

$$\rho_1 = \frac{p_1}{gRT_1} = \frac{100 \times 10^3}{(9.81)(29.2)(273 + 5)} = 1.256 \text{ kg/m}^3$$

and from

$$\frac{\rho_2}{\rho_1} - \frac{V_1}{V_2}$$

$$\rho_2 = \frac{(1.256 \text{ kg/m}^3)(600 \text{ m/s})}{254.8 \text{ m/s}} = 2.957 \text{ kg/m}^3$$

• PROBLEM 12-38

A pitot static tube such as that shown in Fig. 1 is used to measure the speed for an aircraft.

(a) If the free stream temperature is -40°C and the difference between the dynamic and static pressure is $p_2 - p_1 = 14$ kPa, calculate the velocity of the aircraft. The free stream pressure is $p_1 = 69$kPa.

(b) If the aircraft flies at M = 2.7 at an altitude of 20Km on a standard day, determine the final temperature of the air.

(c) If the pitot tube of the aircraft senses a total pressure of 55.3 kPa, determine the Mach number and speed of the aircraft. Assume that the ambient pressure is 4.05kPa and that the aircraft flies at the same altitude as in question (b).

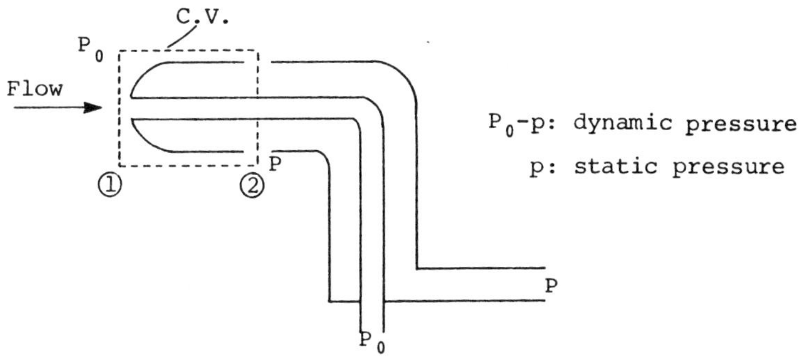

Fig. 1: Pitot-static tube.

Solution: (a) For one-dimensional flow between the points 1 and 2, the first law of thermodynamics can be stated as

$$\dot{Q} + \dot{W} = \frac{\partial}{\partial t}\int e\rho dV + \int_{CS} (u + pv + \frac{V^2}{2} + gz)\rho V dA \quad (1)$$

Assuming an adiabatic steady flow with no work crossing the boundary of the control volume, the above equation becomes

$$\int_{CS} (u + pv + \frac{V^2}{2} + gz)\rho V dA = 0 \quad (2)$$

Since the enthalpy is h = u + pv, the equation (2) can be written as

$$(h_1 - h_2) + g(z_1 - z_2) + \frac{V_1^2 - V_2^2}{2} = 0 \qquad (3)$$

where $V_2 = 0, \; z_1 - z_2 = 0.$

Considering the air as an ideal gas,

$$h_1 - h_2 = C_p \int_{T_2}^{T_1} dT$$

$$= C_p(T_1 - T_2) \qquad (4)$$

From the first law we can obtain the expression relating pressures and temperatures during an isentropic compression, or

$$\frac{T_2}{T_1} = \left(\frac{p_2}{p_1}\right)^{(\gamma - 1)/\gamma}$$

or

$$\frac{T_2 - T_1}{T_1} = \left(\frac{p_2}{p_1}\right)^{(\gamma - 1)/\gamma} - 1$$

or

$$T_2 - T_1 = T_1\left[\left(\frac{p_2}{p_1}\right)^{(\gamma - 1)/\gamma} - 1\right] \qquad (5)$$

where $\gamma = C_p/C_v$ is the ratio of specific heats of the gas.

Combining equations (3), (4) and (5), we obtain the expression which relates the velocity of the aircraft to the temperature of the airstream and the dynamic and static pressure, or

$$V_1 = \left[2C_p T_1 \left[\left(\frac{p_2}{p_1}\right)^{(\gamma-1)/\gamma} - 1\right]\right]^{\frac{1}{2}} \qquad (6)$$

Substituting numerical values into equation (6), we obtain

$$V_1 = \left[2(1.004)(233) \quad \left[\left(\frac{83}{69}\right)^{(1.4-1)/1.4} - 1\right]\right]^{\frac{1}{2}}$$

$$= 5 \text{ m/sec}$$

(b) Because the aircraft flies with M = 2.7, a shock wave similar to the type shown in Fig. 2 is formed in front of the pitot tube.

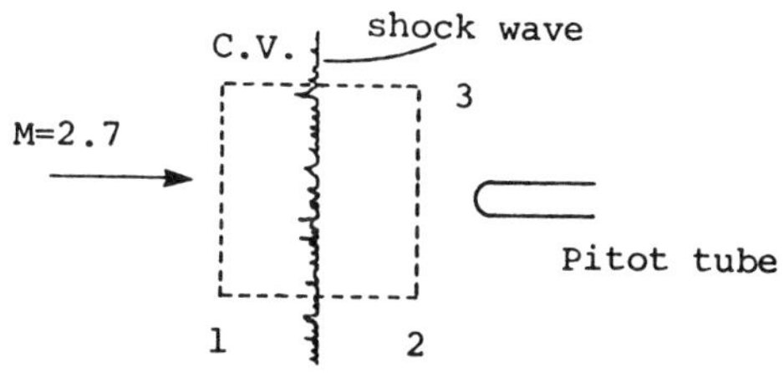

Fig. 2

The compressible flow tables can be used in the solution by making the following assumptions:

a) steady flow
b) air is an ideal gas
c) flow undergoes normal shock from 1 to 2
d) flow is isentropic from 2 to 3

At an altitude of 20km (see appendix, page 922)

$$T = T_1 = 217^0K \quad \text{and} \quad p = p_1 = 5.53kPa$$

From the table of one-dimensional isentropic relations (see appendix, page 938) with M = 2.7

$$\frac{T_1}{T_{01}} = 0.47 \quad \text{or} \quad T_{01} = 533^0K \qquad (7)$$

Because the flow is isentropic from 2 to 3,

$$T_{01} = T_{02} = T_3 \qquad (8)$$

therefore the temperature of the air is

$$T_3 = 533^0K$$

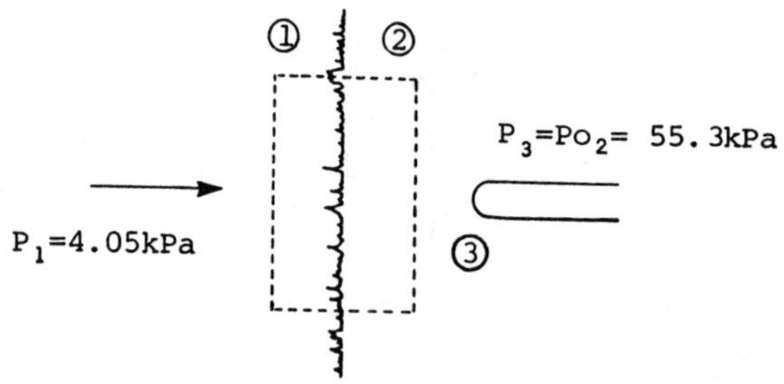

Fig. 3

(c) The compressible flow tables will be used in the solution by making the following assumptions:

a) steady flow
b) air is an ideal gas
c) flow undergoes normal shock from 1 to 2
d) flow is isentropic from 2 to 3

$$\frac{P_{03}}{P_1} = \frac{P_{03}}{P_{01}} \times \frac{P_{01}}{P_1} = \frac{P_{02}}{P_{01}} \times \frac{P_{01}}{P_1} \tag{9}$$

where $\frac{P_{02}}{P_{01}}$ = the stagnation pressures at section 1 as a function of M_1 from the table 1

$\frac{P_{04}}{P_1}$ = the pressure at section 1 as a function of M_1 from the table of one-dimensional isentropic relations (see appendix, page 937-938).

Table 1 Normal shock functions (one-dimensional flow, ideal gas, K = 1.4)

M_1	$\frac{p_{02}}{p_{01}}$
2.82	0.351
2.94	0.345
2.96	0.339
2.98	0.334
3.00	0.328
3.10	0.301
3.20	0.276
3.30	0.253
3.40	0.232
3.50	0.213
3.60	0.195
3.70	0.179
3.80	0.164
3.90	0.151
4.00	0.138

Using these two tables and equation (9) a trial and error solution must be used to determine M_1 such that $\frac{P_{03}}{P_{01}} = \frac{55.3}{4.05} =$ 13.65.

This can be obtained by assuming values for the Mach number M_1. After some trial and error procedures, the following table can be obtained.

M_1	P_{02}/P_{01}	P_1/P_{01}	P_{02}/P_1
3.0	0.328	0.0272	12.06
3.4	0.232	0.0151	15.36
3.2	0.276	0.020	13.65

From the above table we can see that the closest Mach number corresponding to the $P_{03}/P_{01} = 13.65$ is $M_1 = 3.2$. Therefore the Mach number of the aircraft is

$$M_1 = 3.2$$

The speed of the aircraft at an altitude of 20km is given by the relation

$$V_1 = M_1 C_1 = M_1 (KRT_1)^{\frac{1}{2}} \tag{10}$$

where T_1 = the temperature of the air at an altitude of 20km.

Substituting numerical values into equation (10), we obtain

$$V_1 = M_1 (KRT_1)^{\frac{1}{2}} = 3.2(1.4 \times 287 \times 217)^{\frac{1}{2}} = 944.89 \text{ m/sec}$$

• PROBLEM 12-39

(a) An oblique shock inclined at $\beta = 30^0$ to an approaching supersonic stream of $M_1 = 5$ strikes and is reflected from a solid wall as shown in Fig. 1(a). Calculate the final Mach number and the reflected shock position.

(b) If the same supersonic stream ($M_1 = 5$) with $p_1 = 1$ atm, and $T_1 = 250^0$K is deflected by an angle $\theta = 30^0$ due to a compression corner as shown in Fig. 1(b), calculate the shock wave angle β and p_2, T_2, M_2, p_{02} and T_{02} behind the shock wave.

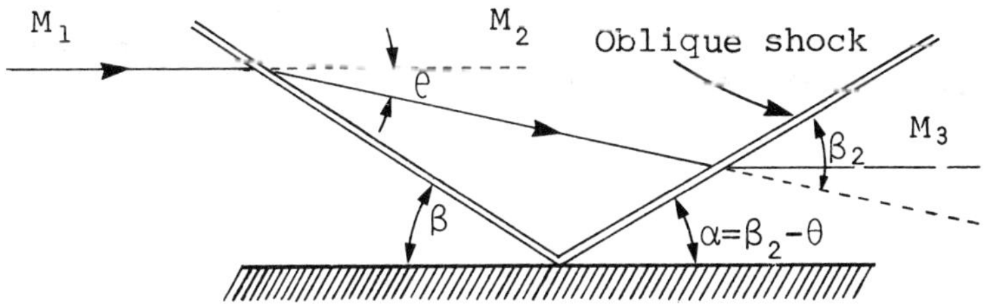

Fig. 1a

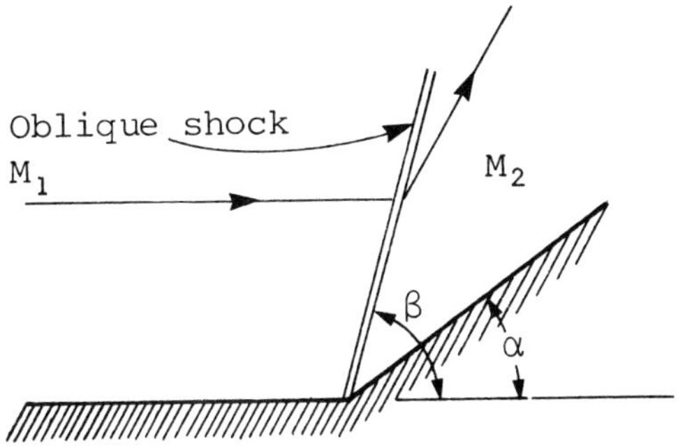

Fig. 1b

Solution: (a) From Fig. 3 for $M_1 = 5$ and $\beta = 30^0$, the angle of flow deflection is $\theta = 20^0$. From the equation for an oblique shock wave,

$$M_2 = \left[\frac{(M^2\sin^2\beta)(\gamma - 1) + 2}{(2\gamma M^2\sin^2\beta + \gamma - 1)\sin(\beta - \theta)}\right]^{\frac{1}{2}} \quad (1)$$

For $M = M_1 = 3$, $\beta = 30^0$, $\theta = 20^0$ and $\gamma = 1.4$, the above equation gives

$$M_2 = 2.07$$

The reflected shock with $M_2 = 2.07$, must return the flow parallel to the wall; that is $\theta = 20^0$ for the reflected shock. Thus, once again from Fig. 3 for $M_1 = 2.07$ and $\theta = 20^0$ we obtain $\beta_2 = 52.5^0$.

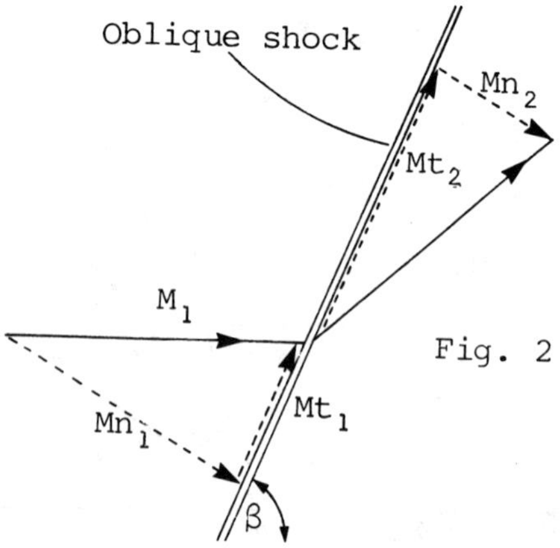

Fig. 2

Now employing equation (1) and substituting the numerical values,

$$M = M_2 = 2.07$$

$$\beta = \beta_2 = 52.5^0$$

$$\gamma = 1.4$$

we obtain

$$M_3 = 0.84$$

The final inclination angle is $\alpha = \beta_2 - \theta = 32.5^0$.

Table 1 Normal Shock Properties

M_1	$\frac{p_2}{p_1}$	$\frac{\rho_2}{\rho_1}$	$\frac{T_2}{T_1}$	$\frac{p_{02}}{p_{01}}$	$\frac{p_{01}}{p_1}$	M_{n_2}
0.3000+01	0.1033+02	0.3857+01	0.2679+01	0.3283+00	0.1206+02	0.4752+00
0.3050+01	0.1069+02	0.3902+01	0.2738+01	0.3145+00	0.1245+02	0.4723+00
0.3100+01	0.1104+02	0.3947+01	0.2799+01	0.3012+00	0.1285+02	0.4695+00
0.3150+01	0.1141+02	0.3990+01	0.2860+01	0.2885+00	0.1325+02	0.4669+00
0.3200+01	0.1178+02	0.4031+01	0.2922+01	0.2762+00	0.1366+02	0.4643+00
0.3250+01	0.1216+02	0.4072+01	0.2985+01	0.2645+00	0.1407+02	0.4619+00
0.3300+01	0.1254+02	0.4112+01	0.3049+01	0.2533+00	0.1449+02	0.4596+00
0.3350+01	0.1293+02	0.4151+01	0.3114+01	0.2425+00	0.1492+02	0.4573+00
0.3400+01	0.1332+02	0.4188+01	0.3180+01	0.2322+00	0.1535+02	0.4552+00
0.3450+01	0.1372+02	0.4225+01	0.3247+01	0.2224+00	0.1579+02	0.4531+00

(b) For $M_1 = 5$ and $\theta = 30^0$, Fig. 3 gives $\beta = 42.5^0$.

From Fig. 2 we can see that

$$M_{n_1} = M_1 \sin\beta = (5)\sin 42.5^0 = 3.37$$

From table 1, for $M_{n_1} = 3.37$,

$$\frac{p_2}{p_1} = 13.01, \quad \frac{T_2}{T_1} = 3.16, \quad M_{n_2} = 0.45$$

$$\frac{p_{02}}{p_{01}} = 0.24$$

Hence $$p_2 = \frac{p_2}{p_1} p_1 = (13.01)(1) = 13.01 \text{ atm}$$

$$T_2 = \frac{T_2}{T_1} T_1 = (3.16)(250) = 790^0 K$$

$$M_2 = \frac{M_{n_2}}{\sin(\beta - \theta)} = \frac{0.45}{\sin(42.5^0 - 30^0)} = 0.473$$

From table 2, for $M_1 = 5$

$$\frac{p_{01}}{p_1} = 529.1 \quad \text{and} \quad \frac{T_{01}}{T_1} = 6$$

Hence

$$p_{02} = \frac{p_{02}}{p_{01}} \frac{p_{01}}{p_1} p_1 = (0.24)(529.1)(1) = 128.30 \text{ atm}$$

$$T_{02} = T_{01} = \frac{T_{01}}{T_1} T_1 = (6)(250) = 1500^0\text{K}$$

Table 2 Isentropic flow properties

M_1	$\frac{p_{01}}{p_1}$	$\frac{\rho_{01}}{\rho_1}$	$\frac{T_{01}}{T_1}$	$\frac{A}{A^*}$
0.4550+01	0.3080+03	0.5991+02	0.5140+01	0.1728+02
0.4600+01	0.3276+03	0.6261+02	0.5232+01	0.1802+02
0.4650+01	0.3483+03	0.6542+02	0.5324+01	0.1879+02
0.4700+01	0.3702+03	0.6833+02	0.5418+01	0.1958+02
0.4750+01	0.3933+03	0.7135+02	0.5512+01	0.2041+02
0.4800+01	0.4177+03	0.7448+02	0.5608+01	0.2126+02
0.4850+01	0.4434+03	0.7772+02	0.5704+01	0.2215+02
0.4900+01	0.4705+03	0.8109+02	0.5802+01	0.2307+02
0.4950+01	0.4990+03	0.8457+02	0.5900+01	0.2402+02
0.5000+01	0.5291+03	0.8818+02	0.6000+01	0.2500+02

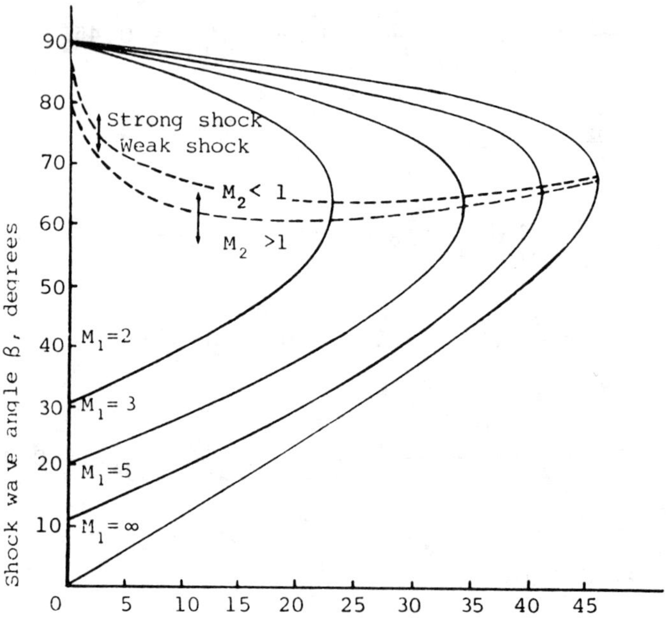

Fig. 3: Deflection angle θ, degrees

CHAPTER 13

FLOW METERS

Basic Attacks and Strategies for Solving Problems in this Chapter. See pages 807 to 846 for step-by-step solutions to problems.

There are two broad categories of devices that measure fluid flows: local-velocity meters and volume flow meters. In addition, liquid U-tube manometers are used to measure pressure in fluid flows as well as in static applications.

Consider first the U-tube manometer as sketched below:

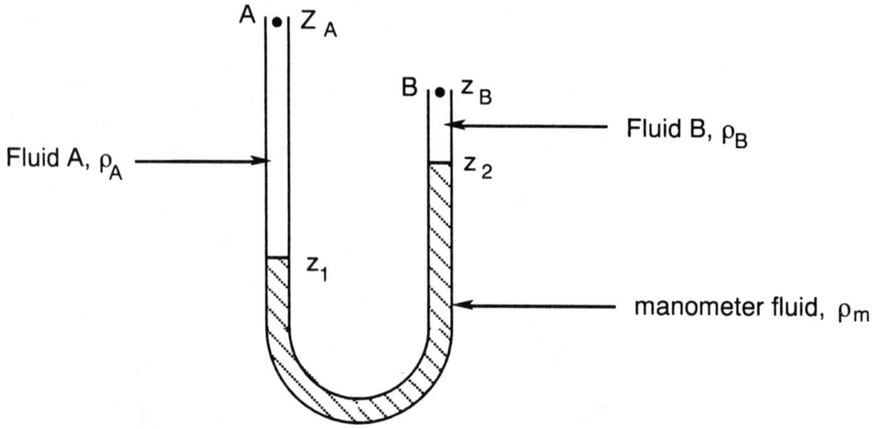

The sketch shows a general application where the pressure difference between point A and point B is desired. The manometer fluid must have a density greater than either ρ_A or ρ_B. The solution involves simple repeated applications of the fundamental hydrostatics equation from Chapter 2,

$$\Delta p = -\rho g \Delta z. \tag{1}$$

Keeping in mind that Equation (1) applies only within one fluid of a constant density, it must be applied in a piecewise fashion. First, write the expression from point A to point 1 in fluid A:

$$p_1 = p_A + \rho_A g(z_A - z_1). \tag{2}$$

Similarly, from point 1 to point 2,

$$p_1 = p_2 + \rho_m g(z_2 - z_1), \tag{3}$$

and from point B to point 2,

$$p_2 = p_B + \rho_B g(z_B - z_2). \tag{4}$$

For more complicated manometers (as in Problem 13.5), it may be necessary to apply Equation (1) several more times. In the above case, Equations (2) – (4) contain four unknowns (p_1, p_2, p_A, and p_B) and there are only three equations. Fortunately, all that is desired is the difference between p_A and p_B. This can be found by eliminating p_1 and p_2 in the equation set. The result is

$$p_A - p_B = \rho_m g(z_2 - z_1) + \rho_B g(z_B - z_1) - \rho_A g(z_A - z_1). \tag{5}$$

In the most common case, the right side of the manometer is open to the atmosphere so that $p_B = p_{atm}$ = known, and the second term on the right-hand side of Equation (5) is negligible since ρ_{air} is very small compared to the density of liquids.

Consider now the pitot tube, which is simply a slender tube with a hole in the front, aligned with the flow. The pressure at the nose of any body in an incompressible flow is the stagnation pressure p_o. Thus, p_o can be measured by connecting the opposite end of the pitot tube to a pressure meter (such as a manometer). The static pressure p can also be measured in a flow, either by a separate pressure tap or with additional holes in the pitot probe itself, such as is shown in the sketch in Problem 13.17. Bernoulli's equation is used to calculate the velocity from the pressure difference between p_o and p as follows (neglecting gravity):

$$p_o = p + {}^1\!/_2\, \rho V^2, \tag{6}$$

hence,

$$V = \left(\frac{2(p_o - p)}{\rho} \right)^{1/2}. \tag{7}$$

The pitot tube and pitot-static tube (which contains holes for both stagnation and static pressure) are local-velocity meters since they can easily be traversed through the fluid flow. Often, it is only necessary to measure the volume flow rate in a pipe flow. The three most common volume flow meters are the orifice meter, the venturi meter, and the flow nozzle. All three work on the principle that pressure in an incompressible flow decreases as the velocity increases through a throat of a smaller area than the pipe's cross-sectional area. This is nothing more than Bernoulli's equation, and so the three devices are called Bernoulli obstruction devices.

In all three devices, a static pressure tap is located near the throat, and a second tap is located just upstream of the device. If there were no losses, the pressure difference between these two would give the average velocity (and hence the volume flow rate) by Bernoulli's equation and the integral conservation of mass. Of course, fluid that flows through any real device will not be inviscid, and frictional losses are taken into account through a *discharge coefficient* C_d which is always less than 1.0. (Empirical formulae for C_d can be obtained for any of the three devices.) The final expression for the volume flow rate is

$$Q = C_d A_t \left[\frac{2(p_1 - p_2)}{\rho(1 - \beta^4)} \right]^{1/2}, \tag{8}$$

where p_1 and p_2 are the pressure upstream of the throat and near the throat respectively,

A_t is the throat (minimum) area, and

β is the ratio of the throat diameter to the upstream pipe diameter.

As seen in Equation (8), only the pressure *difference* is required, which can easily be measured with a U-tube manometer or some other differential pressure meter. Gravity has been neglected in the above analysis, but can easily be included if necessary, as in Problem 13.24.

Step-by-Step Solutions to Problems in this Chapter, "Flow Meters"

MANOMETER

• PROBLEM 13-1

An open-tube mercury manometer (Figure) is connected to a gas tank. The mercury is 39.0 cm higher on the right side than on the left when a barometer nearby reads 75.0 cm-Hg. What is the absolute pressure of the gas? Express the answer in cm-Hg, atm, Pa, and lb/in^2.

The open-tube manometer, as used to measure the pressure in a tank.

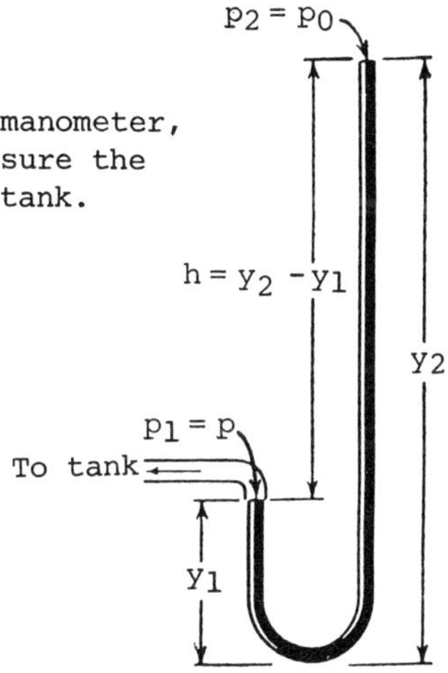

Solution: The gas pressure is the pressure at the top of the left mercury column. This is the same as the pressure at the same horizontal level in the right column. The pressure at this level is the atmospheric pressure (75.0 cm-Hg) plus the pressure exerted by the extra 39.0-cm column of Hg, or (assuming standard values of mercury density and gravity) a total of 114 cm-Hg. Therefore, the absolute pressure of the gas is

$$75 + 39 = 114 \text{ cm-Hg} = \frac{114}{76} \text{ atm} = 1.50 \text{ atm}$$

$$= 1.52 \times 10^5 \text{ Pa.} = (1.50)(14.7) \text{ lb/in.}^2$$

$$= 22.1 \text{ lb/in.}^2.$$

• PROBLEM 13-2

This vertical pipe line with attached gage and manometer contains oil and mercury as shown. The manometer is open to the atmosphere. There is no flow in the pipe. What will be the gage reading, p_x?

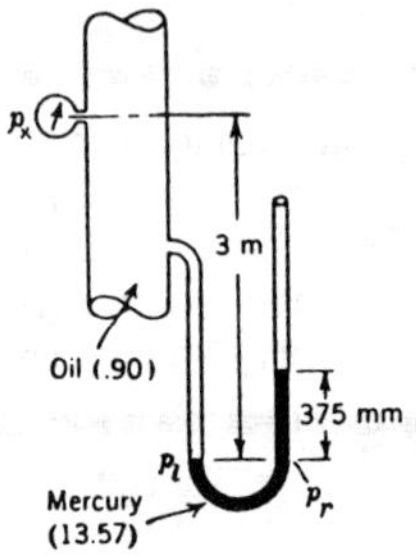

Solution:

$$z_2 - z_1 = \int_{p_1}^{p_2} \frac{dp}{\gamma} \qquad (1)$$

For a fluid of constant density

$$z_2 - z_1 = h = \frac{p_1 - p_2}{\gamma}$$

or (2)

$$p_2 - p_1 = -\gamma h$$

$$p_\ell = p_x + (0.90 \times 9.8 \times 10^3)3 \qquad (2)$$

$$p_r = (13.57 \times 9.8 \times 10^3)0.375. \qquad (2)$$

because

$$p_\ell = p_r$$

$$p_x = 23.4 \text{ kPa}$$

• PROBLEM 13-3

Determine the pressure difference, $p_A - p_B$ for the manometer shown below.

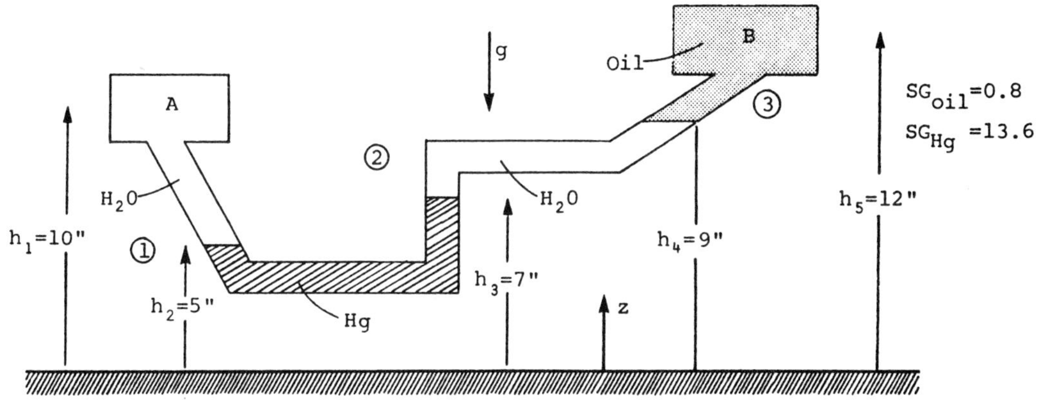

Solution: Employing the pressure-height relation, $\frac{dp}{dz} = -\rho g = -\gamma$ between successive point around the manometer, we have

$$p_1 = p_A + \gamma_{H_2O}(h_2 - h_1) \Rightarrow p_A - p_1 = -\gamma_{H_2O}(h_2 - h_1)$$

$$p_2 = p_1 - \gamma_{Hg}(h_3 - h_2) \Rightarrow p_1 - p_2 = \gamma_{Hg}(h_3 - h_2)$$

$$p_3 = p_2 - \gamma_{H_2O}(h_4 - h_3) \Rightarrow p_2 - p_3 = \gamma_{H_2O}(h_4 - h_3)$$

$$p_B = p_3 - \gamma_{oil}(h_5 - h_4) \Rightarrow p_3 - p_B = \gamma_{oil}(h_5 - h_4)$$

Adding the above equations we have

$$p_A - p_B = -\gamma_{H_2O}(h_2 - h_1) + \gamma_{Hg}(h_3 - h_2) + \gamma_{H_2O}(h_4 - h_3) + \gamma_{oil}(h_5 - h_4)$$

$$= -\gamma_{H_2O}(h_2 - h_1) + 0.8\gamma_{H_2O}(h_3 - h_2) + \gamma_{H_2O}(h_4 - h_3) + 13.6(h_5 - h_4)$$

$$= \gamma_{H_2O}\left[(h_1 - h_2) + 0.8(h_3 - h_2) + (h_4 - h_3) + 13.6(h_5 - h_4)\right]$$

$$= \gamma_{H_2O}\left[(5) + (0.8)(2) + (2) + (13.6)(3)\right]$$

$$= \gamma_{H_2O}(49.4)\text{in} = 62.4\ \frac{\text{lbf}}{\text{ft}^3}\ 4.94\ \frac{1}{12}\ \text{ft}\ \frac{\text{ft}^2}{144\ \text{in}^2}$$

$$\Rightarrow \quad p_A - p_B = 1.78\ \text{lbf/in}^2$$

• PROBLEM 13-4

A sensitive manometer containing water and oil is shown in the figure. What pressure difference is indicated when the liquid levels in the upper chambers were initially at the same level and the manometer deflection was zero when $p_A = p_B$? The level in A is 1/450 cm higher and that in B is 1/450 cm lower than initially.

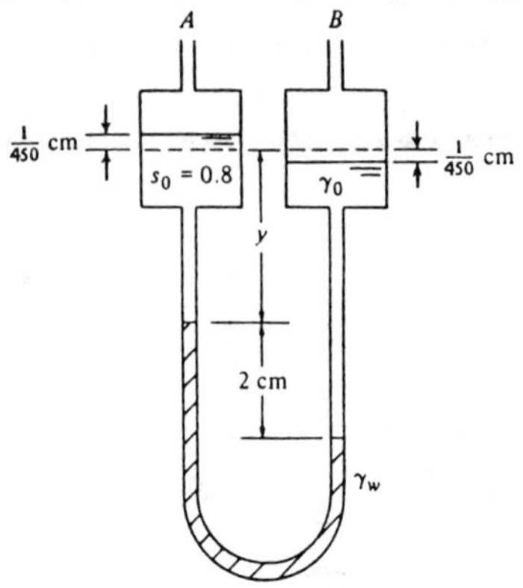

Micromanometer.

Solution: The hydrostatic equations give

$$p_A + \gamma_0[y + (1/450)(0.01)] + \gamma_w(0.02) - \gamma_0[(0.02) + y - (1/450)(0.01)] = p_B$$

$$p_B - p_A = (0.02)(\gamma_w - \gamma_0) + \frac{2\gamma_0}{45,000}$$

$$= 39.6 \text{ Pa}, \quad \text{which is less than } 1/2500 \text{ atm}$$

Note that the relative cross-sectional areas of chambers A and B and the manometer tubing determine the magnitude of the second term in the final expression for $p_B - p_A$, and this term often may be neglected.

• PROBLEM 13-5

Find the pressure difference $p_A - p_B$ in the figure if $z_A = 1.6$ m, $z_1 = 0.7$ m, $z_2 = 2.1$ m, $z_3 = 0.9$ m, $z_B = 1.8$ m, fluids 1 and 3 are water, and fluids 2 and 4 are mercury.

Solution: The specific weights of water and mercury are 9790 and 132,800 N/m^3, respectively. The pressure relations have already been derived. Summing them and substituting numerical values, we have

$$p_A - p_B = -9790(1.6 - 0.7) - 132,800(0.7 - 2.1)$$

$$- 9790(2.1 - 0.9) - 132,800(0.9 - 1.8)$$

$$= -8811 + 185,920 - 11,748 + 119,520$$

$$= 284,881 \text{ Pa} = 285 \text{ kPa}$$

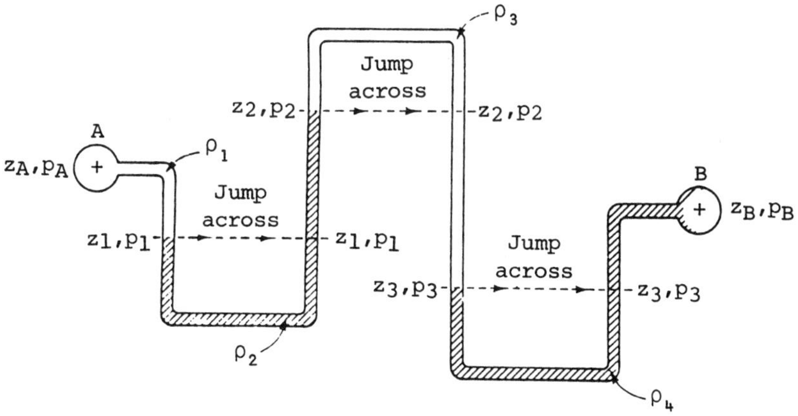

• PROBLEM 13-6

The pressure drop across an orifice in a pipeline conveying oil can be measured with an acetylene tetrabromide manometer as shown in the figure. What is the magnitude of $p_1 - p_2$?

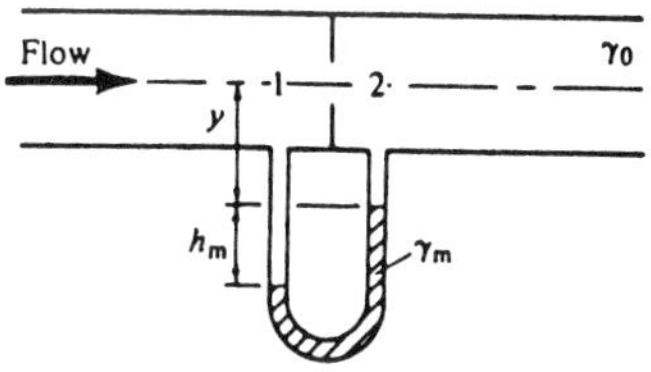

Differential manometer.

Solution: The hydrostatic equations are applied from point 1 to point 2 to obtain

$$p_1 + \gamma_0(y + h_m) - \gamma_m h_m - \gamma_0 y = p_2$$

so that

$$p_1 - p_2 = h_m(\gamma_m - \gamma_0)$$

Let $h_m = 36$ cm and the specific weight of the oil be $s_0 = 0.80$. Then $\gamma_m = (2.96)(9810) = 29.0 \text{ kN/m}^3$

from Table 2, $\gamma_0 = (9810)(0.80) = 7850\ N/m^3$,

and

$$p_1 - p_2 = (0.36)(29{,}000 - 7850) = 7.61\ kPa$$

TABLE 1

DENSITY OF SOME LIQUIDS

Temperature (°C)	Density (kg/m³)			
	water	mercury	benzene	glycerine
0	999.8	13,595		1275
5	999.9	13,583		
10	999.7	13,571		
15	999.1	13,558	880	1270
20	998.2	13,546		
25	997.0	13,534		
30	995.6	13,521		
35	994.0	13,509		
40	992.2	13,497	866	1250

TABLE 2

TYPICAL SPECIFIC GRAVITY OF SOME LIQUIDS AT 20°C

liquid	Specific Gravity
Gasoline	0.66-0.69
Denatured alcohol	0.80
Kerosene	0.80-0.84
Crude oil	0.80-0.92
Castor oil	0.97
Sea water	1.025
Carbon tetrachloride	1.594
Acelylene tetrabromide	2.962
Mercury	13.546

The piezometric head difference between points 1 and 2 is

$$h_1 - h_2 = (p_1 - p_2)/\gamma_0 = 0.97\ \text{m of oil}$$

Note that the distance y does not appear in the solution and thus a differential manometer of this type may be placed at any elevation with respect to the pipe, whereas in other problems, the position of the manometer is important.

Figure 1 is a schematic drawing of a mercury manometer connected between two parallel pipes, A filled with water and B filled with carbon tetrachloride. The manometer fluid in between is mercury. From this schematic drawing determine the pressure difference between the centerlines of the two pipes.

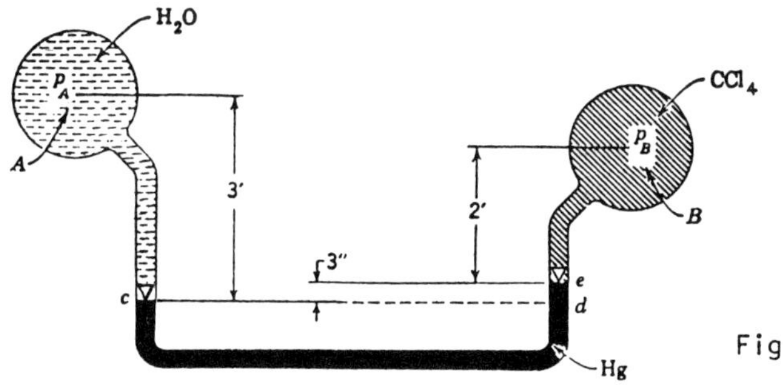

Fig. 1

Solution: Draw a schematic free body diagram (Sketch 1) between points A and c.

$$\Sigma F_V = 0 + \downarrow$$

$$p_A A_1 + 3\gamma_{H_2O} A_1 - p_c A_1 = 0$$

$$p_c = p_A + 3\gamma_{H_2O}$$

In the same horizontal plane are c and d. Since it is possible to go from c to d without leaving the mercury, the pressure at c equals the pressure at d. Therefore, $p_c = p_d$ and $p_A = p_d - 3\gamma_{H_2O}$. Draw a free body diagram of the fluid from d to e (Sketch 2) and another of the fluid from e to B (Sketch 3).

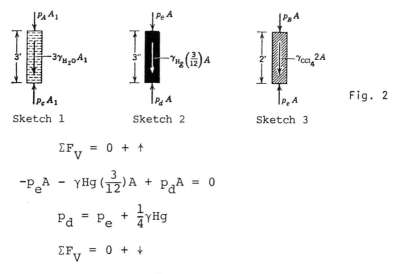

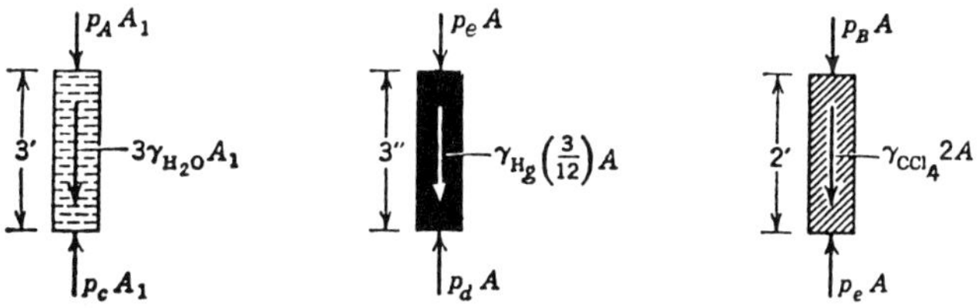

Fig. 2

$$\Sigma F_V = 0 + \uparrow$$

$$-p_e A - \gamma Hg\left(\frac{3}{12}\right)A + p_d A = 0$$

$$p_d = p_e + \frac{1}{4}\gamma Hg$$

$$\Sigma F_V = 0 + \downarrow$$

$$p_B A + 2\gamma_{CCl_4} A - p_e A = 0$$

$$p_e = p_B + 2\gamma_{CCl_4}$$

Substitute these equations for p_e and p_d into

$$p_A = p_d - 3\gamma_{H_2O}$$

which gives $$p_A - p_B = \frac{1}{4}\gamma Hg + 2\gamma_{CCl_4} - 3\gamma_{H_2O}.$$

• PROBLEM 13-8

> The double U-tube configuration shown in the Figure is used to measure the specific gravity of a liquid whose density is less than that of water. Determine the specific gravity of the unknown fluid in terms of the various column heights.

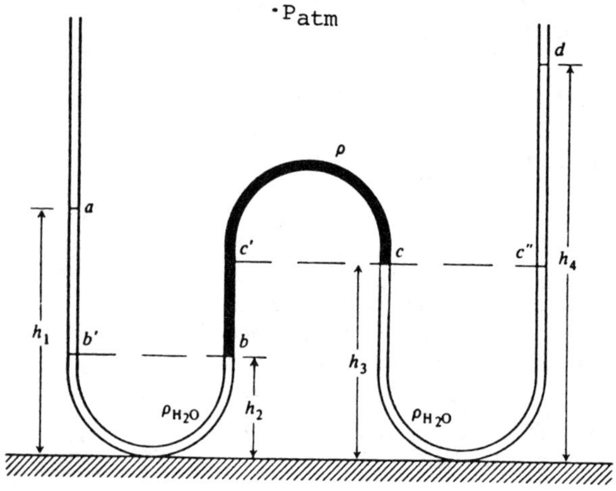

Solution: In multiple-bend manometers it is less confusing to write the overall pressure difference as the sum of the pressure differences across each fluid column. First we label each fluid interface a, b, c, and d. Then we can write

$$p_a - p_d = (p_a - p_b) + (p_b - p_c) + (p_c - p_d)$$

Now we evaluate each pressure-difference term in terms of the fluid density and column height:

$p_a - p_d$. Since both ends of the manometer are open to the atmosphere and the pressure changes due to columns of air are negligibly small compared with pressure changes associated with columns of liquid,

$$p_a = p_{atm} \quad p_d = p_{atm} \quad \text{and thus} \quad p_a - p_d = 0$$

$p_a - p_b$. Any two locations at the same elevation in the same fluid must be at the same pressure. Thus the pressure at b' must be the same as at b. Also, since b' is lower than a, the pressure at b' is greater than at a. Thus

$$p_a - p_b = p_a - p'_b = -\rho_{H_2O}g(h_1 - h_2)$$

$p_b - p_c$. Again the pressure at c' and c must be equal. Since b is lower in elevation than c', its pressure must be greater than that at c'

$$p_b - p_c = p_b - p'_c = \rho g(h_3 - h_2)$$

$p_c - p_d$. The pressure at c" and c must be equal; c" is lower in elevation than d, and its pressure is greater than that at d

$$p_c - p_d = p''_c - p_d = \rho_{H_2O}g(h_4 - h_3)$$

Substituting the various pressure terms back into th pressure difference-equation and solving for the gravity of the unknown fluid SG = ρ/ρ_{H_2O} gives

$$SG = \frac{(h_1 - h_2) - (h_4 - h_3)}{h_3 - h_2}$$

U-TUBE

• PROBLEM 13-9

A device A which measures flow characteristics of fluids is installed in a pipeline transporting water. The device produces a pressure drop in the water flow through the pipe. This pressure drop is measured by a manometer filled with both mercury and a liquid L above the mercury that has a specific gravity half that of mercury. With the column configuration shown, find the pressure drop across the measuring device.

Solution: In the use of a manometer to measure pressure differences, the pressure difference is dependent on the difference in column heights. Accordingly,

$$P_2 - P_1 = h_{1-2}\rho_w \tag{1}$$

where ρ_w is the density of water and the subscript 1-2 represents the column length between pressure levels designated by P_1 and P_2. Similarly,

$$P_2 - P_3 = h_{2-3}\rho_m \tag{2}$$

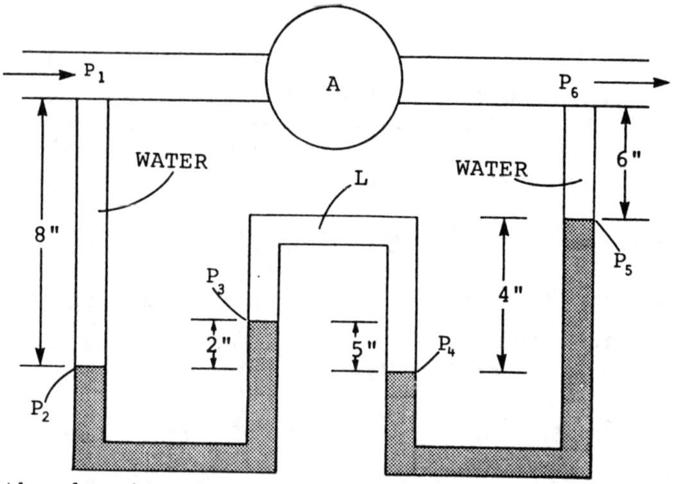

where ρ_m is the density of mercury present in the bottom part (shaded) of the manometer.

Continuing with writing the relationships of pressure differences for the remainder of the manometer,

$$P_4 - P_3 = h_{3-4}\rho_L \tag{3}$$

$$P_4 - P_5 = h_{4-5}\rho_m \tag{4}$$

and

$$P_5 - P_6 = h_{5-6}\rho_w \tag{5}$$

We are seeking to solve for $P_1 - P_6$. In examining the left sides of equations (1) to (5), we see that if some equations are appropriately multiplied by -1, and equations (1) to (5) are all added, then the desired result for $P_1 - P_6$ may be obtained as follows:

Multiplying equations (2), (4), and (5) by -1,

$$P_3 - P_2 = -h_{2-3}\rho_m \tag{6}$$

$$P_5 - P_4 = -h_{4-5}\rho_m \tag{7}$$

$$P_6 - P_5 = -h_{5-6}\rho_w \tag{8}$$

Rewriting equations (1), (6), (3), (7), (8) in orderly sequence,

$$P_2 - P_1 = h_{1-2}\rho_w$$
$$P_3 - P_2 = -h_{2-3}\rho_m$$
$$P_4 - P_3 = h_{3-4}\rho_L$$
$$P_5 - P_4 = -h_{4-5}\rho_m$$
$$P_6 - P_5 = -h_{5-6}\rho_w$$

Adding left and right sides of the preceding sequence results in

$$P_6 - P_1 = h_{1-2}\rho_w + h_{3-4}\rho_L - (h_{2-3}\rho_m + h_{4-5}\rho_m + h_{5-6}\rho_w)$$

Substituting values and expressing the result in lbsf/in^2,

$$P_6 - P_1 = \rho_w(h_{1-2} - h_{5-6}) - \rho_m(h_{2-3} + h_{4-5}) + \rho_L h_{3-4}$$

$$\rho_w = 62.4\ \frac{\text{lb f}}{\text{ft}^3} \times \frac{\text{ft}^3}{1728\ \text{in}^3} = .0361\ \frac{\text{lb f}}{\text{in}^3}$$

$$\rho_m = 13.6\rho_w = .491\ \text{lb f/in}^3$$

$$\rho_L = .246\ \text{lb f/in}^3$$

$$P_6 - P_1 = .0361(8 - 6) - .491(2 + 4) + .246(5)$$
$$= 4.25\ \text{lb f/in}^2$$

• PROBLEM 13-10

A U-tube is partly filled with water. Another liquid, which does not mix with water, is poured into one side until it stands a distance d above the water level on the other side, which has meanwhile risen a distance l (Figure). Find the density of the liquid relative to that of water.

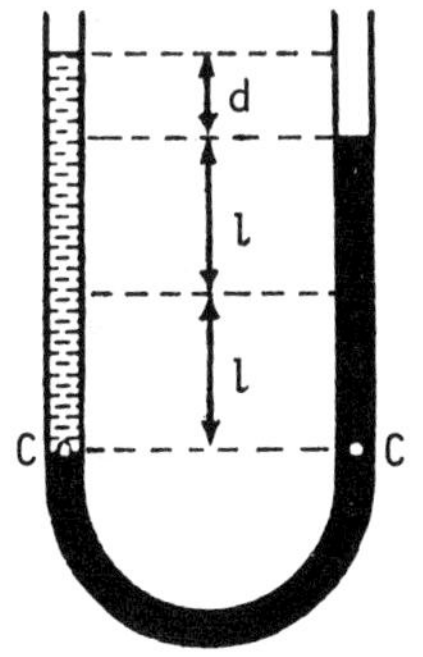

Solution: In the figure points C are both at the same pressure. Hence, the pressure drop from C to each surface is the same, for each surface is at atmospheric pressure.

The pressure drop on the water side is $\rho_w g 2l$; the $2l$ comes from the fact that the water column has risen a distance l on one side and fallen a distance l on the other side, from its initial position. The pressure drop on the other side is $\rho g(d + 2l)$, where ρ is the density of the unknown liquid. Hence,

$$\rho_w g 2l = \rho g(d + 2l)$$

and

$$\frac{\rho}{\rho_w} = \frac{2l}{(2l + d)}$$

The ratio of the density of a substance to the density of water is called the relative density (or the specific gravity) of that substance.

• PROBLEM 13-11

An inverted U-tube of the form shown in the accompanying figure is used to measure the pressure difference between two points A and B in an inclined pipeline through which water is flowing ($\rho H_2O = 10^3 kg\ m^{-3}$). The difference of level h = 0.3 m, a = 0.25 m and b = 0.15 m.

Calculate the pressure difference $p_B - p_A$ if the top of the manometer is filled with (a) air, (b) oil of relative density 0.8.

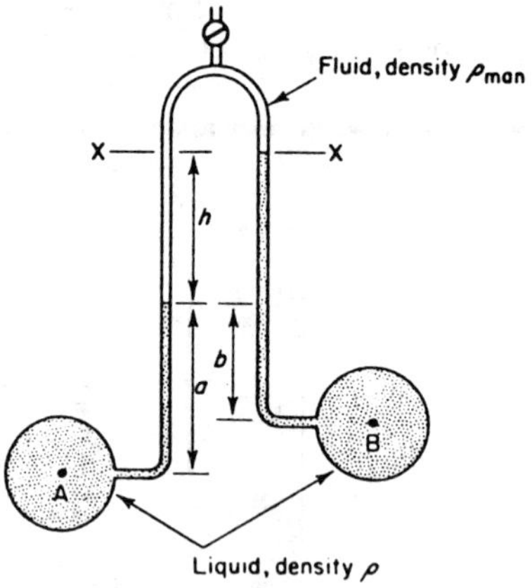

Inverted U-tube manometer

Solution: In either case, the pressure at XX will be the same in both limbs, so that

$$p_{XX} = p_A - \rho ga - \rho_{man} gh = p_B - \rho g(b + h),$$

$$p_B - p_A = \rho g(b - a) + gh(\rho - \rho_{man}).$$

(a) If the top is filled with air ρ_{man} is negligible compared with ρ. Therefore,

$$p_B - p_A = \rho g(b - a) + \rho gh = \rho g(b - a + h)$$

Putting $\rho = \rho H_2O = 10^3 kg\ m^{-3}$, $b = 0.15$ m, $a = 0.25$ m, $h = 0.3$ m:

$$p_B - p_A = 10^3 \times 9.81(0.15 - 0.25 + 0.3)$$

$$= 1.962 \times 10^3 N\ m^{-2}.$$

(b) If the top is filled with oil of relative density 0.8, $\rho_{man} = 0.8\ \rho H_2O = 0.8 \times 10^3 kg\ m^{-3}$.

$$p_B - p_A = \rho g(b - a) + gh(\rho - \rho_{man})$$

$$= 10^3 \times 9.81(0.15 - 0.25) + 9.81 \times 0.3 \times 10^3(1 - 0.8) N\ m^{-2}$$

$$= 10^3 \times 9.81(-0.05 + 0.06)$$

$$= 98.1\ N\ m^{-2}.$$

Compute $p_A - p_B$ for the arrangement and conditions shown in Fig. 1.

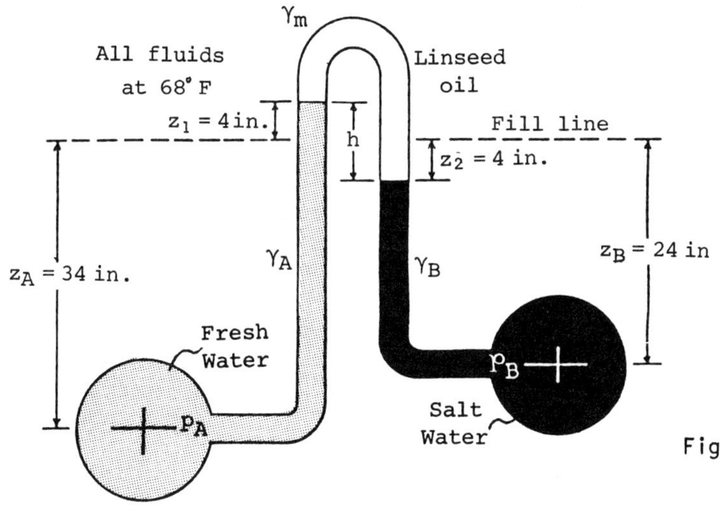

Fig. 1

Solution: Data Reduction

(1) unit conversion

$$z_A = \frac{34}{12} = \frac{17}{6}, \quad z_B = \frac{24}{12} = 2,$$

$$z_1 = z_2 = \frac{4}{12} = \frac{1}{3}, \quad h = \frac{4 + 4}{12} = \frac{2}{3}$$

(2) assume standard gravity

$g = 32.17 \text{ ft/sec}^2$

(3) fluid properties, 68°F

Table fresh water, $\rho_A = 1.937 \text{ slugs/ft}^3$

$\gamma_A = \rho_A g = 1.937 \times 32.17 = 62.31 \text{ lbf/ft}^3$

Table salt water, $\rho_B = 1.988 \text{ slugs/ft}^3$

$\gamma_B = \rho_B g = 1.988 \times 32.17 = 63.95 \text{ lbf/ft}^3$

Table linseed oil, $\rho_m = 1.828 \text{ slugs/ft}^3$

$\gamma_m = \rho_m g = 1.828 \times 32.17 = 58.81 \text{ lbf/ft}^3$

Table *Density of Selected Liquids*

TEMPERATURE °C		0	10	20	30	40	50
TEMPERATURE °F		32	50	68	86	104	122
LIQUID	Pressure, psia	ρ in slugs ft^{-3} or lbf sec^2 ft^{-4}					
Alcohol, ethyl[f]	14.7	1.564	1.548	1.532	1.515	1.498	1.481
Benzene[f]	14.7	1.746	1.726	1.705	1.684	1.663	1.642
Carbon tetrachloride[f]	14.7	3.168	3.130	3.093	3.055	3.017	2.979
Glycerin[f]	14.7	2.469	2.459	2.447	2.436	2.424	2.416
Mercury[f]	14.7	26.38	26.33	26.28	26.24	26.19	26.14
Octane, normal[f]	14.7	1.394	1.378	1.362	1.347	1.331	1.315
Oil, castor[k]	14.7	1.889	1.876	1.863	1.850	1.836	1.832
Oil, linseed[f]	14.7	1.853	1.840	1.828	1.815	1.803	1.791
Turpentine[f]	14.7	1.705	1.689	1.673	1.657	1.641	1.625
Water, fresh[d]	14.7	1.940	1.940	1.937	1.932	1.925	1.917
	10,000	2.005	2.005	1.992	1.986	1.980	1.967
	15,000	2.031	2.031	2.018	2.012	2.005	1.992
Water, sea[h]	14.7	1.995	1.992	1.988	1.982	–	–

Develop an Equation for this Application Based on Equilibrium

$$p_A - \gamma_A(z_A + z_1) = p_B - \gamma_B(z_B - z_2) - \gamma_m h$$

or

$$p_A - p_B = \gamma_A(z_A + z_1) - \gamma_B(z_B - z_2) - \gamma_m h$$

Solve Developed Equation

$$p_A - p_B = 62.31\left(\frac{17}{6} + \frac{1}{3}\right) - (63.95)\left(2 - \frac{1}{3}\right)$$

$$- (58.81)\left(\frac{2}{3}\right) = 51.52 \text{ lbf/ft}^2$$

$$p_A = p_B = \frac{51.52}{144} = 0.3578 \text{ psi}$$

• PROBLEM 13-13

An inverted U-tube is used to measure the differential pressure between two points in a pipe carrying an oil of 55.6 lb/ft^3 density. If the density of the compressed air is 0.4 lb/ft^3, what is the differential pressure for a reading of 10 in.?

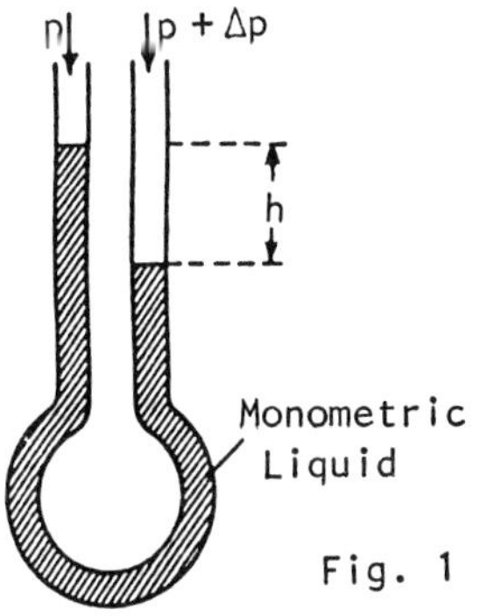

Fig. 1

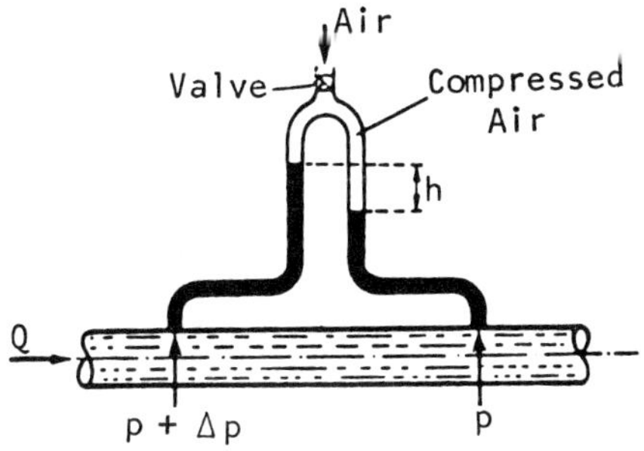

Fig. 2

Solution: Consider a U-tube manometer, the ends of which have been connected to two points of a system where the respective pressures are p and p + Δp, as shown in Fig. 1. Let h be the observed displacement of the manometric liquid due to the pressure Δp in excess of p, then the relationship between Δp and h is obtained from a balance of pressure heads as follows.

Let γ_1 and γ be the densities of the manometric liquid and the fluid producing the differential effect respectively, then expressing the pressure heads in terms of the manometric liquid

$$\frac{p + \Delta p}{\gamma_1} + h\frac{\gamma}{\gamma_1} = \frac{p}{\gamma_1} + h$$

from which

$$\frac{\Delta p}{\gamma_1} = h\left(1 - \frac{\gamma}{\gamma_1}\right) \tag{1}$$

By inverting a U-tube, a type of manometer is obtained as shown in Fig. 2. If the compression is reasonably low, the density of the air above the menisci will be small relative to the density of the liquid, and the relationship between Δp and h may be presented approximately by

$$h = \frac{\Delta p}{\gamma} \tag{2}$$

where γ is the density of the liquid.

With reference to Fig. 2, let γ_a be the density of the compressed air, when using eq. (1)

$$\frac{\Delta p}{\gamma} = h\left(1 - \frac{\gamma_a}{\gamma}\right)$$

in which γ is the density of the oil. Substituting the data in the above equation

$$\frac{\Delta p}{55.6} = \frac{10}{12}\left(1 - \frac{0.4}{55.6}\right)$$

from which

$$\Delta p = 44.0 \text{ psf } (0.306 \text{ psi})$$

Should the term γ_a/γ be ignored, and eq. (2) used instead, the differential pressure would be

$$\Delta p = h\gamma = \frac{10}{12}(55.6)$$

$$\Delta p = 44.3 \text{ psf}$$

The error in using the approximate equation (2) is less than 1.5 per cent, and this is acceptable in most engineering problems.

PITOT TUBE

• PROBLEM 13-14

A two-dimensional flow of liquid discharges from a large reservoir through the sharp-edged opening; a pitot tube at the center of the vena contracta produces the reading indicated. Calculate the velocities at points A, B, C, and D, and the flowrate.

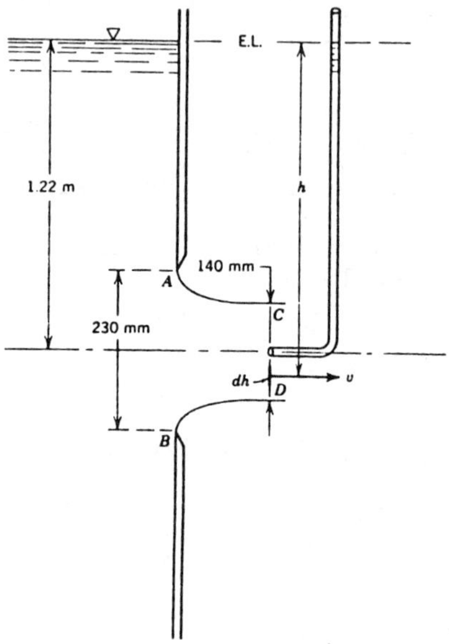

Solution: The pitot tube reading determines the position of the energy line (and also that of the free surface in the reservoir). Since the pressures at points A, B, C, and D are all zero, the respective velocity heads are determined by the vertical distances between the points and the energy line. The velocities at A, B, C, and D are 4.66 m/s, 5.12 m/s, 4.75 m/s, and 5.03 m/s, respectively. The flowrate may be computed by integrating the product vdA over the flow cross section CD, in which $v = \sqrt{2g_n h}$:

$$q = \int_C^D vdA = \int_{h_C}^{h_D} \sqrt{2g_n h}\, dh = \sqrt{2g_n}\, \frac{2}{3}\, h^{3/2} \Big|_{1.15}^{1.29}$$

$$= 0.685 \ m^3/s \cdot m$$

The limits of integration are then the vertical distances (in metres) from energy line to points D and C, respectively. Assuming that the mean velocity at section CD is at the center and thus measured by the pitot tube, $V = \sqrt{2g_n \times 1.22} = 4.89$ m/s, the (approximate) flowrate, q, may be computed from $q = 4.89 \times 0.140 = 0.685 \ m^3/s \cdot m$, giving the same result.

• PROBLEM 13-15

If the speed of an airplane is measured by means of a stagnation tube, what correction must be applied to the constant-density indication when the plane is cruising at 300 mph?

Solution: The problem is reduced to one of steady flow by the principle of relative motion. The velocity of the approaching flow is then $v_0, v = 0$, and according to the energy equation for acceleration under adiabatic conditions,

$$p - p_0 = \frac{\rho_0 v_0^2}{2}\left(1 + \frac{\rho_0 v_0^2}{4kp_0}\right)$$

$$= \frac{0.0024\left(300 \times \frac{5280}{3600}\right)^2}{2}\left[1 + \frac{0.0024\left(300 \times \frac{5280}{3600}\right)^2}{4 \times 1.40 \times 14.7 \times 144}\right]$$

$$= 232 + 9 = 241 \text{ psf}$$

Without the density correction, the resulting pressure rise would indicate

$$v_0 = \sqrt{\frac{2(p - p_0)}{\rho}} = \sqrt{\frac{2 \times 241}{0.0024}} = 446 \text{ fps} = 305 \text{ mph}$$

The correction is therefore 300 - 305 = -5 mph.

A pitot tube is a device that enables one to determine the local velocity in a flow. A sketch of the device is shown in the Fig. Determine an expression for the local velocity in terms of the manometer reading, h.

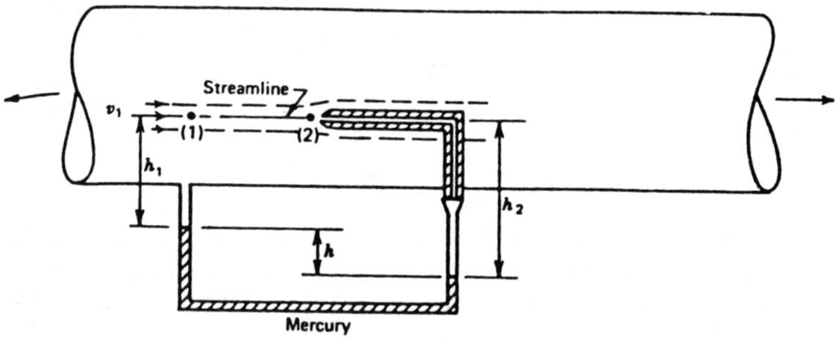

Solution: Observe that points 1 and 2 are on the same streamline and that point 2 is a stagnation point, i.e., a point where the velocity is zero. Since points 1 and 2 are both on the same streamline, the Bernoulli equation is applicable. Then

$$\frac{p_1}{\gamma} + \frac{v_1^2}{2g} + Z_1 = \frac{p_2}{\gamma} + \frac{v_2^2}{2g} + Z_2$$

Since $Z_1 = Z_2$, and v_2 is zero

$$v_1^2 = \frac{2g}{\gamma}(p_2 - p_1)$$

From the manometer reading, we have that

$$p_2 = p_1 + \gamma h_1 + \gamma_{hg} h - \gamma h_2$$

Also

$$h_1 + h = h_2$$

or

$$p_2 - p_1 = (\gamma_{hg} - \gamma)h$$

Therefore,

$$v_1 = \sqrt{2gh\left(\frac{\gamma_{hg}}{\gamma} - 1\right)}$$

It should be noted that in order to obtain an accurate reading for v_1, the longitudinal axis of the pitot tube must be parallel to the flow direction.

• PROBLEM 13-17

An aircraft flying at an altitude of 15,000 ft (ρ = 0.00150 slug/ft^3) has an IAS (indicated airspeed) of 350 mph. What is its TAS (true airspeed)?

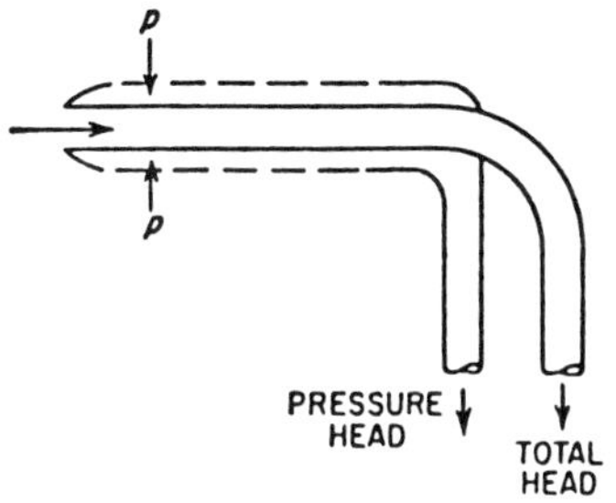

A pitot-static tube.

Solution: The pitot-static tube consists of two tubes, one within the other and sealed at the joints, as shown in the figure. The inner or pitot tube is open at the end and, when pointing directly into the flow, records the total head of the flow, $p + \rho V^2/2$.

The outer or static tube has a number of holes drilled in it at right angles to the flow direction so that this tube records only the local static pressure p.

The difference between the two pressures thus obtained is the dynamic pressure $\rho V^2/2$, which can be converted into velocity if the density ρ is known.

The most common application of the pitot-static tube is aircraft velocity measurement. The two pressures are led into an airspeed indicator (ASI) which is calibrated directly in mph, assuming ρ to have its sea-level value of 0.00238 slug/ft^3. At sea level, then, the indicated airspeed (IAS) will be the same as the true airspeed (TAS); but at altitudes where the air density is less than the sea-level value, the IAS will be less than the TAS.

The ASI is recording the same dynamic pressure that it would record at sea level and 350 mph; therefore

$$\frac{\rho V^2}{2} = \frac{0.00238}{2}\left(350 \times \frac{88}{60}\right)^2 \text{ psf}$$

But at 15,000 ft ρ = 0.00150 slug/ft^3; therefore

$$\frac{1}{2} \times 0.00150\left(v \times \frac{88}{60}\right)^2 = \frac{1}{2} \times 0.00238\left(350 \times \frac{88}{60}\right)^2$$

where v is the TAS in mph. Hence

$$v = \frac{0.00238}{0.00150} \times 350 = 512 \text{ mph}$$

An oil (sp.gr. = 0.88, viscosity = 50 cp) flows in a pipe of 3 in. diameter. The flow is measured by a Pitot tube located centrally, and the differential head is indicated by a U-tube, containing water as the manometric liquid. What is the rate of flow for a reading of 16.4 in.?

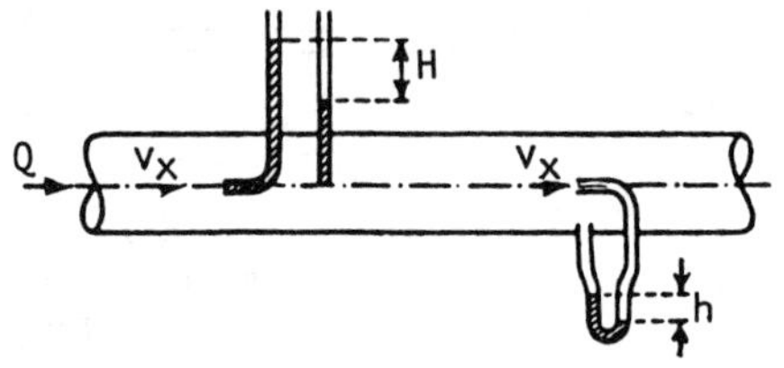

Solution: The Pitot tube is a simple device used for the determination of point velocities in flowing fluids, as shown in the figure.

The differential effect in the Pitot tube is due to the velocity of the fluid, i.e. its kinetic energy, at the point of measurement. Let this point velocity be v_x, then the differential head produced is given by

$$H = \frac{v_x^2}{2g} \qquad (1)$$

The velocity head is usually measured by a manometer, and this makes the Pitot tube a practical instrument for the measurement of point velocities in gases as well as liquids.

From a series of measurements across a stream, the average velocity can be evaluated by a process of integration. This is often a laborious procedure, and in case of circular ducts, such as pipes, it can be simplified by taking only a single measurement at their axis, where the point velocity is a maximum ($v_x = v_{max}$), and calculating the average velocity from the relationship

$$v = \alpha v_{max} \qquad (2)$$

Using the equation

$$H = h\left(\frac{\gamma_1}{\gamma} - 1\right)$$

$$H = \frac{16.4}{12}\left(\frac{1.00}{0.88} - 1\right)$$

$$H = 0.187 \text{ ft}$$

From eq. (1)

$$v_x = v_{max} = \sqrt{(2gH)} = \sqrt{\{(64.4)(0.187)\}}$$

$$v_{max} = 3.46 \text{ fps}$$

Assuming laminar flow, $\alpha = 0.5$, and (from eq. (2)) the average velocity is

$$v = 1.73 \text{ fps}$$

$$Q = vA = (1.73)\left(\frac{3}{12}\right)^2\left(\frac{\pi}{4}\right)$$

$$Q = 0.085 \text{ cfs}$$

This is equivalent to 5.1 cfm, or to

$$(5.1)(0.88)(62.4) = 280 \text{ lb/min}$$

Checking on the Reynolds number

$$Re = \frac{Dv\rho}{\mu} = \frac{(3/12)(1.73)(0.88)(62.4)}{(50)(0.000672)}$$

$$Re = 707$$

This is less than the critical number 2000, thus proving the correctness of the assumption made.

ORIFICE METER

• PROBLEM 13-19

A 12 in. diameter pipe is equipped with a 4 in. diameter orifice plate (see Figure). By calibration the discharge coefficient was determined to be 0.63. What is the discharge in the pipe if the differential pressure registered during the flow equals 12 psi?

Solution: A good approximate formula for orifices in practical applications may be written in the simple form of

$$Q = ca\sqrt{\frac{2g}{\gamma}(\Delta p)}$$

where

$$c = 0.63$$

$$a = \frac{\pi}{4}\left[\frac{4}{12}\right]^2 = 0.0872 \text{ ft}^2$$

$$\gamma = 62.4 \frac{\text{lb}}{\text{ft}^3}$$

$$\Delta p = 12 \times 144 = 1728 \frac{\text{lb}}{\text{ft}^2}$$

Then

$$Q = 0.63 \times 0.0872\sqrt{\frac{2(32.2)(1728)}{62.4}}$$

$$= 2.32 \text{ cfs}$$

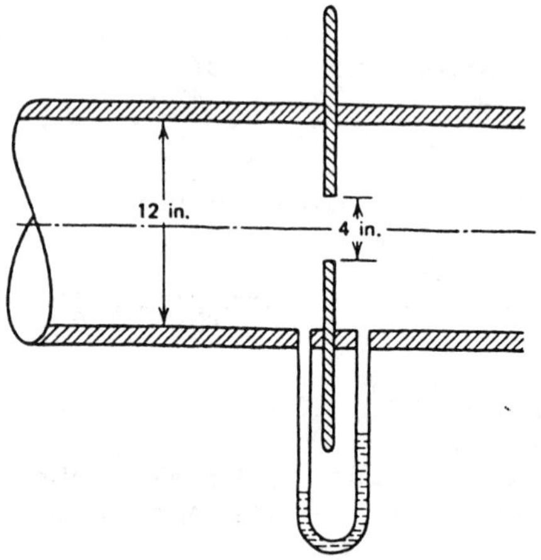

• PROBLEM 13-20

An orifice of area A_0 and velocity coefficient $c_v = 0.8$ is installed in a pipe of area $A_p = 2A_0$. The pipe is attached to a dam as shown and the water level in the dam is 100 ft. higher than the outlet from the reservoir. The contraction coefficient of the orifice is unity. There are no other losses in the pipe nor between the end of the orifice and the pipe. The velocity leaving the pipe V_p is to be determined.

Solution: We shall solve this first by using the velocity coefficient directly. From the Bernoulli equation the velocity V_0 downstream of the orifice is

$$V_0 = c_v\sqrt{2g\left(h_t - \frac{p}{\rho g}\right)}$$

where p is the static gage pressure at the end of the orifice.

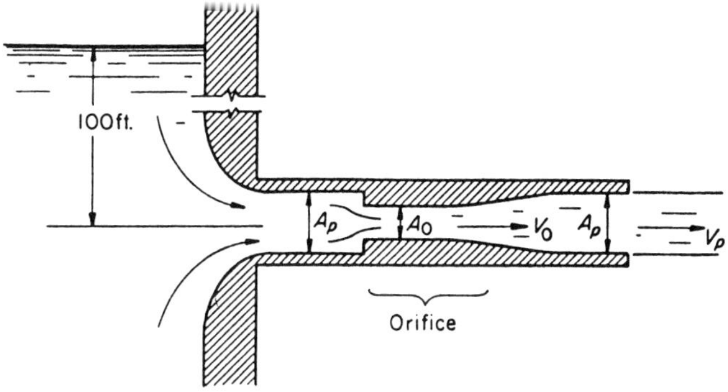

To obtain the pressure at the end of the orifice we write Bernoulli's equation again between the orifice outlet and the end of the pipe:

$$\frac{p}{\rho g} + \frac{V_0^2}{2g} = \frac{V_p^2}{2g} ,$$

or since $A_0 V_0 = A_p V_p$

$$\frac{p}{\rho g} = - \frac{V_p^2}{2g}\left\{\left(\frac{A_p}{A_0}\right)^2 - 1\right\} ,$$

and finally

$$V_0 = \frac{A_p}{A_0} V_p = c_v \sqrt{2g\left\{h_t + \frac{V_p^2}{2g}\left(\frac{A_p^2}{A_0^2} - 1\right)\right\}} .$$

Solving for V_p we have

$$V_p^2 = \frac{2gh_t}{1 + \frac{A_p^2}{A_0^2}\left(\frac{1}{c_v^2} - 1\right)} ,$$

or

$$V_p = \left\{\frac{2 \times 32.2 \times 100}{1 + 4\left(\frac{1}{0.64} - 1\right)}\right\}^{1/2} ,$$

$$V_p = 44.5 \text{ ft./sec.}$$

The other approach consists of writing the Bernoulli equation including the loss term between the pipe outlet and the reservoir surface, i.e.,

$$h_t - k\frac{V_0^2}{2g} = \frac{V_p^2}{2g} .$$

But $$k = \frac{1}{c_v^2} - 1 ,$$

and

$$V_0 = \frac{A_p}{A_0} V_p ,$$

so

$$V_p^2 = \frac{2gh_t}{1 + \frac{A_p^2}{A_0^2}\left(\frac{1}{c_v^2} - 1\right)}$$

as before.

• PROBLEM 13-21

The flow of water is measured by an orifice of 1 in. diameter inserted in a horizontal pipe of 1.61 in. diameter. The differential head is measured by a U-tube manometer containing mercury as the manometric liquid. If the coefficient of discharge is 0.61, what is the manometer reading for a flow of 150 lb/min?

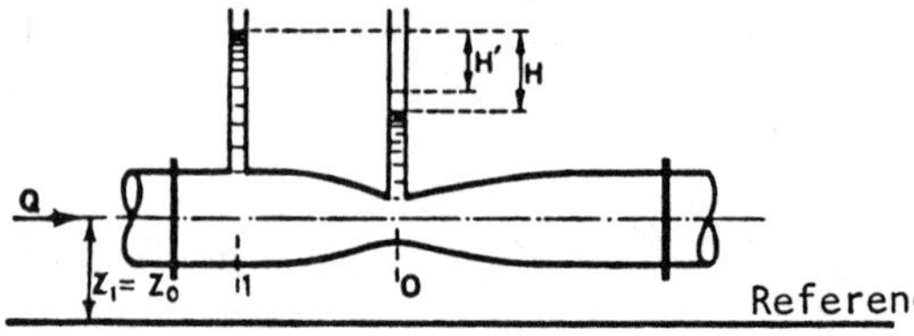

Fig. 1

Solution: Consider in Fig. 1 two sections, 1 and 0, of a Venturi tube in a horizontal arrangement. Let us first assume that the fluid flowing through the tube behaves ideally, then applying Bernoulli's theorem between the two sections considered

$$Z_1 + \frac{p_1}{\gamma} + \frac{v_1^2}{2g} = Z_0 + \frac{p_0}{\gamma} + \frac{v_0^2}{2g}$$

But for a horizontal arrangement $Z_1 = Z_0$, and

$$\frac{p_1 - p_0}{\gamma} = \frac{v_0^2 - v_1^2}{2g}$$

Also from the diagram (Fig. 1), it will be observed that

$$\frac{p_1 - p_0}{\gamma} = H'$$

then

$$H' = \frac{v_0^2 - v_1^2}{2g} \qquad (1)$$

From the law of continuity

$$Q = v_1 D_1^2 \pi/4 = v_0 D_0^2 \pi/4$$

from which

$$v_1 = v_0 \left(\frac{D_0}{D_1}\right)^2$$

where D_1 is the diameter of the pipe carrying the fluid, assumed to be the same as the diameter of the tube at the section 1.

Substituting for v_1 in eq. (1)

$$H' = \frac{v_0^2[1 - (D_0/D_1)^4]}{2g}$$

Let $\beta = D_0/D_1$, then

$$H' = \frac{v_0^2[1 - \beta^4]}{2g}$$

from which

$$v_0 = \sqrt{\left(\frac{2gH'}{1 - \beta^4}\right)} \qquad (2)$$

Now, let Q_0 be the theoretical flow referred to the section 0, then

$$Q_0 = A_0 v_0 = A_0 \sqrt{\left(\frac{2gH'}{1 - \beta^4}\right)}$$

Due to the energy loss in the flow of a real fluid, the observed differential head is greater than the theoratical one, and the actual velocity at the throat is less than that which could be calculated from eq. (2). Let H be the observed differential head, then the actual flow can be obtained from the equation

$$Q = C_D A_0 \sqrt{\left(\frac{2gH}{1 - \beta^4}\right)} \qquad (3)$$

where C_D is the coefficient of discharge.

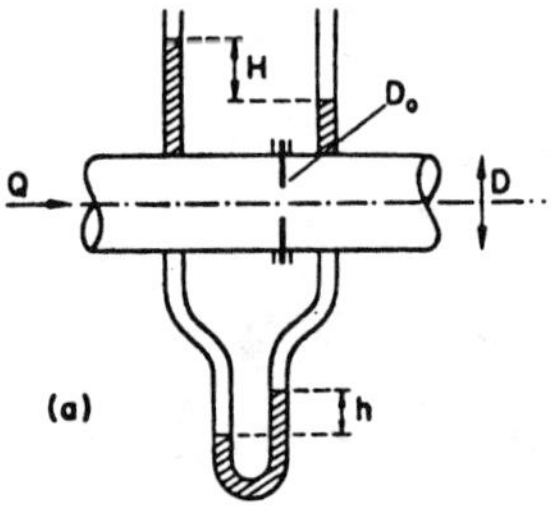

Fig. 2

In this problem (see Fig. 2), Eq. (3) can be applied to the orifice meter to find H, and from H, we can calculate the manometer reading, h. Referring to Fig. 2

$$\beta = D_0/D = 1.00/1.61$$

$$1 - \beta^4 = 1 - 0.144 = 0.856$$

$$A_0 = D_0^2\pi/4 = (1/12)^2\pi/4 = 0.00544 \text{ ft}^2$$

$$Q = \frac{150}{(62.4)(60)} = 0.0400 \text{ cfs}$$

Substituting the data in eq. (3)

$$0.04 = (0.61)(0.00544)\sqrt{\left(\frac{64.4H}{0.856}\right)}$$

from which

$$H = 1.932 \text{ ft}$$

Using

$$H = h\left(\frac{\gamma_1}{\gamma} - 1\right)$$

The mercury-water density ratio is $\gamma_1/\gamma = 13.6/1.0 = 13.6$, then

$$h = \frac{1.932}{13.6 - 1} = 0.153 \text{ ft (1.84 in.)}$$

• PROBLEM 13-22

A tank of 5 m^2 plan area is fitted with a sharp-edged orifice of 5 cm diameter in its base. The coefficient of discharge of the orifice is 0.62. Calculate the time taken for the level in the tank to fall from 2 metres depth to 0.5 metres depth. If water is now admitted to the tank at a rate of 0.01 m^3/s calculate the rate at which the surface will be rising when the depth in the tank is 1 metre and the depth in the tank when the level becomes steady.

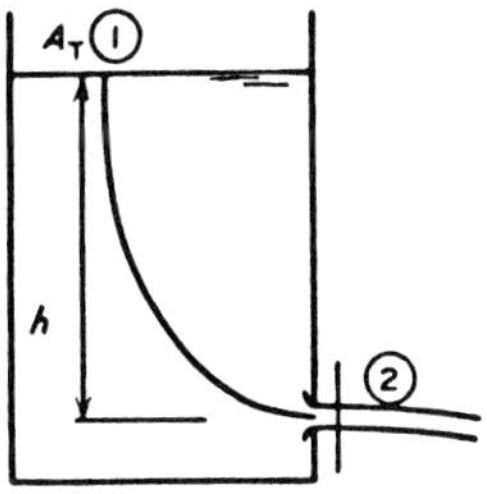

Solution: Energy losses as fluid moves down a streamtube as shown in the figure are very small, so the application of Bernoulli should be appropriate. The centre line of the tube is defined as the line joining ① and ②, so

$$\frac{p_1}{\gamma} + \frac{V_1^2}{2g} + z_1 = \frac{p_2}{\gamma} + \frac{V_2^2}{2g} + z_2 \qquad (1)$$

To solve this equation for one variable, the values of all the others must be known. p_1 is atmospheric pressure, V_1 is very small if the orifice is small and the tank large so $V_1^2/2g$ is negligible, $z_1 - z_2$ is the height of the water surface above point 2. At the first point where the streamlines are parallel in the effluxing jet the pressure is atmospheric so this is the position chosen for point 2. This point where the sides of the jet first became parallel is called the vena contracta and the pressure here must be atmospheric pressure. Then if p_a is the atmospheric pressure equation (1) becomes

$$p_a/\gamma + 0 + h + z_2 = p_a/\gamma + V_2^2/2g + z_2$$

so

$$V_2^2/2g = h$$

so

$$V_2 = \sqrt{(2gh)}$$

This is not perfectly true because there is a small energy loss which was not taken into account. A coefficient can be introduced so that this expression gives an accurate result

$$V = C_v\sqrt{(2gh)}$$

The coefficient C_v is called the coefficient of velocity of the orifice. For a sharp-edged small orifice (that is $d_0 << h$) C_v is approximately 0.98 which shows how well the energy approach works. The flow through the orifice is given by the product of the velocity and the area of the jet at the vena contracta. That is

$$Q = VA_{vc}$$

$$Q = C_v A_{vc} \sqrt{(2gh)}$$

Unfortunately, the area of the vena contracta is not easily calculated by a theoretical method and an element of empiricism must be introduced. The ratio of the area of the vena contracta to the area of the orifice is called the coefficient of contraction of the orifice C_c.

$$C_c = A_{vc}/A_0$$

so

$$Q = C_c A_0 C_v \sqrt{(2gh)} = C_d A_0 \sqrt{(2gh)} \qquad (2)$$

the coefficient C_d is called the coefficient of discharge and is equal to the product $C_v C_c$. Let the cross sectional area of the tank be A_T. The continuity equation is

$$-A_T dh = Q dt$$

(tank volume reduction = outflow rate × time in which the volume reduction occurs). h is the surface elevation above the orifice at time t and the negative sign is introduced because dh is a depth reduction.

$$\therefore \quad -A_T dh = C_d A_0 \sqrt{(2gh)}\, dt \qquad (3)$$

$$\therefore \quad dt = \frac{-A_T dh}{C_d A_0 (2gh)^{1/2}}$$

Integrating

$$t = \frac{2A_T}{C_d A_0 (2g)^{1/2}} [H_0^{1/2} - H_1^{1/2}]$$

H_0 is the initial depth at time t = 0 and H_1 is the depth at time desired.

$$\text{Time} = \frac{2A_T}{C_d A_0 \sqrt{(2g)}} (H_0^{1/2} - H_1^{1/2})$$

$$= \frac{2 \times 5}{0.62 \times \frac{\pi}{4}(0.05)^2 \times \sqrt{19.62}} (2^{1/2} - 0.5^{1/2})$$

$$= 1.31 \times 10^3 \text{s}$$

Rate of rise of surface

From Equation (3)

$$A_T \frac{dh}{dt} = Q - C_d A_0 \sqrt{(2gh)} \quad \text{where } Q = \text{inflow}$$

$$A_T \frac{dh}{dt} = Q - 0.62 \times \frac{\pi}{4} \times (0.05)^2 \sqrt{(19.62 \times 1.0)}$$

$$\frac{dh}{dt} = 0.922 \times 10^{-3} \text{m/s}$$

Level in steady state

From equation (2)

$$Q = C_d A_0 \sqrt{(2gh_s)}$$

$$h_s = \frac{Q^2}{C_d^2 A_0^2 2g}$$

$$= 3.44 \text{ m}$$

VENTURI TUBE

• PROBLEM 13-23

Determine the maximum discharge of water at 32°F that may be carried by a 3-in venturi meter, which has a coefficient of velocity of 0.95 without cavitation when the inlet pressure is 24 psia.

Solution: Since cavitation in a liquid occurs when the pressure in a pipe approaches the vapor pressure of the liquid, the absolute pressure head of water at the throat will be 0.204 ft. The ideal velocity can be determined by using Bernoulli's equation

$$\frac{p_1}{\gamma} + z_1 + \frac{V_1^2}{2g} = \frac{p_2}{\gamma} + z_2 + \frac{V_2^2}{2g}$$

Substituting in the values,

$$\frac{24}{0.433} + 0 + \frac{V_1^2}{64.4} = 0.204 + 0 + \frac{V_2^2}{64.4} \qquad (1)$$

Since from continuity $V_1 A_1 = V_2 A_2$

then $\quad V_1 (6 \text{ in})^2 = V_2 (3 \text{ in})^2$

$$V_2 = 4V_1$$

$$V_2^2 = 16V_1^2$$

Substituting in for V_2^2 in equation (1) yields

$$\frac{15V_1^2}{64.4} = 55.5 - 0.204 = 55.296$$

Therefore

$$V_1 = \left[\frac{55.296(64.4)}{15}\right]^{1/2} = 15.4 \text{ fps (ideal)}$$

To find V_{Actural} multiply V_{Ideal} by the coefficient of velocity.

$$V_{1_{\text{Act.}}} = 0.95(15.4) = 14.6 \text{ fps}$$

The discharge is defined as

$$Q = Va \quad \text{where } a = \text{area of vena contracta}$$

$$= \frac{\pi}{4}(d)^2$$

Thus

$$Q = V_1\frac{\pi}{4}\left(\frac{3}{12}\right)^2 = \frac{\pi}{4}\left(\frac{1}{4}\right)^2 V_1$$

$$= \frac{\pi}{64}(14.6) = 0.7163$$

• PROBLEM 13-24

A venturimeter of throat diameter 4 cm is fitted into a pipeline of 10 cm diameter. The coefficient of discharge is 0.96. Calculate the flow through the meter when the reading on a mercury-water manometer connected across the upstream and throat tapping is 25 cm. If the energy loss in the downstream divergent cone of the meter is $10v_p^2/2g$ per unit weight of fluid, calculate the head loss across the meter. (v_p is the velocity in the pipeline.)

Solution: Consider the venturimeter shown in the Figure. Apply Bernoulli's equation to the entry section 1 and to the throat section 2

$$\frac{p_1}{\gamma} + \frac{v_1^2}{2g} + z_1 = \frac{p_2}{\gamma} + \frac{v_2^2}{2g} + z_2$$

$$\therefore \quad \frac{v_2^2 - v_1^2}{2g} = \frac{p_1 - p_2}{\gamma} + z_1 - z_2$$

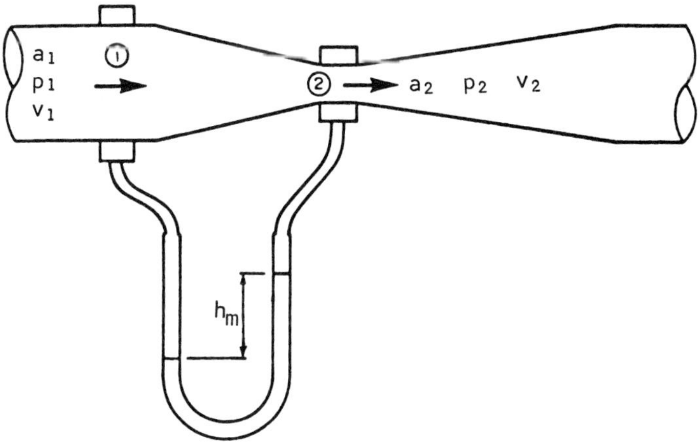

But by continuity $a_1v_1 = a_2v_2$, so

$$v_2 = \frac{a_1}{a_2} v_1$$

$$\therefore \quad \left[\left(\frac{a_1}{a_2}\right)^2 - 1\right] \frac{v_1^2}{2g} = \frac{\Delta p}{\gamma} + z_1 - z_2$$

$$\therefore \quad v_1 = \left[\frac{2g(\Delta p/\gamma + z_1 - z_2)}{(a_1/a_2)^2 - 1}\right]^{1/2}$$

$$\therefore \quad Q = a_1v_1 = a_1a_2\left[\frac{2g(\Delta p/\gamma + z_1 - z_2)}{a_1^2 - a_2^2}\right]^{1/2}$$

Because small energy losses occur in the convergent section a coefficient of discharge must be introduced into Bernoulli's equation.

$$\therefore \quad Q = C_d a_1\left[\frac{2g(\Delta p/\gamma + z_1 - z_2)}{(a_1/a_2)^2 - 1}\right]^{1/2}$$

$$Q = C_d a_p \sqrt{\left\{2g\,\frac{\left(\frac{p_1 - p_2}{\gamma} + z_1 - z_2\right)}{a_p^2 - a_t^2}\right\}}$$

$$Q = 0.96 \times \frac{\pi}{4} \times (0.10)^2 \sqrt{\left\{\frac{19.62 \times 0.25 \times 12.6}{(10/4)^4 - 1}\right\}}$$

$$Q = 9.61 \times 10^{-3}\,\mathrm{m^3/s}$$

The energy loss in the convergent cone can be calculated if the C_d value is known. The Bernoulli equation is written including an allowance for friction loss, that is

$$\frac{p_1}{\gamma} + \frac{v_1^2}{2g} + z_1 = \frac{p_2}{\gamma} + \frac{v_2^2}{2g} + z_2 + h_f$$

where h_f is the head loss due to friction.

Then as before

$$Q = a_1 \left(\frac{2g(\Delta p/\gamma + z_1 - z_2 - h_f)}{(a_1/a_2)^2 - 1} \right)^{1/2}$$

from before

$$Q = C_d a_1 \left(\frac{2g(\Delta p/\gamma + z_1 - z_2)}{(a_1/a_2)^2 - 1} \right)^{1/2}$$

$$\therefore \quad C_d^2(\Delta p/\gamma + z_1 - z_2) = (\Delta p/\gamma + z_1 - z_2 - h_f)$$

$$\therefore \quad h_f = (\Delta p/\gamma + z_1 - z_2)(1 - C_d^2)$$

but

$$\frac{\Delta p}{\gamma} + z_1 - z_2 = \frac{Q^2}{C_d^2 2ga_1^2} \left[\left(\frac{a_1}{a_2} \right)^2 - 1 \right] = \frac{1}{C_d^2} \left(\frac{v_2^2 - v_1^2}{2g} \right)$$

so

$$h_f = \left(\frac{1}{C_d^2} - 1 \right) \frac{v_2^2 - v_1^2}{2g}$$

Energy loss/unit wt across the convergent cone

$$= \left(\frac{1}{C_d^2} - 1 \right) \left(\frac{v_t^2 - v_p^2}{2g} \right)$$

Therefore total energy loss across the meter

$$= \left(\frac{1}{C_d^2} - 1 \right) \left(\frac{v_t^2 - v_p^2}{2g} \right) + 10 \frac{v_p^2}{2g}$$

$$= 1.01 \text{ m}$$

• PROBLEM 13-25

An oil (sp.gr. = 0.92) flows through a vertical tube of 8 in diameter. The flow is measured by a Venturi tube of 4 in. diameter throat with a U-tube manometer, containing mercury, as shown in the figure. If $C_D = 0.98$, what is (a) the flow, for a manometer reading of 9 in? (b) the manometer reading, for a flow of 2 ft³/sec?

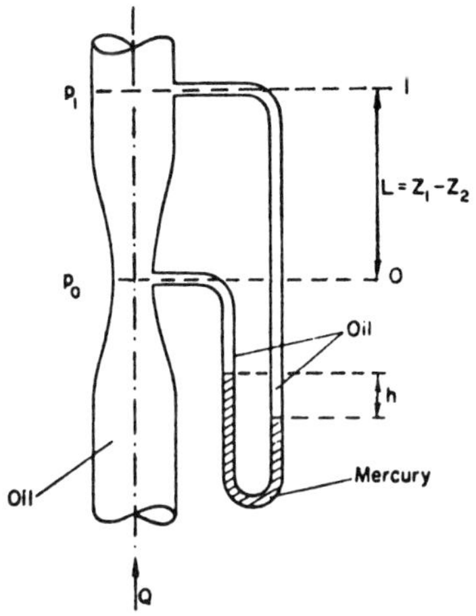

Solution: (a) For a Venturi tube inclined vertically, the Bernoulli equation, with reference to the figure, is given by

$$\frac{p_1 - p_0}{\gamma} = \frac{v_0^2 - v_1^2}{2g} + (Z_0 - Z_1)$$

Since

$$H' = \frac{p_1 - p_0}{\gamma}$$

and

$$\frac{v_0^2 - v_1^2}{2g} = \frac{v_0^2(1 - \beta^4)}{2g}$$

then

$$H' = \frac{v_0^2(1 - \beta^4)}{2g} + (Z_0 - Z_1)$$

or

$$H' = \frac{v_0^2(1 - \beta^4)}{2g} - (Z_1 - Z_0)$$

Also by a balance of pressure heads above the lower meniscus of the mercury in the U-tube

$$\frac{p_0}{\gamma} + h\,\frac{\gamma_1}{\gamma} = \frac{p_1}{\gamma} - (Z_1 - Z_0) + h$$

from which

$$\frac{p_1 - p_0}{\gamma} = h\left(\frac{\gamma_1}{\gamma} - 1\right) - (Z_1 - Z_0)$$

But

$$H' = \frac{p_1 - p_0}{\gamma}$$

then

$$h\left(\frac{\gamma_1}{\gamma} - 1\right) - (Z_1 - Z_0) = \frac{v_0^2(1 - \beta^4)}{2g} - (Z_1 - Z_0)$$

from which

$$v_0 = \sqrt{\left(\frac{2gh(\gamma_1/\gamma - 1)}{1 - \beta^4}\right)}$$

The actual rate of flow is then given by

$$Q = C_D A_0 \sqrt{\left(\frac{2gh(\gamma_1/\gamma - 1)}{1 - \beta^4}\right)}$$

But $\quad 1 - \beta^4 = 1 - (4/8)^4 = 15/16, \quad$ then

$$h(\gamma_1/\gamma - 1) = (9/12)(13.6/0.92 - 1) = 10.35 \text{ ft}$$

$$Q = (0.98)\left[\left(\frac{4}{12}\right)^2 \frac{\pi}{4}\right]\sqrt{\left(\frac{(64.4)(10.35)}{15/16}\right)}$$

$$Q = 2.278 \text{ cfs } (7848 \text{ lb/min})$$

(b) $$2 = (0.98)\left[\left(\frac{4}{12}\right)^2 \frac{\pi}{4}\right]\sqrt{\left(\frac{64.4h(13.6/0.92 - 1)}{15/16}\right)}$$

from which

$$h = 0.5965 \text{ ft } (7.16 \text{ in})$$

• PROBLEM 13-26

The flow of water in a pipe is measured by a Venturi Meter. The diameters of the pipe and throat are 4 in and 2 in respectively. The constant of the meter is 0.98, and the drop of pressure from inlet to throat is 2.000 ft of mercury. Find the discharge.

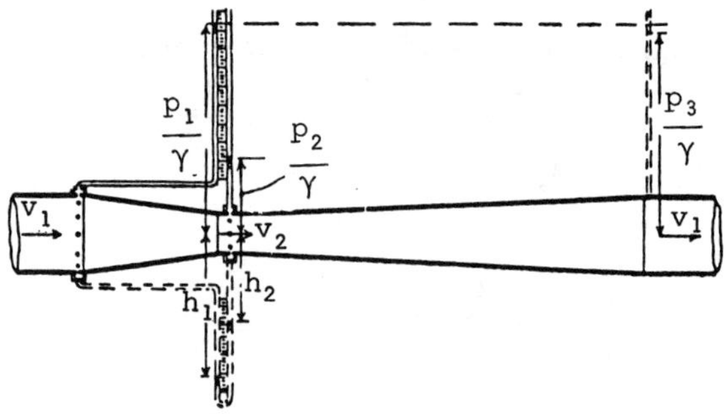

Solution: Here a length of the pipe, of cross-sectional area a_1, is made to taper fairly rapidly so as to reduce the area to a_2 at the throat; a long taper is then provided to enlarge the area very gradually back to a_1. Pressure tubes, or manometers, are led from annular chambers surrounding the inlet and throat respectively, fine holes being drilled from the annular chambers into the pipe around its perimeter. Consequently the pressure tubes, brought side by side in front of a scale, give the values of $\frac{p_1}{\gamma}$ and $\frac{p_2}{\gamma}$ at the centre of the pipe. We thus have

$$\frac{v_1^2}{2g} + \frac{p_1}{\gamma} + z_1 = \frac{v_2^2}{2g} + \frac{p_2}{\gamma} + z_1;$$

if the axis of the pipe is level, or

$$\frac{p_1 - p_2}{\gamma} = \frac{v_2^2 - v_1^2}{2g}.$$

But $a_2 v_2 = a_1 v_1$;

$$\therefore\ v_2 = \frac{a_1 v_1}{a_2}; \quad \therefore\ v_2^2 = \frac{a_1^2 v_1^2}{a_2^2};$$

$$\therefore\ \frac{p_1 - p_2}{\gamma} = \frac{v_1^2}{2g}\left(\frac{a_1^2}{a_2^2} - 1\right);$$

$$\therefore\ v_1^2 = 2g \cdot \frac{p_1 - p_2}{\gamma} \cdot \frac{a_2^2}{a_1^2 - a_2^2};$$

$$\therefore\quad Q = a_1 v_1 = \frac{a_1 a_2}{\sqrt{a_1^2 - a_2^2}}\sqrt{2g \cdot \frac{p_1 - p_2}{\gamma}}$$

$$= \frac{a_2}{\sqrt{1 - \frac{a_2^2}{a_1^2}}}\sqrt{2g\,\frac{p_1 - p_2}{\gamma}} = \text{discharge in the pipe.}$$

Actually, the discharge is rather less than this, owing to fluid resistance, and it is necessary to introduce a "coefficient of velocity," C (usually from 0·96 to 0.99 in value), and write

$$Q = \frac{C \cdot a_2}{\sqrt{1 - \frac{a_2^2}{a_1^2}}}\sqrt{2g\,\frac{p_1 - p_2}{\gamma}},$$

a formula which holds over a considerable range of discharges.

If there were a third pressure tube at the outlet of the meter, we should find $\frac{p_3}{\gamma}$ there only a little less than $\frac{p_1}{\gamma}$, showing that most of the increased velocity head at a_2 had been reconverted into pressure head. The angles where the cones join the parallel parts are rounded off or "streamlined."

Taking the formula

$$Q = 0.98 \frac{a_1 a_2}{\sqrt{a_1^2 - a_2^2}} \sqrt{2g \frac{p_1 - p_2}{\gamma}}$$

we first find $\dfrac{a_1 a_2}{\sqrt{a_1^2 - a_2^2}}$ where $a_1 = 4a_2$

$$\frac{a_1 a_2}{\sqrt{a_1^2 - a_2^2}} = \frac{4a_2^2}{\sqrt{16a_2^2 - a_2^2}} = \frac{4a_2}{\sqrt{15}} = \frac{a_1}{\sqrt{15}} = \frac{\frac{\pi}{4} \cdot \frac{1}{9}}{\sqrt{15}}$$

and $\sqrt{2g} = \sqrt{64.4} = 8.025$. Also

$$\frac{p_1 - p_2}{\gamma} = \frac{\gamma_m - \gamma}{\gamma} \times 2.000 = \frac{13.60 - 1}{1} \times 2.000$$

$$= 12.60 \times 2 = 25.20 \text{ ft of water;}$$

$$\therefore \quad Q = 0.98 \times \frac{\pi}{4} \times \frac{1}{9} \times \frac{1}{\sqrt{15}} \times 8.025 \times \sqrt{25.20} = 0.890 \text{ ft}^3 \text{ per sec.}$$

• PROBLEM 13-27

For the venturi meter shown in the figure, determine the volume flow rate, Q, through the venturi in terms of the manometer reading, h.

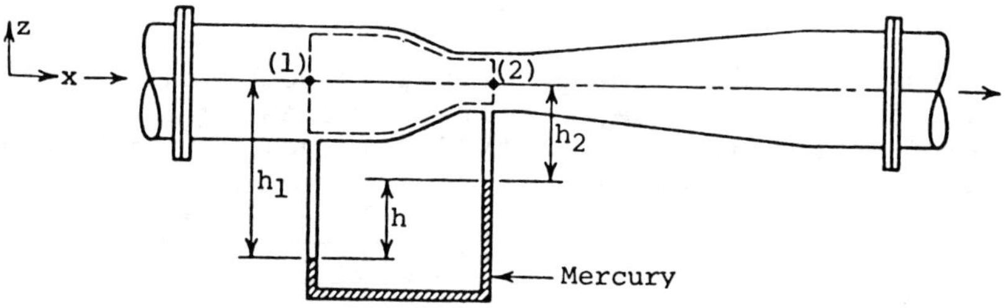

Solution: Applying the Bernoulli equation and the law of conservation of mass and taking $z_1 = z_2$, we obtain

$$\frac{p_1}{\gamma} + \frac{V_1^2}{2g} = \frac{p_2}{\gamma} + \frac{V_2^2}{2g}$$

and

$$A_1 V_1 = A_2 V_2$$

Rearranging the terms, gives

$$V_2^2 - V_1^2 = \frac{2g}{\gamma}(p_1 - p_2)$$

Substituting for V_1, we obtain

$$V_2^2 - \left(\frac{A_2 V_2}{A_1}\right)^2 = V_2^2\left(1 - \frac{A_2^2}{A_1^2}\right) = \frac{2g}{\gamma}(p_1 - p_2)$$

Solving for V_2, we obtain

$$V_2 = \frac{1}{\sqrt{1 - (A_2/A_1)^2}}\sqrt{\frac{2g}{\gamma}(p_1 - p_2)} \qquad (1)$$

Equation (1) gives V_2 in terms of the pressure difference $(p_1 - p_2)$. The ideal volume flow rate is given by

$$Q = A_2 V_2 = \frac{A_2}{\sqrt{1 - (A_2/A_1)^2}}\sqrt{\frac{2g}{\gamma}(p_1 - p_2)}$$

However, viscous effects and other minor losses cause the actual average velocity to be somewhat less than the ideal value, expressed by Eq. 1. An empirical coefficient, c_D, modifies this relationship. Thus,

$$Q = A_2 V_2 = \frac{c_D A_2}{\sqrt{1 - (A_2/A_1)^2}}\sqrt{\frac{2g}{\gamma}(p_1 - p_2)} \qquad (2)$$

For a Herschel-type ventury meter, a typical range for c_D is

$$0.93 < c_D < 0.99$$

Now $(p_1 - p_2)$ can be determined from the manometer readings; that is,

$$p_1 = p_2 + \gamma h_2 + \gamma_{hg} h - \gamma h_1$$

or

$$p_1 - p_2 = \gamma(h_2 - h_1) + \gamma_{hg} h$$

But

$$h_1 - h - h_2 = 0 \quad \text{or} \quad h_2 - h_1 = -h$$

Thus

$$p_1 - p_2 = \gamma h + \gamma_{hg} h = (\gamma_{hg} - \gamma)h \qquad (3)$$

Substituting Eq. 3 into Eq. 2, we obtain

$$Q = \frac{c_D A_2}{\sqrt{1 - (A_2/A_1)^2}}\sqrt{2gh\left(\frac{\gamma_{hg}}{\gamma} - 1\right)}$$

A 4-in. by 1-in. nozzle, shown in the figure, is attached to the end of a 4-in hose line. The velocity of the water leaving the nozzles is 96 fps, the coefficient of velocity, C_v, is 0.96 and the coefficient of contraction, C_c, is 0.80. Determine the necessary pressure at the base of the nozzle.

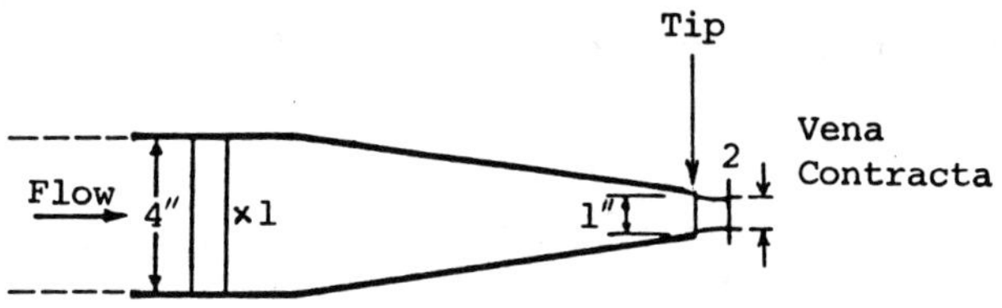

Solution: The figure shows the nozzle attached to the 4-in. hose line. Since the coefficient of velocity, C_v, is $v_{act'l}/v_{ideal}$,

$$v_{ideal} = \frac{96}{0.96} = 100 \text{ fps.}$$

The ratio of the area of the vena contracta to the area of the tip of the nozzle is called the coefficient of contraction.

Incorporated with the continuity equation yields

$$\frac{v_1}{v_2} = \frac{a_2}{a_1} = \frac{d_2^2}{d_1^2} = 0.8\ \frac{(1)^2}{4^2} = \frac{1}{20},$$

$$v_1 = \frac{100}{20} = 5 \text{ fps (ideal).}$$

Substituting these values into the Bernoulli equation, gives

$$\frac{p_1}{0.433} + 0 + \frac{5^2}{64.4} = 0 + 0 + \frac{100^2}{64.4}$$

and

$$\frac{p_1}{0.433} = \frac{10,000 - 25}{64.4} = \frac{9975}{64.4} = 155 \text{ ft.}$$

Therefore,

$$p_1 = 0.433(155) = 67.1 \text{ psi.}$$

• PROBLEM 13-29

A 6-in. by 2-in. nozzle, shown in the figure, is attached to the end of a 6-in. pipe. The pressure at the base of the nozzle (where it joins the pipe) is 60 psi and the coefficient of velocity, C_v, is 0.98. Determine the velocity of the jet.

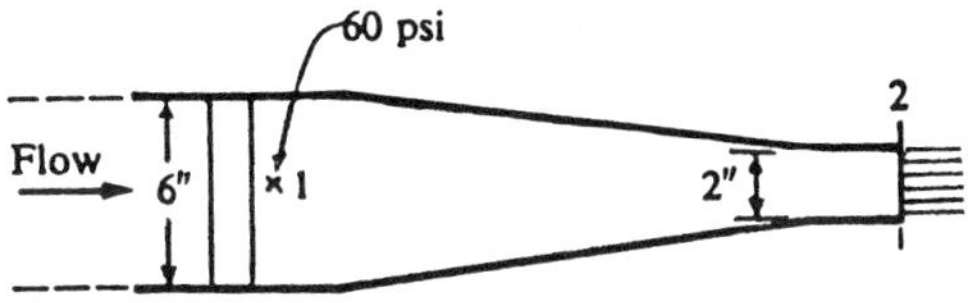

Solution: The figure shows the nozzle attached to the 6-in. pipe. Writing the Bernoulli equation from point 1 to point 2 to determine the ideal velocity gives,

$$\frac{p_1}{\gamma} + z_1 + \frac{V_1^2}{2g} = \frac{p_2}{\gamma} + z_2 + \frac{V_2^2}{2g}$$

where $\frac{62.4}{144} = 0.433$ lb/sq.in-ft $= \gamma$

$$\frac{60}{0.433} + 0 + \frac{v_1^2}{64.4} = 0 + 0 + \frac{v_2^2}{64.4} .$$

Also from the continuity equation $A_1V_1 = A_2V_2$, yields

$$\frac{v_1}{v_2} = \frac{4}{36} = \frac{1}{9} .$$

Therefore,

$$v_2 = 9v_1 \quad \text{and} \quad v_2^2 = 81v_1^2 .$$

Substituting these values into the Bernoulli equation gives

$$\frac{60}{0.433} = \frac{81v_1^2 - v_1^2}{64.4} = \frac{80v_1^2}{64.4} .$$

Therefore,

$$v_1 = \left[\frac{64.4(60)}{80(0.433)}\right]^{1/2} = 10.55 \text{ fps (ideal velocity)}.$$

Since

$$\frac{v_{act'l}}{v_{ideal}} = 0.98, \quad v_{act'l} = 0.98(10.55)$$
$$= 10.34 \text{ fps},$$

and

$$v_2 \text{(the velocity of the jet)} = 9v_1 = 9(10.34)$$
$$= 93.1 \text{ fps}.$$

FLOW NOZZLE

• PROBLEM 13-30

A flow nozzle of 6 in diameter is placed in an 18 in diameter pipe. During calibration the pressure differential was measured to be 10 psi for a flow of 7.2 cfs and 17 psi for a flow of 9.3 cfs. Determine the discharge coefficient of the measuring nozzle (see Figure).

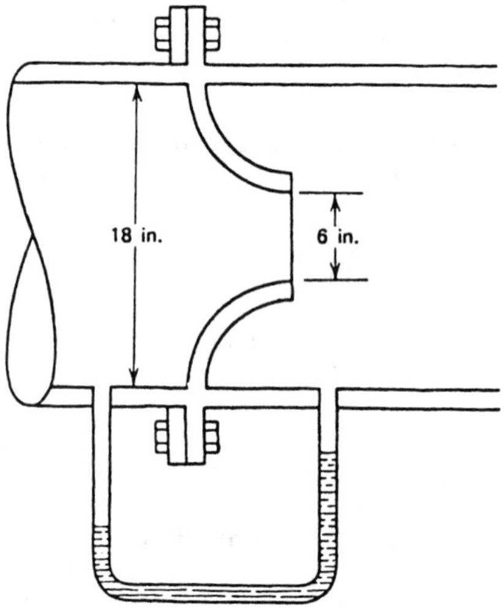

Solution: A good approximate formula for orifices in practical applications may be written in the simple form of

$$Q = ca\sqrt{\frac{2g}{\gamma}\Delta p}$$

$$a = \frac{\pi}{4}\left(\frac{1}{2}\right)^2 = \frac{\pi}{16}\ \text{ft}^2$$

$$c_1 = \frac{7.2}{\frac{\pi}{16}\sqrt{\frac{2\times 32.2}{62.4}\times 10\times 144}}$$

$$= 0.95$$

$$c_2 = \frac{9.3}{\frac{\pi}{16}\sqrt{\frac{2\times 32.2}{62.4}\times 17.0\times 144}}$$

$$= 0.942$$

$$c = \frac{1}{2}(c_1 + c_2) = \frac{1}{2}(0.95 + 0.942) \simeq 0.946$$

CHAPTER 14

FLOW THROUGH HYDRAULIC STRUCTURES

Basic Attacks and Strategies for Solving Problems in this Chapter. See pages 847 to 869 for step-by-step solutions to problems.

A weir is an obstruction in the bottom of a channel over which the flow must deflect. Defining H as the height of the upstream flow above the crest of the weir, the volume flow rate per unit width, q, turns out to be proportional to $H^{3/2}$ for a sharp-crested weir, i.e.,

$$q = {}^2/_3 C_W (2g)^{1/2} H^{3/2}, \tag{1}$$

where C_W is the weir coefficient, which is an empirically determined coefficient that takes into account the losses associated with end effects, friction, etc. Typically,

$$C_W \approx 0.611 + \frac{0.075H}{Y}, \tag{2}$$

where Y is the height of the crest of the weir. Once q is determined, the total discharge (or volume flow rate) Q is then simply q multiplied by the width of the weir.

For a broad-crested weir, i.e., one with a flat top, the surface of the fluid on top of the weir usually sinks to a height h above the crest, where $h \approx 2H/3$. The discharge q per unit width then becomes

$$q = c(2g)^{1/2} h^{3/2}, \tag{3}$$

and c is the discharge coefficient of the weir which must be empirically determined and is usually tabulated as a function of weir geometry (c also varies with h).

Pressure (surge) waves can occur in pipes. A well-known example is the "banging" of water pipes when a valve is suddenly closed. Since the speed of sound c in water is approximately 1,480 m/s, the pressure surge travels down a pipe at this speed. To reduce this effect, the valve must be closed slower than some maximum time (critical time) t_c given by $t_c = 2L/c$ where L is the pipe length.

Other types of problems are included in this chapter, such as the flow through small orifices, where the flow rate Q is given by

$$Q = C_D\, a\, (2gh)^{1/2},$$

where a is the cross-sectional area of the orifice,

h is the difference in the head from one side of the orifice to the other, and

C_D is a discharge coefficient which again must be determined empirically.

Problems 14.6 – 14.8 analyze unsteady flows using Equation (4).

Other problems—such as flows in culverts and over spillways, and seepage flows—involve many empirical relationships. Problems 14.9 – 14.12 treat these special cases.

Step-by-Step Solutions to Problems in this Chapter, "Flow Through Hydraulic Structures"

FLOW OVER WEIRS

• PROBLEM 14-1

The discharge coefficient of a right-angled weir is to be determined over a range of values of upstream head h. To calibrate the weir, the discharge from the weir was collected in a tank and weighed over a clocked time interval. This procedure resulted in the following measurement of volume flow rate versus upstream head.

Q (cfm)	h (inches)
0.30	1
1.73	2
4.73	3
9.70	4
16.0	5
26.2	6
39.2	7
54.2	8
73.1	9
94.8	10

From the preceding data, calculate discharge coefficient versus h.

For a right-angled triangular weir, the ideal discharge rate is

$$Q_i = \frac{8}{15}\sqrt{2g}\ h^{5/2}$$

Solution: The actual discharge rate is

$$Q = CQ_i$$

Therefore the discharge coefficient C is given by

$$C = \frac{Q(\text{measured})}{(8/15)\sqrt{2g}\; h^{5/2}} = \frac{Q(\text{measured})}{4.28h^{5/2}} = \frac{1.945\; Q\; (\text{cfm})}{[h(\text{in.})]^{5/2}}$$

Thus we compute the variation of the discharge coefficient with upstream head.

h (inches)	$h^{5/2}$ (in.)$^{5/2}$	C
1	1.0	0.583
2	5.65	0.597
3	15.6	0.591
4	32.0	0.590
5	53.0	0.588
6	87.0	0.587
7	130	0.586
8	180	0.585
9	243	0.585
10	315	0.586

The discharge coefficient of this triangular weir is therefore 0.585, except for very low head. Hence the relation for weir discharge is

$$Q = 2.50h^{5/2}$$

with Q in cubic feet per second and h in feet.

• PROBLEM 14-2

A rectangular broad crested weir is 30 ft long and is known to have a discharge coefficient of 0.7. Determine the discharge if the upstream water level is 2 ft over the crest.

Solution: The weir formula applies which is given by

$$Q = cb\sqrt{2g}\, h_1^{3/2}$$

in which Q is the total discharge over the rectangular weir of b width under a head of h_1. The parameter c in this formula is the discharge coefficient of the weir.

$$Q = cb\sqrt{2g}h_1^{3/2}$$

$$= 0.7(30)\sqrt{64.4}(2^{3/2})$$

$$= 476 \text{ cfs}$$

• PROBLEM 14-3

Flow is occurring in a rectangular channel at a velocity of 3 fps and depth of 1.0 ft (see Figure). Neglecting the effect of velocity of approach, determine the height of sharp-crested suppressed weir that must be installed to raise the water depth upstream of the weir to 4 ft.

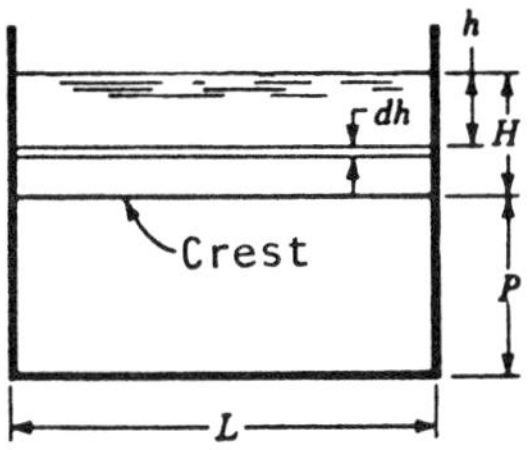

Solution: With reference to the accompanying figure, the flow rate Q for a sharp-crested suppressed weir is given by

$$Q = 3.33\ LH^{3/2}$$

L = length of weir crest = width of channel

$$Q = AV = LyV = L(1)(3) = 3.33LH^{3/2}$$

$$H^{3/2} = \frac{3.0}{3.33} = 0.90 \qquad H = 0.93\text{ ft}$$

$$P = \text{height of weir} = 4.00 - 0.93 = 3.07\text{ ft}$$

• PROBLEM 14-4

A weir 10 m long with a cross-sectional design similar to case number 4 of Table 1 is flowing with an upstream head of 45 cm. Determine the discharge.

Solution: From Table 1 the metric discharge coefficient at 45 cm upstream head is M = 1.77.

Using Figure 1 with b = 10 m, H = 0.45 m, and M = 1.77 the discharge is found to be

$$Q = 4.5\ \frac{m^3}{s}$$

Table 1
Discharge Coefficients for Broad Crested Weirs[a]

	Cross section	Upstream head h [m]							
		0.15	0.30	0.45	0.60	0.75	0.90	1.20	1.50
1	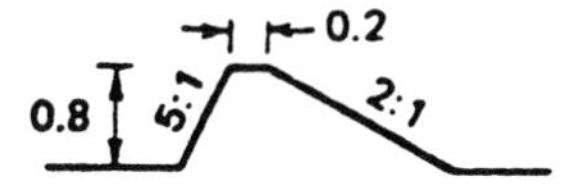	1.61	1.86	1.98					
2	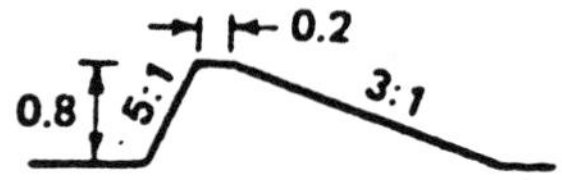	1.60	1.80	1.90					
3	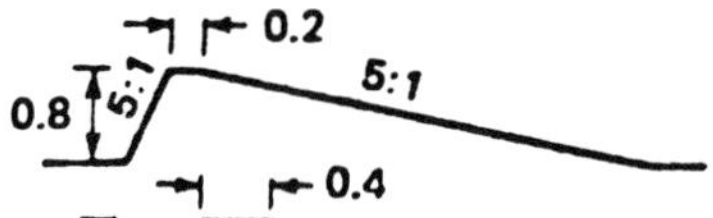	1.58	1.75	1.79					
4	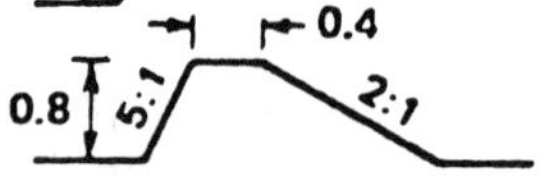	1.53	1.64	1.77					
5	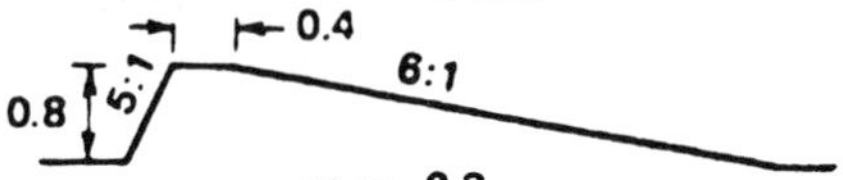	1.54	1.62	1.69					
6	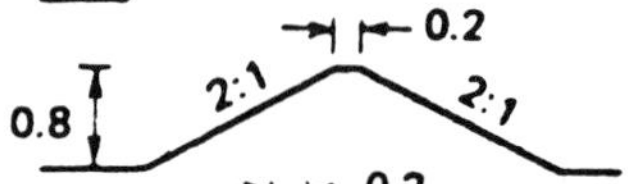	1.72	1.88	1.98					
7	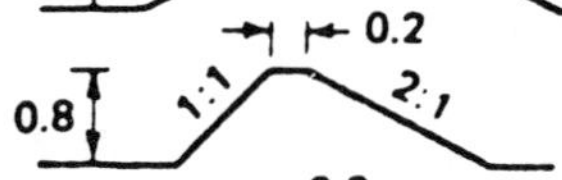	1.65	1.88	2.00					
8	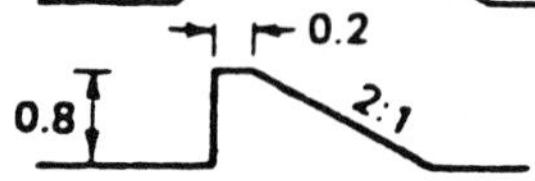	1.53	1.80	1.93					
9	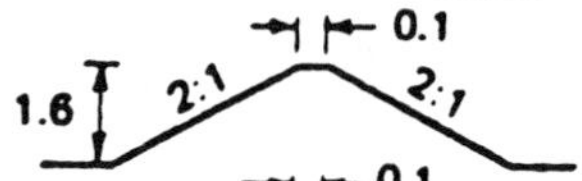				1.96	1.96	1.97	1.99	2.02
10	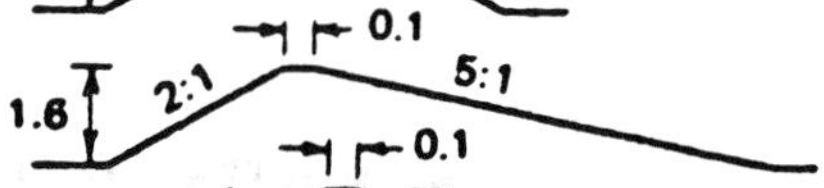				1.94	1.92	1.89	1.92	1.97
11	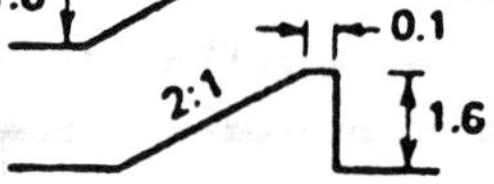		2.12	2.10	2.08	2.08	2.06	2.04	2.00
12	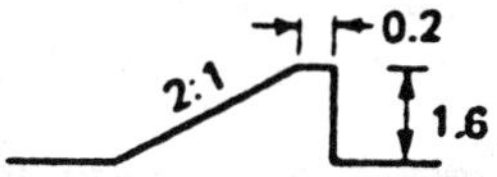		1.88	1.96	2.01	2.04	2.05	2.05	2.05
13	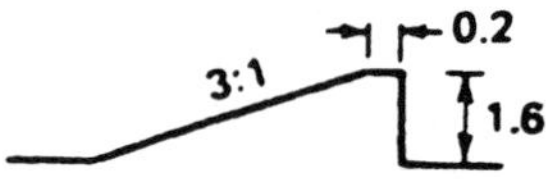				1.96	1.96	1.96	1.96	1.96
14					1.86	1.86	1.86	1.86	1.86

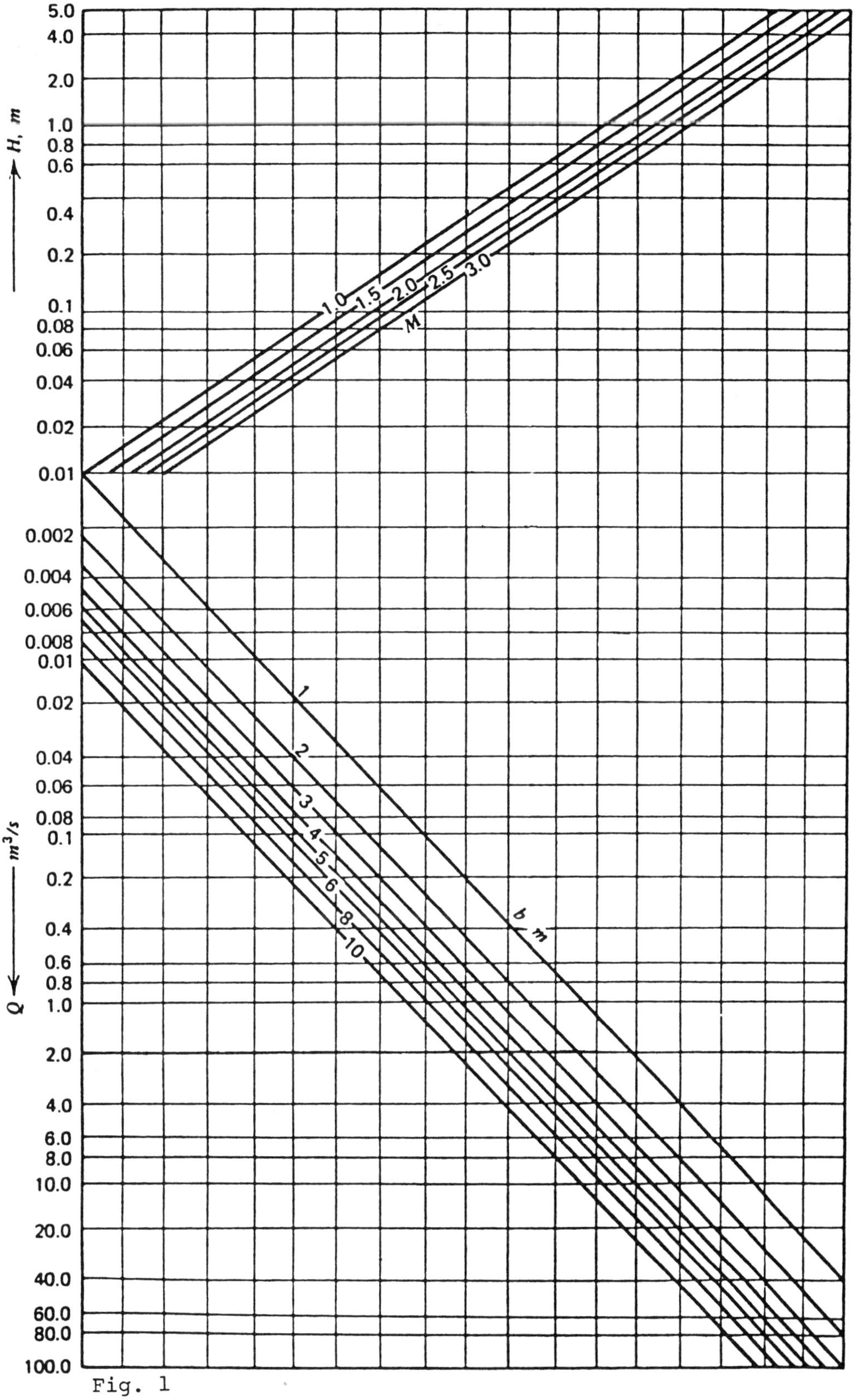

H, m
5.0
4.0
2.0
1.0
0.8
0.6
0.4
0.2
0.1
0.08
0.06
0.04
0.02
0.01
1.0
1.5
2.0
2.5
3.0
M
0.002
0.004
0.006
0.008
0.01
0.02
0.04
0.06
0.08
0.1
0.2
0.4
0.6
0.8
1.0
2.0
4.0
6.0
8.0
10.0
20.0
40.0
60.0
80.0
100.0
Q m^3/s
1
2
3
4
5
6
8
10
b m

Fig. 1

• PROBLEM 14-5

Water from a reservoir flows through a rigid pipe at a velocity of 2.5 m/s. This flow is completely stopped by the closure of a valve situated 1100 m from the reservoir. Determine the maximum rise of pressure in N/m^2 above that corresponding to uniform flow when valve closure occurs in (a) 1 second, (b) 5 seconds.

c = 1430 m/s. In (b), assume that the pressure rises at a uniform rate with time and there is no damping of the pressure wave.

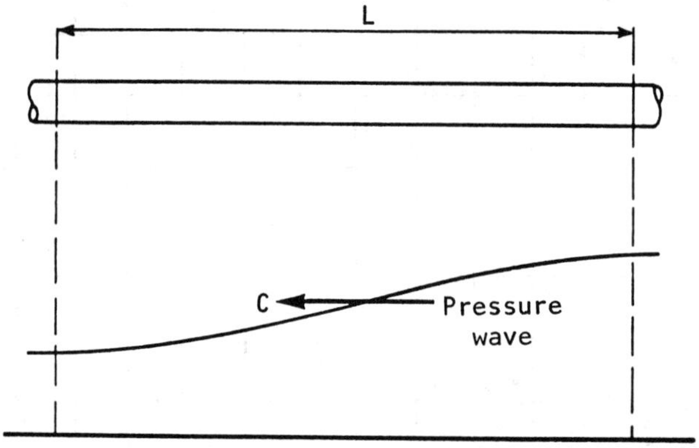

Solution: Consider a wave traversing a length of pipe of length L at a wave speed of magnitude C (see Fig.). The fluid in the pipe length L will be decelerated by an amount Δv in a time Δt

where Δt is the time taken for the wave to traverse the length L that is $\Delta t = L/c$.

$$\Delta h_i = -\frac{L}{g} \times -\frac{\Delta v}{\Delta t} = \frac{L}{g}\frac{\Delta v}{L/c}$$

$$\therefore \quad \Delta h_i = c\Delta v/g$$

This equation will accurately predict the magnitude of the pressure surge if the valve closure occurs in a time less than or equal to the pipe period 2L/c.

The value of $2L/c = 2200/1430$

$$= 1.538 \text{ s}$$

As this is greater than 1 second, (a) represents a sudden closure. The relevant equation is

$$\Delta H = \frac{c\Delta v}{g} = \frac{1430 \times 2.5}{9.81} = 364.4 \text{ m}$$

$$\Delta p = \rho g \Delta H.$$

$$= 9810 \times 364.4$$

$$= 3.575 \times 10^6 \ N/m^2$$

In case (b), the time of closure > 1.538 s so the sudden closure analysis will not be applicable. The question suggests that a rigid pipe--incompressible fluid theory could be used. The assumption that pressure rises at a uniform rate implies this.

Let

$$p = kt$$

but

$$p = \frac{-wL}{g}\frac{dv}{dt} = kt$$

so

$$\frac{dv}{dt} = -\frac{g}{wL}kt$$

$$\therefore \quad v = -\frac{gk}{wL}\frac{t^2}{2} + \text{constant}$$

When

$$t = 5, \ v = 0$$

so

$$\text{constant} = \frac{25gk}{2wL}$$

hence

$$v = \frac{gk}{2wL}(25 - t^2)$$

when

$$t = 0, \ v = 2.5$$

so

$$2.5 = \frac{25gk}{2wL}$$

$$\therefore \quad k = \frac{5wL}{25g} = \frac{5 \times 9810 \times 1100}{25 \times 9.81}$$

$$= 220\ 000$$

so

$$p_{5 \text{ seconds}} = kt = 5 \times 220\ 000$$

$$= 1.1 \ MN/m^2$$

Water is discharged from a large tank to a smaller one through a short pipe, as shown in Fig. 1. The cross-sectional areas of the tanks are 4 ft^2 and 1 ft^2, respectively. What time will be required to bring the water surfaces in the tanks to the same level, if initially the water surface in the larger tank is 9 ft above that in the smaller one? The pipe may be regarded as being equivalent to an orifice of 1 in. diameter and with a coefficient of discharge of 0.5.

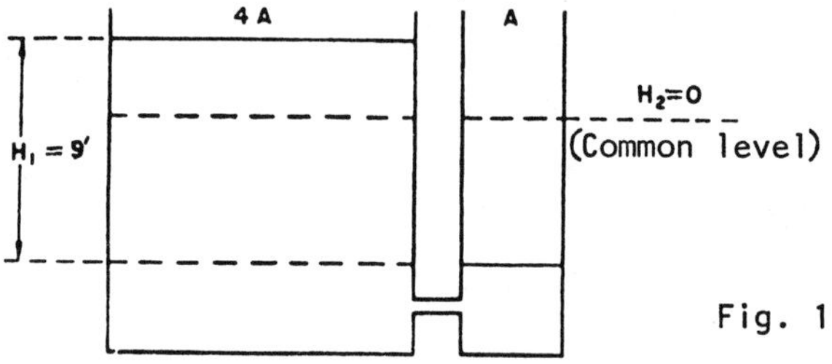

Fig. 1

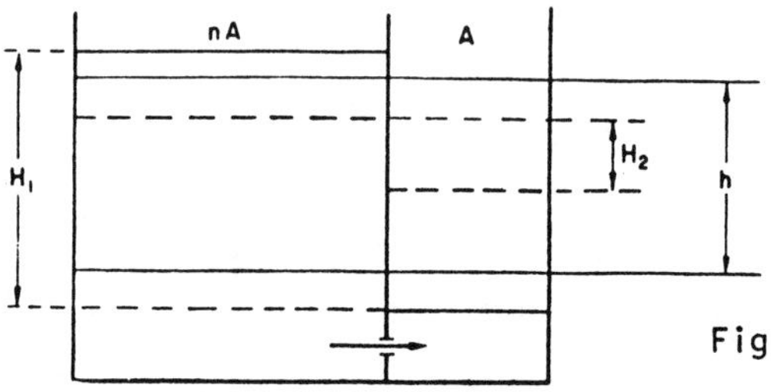

Fig. 2

Solution: Consider, in Fig. 2, a tank of uniform cross-section, partitioned by a vertical wall into two compartments, a smaller one of cross-sectional area A and a larger one of cross-sectional area nA. Assume that the flow takes place between the compartments through a submerged orifice at the base of the partition, and let h be an instantaneous differential head. If Q is the flow at this instant, then in a differential time $d\theta$, the fall of the liquid surface in the larger compartment will be $(Q/nA)d\theta$. At the same time, the liquid surface in the smaller compartment will rise by $(Q/A)d\theta$. The rate of change of the differential head will therefore be

$$\frac{dh}{d\theta} = -\left(\frac{Q}{nA} + \frac{Q}{A}\right) = -\frac{Q(n+1)}{nA}$$

from which

$$Q = -\frac{nA}{1+n}\left(\frac{dh}{d\theta}\right)$$

At the same instant, the rate of flow through the orifice will be

$$Q = C_D a\sqrt{(2gh)}$$

It follows that

$$\frac{dh}{d\theta} = -C_D a\sqrt{(2gh)}\left(\frac{1 + n}{nA}\right)$$

or

$$d\theta = -\frac{nA(dh)}{(1 + n)C_D a\sqrt{(2gh)}}$$

Integrating between the limits $h = H_2$ and $h = H_1$

$$\theta = -\frac{nA}{(1 + n)C_D a\sqrt{(2g)}}\int_{H_1}^{H_2} h^{-1/2}\,(dh)$$

$$\theta = \frac{nA}{(1 + n)C_D a\sqrt{(2g)}}\int_{H_2}^{H_1} h^{-1/2}\,(dh) \tag{1}$$

Referring to Fig. 1

$H_1 = 9$ ft

$H_2 = 0$

$n = 4$

The value of the integral of eq. (1) is

$$\int_{H_2=0}^{H_1=9} h^{-1/2}\,(dh) = [2h^{1/2}]_0^9 = 6$$

Substituting this value in eq. (1)

$$\theta = \frac{(4)(1)(6)}{(1 + 4)(0.5)(1/12)^2\pi/4\sqrt{64.4}}$$

$$\theta = 220 \text{ sec}$$

• PROBLEM 14-7

A water tank, 12 ft high, forms a frustrum, with diameters of 6 ft and 4 ft at the top and bottom, respectively. If water is discharged through an orifice of 2 in. diameter at the base of the tank, what time will be required to lower the water surface from 9 ft to 4 ft, measured above the orifice? The coefficient of discharge is 0.6.

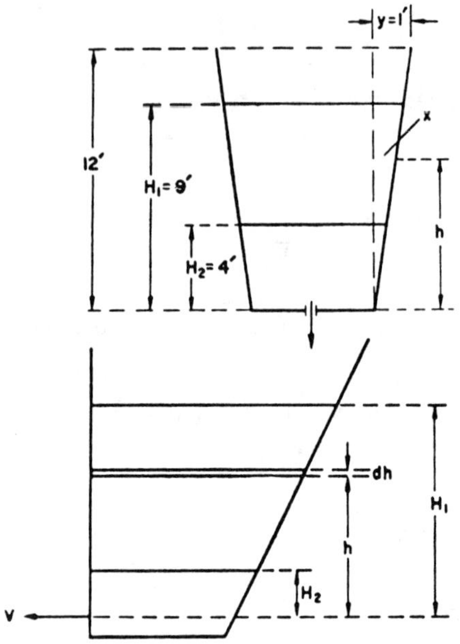

Fig. 1

Fig. 2

Solution: Consider, in Fig. 2, the discharge from a tank of linearly varying section, through an orifice located at the base of the tank. Let H_1 be the initial head, under which the flow takes place, and H_2 be the head after a time θ.

Now, assume an instant when the depth measured above the center of the orifice is h, and let $d\theta$ be the time required to lower this depth by dh, then

$$Q(d\theta) = A(dh)$$

where Q is the instantaneous rate of flow through the orifice, and A is the cross-sectional area of the tank at the depth considered.

For a small orifice, the flow is given by eq.

$$Q = C_D a\sqrt{(2gh)}$$

where a is the cross-sectional area of the orifice, hence

$$\frac{A(dh)}{d\theta} = C_D a\sqrt{(2gh)}$$

from which

$$d\theta = \frac{Ah^{-1/2}\,(dh)}{C_D a\sqrt{(2g)}} \tag{1}$$

Referring to Fig. 1, consider a section of the frustrum at a height h above the orifice, then

$$\frac{x}{y} = \frac{h}{12}$$

then, since y = 1 ft

$$x = \frac{h}{12}$$

Let D be the diameter at this section, then

$$D = 4 + 2x = 4 + \frac{h}{6}$$

The cross-sectional area of the section is therefore

$$A = D^2\frac{\pi}{4} = \left(4 + \frac{h}{6}\right)^2\frac{\pi}{4}$$

Substituting for A in eq. (1) and integrating

$$\theta = \frac{\pi/4}{C_D a\sqrt{(2g)}}\int_{h=4}^{h=9}\left(4 + \frac{h}{6}\right)^2 h^{-1/2}(dh)$$

Evaluating the integral only

$$\int_{h=4}^{h=9}\left(16 + \frac{4}{3}h + \frac{h^2}{36}\right)h^{-1/2}dh = \left[\frac{16h^{1/2}}{0.5} + \frac{8}{9}h^{3/2} + \frac{h^{5/2}}{90}\right]_4^9$$

$$= 51.2$$

$$\theta = \frac{(\pi/4)(51.2)}{(0.6)(2/12)^2\ \pi/4\sqrt{64.4}}$$

$$\theta = 383\ \text{sec.}$$

• PROBLEM 14-8

A cylindrical drum, of 4 ft radius and 12 ft long, is half-full of water. The cylinder lies with its longitudinal axis horizontal. What is the diameter of an orifice, located at the bottom of the cylindrical part of the drum, if emptying the drum takes 50 min 34 sec? The average value of the coefficient of discharge may be assumed to be 0.6.

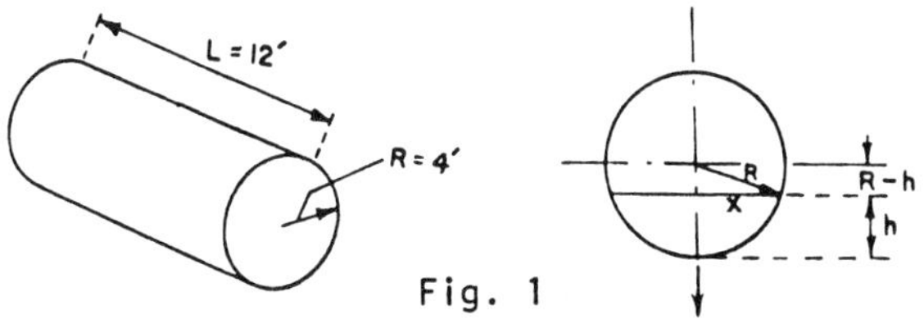

Fig. 1

Solution: Consider, in Fig. 2, the discharge from a tank of linearly varying section, through an orifice located at the base of the tank. Let H_1 be the initial head, under which the flow takes place, and H_2 be the head after a time θ.

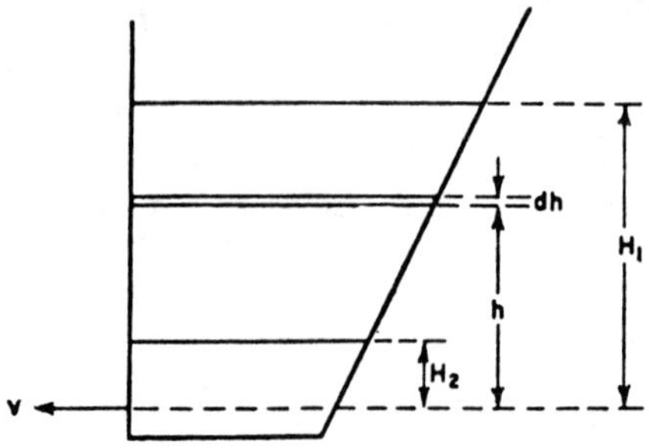

Fig. 2

Now, assume an instant when the depth measured above the center of the orifice is h, and let $d\theta$ be the time required to lower this depth by dh, then

$$Q(d\theta) = A(dh)$$

where Q is the instantaneous rate of flow through the orifice, and A is the cross-sectional area of the tank at the depth considered.

For a small orifice, the flow is given by

$$Q = C_D a\sqrt{(2gh)}$$

where a is the cross-sectional area of the orifice, hence

$$\frac{A(dh)}{d\theta} = C_D a\sqrt{(2gh)}$$

from which

$$d\theta = \frac{Ah^{-1/2}(dh)}{C_D a\sqrt{(2g)}} \qquad (1)$$

Consider, in Fig. 1, an instant when the water surface in the drum is h ft above the orifice, then the free surface area at this instant is

$$A = 2xL$$

But

$$x^2 = R^2 - (R - h)^2 = 2Rh - h^2$$

then

$$A = 2L(2Rh - h^2)^{1/2} = 24(8h - h^2)^{1/2}$$

Substituting the data in eq. (1), and taking $a = d^2\pi/4$, when d is the orifice diameter

$$d\theta = \frac{24(8h - h^2)^{1/2} h^{-1/2} (dh)}{(0.6)(\pi/4)\sqrt{64.4}(d^2)}$$

$$\frac{24}{(0.6)(\pi/4)\sqrt{64.4}} = 6.348$$

$$(8h - h^2)^{1/2} h^{-1/2} = (8 - h)^{1/2}$$

$$d\theta = \frac{6.348}{d^2}(8 - h)^{1/2}\ dh$$

$$\theta = \frac{6.348}{d^2}\int_0^4 (8 - h)^{1/2}\ dh = \frac{6.348}{d^2}\left[-\frac{2}{3}(8 - h)^{3/2}\right]_0^4$$

$$\theta = \frac{6.348}{d^2}\left[\frac{16}{3}(\sqrt{(8)} - 1)\right]$$

$$\theta = 3034 \text{ sec (given)}$$

$$3034 = \frac{6.348}{d^2}\left(\frac{16}{3}\right)(1.828)$$

from which

$$d = 0.143 \text{ ft } (1.72 \text{ in.})$$

CULVERT AND SPILLWAY

• PROBLEM 14-9

A 60 in. diameter concrete culvert is 100 ft long and its entrance loss coefficient is $k_e = 0.2$ (see Figure 1).

Determine the head loss in the culvert if the discharge is 140 cfs and the pipe is flowing such that both ends are submerged. Assume concrete roughness n = .012 and entrance loss coefficient $k_e = 0.2$.

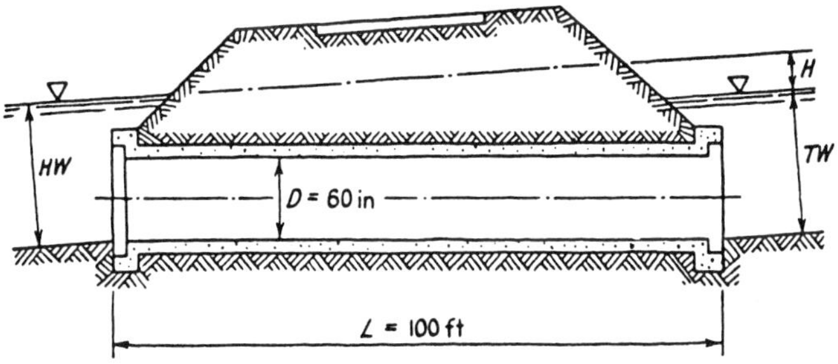

Fig. 1

Solution: The culvert is flowing.

$$n = 0.012$$

$$k_e = 0.2$$

$$L = 100 \text{ ft.}$$

$$Q = 140 \text{ cfs}$$

$$D = \frac{60}{12} = 5 \text{ ft}$$

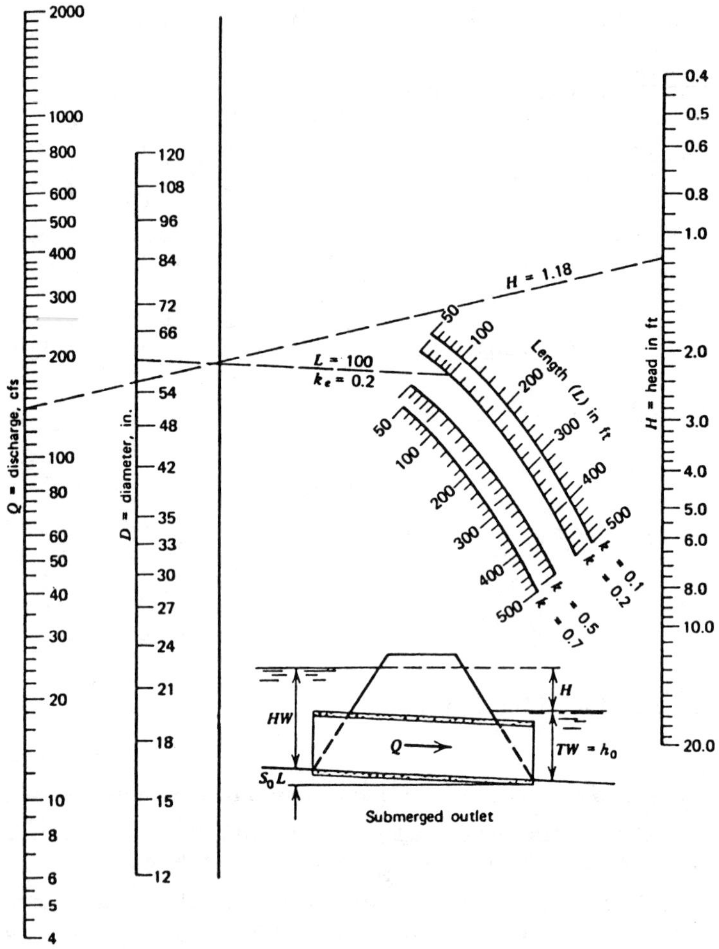

Fig. 2

Using the equation for culvert head loss

$$H = \left[\frac{2.5204(1 + k_e)}{D^4} + \frac{466.18n^2L}{D^{16/3}}\right]\left(\frac{Q}{10}\right)^2$$

H = head in ft.

k_e = entrance loss coefficient

D = diameter of pipe in ft.

n = Kutter's roughness coefficient

L = length of culvert in ft.

Q = design discharge in cfs.

$$H = \left[\frac{2.5204(1 + k_e)}{D^4} + \frac{466.18n^2L}{D^{16/3}}\right](Q/10)^2$$

$$= \left[\frac{2.5204(1.2)}{5^4} + \frac{466.18(0.012)^2 \times 100}{5^{16/3}}\right]\left(\frac{140}{10}\right)^2$$

$= [0.0048 + 0.0013]196$

$= 1.187$ ft

or from nomograph in Figure 2, we find H = 1.18 ft.

• PROBLEM 14-10

Refer to the Figure. At section 2 the water surface is at elevation 30.5, and the 60° spillway surface at elevation 30.0. The velocity in the water surface V_{s_2} at section 2 is 6.1 m/s. Calculate the pressure and velocity on the spillway face at section 2. If the bottom of the approach channel is at elevation 29.0, calculate the depth and velocity in the approach channel.

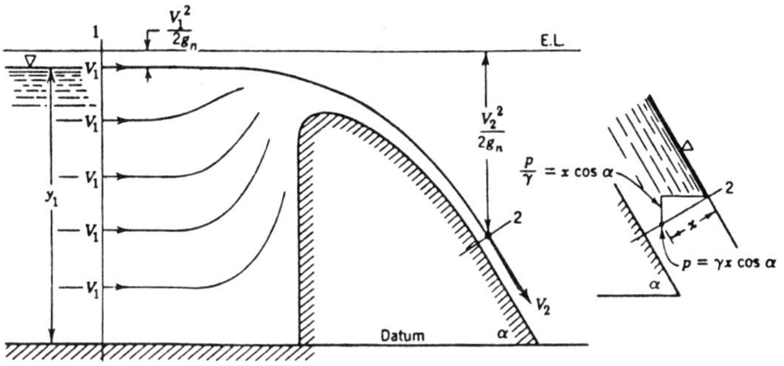

Solution: Thickness of sheet of water at section 2 = (30.5 - 30.0)/cos 60° = 1 m. Pressure on spillway face at section 2 = 1 x 9.8 x 10^3 x cos 60° = 4.9 kPa. Elevation of energy line = 30.5 + $(6.1)^2/2g_n$ = 32.4m

$$32.4 = \frac{4.9}{9.8} + \frac{V_{F2}^{\;2}}{2g_n} + 30.0; \qquad V_{F2} = 6.1 \text{ m/s}$$

which is to be expected from the one-dimensional assumption. Evidently all velocities through section 2 are 6.1 m/s, so

$q = 1 \times 1 \times 6.1 = 6.1$ m³/s per meter of spillway length

At section 1, $y_1 + 29.0 = z_1 + p_1/\gamma$; and, applying the Bernoulli equation,

$$y_1 + \frac{V_1^2}{2g_n} = y_1 + \left(\frac{1}{2g_n}\right)\left(\frac{6.1}{y_1}\right)^2 = 3.4 \text{ m}$$

Solving this (cubic) equation by trial, the roots are y_1 = 3.22 m, 0.85 m, and -0.69 m. The second and third roots are invalid here, so the depth in the approach channel will be 3.22 m. The velocity V_1 may be computed from

$$\frac{V_1^2}{2g_n} = 3.4 - 3.22$$

or from

$$V_1 = \frac{6.1}{3.22}$$

both of which give $V_1 = 1.9$ m/s.

• PROBLEM 14-11

A 6 ft wide vertical gate on the top of a spillway withholds a 4 ft deep water.

(A) Determine the discharge under the gate if it is raised by 1 ft.

(B) Determine the required opening of the gate for a discharge of 20 cfs.

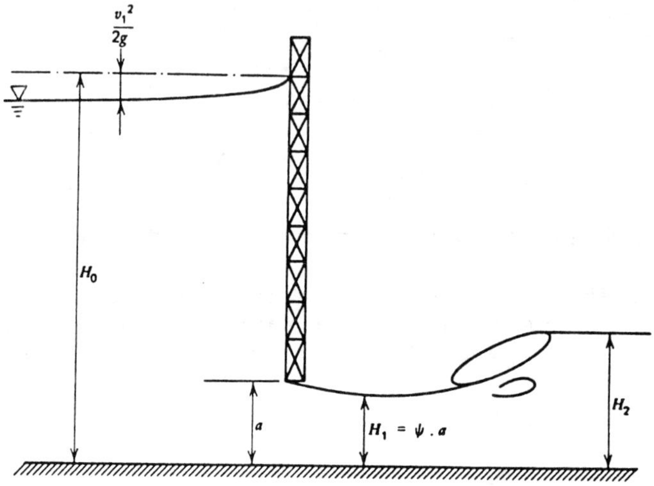

Fig. 1

Solution: (A) Flow under a vertical gate can be defined as a square orifice problem as long as the opening height, a, under the gate is small when compared to the upstream energy level H_0, and the downstream water level H_2 does not influence the flow. Including the discharge coefficient c the discharge formula for flow through openings under pressure is in the form

$$Q = cA\sqrt{2gh} \tag{1}$$

In the case when the jet issuing through the opening exits into the downstream water, the value of h in the above equation is the difference between the two water levels. By Equation 1 we may write

$$Q = bac\sqrt{2g(H_0 - H_1)} \tag{2}$$

in which b is the width of the gate and the other terms are as shown in Figure 1. Direct use of this equation is difficult in practice because of the uncertainty in the determination of H_1, the depth of water at the vena contracta. Since H_1 depends on the opening height a, we may write

$$H_1 = \psi a$$

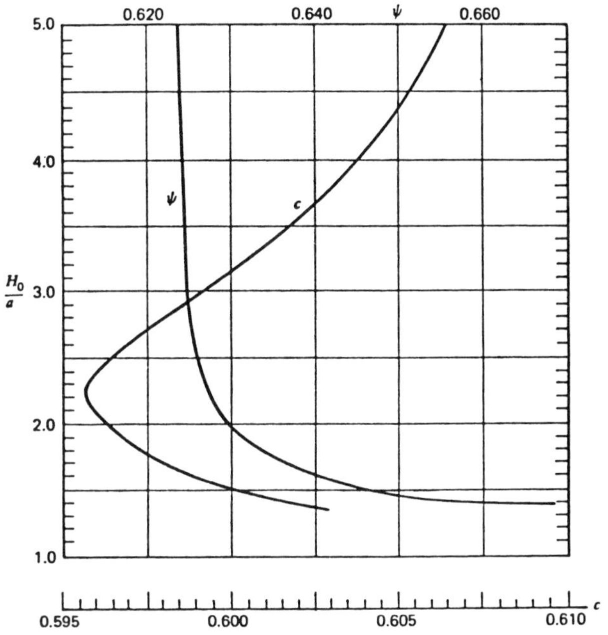

Fig. 2

Discharge coefficient for flow under gates.

Experimental values for ψ were found to depend on H_0/a and are shown in Figure 2.

To eliminate the difficulty in the use of Equation 2, Francke rewrote the discharge formula in the form of

$$Q = bac\sqrt{2g}\ \frac{H_0}{\sqrt{H_0 + \psi \cdot a}} \tag{3}$$

Experimentally determined values for the discharge coefficient c and for the factor ψ are shown in Figure 2. For H_0/a values exceeding the range of the graph, the c value approaches 0.07, and ψ approaches 0.624.

Using Equation 3 with

$$b = 6.0 \text{ ft}$$

$$a = 1.0 \text{ ft}$$

$$H_0 = 4.0 \text{ ft}$$

we have

$$Q = 6(1.0)c\sqrt{64.4}\left(\frac{4}{\sqrt{4 + \psi}}\right)$$

The coefficients c and ψ may be obtained from Figure 2 with

$$\frac{H_0}{a} = \frac{4}{1} = 4$$

obtaining

$$\psi = 0.624$$

$$c = 0.5985$$

Hence

$$Q = 6(0.5985)8.025 \frac{4}{\sqrt{4.624}} = 53.60 \text{ cfs}$$

(B) Entering Equation 3 with our known values, we have

$$20 = 6ac\sqrt{64.4} \frac{4}{\sqrt{4 + \psi a}}$$

Simplifying we have

$$\frac{20}{24\sqrt{64.4}} = 0.103 = \frac{ac}{\sqrt{4 + \psi a}}$$

By observing Figure 2 we note that a trial and error solution is possible by assuming the value of a.

Assuming a = 0.5, $H_0/a = 8$, and $\psi = 0.6245$, $c \simeq 0.665$ approximately. Hence

$$\frac{0.5(0.665)}{\sqrt{4 + 0.5(0.6245)}} = 0.16 > 0.103$$

Therefore a must be further reduced. For our second trial let's take a = 0.4 for this:

$$\frac{H_0}{a} = \frac{4}{0.4} = 10.0$$

and the values of the coefficients may be assumed to be $\psi = 0.6245$ as before and c = 0.7. Therefore

$$\frac{0.4(0.7)}{\sqrt{4 + 0.4(0.6245)}} = 0.136 > 0.103$$

which is still too large. Next, assume a = 0.3. $H_0/a =$ 13.3 a very large number for which the coefficients are as before. Hence

$$\frac{0.3(0.7)}{\sqrt{4 + 0.3(0.6245)}} = 0.1026$$

which nearly equals 0.103. Hence our assumption for a = 0.3 ft was correct. At this opening the gate delivers 20 cfs discharge.

SEEPAGE

• PROBLEM 14-12

A levee built of clay was placed on the top of a 6 in. sand layer. The bottom width of the levee is 60 ft. During flood the water elevation difference between the inside and the outside of the levee is 8 ft. If we assume that the permeability of the sand layer is 1.0 cm/s, calculate the expected seepage discharge for a 200 ft long portion of the levee.

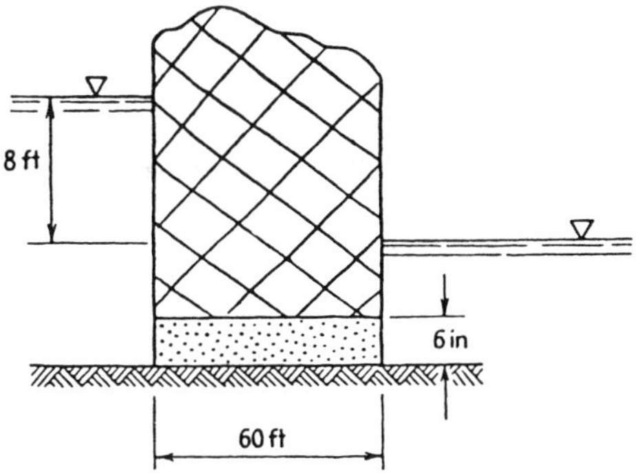

Fig. 1

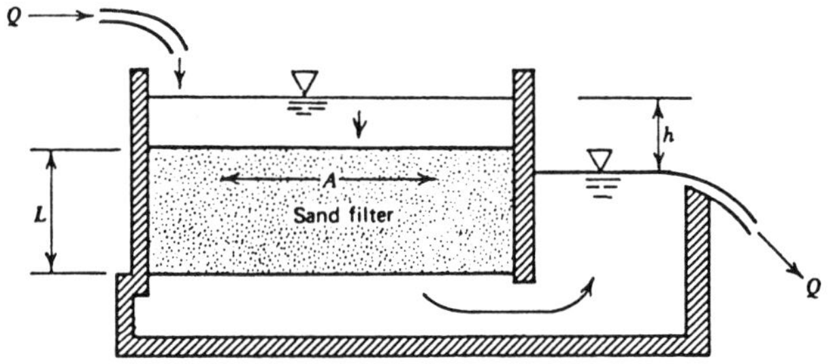

Fig. 2

Darcy's sand filter experiment.

Solution: Well over 100 years ago, Henry Darcy, a French water supply expert, explained the hydraulic behavior of sand filters used in the treatment of water. He postulated that the discharge Q of a sand filter of A surface and L thickness is proportional to the energy loss h that takes place across the filter. Figure 2 shows the arrangement considered by Darcy. This resulted in the formula

$$\frac{Q}{A} = k\,\frac{h}{L}$$

in which k is called the coefficient of permeability with the dimension of velocity.

$$Q = Ak\,\frac{h}{L}$$

$$A = \frac{1}{2} \times 200 = 100 \text{ ft}^2$$

$$k = 1.0 \frac{cm}{s} = \frac{1}{2.54} \times \frac{1}{12} = 0.0328 \frac{ft^2}{sec}$$

$$h = 8 \text{ ft}$$

$$L = 60 \text{ ft}$$

thus

$$Q = 100 \times 0.0328 \times \frac{8}{60} = 0.437 \text{ cfs}$$

PRESSURE (SURGE) WAVES

• PROBLEM 14-13

(a) If the initial water velocity in a rigid pipe is 4 ft/sec (1.22 m/s), and if the initial pressure is 40 psi (p_g = 276 kPa), what maximum pressure will result with rapid closure of a valve in the pipe? Assume that T = 60°F (16°C).

(b) If the pipe of part (a) leads from the reservior and is 4,000 ft long, what is the maximum time of closure for the generation of the maximum pressure of 299 psi, as given in part (a)?

Solution: (a)

$$\Delta p = \rho V c$$

Here

$$\rho = 1.94 \text{ slugs/ft}^3 (1000.0 \text{ kg/m}^3)$$

$$V = 4 \text{ ft/sec } (1.22 \text{ m/s})$$

and

$$c = \sqrt{\frac{E_V}{\rho}} = \sqrt{\frac{320{,}000 \times 144}{1.94}} = 4{,}800 \text{ ft/sec}$$

SI units $E_V = 2.20 \text{ GN/m}^2$

so

$$c = \sqrt{2.20 \times (10^9 \text{ N/m}^2)/(10^3 \text{ kg/m}^3)} = 1.48 \times 10^3 \text{ m/s}$$

Therefore,

$$\Delta p = 1.94 \times 4 \times 4{,}800 = 37{,}200 \text{ psf}$$

$$= 259 \text{ psi}$$

SI units

$$\Delta p = 1000 kg/m^3 \times 1.22\ m/s \times 1.48 \times 10^3\ m/s$$

$$= 1.81\ MN/m^3 = 1.81\ MPa$$

Then the maximum pressure will be the initial pressure plus the pressure change, which is

$$p_{max} = 40 + 259 = 299\ psi$$

SI units

$$p_{max} = 0.276\ MPa + 1.81\ MPa = 2.08\ MPa$$

(b) $t_c = \frac{2L}{c}$

where

t_c = maximum time of closure (critical time) to produce maximum pressure

L = length of pipe = 4,000 ft

c = speed of pressure wave = 4,800 ft/sec

Then

$$t_c = 2 \times \frac{4,000}{4,800} = 1.67\ sec$$

As indicated by part (a), water-hammer pressures can be quite large; therefore, engineers must design piping systems to keep the pressure within acceptable design limits. This is done by installing an accumulator near the valve and/or operating the valve so that rapid closure is prevented. Accumulators may be in the form of air chambers for relatively small systems, or surge tanks (large open tank connected by a branch pipe to the main pipe) for cases such as large hydropower systems.

• PROBLEM 14-14

A rectangular channel 6 m wide with water flow 1.5 m deep, discharging water at 30 m^3s^{-1} suddenly has the discharge reduced by 40 percent due to a gate closure. Calculate the speed and height of the surge wave, assuming the channel is sufficiently deep to accommodate the surge.

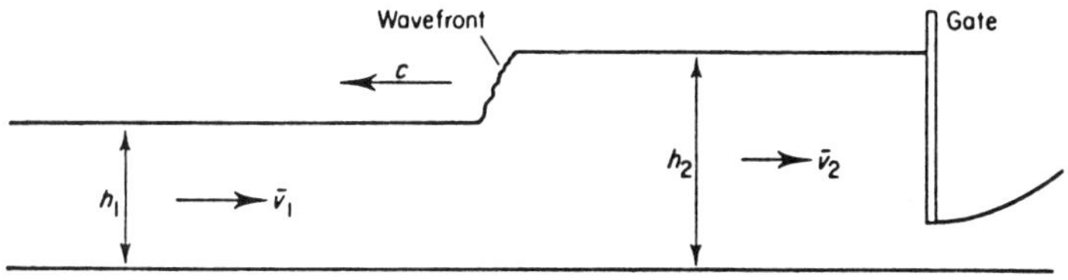

Propagation of a positive surge wave in a rectangular channel

Fig. 1

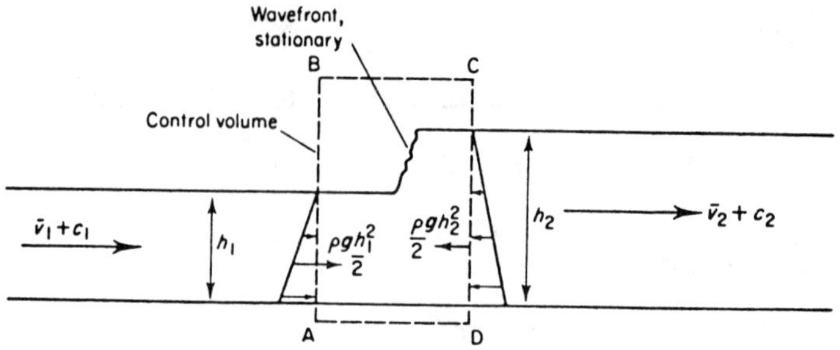

Fig. 2

Reduction of the surge condition to steady state by imposition of the wave speed *c*

Solution: Referring to Figure 1,

$\overline{v}_1$ = mean undisturbed flow velocity

$= 30\ m^3s^{-1}/9m^2 = 3.33\ m\ s^{-1}$

$\overline{v}_2$ = mean flow velocity downstream of the moving surge water

$= 0.6 \times 30\ m^3\ s^{-1}/6 \times h_2$

$= 3/h_2\ m\ s^{-1}$

where h_2 is depth behind wave.

h_1 = undisturbed flow depth = 1.5 m.

By consideration of the control volume ABCD in Fig. 2, the equation of continuity may be written as

$$(\overline{v}_1 + c)h_1 = (\overline{v}_2 + c)h_2 \tag{1}$$

and the equation of motion, force = rate of change of momentum, may be expressed as

$$\rho g h_1^2 - \rho g h_2^2 = \rho h_1(\overline{v}_1 + c)(\overline{v}_2 + c - \overline{v}_1 - c) \tag{2}$$

From equation (1),

$$(\overline{v}_1 + c)h_1 = (\overline{v}_2 + c)h_2,$$

where c is wave speed. Therefore

$$(3.33 + c)1.5 = (3 + ch_2),$$

$$5 = 3 + c(h_2 - 1.5),$$

$$c = 2/(h_2 - 1.5), \tag{3}$$

and from equation (2),

$$h_1^2 - h_2^2 = (h_1/g)(\overline{v}_1 + c)(\overline{v}_2 + c - \overline{v}_1 - c_1)$$

$$2.25 - h_2^2 = (1.5/9.81)(3.33 + c)(3/h_2 - 3.33). \tag{4}$$

A graphical solution obtained by plotting $y_1 = 2.25 - h_2^2$ and $y_2 = 0.154\ (3.33 + c)(3/h_2 - 3.33)$ yields a value of $h_2 = 2.05\ m^2$. Therefore, it follows that

Depth of surge wave = 2.05 m.

Also

$$\overline{v}_2 = 3/h_2 = 3/2.05 = 1.46\ m\ s^{-1}.$$

The speed of propagation of the surge wave is

$$c = 2/(h_2 - 1.5) = 2/0.55 = 3.6\ m\ s^{-1}.$$

CHAPTER 15

PROPULSION AND TURBOMACHINES

Basic Attacks and Strategies for Solving Problems in this Chapter. See pages 870 to 920 for step-by-step solutions to problems.

A *pump* is a turbomachine which adds energy to a fluid, while a *turbine* extracts energy from the fluid. Centrifugal pumps or blowers, for example, are used to push water or air through piping systems or ducts. Huge turbines installed in a hydroelectric dam convert the large head associated with water at high elevations into rotating kinetic energy, which is used to drive an electric generator. Some machinery, like aircraft turbojet engines, employ both pumps (i.e., compressors) and turbines to perform a task.

The fundamental difference between a rocket and a jet engine is that the jet makes use of the surrounding air as its oxidizer, so that a jet aircraft has to carry only the fuel, while the rocket engine must be supplied with both an oxidizer and fuel carried on board the rocket. In either case, thrust is determined by application of the integral (control volume) conservation of momentum law discussed in Chapter 6. For a rocket, where there is no inlet, the thrust is

$$T = (p_e - p_{atm}) A_e + \dot{m}_{total} V_e, \quad (1)$$

where p_e, A_e, and V_e refer to the exit pressure, exit area, and exit velocity respectively.

$\dot{m}_{total}$ is the combined flow rate (mass per second) of the oxidizer and fuel.

In many cases, the pressure term in Equation (1) is negligible, and the thrust is simply

$$T \approx \dot{m}_{total} V_e.$$

If V_e is unknown, the compressible flow concepts discussed in Chapter 12 can be used to find V_e (see Problem 15.4). The power P developed by a rocket is simply the thrust times the velocity of the rocket.

To evaluate the dynamics of a rocket lifting off from Earth, Equation (1) for the thrust must be combined with other forces acting on the rocket drawn as a free-body diagram, specifically aerodynamic drag and the weight of the rocket (which decreases steadily as fuel is burned and exhausted). Using Newton's second law, the sum of all these forces is equated to the mass of the rocket times its acceleration to form a differential equation for the rocket's motion. See Problem 15.10 for an example calculation.

For jet engines, the pressure and mass flux at the *inlet* must be considered as well. Application of the integral conservation of momentum equation yields

$$T = p_2 A_2 - p_1 A_1 + (\dot{m}_a + \dot{m}_f)\, V_2 - \dot{m}_a V_1, \tag{2}$$

where subscripts 1 and 2 refer to the inlet and exit respectively, and $\dot{m}_a$ and $\dot{m}_f$ are the mass flow rates of the air and the fuel respectively. (Note that if the English units of lbm/s are used for $\dot{m}$, the last two terms in Equation (2) must be divided by the gravitational constant g_c.) Again, since the inlet and exit pressures are not very different than atmospheric pressures, the pressure terms are typically negligible, as in Problem 15.11.

Many turbomachinery problems involve the selection of an appropriate pump to supply a specified volume flow rate Q through a piping system with some total head loss H. H is calculated using the methods in Chapter 7 for pipe flows, as illustrated in Problem 15.17. Then, the *specific speed* n_s is computed from

$$n_s = \frac{nQ^{1/2}}{H^{3/4}}, \tag{3}$$

where n is the pump shaft rotation speed in rpm, Q must be in gallons per minute, and H in feet. Specific speed is thus a single parameter that represents the pump speed, flow rate, and head; n_s is useful for selecting a pump with the optimum efficiency for a given job, as illustrated in Problem 15.17.

Often, pump designers or engineers who select pumps need to analyze the performance of a pump at off-design conditions or have to choose a pump size or rotational speed for a new application. For pumps of the same family (i.e., geometrically similar but of various sizes), and for the same pump operated at a faster or slower shaft speed (rpm), certain similarity rules can be applied, namely,

$$\begin{aligned} Q &\propto n D^3 \\ H &\propto n^2 D^2 \\ P &\propto \rho n^3 D^5, \end{aligned} \tag{4}$$

where n is the rpm of the shaft,

D is some characteristic pump dimension (e.g., blade diameter), and

P is the brake horsepower.

In problems where a pump needs to be scaled up or down either in size or speed to provide a different volume flow rate, these scaling laws are applied to solve for the new parameters. Ratios such as

$$\frac{Q_2}{Q_1} = \frac{n_2 D_2^3}{n_1 D_1^3} \quad \text{and} \quad \frac{P_2}{P_1} = \frac{\rho_2 n_2^3 D_2^5}{\rho_1 n_1^3 D_1^5} \tag{5}$$

are used to solve for the unknowns. (See Problem 15.13 for an example.)

Other turbomachining example problems are provided in detail in this chapter, such as propellers (15.5), adiabatic centrifugal turbocompressors (15.16), and impulse turbines and Pelton wheels (15.18, 24, and 25). For the latter, the integral control volume equation for the conservation of momentum is applied to evaluate the force exerted on a blade due to the impingement of a water jet.

Step-by-Step Solutions to Problems in this Chapter, "Propulsion and Turbomachines"

PROPULSION

• PROBLEM 15-1

A rocket engine shown in Fig. 1 has a fuel and oxidizer flow rate of Q = 3.545 kg/s and if the flow leaves the nozzle at V_R = 1900 m/s through an area A_R = 0.0121m^2 with a pressure p_e = 110kPa, what is the thrust, R_x, developed by the rocket motor? The thrust of the rocket engine is attributed to the force developed by the fluid in the rocket thrust chamber. Neglect momentum of entering fluids and the body force due to gravity.

Assume that the velocity of the gas at the nozzle exit remains constant with time.

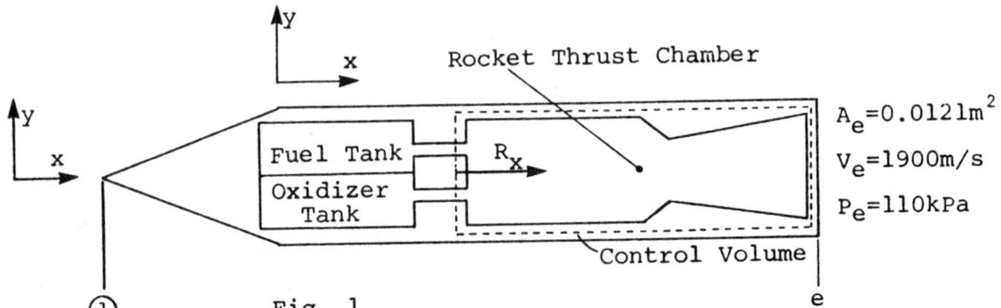

Fig. 1

Solution: Applying the x component of the momentum equation to the control volume shown in Fig. 1, we have,

$$F_{sx} + \cancelto{0}{F_{Bx}} - \int_{CV} \cancelto{0}{a_{rfx}} \rho dV = \frac{\partial}{\partial t}\int_{CV} \cancelto{0}{U_{xy}} \rho dV + \int_{CS} U_{xyz} \rho V_e dA$$

$$\Rightarrow F_{sx} = \int_{CS} U_{xyz} \rho V_e \, dA$$

$$\Rightarrow p_{atm}\cancelto{0}{A} - (p_e - p_{atm})A_1 + R_x = \rho_c V_c^2 A_e$$

$$\Rightarrow R_x = (p_e - p_{atm})A_e + \rho_e V_e^2 A_e$$

$$\Rightarrow R_x = (p_e - p_{atm})A_e + QV_e$$

$$\Rightarrow R_x = (110 - 101.32)(0.0121) + (3.545)(1900) = 6{,}735.60\text{N}$$

• PROBLEM 15-2

A liquid-propellant rocket uses 22 lb of fuel per second and 300 lb of oxidant per second. The exhaust gases leave the rocket at 2,000 fps. Calculate the rocket thrust.

Solution: (a) Consider a rocket which is using a total of M lb of fuel and oxidant per second and ejecting it at a velocity of V fps. In 1 sec an amount of burned fuel and oxidant is ejected with a momentum of MV/g lb-sec and therefore the rate of change of momentum of the rocket is MV/g lb-sec/sec. The rate of change of momentum is equal to the thrust, and so

$$T = \frac{MV}{g}$$

$$= \frac{300 + 22}{32.2} \times 2{,}000 = 20{,}000 \text{ lb}$$

• PROBLEM 15-3

A typical rocket utilizes a liquid fuel-oxidizer combination, which it burns at the rate of 36,000 lb_m/h. The hot gases leave the nozzle with a velocity of 2000 ft/s relative to the rocket body. The static pressure in the rocket exhaust is the same as that in the surrounding atmosphere. The diameter of the exhaust nozzle (and thus the jet) is 8 in., and the rocket carries a payload of 200 lb_m and has (at the moment of concern) a total weight of 400 lb_f. The rocket velocity is 1000 ft/s. Find the engine thrust and the instantaneous horsepower developed.

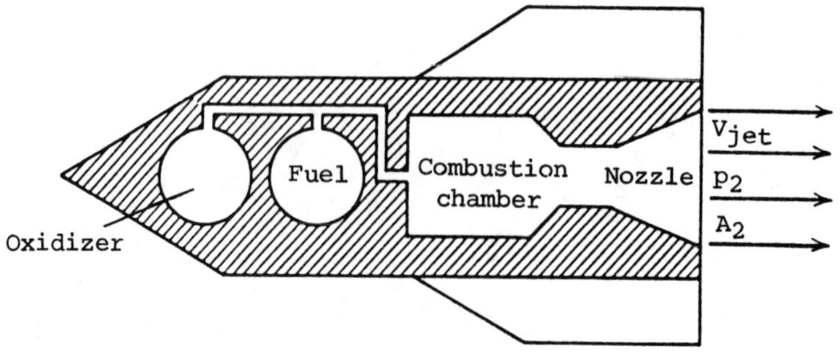

Solution: The engine thrust is given by

$$F_T = \left(\frac{\dot{m}_f}{g_c} V_{jet} + p_2 A_2 \right)$$

Note that $p_2 = 0$ and thus $p_2 A_2 = 0$

$$F_T = \frac{10 \times 2000}{32.2} = 620 \ lb_f$$

then the power developed, P, is

$$P = \frac{F_T V_{rocket}}{550} = \frac{620 \times 1000}{550} = 1127 \ hp$$

• PROBLEM 15-4

A rocket nozzle has a converging exhaust nozzle of exit area 20 in^2. (See Figure 1) If p_c = 100 psia and T_c = 2500 °R, find the rocket thrust for p_b = 20 psia and p_b = 0 psia. Assume the rocket exhaust gases to have γ = 1.4 with a mean molecular weight of 20. Take the flow in the nozzle to be isentropic.

Solution: Select a control volume as shown in Figure 2. The only forces acting on the fluid inside the control volume are pressure forces and the thrust force (see Figure 3). From the momentum equation

$$\sum F_x = \int_{\substack{\text{control} \\ \text{surfaces}}} V_x \frac{\rho V_n dA}{g_c}$$

$$\sum F_x = T - (p_e - p_b)A_e$$

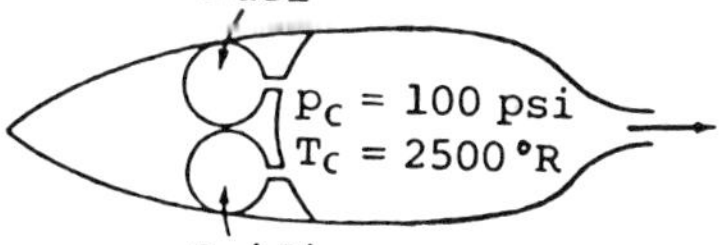

Fig. 1

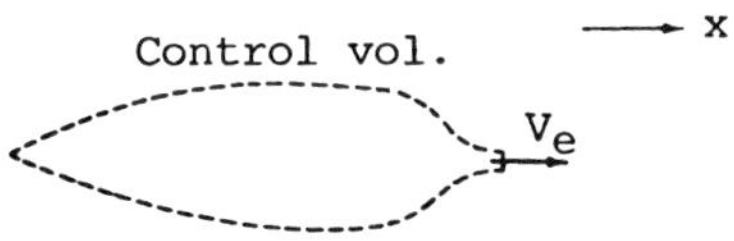

Fig. 2

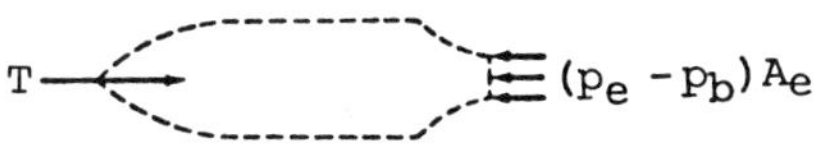

Fig. 3

For both values of p_b,

$$\frac{p_b}{p_e} < \left(\frac{p_b}{p_r}\right)_{crit}$$

so the nozzle is choked and

$$M_e = 1$$

or

$$V_e = \sqrt{\gamma g_c R T_e}$$

where

$$R = \frac{\overline{R}}{MW} = 77.3 \ \frac{\text{ft-lb}_f}{\text{lb}_m\text{-}^\circ R}$$

where $\overline{R}$ the universal gas constant ($\overline{R}$ = 1545 ft-lb_f/mol-deg). Therefore,

$$V_e = \sqrt{1.4 \times 32.2 \times 77.3 \times (0.833 \times 2500)} = 2700 \text{ fps}$$

for $p_b = 0$ and 20 psi

$$p_e = 0.528 p_c = 52.8 \text{ psi}$$

for $\quad M_e = 1$

$$\dot{m} = \rho_e A_e V_e = \frac{p_e}{RT_e} A_e V_e$$

$$= \left(\frac{52.8 \times 20}{77.3 \times 2080}\right) 2700$$

$$= 17.7 \text{ lb}_m/\text{sec}$$

for $\quad p_b = 0 \quad$ and 20 psi

For $p_b = 20$ psi

$$T = (p_e - p_b) A_e + \frac{2700(17.7)}{32.2}$$

$$= 32.8(20) + 1480$$

$$= 656 + 1480$$

$$= 2136 \text{ lb}_f$$

For $p_b = 0$ psi

$$T = 52.8(20) + 1480$$

$$= 1056 + 1480$$

$$= 2536 \text{ lb}_f .$$

The thrust calculated represents the force of the rocket on the fluid; the force of fluid on rocket is equal in magnitude and opposite in direction.

• PROBLEM 15-5

Water-tunnel tests on a 6-in. model ship propeller yield a thrust of 12 lb when the propeller speed is 120 rpm and the water speed is 10 fps. What would be the equivalent thrust of a similar 10-ft propeller when the speed of the ship is 15 knots? To what slipstream velocity would this correspond?

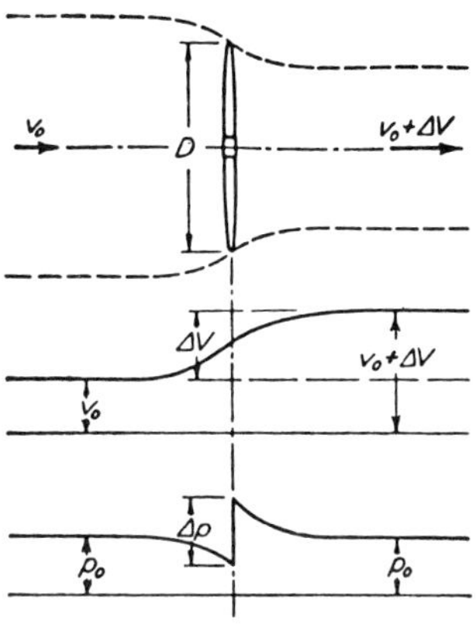

One-dimensional analysis
of flow past a propeller.

Solution: The equation of momentum is thus

$$F_T = Q\rho\Delta V$$

in which the thrust may also be expressed as the product of the pressure change and the area of the propeller section:

$$F_T = \frac{\pi D^2}{4}\Delta p$$

The equation of work and energy for zero loss is

$$\Delta p = \frac{\rho(v_0 + \Delta V)^2}{2} - \frac{\rho v_0^2}{2}$$

Solution of these equations for Q results in

$$Q = \frac{\pi D^2}{4}\left(v_0 + \frac{\Delta V}{2}\right)$$

which indicates (refer to the figure) that half the velocity change occurs before the propeller and half thereafter. Finally, solution for the velocity change (the velocity of the slipstream) is

$$\Delta V = -v_0 + \sqrt{v_0^2 + \frac{8F_T}{\pi D^2\rho}}$$

Since dynamically similar conditions would have to prevail for purposes of comparison, the advance-diameter ratios should first of all be the same. The rotational speed of the prototype propeller is

$$n_p = \left(\frac{v_0}{D}\right)_p \left(\frac{nD}{v_0}\right)_m = \frac{15 \times 1.69}{10} \times \frac{2 \times \frac{1}{2}}{10} = 0.254 \text{ rps}$$

The thrust is then found from the equality of the thrust coefficients:

$$(F_T)_p = (\rho n^2 D^4)_p \left(\frac{F_T}{\rho n^2 D^4}\right)_m = 1.94 \times \overline{0.254}^2 \times \overline{10}^4$$

$$\times \frac{12}{1.94 \times 2^2 \times \overline{0.5}^4}$$

$$= 31{,}000 \text{ lb}$$

The slipstream velocity is finally computed from the equation for the velocity of a slipstream

$$\Delta V = -v_0 + \sqrt{v_0{}^2 + \frac{8F_T}{\pi D^2 \rho}} = -25.4 +$$

$$\sqrt{\overline{25.4}^2 + \frac{8 \times 31{,}000}{\pi \times 10^2 \times 1.94}}$$

$$= -25.4 + 32.4 = 7.0 \text{ fps}$$

• PROBLEM 15-6

An aircraft traveling at 500 mph is propelled by a jet engine developing 8,000 lb thrust and operating with an air/fuel ratio of 25:1. If the exhaust velocity is 1,000 mph, find the required fuel flow rate.

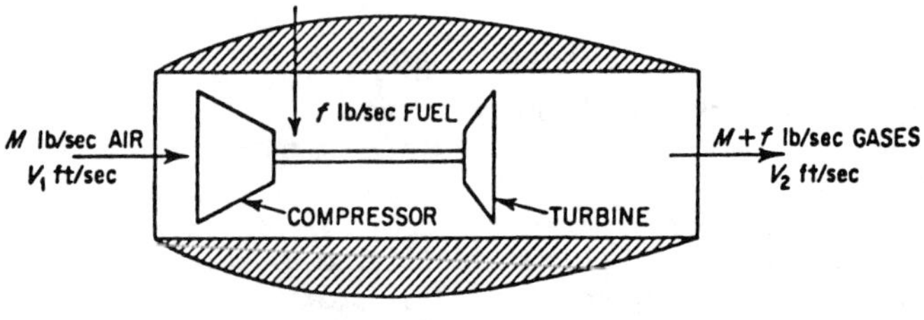

A jet engine.

Solution: Consider the simple jet engine shown in the figure.

In 1 sec an amount of gas is ejected with a momentum of MV/g lb-sec and therefore the rate of change of momentum of the rocket is MV/g lb-sec/sec. The engine is in motion at a speed of V_1 fps, but to simplify the calculations it is considered stationary in an airstream with a velocity of V_1 fps in the opposite direction. The exhaust gas leave the stationary engine with a final velocity of V_2 fps.

For an airflow rate of M lb/sec and a fuel flow rate of f lb/sec, the initial and final rate of momentum flow of the air and gas are

$$\text{intial momentum rate} = \frac{MV_1}{g} \text{ lb-sec/sec}$$

$$\text{final momentum rate} = \frac{M + f}{g} V_2 \text{ lb-sec/sec}$$

Therefore the rate of change of momentum or thrust is

$$T = \frac{M + f}{g} V_2 - \frac{MV_1}{g} \quad \text{lb}$$

$$8{,}000 = \frac{(M + f) \times 1{,}000 \times 88}{g \times 60} - \frac{M \times 500 \times 88}{g \times 60}$$

But M = 25f; therefore

$$8{,}000 = \frac{26 \times f \times 1{,}000 \times 88}{60g} - \frac{25f \times 500 \times 88}{60g}$$

Hence

$$f = 12.95 \text{ lb/sec}$$

• PROBLEM 15-7

A ramjet like that shown in the figure has equal intake and exhaust areas of 2.5 ft^2. In normal steady operation the air at the inlet station, 1, has a velocity of 2500 ft/sec, and a density of 0.0286 lbm/ft^3; at outlet the corresponding values are 3500 ft/sec and 0.0267 lbm/ft^3. The pressure at the outlet is 700 lbf/ft^2, while atmospheric pressure is 628 lbf/ft^3, and the fuel/air ratio is 0.015:1 (by mass). Calculate the gross thrust. Assume $g_c = 32.17\ lb_m\text{-ft}/lb_f\text{-s}^2$

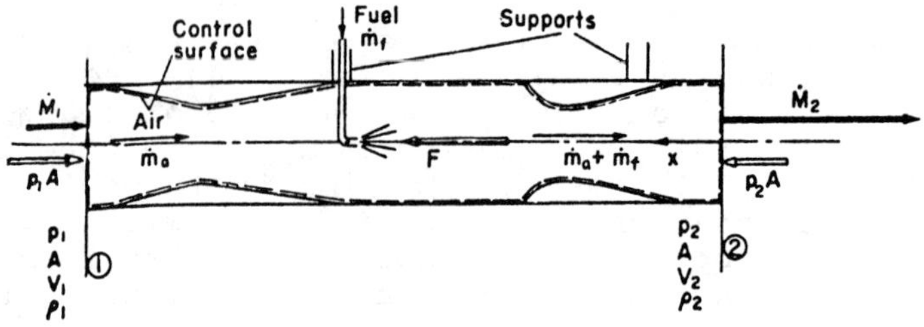

Solution: The apparatus illustrated in the figure is a simple propulsive duct, or ramjet, suitable for driving an aircraft or any other vehicle operating at supersonic speeds within the earth's atmosphere. Air from the atmosphere is forced into the entry, station 1, by the speed of the duct, and after decelerating burns with the injected fuel in the central section; it then accelerates to high velocity in the nozzle. The entry pressure p_1 is uniform and equal to that of the surrounding atmosphere, while in normal operation the outlet pressure p_2 is usually rather greater than atmospheric. In between stations 1 and 2 the duct surface exerts stresses on the fluid stream whose net effect can be regarded as a single force F, in the direction of flow. It is convenient to take the direction of flight as positive, and to define the control volume by the duct surface and stations 1 and 2. Apart from F the only forces acting in the flow direction on the contents of the control volume are $-p_1A_1$ and $+p_2A_2$. The momentum flow rates (in the same diretion) of the entering air and fuel streams are $-\dot{m}_aV_1$ and zero respectively, while that of the leaving stream is $-(\dot{m}_a + \dot{m}_f)V_2$.

In general, the force-momentum relationship is given by

$$g_c(p_1A_1 - p_2A_2 + F) = \dot{m}(V_2 - V_1) \qquad (1)$$

The relation between force and momentum flow rate then enables us to write, for steady operation along a horizontal course,

$$g_c(-p_1A_1 + p_2A_2 + F) = -(\dot{m}_a + \dot{m}_f)\, V_2 - (-\dot{m}_aV_1).$$

$$\therefore \quad F = -p_2A_2 + p_1A_1 - \frac{(\dot{m}_a + \dot{m}_f)^2}{\rho_2 A_2 g_c} + \frac{(\dot{m}_a)^2}{\rho_2 A_1 g_c}$$

or

$$F = -p_2A_2 + p_1A_1 - (\dot{m}_a + \dot{m}_f)V_2/g_c + \dot{m}V_1/g_c \qquad (2)$$

Force F is equal and opposite to the gross thrust of the ramjet, i.e., the thrust experienced by the internal surfaces of the duct.

The air mass flow rate is, from the values given for station 1,

$$\dot{m}_a = \rho_1 V_1 A_1 = 0.0286 \times 2500 \times 2.5 \text{ lbm/sec}$$

$$= 178.75 \text{ lbm/sec}$$

so that the fuel mass flow rate,

$$\dot{m}_f = 0.015\,\dot{m}_a = 2.68 \text{ lbm/sec,}$$

and the flow rate at station 2 is

$$\dot{m}_a + \dot{m}_f = 181.4 \text{ lbm/sec.}$$

The gross thrust (-F) can be calculated from equation 2:

$$(-F) = p_2A_2 - p_1A_1 + \frac{(\dot{m}_a + \dot{m}_f)V_2}{g_c} - \frac{\dot{m}_a V_1}{g_c}$$

$$= \left[(700 - 628)2.5 + \frac{181.4 \times 3500}{32.17} - \frac{178.8 \times 2500}{32.17}\right] \text{ lbf}$$

$$= (180 + 19740 - 13900) \text{ lbf}$$

$$= 6020 \text{ lbf.}$$

• PROBLEM 15-8

Derive a formula for the propulsion efficiency of a jet propelled vessel in still water if u is the absolute velocity of the vessel, v_r the velocity of the jet relative to the vessel when the intake is (a) at the bows facing the direction of motion, (b) amidships at right angles to the direction of motion.

Solution: (a) For intake in the direction of motion,

Mass of fluid entering control volume in unit time $= \rho Q$,

v_{in} = Mean velocity of water at inlet direction of motion relative to control volume $= u$,

v_{out} = Mean velocity of water at outlet in direction of motion relative to control volume $= v_r$.

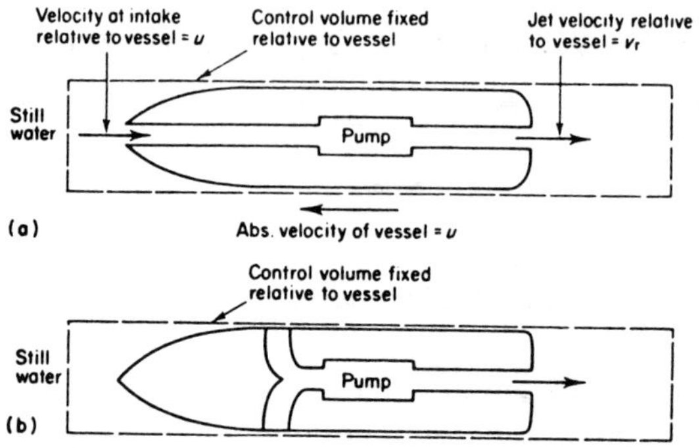

Jet propulsion of vessels. (*a*) Intake in direction of motion. (*b*) Intake in side of vessel

The momentum equation is in the form

$$F = \dot{m}(v_{out} - v_{in})$$

and assuming that the pressure in the water is the same at outlet and inlet,

$$\text{Propelling force} = \rho Q(v_{out} - v_{in}) = \rho Q(v_r - u),$$

$$\text{Work done per unit time} = \text{Propelling force} \times \text{Speed of boat}$$

$$= \rho Q(v_r - u)u$$

In unit time, a mass of water ρQ enters the pump intake with a velocity u and leaves with a velocity v_r.

Kinetic energy per unit time at inlet $= \frac{1}{2}\rho Q u^2$,

Kinetic energy per unit time at outlet $= \frac{1}{2}\rho Q v_r^2$.

Kinetic energy per unit time supplied by pump $= \frac{1}{2}\rho Q(v_r^2 - u^2)$,

$$\text{Hydraulic efficiency} = \frac{\text{Work done per unit time}}{\text{Energy supplied per unit time}}$$

$$= \rho Q(v_r - u)u/\frac{1}{2}\rho Q(v_r^2 - u^2)$$

$$= 2u/(v_r + u).$$

(b) For intake at right angles to the direction of motion (Fig.1 (b)), the control volume used will be the same as in (a), as will the rate of change of momentum through the control volume, and therefore the propelling force. Hence,

$$\text{Work done per unit time} = \rho Q(v_r - u)u.$$

As, however, the intake to the pumps is at right angles to the direction of motion, the forward velocity of the vessel will not assist the intake of water to the pumps, and, therefore, the whole of the energy of the outgoing jet must be provided by the pumps.

$$\text{Energy supplied per unit time} = \frac{1}{2}\ \rho Q v_r^2\ ,$$

$$\text{Hydraulic efficiency} = \frac{\text{Work done per unit time}}{\text{Energy supplied per unit time}}$$

$$= \rho Q(v_r - u)u/\frac{1}{2}\ \rho Q v_r^2$$

$$= 2(v_r - u)u/v_r^2\ .$$

• PROBLEM 15-9

A rocket sled weighs 4 tons including 1 ton of fuel. The motion resistance in the track on which the sled rides and in the air equals KV, where K is 100 lb/fps and V is the velocity of the sled in feet per second. Compute the maximum possible velocity of the sled when the exit velocity u of the rocket exhaust gas relative to the rocket is 10,000 fps and the rocket burns fuel at the rate of 200 lb/sec.

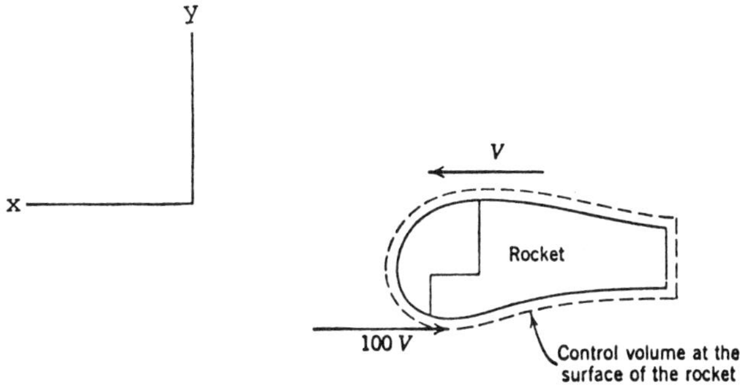

Solution: Consider the axis system as attached to the earth, and measure all velocities relative to the axis system. You cannot consider the axis system as attached to the rocket because the rocket is accelerating and an accelerating axis system is in violation of the restrictions on Newton's second law. Draw a control volume around the rocket which moves with the rocket. (See Sketch.)

$$F_x = \frac{M_{x_2}' - M_{x_1}'}{dt} + \int_A \bar{V}_x \rho V \cos\theta \, dA$$

The resistance term 100V includes all external forces acting on the rocket. Except at the exhaust area, the integration term around the control volume is zero everywhere. Measure the vector velocity $\bar{V}_x$ relative to the axis system, $\bar{V}_x = V - u$. This is constant over the exhaust area and, therefore, may be taken outside the integration. Thus, you get

$$-100V = \frac{M_{x_2}' - M_{x_1}'}{dt} + (V - u) \int_{\substack{\text{exhaust} \\ \text{area}}} \rho V \cos\theta \, dA$$

Over the exhaust area, $\theta = 0$. Therefore, $\cos\theta = 1$. The

$$\int \rho V \, dA$$

is the mass flow rate of fuel leaving the rocket, 200/32.2 slugs/sec. Now the momentum equation is

$$-100V = \frac{M_{x_2}' - M_{x_1}'}{dt} + \left(\frac{200}{32.2}\right)(V - u)$$

The only term left to consider is the $(M_{x_2}' - M_{x_1}')/dt$ term. Take the general case at time t when the rocket and remaining fuel have a total mass M and are traveling at a velocity V in the positive x direction. The momentum of the mass in the control volume, MV, is the momentum of the rocket and remaining fuel. At time t + dt the mass in the control volume is M - dM and the velocity is V + dV. Therefore, the momentum M_{x_2}' of the mass in the control volume at time t + dt is (M - dM)(V + dV). Substitute these values into $(M_{x_2}' - M_{x_1}')/dt$ to get

$$\frac{M_{x_2}' - M_{x_1}'}{dt} = \frac{(M - dM)(V + dV) - MV}{dt}$$

$$= \frac{M\,dV - V\,dM - dM\,dV}{dt}$$

Neglect the dM dV term because it is small compared to the other terms. Thus

$$\frac{M_{x_2}' - M_{x_1}'}{dt} = \frac{M\,dV - V\,dM}{dt} = M\frac{dV}{dt} - V\frac{dM}{dt}$$

where dM/dt is the rate of change of mass in the control volume. The rate of change equals the mass rate at which fuel burns, or dM/dt = 200/32.2 slugs/sec. Therefore, the momentum equation becomes

$$-100V = M\frac{dV}{dt} - \frac{200}{32.2}V + \frac{200}{32.2}(V - u)$$

or

$$-100V = M\frac{dV}{dt} - \frac{200}{32.2}u = M\frac{dV}{dt} - \frac{(200)(10{,}000)}{32.2}$$

You must remember that M is the mass of the rocket and remaining fuel at time t and, therefore, is a function of t. Take the time t = 0 as the time when the rocket first ignites. Then M = [4(2000) - (200)t]/32.2 which is the original mass minus the mass burned in time t. Reduce the momentum equation now to the form

$$-100V = \left(\frac{8000 - 200t}{32.2}\right)\frac{dV}{dt} - \frac{(200)(10{,}000)}{32.2}$$

Solve this differential equation by separating the variables and integrating

$$\int_0^t \frac{(32.2)dt}{8000 - 200t} = \int_0^V \frac{dV}{\left[\frac{(2)(10)^6}{32.2} - 100V\right]}$$

$$\frac{32.2}{200}\ln\frac{8000 - 200t}{8000} = \frac{1}{100}\ln\left(\frac{\frac{2(10)^6}{32.2} - 100V}{\frac{2(10)^6}{32.2}}\right)$$

$$\left(\frac{8000 - 200t}{8000}\right)^{16.1} = \frac{2(10)^6 - 3220V}{2(10)^6}$$

$$-V = \left[(2)(10)^6\left(1 - \frac{200t}{8000}\right)^{16.1} - 2(10)^6\right]\frac{1}{3220}$$

$$V = \frac{2(10)^6}{3220}\left[1 - \left(1 - \frac{t}{40}\right)^{16.1}\right]$$

The rocket attains maximum velocity at the instant that the fuel burns out completely. One ton of fuel burns at the rate of 200 lb/sec; the fuel, then, will burn for 10 sec. Substitute t = 10 sec and solve for V.

$$V = 622\left[1 - \left(1 - \frac{10}{40}\right)\right]^{16.1}$$

$$= 615 \text{ fps}$$

• PROBLEM 15-10

The initial mass of a rocket is to be 100,000 lb, of which 5 per cent is to be structure, the remainder fuel and payload. If the specific impulse of the propellant combination used is 300 $lb_f/lb_m/sec$, calculate the payload mass that would be able to escape from earth for a single-state and for a two-stage rocket. Neglect aerodynamic drag, gravity forces, and the term imvolving $p_e - p_a$ in the thrust equation. Assume escape velocity of 36,800 fps.

Solution: Examine the case of a rocket moving vertically against the earth's gravity field. The momentum equation is given by

$$-D - M_R\frac{g}{g_c} - \frac{M_R}{g_c}\frac{dV_R}{dt} + \text{thrust} = 0$$

where M_R and V_R are the mass and velocity of the rocket, respectively. Since no fluid is ingested in the case of a rocket, that is, $\dot{m} = 0$ and $A_i = 0$, the expression for the rocket's thrust becomes

$$\text{Thrust} = \frac{\dot{m}_f}{g_c}V_e + (p_e - p_a)A_e \qquad (1)$$

Combining the momentum equation above with the expression for the thrust (1), we obtain

$$-D - M_R \frac{g}{g_c} - \frac{M_R}{g_c} \frac{dV_R}{dt} + \frac{\dot{m}_f}{g_c} V_e + (p_e - p_a) A_e = 0$$

If we neglect the effects of drag and gravity, we obtain

$$(p_e - p_a) A_e = \frac{M_R}{g_c} \frac{dV_R}{dt} - \frac{\dot{m}_f}{g_c} V_e$$

Integrating, we obtain for constant $\dot{m}_f$, p_a, and V_e, with $V_R = V_{R_0}$ at $t = 0$ and $M_R = M_0 - \dot{m}_f t$

$$V_R - V_{R_0} = - \left[(p_e - p_a) \frac{A_e g_c}{\dot{m}_f} + V_e \right] \ln \left(1 - \frac{\dot{m}_f t}{M_0} \right)$$

where M_0 is the initial rocket mass. If we choose the time t when the burn of the rocket propellant ceases, $\dot{m}_f t$ represents the total propellant mass M_{pr} consumed during the burn. Hence at propellant burnout we have

$$V_R - V_{R_0} = - \left[\frac{(p_e - p_a)}{\dot{m}_f} A_e g_c + V_e \right] \ln \left(1 - \frac{M_{pr}}{M_0} \right)$$

where M_{pr}/M_0 is called the propellant fraction, that is, that fraction of the total rocket mass that is propellant mass. Since $M_0 - M_{pr}$ is the mass of the rocket at burnout, $M_0/M_0 - M_{pr})$ is called the mass fraction of the rocket, that is, the ratio of initial to burnout mass of the rocket; therefore

$$-\ln \left(1 - \frac{M_{pr}}{M_0} \right) = \ln \frac{M_0}{M_0 - M_{pr}}$$

The term

$$\left[\frac{(p_e - p_a)}{\dot{m}_f} A_e g_c + V_e \right]$$

is called the characteristic velocity, designated by c. It represents the effective exhaust velocity of the rocket engine. When the ambient pressure p_a equals the exhaust pressure p_e, the characteristic velocity equals the exhaust velocity v_e.

The increase in rocket velocity during the burn time is therefore expressed by

$$V_R - V_{R_0} = c \ln \frac{M_0}{M_0 - M_{pr}}$$

If the ratio of initial to burnout mass is the same for each stage, and each stage has the same exhaust velocity, then the increment of velocity provided by each stage will be the same (neglecting drag, gravity, and the pressure term in the thrust equation). In other words, after n stages, the velocity of the rocket will be

$$V_R = nV_e \ln \frac{M_0}{M_0 - M_{pr}}$$

where $M_0/(M_0 - M_{pr})$ is the mass ratio for each stage.

For a single stage rocket

$$V_R = V_e \ln \frac{M_0}{M_0 - M_{pr}}$$

with $V_e = 32.2(300) = 9660$ fps.

Therefore,

$$36{,}800 = 9660 \ln \frac{M_0}{M_0 - M_{pr}}$$

or

$$\frac{M_0}{M_0 - M_{pr}} = 45$$

but

$$M_0 = 100{,}000$$

$$M_0 - M_{pr} = \frac{100{,}000}{45} = 2220 \text{ lb}_m$$

Since 5000 lb_m must be devoted to structure, it follows that a single-stage rocket could not escape.

For a two-stage rocket, with the same mass ratio for each stage,

$$36,800 = 2(9660)\ \ln \frac{M_0}{M_0 - M_{pr}}$$

so that

$$\frac{M_0}{M_0 - M_{pr}} = 6.7$$

or, for each stage

$$M_0 - M_{pr} = 0.149M$$

For the first stage, $M_0 = 100,000\ lb_m$, so that $M_0 - M_{pr} = 14,900\ lb_m$, of which 5000 lb_m is structure. The payload of the first stage is thus 9900 lb_m, which is the initial mass of the second stage. Of this mass, 14.9 per cent of 0.149(9900) = 1470 lb_m is payload and structure. Again, 5 per cent or 495 lb is structure; the remainder, amounting to 1470 - 495 = 975 lb_m, is the payload that could be accelerated to escape velocity.

• PROBLEM 15-11

The air intake duct of a jet engine being tested on a stationary stand takes in air at an average velocity of 600 fps. Air enters the intake duct, and exhaust gases exit the tail pipe at atmospheric pressure. Fuel enters the top of the engine at the rate of 1 slug of fuel to 50 slugs of intake air. The intake duct area is 2 ft^2. The exit velocity of the exhaust gases is 4000 fps. The density of the entering air is 0.00238 $slug/ft^3$. What force is necessary to hold the engine?

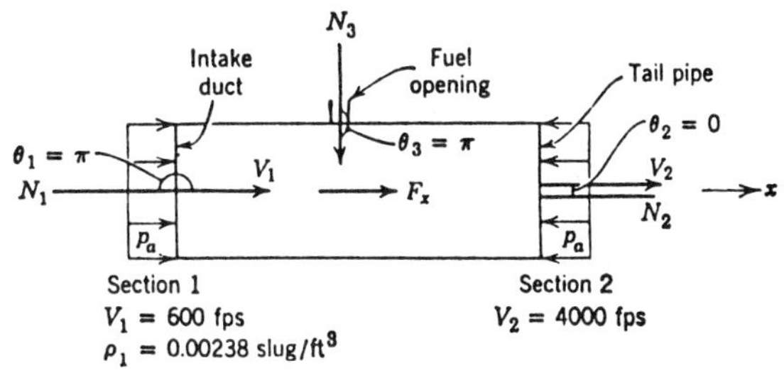

Solution: Sketch a schematic control volume of the engine. Write the momentum equation for the x direction.

$$\sum F_x = \frac{M_{x_2}' - M_{x_1}'}{dt} + \int_A \overline{V}_x \rho V \cos\theta \, dA$$

This is steady flow; therefore, $(M_{x_2}' - M_{x_1}')/dt = 0$

The forces due to the external pressures must be zero because the pressure is contant on all external surfaces of the control volume. Thus, the external forces $\sum f_x$ are

$$\sum F_x = F_x$$

and the momentum equation becomes

$$F_x = \int_A \overline{V}_x \rho V \cos\theta \, dA \qquad (1)$$

Since $\theta = \pi/2$ at all control volume surfaces except the intake duct, the fuel injection opening, and the tail pipe, the value of the integral is zero except at these three surfaces. The fuel has no component V_{x_3} in the x direction so that the above integral in Equation (1) over this opening is zero, which reduces the equation to

$$F_x = \int_{A_1} \overline{V}_{x_1} \rho_1 V_1 \cos\theta_1 \, dA + \int_{A_2} \overline{V}_{x_2} \rho_2 V_2 \cos\theta_2 \, dA$$

Because all quantities are constant over the areas of integration, take them outside the integral sign.

$$F_x = \rho_1 V_1 (-1) \overline{V}_{x_1} \int_{A_1} dA + \rho_2 V_2 (+1) \overline{V}_{x_2} \int_{A_2} dA$$

where $\overline{V}_{x_1}$ is the x component of V_1 and $\overline{V}_{x_2}$ is the x component of V_2. Therefore,

$$\overline{V}_{x_1} = V_1$$

and

$$\bar{V}_{x_2} = V_2$$

Thus,

$$F_x = -\rho_1 A_1 V_1^2 + \rho_2 A_2 V_2^2$$

To find the values of ρ_2 and A_2, try the continuity equation. For steady flow it is

$$0 = \int_A \rho V \cos\theta \, dA$$

Since $\theta = \pi/2$ at all control surfaces except the three mentioned above, the equation becomes

$$0 = \int_{A_1} \rho_1 V_1 \cos\theta_1 \, dA + \int_{A_2} \rho_2 V_2 \cos\theta_2 \, dA +$$

$$\int_{A_3} \rho_3 V_3 \cos\theta_3 \, dA$$

Again these are constant over the integration area and so take them outside the integral sign and carry out the integration.

$$0 = \rho_1 V_1 A_1 \cos\theta_1 + \rho_2 V_2 A_2 \cos\theta_2 + \rho_3 V_3 A_3 \cos\theta_3$$

Substitute the values of θ taken from the sketch of the control volume.

$$0 = -\rho_1 A_1 V_1 + \rho_2 A_2 V_2 - \rho_3 A_3 V_3 \qquad (2)$$

The term $\rho_1 A_1 V_1$ is the mass intake of air, and the term $\rho_3 A_3 V_3$ is the mass intake of fuel. The problem states that 50 slugs of air are used for every slug of fuel. Therefore,

$$\rho_1 A_1 V_1 = 50 \rho_3 A_3 V_3$$

which, when substituted into the continuity equation (2), becomes

$$0 = -\rho_1 A_1 V_1 + \rho_2 A_2 V_2 - \frac{1}{50} \rho_1 A_1 V_1$$

or

$$\frac{51}{50}\,\rho_1 A_1 V_1 = \rho_2 A_2 V_2$$

Rearrange the momentum equation for F_x from

$$F_x = -\rho_1 A_1 V_1{}^2 + \rho_2 A_2 V_2{}^2$$

to

$$F_x = (-\rho_1 A_1 V_1)V_1 + (\rho_2 A_2 V_2)V_2$$

Substitute the above for $\rho_2 A_2 V_2$ to obtain

$$F_x = (-\rho_1 A_1 V_1)V_1 + \frac{51}{50}(\rho_1 A_1 V_1)V_2$$

$$F_x = \rho_1 A_1 V_1\left(\frac{51}{50}V_2 - V_1\right)$$

By using the values given in the problem, calculate F_x as

$$F_x = (0.00238)(2)(600)\left(\frac{51}{50}4000 - 600\right)$$

$$= 9940 \text{ lb}$$

If this force is required to hold the engine in place, the engine develops 9940 lb of thrust under the conditions indicated.

TURBOMACHINES

• PROBLEM 15-12

Air flows in a sheet-metal ventilation duct having a cross-sectional area of 6 ft^2 at a velocity of 40 fps. If a butterfly type of control were suddenly closed, what initial force would be exerted upon it?

Solution: The resulting pressure rise on the upstream side would be

$$\Delta p = \rho V c = 0.0025 \times 40 \times 1100 = 110 \text{ psf}$$

Unless the control were at the end of the duct, there would

be an essentially equal drop in pressure on the downstream side. The total force would then be the differential pressure intensity times the area of one side of the control surface:

$$F = 2 \times 110 \times 6 = 1320 \text{ lb}$$

• PROBLEM 15-13

The axial-flow fan of a laboratory wind tunnel produces an air velocity of 65 fps at the test section when driven by a 3-hp motor at a speed of 1200 rpm. What motor speed and power would be required to increase the velocity to 100 fps?

Solution: Since the cross-sectional area remains constant, it is seen from the discharge coefficient $C_Q = Q/nD^3$ that the velocity and rotational speed are directly proportional:

$$N_{100} = N_{65} \times \frac{100}{65} = 1200 \times \frac{100}{65} = 1846 \text{ rpm}$$

From the power coefficient the required horsepower is computed as

$$\frac{P_{100}}{P_{65}} = \frac{(\rho n^3 D^5)_{100}}{(\rho n^3 D^5)_{65}} = \left(\frac{n_{100}}{n_{65}}\right)^3 = \left(\frac{100}{65}\right)^3 = 3.64$$

$$P_{100} = 3.64 \times 3 = 10.9 \text{ hp}$$

• PROBLEM 15-14

A turbine having an efficiency (including draft tube and valve) of 60% is fed from a reservoir by 150 ft of 10-in. pipe for which f = 0.02. At what rate of flow under a 25-ft head would the turbine deliver 7 hp? (Figure)

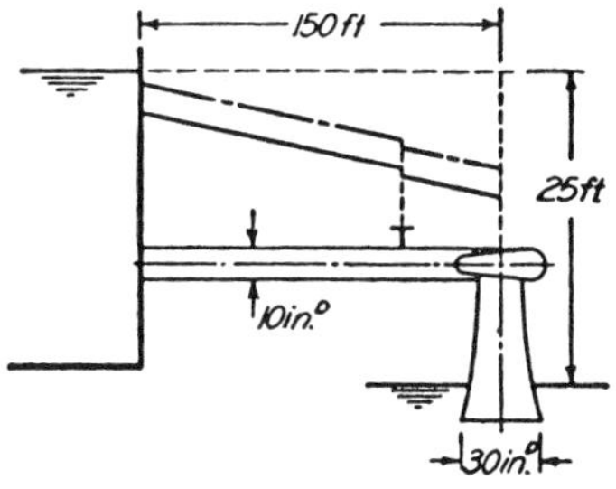

Solution: The head relationship is as follows:

$$25 = \Sigma H_L + \frac{P_T/0.60}{Q\gamma} = \frac{Q^2}{2gA^2}\left[0.5 + 0.02\,\frac{150}{10/12} + \left(\frac{10}{30}\right)^4\right]$$

$$+ \frac{7 \times 550/0.60}{Q \times 62.4}$$

$$= 0.214Q^2 + 103/Q$$

Solution for Q yields three roots, two of which are positive:

$$Q = 5.8 \text{ cfs} \quad \text{or} \quad 6.6 \text{ cfs}$$

This means simply that the lower rate of flow involves less loss, with a correspondingly higher available head, and vice versa.

• PROBLEM 15-15

A pump uses an impeller with radial blades to raise the pressure of water flowing axially into the impeller at 100 gallons/min. A motor drives the impeller at 2800 RPM and the water exits the impeller with a velocity of 8 ft/sec relative to the blades. Calculate the width of the impeller rim at the exit, the motor torque, and required motor horsepower if the impeller diameter is 5".

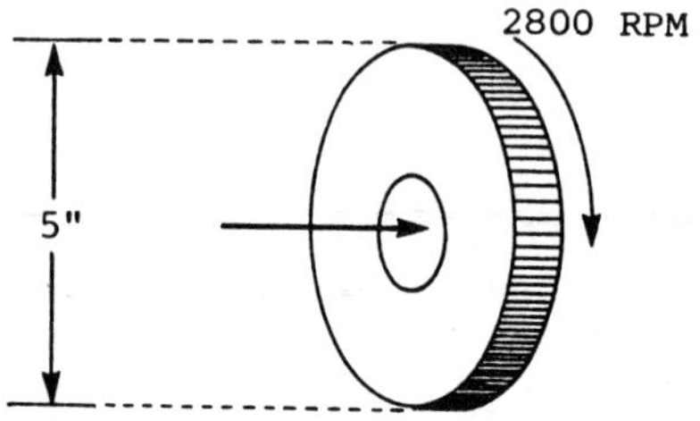

Solution: To calculate the width of the impeller at the exit, it is essential to note that the flow rate Q_2 of the water at the exit is equal to the flow rate Q_1 at the entrance, assuming incompressibility. Now, the flow rate at the entrance is given as 100 gal/min. At the exit the following relationship is applicable:

$$Q_2 = V_2 A_2$$

where V_2 is the exit velocity relative to the blades, and A_2 is the exit area which may be found from

$$A_2 = \pi D_2 W_2$$

where D_2 is the impeller diameter and W_2 is the impeller width at the exit.

Since $Q_2 = Q_1 = 100$ gal/min.

$$V_2 = 8 \text{ ft/sec}$$

$$D_2 = 5 \text{ in.,}$$

we can solve for W_2 as

$$W_2 = \frac{Q_2}{V_2 \pi D_2}$$

and

$$W_2 = 100 \frac{\text{gal}}{\text{min}} \times \frac{\text{ft}^3}{7.48 \text{ gal}} \times \frac{\text{min}}{60 \text{ sec}} \times \frac{\text{sec}}{8 \text{ ft}} \times \frac{1}{\pi} \times \frac{1}{5 \text{ in}} \times \frac{12 \text{ in}}{\text{ft}}$$

$$W_2 = .0213 \text{ ft} \quad \text{or} \quad \simeq 1/4''$$

To calculate the motor torque, we apply the basic equation

$$T = \int \bar{r} \times \bar{V} \rho \bar{V} \cdot d\bar{A}$$

from which

$$T = \frac{D_2}{2} V_2 \rho Q_2$$

since $$V_2 = \omega \frac{D_2}{2}$$

$$T = \frac{D_2}{2}\left(\omega \frac{D_2}{2}\right) \rho Q_2$$

or

$$T = \frac{\omega}{4} D_2^2 \rho Q_2$$

Substituting values,

$$T = \frac{2800}{4}\ \frac{rev}{min} \times 2\pi\ \frac{rad}{rev} \times \frac{min}{60\ sec} \times 5^2\ in^2$$

$$\times\ 1.94\ \frac{slug}{ft^3} \times 150\ \frac{gal}{min} \times \frac{ft^3}{7.48\ gal} \times \frac{min}{60\ sec} \times \frac{ft^2}{12^2\ in^2}$$

$$= 8.25\ ft\ lbs$$

To calculate the required motor horsepower, P,

$$P = T\omega$$

where ω = Angular velocity

$$P = 8.25\ ft\ lbs \times 2800\ \frac{rev}{min} \times 2\pi\ \frac{rad}{rev} \times \frac{min}{60\ sec} \times \frac{HP\ sec}{550\ ft\ lbs}$$

from which

$$P = 4.4\ HP$$

• PROBLEM 15-16

An adiabatic turbocompressor has blades that are radial at the exit of its 150-mm-diameter impeller. It is compressing 0.5 kg/s air at 98.06 kPa abs, t = 15°C, to 294.18 kPa abs. The entrance area is 60 cm^2, and the exit area is 35 cm^2; $\eta = 0.75$; $\eta_m = 0.90$. Determine the rotational speed of the impeller and the actual temperature of air at the exit.

Solution:

Centrifugal compressors operate according to the same principles as turbo-machines for liquids. It is important for the fluid to enter the impeller without shock, i.e., with the relative velocity tangent to the blade. Work is done on the gas by rotation of the vanes, the moment-of-momentum equation relating torque to production of tangential velocity. At the impeller exit the high-velocity gas must have its kinetic energy converted in part to flow energy by suitable expanding flow passages. For adiabatic compression (no cooling of the gas) the actual work w_a of compression per unit mass is compared with the work w_{th} per unit mass to compress the gas to the same pressure isentropically. For cooled compressors the work w_{th} is based on the isothermal work of compression to the same pressure as the actual case. Hence,

$$\eta = \frac{w_{th}}{w_a} \tag{1}$$

is the formula for effiency of a compressor.

The efficiency formula for compression of a perfect gas is developed for the adiabatic compressor, assuming no internal leakage in the machine, i.e., no short-circuiting of high-pressure fluid back to the low-pressure end of the impeller. Centrifugal compressors are usually multistage, with pressure ratios up to 3 across a single stage. From the moment-of-momentum equation with inlet absolute velocity, radial, $\alpha_1 = 90°$, the theoretical torque T_{th} is

$$T_{th} = \dot{m}V_{u_2}r_2 \tag{2}$$

in which $\dot{m}$ is the mass per unit time being compressed, Vu_2 is the tangential component of the absolute velocity leaving the impeller, and r_2 is the impeller radius at exit. The actual applied torque T_a is greater than the theoretical torque by the torque losses due to bearing and packing friction plus disk friction; hence,

$$T_{th} = T_a\eta_m \tag{3}$$

if η_m is the mechanical efficiency of the compressor.

In addition to the torque losses, there are irreversibilities due to flow through the machine. The actual work of compression through the adiabatic machine is obtained from the steady-flow energy equation, neglecting elevation changes and replacing $u + p/\rho$ by h,

$$-w_a = \frac{V_{2a}^2 - V_1^2}{2} + h_2 - h_1 \tag{4}$$

The isentropic work of compression can be obtained from this equation in differential form, neglecting the z terms,

$$-dw_{th} = V\ dV + d\,\frac{p}{\rho} + du = V\ dV + \frac{dp}{\rho} + pd\,\frac{1}{\rho} + du$$

The last two terms are equal to T ds, which is zero for isentropic flow, so that

$$-dw_{th} = V\ dV + \frac{dp}{\rho} \tag{5}$$

By integrating for p/ρ^k = const between sections 1 and 2,

$$-w_{th} = \frac{V_{2th}^2 - V_1^2}{2} + \frac{k}{k-1}\left(\frac{p_2}{\rho_{2th}} - \frac{p_1}{\rho_1}\right)$$

$$= \frac{V_{2th}^2 - V_1^2}{2} + c_p T_1 \left[\left(\frac{p_2}{p_1}\right)^{(k-1)/k} - 1\right] \quad (6)$$

The efficiency may now be written

$$\eta = \frac{-w_{th}}{-w_a} = \frac{(V_{2th}^2 - V_1^2)/2 + c_p T_1 [(p_2/p_1)^{(k-1)/k} - 1]}{(V_{2a}^2 - V_1^2)/2 + c_p (T_{2a} - T_1)} \quad (7)$$

since $h = c_p T$. In terms of Eqs. (2) and (3)

$$-w_a = \frac{T_a \omega}{\dot{m}} = \frac{T_{th}\omega}{\eta_m \dot{m}} = \frac{V_{u2} r_2 \omega}{\eta_m} = \frac{V_{u2} u_2}{\eta_m} \quad (8)$$

then

$$\eta = \frac{\eta_m}{V_{u2} u_2}\left\{c_p T_1 \left[\left(\frac{p_2}{p_1}\right)^{(k-1)/k} - 1\right] + \frac{V_{2th}^2 - V_1^2}{2}\right\} \quad (9)$$

Use of this equation is made in the following.

The density at the inlet is

$$\rho_1 = \frac{p_1}{RT_1} = \frac{9.806 \times 10^4 \text{ N/m}^2}{(287 \text{ J/kg} \cdot \text{K})(273 + 15\text{k})} = 1.186 \text{ kg/m}^3$$

and the velocity at the entrance is

$$V_1 = \frac{\dot{m}}{\rho_1 A_1} = \frac{0.5 \text{ kg/s}}{(1.186 \text{ kg/m}^3)(0.006 \text{ m}^2)} = 70.26 \text{ m/s}$$

The theoretical density at the exit is

$$\rho_{2th} = \rho_1\left(\frac{p_2}{p_1}\right)^{1/k} = 1.186 \times 3^{1/1.4} = 2.60 \text{ kg/m}^3$$

and the theoretical velocity at the exit is

$$V_{2th} = \frac{\dot{m}}{\rho_{2th} A_2} = \frac{0.5}{2.60 \times 0.0035} = 54.945 \text{ m/s}$$

For radial vanes at the exit, $V_{u2} = u_2 = \omega r_2$.

From equation (9)

$$u_2^2 = \frac{\eta_m}{\eta} \left\{ c_p T_1 \left[\left(\frac{p_2}{p_1}\right)^{(k-1)/k} - 1 \right] + \frac{V_{2th}^2 - V_1^2}{2} \right\}$$

$$= \frac{0.90}{0.75} \left[(0.24 \times 4187)(273 + 15)(3^{0.4/1.4} - 1) + \frac{54.945^2 - 70.26^2}{2} \right]$$

and $u_2 = 359.56$ m/s. Then

$$\omega = \frac{u_2}{r_2} = \frac{359.56}{0.075} = 4794 \text{ rad/s}$$

and

$$N = \omega \frac{60}{2\pi} = 4794 \frac{60}{2\pi} = 45{,}781 \text{ rpm}$$

The theoretical work w_{th} is the term in the brackets in the expression for u_2^2. It is $-w_{th} = 0.1058 \times 10^6 \text{m} \cdot \text{N/kg}$. Then from Eq. (1),

$$w_a = \frac{w_{th}}{\eta} = -\frac{1.058 \times 10^5}{0.75} = -1.411 \times 10^5 \text{ m} \cdot \text{N/kg}$$

Since the kinetic-energy term is small, Eq. (4) can be solved for $h_2 - h_1$ and a trial solution effected:

$$h_2 - h_1 = c_p (T_{2a} - T_1) = 1.411 \times 10^5 + \frac{70.26^2 - V_{2a}^2}{2}$$

As a first approximation, let $V_{2a} = V_{2th} = 54.945$; then

$$T_{2a} = 288 + \frac{1}{0.24 \times 4187}\left(1.411 \times 10^5 + \frac{70.26^2 - 54.945^2}{2}\right)$$

$$= 429.4\text{K}$$

For this temperature the density at the exit is 2.387 kg/m^3, and the velocity is 59.85 m/s. Insertion of this value in place of 54.945 reduces the temperature to T_{2a} = 429.1K.

• PROBLEM 15-17

The 500 gpm is to be pumped to a height of 60 ft, overcoming all losses through a 156 ft long cast iron pipe that includes an open globe valve (Figure 1). Assuming that the motor is 3600 rpm, determine the pump size (Q, H, n_s) for the cases when the pipeline has a 6 or 8 in. diameter.

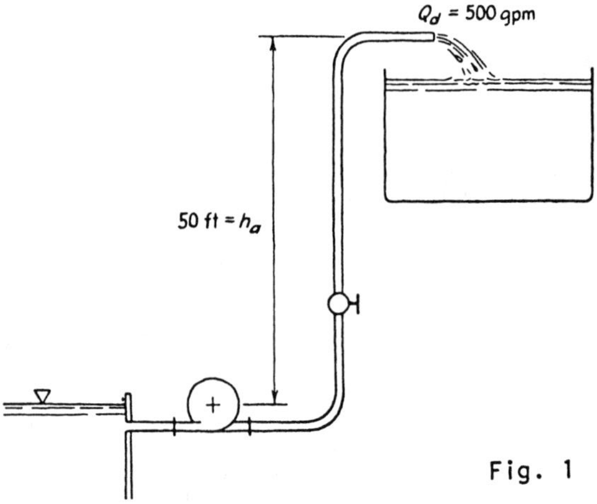

Fig. 1

Solution:

$$500 \text{ gpm} = \frac{500 \text{ gpm}}{450 \text{ gpm/cfs}} = 1.11 \text{ cfs}$$

$$\text{average velocity} = \frac{Q}{A} = V$$

For a 6 in. pipe,

$$\text{average velocity} = \frac{1.11}{(\pi/4)(0.5)^2} = 5.658 \text{ fps}$$

For an 8 in. pipe

$$\text{average velocity} = \frac{1.11}{(\pi/4)(8/12)^2} = 3.183 \text{ fps}$$

The total head loss equals friction loss in pipe plus friction loss in valve plus exit loss.

Consider a 6 in. pipe. Assume the water temperature is 65° F. From the properties of water at 65° F

$$\mu = 2.196 \times 10^{-5} \frac{\text{lbf sec}}{\text{ft}^2}$$

$$= 1.937 \frac{\text{slugs}}{\text{ft}^3}$$

$$\nu = \frac{\mu}{\rho}$$

$$= 1.134 \times 10^{-5} \frac{\text{ft}^2}{\text{sec}}$$

TABLE 1
Viscosity of Water

Temperature in °C	Absolute Viscosity	
	Centipoise	lb sec/ft^2
0	1.792	0.374×10^{-4}
4	1.567	0.327×10^{-4}
10	1.308	0.272×10^{-4}
20	1.005	0.209×10^{-4}
20.2	1.000	0.208×10^{-4}
30	0.8807	0.183×10^{-4}
50	0.5494	0.114×10^{-4}
70	0.4061	0.084×10^{-4}
100	0.2838	0.059×10^{-4}
150	0.184	0.038×10^{-4}

TABLE 2

Commercial Pipe Surface, New	Equivalent Sand Grain Roughness, *e* in ft	e^3 in ft^3
Glass, drawn brass, copper, lead	Smooth	—
Wrought iron, steel	1.5×10^{-4}	3.4×10^{-12}
Asphalted cast iron	4.0×10^{-4}	64×10^{-12}
Galvanized iron	5.0×10^{-4}	124×10^{-12}
Cast iron	8.5×10^{-4}	614×10^{-12}
Concrete	$10\text{–}100 \times 10^{-4}$	$10^{-9}\text{–}10^{-6}$
Riveted steel	$30\text{–}300 \times 10^{-4}$	$10^{-8}\text{–}10^{-5}$

From Table 2 the roughness of the pipe (cast iron) = 0.00085 ft.

$$\frac{e}{D} = \frac{0.00085}{0.5} = 0.0017 \qquad \text{for a 6 in. pipe}$$

$$= \frac{0.00085}{8/12} = 0.00127 \qquad \text{for an 8 in. pipe}$$

Next calculate the Reynolds number as $R_e = VD/\nu$.

$$R_{e6in.\ D} = \frac{5.658 \times 0.5}{1.134 \times 10^{-5}} = 2.49 \times 10^5$$

for flow in a 6 in. pipe

$$R_{e6in.\ D} = \frac{3.183 \times 8/12}{1.134 \times 10^{-5}} = 1.87 \times 10^5$$

for flow in an 8 in. pipe

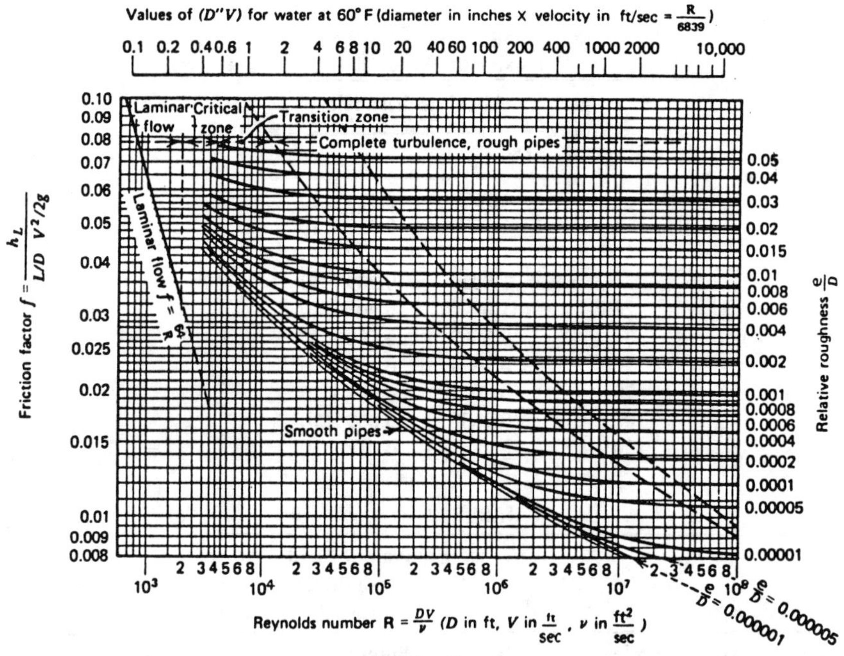

Moody diagram.

Fig. 2

By entering the Moody diagram, we get the following:

For a 6 in. D pipe condition e/D = 0.0017 and $R_e = 2.49 \times 10^{-5}$; we obtain f = 0.023.

For an 8 in. D pipe condition $e/D = 0.0013$ and $R_e = 1.87 \times 10^5$; we obtain $f = 0.022$.

Next calculate the friction loss in pipe h_L.

$$h_L = f \frac{L}{D} \frac{V^2}{2g}$$

For a 6 in. pipe 156 ft long,

$$h_{L_{6\text{ in. D}}} = 0.023 \left(\frac{156}{0.5}\right)\left(\frac{5.658^2}{2g}\right) = 3.57 \text{ ft}$$

and for an 8 in. pipe 156 ft long,

$$h_{L_{8\text{ in. D}}} = 0.022 \left(\frac{156}{8/12}\right)\left(\frac{3.183^2}{2g}\right) = 0.81 \text{ ft}$$

The minor losses equal the loss in valve plus exit loss, which equal

$$k_1 \frac{V^2}{2g} + k_2 \frac{V^2}{2g}$$

or

$$(k_1 + k_2) \frac{V^2}{2g}$$

k_1 for globe valve = 10

k_2 for exit loss = 1

Then, minor losses in the 6 in. pipe =

$$(10 + 1) \left(\frac{5.658^2}{2g}\right) = 5.47 \text{ ft}$$

8 " pipe =

$$(10 + 1) \left(\frac{3.183^2}{2g}\right) = 1.73 \text{ ft}$$

Therefore the total loss in a 6 in. pipe equals friction loss plus minor loss.

$$H_6 = 3.57 + 5.47 = 9.04 \text{ ft}$$

Total loss in an 8 in. pipe is

$$H_8 = 0.81 + 1.73 = 2.54 \text{ ft}$$

Then the total discharge head

$$H_d = 60 + 9.04 = 69.04 \text{ ft for a 6 in. pipeline}$$

$$= 60 + 2.54 = 62.54 \text{ ft for an 8 in. pipeline}$$

Pump selection:

The discharge, head, and speed at optimum performance of various pumps is consolidated into one number called specific speed. The specific speed n, of a pump is computed from

$$n_s = \frac{\text{rpm}\sqrt{\text{gpm}}}{H^{3/4}}$$

where n_s is the specific speed in revolutions per minute, rpm is the rotational speed of the shaft, gpm is the discharge in gallons per minute, and H is the total dynamic head in feet that the pump is expected to develop, all at optimum efficiency.

We get for a 6 in. pump:

$$n_s = \frac{3600\sqrt{500}}{(69.04)^{3/4}}$$

$$= 3361 \text{ rpm}$$

and for an 8 in. pump

$$n_s = \frac{3600\sqrt{500}}{(62.54)^{3/4}}$$

$$= 3619 \text{ rpm}$$

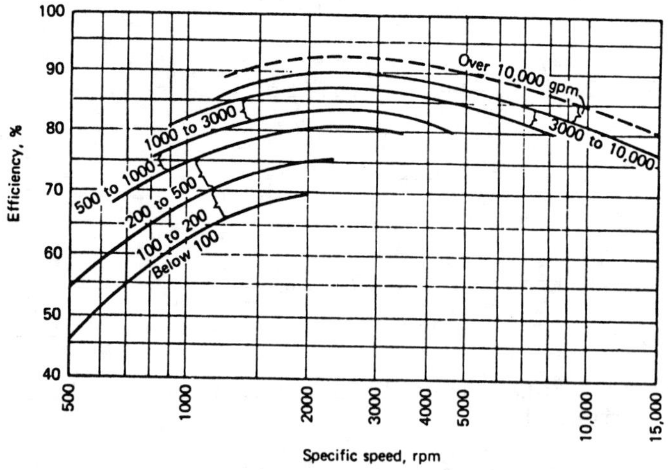

Fig. 3

By entering into Figure 3 for $n_s = 3361$ and 3619 rpm, we note that the same kind of pump will suffice for both applications, a high specific speed radial flow pump with an approximate peak efficiency of 80%. Hence, we need a 6 in. pump having Q = 500 gpm, H = 69 ft, and $n_s = 3361$ rpm, or an 8 in. pump having Q = 500 gpm, H = 62.5 ft, and $n_s = 3619$ rpm.

• PROBLEM 15-18

Describe an impulse turbine.

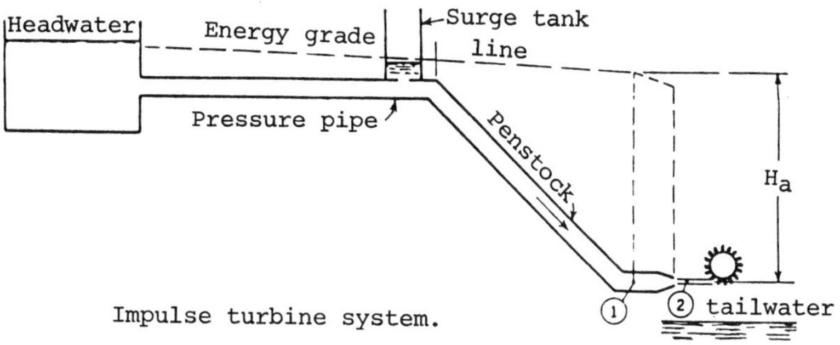

Impulse turbine system.

Fig. 1

Solution: The impulse turbine is one in which all available energy of the flow is converted by a nozzle into kinetic energy at atmospheric pressure before the fluid contacts the moving blades. Losses occur in flow from the reservoir through the pressure pipe(penstock)to the base of the nozzle, which may be computed from pipe friction data. At the base of the nozzle the available energy, or total head, is

$$H_a = \frac{p_1}{\gamma} + \frac{V_1^2}{2g} \tag{1}$$

from Fig. 1. With C_v the nozzle coefficient, the jet velocity V_2 is

$$V_2 = C_v\sqrt{2gH_a} = C_v\sqrt{2g\left(\frac{p_1}{\gamma} + \frac{V_1^2}{2g}\right)} \tag{2}$$

The head lost in the nozzle is

$$H_a - \frac{V_2^2}{2g} = H_a - C_v^2 H_a = H_a(1 - C_v^2) \tag{3}$$

and the efficiency of the nozzle is

$$\frac{V_2^2/2g}{H_a} = \frac{C_v^2 H_a}{H_a} = C_v^2 \tag{4}$$

The jet, with velocity V_2, strikes double-cupped buckets which split the flow and turn the relative velocity through the angle θ (Fig. 2).

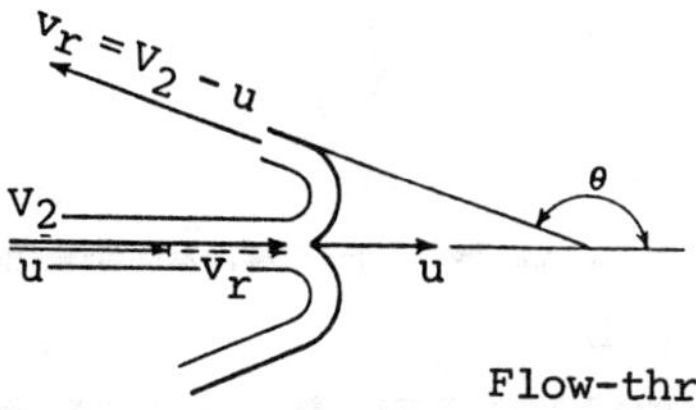

Fig. 2

Flow-through bucket.

The x component of momentum is changed by (Fig. 2)

$$F = \rho Q(v_r - v_r \cos \theta)$$

and the power exerted on the vanes is

$$Fu = \rho Q u v_r (1 - \cos \theta) \tag{5}$$

To maximize the power theoretically, two conditions must be met: $\theta = 180°$ and uv_r must be a maximum; that is, $u(V_2 - u)$ must be a maximum. To determine when $(uv_r)_{max}$ occurs, differentiate with respect to u and equate to zero.

$$(V_2 - u) + u(-1) = 0$$

The condition is met when $u = V_2/2$. After making these substitutions into Eq. (5),

$$Fu = \rho Q \frac{V_2}{2}\left(V_2 - \frac{V_2}{2}\right)[1 - (-1)] = \gamma Q \frac{V_2^2}{2g} \tag{6}$$

which accounts for the total kinetic energy of the jet. The velocity diagram for these values shows that the absolute velocity leaving the vanes is zero.

Practically, when vanes are arranged on the periphery of a wheel the fluid must retain enough velocity to move out of the way of the following bucket. Most of the practical impulse turbines are Pelton wheels. The jet is split in two and turned in a horizontal plane, and half is discharged from each side to avoid any unbalanced thrust on the shaft. There are losses due to the splitter and to friction between jet and bucket surface, which make the most economical speed somewhat less than $V_2/2$. It is expressed in terms of the speed factor

$$\phi = \frac{u}{\sqrt{2gH_a}} \tag{7}$$

For most efficient turbine operation ϕ has been found to depend upon specific speed as shown in Table 1. The angle θ of the bucket is usually 173 to 176°. If the diameter of the jet is d and the diameter of the wheel is D at the centerline of the buckets, it has been found in practice that the diameter ratio D/d should be about $54/N_s$ (ft, hp, rpm), or $206/N_s$ (m, kW, rpm) for maximum efficiency.

Table 1. Dependence of ϕ on specific speed

Specific speed N_s		
(m, kW, rpm)	(ft, hp, rpm)	ϕ
7.62	2	0.47
11.42	3	0.46
15.24	4	0.45
19.05	5	0.44
22.86	6	0.433
26.65	7	0.425

In the majority of installations only one jet is used; it discharges horizontally against the lower periphery of the wheel as shown in Fig. 1. The wheel speed is carefully regulated for the generation of electric power. A governor operates a needle valve that controls the jet discharge by changing its area. So V_2 remains practically constant for a wide range of positions of the needle valve.

The efficiency of the power conversion drops off rapidly with change in head (which changes V_2), as is evident when power is plotted against V_2 for constant u in Eq. (5). The wheel operates in atmospheric air although it is enclosed by a housing. It is therefore essential that the wheel be placed above the maximum floodwater level of the river into which it discharges. The head from nozzle to tail water is wasted. Because of their inefficiency at other than the design head and because of the wasted head, Pelton wheels usually are employed for high heads, eg., from 200 m to more than 1 km. For high heads, the efficiency of the complete installation, from head water to tail water, may be in the high 80s.

Impulse wheels with a single nozzle are most efficient in the specific speed range of 2 to 6, when P is in horsepower, H is in feet, and N is in revolutions per minute. Multiple nozzle units are designed in the specific speed range of 6 to 12.

When the centrifugal pump in the accompanying figure is rotating at 1,650 rpm, the steady-flow rate is 1,600 gpm. If the pump speed is increased instantaneously to 2,000 rpm, determine the flow rate as a function of time. Assume that the coefficient of loss at the pipe entrance to be 0.50 and the head developed by the pump is proportional to the square of the rotative speed.

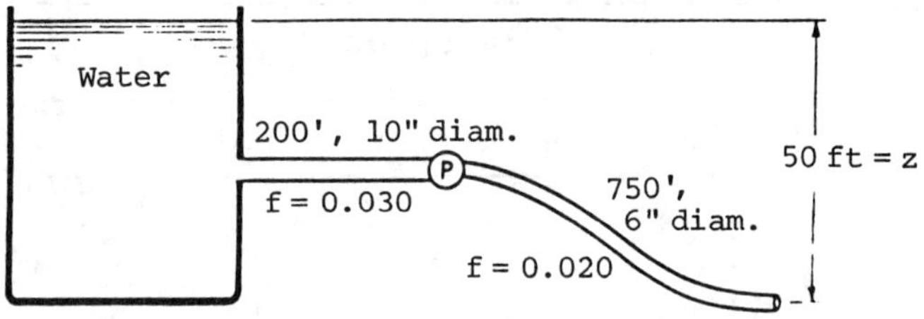

Solution: Writing the unsteady flow energy equation,

$$Z - 0.5\frac{V_1^2}{2g} - f_1\frac{L_1}{D_1}\frac{V_1^2}{2g} + h_p - f_2\frac{L_2}{D_2}\frac{V_2^2}{2g} = \frac{V_2^2}{2g} + \frac{L_1}{g}\frac{dV_1}{dt} + \frac{L_2}{g}\frac{dV_2}{dt} \qquad (1)$$

where the subscripts 1 and 2 refer to the 10- and 6-in.-diameter pipes, respectively. Note that the accelerative head for each pipe depends on the respective L and dV/dt values.

From continuity,

$$V_1 = \frac{A_2V_2}{A_1} = \left(\frac{6}{10}\right)^2 V_2 = 0.36V_2$$

Hence

$$\frac{dV_1}{dt} = \frac{A_2}{A_1}\frac{dV_2}{dt} = 0.36\frac{dV_2}{dt}$$

Thus substituting back into equation (1) yields,

$$50 - 0.5\frac{(0.36V_2)^2}{2g} - 0.030\left(\frac{200}{10/12}\right)\frac{(0.36V_2)^2}{2g} + h_p - 0.020\left(\frac{750}{6/12}\right)\frac{V_2^2}{2g}$$

$$= \frac{V_2^2}{2g} + \frac{200}{g}(0.36)\frac{dV_2}{dt} + \frac{750}{dt}\frac{dV_2}{dt}$$

Evaluating and combining terms,

$$50 + hp = 32.0\frac{V_2^2}{2g} + \frac{822}{g}\frac{dV_2}{dt} \qquad (2)$$

With the original steady-flow conditions $(dV/dt = 0)$,

$$V_2 = \frac{Q}{A_2} = \frac{1{,}600/449}{0.196} = 18.2 \text{ fps}$$

and

$$h_p = 32\frac{V_2^2}{2g} - 50 = 115 \text{ ft}$$

After the speed is increased to 2,000 rpm,

$$h_p = 115(2{,}000/1{,}650)^2 = 169 \text{ ft}$$

Substituting into (2),

$$50 + 169 = 32\frac{V_2^2}{2g} + \frac{822}{g}\frac{dV_2}{dt}$$

Knowing the relationship,

$$V_2^2 = \frac{Q^2}{A^2}$$

we can express the foregoing in terms of Q, yielding

$$219 = \frac{32}{2g}\frac{Q^2}{(0.196)^2} + \frac{822}{g(0.196)}\frac{dQ}{dt}$$

$$219 = 12.9Q^2 + 130\frac{dQ}{dt} \qquad \text{(b)}$$

Solving for dt and integrating, noting that at $t = 0$, $Q = 3.56$ cfs,

$$\int_0^t dt = 130\int_{3.56}^{Q}\frac{dQ}{219 - 12.9Q^2}$$

$$t = 122 \ln \frac{14.8 + 3.6Q}{14.8 - 3.6Q} - 320$$

$$e^{0.0082t + 2.62} = \frac{14.8 + 3.6Q}{14.8 - 3.6Q}$$

Finally,

$$Q = 4.10 \frac{e^{0.0082t + 2.62} - 1}{e^{0.0082t + 2.62} + 1}$$

Note that as $t \to \infty$, $Q \to 4.10$ cfs, the steady-state flow rate for the condition where $h_p = 169$ ft. It should also be noted that the speed of a pump cannot be changed instantaneously from one value to another, as was assumed in this example.

• PROBLEM 15-20

A pump that will deliver 84,500 gpm against a head of 225 ft when operating at 600 rpm is desired. Determine the specific speed of this pump and its approximate dimensions.

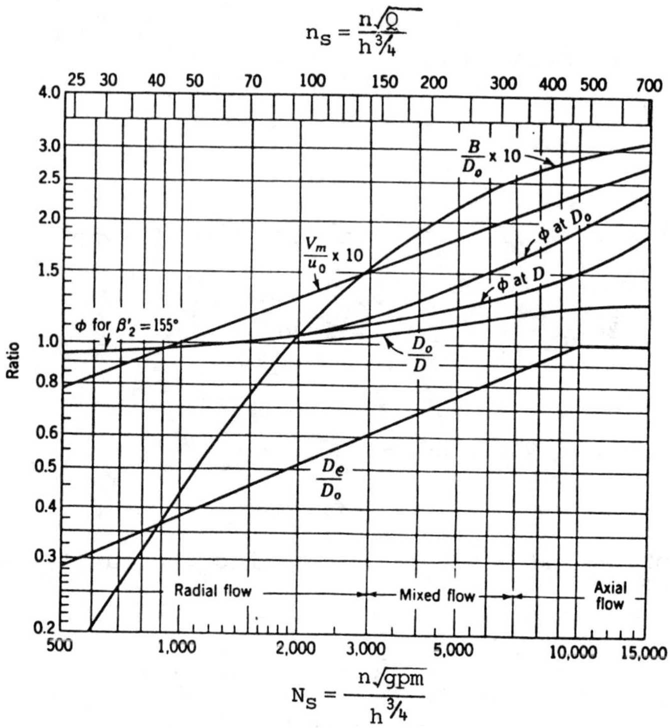

Factors and proportions for pumps Fig. 1

Solution:

$$N_s = \frac{600\sqrt{84,500}}{(225)^{3/4}} = 3,000 \qquad (1)$$

Letting $\beta_2{}^1 = 155°$, the velocity factor can be determined from figure (1)

$\phi = 1.1$ at diameter D. ⁻ence the peripheral velocity is

$$u_2 = \phi\sqrt{2gh} = 1.1\sqrt{2gh} = 132.5 \text{ fps}$$

Then from equation (2) we can determine the diameter,

$$D = \frac{153.2\phi\sqrt{h}}{N} \qquad (2)$$

$$D = 153.2 \times 1.1 \times \frac{\sqrt{225}}{600} = 4.22 \text{ ft} = 50.7 \text{ in.}$$

From Fig. 1,

$$\frac{D_o}{D} = 1.07 \qquad \frac{B}{D_o} = 0.155$$

$$\frac{D_e}{D_O} = 0.6 \qquad \frac{V_m}{u_O} = 0.15$$

Hence

$$D_o = 1.07 \times 50.7 = 54.3 \text{ in.}$$

$$B = 0.155 \times 54.3 = 8.42 \text{ in.}$$

The eye diameter is

$$D_e = 0.6 \times 54.3 = 32.6 \text{ in.}$$

The peripheral velocity at D_O is

$$D_O = 1.07 \times 132.5 = 142 \text{ fps}$$

$$(V_m)_2 = 0.15 \times 142 = 21.3 \text{ fps}$$

$$\text{Cirumferential area} = 0.95\pi \times 8.42 \times \frac{50.7}{144} = 8.85 \text{ ft}^2$$

$$Q = A_{circum}(V_m)_2 = 8.85 \times 21.3 = 188.5 \text{ cfs} = 84,500 \text{ gpm}$$

which checks the initial value.

• PROBLEM 15-21

(a) It is desired to deliver 1,600 gpm at a head of 900 ft with a single-stage pump. What would be the minimum speed that could be used?

(b) For the conditions of (a), how many stages must the pump have if a rotative speed of 600 rpm is to be used?

Solution: The gallons-per-minute basis specific speeds for volute centrifugal pumps range from 500 to 5,000.

(a) Assuming that the minimum practical specific speed is 500, we get

$$n = \frac{N_s h^{3/4}}{\sqrt{gpm}} = \frac{500(900)^{3/4}}{\sqrt{1,600}} = 2,060 \text{ rpm}$$

(b) $$h^{3/4} = \frac{n\sqrt{gpm}}{N_s} = \frac{600\sqrt{1,600}}{500} = 48$$

or

$$h = 175 \text{ ft per stage}$$

Hence

$$900/175 = 5.14 \quad \text{(6 stages are required)}$$

To meet the exact specifications of head and capacity, either the rotative speed or the specific speed or both could be changed slightly.

• PROBLEM 15-22

Determine (a) the specific speed of a pump that is to deliver 2,000 gpm against a head of 150 ft with a rotative speed of 600 rpm. (b) If the rotative speed were doubled, what would be the flow rate and the head developed by the pump? Assume no change in efficiency. (c) The specific speed for the conditions given in (b). And (d) find the required operating speed of a two-stage pump to satisfy the requirements in (a).

Solution: (a)

$$N_s = \frac{n\sqrt{gpm}}{h^{3/4}} = \frac{600\sqrt{2,000}}{(150)^{3/4}} = 625$$

(b) Since the speed rate is directly proportional to the rotative speed,

$Q \alpha\, n$, so $Q = 2 \times 2{,}000 = 4{,}000$ gpm

$h \alpha\, n^2$, so $h = 2^2 \times 150 = 600$ ft

(c) Using equation (1)

$$N_s = \frac{1{,}200\sqrt{4{,}000}}{(600)^{3/4}} = 625$$

This result was expected, for the same impeller was involved in (a) and (b).

(d) Using equation (1) and solving for n,

$$N_s = 625 = \frac{n\sqrt{2{,}000}}{(75)^{3/4}} = 365 \text{ rpm}$$

• PROBLEM 15-23

For the Thomson Jet Pump (Fig. 1), determine the volumetric flow rate for, a) the nozzle, Q_1, the diverge discharge pipe, Q_D and the suction pipe, Q_S. b) find the pressure $\frac{P_d}{w}$ at the throat of the mixing cone and $\frac{P_j}{w}$ in the plane of the orifices. And, c) determine the efficiency of the pump.

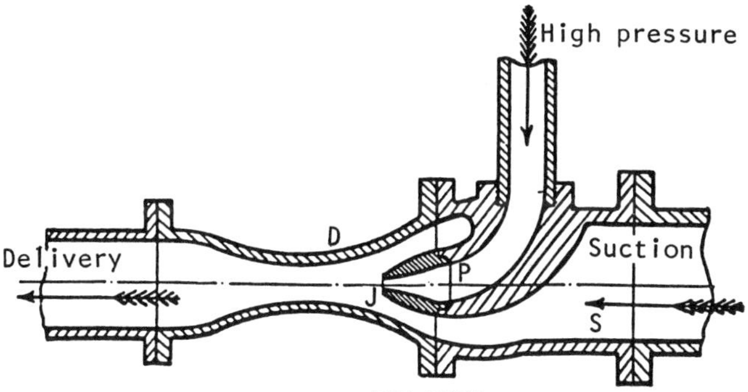

Fig. 1

Solution:

Let h_1 = height of supply head above jet in feet. = 40 feet

" h_s = height of jet above suction supply " " = 15 feet

" h_d = height of delivery head above jet " " = 10 feet

" a_1 = area of nozzle in square feet. = 0.2 sq. ft.

" a_s = area of annular suction pipe in plane of nozzle, square feet. = 0.4 sq. ft.

" a_d = area of mixing chamber at throat, square feet. = 0.6 sq. ft.

" Q = volume passing per second, v = velocity, and p the pressure in lbs. per square foot at the point denoted by a suffix.

a) To solve for the velocities, equation (1) is used.

or

$$(a_1 - a_s)h_1 - 2a_s h_s - a_d h_d = \frac{1}{2g}\{a_d v_d^2 - (a_1 + a_s)v_1^2\} . \quad (1)$$

yielding,

$$-(0.2 \times 40 + 0.8 \times 15 + 0.6 \times 10)\frac{64.4}{0.6} = V_d^2 - V_1^2$$

$$\therefore \quad V_1^2 - V_d^2 = 2,790 \quad (a)$$

A second velocity equation, eq. (2)

$$a_1 v_1 + a_s\sqrt{v_1^2 - 2g(h_1 + h_s)} = a_d v_d \quad (2)$$

yields

$$(0.6V_d - 0.2V_1)^2 = 0.16(V_1^2 - 64.4 \times 55)$$

$$\therefore \quad V_1^2 + 2V_1V_d - 3V_d^2 = 4,718 \quad (b)$$

Substituting for V_d^2 in (b) from (a)

$$V_1^2 + 2V_1\sqrt{V_1^2 - 2,790} - 3(V_1^2 - 2,790) = 4718$$

$$\therefore \quad V_1^2 - 1826 = V_1\sqrt{V_1^2 - 2790}$$

Squaring both sides and combining the V_1 terms:

$$V_1^2 = 3,866$$

$$\therefore \quad V_1 = 62.2 \text{ fps}$$

Substituting this value into (a)

$$V_d = \sqrt{3,866-2790} = 32.8 \text{ fps}$$

From equation (3):

or

$$v_s^2 = v_1^2 - 2g(h_1 + h_s). \tag{3}$$

yields

$$V_s = \sqrt{3,866-3542} = 18 \text{ fps}$$

Having found the velocities, the volumetric flow rate can now be determined.

$$Q_1 = 0.2V_1 = 12.4 \text{ c.f.s.}$$

$$Q_d = 0.6V_d = 19.7 \text{ c.f.s.}$$

$$Q_s = 0.4V_s = 7.2 \text{ c.f.s.}$$

b) The pressures at the throat and in the plane of the orifice are determined by equation (5) and (6).

$$\frac{P_j}{W} + \frac{V_1^2}{2g} = h_1 \tag{5}$$

$$\frac{P_d}{W} + \frac{V_d^2}{2g} = hd \tag{6}$$

yielding,

$$\frac{P_j}{W} = 40 - 60.1 = -20.1 \text{ ft. of water}$$

and

$$\frac{P_d}{W} = 10 - 16.7 = -6.7 \text{ ft. of water}$$

c) The efficiency of the pump is given by:

$$y = \frac{Q_s(H_d + h_s)}{Q_1(h_1 - H_d)}$$

Assume $H_d = 0.7h_d$

Because there is a loss of energy by eddy formation in the diverging discharge pipe, the actual height through which the water may be forced by the pump is less, then,

$$y = \frac{7.2(7 + 15)}{12.4 \times 33} = 0.387$$

This efficiency is low, because the action depends on the mixing of two streams moving with different velocities, and hence involves considerable loss of shock.

• PROBLEM 15-24

A. A Pelton wheel is to be selected to drive a generator at 600 rpm. The water jet is 75 mm in diameter and has a velocity of 100 m/s. With the blade angle at 170°, the ratio of vane speed to initial jet speed at 0.47, and neglecting losses, determine (a) diameter of wheel to centerline of buckets (vanes), (b) power developed, and (c) kinetic energy per newton remaining in the fluid.

B. A small impulse wheel is to be used to drive a generator for 60-Hz power. The head is 100 m, and the discharge is 40 l/s. Determine the diameter of the wheel at the centerline of the buckets and the speed of the wheel, $C_v = 0.98$. Assume efficiency of 80 percent.

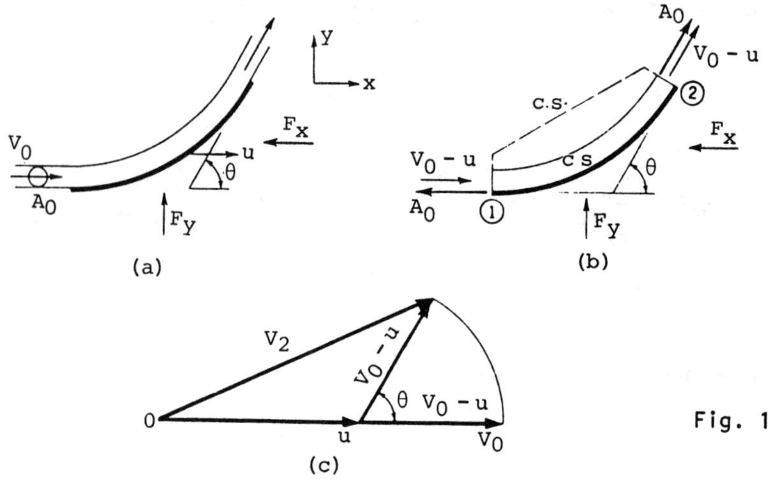

Fig. 1

(a) moving vane. (b) Vane flow viewed as steady-state problem by superposition of velocity u to the left. (c) polar vector diagram.

Solution: A) The peripheral speed of the wheel is

$$u = 0.47 \times 100 = 47 \text{ m/s}$$

Then

$$\frac{600}{60}\left(2\pi \frac{D}{2}\right) = 47 \text{ m/s}$$

or

$$D = 1.495 \text{ m}$$

(b) The power, in kilowatts, is computed to be

$$(1000 kg/m^3)\ \frac{\pi}{4}\ (0.075m)^2 (100m/s)(47\ m/s)(100 - 47\ m/s)$$

$$\times\ [1 - (-0.9848)]\ \frac{1\ kW}{1000\ W} = 2184\ kW$$

(c) From Fig. 1, the absolute velocity components leaving the vane are

$$V_x = (100 - 47)(-0.9848) + 47 = -5.2\ m/s$$

$$V_y = (100 - 47)(0.1736) = 9.2\ m/s$$

The kinetic energy remaining in the jet is

$$\frac{5.2^2 + 9.2^2}{2 \times 9.806} = 5.69\ m \cdot N/N$$

B) The power is

$$P = \gamma Q H_a e = 9806 \times 0.040 \times 100 \times 0.80 = 31.38\ kW$$

By taking a trial value of N_s of 15,

$$N = \frac{N_s H_a^{5/4}}{\sqrt{P}} = \frac{15 \times 100^{5/4}}{\sqrt{31.38}} = 847\ rpm$$

For 60-Hz power the speed must be 3600 divided by the number of pairs of poles in the generator. For five pairs of poles the speed would be 3600/5 = 720 rpm, and for four pairs of poles it would be 3600/4 = 900 rpm. The closer speed 900 is selected. Then

$$N_s = \frac{N\sqrt{P}}{H_a^{5/4}} = \frac{900\sqrt{31.38}}{100^{5/4}} = 15.94$$

For $N_S = 15.94$, $\phi = 0.448$

$$u = \phi\sqrt{2gH_a} = 0.448\sqrt{2 \times 9.806 \times 100} = 19.84\ m/s$$

and

$$\omega = \frac{900}{60}\ 2\pi = 94.25\ rad/s$$

The peripheral speed u and D and ω are related:

$$u = \frac{\omega D}{2} \qquad D = \frac{2u}{\omega} = \frac{2 \times 19.84}{94.25} = 421 \text{ mm}$$

The diameter d of the jet is obtained from the jet velocity V_2; thus

$$V_2 = C_v\sqrt{2gH_a} = 0.98\sqrt{2 \times 9.806 \times 100} = 43.4 \text{ m/s}$$

$$a = \frac{Q}{V_2} = \frac{0.040}{43.4} = 9.22 \text{ cm}^2 \qquad d = \sqrt{\frac{4a}{\pi}} = \sqrt{\frac{0.000922}{0.7854}} = 34.3 \text{ mm}$$

where a is the area of jet. Hence, the diameter ratio D/d is

$$\frac{D}{d} = \frac{421}{34.3} = 12.27$$

The desired diameter ratio for best efficiency is

$$\frac{D}{d} = \frac{206}{N_s} = \frac{206}{15.94} = 12.92$$

so the ratio D/d is satisfactory. The wheel diameter is 421 mm, and the speed is 900 rpm.

• PROBLEM 15-25

It is required to design a Pelton wheel to work under the following specifications:

1. Coefficient of velocity (C_v) =0.985
2. Efficiency Factor =0.85
3. Pitch circle radius (r) = 25.9 ft.
4. Effective head = 500 ft.
5. Velocity of pitch circle (V_{PC}) =0.46 ft./sec.
6. Must develop 800 H.P. at 360 rpm.

Solution: The velocity of efflux of the jet,

$$V_e = C_v \sqrt{2gh} = 0.985 \sqrt{2 \times (32.2) \times 500}$$

$$= \frac{177 \text{ ft}}{\text{sec}}$$

The peripheral velocity is,

$$V_P = V_{PC} \times V_e = 0.46(177) = \frac{81.3 \text{ ft}}{\text{sec}}$$

The radius of pitch is,

$$r_P = \frac{V_P \times 60}{2\pi \times 360} = 2.158 \text{ ft.}$$

i.e., the diameter of pitch = 4.2 ft. Now the energy passing the nozzle per second is

$$\frac{\text{H.P.} \times 550}{0.85} = 518{,}000 \text{ ft. lbs.}$$

and since each cubic foot of water contains

$$\frac{62.4 \times (177)^2}{2g} = 30{,}380 \text{ ft. lbs}$$

in the form of kinetic energy, this requires,

$$\frac{518{,}000}{30{,}380} = \frac{17.06 \text{ ft.}^3}{\text{sec}}$$

The required area of the nozzle is

$$\frac{17.06}{177} = 13.89 \text{ ins.}^2$$

To determine the number of buckets take

$$n = 7.5\sqrt{\frac{r}{t}}$$

where

r = pitch circle radius

t = diameter of pitch

$$n = 7.5\sqrt{\frac{25.9}{4.2}} = 18.6$$

This value may be rounded off to 20 for convenience in balancing.

Next applying the formula and solving for s,

$$n = \frac{\pi}{\sqrt{1 - \left(\frac{r + t/2}{r + s}\right)^2}}$$

yields s = 2.5 in., the amount by which the buckets must project beyond the pitch circle for continuous impact.

• PROBLEM 15-26

A centrifugal pump has the following characteristics:

Tip diameter (D_2) = 5"
Hub diameter (D_1) = 1¼"
Impeller exit width (e_2) = 1/8"
Impeller speed (N) = 3460 rpm
Blade outlet angle (β_2) = 30°
Water enters at 10 ft/sec

Assume: (1) Efficiency (η) = 100%
(2) $V_1 = V_{n_1} = V_{n2} = V_n$
(3) No recirculation and uniform flow

Determine:

(1) Rate of flow (Q)
(2) Total head (H)
(3) Power (Pe)
(4) Velocities: $V_1, V_2, V_{r_1}, V_{r_2}$
(5) Blade inlet angle (β_1)
(6) Specific Speed (N_S)
(7) Torque input to the runner (T_{shaft})

Solution: (1) Rate of flow

$$Q = VA_1 = 10 \text{ ft/sec}\left(\frac{\pi D_1^2}{4}\right) = \frac{10\pi}{4}\left(\frac{1.25}{12}\right)^2$$

$$Q = 0.085 \text{ ft}^3\text{/sec}$$

(2) The total head for a centrifugal pump is defined as

$$H = \frac{u_2}{g}\left(u_2 - \frac{Q \text{ Cot}\beta_2}{A_2}\right)$$

$$u_2 = \pi N D_2 = \pi \frac{3460}{60}\left(\frac{5}{12}\right) = 75.48 \text{ ft/sec}$$

$$A_2 = \pi D_2 e_2 = \pi\left(\frac{5}{12}\right)\left(\frac{1}{8 \times 12}\right) = 0.0136 \text{ ft}^2$$

$$H = \frac{75.48}{32.2}\left(75.48 - \frac{0.085 \text{ Cot } 30^0}{0.0136}\right) = 151.6 \text{ ft}$$

(3) Power

$$Pe = \frac{\gamma QH}{\eta} = \frac{62.4(0.085)(15.8)}{1(550)}$$

$$Pe = 1.46 \text{ HP}$$

(4) Velocities

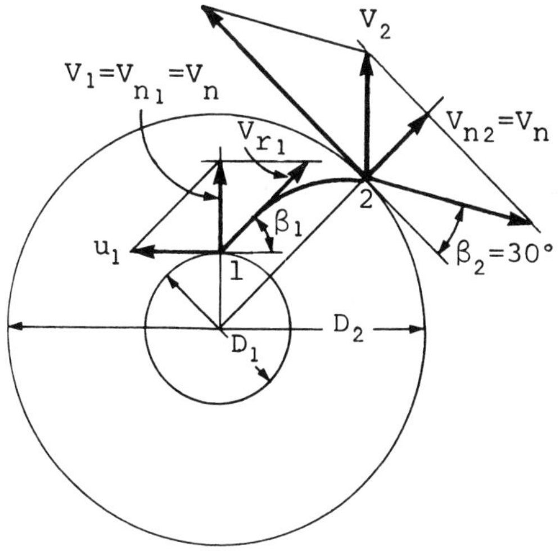

$$V_1 = V_{n_1} = V_{n_2} = \frac{Q}{A_2} = \frac{0.085}{0.0136} = 6.25 \text{ ft/sec}$$

$$u_1 = \pi ND_1 = \pi\left(\frac{3460}{60}\right)\left(\frac{1.25}{12}\right) = 18.87 \text{ ft/sec}$$

$$\tan \beta_1 = \frac{V_1}{u_1} = \frac{6.25}{18.87} = 0.33$$

$$\beta_1 = 18.33^0 \quad \text{(Blade inlet angle)}$$

From the inlet velocity polygon, we have

$$V_{r_1} = \sqrt{u_1^2+V_1^2} = \sqrt{(18.87)^2+(6.25)^2} = 19.87 \text{ ft/sec}$$

from the outlet velocity polygon, we get

$$V_{r_2} = \frac{V_{n_2}}{\sin \beta_2} = \frac{6.25}{\sin 30^0} = 12.5 \text{ ft/sec}$$

$$V_2 = \sqrt{u_2^2 + V_{r_2}^2 - 2u_2 V_{r_2} \cos 30^0}$$

$$V_2 = 64.95 \text{ ft/sec}$$

(6) Specific speed

$$N_s = \frac{N\sqrt{Q}}{H^{3/4}} = \frac{3460(0.085 \times 7.48 \times 60)^{\frac{1}{2}}}{(151.6)^{3/4}}$$

$$N_s = 494.6$$

(7) From the moment of momentum equation, we have

$$T_{shaft} = N\left(\frac{D_2}{2}\right)^2 \rho Q$$

$$= 3460\left(\frac{2\pi}{60}\right)\left(\frac{5}{2 \times 12}\right)^2 (1.94)0.085$$

$$T_{shaft} = 2.6 \text{ ft-lb}_f.$$

APPENDIX

Constants and Conversion Factors

MATHEMATICAL CONSTANTS

$$e = 2.71828\ldots$$
$$\ln 10 = 2.30259\ldots$$
$$\pi = 3.14159\ldots$$

PHYSICAL CONSTANTS

Gas law constant	$R = 1.987_2$ cal g-mole^{-1} °K^{-1}
	$= 82.05_7$ cm^3 atm g-mole^{-1} °K^{-1}
	$= 8.314_4 \times 10^7$ g cm^2 sec^{-2} g-mole^{-1} °K^{-1}
	$= 8.314_4 \times 10^3$ kg m^2 sec^{-2} kg-mole^{-1} °K^{-1}
	$= 4.968_6 \times 10^4$ lb$_m$ ft^2 sec^{-2} lb-mole^{-1} °R^{-1}
	$= 1.544_3 \times 10^3$ ft lb$_f$ lb-mole^{-1} °R^{-1}
Standard acceleration of gravity	$g_0 = 980.665$ cm sec^{-2}
	$= 32.1740$ ft sec^{-2}
Joule's constant (mechanical equivalent of heat)	$J_c = 4.1840 \times 10^7$ erg cal^{-1}
	$= 778.16$ ft lb$_f$ Btu^{-1}
Avogadro's number	$\tilde{N} = 6.02_3 \times 10^{23}$ molecules g-mole^{-1}
Boltzmann's constant	$\kappa = R/\tilde{N} = 1.380_5 \times 10^{-16}$ erg molecule^{-1} °K^{-1}
Faraday's constant	$\mathscr{F} = 9.652 \times 10^4$ abs-coulombs g-equivalent^{-1}
Planck's constant	$h = 6.62_4 \times 10^{-27}$ erg sec

Stefan-Boltzmann constant	$\sigma = 1.355 \times 10^{-12}$ cal sec^{-1} cm^{-2} °K^{-4}
	$= 0.1712 \times 10^{-8}$ Btu hr^{-1} ft^{-2} °R^{-4}
Electronic charge	$e = 1.602 \times 10^{-19}$ abs-coulomb
Speed of light	$c = 2.99793 \times 10^{10}$ cm sec^{-1}

CONVERSION FACTORS

To convert any physical quantity from one set of units into another, multiply by the appropriate table entry. For example, suppose that p is given as 10 lb_f in^{-2} but is desired in units of poundals ft^{-2}. From Table C.3–2, the result is

$$p = 10 \times 4.6330 \times 10^3 = 4.6330 \times 10^4 \text{ poundals ft}^{-2} \quad \text{or} \quad \text{lb}_m \text{ ft}^{-1} \text{ sec}^{-2}$$

In addition to the extended tables, we give a few of the commonly used conversion factors here:

Given a quantity in these units	Multiply by	To get quantity in these units
Pounds	453.59	Grams
Kilograms	2.2046	Pounds
Inches	2.5400	Centimeters
Meters	39.370	Inches
Gallons (U.S.)	3.7853	Liters
Gallons (U.S.)	231.00	Cubic inches
Gallons (U.S.)	0.13368	Cubic feet
Cubic feet	28.316	Liters

Standard-atmosphere Table

Altitude, ft	Temperature, °F	Pressure in. Hg	Pressure psf	ρ/ρ_0†	c, fps
0	59.0	29.92	2,116	1.000	1,117
1,000	55.4	28.86	2,041	0.9711	1,113
2,000	51.9	27.82	1,968	0.9428	1,109
3,000	48.3	26.82	1,897	0.9151	1,105
4,000	44.7	25.84	1,828	0.8881	1,101
5,000	41.2	24.90	1,761	0.8617	1,098
6,000	37.6	23.98	1,696	0.8359	1,094
7,000	34.0	23.09	1,633	0.8106	1,090
8,000	30.5	22.22	1,572	0.7860	1,086
9,000	26.9	21.39	1,513	0.7620	1,082
10,000	23.3	20.58	1,455	0.7385	1,078
11,000	19.8	19.79	1,400	0.7156	1,074
12,000	16.2	19.03	1,346	0.6932	1,070
13,000	12.6	18.29	1,294	0.6713	1,066
14,000	9.1	17.58	1,243	0.6500	1,062
15,000	5.5	16.89	1,194	0.6292	1,058
16,000	1.9	16.22	1,147	0.6090	1,054
17,000	− 1.6	15.57	1,101	0.5892	1,050
18,000	− 5.2	14.94	1,057	0.5699	1,045
19,000	− 8.8	14.34	1,014	0.5511	1,041
20,000	−12.3	13.75	972.5	0.5328	1,037
21,000	−15.9	13.18	932.4	0.5150	1,033
22,000	−19.5	12.64	893.7	0.4976	1,029
23,000	−23.0	12.11	856.3	0.4806	1,025
24,000	−26.6	11.60	820.2	0.4642	1,021
25,000	−30.2	11.10	785.3	0.4481	1,016
26,000	−33.7	10.63	751.6	0.4325	1,012
27,000	−37.3	10.17	719.1	0.4173	1,008
28,000	−40.9	9.725	687.8	0.4025	1,004
29,000	−44.4	9.297	657.6	0.3881	999
30,000	−48.0	8.885	628.4	0.3741	995
31,000	−51.6	8.488	600.3	0.3605	991
32,000	−55.1	8.106	573.3	0.3473	986
33,000	−58.7	7.737	547.2	0.3345	982
34,000	−62.2	7.382	522.1	0.3220	978
35,000	−65.8	7.041	498.0	0.3099	973
36,000	−67.6	6.702	474.8	0.2971	971
37,000	−67.6	6.397	452.5	0.2844	971

† $\rho_0 = 0.002378$ slug/ft³.

Physical Properties of Water

1. Latent Heat of Water at 273. 15 K (0°C)

Latent heat of fusion = 1436.3 cal/g mol

= 79.724 cal/g

= 2585.3 btu/lb mol

= 6013.4 kJ/kg mol

Latent heat of vaporization at 298.15 K (25°C)

Pressure (mm Hg)	*Latent Heat*
23.75	44,020 kJ/kg mol, 10.514 kcal/g mol, 18,925 btu/lb mol
760	44,045 kJ/kg mol, 10.520 kcal/g mol, 18,936 btu/lb mol

2. Vapor Pressure of Water

Temperature		*Vapor Pressure*		*Temperature*		*Vapor Pressure*	
K	*°C*	*kPa*	*mm Hg*	*K*	*°C*	*kPa*	*mm Hg*
273.15	0	0.611	4.58	323.15	50	12.333	92.51
283.15	10	1.228	9.21	333.15	60	19.92	149.4
293.15	20	2.338	17.54	343.15	70	31.16	233.7
298.15	25	3.168	23.76	353.15	80	47.34	355.1
303.15	30	4.242	31.82	363.15	90	70.10	525.8
313.15	40	7.375	55.32	373.15	100	101.325	760.0

3. Density of Liquid Water

Temperature K	Temperature °C	Density g/cm^3	Density kg/m^3	Temperature K	Temperature °C	Density g/cm^3	Density kg/m^3
273.15	0	0.99987	999.87	323.15	50	0.98807	988.07
277.15	4	1.00000	1000.00	333.15	60	0.98324	983.24
283.15	10	0.99973	999.73	343.15	70	0.97781	977.81
293.15	20	0.99823	998.23	353.15	80	0.97183	971.83
303.15	30	0.99568	995.68	363.15	90	0.96534	965.34
313.15	40	0.99225	992.25	373.15	100	0.95838	958.38

4. Viscosity of Liquid Water

Temperature K	Temperature °C	Viscosity [$(Pa \cdot s)\,10^3$, $(kg/m \cdot s)\,10^3$, or cp]	K	°C	Viscosity [$(Pa \cdot s)\,10^3$, $(kg/m \cdot s)\,10^3$, or cp]
273.15	0	1.7921	323.15	50	0.5494
275.15	2	1.6728	325.15	52	0.5315
277.15	4	1.5674	327.15	54	0.5146
279.15	6	1.4728	329.15	56	0.4985
281.15	8	1.3860	331.15	58	0.4832
283.15	10	1.3077	333.15	60	0.4688
285.15	12	1.2363	335.15	62	0.4550
287.15	14	1.1709	337.15	64	0.4418
289.15	16	1.1111	339.15	66	0.4293
291.15	18	1.0559	341.15	68	0.4174
293.15	20	1.0050	343.15	70	0.4061
293.35	20.2	1.0000	345.15	72	0.3952
295.15	22	0.9579	347.15	74	0.3849
297.15	24	0.9142	349.15	76	0.3750
299.15	26	0.8737	351.15	78	0.3655
301.15	28	0.8360	353.15	80	0.3565
303.15	30	0.8007	355.15	82	0.3478
305.15	32	0.7679	357.15	84	0.3395
307.15	34	0.7371	359.15	86	0.3315
309.15	36	0.7085	361.15	88	0.3239
311.15	38	0.6814	363.15	90	0.3165
313.15	40	0.6560	365.15	92	0.3095
315.15	42	0.6321	367.15	94	0.3027
317.15	44	0.6097	369.15	96	0.2962
319.15	46	0.5883	371.15	98	0.2899
321.15	48	0.5683	373.15	100	0.2838

5. Heat Capacity of Liquid Water at 101. 325 kPa (1 Atm)

Temperature		Heat Capacity, c_p		Temperature		Heat Capacity, c_p	
°C	K	cal/g·°C	kJ/kg·K	°C	K	cal/g·°C	kJ/kg·K
0	273.15	1.0080	4.220	60	333.15	1.0001	4.187
10	283.15	1.0019	4.195	70	343.15	1.0013	4.192
20	293.15	0.9995	4.158	80	353.15	1.0029	4.199
30	303.15	0.9987	4.181	90	363.15	1.0050	4.208
40	313.15	0.9987	4.181	100	373.15	1.0076	4.219
50	323.15	0.9992	4.183				

6. Thermal Conductivity of Liquid Water

Temperature			Thermal Conductivity	
°C	°F	K	btu/h·ft·°F	W/m·K
0	32	273.15	0.329	0.569
37.8	100	311.0	0.363	0.628
93.3	200	366.5	0.393	0.680
148.9	300	422.1	0.395	0.684
215.6	420	588.8	0.376	0.651
326.7	620	599.9	0.275	0.476

7. Vapor Pressure of Saturated Ice-Water Vapor and Heat of Sublimation

Temperature			Vapor Pressure			Heat of Sublimation	
K	°F	°C	kPa	psia	mm Hg	btu/lb_m	kJ/kg
273.2	32	0	6.107×10^{-1}	8.858×10^{-2}	4.581	1218.6	2834.5
266.5	20	−6.7	3.478×10^{-1}	5.045×10^{-2}	2.609	1219.3	2836.1
261.0	10	−12.2	2.128×10^{-1}	3.087×10^{-2}	1.596	1219.7	2837.0
255.4	0	−17.8	1.275×10^{-1}	1.849×10^{-2}	0.9562	1220.1	2838.0
249.9	−10	−23.3	7.411×10^{-2}	1.082×10^{-2}	0.5596	1220.3	2838.4
244.3	−20	−28.9	3.820×10^{-2}	6.181×10^{-3}	0.3197	1220.5	2838.9
238.8	−30	−34.4	2.372×10^{-2}	3.440×10^{-3}	0.1779	1220.5	2838.9
233.2	−40	−40.0	1.283×10^{-2}	1.861×10^{-3}	0.09624	1220.5	2838.9

8. Heat Capacity of Ice

Temperature		c_p		*Temperature*		c_p	
°F	K	btu/lb$_m$·°F	kJ/kg·K	°F	K	btu/lb$_m$·°F	kJ/kg·K
32	273.15	0.500	2.093	−10	249.85	0.461	1.930
20	266.45	0.490	2.052	−20	244.25	0.452	1.892
10	260.95	0.481	2.014	−30	238.75	0.442	1.850
0	255.35	0.472	1.976	−40	233.15	0.433	1.813

9. Properties of Saturated Steam and Water (Steam Table), SI Units

Temperature (°C)	*Vapor Pressure (kPa)*	*Specific Volume (m³/kg)*		*Enthalpy (kJ/kg)*		*Entropy (kJ/kg·K)*	
		Liquid	*Sat'd Vapor*	*Liquid*	*Sat'd Vapor*	*Liquid*	*Sat'd Vapor*
0.01	0.6113	0.0010002	206.136	0.00	2501.4	0.0000	9.1562
3	0.7577	0.0010001	168.132	12.57	2506.9	0.0457	9.0773
6	0.9349	0.0010001	137.734	25.20	2512.4	0.0912	9.0003
9	1.1477	0.0010003	113.386	37.80	2517.9	0.1362	8.9253
12	1.4022	0.0010005	93.784	50.41	2523.4	0.1806	8.8524
15	1.7051	0.0010009	77.926	62.99	2528.9	0.2245	8.7814
18	2.0640	0.0010014	65.038	75.58	2534.4	0.2679	8.7123
21	2.487	0.0010020	54.514	88.14	2539.9	0.3109	8.6450
24	2.985	0.0010027	45.883	100.70	2545.4	0.3534	8.5794
27	3.567	0.0010035	38.774	113.25	2550.8	0.3954	8.5156
30	4.246	0.0010043	32.894	125.79	2556.3	0.4369	8.4533
33	5.034	0.0010053	28.011	138.33	2561.7	0.4781	8.3927
36	5.947	0.0010063	23.940	150.86	2567.1	0.5188	8.3336
40	7.384	0.0010078	19.523	167.57	2574.3	0.5725	8.2570
45	9.593	0.0010099	15.258	188.45	2583.2	0.6387	8.1648
50	12.349	0.0010121	12.032	209.33	2592.1	0.7038	8.0763
55	15.758	0.0010146	9.568	230.23	2600.9	0.7679	7.9913
60	19.940	0.0010172	7.671	251.13	2609.6	0.8312	7.9096
65	25.03	0.0010199	6.197	272.06	2618.3	0.8935	7.8310
70	31.19	0.0010228	5.042	292.98	2626.8	0.9549	7.7553
75	38.58	0.0010259	4.131	313.93	2635.3	1.0155	7.6824
80	47.39	0.0010291	3.407	334.91	2643.7	1.0753	7.6122
85	57.83	0.0010325	2.828	355.90	2651.9	1.1343	7.5445
90	70.14	0.0010360	2.361	376.92	2660.1	1.1925	7.4791
95	84.55	0.0010397	1.9819	397.96	2668.1	1.2500	7.4159
100	101.35	0.0010435	1.6729	419.04	2676.1	1.3069	7.3549
105	120.82	0.0010475	1.4194	440.15	2683.8	1.3630	7.2958
110	143.27	0.0010516	1.2102	461.30	2691.5	1.4185	7.2387

(continued)

Temperature (C)	Vapor Pressure (kPa)	Specific Volume (m^3/kg)		Enthalpy (kJ/kg)		Entropy (kJ/kg · K)	
		Liquid	Sat'd Vapor	Liquid	Sat'd Vapor	Liquid	Sat'd Vapor
115	169.06	0.0010559	1.0366	482.48	2699.0	1.4734	7.1833
120	198.53	0.0010603	0.8919	503.71	2706.3	1.5276	7.1296
125	232.1	0.0010649	0.7706	524.99	2713.5	1.5813	7.0775
130	270.1	0.0010697	0.6685	546.31	2720.5	1.6344	7.0269
135	313.0	0.0010746	0.5822	567.69	2727.3	1.6870	6.9777
140	316.3	0.0010797	0.5089	589.13	2733.9	1.7391	6.9299
145	415.4	0.0010850	0.4463	610.63	2740.3	1.7907	6.8833
150	475.8	0.0010905	0.3928	632.20	2746.5	1.8418	6.8379
155	543.1	0.0010961	0.3468	653.84	2752.4	1.8925	6.7935
160	617.8	0.0011020	0.3071	675.55	2758.1	1.9427	6.7502
165	700.5	0.0011080	0.2727	697.34	2763.5	1.9925	6.7078
170	791.7	0.0011143	0.2428	719.21	2768.7	2.0419	6.6663
175	892.0	0.0011207	0.2168	741.17	2773.6	2.0909	6.6256
180	1002.1	0.0011274	0.19405	763.22	2778.2	2.1396	6.5857
190	1254.4	0.0011414	0.15654	807.62	2786.4	2.2359	6.5079
200	1553.8	0.0011565	0.12736	852.45	2793.2	2.3309	6.4323
225	2548	0.0011992	0.07849	966.78	2803.3	2.5639	6.2503
250	3973	0.0012512	0.05013	1085.36	2801.5	2.7927	6.0730
275	5942	0.0013168	0.03279	1210.07	2785.0	3.0208	5.8938
300	8581	0.0010436	0.02167	1344.0	2749.0	3.2534	5.7045

10. Properties of Saturated Steam and Water (Steam Table), English Units

Temperature (°F)	Vapor Pressure (psia)	Specific Volume (ft^3/lb_m)		Enthalpy (btu/lb_m)		Entropy ($btu/lb_m \cdot °F$)	
		Liquid	Sat'd Vapor	Liquid	Sat'd Vapor	Liquid	Sat'd Vapor
32.02	0.08866	0.016022	3302	0.00	1075.4	0.000	2.1869
35	0.09992	0.016021	2948	3.00	1076.7	0.00607	2.1764
40	0.12166	0.016020	2445	8.02	1078.9	0.01617	2.1592
45	0.14748	0.016021	2037	13.04	1081.1	0.02618	2.1423
50	0.17803	0.016024	1704.2	18.06	1083.3	0.03607	2.1259
55	0.2140	0.016029	1431.4	23.07	1085.5	0.04586	2.1099
60	0.2563	0.016035	1206.9	28.08	1087.7	0.05555	2.0943
65	0.3057	0.016042	1021.5	33.09	1089.9	0.06514	2.0791

Temperature (°F)	Vapor Pressure (psia)	Specific Volume (ft^3/lb_m)		Enthalpy (btu/lb_m)		Entropy ($btu/lb_m \cdot °F$)	
		Liquid	Sat'd Vapor	Liquid	Sat'd Vapor	Liquid	Sat'd Vapor
70	0.3622	0.016051	867.7	38.09	1092.0	0.07463	2.0642
75	0.4300	0.016061	739.7	43.09	1094.2	0.08402	2.0497
80	0.5073	0.016073	632.8	48.09	1096.4	0.09332	2.0356
85	0.5964	0.016085	543.1	53.08	1098.6	0.10252	2.0218
90	0.6988	0.016099	467.7	58.07	1100.7	0.11165	2.0083
95	0.8162	0.016114	404.0	63.06	1102.9	0.12068	1.9951
100	0.9503	0.016130	350.0	68.05	1105.0	0.12963	1.9822
110	1.2763	0.016166	265.1	78.02	1109.3	0.14730	1.9574
120	1.6945	0.016205	203.0	88.00	1113.5	0.16465	1.9336
130	2.225	0.016247	157.17	97.98	1117.8	0.18172	1.9109
140	2.892	0.016293	122.88	107.96	1121.9	0.19851	1.8892
150	3.722	0.016343	96.99	117.96	1126.1	0.21503	1.8684
160	4.745	0.016395	77.23	127.96	1130.1	0.23130	1.8484
170	5.996	0.016450	62.02	137.97	1134.2	0.24732	1.8293
180	7.515	0.016509	50.20	147.99	1138.2	0.26311	1.8109
190	9.343	0.016570	40.95	158.03	1142.1	0.27866	1.7932
200	11.529	0.016634	33.63	168.07	1145.9	0.29400	1.7762
210	14.125	0.016702	27.82	178.14	1149.7	0.30913	1.7599
212	14.698	0.016716	26.80	180.16	1150.5	0.31213	1.7567
220	17.188	0.016772	23.15	188.22	1153.5	0.32406	1.7441
230	20.78	0.016845	19.386	198.32	1157.1	0.33880	1.7289
240	24.97	0.016922	16.327	208.44	1160.7	0.35335	1.7143
250	29.82	0.017001	13.826	218.59	1164.2	0.36772	1.7001
260	35.42	0.017084	11.768	228.76	1167.6	0.38193	1.6864
270	41.85	0.017170	10.066	238.95	1170.9	0.39597	1.6731
280	49.18	0.017259	8.650	249.18	1174.1	0.40986	1.6602
290	57.33	0.017352	7.467	259.44	1177.2	0.42360	1.6477
300	66.98	0.017448	6.472	269.73	1180.2	0.43720	1.6356
310	77.64	0.017548	5.632	280.06	1183.0	0.45067	1.6238
320	89.60	0.017652	4.919	290.43	1185.8	0.46400	1.6123
330	103.00	0.017760	4.312	300.84	1188.4	0.47722	1.6010
340	117.93	0.017872	3.792	311.30	1190.8	0.49031	1.5901
350	134.53	0.017988	3.346	321.80	1193.1	0.50329	1.5793
360	152.92	0.018108	2.961	332.35	1195.2	0.51617	1.5688
370	173.23	0.018233	2.628	342.96	1197.2	0.52894	1.5585
380	195.60	0.018363	2.339	353.62	1199.0	0.54163	1.5483
390	220.2	0.018498	2.087	364.34	1200.6	0.55422	1.5383
400	247.1	0.018638	1.8661	375.12	1202.0	0.56672	1.5284
410	276.5	0.018784	1.6726	385.97	1203.1	0.57916	1.5187
450	422.1	0.019433	1.1011	430.2	1205.6	0.6282	1.4806

11. Properties of Superheated Steam (Steam Table), SI Units (v, specific volume, m^3/kg; H, enthalpy, kJ/kg; s, entropy, kJ/kg·K)

Absolute Pressure, kPa (Sat. Temp., °C)		*Temperature (°C)*							
		100	*150*	*200*	*250*	*300*	*360*	*420*	*500*
	v	17.196	19.512	21.825	24.136	26.445	29.216	31.986	35.679
10	H	2687.5	2783.0	2879.5	2977.3	3076.5	3197.6	3320.9	3489.1
(45.81)	s	8.4479	8.6882	8.9038	9.1002	9.2813	9.4821	9.6682	9.8978
	v	3.418	3.889	4.356	4.820	5.284	5.839	6.394	7.134
50	H	2682.5	2780.1	2877.7	2976.0	3075.5	3196.8	3320.4	3488.7
(81.33)	s	7.6947	7.9401	8.1580	8.3556	8.5373	8.7385	8.9249	9.1546
	v	2.270	2.587	2.900	3.211	3.520	3.891	4.262	4.755
75	H	2679.4	2778.2	2876.5	2975.2	3074.9	3196.4	3320.0	3488.4
(91.78)	s	7.5009	7.7496	7.9690	8.1673	8.3493	8.5508	8.7374	8.9672
	v	1.6958	1.9364	2.172	2.406	2.639	2.917	3.195	3.565
100	H	2676.2	2776.4	2875.3	2974.3	3074.3	3195.9	3319.6	3488.1
(99.63)	s	7.3614	7.6134	7.8343	8.0333	8.2158	8.4175	8.6042	8.8342
	v		1.2853	1.4443	1.6012	1.7570	1.9432	2.129	2.376
150	H		2772.6	2872.9	2972.7	3073.1	3195.0	3318.9	3487.6
(111.37)	s		7.4193	7.6433	7.8438	8.0720	8.2293	8.4163	8.6466
	v		0.4708	0.5342	0.5951	0.6548	0.7257	0.7960	0.8893
400	H		2752.8	2860.5	2964.2	3066.8	3190.3	3315.3	3484.9
(143.63)	s		6.9299	7.1706	7.3789	7.5662	7.7712	7.9598	8.1913
	v			0.2999	0.3363	0.3714	0.4126	0.4533	0.5070
700	H			2844.8	2953.6	3059.1	3184.7	3310.9	3481.7
(164.97)	s			6.8865	7.1053	7.2979	7.5063	7.6968	7.9299
	v			0.2060	0.2327	0.2579	0.2873	0.3162	0.3541
1000	H			2827.9	2942.6	3051.2	3178.9	3306.5	3478.5
(179.91)	s			6.6940	6.9247	7.1229	7.3349	7.5275	7.7622
	v			0.13248	0.15195	0.16966	0.18988	0.2095	0.2352
1500	H			2796.8	2923.3	3037.6	3.1692	3299.1	3473.1
(198.32)	s			6.4546	6.7090	6.9179	7.1363	7.3323	7.5698
	v				0.11144	0.12547	0.14113	0.15616	0.17568
2000	H				2902.5	3023.5	3159.3	3291.6	3467.6
(212.42)	s				6.5453	6.7664	6.9917	7.1915	7.4317
	v				0.08700	0.09890	0.11186	0.12414	0.13998
2500	H				2880.1	3008.8	3149.1	3284.0	3462.1
(223.99)	s				6.4085	6.6438	6.8767	7.0803	7.3234
	v				0.07058	0.08114	0.09233	0.10279	0.11619
3000	H				2855.8	2993.5	3138.7	3276.3	3456.5
(233.90)	s				6.2872	6.5390	6.7801	6.9878	7.2338

12. Properties of Superheated Steam (Steam Table), English Units (v, specific volume, ft^3/lb_m; H, enthalpy, btu/lb_m; s, entropy, $btu/lb_m \cdot {}^\circ F$)

Absolute Pressure, psia (Sat. Temp., °F)		*Temperature (°F)* 200	300	400	500	600	700	800	900	1000
	v	392.5	452.3	511.9	571.5	631.1	690.7	750.3	809.9	869.5
1.0	*H*	1150.1	1195.7	1241.8	1288.5	1336.1	1384.5	1433.7	1483.8	1534.8
(101.70)	*s*	2.0508	2.1150	2.1720	2.2235	2.2706	2.3142	2.3550	2.3932	2.4294
	v	78.15	90.24	102.24	114.20	126.15	138.08	150.01	161.94	173.86
5.0	*H*	1148.6	1194.8	1241.2	1288.2	1335.8	1384.3	1433.5	1483.7	1534.7
(162.21)	*s*	1.8715	1.9367	1.9941	2.0458	2.0930	2.1367	2.1775	2.2158	2.2520
	v	38.85	44.99	51.03	57.04	63.03	69.01	74.98	80.95	86.91
10.0	*H*	1146.6	1193.7	1240.5	1287.7	1335.5	1384.0	1433.3	1483.5	1534.6
(193.19)	*s*	1.7927	1.8592	1.9171	1.9690	2.0164	2.0601	2.1009	2.1393	2.1755
	v		30.52	34.67	38.77	42.86	46.93	51.00	55.07	59.13
14.696	*H*		1192.6	1239.9	1287.3	1335.2	1383.8	1433.1	1483.4	1534.5
(211.99)	*s*		1.8157	1.8741	1.9263	1.9737	2.0175	2.0584	2.0967	2.1330
	v		22.36	25.43	28.46	31.47	34.77	37.46	40.45	43.44
20.0	*H*		1191.5	1239.2	1286.8	1334.8	1383.5	1432.9	1483.2	1534.3
(227.96)	*s*		1.7805	1.8395	1.8919	1.9395	1.9834	2.0243	2.0627	2.0989
	v		7.260	8.353	9.399	10.425	11.440	12.448	13.452	14.454
60.0	*H*		1181.9	1233.5	1283.0	1332.1	1381.4	1431.2	1481.8	1533.2
(292.73)	*s*		1.6496	1.7134	1.7678	1.8165	1.8609	1.9022	1.9408	1.9773
	v			4.934	5.587	6.216	6.834	7.445	8.053	8.657
100.0	*H*			1227.5	1279.1	1329.3	1379.2	1429.6	1480.5	1532.1
(327.86)	*s*			1.6517	1.7085	1.7582	1.8033	1.8449	1.8838	1.9204
	v			3.221	3.679	4.111	4.531	4.944	5.353	5.759
150.0	*H*			1219.5	1274.1	1325.7	1376.6	1427.5	1478.8	1530.7
(358.48)	*s*			1.5997	1.6598	1.7110	1.7568	1.7989	1.8381	1.8750
	v			2.361	2.724	3.058	3.379	3.693	4.003	4.310
200.0	*H*			1210.8	1268.8	1322.1	1373.8	1425.3	1477.1	1529.3
(381.86)	*s*			1.5600	1.6239	1.6767	1.7234	1.7660	1.8055	1.8425
	v				2.150	2.426	2.688	2.943	3.193	3.440
250.0	*H*				1263.3	1318.3	1371.1	1423.2	1475.3	1527.9
(401.04)	*s*				1.5948	1.6494	1.6970	1.7401	1.7799	1.8172
	v				1.766	2.004	2.227	2.442	2.653	2.860
300.0	*H*				1257.5	1314.5	1368.3	1421.0	1473.6	1526.5
(417.43)	*s*				1.5701	1.6266	1.6751	1.7187	1.7589	1.7964
	v				1.2843	1.4760	1.6503	1.8163	1.9776	2.136
400	*H*				1245.2	1306.6	1362.5	1416.6	1470.1	1523.6
(444.70)	*s*				1.5282	1.5892	1.6397	1.6884	1.7252	1.7632

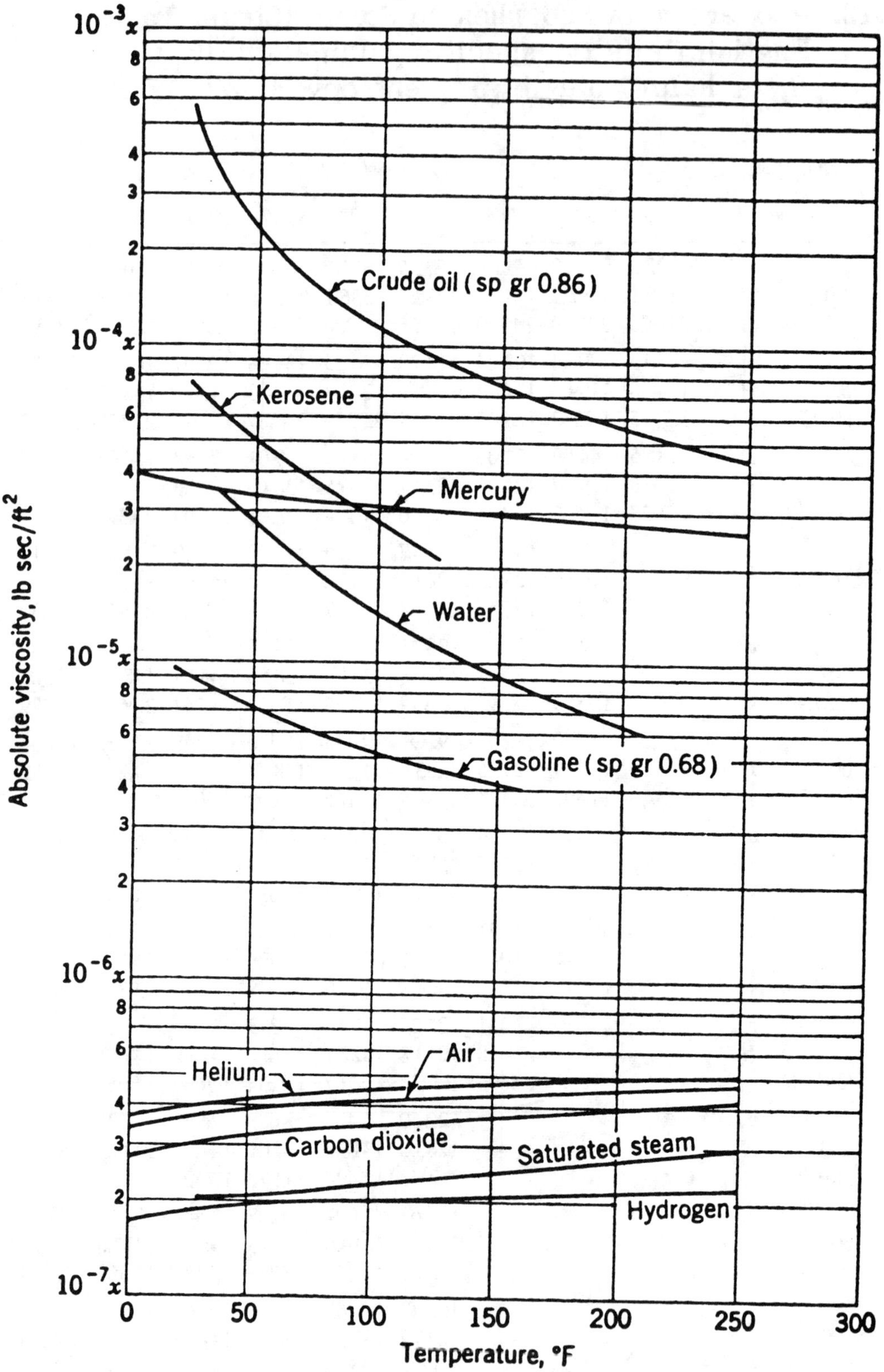

Absolute-viscosity curves.

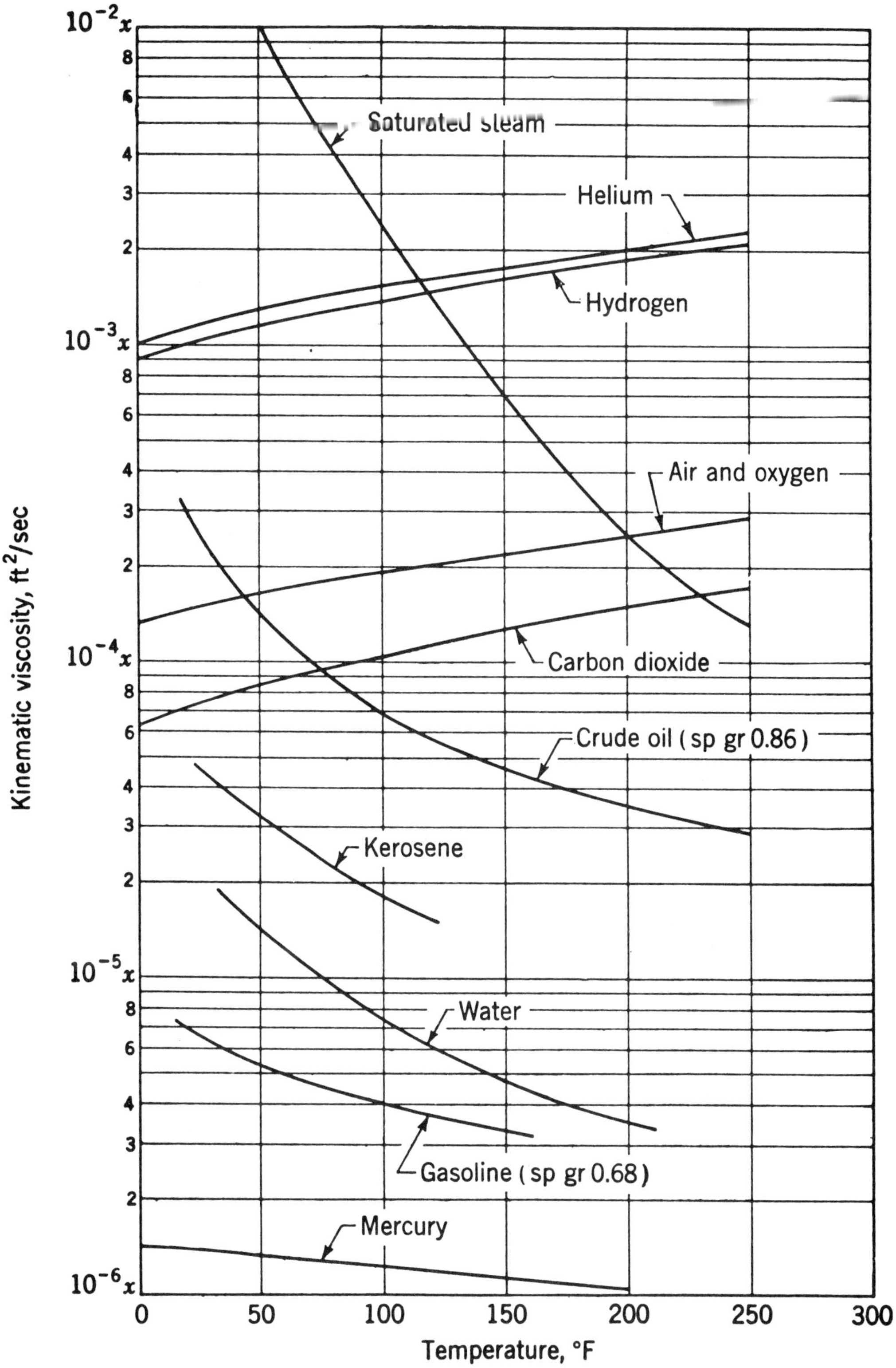

Kinematic-viscosity curves for standard atmospheric pressure.

TABLE: FRICTION FACTOR FOR A PERFECT GAS K = 1.4

M	$\frac{fK}{H}$ L
0	∞
.10	93.6903
.20	20.3417
.30	7.4190
.40	3.2312
.50	1.4967
.60	0.6872
.70	0.2914
.80	0.1012
.90	0.0203
1	0
1.10	0.0139
1.20	0.0471
1.30	0.0908
1.40	0.1396
1.50	0.1905
1.60	0.2413
1.70	0.2909
1.80	0.3386
1.90	0.3841
2	0.4270
2.10	0.4674
2.20	0.5053
2.30	0.5407
2.40	0.5738
2.50	0.6048
2.60	0.6336
2.70	0.6605
2.80	0.6857
2.90	0.7091
3	0.7310
3.10	0.7515
3.20	0.7706
3.30	0.7885
3.40	0.8053
3.50	0.8210
3.60	0.8357
3.70	0.8496
3.80	0.8626
3.90	0.8748
4	0.8863

Thermal Conductivity of Gases at 101.325 kPa (1 Atm Abs)

Temperature			H_2		O_2		N_2		CO		CO_2	
K	°C	°F	W/m·K	btu/h·ft·°F	W/m·K	btu/h·ft·°F	W/m·K	btu/h·ft·°F	W/m·K	btu/h·ft·°F	W/m·K	btu/h·ft·°F
255.4	−17.8	0	0.1592	0.0920	0.0228	0.0132	0.0228	0.0132	0.0222	0.0128	0.0[illegible]32	0.0076
273.2	0	32	0.1667	0.0963	0.0246	0.0142	0.0239	0.0138	0.0233	0.0135	0.0[illegible]45	0.0084
283.2	10.0	50	0.1720	0.0994	0.0253	0.0146	0.0248	0.0143	0.0239	0.0138	0.0[illegible]52	0.0088
311.0	37.8	100	0.1852	0.107	0.0277	0.0160	0.0267	0.0154	0.0260	0.0150	0.0[illegible]73	0.0100
338.8	65.6	150	0.1990	0.115	0.0299	0.0173	0.0287	0.0166	0.0279	0.0161	0.0190	0.0110
366.5	93.3	200	0.2111	0.122	0.0320	0.0185	0.0303	0.0175	0.0296	0.0171	0.0216	0.0125
394.3	121.1	250	0.2233	0.129	0.0343	0.0198	0.0329	0.0190	0.0318	0.0184	0.0239	0.0138
422.1	148.9	300	0.2353	0.136	0.0363	0.0210	0.0348	0.0201	0.0338	0.0195	0.0260	0.0150
449.9	176.7	350	0.2458	0.142	0.0382	0.0221	0.0365	0.0211	0.0355	0.0205	0.0286	0.0165
477.6	204.4	400	0.2579	0.149	0.0398	0.0230	0.0382	0.0221	0.0369	0.0213	0.0308	0.0178
505.4	232.2	450	0.2683	0.155	0.0422	0.0244	0.0400	0.0231	0.0384	0.0222	0.0334	0.0193
533.2	260.0	500	0.2786	0.161	0.0438	0.0253	0.0419	0.0242	0.0407	0.0235	0.0355	0.0205

Heat Capacity of Gases at Constant Pressure at 101.325 kPa (1 Atm Abs)

Temperature			H_2		O_2		N_2		CO		CO_2	
K	°C	°F	kJ/kg·K	btu/lb_m·°F	kJ/kg·K	btu/lb_m·°F	kJ/kg·K	btu/lb_m·°F	kJ/kg·K	btu/lb_m·°F	kJ/kg·K	btu/lb_m·°F
255.4	−17.8	0	14.07	3.36	0.909	0.217	1.034	0.247	1.034	0.247	0.800	0.191
273.2	0	32	14.19	3.39	0.913	0.218	1.038	0.248	1.038	0.248	0.816	0.195
283.2	10.0	50	14.19	3.39	0.917	0.219	1.038	0.248	1.038	0.248	0.825	0.197
311.0	37.8	100	14.32	3.42	0.921	0.220	1.038	0.248	1.043	0.249	0.854	0.204
338.8	65.6	150	14.36	3.43	0.925	0.221	1.038	0.248	1.043	0.249	0.883	0.211
366.5	93.3	200	14.40	3.44	0.929	0.222	1.043	0.249	1.047	0.250	0.904	0.216
394.3	121.1	250	14.44	3.45	0.938	0.224	1.043	0.249	1.047	0.250	0.929	0.222
422.1	148.9	300	14.49	3.46	0.946	0.226	1.047	0.250	1.051	0.251	0.950	0.227
449.9	176.7	350	14.49	3.46	0.955	0.228	1.047	0.250	1.055	0.252	0.976	0.233
477.6	204.4	400	14.49	3.46	0.963	0.230	1.051	0.251	1.059	0.253	0.996	0.238
505.4	232.2	450	14.52	3.47	0.971	0.232	1.055	0.252	1.063	0.254	1.017	0.243
533.2	260.0	500	14.52	3.47	0.976	0.233	1.059	0.253	1.068	0.255	1.030	0.246

One-dimensional Isentropic Relations†

M	A/A^*	p/p_0	ρ/ρ_0	T/T_0	M	A/A^*	p/p_0	ρ/ρ_0	T/T_0
0.00	.....	1.000	1.000	1.000	0.78	1.05	0.669	0.750	0.891
0.01	57.87	0.9999	0.9999	0.9999	0.80	1.04	0.656	0.740	0.886
0.02	28.94	0.9997	0.9999	0.9999	0.82	1.03	0.643	0.729	0.881
0.04	14.48	0.999	0.999	0.9996	0.84	1.02	0.630	0.719	0.876
0.06	9.67	0.997	0.998	0.999	0.86	1.02	0.617	0.708	0.871
0.08	7.26	0.996	0.997	0.999	0.88	1.01	0.604	0.698	0.865
0.10	5.82	0.993	0.995	0.998	0.90	1.01	0.591	0.687	0.860
0.12	4.86	0.990	0.993	0.997	0.92	1.01	0.578	0.676	0.855
0.14	4.18	0.986	0.990	0.996	0.94	1.00	0.566	0.666	0.850
0.16	3.67	0.982	0.987	0.995	0.96	1.00	0.553	0.655	0.844
0.18	3.28	0.978	0.984	0.994	0.98	1.00	0.541	0.645	0.839
0.20	2.96	0.973	0.980	0.992	1.00	1.00	0.528	0.632	0.833
0.22	2.71	0.967	0.976	0.990	1.02	1.00	0.516	0.623	0.828
0.24	2.50	0.961	0.972	0.989	1.04	1.00	0.504	0.613	0.822
0.26	2.32	0.954	0.967	0.987	1.06	1.00	0.492	0.602	0.817
0.28	2.17	0.947	0.962	0.985	1.08	1.01	0.480	0.592	0.810
0.30	2.04	0.939	0.956	0.982	1.10	1.01	0.468	0.582	0.805
0.32	1.92	0.932	0.951	0.980	1.12	1.01	0.457	0.571	0.799
0.34	1.82	0.923	0.944	0.977	1.14	1.02	0.445	0.561	0.794
0.36	1.74	0.914	0.938	0.975	1.16	1.02	0.434	0.551	0.788
0.38	1.66	0.905	0.931	0.972	1.18	1.02	0.423	0.541	0.782
0.40	1.59	0.896	0.924	0.969	1.20	1.03	0.412	0.531	0.776
0.42	1.53	0.886	0.917	0.966	1.22	1.04	0.402	0.521	0.771
0.44	1.47	0.876	0.909	0.963	1.24	1.04	0.391	0.512	0.765
0.46	1.42	0.865	0.902	0.959	1.26	1.05	0.381	0.502	0.759
0.48	1.38	0.854	0.893	0.956	1.28	1.06	0.371	0.492	0.753
0.50	1.34	0.843	0.885	0.952	1.30	1.07	0.361	0.483	0.747
0.52	1.30	0.832	0.877	0.949	1.32	1.08	0.351	0.474	0.742
0.54	1.27	0.820	0.868	0.945	1.34	1.08	0.342	0.464	0.736
0.56	1.24	0.808	0.859	0.941	1.36	1.09	0.332	0.455	0.730
0.58	1.21	0.796	0.850	0.937	1.38	1.10	0.323	0.446	0.724
0.60	1.19	0.784	0.840	0.933	1.40	1.11	0.314	0.437	0.718
0.62	1.17	0.772	0.831	0.929	1.42	1.13	0.305	0.429	0.713
0.64	1.16	0.759	0.821	0.924	1.44	1.14	0.297	0.420	0.707
0.66	1.13	0.747	0.812	0.920	1.46	1.15	0.289	0.412	0.701
0.68	1.12	0.734	0.802	0.915	1.48	1.16	0.280	0.403	0.695
0.70	1.09	0.721	0.792	0.911	1.50	1.18	0.272	0.395	0.690
0.72	1.08	0.708	0.781	0.906	1.52	1.19	0.265	0.387	0.684
0.74	1.07	0.695	0.771	0.901	1.54	1.20	0.257	0.379	0.678
0.76	1.06	0.682	0.761	0.896	1.56	1.22	0.250	0.371	0.672

† For a perfect gas with constant specific heat. $k = 1.4$.

M	A/A^*	p/p_0	ρ/ρ_0	T/T_0	M	A/A^*	p/p_0	ρ/ρ_0	T/T_0
1.58	1.23	0.242	0.363	0.667	2.30	2.19	0.080	0.165	0.486
1.60	1.25	0.235	0.356	0.661	2.32	2.23	0.078	0.161	0.482
1.62	1.27	0.228	0.348	0.656	2.34	2.27	0.075	0.157	0.477
1.64	1.28	0.222	0.341	0.650	2.36	2.32	0.073	0.154	0.473
1.66	1.30	0.215	0.334	0.645	2.38	2.36	0.071	0.150	0.469
1.68	1.32	0.209	0.327	0.639	2.40	2.40	0.068	0.147	0.465
1.70	1.34	0.203	0.320	0.634	2.42	2.45	0.066	0.144	0.461
1.72	1.36	0.197	0.313	0.628	2.44	2.49	0.064	0.141	0.456
1.74	1.38	0.191	0.306	0.623	2.46	2.54	0.062	0.138	0.452
1.76	1.40	0.185	0.300	0.617	2.48	2.59	0.060	0.135	0.448
1.78	1.42	0.179	0.293	0.612	2.50	2.64	0.059	0.132	0.444
1.80	1.44	0.174	0.287	0.607	2.52	2.69	0.057	0.129	0.441
1.82	1.46	0.169	0.281	0.602	2.54	2.74	0.055	0.126	0.437
1.84	1.48	0.164	0.275	0.596	2.56	2.79	0.053	0.123	0.433
1.86	1.51	0.159	0.269	0.591	2.58	2.84	0.052	0.121	0.429
1.88	1.53	0.154	0.263	0.586	2.60	2.90	0.050	0.118	0.425
1.90	1.56	0.149	0.257	0.581	2.62	2.95	0.049	0.115	0.421
1.92	1.58	0.145	0.251	0.576	2.64	3.01	0.047	0.113	0.418
1.94	1.61	0.140	0.246	0.571	2.66	3.06	0.046	0.110	0.414
1.96	1.63	0.136	0.240	0.566	2.68	3.12	0.044	0.108	0.410
1.98	1.66	0.132	0.235	0.561	2.70	3.18	0.043	0.106	0.407
2.00	1.69	0.128	0.230	0.556	2.72	3.24	0.042	0.103	0.403
2.02	1.72	0.124	0.225	0.551	2.74	3.31	0.040	0.101	0.400
2.04	1.75	0.120	0.220	0.546	2.76	3.37	0.039	0.099	0.396
2.06	1.78	0.116	0.215	0.541	2.78	3.43	0.038	0.097	0.393
2.08	1.81	0.113	0.210	0.536	2.80	3.50	0.037	0.095	0.389
2.10	1.84	0.109	0.206	0.531	2.82	3.57	0.036	0.093	0.386
2.12	1.87	0.106	0.201	0.526	2.84	3.64	0.035	0.091	0.383
2.14	1.90	0.103	0.197	0.522	2.86	3.71	0.034	0.089	0.379
2.16	1.94	0.100	0.192	0.517	2.88	3.78	0.033	0.087	0.376
2.18	1.97	0.097	0.188	0.513	2.90	3.85	0.032	0.085	0.373
2.20	2.01	0.094	0.184	0.508	2.92	3.92	0.031	0.083	0.370
2.22	2.04	0.091	0.180	0.504	2.94	4.00	0.030	0.081	0.366
2.24	2.08	0.088	0.176	0.499	2.96	4.08	0.029	0.080	0.363
2.26	2.12	0.085	0.172	0.495	2.98	4.15	0.028	0.078	0.360
2.28	2.15	0.083	0.168	0.490	3.00	4.23	0.027	0.076	0.357

INDEX

Numbers on this page refer to PROBLEM NUMBERS, not page numbers

Numbers on this page refer to PROBLEM NUMBERS, not page numbers

Numbers on this page refer to PROBLEM NUMBERS, not page numbers

Numbers on this page refer to PROBLEM NUMBERS, not page numbers

Numbers on this page refer to PROBLEM NUMBERS, not page numbers

Numbers on this page refer to PROBLEM NUMBERS, not page numbers

Numbers on this page refer to PROBLEM NUMBERS, not page numbers